Functional Nanomaterials

Functional Nanomaterials

Synthesis, Properties, and Applications

Edited by Wai-Yeung Wong and Qingchen Dong

Editors

Prof. Wai-Yeung Wong
The Hong Kong Polytechnic University
Department of Applied Biology and
Chemical Technology
Hung Hom, Kowloon
Hong Kong
China

Prof. Qingchen Dong
Shanghai University
MOE Key Laboratory of Advanced
Display and System Applications
149 Yanchang Road
Jingan District
200072 Shanghai
China

Cover Image: © SEBASTIAN KAULITZKI/SCIENCE PHOTO LIBRARY/Getty Images

All books published by **WILEY-VCH** are carefully produced. Nevertheless, authors, editors, and publisher do not warrant the information contained in these books, including this book, to be free of errors. Readers are advised to keep in mind that statements, data, illustrations, procedural details or other items may inadvertently be inaccurate.

Library of Congress Card No.: applied for

British Library Cataloguing-in-Publication Data
A catalogue record for this book is available from the British Library.

Bibliographic information published by the Deutsche Nationalbibliothek
The Deutsche Nationalbibliothek lists this publication in the Deutsche Nationalbibliografie; detailed bibliographic data are available on the Internet at <http://dnb.d-nb.de>.

© 2022 WILEY-VCH GmbH, Boschstr. 12, 69469 Weinheim, Germany

All rights reserved (including those of translation into other languages). No part of this book may be reproduced in any form – by photoprinting, microfilm, or any other means – nor transmitted or translated into a machine language without written permission from the publishers. Registered names, trademarks, etc. used in this book, even when not specifically marked as such, are not to be considered unprotected by law.

Print ISBN: 978-3-527-34797-1
ePDF ISBN: 978-3-527-82854-8
ePub ISBN: 978-3-527-82855-5
oBook ISBN: 978-3-527-82856-2

Typesetting Straive, Chennai, India
Printing and Binding CPI Group (UK) Ltd, Croydon, CR0 4YY

Printed on acid-free paper

Contents

Preface *xi*
About the Editor *xiii*

1 Earth-Abundant Metal-Based Nanomaterials for Electrochemical Water Splitting *1*
Weiran Zheng, Yong Li, and Lawrence Yoon Suk Lee
1.1 Electrochemical Water Splitting *1*
1.1.1 General Principle *1*
1.1.2 Overpotential and Tafel Slope *3*
1.1.3 Current Techniques *4*
1.2 Earth-Abundant Metallic Nanomaterials *5*
1.2.1 Hydrogen Evolution Reaction (HER) *6*
1.2.1.1 Mechanism *6*
1.2.1.2 Metal (M^0) Nanoparticles *7*
1.2.1.3 Metal (M^0) Single-Atom Catalysts *8*
1.2.1.4 Metal Phosphides *11*
1.2.1.5 Metal Chalcogenides *12*
1.2.1.6 Metal Nitrides *14*
1.2.1.7 Metal Carbides *15*
1.2.1.8 Metal Oxides/(Oxy)hydroxides *15*
1.2.2 Oxygen Evolution Reaction *16*
1.2.2.1 Mechanism *16*
1.2.2.2 Metal Oxides/Hydroxides *19*
1.2.2.3 Metal (M^{n+}) Single-Atom Catalysts *25*
1.2.2.4 Metal Chalcogenides/Nitrides/Phosphides and Others *26*
1.3 Computer-Assisted Materials Discovery *28*
1.4 Challenge and Outlook *29*
1.4.1 Reliability Comparison Between Results *29*
1.4.2 Gap Between Industrial and Laboratory Research *30*
1.4.3 Outlook *30*
References *31*

2 Studies on Cerium-Based Nanostructured Materials for Electrocatalysis *41*
Xuemei Zhou, Mingkai Zhang, Yuwei Jin, and Yongquan Qu
2.1 Introduction *41*
2.2 Cerium-Based Nanostructure Materials *42*

2.3	Cerium-Based Electrocatalysts for HER	44
2.3.1	Cerium-Doped Electrocatalysts for HER	44
2.3.2	Composites with CeO_2 for HER	45
2.4	Cerium-Based Electrocatalysts for OER	49
2.4.1	Cerium-Doped Electrocatalysts for OER	50
2.4.2	Composites with CeO_2 for OER	50
2.5	Cerium-Based Electrocatalysts for ORR	57
2.5.1	Noble Metals with Ce/Ceria for ORR	58
2.5.2	Doping Ce Element into Earth-Abundant Electrocatalysts for ORR	59
2.5.3	CeO_2-Based Electrocatalysts for ORR	60
2.6	Cerium-Based Electrocatalysts for Other Electrochemical Reactions	63
2.7	Conclusions and Outlooks	65
	Acknowledgment	67
	References	67
3	**Metal-Free Carbon-Based Nanomaterials: Fuel Cell Applications as Electrocatalysts**	**73**
	Lai-Hon Chung, Zhi-Qing Lin, and Jun He	
3.1	Introduction	73
3.2	Heteroatom-Doped Carbon Nanomaterials	75
3.2.1	Heteroatom-Doped Carbon Nanotubes	76
3.2.2	Heteroatom-Doped Graphenes	80
3.2.3	Heteroatom-Doped Graphdiyne	94
3.2.4	Heteroatom-Doped Porous Carbon Nanomaterials	97
3.2.5	Heteroatom-Doped Composite Materials	105
3.2.6	Better ORR Performance in Acidic Medium	108
3.3	Undoped Carbon Nanomaterials	111
3.3.1	Edge as Defect	112
3.3.2	Intrinsic/Topological Defects	114
3.4	Carbon-Based Organic Framework	117
3.5	Application in Fuel Cells	120
3.5.1	Application in Alkaline Fuel Cell and PEMFC	121
3.5.2	Application in Zinc–Air Battery	124
3.6	Conclusion	129
	References	130
4	**Rare Earth Luminescent Nanomaterials and Their Applications**	**141**
	Jianle Zhuang and Xuejie Zhang	
4.1	Introduction	141
4.2	Rare Earth Based UCNPs	142
4.2.1	Development of Upconversion Materials	142
4.2.2	Upconversion Mechanism	143
4.2.2.1	Excited-State Absorption (ESA)	143
4.2.2.2	Energy Transfer Upconversion (ETU)	143
4.2.2.3	Cooperative Upconversion (CUC)	144
4.2.2.4	Cross Relaxation (CR)	144
4.2.2.5	Photon Avalanche (PA)	144
4.2.2.6	Energy Migration-Mediated Upconversion (EMU)	145
4.2.3	Composition of UCNPs	145
4.2.3.1	Host	145
4.2.3.2	Activator	145
4.2.3.3	Sensitizer	146

4.2.4	Synthesis of UCNPs	146
4.2.4.1	Thermal Decomposition	146
4.2.4.2	Hydro/Solvothermal Synthesis	149
4.2.4.3	Coprecipitation	149
4.2.4.4	Sol–Gel Synthesis	150
4.2.4.5	Microwave-Assisted Synthesis	150
4.2.5	Characterization of UCNPs	150
4.2.5.1	Identification of Crystal Structures	151
4.2.5.2	Determination of Size and Morphology	152
4.2.5.3	Characterization of Surface Moieties	153
4.2.5.4	Composition Determination	154
4.2.5.5	Measurement of Optical Properties	155
4.2.5.6	Evaluation of Magnetic Properties	156
4.2.6	Tuning of Upconversion Emission	156
4.2.6.1	Tuning UC Emission Changing by the Chemical Composition and Varying Dopant Concentration	156
4.2.6.2	Tuning UC Emission by Host Matrix Screening	157
4.2.6.3	Tuning UC Emission by Interparticle Energy Transfer or Antenna Effect	158
4.2.6.4	Tuning UC Emission Through Energy Migration	158
4.2.6.5	Tuning UC Emission Using Cross-Relaxation Processes	160
4.2.6.6	Tuning UC Emission Using Core/Shell Structures	160
4.2.6.7	Tuning UC Emission Using Size- and Shape-Induced Surface Effects	160
4.2.6.8	Tuning UC Emission Using FRET or RET	162
4.2.6.9	Tuning Upconversion Emission Through External Stimulus	165
4.2.7	Applications of UCNPs	165
4.2.7.1	Bioimaging	165
4.2.7.2	Therapy	168
4.2.7.3	Optogenetics	170
4.2.7.4	Sensing and Detection	171
4.2.7.5	Photocatalysis	173
4.2.7.6	UCNPs-Mediated Molecular Switches	175
4.2.7.7	Other Technological Applications	176
4.3	Rare Earth Based DCNPs	178
4.3.1	$Y_3Al_5O_{12}$:RE (RE = Ce^{3+}, Tb^{3+})	178
4.3.1.1	Coprecipitation Approach	178
4.3.1.2	Sol–Gel Method	179
4.3.1.3	Solvothermal Method	181
4.3.2	$SrAl_2O_4$:Eu^{2+}, Dy^{3+}	182
4.3.2.1	Hydrothermal Method	182
4.3.2.2	Sol–Gel Method	183
4.3.2.3	Microwave Method	183
4.3.2.4	Electrospinning	183
4.3.3	Y_2O_3:Eu^{3+}	184
4.3.4	$LnVO_4$:Ln^{3+} (Ln = La, Gd, Y; Ln^{3+} = Eu^{3+}, Dy^{3+}, Sm^{3+})	186
4.3.5	$LaPO_4$:Ce^{3+},Tb^{3+}	187
4.3.6	Applications	189
4.3.6.1	Biological Imaging	189
4.3.6.2	Tumor Treatment	190
4.3.6.3	Fluorescent Ink	191
4.4	Summary and Outlook	192
	References	193

5 Metal Complex Nanosheets: Preparation, Property, and Application *207*
Ryota Sakamoto
5.1 Introduction *207*
5.2 Preparation of Metal Complex Nanosheets *208*
5.2.1 Vacuum Phase Fabrication *208*
5.2.2 Mechanical Exfoliation *208*
5.2.3 Liquid-Phase Exfoliation *209*
5.2.4 Liquid/Liquid Interfacial Synthesis *211*
5.2.5 Gas/Liquid Interfacial Synthesis *213*
5.3 Properties of Metal Complex Nanosheets *215*
5.3.1 Electroproperties *215*
5.3.2 Photoproperties *217*
5.3.3 Magnetoproperties *219*
5.4 Outlook on Metal Complex Nanosheets *221*
References *221*

6 Synthesis, Properties, and Applications of Metal Halide Perovskite-Based Nanomaterials *225*
Mei-Li Sun, Cai-Xiang Zhao, Jun-Feng Shu, and Xiong Yin
6.1 Introduction *225*
6.1.1 Crystal Structure and Phase of Metal Halide Perovskites *225*
6.1.2 Classification of Metal Halide Perovskite-Based Nanomaterials *227*
6.1.2.1 Organic–Inorganic Hybrid Perovskite Materials *228*
6.1.2.2 All-Inorganic Perovskite Materials *232*
6.1.2.3 Lead-Free Perovskite Materials and Low-Lead Perovskite Material *234*
6.2 Properties of Metal Halide Perovskite Materials *238*
6.2.1 Tunable Bandgap *238*
6.2.2 High Absorption Coefficient *239*
6.2.3 Excellent Charge Transport Performance *240*
6.2.4 Photoluminescence Properties *240*
6.3 Synthesis of Metal Halide Perovskite-Based Nanomaterials *242*
6.3.1 Hot Injection Method *243*
6.3.2 Ligand-Assisted Reprecipitation Method *244*
6.3.3 Solution Deposition Methods *244*
6.3.3.1 One-Step Method *245*
6.3.3.2 Two-Step Method *247*
6.3.3.3 Other Solution-Processing Methods *249*
6.4 Application of Metal Halide Perovskite-Based Nanomaterials *251*
6.4.1 Perovskite Solar Cells *251*
6.4.2 Perovskite Light-Emitting Diode *254*
6.4.3 Sensing *256*
6.4.4 Other Devices *257*
References *259*

7 Progress in Piezo-Phototronic Effect on 2D Nanomaterial-Based Heterostructure Photodetectors *275*
Yuqian Zhao, Ran Ding, Feng Guo, Zehan Wu, and Jianhua Hao
7.1 Introduction *275*
7.2 Piezo-Phototronic Effect on the Junctions *277*
7.2.1 Fundamental Physics of Piezo-Phototronics *277*

7.2.2	Piezo-Phototronic Effect on P–N Junction *278*
7.2.3	Piezo-Phototronic Effect on Metal–Semiconductor Junction *282*
7.3	Piezo-Phototronic Effect on the Performance of P–N Junction Photodetectors *284*
7.3.1	Photodetector Based on 2D Homojunction *285*
7.3.2	Photodetectors Based on 1D–2D Heterostructure *286*
7.3.3	Photodetectors Based on 2D–2D Heterostructure *289*
7.3.4	Photodetectors Based on 3D–2D Heterostructure *293*
7.4	Conclusion and Future Perspectives *295*
	Acknowledgments *297*
	References *297*

8	**Synthesis and Properties of Conducting Polymer Nanomaterials** *303*
	Ziyan Zhang, Tianyu Sun, Mingda Shao, and Ying Zhu
8.1	Introduction *303*
8.2	Synthesis and Properties *305*
8.2.1	Chemical Synthesis and Properties *306*
8.2.2	Electrochemical Synthesis and Properties *314*
8.3	Summary *329*
	References *329*

9	**Conducting Polymer Nanomaterials for Electrochemical Energy Storage and Electrocatalysis** *337*
	Mingwei Fang, Xingpu Wang, Xueyan Li, and Ying Zhu
9.1	Introduction *337*
9.2	Electrode Materials of Batteries *337*
9.2.1	Electrodes for Metal-Ion Batteries *338*
9.2.1.1	Electrodes for Lithium-Ion Batteries *338*
9.2.1.2	Electrodes for Other Metal-Ion Batteries *345*
9.2.2	Electrodes for Lithium–Sulfur Batteries *348*
9.2.3	Electrodes for All-Polymer Batteries *350*
9.2.4	Electrodes for Dye-Sensitized Solar Cell *352*
9.2.5	Electrodes for Bioelectric Batteries *352*
9.3	Electrocatalysis *355*
9.3.1	Oxygen Evolution Reaction (OER) *356*
9.3.2	Hydrogen Evolution Reaction (HER) *357*
9.3.3	Carbon Dioxide Reduction Reaction (CO_2RR) *361*
9.4	Supercapacitors *363*
9.4.1	CP as the Active Material *364*
9.4.2	CP Composites as the Active Materials *370*
9.5	Summary and Perspective *386*
	References *386*

10	**Conducting Polymer Nanomaterials for Bioengineering Applications** *399*
	Xiang Sun, Meiling Wang, You Liu, Xin Zhang, Yalan Chen, Shiying Li, and Ying Zhu
10.1	Introduction *399*
10.2	Electronic Skin *399*

10.2.1	Wearable Electronic Devices *400*
10.2.2	Self-Healing E-Skin *403*
10.2.3	Energy-Saving E-Skin *405*
10.3	Bioengineering *406*
10.3.1	Tissue Regeneration Engineering *406*
10.3.2	Drug Delivery *414*
10.3.3	Actuators *422*
10.4	Chemical Sensors and Biosensors *424*
10.4.1	Chemical Sensors *424*
10.4.2	Biosensors *427*
10.5	Summary and Perspective *436*
	References *436*

11	**Methods for Synthesizing Polymer Nanocomposites and Their Applications** *447*
	Muwei Ji, Jintao Huang, and Caizhen Zhu
11.1	Factors for Synthesizing Polymer Nanocomposites *448*
11.2	Solution Mixing *451*
11.3	Emulsion Polymerization *456*
11.4	Dispersion Polymerization and Dispersion Copolymerization *458*
11.5	Self-Assembly *461*
11.6	Melting *463*
11.7	*In situ* Polymerization *466*
11.8	Tailoring of Polymers Nanocomposite *471*
11.9	Application of Polymer Nanocomposites *474*
11.10	Outlook *481*
	List of Abbreviations *481*
	References *483*

12	**Spin-Related Electrode Reactions in Nanomaterials** *491*
	Shengnan Sun and Yanglong Hou
12.1	Introduction *491*
12.2	Factors Influencing the Electrochemical System *492*
12.2.1	Forces Caused by Magnetic Fields in Aqueous Solution *492*
12.2.2	Spin States of Electrocatalysts *495*
12.3	Spin-Related Electrode Reactions *496*
12.3.1	Electrodeposition of Metals or Alloys *496*
12.3.2	Hydrogen Evolution Reaction *498*
12.3.3	Oxygen Evolution Reaction *504*
12.3.4	Oxygen Reduction Reaction *513*
12.3.5	Other Catalytic Reactions *517*
12.3.6	Battery *518*
12.3.7	Others *522*
12.4	Conclusion and Outlook *523*
	References *523*

Index *533*

Preface

Since Feynman gave the speech "There's Plenty of Room at the Bottom" at the annual conference of American Physical Society at California Institute of Technology in 1959, nanoscience and nanotechnology emerged as times require, and the field has developed for more than half century. This has achieved leap forward development in the past two decades. Nanoscience is a newly emerging discipline for exploring the micro-world. Due to the specific small size effect, macroscopic quantum tunneling effect, surface and interface effect, etc., nanomaterials have displayed extensive and attractive application perspectives in the fields of optics, electronics, magnetism, catalysis, sensing, and biomedicine. With the development of nanotechnology, some interdisciplinary research works such as nanoelectronics, nanobiology, nanomaterials, nanomedicine, etc. have also been successively established, which also promotes the mutual penetration and integration of multidisciplinary researches. By changing the size, morphology, chemical structure, and composition, a number of advanced functional nanomaterials have been designed, synthesized, and applied in various areas.

Therefore, to bring a focus on the recent research developments of some novel functional nanomaterials such as metal nanoparticles, metal oxide nanomaterials, metal-free carbon-based nanomaterials, polymer nanomaterials, two-dimensional nanomaterials, and perovskite-based nanomaterials and their applications in catalysis, optoelectronics, sensing, biomedicine, etc., a collection of chapters from leading scientists is presented in this book. The fundamental concepts and theories, synthetic strategies, properties characterization, device fabrication, and performance evaluation of the above nanomaterials and devices are covered, which provide readers with a good source of information in understanding the development trend of advanced nanoscience and nanotechnology and exploring more functional nanomaterials. These include the investigation of functional earth-abundant metal-based nanomaterials (by Weiran Zheng, Yong Li, Lawrence Yoon Suk Lee), metal oxide nanomaterials (by Xuemei Zhou, Mingkai Zhang, Yuwei Jin, Yongquan Qu), and metal-free carbon-based nanomaterials (by Lai-Hon Chung, Zhi-Qing Lin, Jun He) for catalysis and the spin-related catalytic reaction in nanomaterials (by Shengnan Sun, Yanglong Hou); the studies on rare-earth luminescent nanomaterials (by Jianle Zhuang, Xuejie Zhang), metal complex nanosheets (by Ryota Sakamoto), metal halide perovskite-based nanomaterials (by Mei-Li Sun, Cai-Xiang Zhao, Jun-Feng Shu, Xiong Yin), and two-dimensional nanomaterials (by Yuqian Zhao, Ran Ding, Feng Guo, Zehan Wu, Jianhua Hao) for applications in optoelectronics and sensing; and the comprehensive introduction of the synthesis, electrochemistry, and bioengineering applications of conducting polymer-based nanomaterials (by Ziyan Zhang, Tianyu Sun, Mingda Shao, Mingwei Fang, Xingpu Wang, Xueyan Li, Xiang Sun, Meiling Wang, You Liu, Xin Zhang, Yalan Chen, Shiying Li, Ying Zhu) and polymers-based nanocomposites (by Muwei Ji, Jintao Huang, Caizhen Zhu).

Nanotechnology has extended the means and ability of human beings to understand and transform the material world to atoms and molecules and provided us with a new concept of designing nanomaterials with unique physical and chemical properties different from the traditional materials. We believe that nanotechnology and nanomaterials with excellent physical and chemical properties will help us address some challenging issues such as energy and climate crisis, healthy problems, information storage, and transmission and display that are confronted by human society in the twenty-first century. We are grateful to all the contributors who have participated in the preparation of this book. Finally, we deeply thank Dr. Lifen Yang, Program Manager, and Ms. Katherine Wong, Senior Managing Editor, of John Wiley & Sons Inc. for their helpful suggestions and discussion on the organization of this book. Without their support, this huge work would not have been possible.

We thank the financial support from the Science, Technology and Innovation Committee of Shenzhen Municipality (JCYJ20180507183413211), National Natural Science Foundation of China (52073242, 62174116, 61774109), the Hong Kong Research Grants Council (PolyU 153062/18P), the RGC Senior Research Fellowship Scheme (SRFS2021-5S01), The Hong Kong Polytechnic University (1-ZE1C), Research Institute for Smart Energy (RISE), Miss Clarea Au for the Endowed Professorship in Energy (847S) and the startup fund of Shanghai University.

Hong Kong   *Wai-Yeung Wong and Qingchen Dong*
05 January 2022

About the Editor

Wai-Yeung Wong is currently Chair Professor and Dean of Faculty of Applied Science and Textiles at The Hong Kong Polytechnic University (PolyU). He is also Professor at the PolyU Shenzhen Research Institute. He received his BSc (1992) and PhD (1995) degrees from the University of Hong Kong. After postdoctoral training with Prof. F. Albert Cotton at Texas A&M University in 1996 and Profs. The Lord Lewis and Paul R. Raithby at the University of Cambridge in 1997, he joined Hong Kong Baptist University (HKBU) from 1998 to 2016. His research interests lie in the areas of metallopolymers and metallo-organic molecules with energy functions and photofunctional properties. His research activities are documented in more than 700 scientific articles, 4 books, 18 book chapters, and 2 US patents. Professor Wong has been named in the list of Highly Cited Researchers from 2014 to 2020 published by Thomson Reuters/Clarivate Analytics. He becomes the first Chinese scientist to be presented with the Chemistry of the Transition Metals Award by the Royal Society of Chemistry in 2010. He has also won the Federation of Asian Chemical Societies (FACS) Distinguished Young Chemist Award from the Federation of Asian Chemical Societies in 2011, Ho Leung Ho Lee Foundation Prize for Scientific and Technological Innovation in 2012, State Natural Science Award (Second Class) of China in 2013, Japanese Photochemistry Association Lectureship Award for Asian and Oceanian Photochemist (Eikohsha Award) in 2014, and Research Grants Council (RGC) Senior Research Fellow Award in 2020. Professor Wong is currently the Editor-in-Chief of *Topics in Current Chemistry*, Editor of *Journal of Organometallic Chemistry*, and Associate Editor of *Journal of Materials Chemistry C and Materials Advances*. At present, he is the Chairman of the Hong Kong Chemical Society.

Qingchen Dong obtained her PhD degree under the tutelage of Prof. Wai-Yeung Wong in 2012 at HKBU. She also worked at Caltech with Prof. H. B. Gray from 2010 to 2011. She joined Taiyuan University of Technology (TYUT) from 2012 to 2021 and now works at Shanghai University (SHU) as a full-time professor. Her research involves the design and synthesis of functional organic and metallo-organic compounds and carbon-based materials for applications in data storage, artificial synapse and neuromorphic computing, optoelectronics, etc. Prof. Dong has published more than 50 papers in *Chemical Society Reviews, Advanced Materials, Advanced Functional Materials, Angewandte Chemie International Edition, Advanced Electronic Materials, Journal of Materials Chemistry* A, etc., edited 2 book chapters, and awarded 3 CN patents. She won the Natural Science Award (Second Class) of Shanxi Province, China, in 2019.

1

Earth-Abundant Metal-Based Nanomaterials for Electrochemical Water Splitting

Weiran Zheng, Yong Li, and Lawrence Yoon Suk Lee

The Hong Kong Polytechnic University, Department of Applied Biology and Chemical Technology, Hung Hom, Kowloon, Hong Kong SAR, China

1.1 Electrochemical Water Splitting

Since the oil crisis in the 1970s and 1980s, hydrogen has been widely recognized to be an efficient and promising energy carrier for our future. Yet, as of 2019, about 95% of global hydrogen production relies on fossil fuels, emitting an excessive amount of carbon into the atmosphere [1]. Typically, for every ton of hydrogen produced by steam reforming of natural gas, around 9–12 tons of CO_2 is released and wasted [2, 3]. One of the most attractive options to produce hydrogen sustainably is to split water molecules using electrical energy generated by sustainable sources, such as wind, hydropower, nuclear energy, etc.

1.1.1 General Principle

In general, the electrochemical water splitting process requires passing electricity through two electrodes in water (Figure 1.1), where the oxidation occurs on the anode to generate oxygen (oxygen evolution reaction [OER]) and the reduction occurs on the cathode to produce hydrogen (hydrogen evolution reaction [HER]). The overall reaction can be simplified as Eq. (1.1):

$$\text{Overall reaction:} \quad H_2O\,(l) \rightarrow \frac{1}{2}O_2(g) + H_2(g) \quad (1.1)$$

Water splitting to hydrogen and oxygen is a thermodynamically uphill process, which requires a Gibbs free energy of $\Delta G° = 237.22\,\text{kJ}\,\text{mol}^{-1}$ or enthalpy of $\Delta H° = 285.84\,\text{kJ}\,\text{mol}^{-1}$ at standard conditions of temperature and pressure (298 K, 1 bar). When converting electrical energy to chemical energy, the equation of $\Delta G° = nFE^0$ applies, where n is the number of transferred electrons (two electrons exchanged for the splitting of one water molecule), F is the Faraday's constant, and E^0 is the standard cell voltage required. Therefore, the thermodynamically required voltage for water splitting is 1.229 V. It should be noted that the voltage value depends on the temperature and water status. For example, the electrochemical

Functional Nanomaterials: Synthesis, Properties, and Applications, First Edition.
Edited by Wai-Yeung Wong and Qingchen Dong.
© 2022 WILEY-VCH GmbH. Published 2022 by WILEY-VCH GmbH.

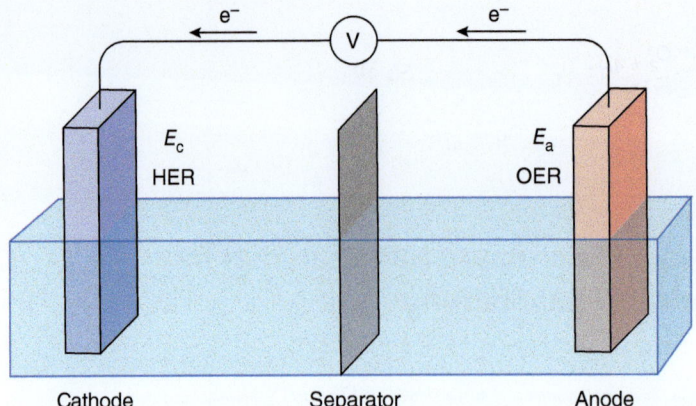

Figure 1.1 Simplified illustration of an electrolyzer for water splitting driven by a power source. The electrons travel through the external circuit and promote the HER at cathode and OER at the anode. A separator, often semipermeable membrane, is used for proton transfer and product separation.

dissociation of water vapor needs only 1.18 V. Since the electrolysis reaction is endothermic, if the reaction is performed without an external heat source, the extra voltage is needed to compensate the temperature factor in the enthalpy. In this case, the equation of $\Delta H° = nFE^0$ applies, producing a value of 1.481 V at standard conditions, known as the thermoneutral potential.

From an electrochemical perspective, the cell voltage needs to drive the two half-reactions at the electrodes:

$$E^0_{cell} = E^0_c - E^0_a$$

where E^0_{cell}, E^0_c, and E^0_a represent the standard cell, cathodic, and anodic potential, respectively.

Although the overall reaction is irrelevant to electrolyte conditions, the two half-reactions follow two routes depending on the proton concentration of the electrolyte (Figure 1.2).

Under acidic conditions:

Cathodic reaction $\quad 2H^+ + 2e^- \rightarrow H_2$ \hfill (1.2)

Anodic reaction $\quad 2H_2O \rightarrow O_2 + 4H^+ + 4e^-$ \hfill (1.3)

Under alkaline conditions:

Cathodic reaction $\quad 2H_2O + 2e^- \rightarrow H_2 + 2OH^-$ \hfill (1.4)

Anodic reaction $\quad 4OH^- \rightarrow O_2 + 2H_2O + 4e^-$ \hfill (1.5)

The Nernst equation can express the thermodynamical potential to drive the anodic side in acidic conditions:

$$E^0_a = E^0_{H_2O/O_2} + \frac{RT}{nF} \ln \frac{\left(a^2_{H^+}\right)\left(f^{1/2}_{O_2}\right)}{a_{H_2O}}$$

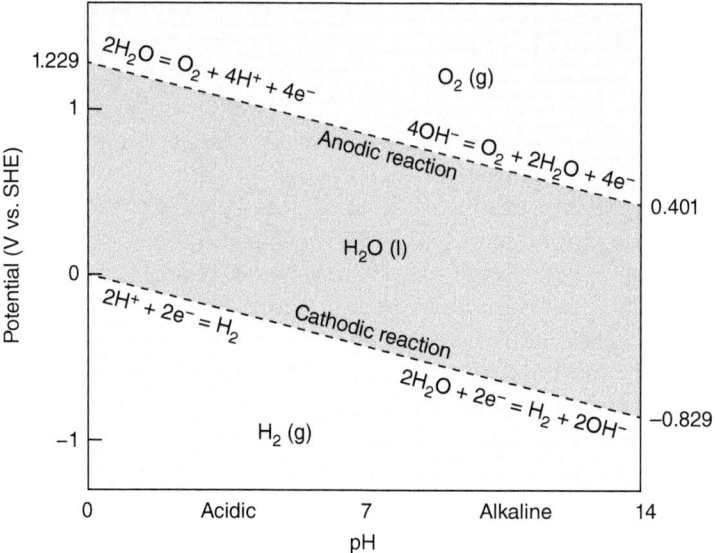

Figure 1.2 Pourbaix diagram (pH potential) of water under standard temperature and pressure (298 K, 1 bar).

where a_{H^+} and a_{H_2O} are the activity of proton and water and f_{O_2} is the fugacity of oxygen in the anodic compartment. In a simplified case when the activity coefficient of the proton is unity and the water activity is the same at all concentration,

$$E_a^0 = 1.229 - 0.059\, pH$$

Similarly, for the cathodic side,

$$E_c^0 = E_{H_2/H^+}^0 + \frac{RT}{nF} \ln \frac{a_{H^+}^2}{f_{H_2}} \approx -0.059\, pH$$

Such expressions also apply for the alkaline conditions providing the electrolyte shares the same proton activity coefficient and water activity. Therefore, the standard potential for both anodic and cathodic reactions depends on pH.

1.1.2 Overpotential and Tafel Slope

Even if the desired potential is met, the reaction may not proceed, and extra potential (overpotential, η) beyond the thermodynamic value (E^0) is commonly required to overcome the reaction energy barriers caused by many factors:

$$\eta = E_{applied} - E^0$$

The causes of overpotential can be divided mainly into three categories: the resistance overpotential due to the ohmic losses in the electric circuit, the concentration overpotential caused by the concentration gradient in the double layer region of electrolyte, and the kinetic overpotential to drive the surface reaction [4]. Both electrode and electrolyte contribute to the ohmic loss. In a laboratory-scale reaction, the potential drop due to ohmic loss is generally small and often not taken into account because the resistance of both electrode and electrolyte is negligible. However, industrial water electrolysis often suffers from the joule heat due to the ohmic resistance and large current. The concentration overpotential is the direct result of reactants consumption in the double-layer region near the electrode, therefore mainly controlled by the diffusion rate of the proton for water electrolysis. Conventionally, using electrolyte with highly acidic/alkaline conditions (extreme pH values) can offset such concentration effect and maximize efficiency. Accordingly, most researchers often employ electrolytes with pH = 1 or 14, and the current understanding of water splitting is also heavily based on the results from extreme pH levels. Industrial alkaline water electrolysis uses 20–40% NaOH or KOH aqueous solutions.

Like the concepts in heterogeneous catalysis, electrocatalysis also requires the reactants to be adsorbed first on the electrode to conduct further bond breaking and/or formation processes. Moreover, the electron transfer from/to the adsorbed reactants suffers resistance. As Eqs. (1.2)–(1.5) indicate, both anodic and cathodic processes demand an overpotential for proton/hydroxide and charge transfer for water splitting. The kinetic overpotential is referred to as the energy required to make the reactions proceed at appreciable rates. The faster the speed of water splitting is (measured as normalized charge flowing in the circuit, or current density), the higher overpotential must be supplied. The kinetic parameter used to describe such dependency is the Tafel slope (unit: $mV\ dec^{-1}$), defined as the overpotential needed to increase the current density by a factor of 10.

Due to the existence of overpotential, the energy efficiency of an electrolysis process is not 100%. When considering the efficiency in the lab, the Faradaic efficiency (ratio of the electrons used for hydrogen production vs. the total charge passed) is commonly used. The experimental value may approach 100% but is always lower than 100% because of some parasitic processes such as the conversion of the electrocatalyst. The energy loss due to resistance, however, cannot be revealed by the Faradaic efficiency. Industrial water electrolyzers often use a more practical way to evaluate the energy efficiency of water splitting by dividing the energy available from the produced hydrogen by the total energy consumed by the cell. Such value reflects the energy loss due to the overpotentials and uncovers the commercial viability of the system.

To achieve the highest energy efficiency, the energy spent on overcoming the extra barriers needs to be minimized. The role of electrocatalysts is to reduce the kinetic overpotential as much as possible. Over the past decades, electrochemists have been working on finding the best electrocatalysts for both HER and OER under acidic and alkaline conditions. Two indicators are generally used in literature for comparison: the overpotential to achieve a current density of $10\ mA\ cm^{-2}$ and the Tafel slope within a specific current range.

1.1.3 Current Techniques

Two main techniques have been commercialized for electrochemical water splitting, including alkaline electrolysis and proton exchange membrane (PEM) electrolysis. The alkaline electrolysis follows the alkaline pathway described in reactions

(1.4) and (1.5) and remains the dominating commercial approach. Both cathode and anode are often made of nickel-based materials, and the separator is a polymer that allows hydroxide ions and water molecules to pass. The PEM electrolysis that has emerged more recently follows the acidic pathway (reactions (1.2) and (1.3)) with a PEM separator placed between the electrodes to allow proton transfer. Considering the highly acidic conditions, the electrocatalysts engaged in the PEM electrolysis need to be stable under the operating environment, leaving far less choices comparing with alkaline electrolysis. The current state-of-the-art catalysts in PEM electrolysis is platinum and iridium/ruthenium oxides for HER and OER, respectively.

Although the alkaline electrolysis is more technologically mature and relatively cost-effective compared with the PEM electrolysis, it has some drawbacks such as low current density, low partial load range, low operational pressures, and hydrogen crossover through the separator. The high efficiency of PEM electrolysis, on the other hand, cannot redeem the expensive noble metal-based electrode materials. Based on the calculation by Chatzitakis and coworkers [5] a typical PEM electrolyzer requires 0.4 mg cm^{-2} Pt on the cathode and 1.54 mg cm^{-2} Ir and 0.54 mg cm^{-2} Ru on the anode, respectively. Yet, for a power density of 1.18 W cm^{-2}, 1.5, 180, and 12 years of the annual production of Pt, Ir, and Ru, respectively, are demanded.

To further improve both techniques, the development of electrocatalyst for alkaline water electrolysis needs to focus on materials bearing high current at relatively low overpotential. As to the PEM electrolysis, finding suitable alternatives that are stable under extremely acidic conditions while showing similar activities to the noble metal-based ones is essential.

1.2 Earth-Abundant Metallic Nanomaterials

The key to large-scale commercialization is to lower the cost of water electrolysis systems. Apart from increasing the activity of noble metal atoms, using earth-abundant metal-based materials as the electrocatalysts is generally appreciated by the research community. The elements currently of particular interest are nickel (Ni), cobalt (Co), iron (Fe), manganese (Mn), titanium (Ti), vanadium (V), and molybdenum (Mo), as shown in Figure 1.3.

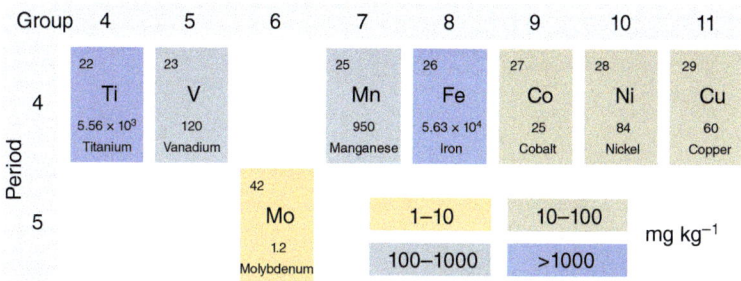

Figure 1.3 Earth-abundant metals that are currently used for water electrolysis and their relative positions in the periodic table. Their abundance in Earth's crust [6] is shown in the unit of mg kg^{-1}.

In the following contents, the current research progress of individual metal-based materials in different forms (metal, metal oxide, metal hydroxide, metal chalcogenide, metal sulfide, etc.) will be introduced based on their application: HER and OER. In many cases, the electrocatalytic materials involve more than one metal element. For clarity, their descriptions are listed under the major element responsible for active sites.

1.2.1 Hydrogen Evolution Reaction (HER)

1.2.1.1 Mechanism

The well-established mechanism for HER (heavily based on Pt electrode) can be generally described as the electrochemical hydrogen adsorption followed by hydrogen desorption reactions and/or chemical desorption [7].

Under acidic conditions (M represents the HER active sites on electrocatalysts):

$$\text{Volmer step} \quad H^+ + M + e^- \rightarrow M - H_{ads} \tag{1.6}$$

$$\text{Heyrovsky step} \quad M - H_{ads} + H^+ + e^- \rightarrow M + H_2 \tag{1.7}$$

$$\text{Tafel step} \quad 2M - H_{ads} \rightarrow 2M + H_2 \tag{1.8}$$

In acidic media, the proton first gains an electron at the active site to form $M - H_{ads}$ intermediate (reaction (1.6)). In alkaline media, instead of proton adsorption, the water molecule is involved in the elemental reactions (reactions (1.9) and (1.10)).

Under alkaline conditions:

$$\text{Volmer step} \quad H_2O + M + e^- \rightarrow M - H_{ads} + OH^- \tag{1.9}$$

$$\text{Heyrovsky step} \quad M - H_{ads} + H_2O + e^- \rightarrow M + H_2 + OH^- \tag{1.10}$$

$$\text{Tafel step} \quad 2M - H_{ads} \rightarrow 2M + H_2 \tag{1.11}$$

Most HER catalytic systems adopt either the Volmer–Heyrovsky or Volmer–Tafel pathway for mechanism understanding. An approximate way to determine the mechanism is to use the Tafel slope, derived from the HER polarization curve. If the Volmer step is the rate-determining step (RDS), the slope is 120 mV dec^{-1}, and when Heyrovsky and Tafel steps are the RDS, the Tafel slopes of 40 and 30 mV dec^{-1} result, respectively.

Both the bonding strength of the $M - H_{ads}$ intermediate and the free energy of hydrogen adsorption (ΔG_H) on the cathode can be used to describe the interaction between the hydrogen atom and active sites. When plotting the two descriptors with the experimentally measured exchange current densities from polycrystalline metal electrodes, a typical volcano relationship emerges (Figure 1.4), demonstrating the so-called Sabatier principle. Neither too strong nor too weak bonding benefits the HER. Only a moderate value of ΔG_H helps the hydrogen gas discharge following the Heyrovsky step (reaction (1.7)) and/or Tafel step (reaction (1.8)).

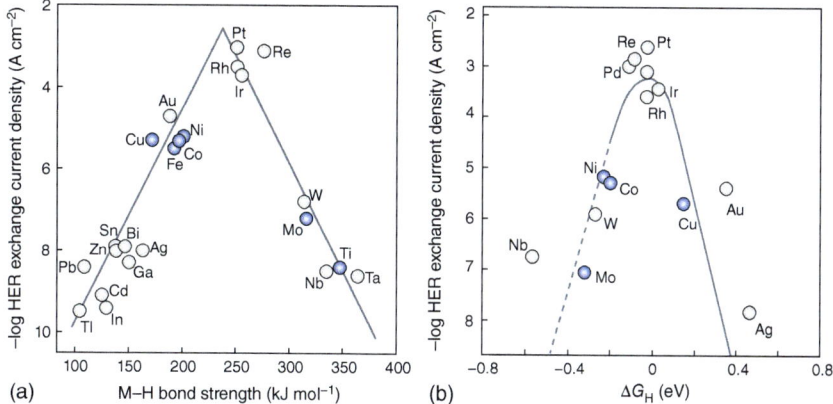

Figure 1.4 HER volcano plots with measured exchange current density of polycrystalline metal vs. (a) the energy of the intermediate metal–hydrogen bond formation. Source: Trasatti [8]. Reproduced with permission of Elsevier. (b) The calculated free energy of H adsorption. Source: Skúlason et al. [9]. Reproduced with permission of American Chemical Society. The earth-abundant metallic elements are marked in blue.

1.2.1.2 Metal (M⁰) Nanoparticles

The metallic form of earth-abundant metals usually shows limited HER activity and stability in both acidic and alkaline media [10, 11]. Major efforts have been devoted to increasing metal sites' activity and stability. Decreasing the size of metals to the nanoscale is a proven strategy to alter the electronic properties of metal sites (Figure 1.5) [13, 14]. The high surface energy of metallic nanoparticles, however, renders them instable. The most straightforward solution is to stabilize the nanoparticles using supports, which can be structurally engineered to affect the electronic properties of the metal nanoparticles to benefit HER activity.

One typical example is Co nanoparticles. A recent report shows that Co nanoparticles encapsulated in nitrogen-enriched carbon material can deliver a reasonable overpotential (at a current density of $10\,\text{mA}\,\text{cm}^{-2}$, or η_{10}) of 265 mV in acid and 337 mV in base [15]. It was claimed that the synergistic effects between Co nanoparticles and N-doped carbon can significantly enhance the activity of Co sites.

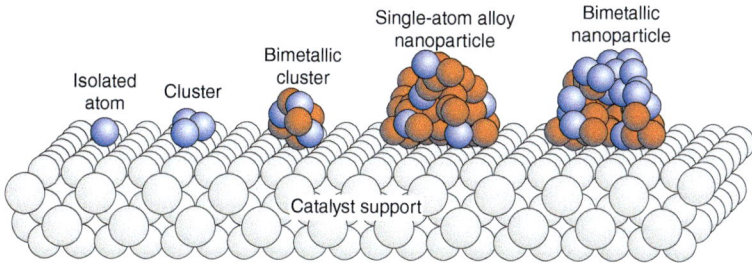

Figure 1.5 Illustration of various forms of metallic electrocatalysts, from isolated metal atoms to bimetallic nanoparticles supported on conventional solid carriers. Source: Liu and Corma [12]. Reproduced with permission of Springer Nature.

Similarly, by decorating TiO$_2$ nanoparticles on the surface of Co nanostructure, the hydrogen adsorption free energy at the material junction can be optimized to achieve a η_{10} of 229 mV in 1.0 M KOH, which is highly enhanced compared with 356 mV using unmodified Co nanomaterial [16]. Similar strategies are frequently applied to other earth-abundant metal nanoparticles. The commonly used supports include the following: *for Co nanoparticles*, carbon-nanotube (CNT)-grafted graphene sheet [17]; *for Ni nanoparticles*, nitrogen-doped graphitized carbon driven by Ni(II)–dimeric complex [18], nitrogen-doped CNT [19, 20], carbon fiber cloth [21], hydrophilic graphene [22], and graphene [23]; and *for Fe nanoparticles*, carbon shell encapsulation [24] and nitrogen-doped carbon [25]. Notably, the formation of metal–N local environment has been proven an efficient method to promote the HER activity of metal nanoparticles. Meanwhile, the employment of support is essential for small metal clusters, especially for the stability during reactions.

Another commonly employed approach is to change the shape of the metal nanoparticle, which exposes highly active sites or desired surface features, such as corners and edges, as demonstrated in the HER study of Cu nanoparticles with different shape [26]. However, the tunability of the single element metal nanoparticle is limited by the intrinsic properties of the element.

Involving two or more earth-abundant metals for the construction of HER electrocatalyst is more popular as more flexibility is possible. The introduction of another/multiple metal(s) into a hosting metal lattice can modify the original electronic structure via strain effect and the bonding strength of adsorbed hydrogen atoms. By combining metals on opposite slopes of the volcano-type plots (Figure 1.4), new materials systems are proposed to produce suitable intermediate hydrogen binding energy and improved HER activity. For instance, although both Cu and Ti are poor HER electrocatalysts, their combination allows the creation of Cu–Cu–Ti hollow sites with hydrogen binding energy close to that of Pt [27]. The various combinations of metals, tunable ratios, and multiple morphologies (core–shell, alloy, etc.) allow nearly infinite possibility to be explored. Some of the most exciting systems include Cu–Ti [27], Cu–Co [28], Cu–Ni [29], Ni–Mo [30], Fe–Co [31], and Ni–Co–Fe [32]. Depending on the mechanism, one element may serve as the primary active sites for HER and the heteroatoms can aid either proton adsorption or electron transfer [30]. In other cases, the active sites are identified as the hollow sites surrounded by different atoms [27].

1.2.1.3 Metal (M⁰) Single-Atom Catalysts

Decreasing the size of metal nanoparticle to the extreme leads to the development of single-atom electrocatalysts (Figure 1.5). Since its first appearance in 2011 [33], the concept of using single-atom catalysts (SACs) has been drawing wide attention due to the maximum atomic utilization close to 100%. The initial intention for SACs as HER catalysts is to minimize the Pt loading without sacrificing HER activity [34]. However, when the single atoms are isolated and supported on different materials, their properties change significantly from their bulk nanoparticles [12]. In addition to the noble metals, several earth-abundant metals, including Co [35], Fe [36], Ni [36, 37], Mo [38], and W [39], have shown unexpected HER activities (selected

Table 1.1 Selected earth-abundant metal single-atom electrocatalyst for HER.

Entry	Metal	Coordination environment	Overpotential at 10 mA cm^{-2} (mV)	Tafel slope (mV dec^{-1})	Electrolyte	References
1	Co	N-doped graphene	147	82	0.5 M H$_2$SO$_4$	[35]
2	Co	Phosphorized carbon nitride	89	38	1 M KOH	[40]
3	Fe	Graphdiyne	66	37.8	0.5 M H$_2$SO$_4$	[36]
4	Fe	N-doped carbon	111	86.1	1 M KOH	[41]
5	Ni	Graphdiyne	88	45.8	0.5 M H$_2$SO$_4$	[36]
6	Ni	Graphene	180	45	0.5 M H$_2$SO$_4$	[37]
7	Mo	N-doped carbon	132	90	0.1 KOH	[42]
8	W	N-doped carbon	85	53	0.1 KOH	[39]
9	W, Mo	N-doped graphene	24	30	0.5 M H$_2$SO$_4$	[43]
10	W, Mo	N-doped graphene	67	45	1 M KOH	[43]

results shown in Table 1.1). Moreover, a series of literature reviews on the topic of SACs were published, covering from their preparation methods to applications [44–47].

The synthesis of SACs with robust structures and reasonable performance is the major challenge since the highly active single atoms tend to aggregate during the catalytic reaction and result in a lower activity. With the advance of experimental technics, current strategies include wet chemistry method, atomic layer deposition (ALD), template-driven decomposition (mainly metal–organic frameworks), and electrochemical deposition [47, 48]. Typically, such methods produce nanocomposites with isolated metal atoms decorated on a support to prevent migration and aggregation of single atoms. Therefore, the nature of the support and the metal coordination sites can impact the activity of SACs for HER significantly. As revealed by Gao et al. using *ab initio* calculation, the hydrogen bonding free energy on the same metal species at different coordination sites of graphene varies largely to show versatile HER activities [49] (Figure 1.6).

Carbon supports, including graphene and CNTs, are the most popular choice for SACs. Fan et al. demonstrated a Ni–C electrocatalyst with carbon nanoparticles dotted with isolated nickel atoms. Under acidic conditions, the Ni–C electrocatalyst exhibited a performance comparable with the commercial Pt/C electrocatalyst even when operating at a current density of 100 mA cm^{-2}. Remarkably, no significant activity loss was observed over 25 hours of continuous operation [50]. Density function theory (DFT) calculations study by He et al. suggested that Fe, Co, Ni, and V supported on carbon are all promising SACs for HER. Especially, V atom on carbon shows a Gibbs free energy of −0.01 eV of hydrogen adsorption, which is close to that of Pt [51].

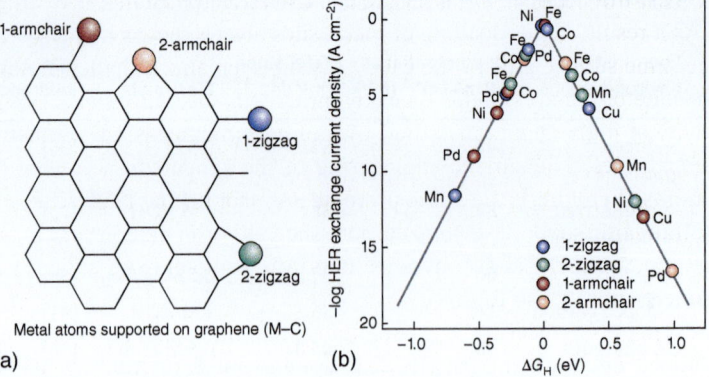

Figure 1.6 (a) Model of metal atoms supported on graphene via metal–carbon coordination. (b) Calculated volcano curve of exchange current density as the function of Gibbs free energy of hydrogen adsorption on coordinated metal sites. Source: Gao et al. [49]. Reproduced with permission of Royal Society of Chemistry.

Functionalization/doping of the carbon supports with other non-metallic atoms allows fine-tuning of the metal sites' properties. One of the most commonly reported cases is nitrogen-doped carbon materials. The N atoms can form strong bondings with the metal sites for better stability and assist electron transfer to the d orbitals of the metal sites [39, 42, 52]. Cao et al. reported a SAC system where Co atoms were anchored by N atoms in graphitic carbon nitride to form Co–N_4 sites (Co atoms are covalently grafted by four N atoms) [53]. The coordinated nitrogen sites can donate electrons to the Co sites to lower the formation barrier of Co–H. Later, by using *operando* spectroscopy, they revealed the formation of a high-valence HO–Co–N environment from Co–N_4 sites in the alkaline reaction conditions, which enabled the water adsorption to assist water splitting (Eq. (1.9)) [40]. With a low Tafel slope of 52 mV dec^{-1}, a Volmer–Heyrovsky mechanism was proposed (Figure 1.7).

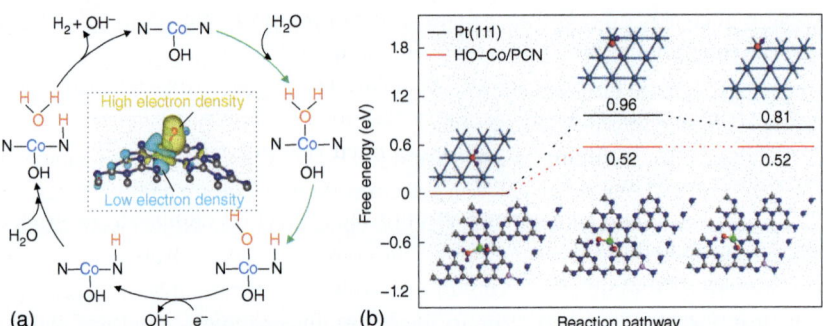

Figure 1.7 (a) Alkaline HER mechanism on Co–N SAC. Inset shows the electron density at different regions of the HO–Co–N environment. (b) Calculated adsorption energies of H_2O and H on the surface of HO–Co–N and Pt(111). Source: Cao et al. [40]. Reproduced with permission of Springer Nature.

For most SACs, the metal active sites are exclusively responsible for hydrogen production. As a result, higher loading of metal sites always means higher overall activity at the same surface area of the catalyst, which remains a challenge nowadays. Current state-of-the-art systems typically have metal loading <3 wt%. Despite the high activity of individual atoms, their overall performance is not compatible with commercial ones. The surface engineering of the support to accommodate more metal atoms, therefore, is the key to promote SACs for further large-scale HER applications. Inorganic supports beyond carbon, such as boron monolayer [54] and boron nitride [55], are under active investigations. More suitable SAC supports are expected to emerge in the near future.

Other than support engineering, engaging more than one metal atoms can offer a more optimized environment for HER, as demonstrated by Yang et al. who demonstrated that W and Mo dual-atom catalyst is a better HER catalyst than W-/Mo-SACs [43].

1.2.1.4 Metal Phosphides

Transition metal phosphides (TMPs, e.g. CoP, Ni_5P_4, MoP, Cu_3P, and FeP) have attracted increasing attention for HER owing to their good hydrogen adsorption property. As the P atom possesses a high electronegativity, the transition metal sites (TM) are usually positively charged in TMPs, and the TM—P bonds can serve as a proton acceptor to promote HER activity [56]. Based on previous reports, the increased P content in TMPs can improve the HER activity. For examples, nickel phosphide and molybdenum phosphide exhibit HER activity trend of $Ni_5P_4 > Ni_2P > Ni_{12}P_5$ [57] and $MoP > Mo_3P > Mo$ [58], respectively. In contrast, the excessive P content may reduce the performance owing to the reduced electronic conductivity. Therefore, the P content in TMPs should be adjusted to optimize the activity. To further improve the HER activity, metal doping, non-metal doping, hybridization, and coupling with carbon materials have been adopted.

Up to date, various metallic elements, such as Fe, Co, Zn, Mn, and Mo, have been doped into TMPs to improve HER activity (Figure 1.8) [59, 60]. Guan et al. synthesized hollow Mo-doped CoP nanoarrays on carbon cloth by low-temperature annealing with NaH_2PO_2 [61]. The Mo-doped CoP showed a superior HER activity with a low overpotential of 40 mV in 1.0 M KOH, and the P sites were considered as the HER active sites. According to the DFT calculations, the hydrogen adsorption energy on the P sites was only 0.07 eV, which is close to zero and thus enhanced the HER activity. Similarly, Mn-doped Ni_2P also exhibited an excellent HER activity with an overpotential of 84 and 122 mV in 0.5 M H_2SO_4 and 1 M KOH, respectively [62]. The Mn dopants were claimed to provide electrons to adjacent Ni atoms and weaken the Ni—H_{ad} bonds to improve the HER activity. Besides, Fe and Co were doped to Ni_2P, and Zn was doped to CoP for HER enhancement [63, 64].

The N and S elements were widely reported as the non-metal dopants to improve the HER activity of TMPs. Zhang's group reported N-doped CoP as an excellent HER catalyst [65]. Because of the higher electronegativity of N than P, the positively charged Co could enhance the Co–Co interaction and lower the d-band. Therefore, the Co—H_{ad} bonds were weakened with an optimized hydrogen adsorption energy

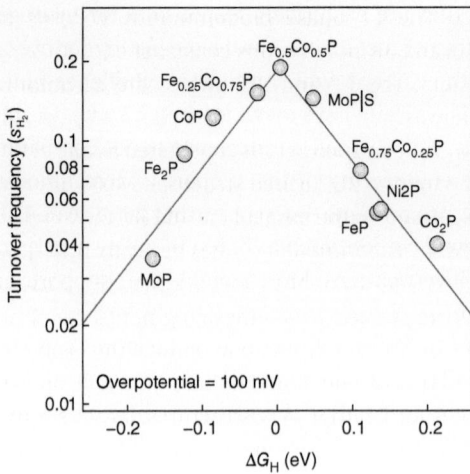

Figure 1.8 HER volcano plot of TMPs showing the average H_2 turnover frequency at an overpotential of 100 mV as a function of ΔG_H. Source: Kibsgaard et al. [59]. Reproduced with permission of Royal Society of Chemistry.

of −0.14 eV, which is higher than the undoped CoP (−0.52 eV) and closer to zero. Recently, S-doped Ni_5P_4 nanoplate arrays and S-doped MoP nanoporous layers were also reported as efficient HER catalysts [66, 67].

Hybridizing the TMPs with other compositions can construct an active interface and induce a synergistic effect, beneficial for HER enhancement. The TMPs hybridized with metal oxides [68, 69], hydroxides [70], sulfides [71], or even another TMP [72] were reported. Luo et al. prepared Mo-doped Ni_3S_2/Ni_xP_y hollow nanorods that delivered 10 mA cm^{-2} at an overpotential of 109 mV for HER in 1.0 M KOH [73]. The heterostructure not only enhanced the adsorption of the water molecule but also decreased the value of hydrogen adsorption energy to improve HER activity.

Coupling with carbon materials is another viable strategy to improve HER activity by enhancing the electronic conductivity and number of active sites. With this respect, carbon materials have gained large attention due to their high conductivity, structural stability, and corrosion resistance. The large surface area of carbon materials can also restrain the aggregation of nanosized TMPs. For examples, MoP nanoparticles decorated on CNTs or anchored on reduced graphene oxide (RGO) exhibited suitable HER activities in both acidic and basic media [74, 75]. Other TMPs were also reported to be modified on carbon materials, such as CoP/N,P co-doped carbon frameworks [76] and CoP/CNT [77], for enhanced HER activities. TMPs were also encapsulated by carbon to form core–shell structures, for example, Co_xP@NC [78] and Ni_2P@graphene [79]. Unlike the usual TMPs exposed to substrates, the encapsulated TMPs cannot provide active sites for HER, and it is generally considered that HER takes place on the exposed carbon sites or N dopants in the carbon shell.

1.2.1.5 Metal Chalcogenides

Transition metal chalcogenides (TMCs) can be represented with a formula MX_2, where M = Mo, W, or V and X = S or Se. It is well accepted that the two-dimensional (2D) layered TMCs exhibit three types of structures: 1T (single X–M–X layer),

2H, and 3R (several X–M–X layers). The 1T phase predominantly consists of edge-sharing MX_6 octahedra, while 2H and 3R polymorphs consist of edge-sharing MX_6 trigonal prisms [80]. Among them, the 1T and 2H phases are commonly employed in water splitting electrocatalysis.

MoS_2, first investigated by Nørskov's group in 2005 [81], is considered as one of the promising HER catalysts. Owing to the unique 2D layered structure, MoS_2 has two distinct types of surfaces: (i) the S sites located on the basal plane and (ii) the exposed Mo and S sites on the edge plane [82]. Many experimental studies have attributed the HER activity of MoS_2 to its edge sites [83, 84]. To further improve the HER activity of MoS_2, abundant edge active sites were created by synthesizing nanostructures. Cui's group reported MoS_2 with vertically aligned layers to maximize the exposure of active edge sites and found that the HER activity could be modulated by adjusting the density of the exposed edge sites [85]. Other nanostructures, such as MoS_2 nanoparticles, core–shell MoO_3–MoS_2 nanowires, and molecular clusters, have also been reported [86, 87].

It was reported that the phase of MoS_2 can affect HER activity: 1T MoS_2 exhibits a higher catalytic activity than 2H MoS_2. The 2H phase is generally considered as an inactive toward HER catalysis because of semiconductive property that inhibits the charge transfer. On the other hand, the 1T phase has a good hydrogen adsorption property and conductivity to improve the charge transfer kinetics [88]. Voiry et al. prepared metallic 1T phase from 2H MoS_2 by a solvent-free intercalation method, which showed an enhanced HER activity [89]. Meanwhile, when a small amount of CNTs was modified on the surface of 2H MoS_2 phase, the catalytic activity was also improved, confirming that the importance of conductivity for HER. Aiming to enhance the conductivity, the integration of MoS_2 and carbon materials was recently reported for HER enhancement, for example, MoS_2/Co–N-doped carbon nanocages [90] and CNTs/MoS_2 nanoflake [91].

The first-row transition metals (Ni, Fe, Co) are also useful for TMCs to form highly active HER catalysts. The crystalline structures of MX_2 (Ni, Fe, Co) are slightly different from MoS_2, which consists of corner-sharing or edge-sharing MX_6 octahedra that forms pyrite or marcasite structures, respectively [92]. Among the nickel chalcogenides, NiS, NiS_2, and Ni_3S_2 have been reported for highly active HER activities. Jiang et al. found that Ni_3S_2 exhibited better HER activity than NiS and NiS_2, owing to the larger active surface area and higher intrinsic conductivity [93]. However, the bonds of adsorbed hydrogen intermediates on Ni_3S_2 are still too strong, making it challenging to generate H_2. To optimize the hydrogen adsorption property, interface engineering has been widely adopted by hybridizing Ni_3S_2 with other compositions. Cu nanodots were decorated on Ni_3S_2 nanotubes by Feng et al. to work as electrons donors [94]. The positively charged Cu in Cu/Ni_3S_2 optimized the hydrogen adsorption energy and thus facilitated the water dissociation. Based on the S–H_{ad} peak observed from the *in situ* Raman spectra, the authors claimed that S sites were responsible for HER activity. An Ni–Ni_3S_2 hybrid structure was also reported to improve the HER activity [95]. According to the first-principles calculations, the metallic Ni is believed to work as the active material for HER, and the Ni/Ni_3S_2 interface facilitates the water adsorption and dissociation. Ni selenides are another type of nickel chalcogenides and have been also reported for HER activity recently. Anantharaj et al. compared different structures of Ni selenides by controlling the ratio of Ni to Se, whose HER activity followed the order of $NiSe_2 > Ni_3Se_4 > Ni_{0.85}Se$ [96]. The HER activity was improved by increasing the content of Se, indicating that Se sites might be responsible for the proton adsorption [97]. Wang et al. synthesized Se-enriched $NiSe_2$ and confirmed that the hydrogen adsorption energy on Se sites is much lower than that of Ni sites, which endowed an excellent HER activity with an ultralow Tafel slope of 32 mV dec^{-1} [98].

Cobalt chalcogenides have attracted attention as efficient HER catalysts of low overpotential and small Tafel slope. The HER activity of cobalt chalcogenides (CoS_2 and $CoSe_2$) can be further improved by doping heteroatoms. Ternary pyrite-type CoPS nanostructures (film, nanowires, and nanoplates) were synthesized by Jin's group and showed excellent HER activities [99]. Compared with CoS_2, the existence of adjacent P—H_{ad} bonds in CoPS reduced the oxidation state of Co^{3+} to Co^{2+}. The Co^{2+} sites were believed to optimize hydrogen adsorption sites and would be oxidized back to Co^{3+} after hydrogen adsorption. Dutta et al. also synthesized CoSSe microspheres to improve the HER activity of CoS_2 by optimizing the hydrogen adsorption energy of Co sites [100]. Ni atoms were doped to $CoSe_2$ to generate $Ni_{0.33}Co_{0.67}Se_2$ nanostructure by Xia et al. [101] The $Ni_{0.33}Co_{0.67}Se_2$ exhibited an excellent HER activity with an overpotential of 65 mV in 0.5 M H_2SO_4, which was attributed to the improved conductivity and increased active surface area.

There are only a few reports on the HER activity of Fe chalcogenides, owing to the unfavorable intrinsic activity of Fe. Recently, Jasion et al. tried to improve the HER activity of FeS_2 by controlling the ratio of Fe to S in its 0, 1, and 2D nanostructures (cubes, wires, and disks, respectively) [102]. Among them, the 2D FeS_2 exhibited excellent HER activity and stability. Inspired by this work, Miao et al. synthesized mesoporous FeS_2 nanoparticles containing exposed (210) facets [103], which demonstrated excellent HER activity in alkaline media. Compared with the (100) surface, a higher water adsorption energy and a lower activation barrier energy were obtained on the (210) surface, resulting in the promotion of HER activity. Similarly, $FeSe_2$ is also known to have an inactive HER activity. Even though coupling Co or Mo could improve the performance, the prepared $Fe_xCo_{1-x}Se_2$ still suffered from an insufficient catalytic activity, much lower than that of Pt. [104] The incorporation of Mo caused the phase separation to form 1T $MoSe_2/FeSe_2$ heterostructure, and the improved HER activity might be attributed to the 1T $MoSe_2$ phase near the interface [105].

1.2.1.6 Metal Nitrides

Transition metal nitrides have been also reported as earth-abundant catalysts for HER. A nitrogen-rich 2D Mo_5N_6 nanosheet was synthesized by Qiao's group and demonstrated superior HER activity [106]. The higher valence state of Mo (+4) in Mo_5N_6 than that of MoN (+3) facilitated the water dissociation in alkaline media and induced the downshift of the d-band center of Mo, which resulted in optimizing the hydrogen adsorption energy. Liu et al. introduced N vacancies into Ni_3N to further improve HER activity and achieved an ultralow overpotential of 55 mV in

1.0 M KOH media [107]. The N vacancies enriched in Ni_3N_{1-x} were also beneficial for hydrogen adsorption due to the downshift of d-band center. Similarly, Chen et al. doped V atoms into Co_4N to tailor the d-band center of V-Co_4N, which also exhibited an excellent HER activity closed to the Pt/C catalyst [108]. Besides, constructing an active interface in nitride heterostructures was also reported to improve the HER activity, for instances, Ni/Co_2N [109] and Mo_2N/CeO_2 [110].

1.2.1.7 Metal Carbides

The HER catalysts based on transition metal carbides are mainly focus on molybdenum carbides, including α-MoC_{1-x}, β-Mo_2C, and η-MoC [111]. Recent works showed that the catalytic performance of molybdenum carbides can be further improved by doping heteroatoms or constructing an interface. Ma et al. claimed that the d-band center and the strength of Mo—H bonds of β-Mo_2C spheres could be modulated by doping Co atoms [112]. Such modulated Co-Mo_2C exhibited a low onset potential of 27 mV. Huang et al. compared the transition metal (Fe, Co, Ni, and Mn)-modified Mo_2C and found that the Fe-Mo_2C exhibited the best HER activity with an overpotential of 65 mV in 1.0 M KOH media, owing to the optimized hydrogen adsorption energy [113]. The α-MoC_{1-x} decorated with ultrafine Pt nanoparticles was reported to exhibit a higher intrinsic HER performance than commercial Pt/C [114].

1.2.1.8 Metal Oxides/(Oxy)hydroxides

The pristine transition metal oxides (NiO, Co_3O_4, TiO_2, MoO_2, MnO_2, Fe_2O_3, V_2O_5, etc.) are normally considered to be inactive HER catalysts owing to their semiconductive property and unfavorable hydrogen adsorption and desorption processes [115]. To improve their HER activities, some strategies have been widely discussed, such as morphology engineering, oxygen vacancies (OVs), and heteroatoms doping.

Morphology engineering is an effective way to create active sites on the surface of catalysts. The common strategy is to synthesize nanoscale metal oxides with abundant active sites exposed like growing nanostructure arrays on conductive substrates and introduce porous or hollow structures. Zhang's group synthesized ultrathin δ-MoO_2 nanosheets on Ni foam by hydrothermal method [116]. The MoO_2 nanosheets were only two monolayer thick and exhibited an excellent HER activity in alkaline media due to abundant OVs and active sites. Other similar works were reported, such as porous MoO_2 nanosheets [117], porous WO_2 hexahedral networks [118], and $NiCo_2O_4$ hollow microcuboids [119].

In addition to synthesizing nanostructures, improving the intrinsic HER activity of metal oxides has been also well investigated. It is widely reported that creating OVs and heteroatom doping are promising strategies to enhance the unfavorable HER activity of pure metal oxides [120]. According to the first-principles calculations, the electrons near the defect are easier to be excited, thus improving the conductivity of the materials and optimizing the hydrogen adsorption energy [121]. For example, VO-enriched MoO_x and TiO_2 were applied as HER catalysts in an acidic solution [122, 123], while VO-enriched NiO and CoO can usually catalyze HER in alkaline media [124, 125]. It is worth noting that excessive OVs might reduce catalytic activity due to structural instability and decreased electronic conductivity [126]. Therefore, it is important to adjust the OV concentration in catalysts to obtain the highest conductivity and HER activity.

The heteroatom doping can be divided into non-metal doping and metal doping. Non-metal doping is considered as an effective way to induce OVs in metal oxides. For instance, Zhang et al. reported that N-doped NiO could modulate the surface charge redistribution and induce OVs to facilitate the water dissociation, adsorption of H_{ad}, and desorption of OH_{ad} [127]. Other dopants were also reported, including P dopants in MoO_{3-x} nanosheets [128] and Co_3O_4 [129] and S dopants in CoO_x [130] and MoO_2 nanosheets [131]. Metal doping is another way to tune the electronic structure of metal oxides to optimize the hydrogen adsorption energy, which has also been widely engaged in binary metal oxides (Ni-doped Co_3O_4 nanosheets [132], $NiFe_2O_4$ [133], $ZnCo_2O_4$ [134], and $CuCo_2O_4$ [135]) and ternary metal oxides (Ce-$MnCo_2O_4$ [136], Mo-doped NiFe oxide nanowires [137], and Ni/Zn co-doped CoO nanorods [138]).

Metal (oxy)hydroxides are usually poor HER catalysts because of the unfavorable hydrogen adsorption energies. Considering that the metal (oxy)hydroxides are normally the active materials for OER and exhibit excellent water adsorption and dissociation properties, the metal (oxy)hydroxides can be coupled with other HER active materials to achieve excellent HER activity. Zhang et al. synthesized a three-dimensional (3D) hierarchical heterostructured NiFe layered double hydroxide (LDH) on NiCoP, which required a low overpotential of 120 mV for HER [139]. The interface between LDH and NiCoP increased the electrochemical surface area (ECSA) and optimized the reaction kinetics. Chen et al. prepared Ru-doped NiFe–LDH, which exhibited an ultralow overpotential of 29 mV in 1.0 M KOH [140]. The Ru dopants were beneficial in lowering the kinetic energy barrier of the Volmer step. Recently, other structures, such as NiCoP@NiMn LDH [141], NiCoP nanowiere@NiCo LDH [142], CoSe/NiFe LDH [143], and CuO@CoFe LDH [144], were also reported to show good HER activities.

1.2.2 Oxygen Evolution Reaction

1.2.2.1 Mechanism

Compared with HER that involves two electrons, OER, a four-electron transfer reaction, is kinetically sluggish, making it the rate-limiting process of overall water splitting. Despite its importance, the mechanism of OER is not as well-studied as HER, mainly due to its complexity in different catalytic systems, as well as various reaction conditions.

In recent decades, several reaction mechanisms have been developed based on the results from different materials. Two of them are generally accepted: adsorbate evolution mechanism (AEM) and lattice oxygen mechanism (LOM), which are shown in Figure 1.9 [146]. It should be noted that the mechanism depends heavily on the reaction conditions and surface features of electrocatalysts, thus not fixed.

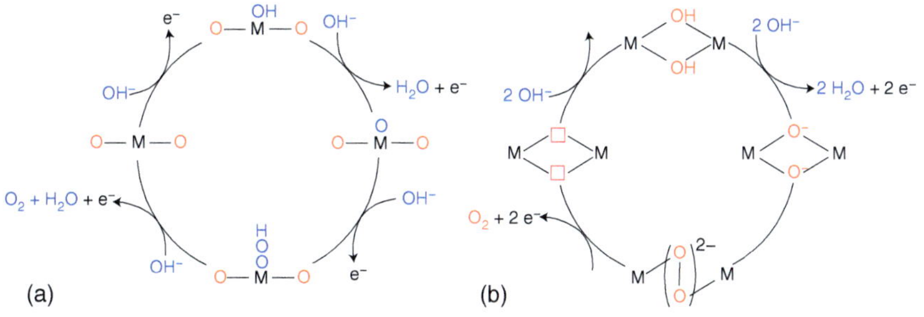

Figure 1.9 Mechanism of OER: (a) adsorbate evolution mechanism and (b) lattice oxygen mechanism. Source: Grimaud et al. [145]. Reproduced with permission of Springer Nature.

The AEM involves four concerted proton–electron transfer (CPET) processes occurring on the metal surface as follows:

$$M + OH^- \rightarrow M - OH + e^- \tag{1.12}$$

$$M - OH \rightarrow M - O + H^+ + e^- \tag{1.13}$$

$$M - O + OH^- \rightarrow M - OOH + e^- \tag{1.14}$$

$$M - OOH \rightarrow M + H^+ + O_2 + e^- \tag{1.15}$$

After the adsorption of hydroxide anion (OH^-) on the active site M (Eq. (1.12)), the adsorbed OH undergoes subsequent deprotonation to form O species (Eq. (1.13)), which can react with another OH^- to form M–OOH intermediate (Eq. (1.14)). Such intermediate is not stable and eventually releases O_2 and regenerates the active sites (Eq. (1.15)). The overall OER activity, therefore, is determined by all four steps. The ideal situation where all reactions occur at exactly 1.229 V (pH = 0) or 0.401 (pH = 14), the thermodynamic potential of OER at standard conditions, is almost impossible because the adsorption energy for the intermediates, from OH to O to OOH, is linearly correlated (scaling relation) (Figure 1.10a) [147]. For the binding energy of OOH and OH on catalytic sites, regardless of metals or metal oxide surfaces, both species involve a M—O single bond structure with a constant difference of Gibbs free energy ($\Delta G_{OOH} - \Delta G_{OH}$) of 3.2 ± 0.2 eV. For most catalytic systems, the reactions (1.13) and (1.14) are the RDS. Therefore, the free energy difference between O and OH bindings ($\Delta G_O - \Delta G_{OH}$) can be used as a universal descriptor to interpret and predict the OER activity of various materials [148].

Similar to the volcano-shaped relationship for HER (Figure 1.4), the $\Delta G_O - \Delta G_{OH}$ value also exhibits a volcano trend for various materials (Figure 1.10b). Regardless of the type of catalysts, OER only occurs when the species have neither too strong nor too weak adsorption strength. Thus, tuning the electrocatalyst surface for suitable adsorption energy to minimize the potential required for reactions (1.13) and (1.14) is the main target for rational OER catalyst design.

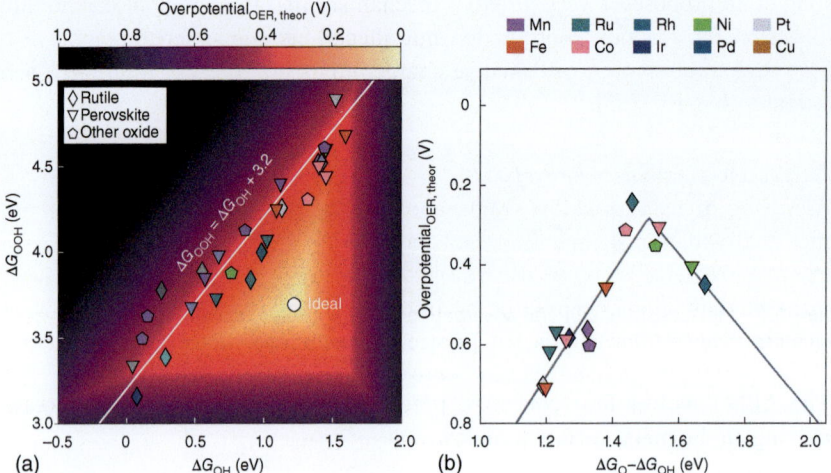

Figure 1.10 (a) Linear scaling relation of **M**–OOH and **M**–OH Gibbs free energy on a heat map of OER overpotential. Source: Man et al. [147]. Reproduced with permission of Springer Nature. (b) OER volcano plot for rutile, perovskite, and other metal oxides. Source: Montoya et al. [148]. Reproduced with permission of Wiley.

Another often referred OER mechanism is LOM. Different from AEM that only one metal site is considered, LOM engages two neighboring metal sites (Figure 1.9b). The two OH species on the metal sites firstly release proton to form **M**–O–**M** species, followed by direct coupling to establish O—O bonding other than **M**–OOH formation in AEM. The discharge of O_2 leaves two vacancies for two OH^- anions to adsorb. Since the LOM mechanism does not produce **M**–OOH species, the scaling relation shown in Figure 1.10a does not exist.

Compared with AEM, the role of lattice oxygen in LOM still lacks a full understanding. In 2017, Grimaud et al. showed direct experimental evidence that the O_2 generated during OER on some highly active sites of perovskites were actually from the material lattice [149]. Further evidence indicated that the switching of reaction pathway from AEM to LOM was dependent on the metal–oxygen covalency. Moreover, some conventional materials, such as $Co_3(PO_4)_2$ [150] and IrO_2 [151], which was believed to follow AEM mechanism, have been proven to adopt LOM mechanism under some specific conditions.

The understanding of OER mechanism benefits and guides the rational design of OER catalysts. Based on AEM, reducing the $\Delta G_O - \Delta G_{OH}$ value is the key for high performance, and methods such as surface doping, vacancy creation, lattice strain engineering, and interfacial engineering have been widely adopted [146]. However, limited by the scaling relation of AEM, a minimal theoretical overpotential of 0.37 eV was predicted [152] while such limitation does not apply to LOM, allowing more freedom for materials discovery, such as bimetallic sites modulation to tune the metal–oxygen covalency.

Still, it is of great importance to point out that OER mechanism is currently under debate. Even for the most well-studied systems, such as IrO_2, a recent study by

Nong et al. proposed a very different mechanism that the potential has no direct impact on the reaction coordination but affects the charge accumulation in the catalyst. As the amount of oxidative charge builds up, the activation free energy decreases linearly [153].

1.2.2.2 Metal Oxides/Hydroxides

Discussing or using metallic (M^0) nanoparticles for OER is not so meaningful since the oxidation of metal species generally occurs at a potential lower than OER requirement. Metal oxides and hydroxides are the most common forms of earth-abundant elements as OER electrocatalysts, and thousands of OER catalysts have been reported and proposed [154]. Burke et al. investigated the activity trend of a series of ultrathin metal oxides/hydroxides coated on Au electrode (Figure 1.11) and showed that bimetallic $NiFeO_xH_y$ is the most promising system among the studied with an overpotential of 336 mV at 10 mA cm^{-2} and a Tafel slope of 30 mV dec^{-1} [155].

In the following contents, current systems showing significant potential for large-scale application and insights for fundamental understanding, such as doped Ni-/Co-based oxides/hydroxides, are introduced briefly.

Ni-Based Oxides/Hydroxides Ni-based oxides and hydroxides are the most popular catalysts so far, and they are the current commercial alkaline water electrolysis catalysts. NiO_x and $Ni(OH)_2$ are as active as noble metal-based ones (e.g. IrO_2) for OER: electrochemical deposited NiO_x shows an overpotential of 420 mV for 10 mA cm^{-2} in 1.0 M NaOH electrolyte (Figure 1.12) [157]. Reducing the size of catalysts can significantly improve the catalytic performance due to the enlarged surface area and enhanced conductivity: nano-$Ni(OH)_2$ (2.3 nm) and NiO_x (6 nm) can deliver

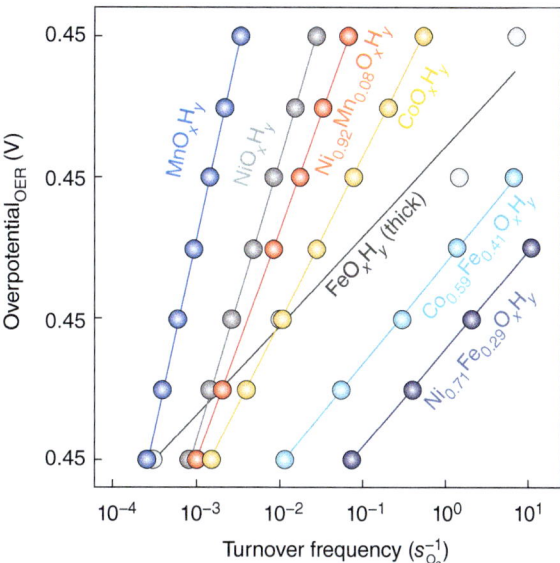

Figure 1.11 Experimental OER activity vs. turnover frequencies (TOF, O_2 generated per second per metal cation) of electrodeposited (oxy)hydroxides on Au microbalance electrodes. Source: Burke et al. [155]. Reproduced with permission of American Chemical Society.

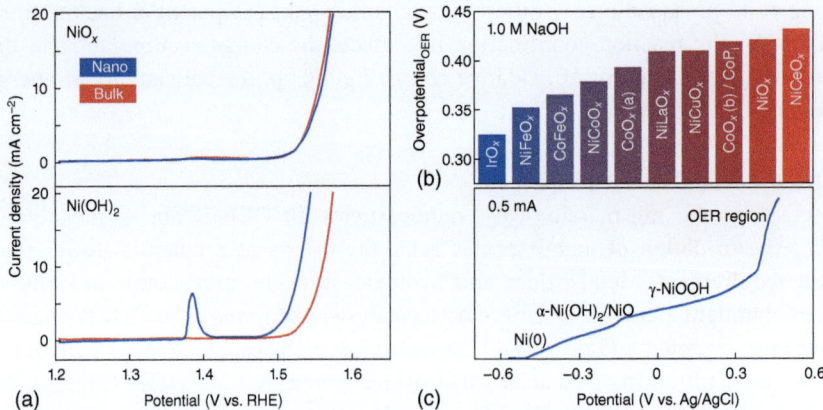

Figure 1.12 (a) Comparison of the OER activity of bulk NiO_x and $Ni(OH)_2$ with the nanostructured ones. Source: Stern and Hu [156]. Reproduced with permission of Royal Society of Chemistry. (b) Overview of the OER overpotentials (at $10\,mA\,cm^{-2}$) of a few metal oxides in 1.0 M NaOH. Source: McCrory et al. [157]. Reproduced with permission of American Chemical Society. (c) Chronopotentiometry measurement of Ni nanoparticle oxidation at a low current of 0.5 mA. Source: Huang et al. [158]. Reproduced with permission of Wiley.

much lower overpotential of 300 and 330 mV under the same conditions, respectively (Figure 1.12a) [156]. During the potential-driven polarization, it is generally accepted that the Ni(II) species first convert to Ni(III) species, in the form of NiOOH (or NiO(OH)) (Figure 1.12c) [158, 159]. By directly engaging NiOOH for OER, higher activities were demonstrated, especially in non-extreme pH ranges, such as the neutral electrolyte. A recent work by Li et al. showed that nanosized NiOOH owns a high concentration of Ni cationic vacancy, providing a large number of active sites for Ni–OOH intermediate formation [160].

Recently, it was noticed that the iron impurities, even in the level of ppm, in the electrolyte and Ni-based oxide/hydroxide can significantly impact OER activity (Figure 1.11) [161, 162]. Most of the earlier studies of Ni-based OER catalysts were likely impacted by such impurity since iron is the most common impurity during Ni refine. Experimental evidence suggested that Fe atoms can gradually dope into the phase of formed γ-NiOOH during polarization, promoting the phase transition to β-NiOOH. Such Fe doping can dramatically increase the electronic conductivity. It is also believed that Fe atoms exert a partial-charge-transfer activation effect on Ni atoms, resulting in high OER performance [163]. Recent DFT calculations showed that the formation energy of Ni–O can be reduced via Fe doping [164]. For Ni–Fe systems, an overpotential trend of Fe-doped β-NiOOH (0.26 V) < $NiFe_2O_4$ (0.42 V) < β-NiOOH (0.46 V) < Fe-doped γ-NiOOH (0.48 V) < γ-NiOOH (0.52 V) < Fe_3O_4 (0.70 V) was predicted by Li and Selloni [161]. Among other doped Ni-based oxides/hydroxides, such as Mo [165], W [166], and Co [167], the Fe–Ni system has shown by far the best efficiency. A guideline work for the surface doping on NiOOH (Figure 1.13) by Oscar Diaz-Morales et al. shows that, when Mn, Fe, Co, Cu, and Zn are engaged on the surface of NiOOH, the overpotential of Ni sites increases slightly while Cr causes decrement. However, the Fe and Mn sites themselves are much active for OER than the Ni sites [168].

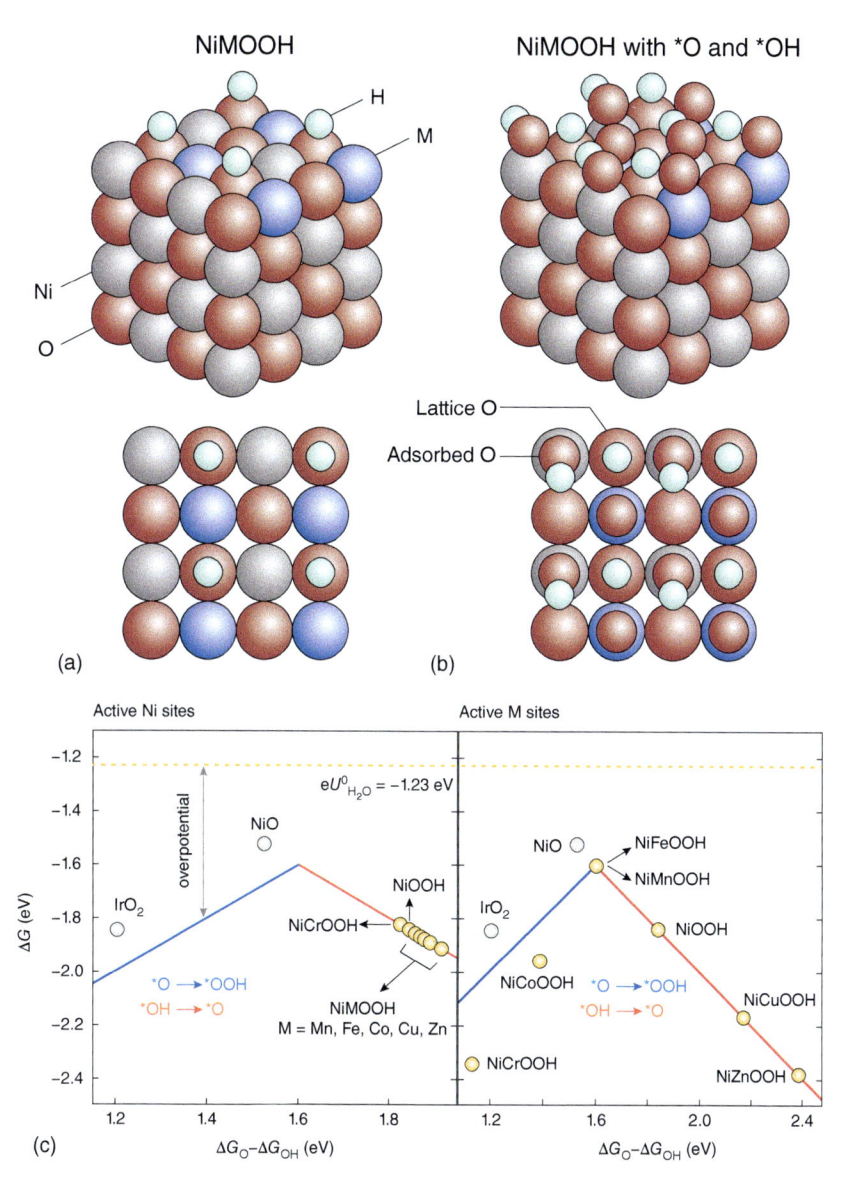

Figure 1.13 Perspective and top views of M-doped NiOOH (001) surface: (a) clear and (b) with *O and *OH adsorbed on M and Ni, respectively. (c) OER volcano plots of M-doped NiOOH with Ni sites (left) and M sites (right) as the active sites. Blue and red lines indicate the potential limiting steps: the blue part is limited by the transformation of *O to *OOH, and the red part is limited by the *OH to *O transition. M = Cr, Mn, Fe, Co, Ni, Cu, and Zn. Source: Diaz-Morales et al. [168]. Reproduced with permission of American Chemical Society.

Further surface engineering of Fe-doped Ni oxides/hydroxides provides extra tunability of OER activity. For example, Fe and V were co-doped to Ni oxide to modulate the local coordination sites of Fe/V/Ni cations. The V sites, surprisingly, showed near-ideal binding energy for OER intermediates and the lowest overpotential compared with nearby Ni and Fe sites [169]. Similar trimetallic systems, such as MoFeNi oxides/hydroxides [165], also showed superior OER performance than their mono/bimetallic siblings. Another extreme case is NiFeCoCeO$_x$ catalytic system where Ni, Fe, and Co are the active sites [170]. Nearly all doped systems attributed the high activity to the synergistic effect between metal cations, increasing the interaction strength of active sites with OH species. Other non-metallic doping examples are also proven effective [171].

Doping strategy focuses on tuning the electronic properties of metal sites. Another commonly applied method is altering the shape to expose more active sites. One of the most efficient ways to achieve high surface area is to increase the porosity of materials. For instance, Fe-doped β-Ni(OH)$_2$ porous structure showed a low overpotential of 218 mV in alkaline media [172]. Making the metal hydroxide into 2D forms to enlarge the number of facial atoms is also a popular approach. Luan et al. studied the morphologic effect of Ni(OH)$_2$ in OER using layer-stacked bud-like, flower-like, petallike, and ultralarge sheetlike Ni(OH)$_2$. The petallike Ni(OH)$_2$ showed the best performance with an overpotential of 260 mV and Tafel slope of 78.6 mV dec^{-1} among all shapes. Although the sheetlike Ni(OH)$_2$ owns a high diffusion rate, the petal-like Ni(OH)$_2$ with small particle size offers more boundary sites for OER [173].

In addition to modifying the electronic and structural properties of Ni-based oxides/hydroxides, engineering the electrocatalyst support for better electronic transfer is equally crucial since most metal oxides/hydroxides are semiconducting. Supports that can improve the interfacial resistance between catalyst and electrode have been explored, such as CNTs [174], graphene [175], and carbon nitride [176]. The support, on the other hand, offers another dimension for the electronic tuning of Ni-based materials. In the case of C$_3$N$_4$, the formation of Ni—N bonding between the NiO and C–N units can lower the Gibbs free energy for OER intermediate adsorption, improve the charge transfer rate, and promote mass diffusion rate at the same time [176].

Co-Based Oxides/Hydroxides Cobalt oxides are highly active OER electrocatalysts, which can offer comparable performance to Ni-based ones. Regardless of the valent states of the catalysts before reaction, such as metallic Co(0) nanoparticles, Co(II) oxide, and Co(III) oxides, it is commonly accepted that the oxidation of Co(III) to Co(IV) species (CoO$_2$) is the precatalytic conversion before O$_2$ discharge. However, the full details of OER on Co oxides and hydroxides are still under debate [177]. *In situ* (or *operando*) methods are heavily relied on for understanding the reaction mechanism of Co-based materials. Favaro et al. pointed out that full conversion of Co(OH)$_2$ and partial conversion of Co$_3$O$_4$ to CoOOH is inevitable, starting at a potential lower than that of thermodynamic OER. Such CoOOH species, as suggested by the *operando* photoelectron spectra, are responsible for generating highly active Co(IV) center for OER [177]. Bergmann et al. added that Co-based catalytic systems, independent of their initial Co or coordination states, transform into a universal structure of HO-bridged Co(II)/Co(III) ion clusters, where the reducible Co(III)–O sites are the active sites for general OER activity [178]. However, Moysiadou et al. recently argued that, based on their *operando* results, the RDS of OER on amorphous CoOOH is the release of O_2 from the superoxide intermediate produced during the Co(III) to Co(IV) oxidation [179]. Based on such findings, any OER performance differences of electrocatalysts based on different valent states of Co can be addressed by the surface area of *in situ* generated Co(III)/Co(IV), as confirmed experimentally by Chou et al. using Co, CoO, and Co_3O_4 nanoparticles with similar surface areas [180].

Currently, the most studied materials among all are spinel cobalt oxide (Co_3O_4), cobaltite (MCo_2O_4, M = Ni, Cu, Mn, etc.), cobalt hydroxide ($Co(OH)_2$), and cobalt oxyhydroxide (CoOOH) [181]. In general, two fundamental issues of the Co-based electrocatalysts are targeted: active site engineering and conductivity improvement.

Active site engineering includes increasing the total population of Co sites and boosting the activity of Co sites. For the former aspect, typical methods include decreasing particle size and controlling the morphology and crystalline phases. For example, Sidhureddy et al. compared the OER performance of shape-controlled Co_3O_4, from one-dimensional (1D) nanorod to 3D nanocube, and reported that the 2D nanosheets showed the best performance due to its high surface area and abundant oxygen defects [182]. For the latter aspect, tuning the activity Co sites requires electronic interaction, and the most frequently visited method is metal doping. The involvement of another (or more) cation(s) in Co oxides/hydroxides often leads to a direct impact of the Co(IV) formation and phase conversion, as well as the creation of oxygen vacancy. Fe [183, 184], Cu [185], and Ni [186] are the most popular choices while cases using Mn [187], Ca [188], and Ti [189] are also reported. Direct doping on CoOOH is also a popular choice [190]. Recently, doping of non-metallic elements, such as S [191] and Cl [192], is gaining attention as an effective way of forming direct bonding between Co and the anions, unlike the bimetallic –Co–O–M– formation using metal doping. Chen et al. introduced fluorine anions (F^-) on the surface of CoOOH, creating a more hydrophilic surface of electrocatalysts to benefit the OH^- adsorption [193].

Extensive doping leads to the formation of bimetallic (or trimetallic) oxides/hydroxides. The most promising systems include $NiCoO_xH_y$ [194] and $FeCoO_xH_y$ [195, 196], which commonly deliver much lower overpotential than the monometallic compounds. For instance, $FeCoO_xH_y$ 2D nanosheets developed by Zhuang et al. exhibited a high current density of 54.9 A g^{-1} with an overpotential of 350 mV [195]. The low Tafel slope of 36.8 mV dec^{-1} makes it one of the best electrocatalytic systems reported so far. Such work is a typical example of combining both morphologic and electronic engineering for OER catalyst.

Metal doping can also significantly improve the conductivity of metal oxides/hydroxides due to the creation of vacancies in the semiconducting structure. Another method to reduce the charge transfer resistance is the support engineering. Similar to previously mentioned Ni-based systems, CNTs [197] and functionalized carbon supports [198] are generally engaged.

Fe-Based Oxides/Hydroxides Iron is the most abundant transition metal element on earth crust and is much cheaper than cobalt and nickel. However, in their bulk forms, such as FeO and $Fe(OH)_2$, the electronic conductivities are generally low [199], limiting the application of most Fe-based oxides/hydroxides for OER electrocatalysis.

Comparing with Ni and Co, the fundamental studies on Fe oxides/hydroxides are not broad. The doping of Fe in Ni/Co-based materials is proven effective and discussed earlier in this section. Lyons and Brandon proposed the redox couple of Fe(III)/Fe(VI) to accompany the OER on Fe oxyhydroxide electrode [200]. Systematic work by Boettcher's group showed that the OER behavior of FeOOH depends on the applied overpotential. At a low overpotential, the activity of FeOOH is impacted by the supporting electrode (Au/AuO_x in their study) where the interfacial Fe cations are active. At a high overpotential, on the other hand, the electrical conductivity of FeOOH increases, benefitting the OER process [199].

Current research focus, therefore, is to improve the conductivity of Fe-based oxides/hydroxides. Once again, metal doping and catalyst support engineering are preferred pathway. Some state-of-the-art systems include Co-doped FeOOH [201], Cu-doped FeOOH [202], Se-doped FeOOH [203], Ti-doped FeOOH [204], and a series of transition metal-doped Fe_2O_3 [205] and Fe_3O_4 [206].

Other Oxides/Hydroxides Studies on other earth-abundant metal element-based oxides and hydroxides, including Cu, Ti, Mn, Mo, and V, are not as booming as Ni/Co/Fe-based ones due to their low activities. A brief introduction is given below with the selected cases:

Cu based: Liu et al. demonstrated that CuO electrocatalyst could perform OER at a rate of $1.0\,mA\,cm^{-2}$ with an overpotential of 430 mV. Although outperformed by Ni/Co/Fe, it is one of the best results achieved for Cu-based system [207]. Making the bulk nanoparticle into 2D shape benefits the mass transfer and surface enlargement, as suggested by the catalytic results of 2D CuO electrocatalyst that delivered an overpotential of 350 mV at $10\,mA\,cm^{-2}$ in 1.0 M KOH [208]. Metal doping improves the catalytic performance significantly: Co-doped CuO with Co atomic ratio of c. 10% recorded an overpotential of 330 mV to achieve $100\,mA\,cm^{-2}$ [209]. An example of morphologic effect was reported by Huan et al. using dendritic nanostructured CuO, which demonstrated an even lower overpotential of 290 mV at $10\,mA\,cm^{-2}$ [210].

Ti based: TiO_2 is well known for its photo(electro)catalytic activity, which is beyond our scope in this chapter. Pure TiO_2 has low electrical conductivity and is predicted to have a very high overpotential for OER [211]. However, theoretical calculations suggested that earth-abundant metal-doped (Cr, Mo, and Mn) TiO_2 could share OER activity similar to RuO_2 [212].

Mn based: Mn_xO_y family (e.g. MnO_2, Mn_2O_3, and Mn_3O_4) is considered as a promising OER catalyst due to its stability under extreme pH conditions (especially in acidic conditions where Ni/Co/Fe-based electrocatalysts are not stable) and abundance [213]. Nocera's group has focused on understanding the OER mechanism on Mn_xO_y [214, 215] and showed that the stabilization of Mn(III) species is the key to promote O_2 formation [216]. The best performance so far was obtained using Mn_2O_3 nanotube arrays coated on conductive Ni foam, which showed an overpotential of 270 mV at 10 mA cm^{-2} and a Tafel slope of 85 mV dec^{-1} [217]. Current studies focus on the morphologic tuning of Mn_xO_y for better conductivity. For example, using sub-10 nm MnO nanocrystals can facilitate the generation of surface Mn(III) species during polarization for improved OER [218]. Still, the major challenge lies in the understanding of Mn-based catalytic system and developing synthesis methods to achieve different morphology.

Mo based: The application of pure Mo oxides/hydroxides for electrochemical OER is rarely reported. One recent literature showed that, regardless of morphology, MoO_3 has a much higher overpotential than most other metal oxides/hydroxides discussed. Even with the help of graphene to reduce the resistance, an overpotential of c. 540 mV was required to deliver merely 0.12 mA cm^{-2} [219]. Currently, Mo is used only as doping elements to modify the properties of hosting materials, such as $Ni(OH)_2$ and $Co(OH)_2$.

V based: V-based oxides/hydroxides are widely used for electrochemical applications. The valence of V ranges from V(II) to V(III) to V(IV) to V(V) in vanadium oxides, offering various oxygen coordination environments for OER. V cations themselves, however, are not active for OER. The most frequently referred OER catalytic systems are MVO_xH_y (M = Co, Ni) [220, 221]. For example, Jiang et al. developed a $Ni_3Fe_{1-x}V_x$ hydroxide material, in which the V sites outperform Ni and Fe sites for OER due to the synergistic electronic effect [169].

1.2.2.3 Metal (M^{n+}) Single-Atom Catalysts

Similar to the original intention of HER SACs, the development of OER SACs initially focused on lowering the loading of noble metals (Ru and Ir) and gradually expanded to earth-abundant metals [222]. For both HER and OER, the study of SACs shares the same challenge of preparing stable and efficient SACs, making the engineering of catalyst support a major topic [223]. Especially, the understanding of reaction mechanism on OER SACs is much more complicated than HER, given the limited technologies with atomic resolution and more complex OER pathway.

Currently, Fe, Ni, and Co SACs are among the most reported OER electrocatalysts. Chen et al. developed a Fe SAC with abundant Fe atoms decorated on N- and S-enriched carbon layer, which exhibited an overpotential of 370 mV at 10 mA cm^{-2} and Tafel slope of 82 mV dec^{-1}. The authors argued that the Fe 3d-electrons are affected by the neighboring N and S atoms, resulting in largely improved electrical conductivity [224]. Zheng et al. demonstrated a series of M–C_3N_4 catalyst (M = Fe, Co, Ni) of OER activity, and Co–C_3N_4 could deliver similar activities with noble metal-based ones in alkaline media. They further identified that the *M*–OOH formation governs the reaction rate, and the hosting N-rich environment for metal atoms is the key to stabilize isolated metal atoms. As the applied potential increases, the N-coordinated Co atoms are oxidized to Co^{3+} and higher states to catalyze OER [225]. Other SACs, beyond Fe, Ni, and Co, are rarely reported.

The coordinating atoms interacting with the metal sites define the catalytic performance of SACs. Other than N atoms, S [226] and O [227] coordinating atoms are also investigated. In some case, the coordination via S or O atoms shows more functionality in OER than electronically affecting the metal sites. For example, Huang and coworkers identified that the C atoms of a Ni–N_4–C_4 SAC could act as active sites for a dual reaction pathway OER [228].

Involving two or more metal atoms for SAC design is a proven strategy for HER, which allows fine-tuning of the activity of selected sites [229]. The same method also works for OER SACs. A Co–Fe double-atom electrocatalyst was prepared by *in situ* electrochemical treatment of Co SACs by Bai et al., and the Co–Fe dimeric unit was found to act as the active site [230]. Other combinations include Co–Mo (N–Co–Mo supported by graphene nanoflake–CNT composite) [231], Co–Ni (Co–Ni sites embedded in N-doped hollow carbon nanocubes) [232], and Fe–Ni (Fe–Ni in N-doped carbon hollow spheres) [233].

Overall, OER SACs are promising candidates for both high-efficient water splitting and theoretical understanding of the OER mechanism. Different combinations and various support choices make the study of SACs one of the fastest developing topics in both electrochemistry and energy-related catalysis. Yet, the durability of SACs should be proven before claiming any superiority over conventional materials. In this regard, the results from *in situ* techniques that can monitor the isolated active sites are always preferred for performance interpretation.

1.2.2.4 Metal Chalcogenides/Nitrides/Phosphides and Others

Materials other than typical metal and metal oxides are often proposed for OER, including metal chalcogenide (MX_2, M = Ti, Mo, or W, X = S, Se, or Te) [234], metal nitrides (M_xN, M = Co, Ni, Fe, etc.) [235], and metal phosphides (M_xP, M = Co, Ni, Fe, etc.) [236]. Many OER catalysts have reported even higher activity than commercial noble metal-based ones, as well as outstanding stability for hours. As discussed in the HER section (Sections 1.2.1.4–1.2.1.6), these materials are also active catalysts for HER. Based on their performances in both HER and OER, they are frequently claimed "bifunctional" despite being neither necessary nor efficient for real devices where the HER and OER are performed separately (Figure 1.1). However, based on the mechanism for OER, the establishment of metal–oxygen bonding is essential for ***M***–OH formation. As pointed out by Song Jin [237], metal sulfides, as well as metal nitrides and phosphides, are thermodynamically less stable than metal oxides when subjected to external potential. Therefore, the formation of metal oxides/hydroxides is inevitable for all metal chalcogenides, metal nitrides, and metal phosphides as soon as they are placed in the electrolyte of extreme pH. Arguably, the surface metal oxides/hydroxides are the real OER catalysts, while the initial metal compounds are the precatalysts [238] (materials that produce real catalysts during the electrochemical process). Such opinion is confirmed by many literatures. For example, Ni_2P, one of the earliest metal phosphide OER electrocatalysts, generates a shell of NiO_x during OER process, explaining its ability to produce O_2 [239]. The real OER catalyst should be expressed, at least, as NiO_x/Ni_2P, other than just Ni_2P. Similarly, a report by Chen et al. on using metallic

Co₄N porous nanowire arrays for OER also acknowledged the existence of a CoO_x layer formed during the electrochemical test [240]. Due to the known high activity of CoO_x, it is unclear whether Co₄N contributes to OER activity or not.

Nevertheless, experimental results also proved that the coexisting M–S/N/P structures can affect the OER activity of outer metal oxides/hydroxides. Xu et al. studied the trend in OER activity in metal phosphides and revealed the activity order of FeP < NiP < CoP < FeNiP < FeCoP < CoNiP < FeCoNiP, which also agrees with the electronegativity of metal sites [241]. Although the surface is converted to metal oxides/hydroxides, the sublayer coordination environments still affect the OER sites.

Since the concept of precatalyst is now generally accepted in the research community, the rational design and understanding of metal chalcogenides/nitrides/phosphides in OER conditions require more carefully designed experiments and *in situ* techniques to uncover the proper reaction mechanism [242].

Other emerging materials, such as metal–organic frameworks, were also reported suitable for OER but shared the same problems discussed above. For example, Ghoshal et al. suggested ZIF-67 (zeolitic imidazolate framework with Co sites) as a highly active OER catalyst [243]. Later, by using *in situ* spectroelectrochemistry, Zheng et al. found that the structure of ZIF-67 is not stable during electrochemical studies and α/β-Co(OH)₂ are the real active species for OER, not the Co sites (Figure 1.14) [244]. Moreover, even without a redox reaction occurring on the Co sites, the sweeping potential can sabotage the weak Co–N coordination sites within a few seconds.

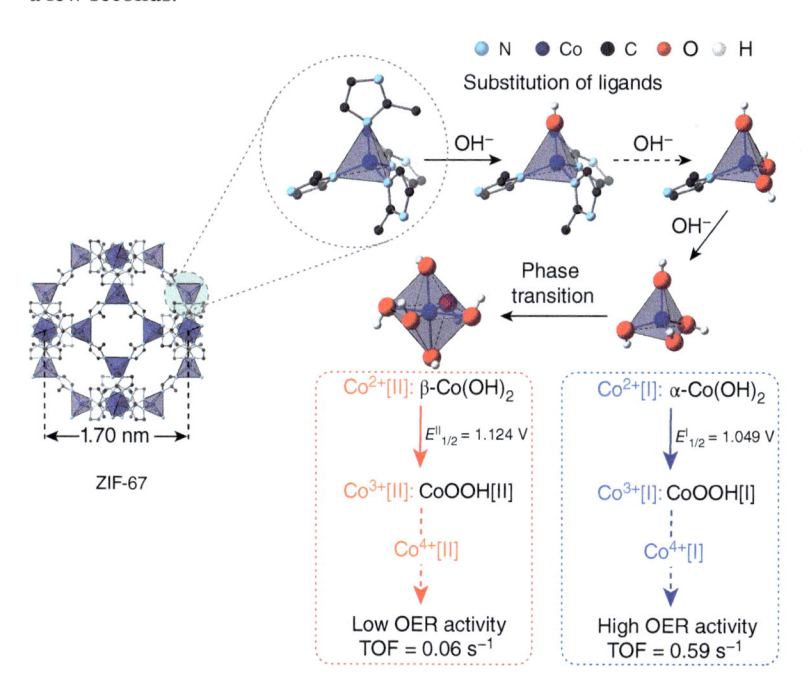

Figure 1.14 Precatalytic conversion of ZIF-67 to α/β-Co(OH)₂ and the experimental OER activity of both Co(OH)₂. Source: Zheng et al. [244]. Reproduced with permission of American Chemical Society.

1.3 Computer-Assisted Materials Discovery

The prosperous development of material science in the last decade has led to the explosive increment in numbers and categories of materials for water splitting. Apparently, it is impossible for any research group in the world to prepare all the materials and benchmark them under the identical condition as before. For example, the bimetallic/trimetallic (or even more) doping on metal oxides/hydroxides OER catalyst can have a doping level from ppm to 100% surface coverage. Besides, metal sites can be doped on different crystalline facets with different ratios. Such factors in practical experimental conditions provide nearly infinite combinations for potentially high OER activity. Experimentally, it can be challenging to prepare the material and confirm the structure–activity relationship.

Computer-assisted materials discovery, including DFT prediction and machine learning, are the future for material design and discovery of electrocatalysts, including HER and OER [245, 246]. Using DFT calculations for understanding the reaction pathway has been a standard technology in the last couple of decades [247]. Meanwhile, the high-throughput screening of water splitting electrocatalysts also has gained wide attention: not only the activity but also stability can be readily predicted. In 2006, Greeley et al. presented a DFT-based screening scheme for HER electrocatalyst involving over 700 binary surface alloys (Figure 1.15), where the superiority of BiPt over Pt was predicted and experimentally confirmed [248]. Later, Björketun et al. exploited a collection of theoretical and experimental databases and

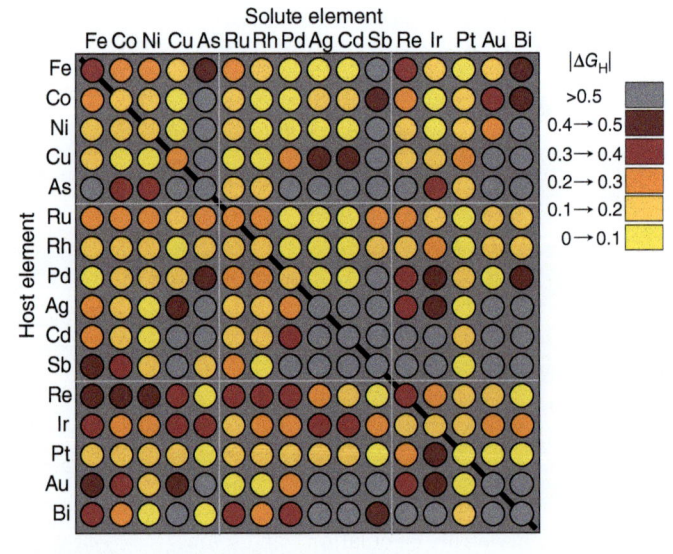

Figure 1.15 Computational high-throughput screening of HER catalyst for ΔG_H on 256 combinations. Source: Greeley et al. [248]. Reproduced with permission of Springer Nature.

found that a binary "substrate–overlayer" of Cu–W is HER active, confirmed by their experiments [249]. Similar predictions were also made for OER catalyst design. For instance, a family of Ni–Fe–Co–Ce oxides was screened by Haber et al., and the compositions of $Ni_{0.5}Fe_{0.3}Co_{0.17}Ce_{0.03}O_x$ and $Ni_{0.3}Fe_{0.07}Co_{0.2}Ce_{0.43}O_x$ were proposed with excellent OER performance and stability [250]. Obviously, it would take years of material synthesis and testing to confirm such a combination without such prediction. Even for systems that are not available due to experimental difficulty, their potential activity can be studied conveniently such as $MoTe_2/WTe_2$ layered material with an interlayer rotation angle of 300° and was predicted to have an overpotential of 30 mV for HER and 170 mV for OER [251].

Current experimental techniques do not limit the computer-assisted materials discovery, but the calculation power does. The rapid advances in computer science have significantly shortened the calculation time in complex systems and improved accuracy. For researchers not familiar with computational chemistry, a few public/open calculation projects/databases are currently available for preliminary material screening, calculation, and property prediction, which mainly include the following:

Open Catalyst Project: https://opencatalystproject.org
Materials Cloud: https://www.materialscloud.org
The Materials Project: https://materialsproject.org
Novel Materials Discovery (NOMAD): https://nomad-lab.eu
Automatic-Flow for Materials Discovery (AFLOW): http://aflowlib.org
The Open Quantum Materials Database (OQMD): http://oqmd.org

1.4 Challenge and Outlook

Hydrogen production by electrochemical water splitting is one of the most promising and feasible ways to realize the large-scale application of hydrogen fuel but currently has few challenges.

1.4.1 Reliability Comparison Between Results

Numerous publications on HER and OER electrocatalysts that appear every day make it increasingly challenging to benchmark the electrocatalysts among various research groups, mainly due to the difference in reaction conditions (scan rate, electrolyte, electrocatalyst pretreatment, electrocatalyst supporting electrode), experimental setups (electrolyte, reference/counter electrode, potential/current sequence), data treatment (current density calculation), and so on.

To confront such problems in the electrochemical community, various guideline papers have been published recently trying to address the experimental issues.

The most important ones are introduced here to benefit the junior researchers to correctly evaluate the HER/OER activity of materials:

For general water electrolyzer: Essential parameters required for material characterization and water splitting performance evaluation [252].

For heterogeneous electrocatalyst: Benchmarking catalyst, electrochemical surface area calculation, and counter electrode selection [253, 254].

For HER/OER electrocatalyst: Electrochemical cell, selection of counter electrode, and contamination of working electrodes [255], essential parameters and ways to report [256]; and problems associated with using conductive foam as supporting electrode [257].

1.4.2 Gap Between Industrial and Laboratorial Research

Currently, many electrocatalytic systems in literature have demonstrated excellent HER and OER activities that outperform the costly commercial noble metal-based electrocatalysts. Yet, they are not quickly commercialized as expected. One of the major issues is the different operating conditions between industrial and laboratorial researches. Most published materials are tested under standard laboratory conditions: small scale, electrolyte of 0.5 M H_2SO_4/1.0 M KOH, room temperature, low current, small working area, etc. Conventional industrial water electrolysis, however, operates at higher temperature (>60 °C) with extreme pH (30% KOH electrolyte for alkaline water electrolysis). Much higher durability is also required since it is impractical to change the electrocatalyst every a few hours. Most importantly, the "novel nanomaterials," despite having low cost of ingredients, are difficult and expensive to be prepared in large quantities. Fancy and fragile nanostructures, although sometimes show attractive performance, are not actually cheaper than the noble metal-based ones due to their high manufacturing cost and poor stability under industrial conditions.

Currently, PEM process for water electrolysis has been proven more efficient than alkaline water splitting, but most earth-abundant metal-based HER/OER catalysts are not suitable for application under acidic conditions, especially at high temperature. The development of such catalytic systems is urgently required for commercial PEM devices.

1.4.3 Outlook

Today, we are witnessing the key transition of the thousand-year-old carbon-based energy economy to the sustainable hydrogen-based energy economy. Due to the increasing environmental concerns, many countries and organizations, including China, European Union, Japan, South Korea, and Unite Kingdom, have committed in law to achieve carbon neutrality between 2050 and 2060. Compared with the last century, remarkable progresses have been achieved in the past few years on water splitting, and more commercial HER/OER electrocatalysts are expected within next 10 years. The rational design of durable and efficient HER/OER electrocatalysts based on earth-abundant metal elements will be the key topic in energy-related electrochemistry.

References

1. IRENA (2019). *Hydrogen: A Renewable Energy Perspective*. Abu Dhabi.
2. Levin, D.B. and Chahine, R. (2010). *Int. J. Hydrogen Energy* 35: 4962–4969.
3. Valente, A., Iribarren, D., and Dufour, J. (2020). *Sci. Total Environ.* 728: 138212.
4. Shinagawa, T. and Takanabe, K. (2017). *ChemSusChem* 10: 1318–1336.
5. Sun, X., Xu, K., Fleischer, C. et al. (2018). *Catalysts* 8: 657–697.
6. Rumble, J. (2020). *CRC Handbook of Chemistry and Physics*, 101e. CRC Press.
7. Lasia, A. (2019). *Int. J. Hydrogen Energy* 44: 19484–19518.
8. Trasatti, S. (1972). *J. Electroanal. Chem. Interfacial Electrochem.* 39: 163–184.
9. Skúlason, E., Tripkovic, V., Björketun, M.E. et al. (2010). *J. Phys. Chem. C* 114: 18182–18197.
10. Rojas, M., Fan, C.L., Miao, H.J., and Piron, D.L. (1992). *J. Appl. Electrochem.* 22: 1135–1141.
11. Behzadian, B., Piron, D., Fan, C., and Lessard, J. (1991). *Int. J. Hydrogen Energy* 16: 791–796.
12. Liu, L. and Corma, A. (2020). *Nat. Rev. Mater.* 6: 244–263.
13. Ouyang, G., Wang, C.X., and Yang, G.W. (2009). *Chem. Rev.* 109: 4221–4247.
14. Mayrhofer, K.J., Blizanac, B.B., Arenz, M. et al. (2005). *J. Phys. Chem. B* 109: 14433–14440.
15. Fei, H., Yang, Y., Peng, Z. et al. (2015). *ACS Appl. Mater. Interfaces* 7: 8083–8087.
16. Yu, H.Z., Wang, Y., Ying, J. et al. (2019). *ACS Appl. Mater. Interfaces* 11: 27641–27647.
17. Chen, Z., Wu, R., Liu, Y. et al. (2018). *Adv. Mater.* 30: e1802011.
18. Devi, B., Koner, R.R., and Halder, A. (2018). *ACS Sustainable Chem. Eng.* 7: 2187–2199.
19. Yan, X., Gu, M., Wang, Y. et al. (2020). *Nano Res.* 13: 975–982.
20. Gong, Y., Wang, L., Xiong, H. et al. (2019). *J. Mater. Chem. A* 7: 13671–13678.
21. Yang, L., Zhou, W., Jia, J. et al. (2017). *Carbon* 122: 710–717.
22. Du, J., Wang, L., Bai, L. et al. (2018). *ACS Sustainable Chem. Eng.* 6: 10335–10343.
23. Wang, L., Li, Y., Xia, M. et al. (2017). *J. Power Sources* 347: 220–228.
24. Tavakkoli, M., Kallio, T., Reynaud, O. et al. (2015). *Angew. Chem. Int. Ed.* 54: 4535–4538.
25. Wang, J., Wang, G., Miao, S. et al. (2014). *Faraday Discuss.* 176: 135–151.

26 Ahmed, J., Trinh, P., Mugweru, A.M., and Ganguli, A.K. (2011). *Solid State Sci.* 13: 855–861.
27 Lu, Q., Hutchings, G.S., Yu, W. et al. (2015). *Nat. Commun.* 6: 6567.
28 Kuang, M., Wang, Q., Han, P., and Zheng, G. (2017). *Adv. Energy Mater.* 7: 1700193–1700200.
29 Ahsan, M.A., Puente Santiago, A.R., Hong, Y. et al. (2020). *J. Am. Chem. Soc.* 142: 14688–14701.
30 Wang, T., Guo, Y., Zhou, Z. et al. (2016). *ACS Nano* 10: 10397–10403.
31 Ahmed, J., Kumar, B., Mugweru, A.M. et al. (2010). *J. Phys. Chem. C* 114: 18779–18784.
32 Khani, H., Grundish, N.S., Wipf, D.O., and Goodenough, J.B. (2019). *Adv. Energy Mater.* 10: 1903215–1903224.
33 Qiao, B., Wang, A., Yang, X. et al. (2011). *Nat. Chem.* 3: 634–641.
34 Hunt, S.T., Milina, M., Wang, Z., and Román-Leshkov, Y. (2016). *Energy Environ. Sci.* 9: 3290–3301.
35 Fei, H., Dong, J., Arellano-Jimenez, M.J. et al. (2015). *Nat. Commun.* 6: 8668.
36 Xue, Y., Huang, B., Yi, Y. et al. (2018). *Nat. Commun.* 9: 1460.
37 Qiu, H.J., Ito, Y., Cong, W. et al. (2015). *Angew. Chem. Int. Ed.* 54: 14031–14035.
38 Chen, W., Pei, J., He, C.-T. et al. (2017). *Angew. Chem. Int. Ed.* 129: 16302–16306.
39 Chen, W., Pei, J., He, C.T. et al. (2018). *Adv. Mater.* 30: e1800396.
40 Cao, L., Luo, Q., Liu, W. et al. (2018). *Nat. Catal.* 2: 134–141.
41 Wang, L., Liu, X., Cao, L. et al. (2020). *J. Phys. Chem. Lett.* 11: 6691–6696.
42 Chen, W., Pei, J., He, C.T. et al. (2017). *Angew. Chem. Int. Ed.* 56: 16086–16090.
43 Yang, Y., Qian, Y., Li, H. et al. (2020). *Sci. Adv.* 6: eaba6586.
44 Zhang, Q. and Guan, J. (2020). *Adv. Funct. Mater.* 30: 2000768–2000820.
45 Gutić, S.J., Dobrota, A.S., Fako, E. et al. (2020). *Catalysts* 10: 290–327.
46 Liu, H., Peng, X., and Liu, X. (2018). *ChemElectroChem* 5: 2963–2974.
47 Pu, Z.H., Amiinu, I.S., Cheng, R.L. et al. (2020). *Nano-Micro Lett.* 12: 21–49.
48 Li, X., Bi, W., Chen, M. et al. (2017). *J. Am. Chem. Soc.* 139: 14889–14892.
49 Gao, G., Bottle, S., and Du, A. (2018). *Catal. Sci. Technol.* 8: 996–1001.
50 Fan, L., Liu, P.F., Yan, X. et al. (2016). *Nat. Commun.* 7: 10667.
51 He, T., Zhang, C., and Du, A. (2019). *Chem. Eng. Sci.* 194: 58–63.
52 Zhang, Y., Li, W., Lu, L. et al. (2018). *Electrochim. Acta* 265: 497–506.
53 Cao, Y., Chen, S., Luo, Q. et al. (2017). *Angew. Chem. Int. Ed.* 56: 12191–12196.
54 Ling, C., Shi, L., Ouyang, Y. et al. (2017). *Nano Lett.* 17: 5133–5139.
55 Sredojević, D.N., Belić, M.R., and Šljivančanin, Ž. (2020). *J. Phys. Chem. C* 124: 16860–16867.
56 Weng, C.C., Ren, J.T., and Yuan, Z.Y. (2020). *ChemSusChem* 13: 3357–3375.
57 Li, H., Lu, S., Sun, J. et al. (2018). *Chemistry* 24: 11748–11754.
58 Xiao, P., Sk, M.A., Thia, L. et al. (2014). *Energy Environ. Sci.* 7: 2624–2629.
59 Kibsgaard, J., Tsai, C., Chan, K. et al. (2015). *Energy Environ. Sci.* 8: 3022–3029.
60 Du, H., Kong, R.M., Guo, X. et al. (2018). *Nanoscale* 10: 21617–21624.
61 Guan, C., Xiao, W., Wu, H. et al. (2018). *Nano Energy* 48: 73–80.
62 Wang, X., Zhou, H., Zhang, D. et al. (2018). *J. Power Sources* 387: 1–8.

References

63 Man, H.-W., Tsang, C.-S., Li, M.M.-J. et al. (2019). *Appl. Catal., B.* 242: 186–193.
64 Liu, T., Liu, D., Qu, F. et al. (2017). *Adv. Energy Mater.* 7: 1700020.
65 Zhou, Q., Shen, Z., Zhu, C. et al. (2018). *Adv. Mater.* 30: 1800140.
66 Chang, J., Li, K., Wu, Z. et al. (2018). *ACS Appl. Mater. Interfaces* 10: 26303–26311.
67 Liang, K., Pakhira, S., Yang, Z. et al. (2019). *ACS Catal.* 9: 651–659.
68 Zhang, T., Yang, K., Wang, C. et al. (2018). *Adv. Energy Mater.* 8: 1801690.
69 Wang, Z., Du, H., Liu, Z. et al. (2018). *Nanoscale* 10: 2213–2217.
70 Zhang, X., Zhu, S., Xia, L. et al. (2018). *Chem. Commun.* 54: 1201–1204.
71 Wu, Z., Wang, J., Xia, K. et al. (2018). *J. Mater. Chem. A* 6: 616–622.
72 Wang, A.-L., Lin, J., Xu, H. et al. (2016). *J. Mater. Chem. A* 4: 16992–16999.
73 Luo, X., Ji, P., Wang, P. et al. (2020). *Adv. Energy Mater.* 10: 1903891.
74 Adam, A., Suliman, M.H., Dafalla, H. et al. (2018). *ACS Sustainable Chem. Eng.* 6: 11414–11423.
75 Zhang, Y., Yang, J., Dong, Q. et al. (2018). *ACS Appl. Mater. Interfaces* 10: 26258–26263.
76 Zhu, Y.-P., Xu, X., Su, H. et al. (2015). *ACS Appl. Mater. Interfaces* 7: 28369–28376.
77 Adam, A., Suliman, M.H., Siddiqui, M.N. et al. (2018). *ACS Appl. Mater. Interfaces* 10: 29407–29416.
78 Lv, X., Ren, J., Wang, Y. et al. (2019). *ACS Sustainable Chem. Eng.* 7: 8993–9001.
79 Miao, M., Hou, R., Liang, Z. et al. (2018). *J. Mater. Chem. A* 6: 24107–24113.
80 Heine, T. (2015). *Acc. Chem. Res.* 48: 65–72.
81 Hinnemann, B., Moses, P.G., Bonde, J. et al. (2005). *J. Am. Chem. Soc.* 127: 5308–5309.
82 Li, W., Liu, G., Li, J. et al. (2019). *Appl. Surf. Sci.* 498: 143869.
83 Jaramillo, T.F., Jørgensen, K.P., Bonde, J. et al. (2007). *Science* 317: 100.
84 An, Y.-R., Fan, X.-L., Luo, Z.-F., and Lau, W.-M. (2017). *Nano Lett.* 17: 368–376.
85 Kong, D., Wang, H., Cha, J.J. et al. (2013). *Nano Lett.* 13: 1341–1347.
86 Chen, Z., Cummins, D., Reinecke, B.N. et al. (2011). *Nano Lett.* 11: 4168–4175.
87 Li, Y., Wang, H., Xie, L. et al. (2011). *J. Am. Chem. Soc.* 133: 7296–7299.
88 Tang, Q. and Jiang, D.-e. (2016). *ACS Catal.* 6: 4953–4961.
89 Voiry, D., Salehi, M., Silva, R. et al. (2013). *Nano Lett.* 13: 6222–6227.
90 Hou, X., Zhou, H., Zhao, M. et al. (2020). *ACS Sustainable Chem. Eng.* 8: 5724–5733.
91 Huang, H., Huang, W., Yang, Z. et al. (2017). *J. Mater. Chem. A* 5: 1558–1566.
92 Kong, D., Cha, J.J., Wang, H. et al. (2013). *Energy Environ. Sci.* 6: 3553–3558.
93 Jiang, N., Tang, Q., Sheng, M. et al. (2016). *Catal. Sci. Technol.* 6: 1077–1084.
94 Feng, J.X., Wu, J.Q., Tong, Y.X., and Li, G.R. (2018). *J. Am. Chem. Soc.* 140: 610–617.
95 An, Y., Huang, B., Wang, Z. et al. (2017). *Dalton Trans.* 46: 10700–10706.
96 Anantharaj, S., Subhashini, E., Swaathini, K.C. et al. (2019). *Appl. Surf. Sci.* 487: 1152–1158.
97 Anantharaj, S., Kundu, S., and Noda, S. (2020). *J. Mater. Chem. A* 8: 4174–4192.

98 Wang, F., Li, Y., Shifa, T.A. et al. (2016). *Angew. Chem. Int. Ed.* 55: 6919–6924.
99 Cabán-Acevedo, M., Stone, M.L., Schmidt, J.R. et al. (2015). *Nat. Mater.* 14: 1245–1251.
100 Dutta, B., Wu, Y., Chen, J. et al. (2019). *ACS Catal.* 9: 456–465.
101 Xia, C., Liang, H., Zhu, J. et al. (2017). *Adv. Energy Mater.* 7: 1602089.
102 Jasion, D., Barforoush, J.M., Qiao, Q. et al. (2015). *ACS Catal.* 5: 6653–6657.
103 Miao, R., Dutta, B., Sahoo, S. et al. (2017). *J. Am. Chem. Soc.* 139: 13604–13607.
104 Xu, X., Ge, Y., Wang, M. et al. (2016). *ACS Appl. Mater. Interfaces* 8: 18036–18042.
105 Chen, Y., Zhang, J., Guo, P. et al. (2018). *ACS Appl. Mater. Interfaces* 10: 27787–27794.
106 Jin, H., Liu, X., Vasileff, A. et al. (2018). *ACS Nano* 12: 12761–12769.
107 Liu, B., He, B., Peng, H.-Q. et al. (2018). *Adv. Sci.* 5: 1800406.
108 Chen, Z., Song, Y., Cai, J. et al. (2018). *Angew. Chem. Int. Ed.* 57: 5076–5080.
109 Sun, K., Zhang, T., Tan, L. et al. (2020). *ACS Appl. Mater. Interfaces* 12: 29357–29364.
110 Wang, C., Lv, X., Zhou, P. et al. (2020). *ACS Appl. Mater. Interfaces* 12: 29153–29161.
111 Tang, C., Zhang, H., Xu, K. et al. (2019). *J. Mater. Chem. A* 7: 18030–18038.
112 Ma, Y., Chen, M., Geng, H. et al. (2020). *Adv. Funct. Mater.* 30: 2000561.
113 Huang, J., Wang, J., Xie, R. et al. (2020). *J. Mater. Chem. A* 8: 19879–19886.
114 Song, H.J., Sung, M.-C., Yoon, H. et al. (2019). *Adv. Sci.* 6: 1802135.
115 Gong, M., Wang, D.-Y., Chen, C.-C. et al. (2016). *Nano Res.* 9: 28–46.
116 Zhao, Y., Chang, C., Teng, F. et al. (2017). *Adv. Energy Mater.* 7: 1700005.
117 Jin, Y., Wang, H., Li, J. et al. (2016). *Adv. Mater.* 28: 3785–3790.
118 Shu, C., Kang, S., Jin, Y. et al. (2017). *J. Mater. Chem. A* 5: 9655–9660.
119 Gao, X., Zhang, H., Li, Q. et al. (2016). *Angew. Chem. Int. Ed.* 55: 6290–6294.
120 Zhu, Y., Lin, Q., Zhong, Y. et al. (2020). *Energy Environ. Sci.* 13: 3361–3392.
121 Zheng, T., Sang, W., He, Z. et al. (2017). *Nano Lett.* 17: 7968–7973.
122 Swaminathan, J., Subbiah, R., and Singaram, V. (2016). *ACS Catal.* 6: 2222–2229.
123 Li, Y.H., Liu, P.F., Pan, L.F. et al. (2015). *Nat. Commun.* 6: 8064.
124 Zhang, T., Wu, M.-Y., Yan, D.-Y. et al. (2018). *Nano Energy* 43: 103–109.
125 Ling, T., Yan, D.-Y., Wang, H. et al. (2017). *Nat. Commun.* 8: 1509.
126 Li, L., Feng, X., Nie, Y. et al. (2015). *ACS Catal.* 5: 4825–4832.
127 Zhang, L., Liu, P.F., Li, Y.H. et al. (2018). *ChemSusChem* 11: 1020–1024.
128 Li, L., Zhang, T., Yan, J. et al. (2017). *Small* 13: 1700441–1700447.
129 Xiao, Z., Wang, Y., Huang, Y.-C. et al. (2017). *Energy Environ. Sci.* 10: 2563–2569.
130 Yu, X., Yu, Z.-Y., Zhang, X.-L. et al. (2020). *Nano Energy* 71: 104652–104659.
131 Geng, S., Liu, Y., Yu, Y.S. et al. (2019). *Nano Res.* 13: 121–126.

132 Zeng, L., Zhou, K., Yang, L. et al. (2018). *ACS Appl. Energy Mater.* 1: 6279–6287.

133 Dalai, N., Mohanty, B., Mitra, A., and Jena, B. (2019). *ChemistrySelect* 4: 7791–7796.

134 Gu, J., Zhang, C., Du, Z., and Yang, S. (2019). *Small* 15: e1904587.

135 Aqueel Ahmed, A.T., Pawar, S.M., Inamdar, A.I. et al. (2019). *Adv. Mater. Interfaces* 7: 1901515–1901525.

136 Huang, X., Zheng, H., Lu, G. et al. (2018). *ACS Sustainable Chem. Eng.* 7: 1169–1177.

137 Chen, Y., Dong, C., Zhang, J. et al. (2018). *J. Mater. Chem. A* 6: 8430–8440.

138 Ling, T., Zhang, T., Ge, B. et al. (2019). *Adv. Mater.* 31: e1807771.

139 Zhang, H., Li, X., Hähnel, A. et al. (2018). *Adv. Funct. Mater.* 28: 1706847–1706856.

140 Chen, G., Wang, T., Zhang, J. et al. (2018). *Adv. Mater.* 30: 1803694–1809700.

141 Wang, P., Qi, J., Chen, X. et al. (2020). *ACS Appl. Mater. Interfaces* 12: 4385–4395.

142 Gao, X., Zhao, Y., Dai, K. et al. (2020). *Chem. Eng. J.* 384: 123373.

143 Sun, H., Li, J.-G., Lv, L. et al. (2019). *J. Power Sources* 425: 138–146.

144 Wang, Y., Wang, T., Zhang, R. et al. (2020). *Inorg. Chem.* 59: 9491–9495.

145 Grimaud, A., Hong, W.T., Shao-Horn, Y., and Tarascon, J.M. (2016). *Nat. Mater.* 15: 121–126.

146 Song, J., Wei, C., Huang, Z.F. et al. (2020). *Chem. Soc. Rev.* 49: 2196–2214.

147 Man, I.C., Su, H.Y., Calle-Vallejo, F. et al. (2011). *ChemCatChem* 3: 1159–1165.

148 Montoya, J.H., Seitz, L.C., Chakthranont, P. et al. (2016). *Nat. Mater.* 16: 70–81.

149 Grimaud, A., Diaz-Morales, O., Han, B. et al. (2017). *Nat. Chem.* 9: 457–465.

150 Surendranath, Y., Kanan, M.W., and Nocera, D.G. (2010). *J. Am. Chem. Soc.* 132: 16501–16509.

151 Fierro, S., Nagel, T., Baltruschat, H., and Comninellis, C. (2007). *Electrochem. Commun.* 9: 1969–1974.

152 Kulkarni, A., Siahrostami, S., Patel, A., and Norskov, J.K. (2018). *Chem. Rev.* 118: 2302–2312.

153 Nong, H.N., Falling, L.J., Bergmann, A. et al. (2020). *Nature* 587: 408–413.

154 Burke, M.S., Enman, L.J., Batchellor, A.S. et al. (2015). *Chem. Mater.* 27: 7549–7558.

155 Burke, M.S., Zou, S., Enman, L.J. et al. (2015). *J. Phys. Chem. Lett.* 6: 3737–3742.

156 Stern, L.A. and Hu, X. (2014). *Faraday Discuss.* 176: 363–379.

157 McCrory, C.C., Jung, S., Peters, J.C., and Jaramillo, T.F. (2013). *J. Am. Chem. Soc.* 135: 16977–16987.

158 Huang, J., Li, Y., Zhang, Y. et al. (2019). *Angew. Chem. Int. Ed.* 58: 17458–17464.

159 Diaz-Morales, O., Ferrus-Suspedra, D., and Koper, M.T.M. (2016). *Chem. Sci.* 7: 2639–2645.
160 Li, L.-F., Li, Y.-F., and Liu, Z.-P. (2020). *ACS Catal.* 10: 2581–2590.
161 Li, Y.-F. and Selloni, A. (2014). *ACS Catal.* 4: 1148–1153.
162 Klaus, S., Cai, Y., Louie, M.W. et al. (2015). *J. Phys. Chem. C* 119: 7243–7254.
163 Trotochaud, L., Young, S.L., Ranney, J.K., and Boettcher, S.W. (2014). *J. Am. Chem. Soc.* 136: 6744–6753.
164 Zhang, F., Wang, W., Li, Z. et al. (2020). *Mol. Catal.* 493: 111082–111089.
165 Jin, Y., Huang, S., Yue, X. et al. (2018). *ACS Catal.* 8: 2359–2363.
166 Yan, J., Kong, L., Ji, Y. et al. (2019). *Nat. Commun.* 10: 2149.
167 Dou, Y., He, C.-T., Zhang, L. et al. (2020). *Cell Rep. Phys. Sci.* 1: 100077–100090.
168 Diaz-Morales, O., Ledezma-Yanez, I., Koper, M.T.M., and Calle-Vallejo, F. (2015). *ACS Catal.* 5: 5380–5387.
169 Jiang, J., Sun, F., Zhou, S. et al. (2018). *Nat. Commun.* 9: 2885.
170 Favaro, M., Drisdell, W.S., Marcus, M.A. et al. (2017). *ACS Catal.* 7: 1248–1258.
171 Zhang, Z., Zhang, T., and Lee, J.Y. (2018). *ACS Appl. Nano Mater.* 1: 751–758.
172 Qiao, X., Kang, H., Li, Y. et al. (2020). *ACS Appl. Mater. Interfaces* 12: 36208–36219.
173 Luan, C., Liu, G., Liu, Y. et al. (2018). *ACS Nano* 12: 3875–3885.
174 Cheng, Y. and Jiang, S.P. (2015). *Prog. Nat. Sci.: Mater. Int.* 25: 545–553.
175 Yao, Y., Xu, Z., Cheng, F. et al. (2018). *Energy Environ. Sci.* 11: 407–416.
176 Liao, C., Yang, B., Zhang, N. et al. (2019). *Adv. Funct. Mater.* 29: 1904020–1904030.
177 Favaro, M., Yang, J., Nappini, S. et al. (2017). *J. Am. Chem. Soc.* 139: 8960–8970.
178 Bergmann, A., Jones, T.E., Martinez Moreno, E. et al. (2018). *Nat. Catal.* 1: 711–719.
179 Moysiadou, A., Lee, S., Hsu, C.S. et al. (2020). *J. Am. Chem. Soc.* 142: 11901–11914.
180 Chou, N.H., Ross, P.N., Bell, A.T., and Tilley, T.D. (2011). *ChemSusChem* 4: 1566–1569.
181 Han, L., Dong, S., and Wang, E. (2016). *Adv. Mater.* 28: 9266–9291.
182 Sidhureddy, B., Dondapati, J.S., and Chen, A. (2019). *Chem. Commun.* 55: 3626–3629.
183 Pan, S., Mao, X., Yu, J. et al. (2020). *Inorg. Chem. Front.* 7: 3327–3339.
184 Jin, H., Mao, S., Zhan, G. et al. (2017). *J. Mater. Chem. A* 5: 1078–1084.
185 Chen, L., Zhang, H., Chen, L. et al. (2017). *J. Mater. Chem. A* 5: 22568–22575.
186 Zhou, Q., Chen, Y., Zhao, G. et al. (2018). *ACS Catal.* 8: 5382–5390.
187 Raj, S., Anantharaj, S., Kundu, S., and Roy, P. (2019). *ACS Sustainable Chem. Eng.* 7: 9690–9698.
188 Su, P., Ma, S., Huang, W. et al. (2019). *J. Mater. Chem. A* 7: 19415–19422.
189 He, L., Liu, J., Hu, B. et al. (2019). *J. Power Sources* 414: 333–344.
190 Zhang, S., Yu, T., Wen, H. et al. (2020). *Chem. Commun.* 56: 15387–15405.
191 Al-Mamun, M., Zhu, Z., Yin, H. et al. (2016). *Chem. Commun.* 52: 9450–9453.
192 Kou, Y., Liu, J., Li, Y. et al. (2018). *ACS Appl. Mater. Interfaces* 10: 796–805.

193 Chen, P., Zhou, T., Wang, S. et al. (2018). *Angew. Chem. Int. Ed.* 57: 15471–15475.
194 Wang, L., Lin, C., Huang, D. et al. (2014). *ACS Appl. Mater. Interfaces* 6: 10172–10180.
195 Zhuang, L., Ge, L., Yang, Y. et al. (2017). *Adv. Mater.* 29: 1606793–1160679.
196 Zhuang, L., Jia, Y., He, T. et al. (2018). *Nano Res.* 11: 3509–3518.
197 Lu, X. and Zhao, C. (2013). *J. Mater. Chem. A* 1: 12053–12059.
198 Mallón, L., Romero, N., Jiménez, A. et al. (2020). *Catal. Sci. Technol.* 10: 4513–4521.
199 Zou, S., Burke, M.S., Kast, M.G. et al. (2015). *Chem. Mater.* 27: 8011–8020.
200 Lyons, M.E. and Brandon, M.P. (2009). *Phys. Chem. Chem. Phys.* 11: 2203–2217.
201 Zhang, X., An, L., Yin, J. et al. (2017). *Sci. Rep.* 7: 43590.
202 Zhang, S., Gao, H., Huang, Y. et al. (2018). *Environ. Sci.: Nano* 5: 1179–1190.
203 Niu, S., Jiang, W.J., Wei, Z. et al. (2019). *J. Am. Chem. Soc.* 141: 7005–7013.
204 Yoon, K.-Y., Ahn, H.-J., Kwak, M.-J. et al. (2016). *J. Mater. Chem. A* 4: 18730–18736.
205 Yin, Y., Zhang, X., and Sun, C. (2018). *Prog. Nat. Sci.: Mater. Int.* 28: 430–436.
206 Han, S., Liu, S., Yin, S. et al. (2016). *Electrochim. Acta* 210: 942–949.
207 Liu, X., Cui, S., Sun, Z. et al. (2016). *J. Phys. Chem. C* 120: 831–840.
208 Pawar, S.M., Pawar, B.S., Hou, B. et al. (2017). *J. Mater. Chem. A* 5: 12747–12751.
209 Xiong, X., You, C., Liu, Z. et al. (2018). *ACS Sustainable Chem. Eng.* 6: 2883–2887.
210 Huan, T.N., Rousse, G., Zanna, S. et al. (2017). *Angew. Chem. Int. Ed.* 56: 4792–4796.
211 Valdés, Á., Qu, Z.W., Kroes, G.J. et al. (2008). *J. Phys. Chem. C* 112: 9872–9879.
212 García-Mota, M., Vojvodic, A., Metiu, H. et al. (2011). *ChemCatChem* 3: 1607–1611.
213 Kumbhar, V.S., Lee, H., Lee, J., and Lee, K. (2019). *Carbon Resour. Convers.* 2: 242–255.
214 Huynh, M., Bediako, D.K., and Nocera, D.G. (2014). *J. Am. Chem. Soc.* 136: 6002–6010.
215 Huynh, M., Shi, C., Billinge, S.J., and Nocera, D.G. (2015). *J. Am. Chem. Soc.* 137: 14887–14904.
216 Morgan Chan, Z., Kitchaev, D.A., Nelson Weker, J. et al. (2018). *Proc. Natl. Acad. Sci. U.S.A.* 115: E5261–E5268.
217 Liu, P.-P., Zheng, Y.-Q., Zhu, H.-L., and Li, T.-T. (2019). *ACS Appl. Nano Mater.* 2: 744–749.
218 Jin, K., Chu, A., Park, J. et al. (2015). *Sci. Rep.* 5: 10279.
219 Chandrasekaran, S., Kim, E.J., Chung, J.S. et al. (2016). *J. Mater. Chem. A* 4: 13271–13279.
220 Gonçalves, J.M., Ireno da Silva, M., Angnes, L., and Araki, K. (2020). *J. Mater. Chem. A* 8: 2171–2206.
221 Cui, Y., Xue, Y., Zhang, R. et al. (2019). *J. Mater. Chem. A* 7: 21911–21917.

222 Lee, W.H., Ko, Y.J., Kim, J.Y. et al. (2020). *Chem. Commun.* 56: 12687–12697.
223 Zhu, C., Shi, Q., Feng, S. et al. (2018). *ACS Energy Lett.* 3: 1713–1721.
224 Chen, P., Zhou, T., Xing, L. et al. (2017). *Angew. Chem. Int. Ed.* 56: 610–614.
225 Zheng, Y., Jiao, Y., Zhu, Y. et al. (2017). *J. Am. Chem. Soc.* 139: 3336–3339.
226 Hou, Y., Qiu, M., Kim, M.G. et al. (2019). *Nat. Commun.* 10: 1392.
227 Li, Y., Wu, Z.S., Lu, P. et al. (2020). *Adv. Sci.* 7: 1903089.
228 Fei, H., Dong, J., Feng, Y. et al. (2018). *Nat. Catal.* 1: 63–72.
229 Ying, Y., Luo, X., Qiao, J., and Huang, H. (2020). *Adv. Funct. Mater.* 31: 2007423–2007447.
230 Bai, L., Hsu, C.S., Alexander, D.T.L. et al. (2019). *J. Am. Chem. Soc.* 141: 14190–14199.
231 Tavakkoli, M., Flahaut, E., Peljo, P. et al. (2020). *ACS Catal.* 10: 4647–4658.
232 Han, X., Ling, X., Yu, D. et al. (2019). *Adv. Mater.* 31: e1905622.
233 Zhu, X.F., Zhang, D.T., Chen, C.J. et al. (2020). *Nano Energy.* 71: 104597–104607.
234 Yin, J., Jin, J., Lin, H. et al. (2020). *Adv. Sci.* 7: 1903070.
235 Peng, X., Pi, C., Zhang, X. et al. (2019). *Sustain. Energy Fuels* 3: 366–381.
236 Dutta, A. and Pradhan, N. (2017). *J. Phys. Chem. Lett.* 8: 144–152.
237 Jin, S. (2017). *ACS Energy Lett.* 2: 1937–1938.
238 Wygant, B.R., Kawashima, K., and Mullins, C.B. (2018). *ACS Energy Lett.* 3: 2956–2966.
239 Stern, L.-A., Feng, L., Song, F., and Hu, X. (2015). *Energy Environ. Sci.* 8: 2347–2351.
240 Chen, P., Xu, K., Fang, Z. et al. (2015). *Angew. Chem. Int. Ed.* 54: 14710–14714.
241 Xu, J., Li, J., Xiong, D. et al. (2018). *Chem. Sci.* 9: 3470–3476.
242 Li, W., Xiong, D., Gao, X., and Liu, L. (2019). *Chem. Commun.* 55: 8744–8763.
243 Ghoshal, S., Zaccarine, S., Anderson, G.C. et al. (2019). *ACS Appl. Energy Mater.* 2: 5568–5576.
244 Zheng, W., Liu, M., and Lee, L.Y.S. (2020). *ACS Catal.* 10: 81–92.
245 Asthagiri, A. and Janik, M.J. (2014). *Computational Catalysis.* RSC Publishing.
246 Soriano-López, J., Schmitt, W., and García-Melchor, M. (2018). *Curr. Opin. Electrochem.* 7: 22–30.
247 Abild-Pedersen, F. (2016). *Catal. Today* 272: 6–13.
248 Greeley, J., Jaramillo, T.F., Bonde, J. et al. (2006). *Nat. Mater.* 5: 909–913.
249 Bjorketun, M.E., Bondarenko, A.S., Abrams, B.L. et al. (2010). *Phys. Chem. Chem. Phys.* 12: 10536–10541.
250 Haber, J.A., Cai, Y., Jung, S. et al. (2014). *Energy Environ. Sci.* 7: 682–688.
251 Ge, L., Yuan, H., Min, Y. et al. (2020). *J. Phys. Chem. Lett.* 11: 869–876.
252 Chatenet, M., Benziger, J., Inaba, M. et al. (2020). *J. Power Sources* 451: 227635–227638.
253 Chen, J.G., Jones, C.W., Linic, S., and Stamenkovic, V.R. (2017). *ACS Catal.* 7: 6392–6393.

254 Voiry, D., Chhowalla, M., Gogotsi, Y. et al. (2018). *ACS Nano* 12: 9635–9638.
255 Bird, M.A., Goodwin, S.E., and Walsh, D.A. (2020). *ACS Appl. Mater. Interfaces* 12: 20500–20506.
256 Anantharaj, S. and Kundu, S. (2019). *ACS Energy Lett.* 4: 1260–1264.
257 Zheng, W., Liu, M., and Lee, L.Y.S. (2020). *ACS Energy Lett.* 5: 3260–3264.

2

Studies on Cerium-Based Nanostructured Materials for Electrocatalysis

Xuemei Zhou[1], Mingkai Zhang[2], Yuwei Jin[1], and Yongquan Qu[3]

[1] Wenzhou University, Key Laboratory of Carbon Materials of Zhejiang Province, College of Chemistry and Materials Engineering, Chashan University Town, Wenzhou City, Zhejiang Province 325027, China
[2] Xi'an Jiaotong University, Frontier Institute of Science and Technology, No.1 West building, 99 Yanxiang Road, Yanta District, Xi'an, Shaanxi Province 710049, China
[3] Northwestern Polytechnical University, Key Laboratory of Special Functional and Smart Polymer Materials of Ministry of Industry and Information Technology, School of Chemistry and Chemical Engineering, 1 Dongxiang Road, Chang'an District, Xi'an, Shaanxi Province 710072, China

2.1 Introduction

Environment-friendly energy storage technologies with high capacity and efficient catalytic conversion techniques are important in producing sustainable energy by converting renewable and intermittent energy resources (solar, wind, hydroelectric power, etc.) into storable and green energy [1]. Electrocatalysis, including hydrogen evolution reaction (HER), oxygen evolution reaction (OER), oxygen reduction reaction (ORR), N_2 reduction reaction (NRR), and CO_2 reduction reaction (CO_2RR), promotes the transformation of universal and economical sources (e.g. water, N_2, CO_2) into various value-added chemicals used in daily human life and industrial manufacturing [2].

Noble metals (e.g. Pt, Ru) or noble metal oxides (e.g. IrO_2, RuO_2) are still recognized as superior electrocatalysts for HER, OER, ORR, etc. [3, 4]. However, precious metals are scarce and expensive. Efforts have been made to develop affordable electrocatalysts in the last decades. Electrocatalytic reactions usually occur at the triphase and interface, where the catalysts and electrode–electrolyte interfaces develop fast electron and efficient mass transport. There are two main strategies to improve the electrocatalytic activity: (i) increasing the number of active sites through the nano-architecture design of the catalysts or the electrodes; and (ii) enhancing the intrinsic activity of each active site via doping or heterostructure engineering to regulate the electronic structure. Various nanostructures with porous, hollow, and two-dimensional structures with more exposed surfaces facilitate mass transport improving electrocatalytic performance. Conductivity is another factor for enhancing electrocatalytic performance.

Recent advances provided good results when additives were integrated with electrocatalysts to enhance electrocatalytic activity and long-term stability [5–8].

Functional Nanomaterials: Synthesis, Properties, and Applications, First Edition.
Edited by Wai-Yeung Wong and Qingchen Dong.
© 2022 WILEY-VCH GmbH. Published 2022 by WILEY-VCH GmbH.

Ceria (CeO_2) with abundant surface defects, relatively cheap, flexible transformation between Ce^{3+} and Ce^{4+}, and high oxygen storage capacity were effective as a catalyst support or hybrid with other materials, even electroactive catalysts [9]. Many CeO_2-based materials have shown superior electrocatalytic performance, such as noble metal species [10], transition metal oxides [11], hydroxides [12], phosphides [13], and sulfides [14].

CeO_2-based nanomaterials with special morphologies and exposed surfaces are also ideal catalysts for comprehensive structure-activity relationship studies. There are developments in characterization techniques, especially in advanced *in situ*/operation spectroscopy and microscopic studies. Active sites can be experimentally identified, and changes in their structure and composition can be accurately monitored, promoting the fundamental understandings of catalytic reaction mechanisms and dynamic evolution of the active site [15–18]. Moreover, the theoretical investigation of thermodynamics and kinetics of the electrocatalytic reaction process is also important in designing and analyzing CeO_2-based electrocatalysts [19, 20]. The experimental results and density functional theory (DFT) calculations obtained the information of electronic interaction and energy barrier in heterojunction interfaces. The applications of CeO_2-based nanomaterials in thermal catalysis (such as CO oxidation, oxidation of volatile organic compounds (VOCs), and low-temperature water–gas shift reactions) [15, 21, 22] and photocatalysis have been widely investigated over the past decades [10, 23]. However, few have focused on CeO_2-based materials for many kinds of electrocatalysis [9, 24, 25].

This chapter discusses the application of CeO_2-based nanomaterials to various electrocatalytic reactions (HER, OER, ORR, etc.) and provides a comprehensive analysis of recent progress, including the catalytic mechanisms for special electrochemical reactions and the strategies for enhancing the catalytic performance of CeO_2-based electrocatalysts. After thorough investigation, a solution will be proposed to guide and enlighten future studies of CeO_2-based materials as efficient and economic electrocatalysts for clean energy conversion.

2.2 Cerium-Based Nanostructure Materials

Cerium in the F block of the periodic table is the most abundant rare earth alkali element, comparable to or higher than the transition metals Co, Ni, and Cu. Cerium exhibits a property of cycling between two oxidation states (Ce^{3+} and Ce^{4+}) due to its ground-state electron in the 4f (Xe $4f^1 5d^1 6s^2$) orbital [26]. Generally, CeO_2 crystallizes in the fluorite crystal structure with space group *Fm3m* containing a cubic close-packed array of metal atoms with eight coordinate Ce^{4+} and four coordinate O^{2-}, as given in Figure 2.1a. The most stable (111) crystal surface is enclosed by threefold-coordinated oxygen atoms and sevenfold-coordinated cerium atoms [27, 28]. The CeO_2(110) crystal plane is ended by threefold oxygen and sixfold cerium atoms; the CeO_2(100) plane is dominated by twofold-coordinated oxygen atoms [27, 28]. The nonstoichiometric CeO_2 is formed by releasing the lattice oxygen to generate oxygen vacancies without destructing its fluorite crystal

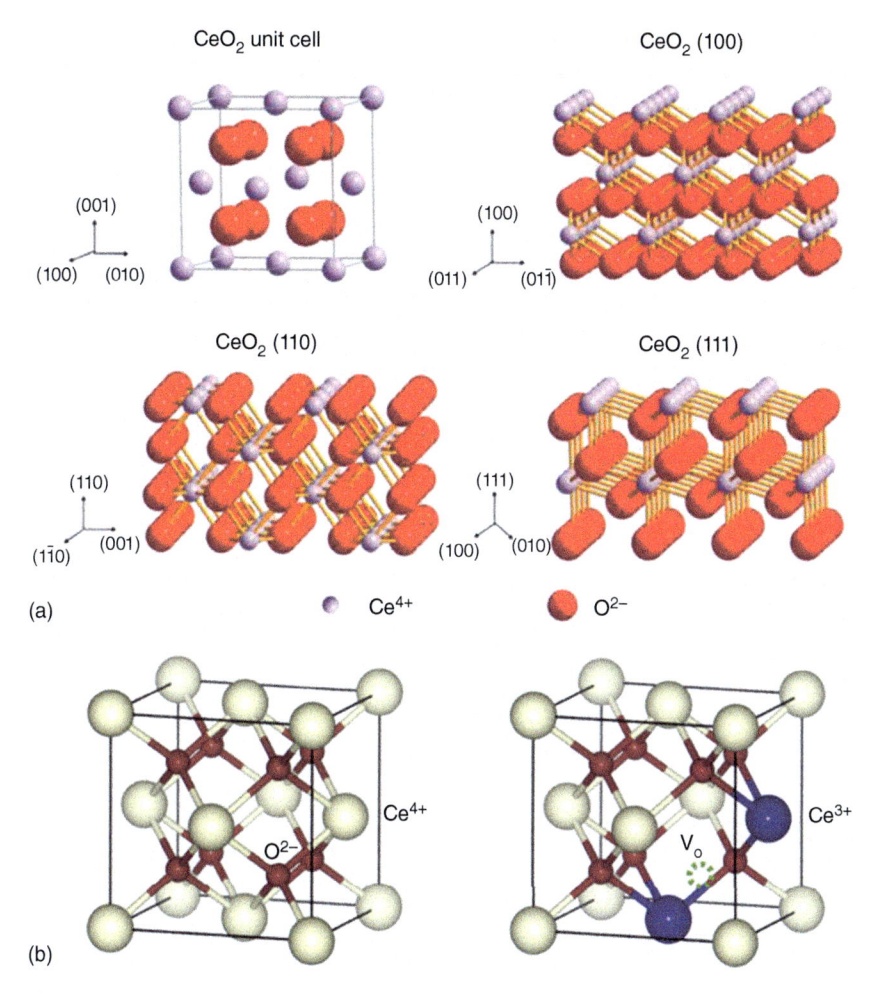

Figure 2.1 (a) Atomic configurations of the unit cell and the (100), (110), and (111) facets of CeO_2. Source: Reprinted with permission from Li and Shen. [27]. Copyright 2014 Royal Chemical Society. (b) Crystal structure of CeO_2 in the presence of one oxygen vacancy accompanied with two generated Ce^{3+} species. Source: Reprinted with permission from Ma et al. [21]. Copyright 2018 Elsevier B. V.

structure (Figure 2.1b) [21]. With the formation of an oxygen vacancy in the lattice, the oxygen atom removal results in the localization of two electrons in the two cerium cations that changes their oxidation states from Ce^{4+} to Ce^{3+}, as presented in Eq. (2.1) [10, 21].

$$2Ce_{Ce}^x + O_O^x \rightarrow V_O^{\bullet\bullet} + 2Ce'_{Ce} + 1/2 O_2(g) \tag{2.1}$$

Moreover, the formation of oxygen vacancies is more significant by decreasing the size of CeO_2 crystals to the nanoscale. Surface defects originating from the oxygen vacancies can bring the unique and attractive surface physicochemical properties of nanostructured CeO_2. They include high surface areas, tunable pore sizes, abundant defects, reversible valence of Ce^{4+}/Ce^{3+}, and tailorable surface chemistry [29].

2.3 Cerium-Based Electrocatalysts for HER

HER is an important half-reaction for water electrolysis. Although Pt-based electrocatalysts are still considered the best catalysts for HER, the development of affordable and high catalytic alternatives is needed [30–32]. Cerium is one of the most abundant and economically rare earth elements, with properties that improve electrocatalytic activity. It is also considered a crucial functional material due to its excellent redox ability and the flexible transformation between Ce^{4+} and Ce^{3+} states. The rich oxygen defects in CeO_2 are conducive to the absorption of OH^- species, fast electron transfer [33], water dissociation, hydrogen adsorption free energy [34], and formation of dangling bonds and gap states, resulting in the electron rearrangement of the interfaces [35]. CeO_2-based materials have the potential to improve electrocatalytic HER activity of various catalysts feasible for large-scale applications. Multiple strategies have been adopted to enhance the performance of CeO_2-based electrocatalysts for HER, and remarkable results have been achieved.

2.3.1 Cerium-Doped Electrocatalysts for HER

Doping cerium into electrocatalysts (such as phosphides, sulfides, and transition metal-layered double hydroxide (LDH)) can effectively tune their intrinsic properties to modulate the adsorption of active hydrogen species, which has been used in electrocatalytic hydrogen production [36–39]. Recently, our group successfully modulated the electronic structure of CoP by doping Ce^{3+} ions, which delivered superior HER catalytic performance in both acidic and alkaline solutions [36, 40]. The lower hydrogen adsorption free energy on different CoP facets after introducing Ce was confirmed by DFT calculations (Figure 2.2a). In contrast, the Bader charge analyses showed that the electronic structure of Co was tuned to a negatively charged state by the introduction of Ce. The experimental results confirmed the successful doping of Ce into CoP (Figure 2.2b).

Meanwhile, according to the polarization curves, the overpotentials and Tafel slopes significantly decreased in both acidic and alkaline electrolytes for HER, which was consistent with the theoretical analyses (Figure 2.2c,d). The results further indicate that doping Ce can optimize the electronic structure of supporting materials to promote HER activity [36, 40]. Recently, chemical doping of Ce into CoS_2 nanowire also significantly improved the HER catalytic stability of the electrodes, exhibiting an activity-durability factor of ~31 times compared with that of CoS_2 alone. Catalytic mechanism and control experiments demonstrated that the Ce-doped CoS_2 significantly weakened the interfacial O_2 adsorption, minimizing the ambient oxidation and electrochemical oxidation that leads to the suppressed oxidation- and corrosion-induced catalysts leaching [38].

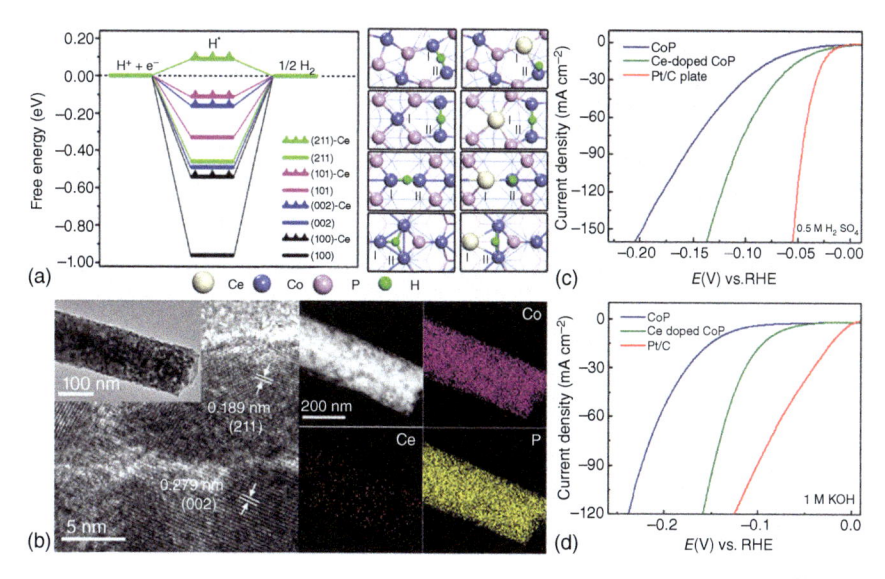

Figure 2.2 (a) Calculated free energy diagram for HER on different facets of CoP and Ce-doped CoP with optimized adsorption structures of H*. (b) Transmission electron microscopy (TEM) and high-resolution TEM (HRTEM) images of Ce-doped CoP nanowires, dark field-TEM images of Ce-doped CoP, and element mapping images of Co, Ce, and P. Polarization curves (with iR corrections) of as-grown catalysts in (c) 0.5 M H_2SO_4 and (d) 1 M KOH solutions with a scan rate of 5 mV s^{-1}. Source: Gao et al. [36]/Reproduced with permission of Elsevier.

Ce-doped NiFe-LDH electrocatalyst was grown on a nickel foam (NF) substrate at room temperature via an electrodeposition method in 1 M KOH. It exhibited excellent catalytic activity for overall water splitting and long-term stability. The Tafel slope value of the optimal Ce-doped NiFe-LDH (NiFe$_{0.8}$Ce$_{0.2}$) was 112 mV dec^{-1}, indicating the Volmer–Heyrovsky mechanism was employed toward HER on NiFe$_{0.8}$Ce$_{0.2}$, and the Volmer step ($H_2O + e^- \rightarrow H_{ads} + OH^-$) was the rate-determining step (RDS) [39].

2.3.2 Composites with CeO_2 for HER

Fundamental structural studies show that ceria is widely used as special support incorporated with other materials, such as metals and their alloys [41–43], hydroxides [44], phosphides [13], and nitrides [45]. These materials enhance the catalytic performance of the electrochemical field through strong interface interactions and tunable electronic coupling effects [46].

Demir et al. reported that the Ru^0/CeO$_2$ (1.86 wt% Ru) catalysts showed superior electrocatalytic HER activity with low overpotential, high exchange current density, small Tafel slope, and long-term stability in acidic electrolytes [41]. In another study, Pd/CeO$_2$/C hybrids, prepared using an impregnation and reduction method, exhibited a superior HER electrocatalytic activity in basic solution. Based on the DFT computations and experiments, the improved HER activity could be attributed to

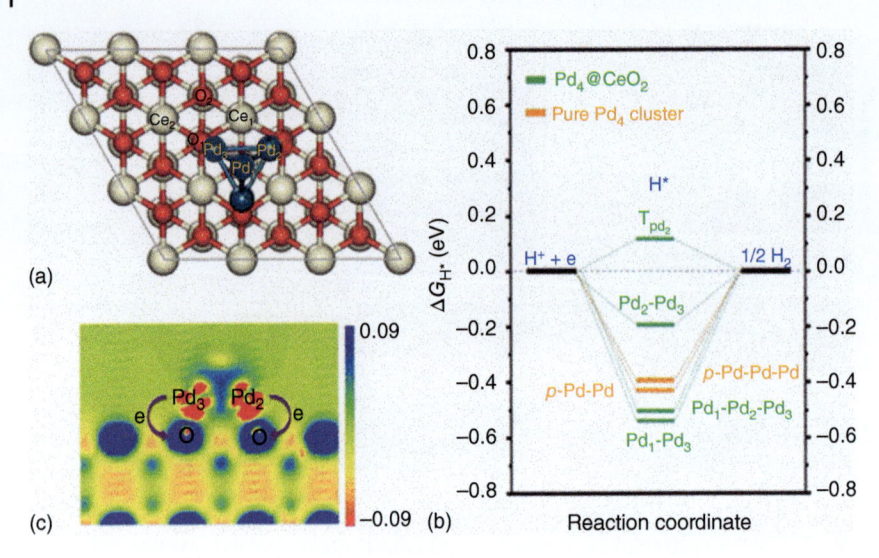

Figure 2.3 (a) Top view of $Pd_4@CeO_2$ model and the possible adsorption sites for H* on the $Pd_4@CeO_2$. (b) The calculated free-energy diagram of HER for the $Pd_4@CeO_2$ and pure Pd_4 cluster at the equilibrium potential. (c) The charge density difference $\Delta\rho$ of the $Pd_4@CeO_2$ model for the slice (vertical to the surface of substrate CeO_2) cut along the Pd2–Pd3 bond, where the red and blue areas denote the decreased and increased electron densities, respectively. Source: Gao et al. [43]/Reproduced with permission from American Chemical Society.

the presence of CeO_2, which acted as a support for anchoring and stabilizing Pd nanoparticles (NPs) that encouraged strong interactions between Pd and CeO_2 [42]. The subunit Pd nanoclusters (NCs) were further confirmed to be the active centers based on the free energy of H* (ΔG_{H^*}) (Figure 2.3) [43]. Meanwhile, the electron transfer between Pd and O atoms in Pd NCs@CeO_2 could effectively regulate the H* adsorption state on the subunit of the Pd cluster compared to the pure Pd NCs. This co-joint function could effectively improve the HER catalytic performance of the hybrid electrocatalysts. This work can guide the application of metal NC-transition metal oxide composites in catalysis.

The non-noble metals (such as Ni [47], Co [48, 49], alloys [48]) were considered alternative HER electrocatalysts, with advantages such as relatively low overpotential and good corrosion resistance in alkaline medium. Sun et al. reported 3D porous Ni–CeO_2 nanosheets grown on Ti mesh (Ni–CeO_2/TM) (Figure 2.4a) as efficient electrocatalysts for HER with a low overpotential of 67 mV at a current density of 10 mA cm^{-2}. The electron-rich Ce^{3+} in the complexes and abundant exposed Ni active sites favored the adsorption of water molecules that facilitated dissociation of the O—H bond, resulting in the improvement of HER performance [50]. Additionally, the possible synergetic effects between CeO_2 particles and the Ni matrix could facilitate the migration of H_{ads} from the Ni to CeO_2 particles [47, 53]. Based on the "volcano curve," integrating NiCo alloy and CeO_2 further enhanced the electrocatalytic performance of alloyed metals toward the HER with low overpotentials of 34 mV at 10 mA cm^{-2}. It exhibited excellent durability in

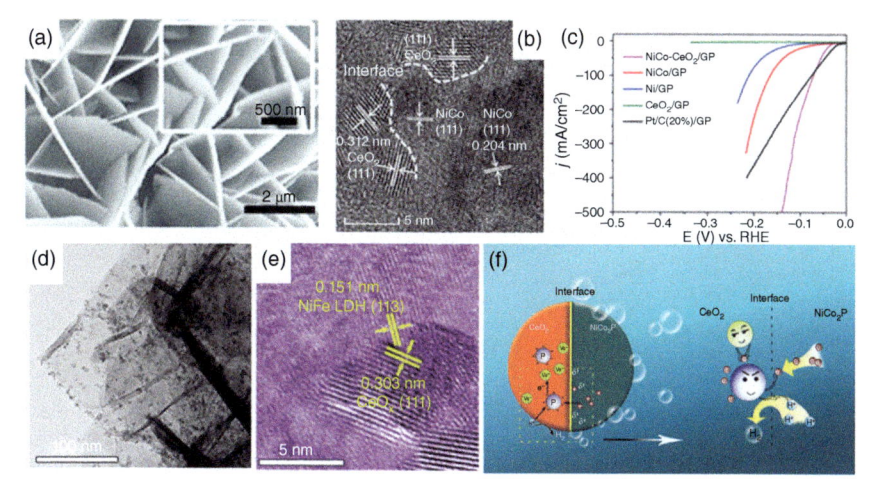

Figure 2.4 (a) Scanning electron microscopy (SEM) images of Ni–CeO$_2$/TM. Source: Sun et al. [50]/Reproduced with permission from Royal Society of Chemistry. (b) HRTEM image of NiCo–CeO$_2$/GP. Source: Sun et al. [50]/Reproduced with permission from Wiley-VCH. (c) Linear sweep voltammetry (LSV) curves of NiCo–CeO$_2$/GP, NiCo/GP, Ni/GP, CeO$_2$/GP, and Pt/C(20%)/GP in 1 M KOH purged with Ar. Source: Reproduced with permission from Sun et al. [48]. Copyright 2020 Wiley-VCH. (d) TEM image and (e) HRTEM image of NF@NiFe LDH/CeO$_x$. Source: Wang et al. [51]/Reproduced with permission from American Chemical Society. (f) Schematic diagram of the interface interaction between NiCo$_2$P and CeO$_2$ in which oxygen vacancies and P atoms coexist in the electrocatalyst of NiCo$_2$P^x/P^{x}FVO–CeO$_2$. Source: Reproduced with permission from Wang et al. [52]. Copyright 2019 American Chemical Society.

alkaline solution (Figure 2.4b,c) [48]. Similarly, the Co–W/CeO$_2$ composite coatings displayed improvement in the catalytic HER in alkaline solution, which could be attributed to the interaction between CeO$_2$ particles and Co–W matrix [49].

Integrating transition metal-LDH with nanostructured CeO$_2$ can facilitate the electron/ion migration, enhance the charge transfer and increase the electrochemical active sites, improving electrocatalytic performance. The strong electronic effects between NiFe LDH and CeO$_x$ in NiFe LDH/CeO$_x$ hybrids (Figure 2.4d,e) resulted in a rearrangement of Ni and Fe valence states, improving electrocatalytic activities of HER and overall water splitting [51]. Based on the DFT results, the introduction of CeO$_x$ could generate abundantly exposed oxygen vacancies, enhancing electrocatalytic performance. Similarly, the Co(OH)$_2$/CeO$_2$ nanoplates displayed an improvement of HER activity due to the increased ratio of Ce^{3+} and the concentration of oxygen defects, which caused the charge interactions between CeO$_2$ nanoplate and Co(OH)$_2$ [44].

Transition-metal phosphides (such as CoP [13, 54–56], Ni$_2$P [57], and Cu$_3$P [58]) have also been used as electrocatalysts for HER in acid or alkaline media. Wang et al. reported CeO$_2$ coated NiCo$_2$P electrocatalysts had improved HER activity. The oxygen vacancies of CeO$_2$ were filled by P, elevating chemical valence states of Ni^{2+} and Co^{3+}, increasing the hydride acceptors, and facilitating the electrons and ions transfer (Figure 2.4f) [52]. In essence, NiCo$_2$P$_{0.3}$/P$_{0.3}$FVO–CeO$_2$ only required an

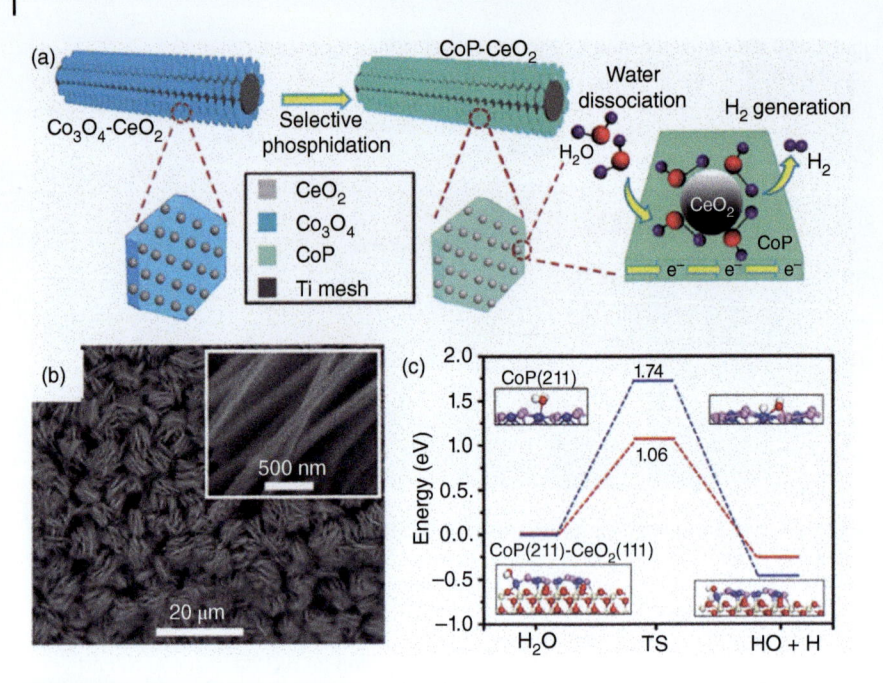

Figure 2.5 (a) Schematic illustration of the selective phosphidation in the catalyst synthesis and the catalytic process. (b) SEM images of CoP–CeO$_2$/Ti. (c) DFT calculated reaction energy diagram of water dissociation for CoP (211) and CoP (211)/CeO$_2$(111). Source: Zhang et al. [56]/Reproduced with permission from Royal Chemical Society.

overpotential of 33.6 mV to deliver a current density of 10 mA cm^{-2} in base media [52]. Meanwhile, a series of MP–CeO$_2$ (M = Ni, Cu, Co) hybrid catalysts directly grown on different substrates (Figure 2.5a,b) served as efficient electrocatalysts for HER [56–58]. DFT calculations confirmed that the charge rearrangement in CoP–CeO$_2$ could decrease water dissociation free energy and adjust hydrogen adsorption free energy (Figure 2.5c), resulting in improved HER. Carbon-based materials with high surface area and high electrical conductivity were considered ideal scaffolds to support catalysts [59]. Carbon-based materials were employed as supporting substrates for CeO$_2$ and nitrogen co-doped CoP nanorod arrays to explore pH-universal metal phosphide electrocatalysts with high activity and stability [13]. Apart from phosphides, metal nitrides (e.g. Ni$_3$N) incorporated with CeO$_2$ are also potential electrocatalysts for HER [45].

The recent studies of CeO$_2$-based electrocatalysts for HER in acid and alkaline solutions are summarized in Table 2.1. Several methods have been applied to improve the stability and activities of CeO$_2$-based electrocatalysts. However, the development of nonprecious electrocatalysts with low overpotential is a challenge. Low overpotential is usually required to achieve excellent HER. The free energy of hydrogen adsorption is the main factor in determining the rate of the whole reaction, especially under alkaline media. The overall activity of CeO$_2$-based electrocatalysts under alkaline conditions could be improved by constructing the interfaces or optimizing binding energy for the intermediates. Furthermore, the mechanisms of CeO$_2$ to enhance HER are indefinite. Therefore, it requires deeper investigation and further development.

Table 2.1 Summarized overpotentials and Tafel slopes for the HER performance of reported electrocatalysts in literature.

Synthesis method	Catalysts	η_{10} (mV)	Tafel slope (mV dec^{-1})	Electrolyte	References
Hydrothermal	Ce-doped CoP	54	59.3	0.5 M H_2SO_4	[36]
Thermal treatment	$Ce_{0.05}$-doped CoP CNTs	146	78	0.5 M H_2SO_4	[37]
Hydrothermal	Ce-doped CoS_2	82	59.4	1 M KOH	[38]
Electrodeposition	Ce-doped NiFe-LDH	147	112	1 M KOH	[39]
Reduction	Ru^0/CeO_2	47	41	0.5 M H_2SO_4	[41]
Chemical deposition, reduction	$Pd/CeO_2/C$	109	75	1 M KOH	[42]
Hydrothermal	Pd NCs@CeO_2	36	235	0.5 M H_2SO_4	[43]
Thermal reduction	Ni–CeO_2/TM	67	111	1 M KOH	[50]
Electrodeposition	Ni–CeO_2	/	131	1 M NaOH	[53]
Solvothermal	Ni_xCeO_{2+x}–CNT	180	/	1 M NaOH	[47]
Electrodeposition	NiCo–CeO_2/GP	34	49.1	1 M KOH	[48]
Hydrothermal, electrodeposition	NF@NiFe LDH/CeO_x	154	101	1 M KOH	[51]
Hydrothermal, reduction	$CeO_2/Co(OH)_2$	317	140	1 M KOH	[44]
Hydrothermal	Ce, N–CoP/CC	41	43	1 M KOH	[13]
Thermal treatment	CeO_2-CoP-C	71	53	0.5 M H_2SO_4	[54]
Thermal treatment	CeO_x/CoP@C-2	127	35	1 M KOH	[55]
Hydrothermal, annealing	Ni_2P–CeO_2	84@20	87	1 M KOH	[57]
Hydrothermal, annealing	CeO_2-Cu_3P	148@20	132	1 M KOH	[58]
Hydrothermal, electrodeposition	$NiCo_2P^x/P^xFVo$–CeO_2	33.6	61.4	1 M KOH	[52]
Hydrothermal, annealing	Ni_3N–CeO_2/TM	80	122	1 M KOH	[45]
Hydrothermal	3D-rGO-CeO_2	340	112.8	1 M KOH	[59]

GP: graphite plate.
η_{10} refers to the overpotential at the current density of 10 mA cm^{-2}.

2.4 Cerium-Based Electrocatalysts for OER

OER, a half-reaction with a two-electron or four-electron process to convert water into oxygen molecules, has always been an obstacle for its sluggish reaction kinetics. During a basic OER process, the O and OH are first adsorbed on the surfaces of the electrocatalysts. Then, these intermediates are oxidized to generate O_2 via two pathways: one is the direct recombination intermediate O to produce O_2 that is usually considered as a rate-determining step because of its higher thermodynamic barrier [60], and the other uses the reaction between the intermediate O and OH⁻ to form hydrogen peroxide and is further oxidized to O_2 and H_2O, which has been commonly accepted [61, 62]. In OER, the formation of the intermediate, hydrogen peroxide, is the key process, and efficient electrocatalysts usually facilitate the formation of this species.

2.4.1 Cerium-Doped Electrocatalysts for OER

Doping Ce ions into transition metal-based electrocatalysts enhance surface properties (such as flexible variation between Ce^{3+} and Ce^{4+} oxidation states) for OER by increasing the mobility and formation of oxygen species [62, 63]. Jaramillo and coworkers reported an improvement of the OER of electrodeposited NiO_x using cerium as a dopant and gold as conductive metal support ($NiCeO_x$–Au) [64]. DFT calculations revealed that the coexistence of both Ce and Au near the under-coordinated Ni edge on NiOOH ($01\bar{1}2$) resulted in a lower theoretical overpotential (Figure 2.6a). Good oxophilicity of Ce and the local active site was observed. The OER energy descriptor $\Delta G_O - \Delta G_{OH}$ was also lowered to an optimal value for OER because of the strong electronic interactions in the $NiCeO_x$–Au systems.

Subsequently, a series of Ce-doped transition metal (TM)-LDH electrocatalysts were developed by different preparation methods (Figure 2.6b,c) [65–69]. By regulating the Ce ratio to less than 40%, the as-grown catalysts displayed lower overpotentials and higher catalytic current densities than the corresponding counterparts without the doped Ce. For example, the 30% CeNiFe-LDH required an overpotential of 242 mV to reach 10 mA cm⁻² for the OER process (Figure 2.6d) [66]. The improvement on the OER activity for these systems by introducing Ce can be summarized as the following factors. First, doping Ce into TM-LDH mainly changed the layer composition and coordination structure of LDH. Thus, it significantly improved the electrical conductivity of the LDH. Second, the intrinsic Ce^{3+}/Ce^{4+} redox transformation contributed to a fast oxygen diffusion via short ion diffusion paths and kinetics of the released oxygen in the lattice due to more oxygen vacancies generated with the increased structural defects. Third, synergic effects derived from Ce-doped TM-LDH led to stronger adsorption of *OH (Figure 2.6e) [66–68]. Therefore, introducing Ce to the host electrocatalysts can promote electronic interaction and create a more defective structure for OER.

2.4.2 Composites with CeO₂ for OER

Compared to Ce ions, CeO_2 has been widely used as an additive to promote activity and stability, facilitating rich defects, abundant active sites, and favorable oxygen storage capacity. The combination with CeO_2 could facilitate the formation of hydroperoxy species (OOH_{ad}) in the composites, improving electrochemical performance [70]. Many electrocatalysts, including metals [70, 71], metal oxides [72, 73],

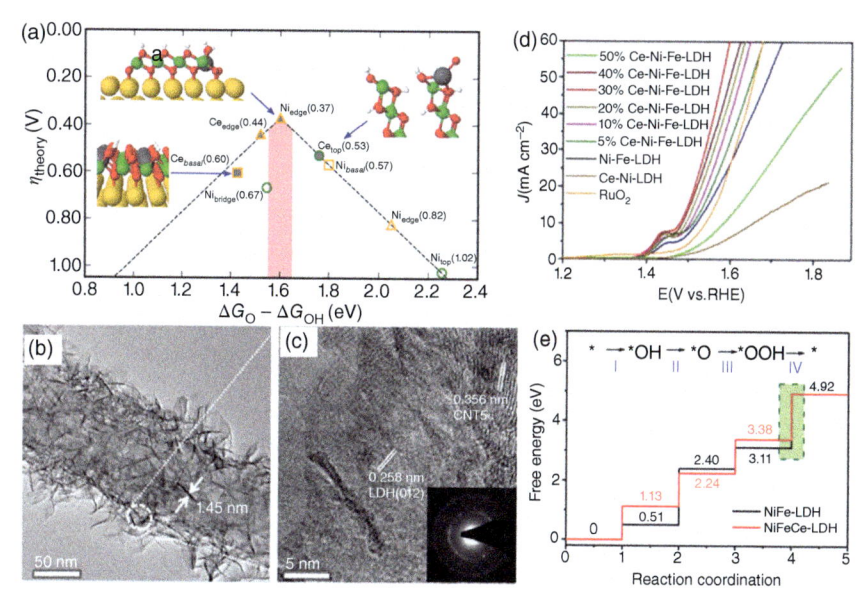

Figure 2.6 (a) Representation of the theoretical overpotential as a function of the difference in O* and HO* adsorption Gibbs energies. Green circles, gold squares, and gold triangles refer to the NiOOH ($01\bar{1}2$) surface, infinite NiO_x–Au thin films, and finite NiO_x–Au thin films, respectively. Grey-filled markers indicate Ce-doped structures. Source: Reproduced with permission from Ng et al. [64]. Copyright 2016, Macmillan Publishers Limited, part of Springer Nature. (b) TEM images and (c) HRTEM image of 5.0% Ce-NiFe-LDH/CNT, and the corresponding selected area electron diffraction (SAED) patterns of 5.0% Ce-NiFe-LDH/CNT (inset). Source: Xu et al. [65]/Reproduced with permission from American Chemical Society. (d) Polarization curves of Ni-Fe-Ce-LDH with different Ce/Fe ratio. Source: Reprinted with permission from Xu et al. [66]. Copyright 2020 Royal Chemical Society. (e) Gibbs free energy diagram for the four steps of OER on NiFe-LDH (black line) and NiFeCe-LDH (red line). The top is the brief four-electron mechanism of OER. Source: Reprinted with permission from Wen et al. [67]. Copyright 2021 Elsevier B.

metal hydroxides, sulfides, etc., hybrid with CeO_2 to achieve high OER activity in alkaline solution.

Co_3O_4 (spinel oxides) is considered a promising OER electrocatalyst in an alkaline medium because of its mixed valances of Co cations [74, 75]. Constructing Co_3O_4/CeO_2 nanohybrids (NHs) could further promote the OER performance of Co_3O_4. It could be attributed to the formation of strongly coupled interfaces that induce changes in the Co^{3+} octahedral environment [76]. For example, Co_3O_4/CeO_2 NHs were developed using Co_3O_4 nanosheets to support CeO_2 nanocubes (Figure 2.7a,b). It exhibited high OER activity with a low overpotential of 270 mV at a current density of 10 mA cm^{-2} and a high turnover frequency (TOF) (0.25 s^{-1}) [77]. The enhanced OER performance resulted from the specific structure of NHs with high exposed oxygen vacancy concentration. The interfaces in the NHs also offered high electron mobility and good conductivity (Figure 2.7c). Porous Co_3O_4@Z67 uniformly coated by CeO_2 shells (Co_3O_4@Z67@CeO_2) displayed remarkable OER behavior with an overpotential of 350 mV at a current density

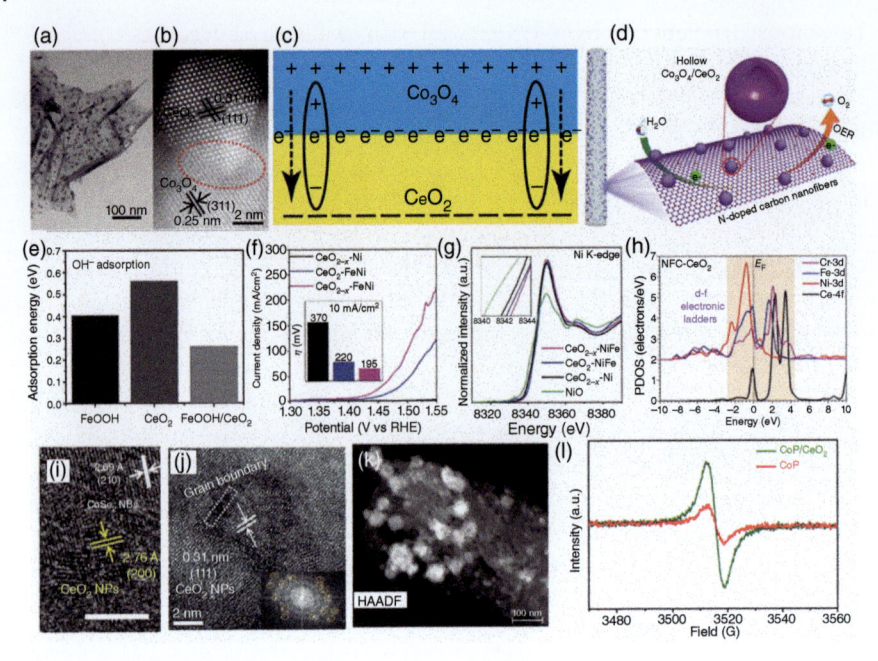

Figure 2.7 (a) TEM and (b) Scanning transmission electron microscopy (STEM) images of the Co_3O_4/CeO_2 NHs. (c) Schematic diagram of the interfacial electronic structure for Co_3O_4/CeO_2 NHs. Source: Liu et al. [77]/Reproduced with permission from WILEY-VCH. (d) Structure and composition advantages of Co_3O_4/CeO_2@N-CNFs for the OER. Source: Reproduced with permission from Li et al. [73]. Copyright 2019 American Chemical Society. (e) The calculated adsorption free energy changes of OH^- on FeOOH, CeO_2, and FeOOH/CeO_2. Source: Reproduced with permission from Feng et al. [78]. Copyright 2016 Wiley-VCH. (f) Polarization curves for as-grown catalysts [79]. (g) XANES spectra Ni K-edge of the samples. The inset of panel shows the enlarged absorption edge, demonstrating the Ni valance charge. Source: Reproduced with permission from Yu et al. [79]. Copyright 2021 Wiley-VCH. (h) The partial density of states (PDOS) of NFC-CeO_2. Source: Reproduced with permission from Xia et al. [80]. Copyright 2020 Wiley-VCH. (i) HRTEM images of the $CeO_2/CoSe_2$ nanobelt composite. Scale bar: 5 nm. Source: Zheng et al. [61]/Reproduced with permission from WILEY-VCH. (j) The HRTEM images and corresponding SAED pattern (the inset) of 14.6% CeO_x/CoS. Source: Xu et al. [14]/Reproduced with permission from WILEY-VCH. (k) STEM images of CoP/CeO_2 heterostructure. (l) ESR spectra of CoP/CeO_2 heterostructure. Source: Li et al. [81]/Reproduced with permission of Elsevier.

of $10\,mA\,cm^{-2}$, resulting from the interface effects between CeO_2 (Ce^{3+}/Ce^{4+} and oxygen defects) and Co_3O_4 (Co^{2+}/Co^{3+} and CoOOH) [82]. High surface area and porosity of N-doped carbon-based materials were employed to improve conductivity and optimize the morphologies and electron environment of the electrocatalysts [72, 73]. Li et al. used the electrospinning strategy to fabricate Co_3O_4/CeO_2 heterostructures *in situ* embedded in N-doped carbon nanofibers (h-Co_3O_4/CeO_2@N-CNFs) as an electrocatalyst with high efficiency and good stability, better than commercial RuO_2 catalysts. Its high catalytic performance resulted from synergistic interactions of the Co_3O_4/CeO_2 heterostructures and

2.4 Cerium-Based Electrocatalysts for OER

the porous 3D network of N-CNFs with rich available active sites, increased mass/charge mobility, and robust structure stability (Figure 2.7d) [73].

Apart from single metal oxides, spinel oxides with different transition metal elements were also considered as OER electrocatalysts due to their superior redox ability and variable valence states [83]. The $FeCo_2O_4/CeO_2$ heterostructures by direct annealing of metal-organic frameworks (MOFs) were constructed to deliver high OER performance [74]. The synergistic effects between $FeCo_2O_4$ and CeO_2 provided more exposed active sites and improved charge transfer and formation of intermediates. The existence of nanopores was beneficial to the diffusion of O_2 and the transport of electrolytes, affording an effective triphase region for OER [83, 84].

During the OER process, the formed hydroxides are the active sites determining the OER activity. Many reports have confirmed that appropriate introduction of CeO_2 in hydroxides like $FeOOH/CeO_2$ [78, 79], $CeO_2/Ni(OH)_2$ [85, 86], $CeO_2/Co(OH)_2$ [44], and $CeO_2/Cu(OH)_2$ [80] can result in a facile absorption of the generated oxygen during the OER that improves the OER activity due to the strong electronic interactions between CeO_2 and hydroxides. The hetero-layered $FeOOH/CeO_2$ nanotube arrays exhibited an efficient OER activity [78]. The X-ray photoelectron spectroscopy (XPS) profiles and DFT calculations showed a strong electronic interaction between FeOOH and CeO_2, which could reduce the energy barriers for various intermediate formations promoting the OER process (Figure 2.7e) [78]. Recently, high-valence nickel-doped CeO_{2-x} covered with FeOOH nanosheets were fabricated. It enhanced OER activity with a low overpotential of 195 mV at 10 mA cm^{-2} and a high TOF value of 0.99 s^{-1} (Figure 2.7f) [79]. Based on Ni K-edge X-ray absorption near edge structure (XANES) spectra, the highest binding energy for Ni in CeO_{2-x}–FeNi could be assigned to the highest Ni valence state (Ni^{3+}/Ni^{4+}) (Figure 2.7g), which could be the active sites in the OER process that could enhance electrocatalytic activity [79, 84]. Furthermore, the coupling between Ce-4f and 3d units (Ni, Fe, and Cr) provided active sites for fast charge transfer to promote the OER (Figure 2.7h) [80].

More studies have been done on metal sulfides [14], selenides [61], and phosphides [81] due to their unique characteristics, such as high surface areas, tunable porosity, and architectures. The decoration of CeO_2 onto $CoSe_2$ nanobelts (Figure 2.7i) could be beneficial in forming hydroperoxy species (OOH_{ad}) on the surfaces of the composites due to the surface structure that has high mobility of oxygen vacancies [61]. The synergistic chemical effects between $CoSe_2$ and CeO_2 would also improve the OER catalytic activity [61]. Another study also confirmed that regulation of the amount of CeO_2 on the surface of hollow CoS (Figure 2.7j) could increase the accessible sites, tailor the electronic states of CoS, and inhibit the corrosion of CoS, resulting in the enhancement of the OER catalytic performance and stability [14]. Recently, the $NiCo_2S_4$ hollow nanotube arrays decorated with CeO_x NPs assembled on a flexible support carbon cloth (CC) were reported to enhance OER stability by keeping horizontal for 20 hours to deliver 10 mA cm^{-2}. It was possibly ascribed to the formed protection film of CeO_x on the surface of $NiCo_2S_4$ [87]. A new type of CoP/CeO_2 heterostructures composed of CoP nanosheets decorated with CeO_2 NPs with the improved OER performance

(Figure 2.7k) was constructed by interface engineering [81]. In this study, the concentration of oxygen vacancies in the CoP/CeO$_2$ heterostructures was much more than that of pure CoP, according to the electron spin resonance (ESR) spectra (Figure 2.7l). Abundant oxygen vacancies were beneficial to the absorption of the oxygen species, leading to the improved OER activities of catalysts [81]. Meanwhile, strong electronic interactions between vanadium (V) and CeO$_2$ could adjust the electron density of Co atoms with increased catalytic active sites and intrinsic activity. These interactions optimize the Gibbs free energy values of H* adsorption energy (ΔG_{H*}) and enable a rapid OER process in alkaline electrolytes [88]. The concentration of oxygen vacancies and the oxygen adsorption capacity were also evaluated via the iodometric titration method and oxygen temperature-programmed desorption (O$_2$-TPD) method [89, 90].

The improvement in alkaline OER activity with the introduction of Ce^{3+} or the hybridization of CeO$_2$ could be attributed to synergistic effects. However, the mechanism of heterostructure electrocatalysts activation via cerium or ceria is complicated. Recently, our group built and estimated two CeO$_2$-incorporated NiO catalysts with different distributions and configurations, namely CeO$_2$-embedded NiO and CeO$_2$-surface-loaded NiO (Figure 2.8a) [90]. The catalysts with embedded

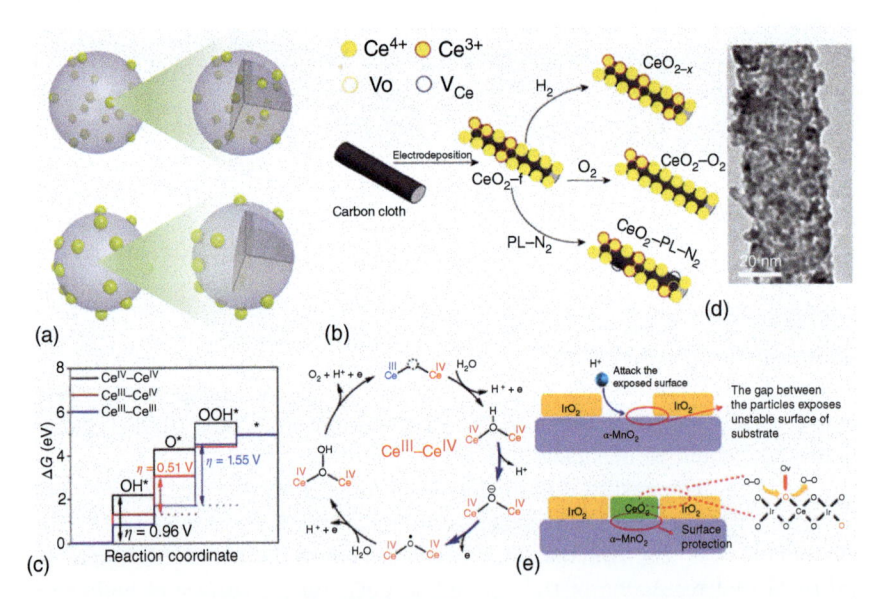

Figure 2.8 (a) Diagrams of CeO$_2$-embedded NiO (Ce–NiO–E) (up) and CeO$_2$-surface loaded NiO (Ce–NiO–L). The gray balls represent NiO and the yellow ones represent CeO$_2$ species. Source: Reproduced with permission from Gao et al. [90]. Copyright 2018 Wiley-VCH. (b) Schematic illustration of the synthesis procedures for the various catalysts of CeO$_2$–f, CeO$_{2-x}$, CeO$_2$–O$_2$, and CeO$_2$–PL–N$_2$. (c) Standard free-energy diagrams for OER and catalytic cycles for Ce^{III}–Ce^{IV}. Source: Reproduced with permission from Yu et al. [91]. Copyright 2020 Wiley-VCH. (d) HRTEM image of CeIr-ox/Mn. Source: Sun et al. [92]/Reproduced with permission from American Chemical Society. (e) Brief description of the effect of additional CeO$_2$ particles on the stability of the catalysts. Source: Reproduced with permission from Sun et al. [92]. Copyright 2020 American Chemical Society.

CeO$_2$ clusters into NiO displayed better OER activities than the surface-loading structure of CeO$_2$, which was mainly ascribed to the increased available active sites, low coordination of Ce–Ce and Ce–O, and the adjusted electronic structures of the intrinsic property of NiO.

Meanwhile, Yang's group designed and synthesized ceria with varying oxidation states of Ce ions to investigate the underlying causes of improved OER of ceria-based hybrid electrocatalysts (Figure 2.8b) [91]. They found that creating Ce^{3+} species played an important role in enhancing the OER performance of ceria and ceria-based catalysts. Furthermore, controlling the modification of the oxidation states of Ce in ceria could lower charge-transfer resistance and optimize the binding energy of OH* intermediates during the OER process. Furthermore, the Ce^{III}–Ce^{IV} pair with the optimal oxygen-binding free energy was found to be the active site that reduced the energy barrier of the reaction steps, making it the crucial index for estimating OER activities of ceria-based electrocatalysts (Figure 2.8c) [91]. These results provide insight into the design of hybrid catalysts with improved OER catalytic activity.

Combining CeO$_2$ with other active species (such as Ir [92, 93], Ru [94]) is an effective approach to adjust the electronic structure and improve OER activity and stability in acid medium. For example, the cerium-decorated iridium dioxide (IrO$_2$) grown on the surface of α-MnO$_2$ nanorods (Ce-Ir-ox/Mn) (Figure 2.8d) via a two-step hydrothermal process delivered a high catalytic activity in acidic electrolyte. The specific mass activity (275 A g$_{Ir}$$^{-1}$ at the overpotential of 300 mV) of the optimized hybrids was 5.7 times higher than that of IrO$_2$, along with excellent stability influenced by the CeO$_2$ layer on the surface (Figure 2.8e). X-ray adsorption spectroscopy (XAS) and XPS analyses suggested Ce decoration caused a lattice distortion on the IrO$_6$ octahedron and a partial occupation of e$_g$ orbitals, resulting in enhanced OER activity [92]. Another study also demonstrated that optimal Ce$_{0.2}$-IrO$_2$@N-doped porous carbon (NPC) electrocatalysts delivered a low overpotential of 224 mV and remarkable durability of 100 hours at a current density of 10 mA cm^{-2} in 0.5 M H$_2$SO$_4$ [93]. This good performance resulted from the addition of Ce that reduced the energy barrier of the OER RDS and the large surface area of NPC structure [93].

The effect of ceria on the OER performance of several types of electrode materials (metal oxides, hydroxides, phosphides, sulfides, selenides) has been discussed. The OER activities of some representative CeO$_2$-based electrocatalysts are listed in Table 2.2. Several factors contributing to OER performance have been investigated using ceria. Studies include the strong electronic effects between CeO$_2$ and other components, the increased oxygen vacancies concentration in CeO$_2$, the tailorable electronic structures, the increased porosity, and the enhanced electron transfer ability with the support of carbon-based materials. The interfaces between CeO$_2$ and other components are beneficial to the electron transfer and OER process. However, the detailed interaction mechanisms still lack and need to be uncovered through *in situ* and operando measurements to design efficient CeO$_2$-based electrocatalysts.

Further insight is needed to understand the reaction mechanism. It is important to reveal the exact changes in the CeO$_2$-based electrocatalysts after the OER. In addition, CeO$_2$-based electrocatalysts displayed superior OER or HER activities, whereas

Table 2.2 Summarized overpotentials and Tafel slopes for the OER performance of reported electrocatalysts in literature.

Synthesis method	Catalysts	η_{10} (mV)	Tafel slope (mV dec^{-1})	Electrolyte	References
Electrodeposition	NiCeO$_x$–Au	271	/	1 M NaOH	[64]
Coprecipitation	5% Ce-NiFe-LDH/CNT	227	33	1 M KOH	[65]
Green synthesis	Ni$_3$Fe/Ce-6	310	49	1 M KOH	[69]
Solvothermal	10% Ce–NiCoLDH/CNT	236	56	1 M KOH	[68]
Template	NiFeCe–LDH/MXene	260	42.8	1 M KOH	[67]
Hydrothermal	30% Ce-Ni-Fe-LDH	242	34	1 M KOH	[66]
ESD	Amorphous CoO$_x$(Ce)	261@20	65.67	1 M KOH	[95]
Thermal treatment	Ce-Co$_3$O$_4$	369	56	1 M KOH	[96]
Coprecipitation	Ce-MnCo$_2$O$_4$-3%	390	125	1 M KOH	[97]
Electrodeposition	N, Ce–CoS$_2$	190	78	1 M KOH	[98]
Reduction	Ru^0/CeO$_2$	420	122	0.5 M KOH	[71]
Hydrothermal	1-RuO$_2$/CeO$_2$	350	74	1 M KOH	[99]
Solvothermal	Co$_3$O$_4$/CeO$_2$	270	60	1 M KOH	[77]
Templates	CeO$_2$/Co$_3$O$_4$	265	68.1	1 M KOH	[76]
Hydrothermal	Co$_3$O$_4$@Z67@CeO$_2$	350	80.7	0.1 M KOH	[82]
Solvothermal	CeO$_2$/Co$_3$O$_4$@NC	274	58.3	0.1 M KOH	[72]
Electrospinning	h-Co$_3$O$_4$/CeO$_2$@N-CNFs	310	85	0.1 M KOH	[73]
Pyrolysis	CeO$_2$/Co@NC	474	/	0.1 M KOH	[100]
Annealing	FeCo$_2$O$_4$/CeO$_2$	/	63	0.1 M KOH	[83]
Electrodeposition	NiCe@NiFe/NF–N	254@100	59.9	1 M KOH	[84]
Hydrolysis	CeO$_{2-x}$-FeNi	195	43	1 M KOH	[79]
Templates	CeO$_2$/FeOOH HLNTs-NF	210	92.3	1 M NaOH	[78]
Hydrothermal	CeO$_2$/Co(OH)$_2$	410	66	1 M KOH	[44]
Solvothermal	Ni$_4$Ce$_1$@CP	220	81.9	1 M KOH	[85]
Electrodeposition	Cu@CeO$_2$@NFC-0.25	230.8	32.7	1 M KOH	[80]
Electrophoretic deposition	CeO$_2$/Ni(OH)$_2$/NOSCF	240	57	1 M KOH	[86]
Thermal treatment	CeO$_2$/CoSe$_2$	288	44	0.1 M KOH	[61]
Solvothermal	14.6% CeO$_x$/CoS	269	50	1 M KOH	[14]
Hydrothermal	CeO$_x$/NiCo$_2$S$_4$/CC	270	126	1 M KOH	[87]
Solvothermal	CoP/CeO$_2$	224	90.3	1 M KOH	[81]

Table 2.2 (Continued)

Synthesis method	Catalysts	η_{10} (mV)	Tafel slope (mV dec^{-1})	Electrolyte	References
Hydrothermal electrodeposition	V-CoP@ a-CeO$_2$	225	58	1 M KOH	[88]
Sol–gel	Ce-NiO-E	382	118.7	1 M KOH	[90]
	Ce-NiO-L	426	131.6		
Electrodeposition	CeO$_{2-x}$	NA	76	1 M KOH	[91]
Solution-precipitation	20CeO$_2$/Co–Bi	453	120	0.1 M PBS	[101]
Coprecipitation	Ru$_{0.80}$Ir$_{0.15}$Ce$_{0.05}$O$_2$	/	45.84	0.5 H$_2$SO$_4$	[94]
Hydrothermal	Ce-Ir-ox/Mn-4	/	52	0.1 M HClO$_4$	[92]
Thermal	Ce$_{0.2}$–IrO$_2$@NPC	224	55.9	0.5 H$_2$SO$_4$	[93]

"a" in V-CoP@a-CeO$_2$: amorphous; CC: carbon cloth; CeO$_2$/FeOOH HLNTs-NF: ; Co–Bi: cobalt borate; ESD: electrostatic spray deposition; NC: N-doped carbon; NFC: nickel-iron-chromium hydroxide; Ni$_4$Ce$_1$@CP: Ni(OH)$_2$–CeO$_2$ supported on carbon paper; NiCe@NiFe/NF-N: porous network NiCeO$_x$@NiFeO$_x$ electrocatalyst on Ni foam; PBS: phosphate-buffered saline.

their investigations of the overall water splitting reaction are rarely reported. Designing and developing OER electrocatalysts with high activity and excellent stability in acidic media is still difficult.

2.5 Cerium-Based Electrocatalysts for ORR

Various clean and sustainable energy technologies, such as proton-exchange membrane fuel cells (PEMFC), anion exchange membrane fuel cells (AEMFCs), and metal-air battery, have been studied and employed for the potential practical use in transportation and portable power generation [102–104]. ORR is significant for those technologies. Depending on the type of electrolytes, two catalytic pathways for ORR at the cathode of fuel cells and metal-air batteries are proposed via a two-electron or four-electron reaction process with protons or hydroxide ions and electrons to generate water or OH$^-$. For the two-electron pathway, O$_2$ molecule is adsorbed on the surface of the catalyst and combined with two hydrogen to give hydrogen peroxide intermediates, which could be further reduced to two water molecules or simply decomposed to a free hydrogen peroxide molecule. For the four-electron pathway, also known as full reduction reaction, direct reduction occurs from the adsorption of the O$_2$ molecule to water without hydrogen peroxide generation [9].

Noble metals and their alloys (such as Pt) have been the most active ORR electrocatalysts. However, their price and durability limit their applications. Therefore, efforts have been made to develop affordable, high-efficiency earth-abundant ORR catalysts. The special 4f shell electron state of ceria that allows electron sharing and bonding inspired many methods. The flexible transition between Ce^{3+} and Ce^{4+}

redox couple, promoting O_2 adsorption, employing ceria to enhance the activity and stability of ORR electrocatalysts is realized by various strategies, including alloying, doping, interface engineering, etc.

2.5.1 Noble Metals with Ce/Ceria for ORR

The 4f electrons of ceria have unique properties. It allows the transformation between Ce^{3+} and Ce^{4+} redox couple and provides excellent oxygen storage capacity. Alloying noble metal with Ce is considered a critical step to improve electrocatalytic ORR activity, regulating the oxygen density on catalyst surfaces. Recently, Chorkendorff and coworkers employed Pt-lanthanide alloys (including La, Ce, Sm, Gd, Tb, and Dy) to investigate the influence of periodic changes in the lanthanides radius with the increase in 4f electrons [105–108]. The thickness of the electrochemical formation, lattice parameter, catalytic activity, and compressive strain of the Pt overlayer were closely related to the radius of the lanthanide elements (Figure 2.9a–d) [105–108]. The alloying-induced strain of catalysts, an important structure-activity descriptor, enhanced the catalytic activity of Pt with the tailored lattice parameter and facilitated the dissolution of the surface residual lanthanides to give the active surface layer for ORR [105]. The catalytic performance delivered a Sabatier volcano behavior as the function of the lattice parameter of Pt alloys. The alloys on the left exhibited a weak binding with OH, while those on the right exhibited otherwise. When the bulk strain was above a certain level, the overlayer became unstable. It affected the Pt–Pt parameter of the overlayer and ORR activity of alloy to relax toward a lower level of surface strain [105–108].

The Pt–Ce nanoalloy electrocatalysts were prepared by using hydrogen reduction technique. The optimal alloyed catalysts exhibited a half-wave potential of 0.5 V (vs. Ag/AgCl) and a current density of 209 A g_{Pt}^{-1} for ORR [109]. The catalytic activity increased with the increase of Ce concentrations. With non-Pt catalysts, Pd-Ir-Ce catalysts with 66 different ternary compositions were prepared through a combination of high-throughput methods to find the optimal ternary compositions with the highest performance toward ORR in acidic solution [110].

Ceria has also been employed as an additive enhancing the activity of noble metal cathode catalysts for ORR. CeO_2 could stabilize the nanosized noble metals by preventing them from aggregation and modulating their structural and electronic features with high active surface areas for catalysis [40]. Fugane et al. revealed that the large defect clusters (i.e. 12 $\left(Pt''_{Ce}-V_O^{\cdot\cdot}\right) + 2\left(Pt''_{Ce}-2V_O^{\cdot\cdot}-2Ce'_{Ce}\right)$) generated by the strong interactions between Pt and CeO_x could accelerate their charge transfer and maximize the ORR performance on metallic Pt electrode [111]. Another study showed that Frenkel defect cluster (i.e. $2Pt_i^{\cdot\cdot}-4O''_i - 4V_O^{\cdot\cdot}-V''''_{Ce}$) in the surface/interface of the Pt–CeO_x nanowires, and Schottky defect clusters (i.e. $\left(Pt''_{Ce}-2V_O^{\cdot\cdot} - 2Ce'_{Ce}\right)$ and $\left(Pt''_{Ce}V_O^{\cdot\cdot}\right)$) in the bulk of Pt–$CeO_x$ nanowires improved the ORR performance [112]. Moreover, the Pt/CeO_x/C electrocatalysts derived from Ce-based MOFs delivered an excellent performance for ORR and H_2/O_2 fuel cell system in acidic solution [108]. The results indicated that incorporating CeO_x could effectively protect Pt NPs against agglomeration. In addition, the stabilized Pt^0 and

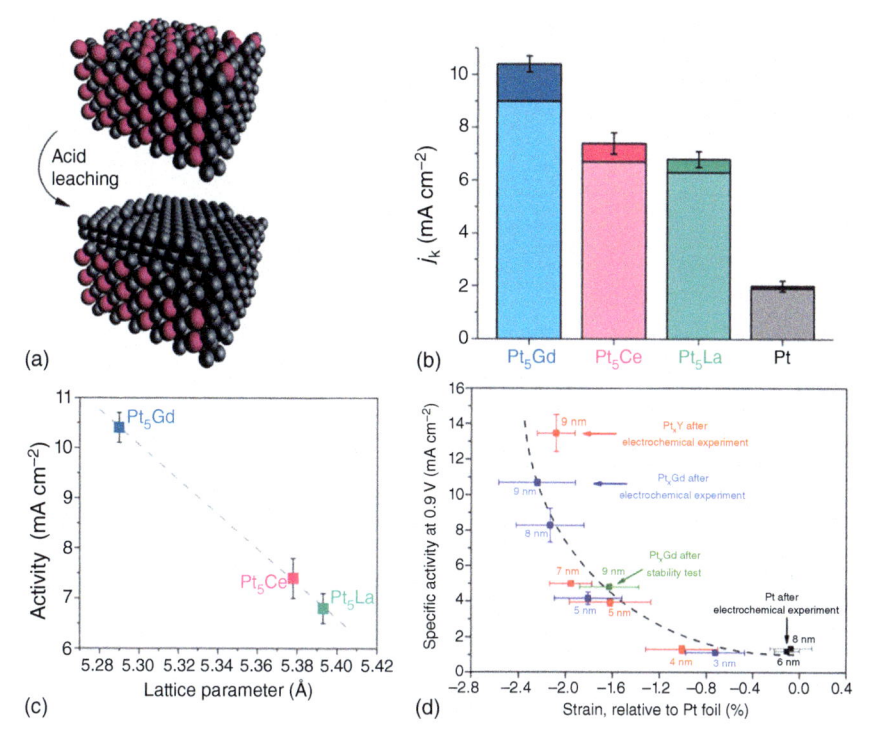

Figure 2.9 (a) Three-dimensional view of the PtCe alloy during sputter-cleaning and after electrochemistry, and gray and red spheres represent Pt and Ce atoms, respectively. (b) Kinetic current density, j_k, of Pt$_5$M and Pt at 1600 revolutions per minute in O$_2$-saturated 0.1 M HClO$_4$. (c) Experimental ORR activity (kinetic current density per geometric surface area at 0.9 V vs. RHE) of polycrystalline Pt$_5$La, Pt$_5$Ce, Pt$_5$Gd$_{33}$ vs. the lattice parameter a of the alloys as determined from XRD measurements. (d) Specific activity results for the ORR on Pt$_x$Gd nanoparticles as a function of the strain, relative to bulk Pt, determined through EXAFS after electrochemical experiment (blue) and after the stability test (green). Source: Reprinted with permission from Ma et al. [21]. Copyright 2018 Elsevier B. V.

Ce^{3+} species in the nanocomposite promoted the H and CO adsorption, resulting in higher activity and durability for ORR than commercial benchmark Pt/C. Similar results were also observed for the cases of Pt$_3$Pd$_1$/CeO$_2$/C [113], Pt/CeO$_2$/C [114], Pt/CeO$_x$/C [115], Pt/CeO$_2$/MWCNTs [116], and Pd-based/ceria [117, 118] in improving and preserving activity through to the robust interactions between noble metals and CeO$_2$, stabilizing noble metal NPs. *In situ* electrochemical X-ray absorption fine structure (XAFS) measurements revealed that improving the ORR activity was ascribed to inhibiting the formation of Pt oxide by the CeO$_x$ layer, where Ce^{3+} was oxidized to Ce^{4+} instead of Pt at the formation potential of Pt oxide [115].

2.5.2 Doping Ce Element into Earth-Abundant Electrocatalysts for ORR

An effective strategy to enhance the performance and stability of ORR is to introduce Ce into earth-abundant electrocatalysts. It promotes the reducibility and dispersion of the active components [119]. An anion exchange process developed a

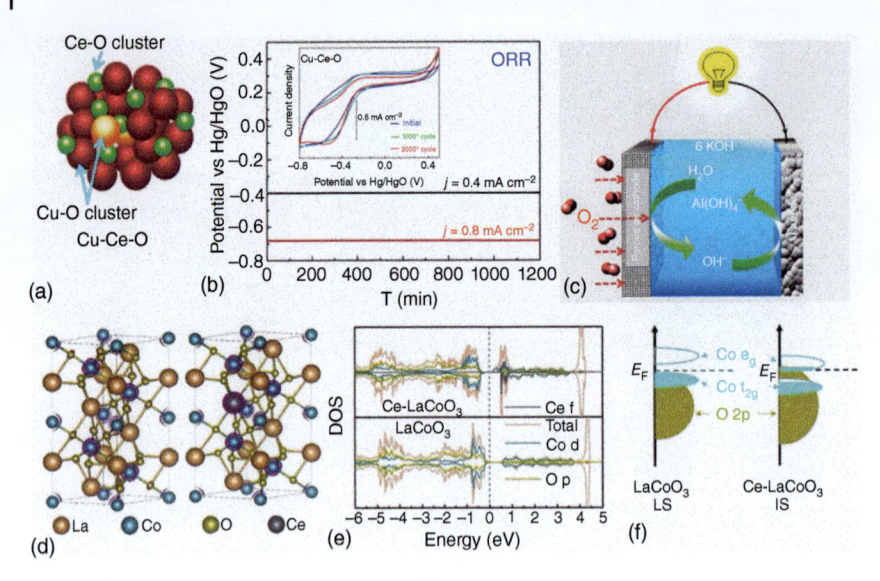

Figure 2.10 (a) Schematic illustration of the resulting Cu–Ce–O oxide. (b) Chronopotentiometry (i–t) measurements of Cu–Ce–O in O_2-saturated 0.1 M KOH on catalyst-modified GC electrode. Insets: CVs for ORR before cycling and after 1000 and 2000 cycles. (c) A schematic of the primary Al-air battery. Source: Hong et al. [120]/Reproduced with permission from American Chemical Society. (d) The crystal models of $LaCoO_3$ before and after Ce substitution on La sites. (e) DOS results of two samples. (f) Schematic representation of Co 3d-O 2p overlap for $LaCoO_3$ and Ce-doped $LaCoO_3$ [121]. Source: Reprinted with permission from Qian et al. [121]. Copyright 2020, Elsevier B. V.

novel Ce-doped CuO nanoplatelets (Cu-Ce-O) catalyst that exhibited comparable electrocatalytic performance and durability for ORR. It had a higher discharge voltage plateau than the benchmark Pt/C catalysts when used in Al-air batteries (Figure 2.10a–c) due to the synergistic effects between Cu and Ce [120].

Recently, perovskite oxides (ABO_3) have emerged as promising alternative ECs for OER/ORR because of their cost-effectiveness, tailorable specific surface area, and unique electronic structure (Figure 2.10d) [121]. Studies have been done to modulate the electronic structure and spin state of perovskites for ORR by introducing Ce [121–124]. Qian et al. reported the improved ORR activity on the Ce-doped $LaCoO_3$ (Ce-5.6%) ECs with the largest limiting current density of 4.91 mA cm^{-2}, a positive half-wave potential of 0.72 V compared with other Ce-doped $LaCoO_3$ ECs [121]. Both theoretical and experimental results confirmed that the improved electrocatalytic activity could be attributed to the low-spin state (LS) of the Co^{3+} to the intermediate-spin state (IS) with a configuration of $e_g = 1$. The introduction of Ce resulted in a synergistic interaction between the enlarged Co 3d-O 2p covalency and boosted electrical conductivity (Figure 2.10e,f) [121].

2.5.3 CeO_2-Based Electrocatalysts for ORR

Pure CeO_2 can be used as a cathode for ORR in Li–O_2 batteries. Oxygen is rapidly adsorbed and released on the surface of CeO_2 [125]. Several reports found that

CeO$_2$ NPs or nanorods with high concentrations of surface oxygen vacancies could display superior ORR performance for Li–O$_2$ batteries even with lower surface areas [126–128]. In increasing the surface oxygen defect concentrations, doping other elements (such as Zr [129], Gd [130], Sm [131], Mn [132], Co [133], S [134]) into CeO$_2$ or surface modification with noble metals is done [129, 135]. Carbon-based materials have also been employed as supports of CeO$_2$ nanomaterials due to their high electrical conductivity, high surface area, and the tunable nanostructures [136–139]. For example, porous nanostructured Zr-doped CeO$_2$ (ZDC) with high active defect sites, prepared as a cathode for ORR in nonaqueous Li–O$_2$ cells, exhibited the maximum discharge capacity of 8435 mA h g^{-1} at 0.1 mA cm^{-2}. The remarkable electrochemical properties of ZDC/carbon were attributed to the fast electron transport provided by the carbon support and the high oxygen reduction activity of the ZDC [129]. Similarly, nano Ce$_2$O$_2$S with enriched oxygen-deficient Ce^{3+} sites supported by N and S dual-doped carbon (NSC) as an active oxygen-supply catalyst displayed superior ORR compared to commercial Pt/C [134]. The ORR performance of the Ce$_2$O$_2$S/NSC-950 was attributed to its moderate binding energy with the absorbed O-species, reducing the reaction obstacle during reduction. The well-ordered mesoporous structure and abundant active sites derived from the N, S dual-doping and rich oxygen vacancies also improved the ORR activity of the hybrids [134]. The active material electrodes had excellent conductivity of the inverse opal carbon (IOC) matrix that supported the CeO$_2$ nanocubes, increasing the cycle life of the Li–O$_2$ battery to 440 cycles [140]. The CeO$_2$ nanocubes with controllable sizes and morphologies, the large amount of active Ce^{3+}, and the IOC matrix with large pores were beneficial to the reversible electrochemical adsorption and dissociation of loosely stacked Li$_2$O$_2$ films, improving the electronic conductivity and postponement of electrode deactivation during the discharge/charge process (Figure 2.11a) [140].

The effects of the interface on the noble metals and ceria were fully discussed in Section 2.5.1. Metals ceria/nonprecious transition metal species (such as single atom [141], metals and metal alloys [144–146], oxides [72, 82, 142, 147–150], and other metal compounds [151]) ECs with various interfaces have also been used for ORR in alkaline media. CeO$_2$ supported metallic Co NPs in N-doped carbon [100, 152], MnO$_2$-based/CeO$_2$ composite [153, 154], porous hexagonal FeCo$_2$O$_4$/CeO$_2$ heterostructure derived from MOFs [83], La0$_7$Sr$_{0.3}$MnO$_3$ particles/flower-like CeO$_2$ [155], and hollow CeO$_2$/CePO$_4$@N. P co-doped carbon composites were applied similarly [151], with intimate interactions and synergistic effects between the CeO$_2$ with an oxygen buffer and the transition metal-based species or the N-doped carbon support.

Recently, Li et al. suggested a "ceria-assisted" strategy to effectively inhibit agglomeration of the isolated Fe single atoms and generate the active single-atom Fe sites by bonding to O in the lattice of CeO$_2$ through spatial confinement and strong trapping for Fe atoms (Figure 2.11b) [141]. The resulting Ce/Fe-N-C materials displayed excellent ORR catalytic activity with a positive half-wave potential of 0.915 V and a kinetic current density of 7.15 mA cm^{-2} at 0.9 V (vs. RHE) [141]. Another study demonstrated CuO$_x$–CeO$_2$/C EC with rich oxide-oxide and oxide-carbon interfaces exhibiting superior ORR performance (a small Tafel slope of 65 mV dec^{-1} and a high

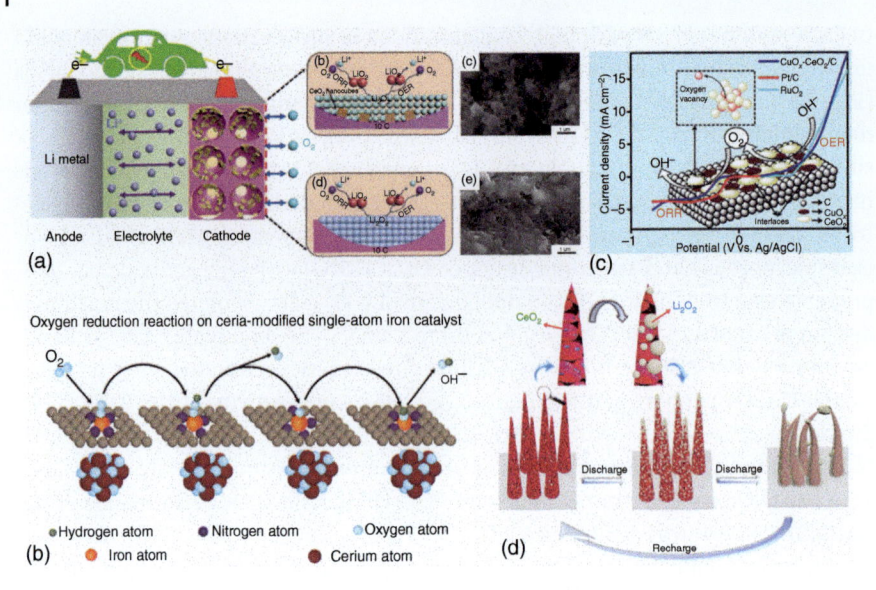

Figure 2.11 (a) Schematic illustration of the reaction mechanism during cycling and the corresponding experimental results. Source: Hou et al. [140]/Reproduced with permission from Wiley-VCH. (b) ORR on ceria-modified single-atom Fe catalyst [141]. Source: Reproduced with the permission from Li et al. [141]. Copyright 2020 American Chemical Society. (c) ORR and OER on CuO_x–CeO_2/C electrodes in alkaline solution [142]. Source: Reproduced with the permission from Goswami et al. [142]. Copyright 2020 American Chemical Society. (d) Schematic illustration for the formation and decomposition of Li_2O_2 on CeO_2/Co_3O_4 NWAs cathodes during the discharge and charge process [143]. Source: Reprinted with permission from Wang et al. [143]. Copyright 2020 Elsevier B. V.

limiting current density of 5.03 mA cm^{-2}) over the benchmark catalyst Pt/C. This result could be attributed to the synergistic effects and the abundant surface oxygen vacancies from the Ce^{4+}/Ce^{3+} redox pair, robust heterointerfaces with oxide-oxide and oxide-carbon, and homogeneous dispersion of oxides over the carbon support (Figure 2.11c) [142].

The free-standing CeO_2/Co_3O_4 nanowires arrays (NWAs) electrodes promoted OER and ORR during charge and discharge processes. The wire-like structures and mesopores with large surface areas and exposed active sites facilitate O_2 and electrolyte diffusion and ion and electron transformation [143]. Thus, Li_2O_2/$Li_{2-x}O_2$ lumps rapidly nucleated and formed at Ce^{3+} sites with higher reaction dynamics, generating into a film along the nanowires succeeded by Co^{2+} catalysis (Figure 2.11d) [143].

Introducing CeO_2 in noble metal NPs can boost their electrocatalytic ORR activities due to the protection of NPs against agglomeration and increased metal dispersion. Moreover, the intrinsic activity and stability of CeO_2 for ORR can be regulated by changing particle sizes, morphologies, and doping species, enhancing the ORR performance of ECs.

Meanwhile, the poor electrical conductivity of CeO_2 can be enhanced by adding carbon-based materials as conductive supports. The integration of CeO_2 with an oxygen buffer into other ORR ECs (e.g. transition metals, transition metal oxides, transition metal compounds) promote intimate interface interactions and synergistic effects that improve the ORR performance in fuel cells and metal-air batteries. Ce^{3+} modulates the binding energy of the adsorbed oxygen species during adsorption, dissociation, and reduction of the oxygen molecules. However, experimental and theoretical evidence is incomplete. Based on previous findings, the facile and dedicated construction of the novel CeO_2-based ORR nanomaterials might provide a feasible approach in designing more efficient ORR ECs.

2.6 Cerium-Based Electrocatalysts for Other Electrochemical Reactions

Ceria with the reversible Ce^{3+}/Ce^{4+} redox pair under mild conditions and high oxygen storage capacity is important for catalysis. It is also explored in other electrocatalytic reactions. Ce/ceria hybrids and host ECs are efficient in the electrochemical oxidation of ethanol [156, 157], formic acid [158], hydrazine [159], N_2 fixation to synthesize NH_3 [160, 161] and the electrochemical CO_2RR, etc. [162–164] except for applications in HER, OER, and ORR.

The Haber–Bosch method of reducing N_2 to ammonia expends large amounts of energy and induces severe environmental problems [165]. Electrochemical reduction of N_2 to NH_3 is a cleaner production of NH_3 by eliminating carbon dioxide emission. In obtaining high catalytic activity for NRR, the ECs should follow the general principles for optimized mass transport, increased oxygen vacancies for chemical adsorption of the reactants, efficient protons and electrons transfers, etc. CeO_2-based catalysts can provide the surface-active catalytic sites for the N_2 chemisorption or catalytic intermediates using their tailorable surface oxygen vacancies and flexible Ce^{3+} content [166]. A similar strategy to improve ORR is also suitable for NRR. For example, CeO_x-induced amorphization of Au NPs anchored on reduced graphite oxide (a-Au/CeO_x–RGO) yielded a superior performance (i.e. Faradaic efficiencies (FEs) of 10.10% and NH_3 yields of $8.3\,\mu g\,h^{-1}\,mg^{-1}_{cat.}$), compared with the crystalline counterpart (c-Au/RGO) at -0.2 V vs. reversible hydrogen electrode (RHE), as given in Figure 2.12a [167]. CeO_2 induced amorphous structure formations and established an intimate interface for fast interfacial charge transfer [167, 169]. The catalysts consisted of Ru NCs-doped, rich oxygen vacancies of CeO_2 nanorods exhibited efficient catalytic activity with FEs of 11.7% at -0.25 V vs. RHE in 0.05 M H_2SO_4 solution. The result could be attributed to the synergistic effects between Ru and CeO_2 with rich oxygen vacancies (Figure 2.12b) [160]. The defective CeO_2 enhanced unsaturated sites of the surface's low valent Ce^{3+} species for N_2 adsorption and Ru NCs for N_2 activation. In another study, the morphologies of CeO_2 were controlled by modulating Fe-doping from crystalline NPs to partial-amorphous nanosheets (Figure 2.12c) [165]. This phase evolution was accompanied by forming more oxygen vacancies, more exposed active sites and faster electron transport, significantly improving NRR activities with FEs of 14.7% (-0.4 V vs. RHE) and NH_3 yields of $26.2\,\mu g\,h^{-1}\,mg^{-1}$ (-0.5 V vs. RHE) [168].

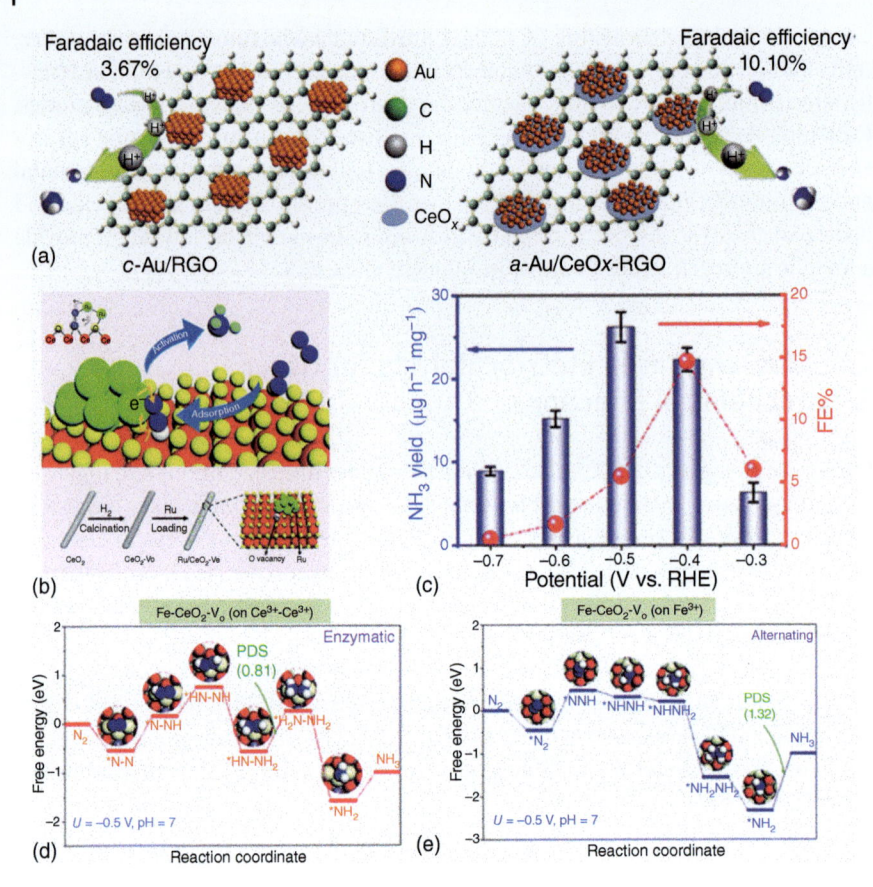

Figure 2.12 (a) Schematic illustration for the electrochemical NRR by catalysts of a-Au/CeO$_x$–RGO and c-Au/RGO under ambient conditions [167]. Source: Reproduced with permission from Li et al. [167]. Copyright 2017 Wiley-VCH. (b) Mechanism of the Ru/CeO$_2$-V$_O$ nanorod electrocatalyst in N$_2$RR [160]. Source: Reprinted with permission from Ding et al. [160]. Copyright 2020 Royal Chemical Society. (c) obtained NH$_3$ yields and FEs; Free energy diagrams of the NRR pathway on (d) Fe-CeO$_2$-V$_O$ (on Ce^{3+}–Ce^{3+}) and (e) Fe-CeO$_2$-V$_O$ (on Fe^{3+}) [168]. Source: Reprinted with permission from Chu et al. [168]. Copyright 2020 Royal Chemical Society.

DFT calculations further indicated that the Ce^{3+}–Ce^{4+} pairs and enriched oxygen vacancies improved the intrinsic NRR activity of active sites and suppressed the competitive reaction of HER. (Figure 2.12d,e). Although these reports demonstrated their potentials for NRR, the applications of CeO$_2$-based nanomaterials in the electrochemical reduction of N$_2$ are in the early stages [170].

Electrochemical CO$_2$RR is the efficient conversion of CO$_2$ to other fine chemicals or fuels, encouraging environmental protection and energy conservation [171, 172]. Ceria can be a promotor of catalytically active metals for CO$_2$RR. An Au–CeO$_x$ interface was constructed to improve the activity and Faradaic efficiency of CO$_2$RR to CO production (Figure 2.13a,b) [158]. The Au–CeO$_x$ interface efficiently enhanced the activation and adsorption of CO$_2$ at interfacial sites on ceria terraces.

The strong interactions among Au, CeO_x, and the hydroxyl groups from water dissociation enhanced the carboxyl intermediate (*COOH) stability on the catalyst's surfaces, compared to the individual Au or CeO_x for the CO_2RR (Figure 2.13c). Similar interface-enhanced CO_2RR performance was also found on the Ag–CeO_x/C interfaces [162]. Apart from the noble metals (e.g. Au, Ag) supported on CeO_x, transition metals or metal oxides on CeO_{2-x} also exhibited excellent activities for the electrochemical CO_2RR [174, 175]. Wang et al. reported single-atomic Cu-doped mesoporous CeO_2 nanorods with multiple oxygen vacancy-bound for selective CO_2 electrocatalytic reduction to CH_4 [173]. They suggested that the strong electronic effect between Cu and CeO_2 resulted in single-atom dispersed Cu species with a low concentration (<5%) and enriched multiple (~3) oxygen vacancies around each Cu site (Figure 2.13d). It supplied many effective catalytic sites for CO_2 molecular reduction to CH_4 with high FEs of 58% [173]. Co_3O_4-CeO_2/low graphitic carbon (LGC)) with oxygen vacancies was fabricated to reduce the overpotential of the electrochemical CO_2RR in formates with the maximum FEs of 76.4% at −0.75 V vs. RHE and the long catalytic stability. The experimental results and DFT calculations confirmed that a high concentration of oxygen vacancies generated from introducing Co_3O_4 into CeO_2/LGC played a crucial role in enhancing activity and formate selectivity. The oxygen vacancies also promoted CO_2 activation and facilitated the electrochemical CO_2RR along $CO_2 \rightarrow CO_2^* \rightarrow COOH^* \rightarrow HCOOH$ path [164].

2.7 Conclusions and Outlooks

This chapter summarizes the progress on the CeO_2-based ECs and their applications in various electrocatalytic reactions, such as HER, OER, ORR, NRR, and CO_2RR. CeO_2-based composite materials are widely studied and employed for their unique structural features such as reversible Ce^{3+}/Ce^{4+} redox pair, abundant surface oxygen vacancies, high oxygen storage capacity, and good chemical and mechanical stability. Thus, they are ECs that improve electrocatalytic performance. Various strategies have been explored to enhance intrinsic catalytic activity, including creating oxygen defects, fabricating heterostructures, doping ions, and modulating electronic structures. The cases have been discussed and summarized. Based on previous discussions, the integration of Ce/ceria into ECs boosts their catalytic properties for different electrochemical reactions. However, the accurate structural design of these composites and the explicit understanding of catalytic reaction mechanisms are still changeable.

(1) Enhancing conductivity. For ECs, excellent conductivity effectively transfers charge carriers from the metal electrode to the adsorption intermediates. It is necessary to mitigate the disadvantage of poor conductivity of pure CeO_2. Introducing other elements as dopants, combining with conductive supports (such as carbon-based materials), designing various morphologies with high surface areas and porosity, and generating abundant oxygen vacancies can enhance the electrical conductivity and promote better catalytic performance.

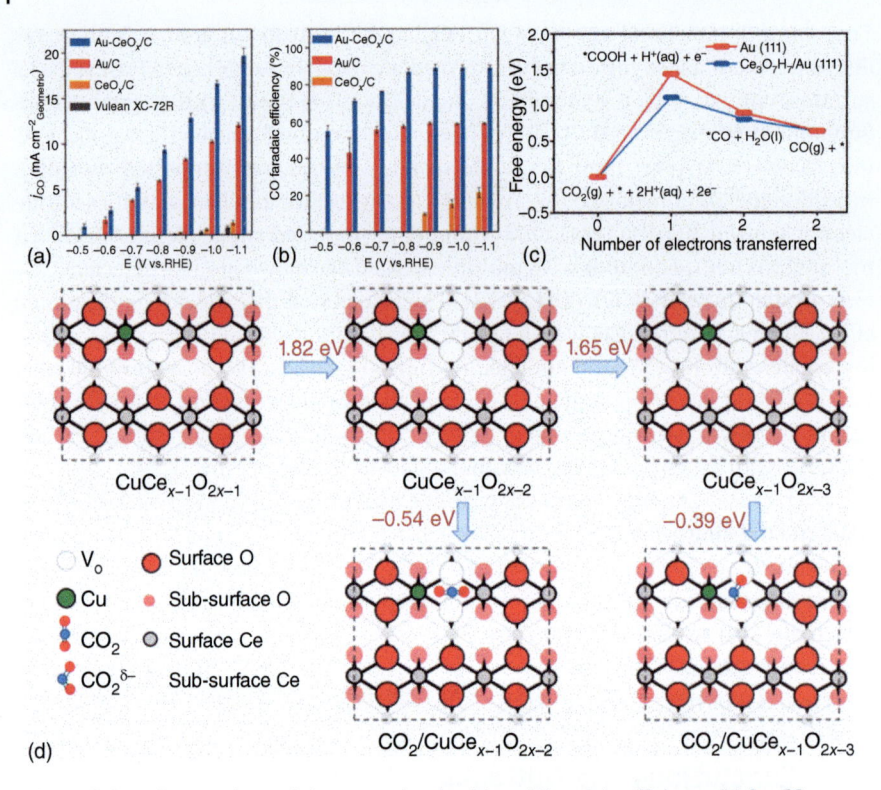

Figure 2.13 Geometric partial current density (a) and Faradaic efficiency (b) for CO production over Au/C, CeO$_x$/C and Au–CeO$_x$/C catalysts in CO$_2$-saturated 0.1 M KHCO$_3$ solution and their dependence on the applied potentials. (c) Calculated free energy diagram. "*X" indicates an adsorbed surface species [162]. Source: Reproduced with permission from Gao et al. [162]. Copyright 2017 American Chemical Society. (d) Theoretical calculations of the most stable structures of Cu-doped CeO$_2$(110) and their effects on CO$_2$ activation [173]. Source: Reproduced with permission from Wang et al. [173]. Copyright 2018 American Chemical Society.

(2) Exploring new methods to synthesize CeO$_2$ and their composites. The synthetic methods of CeO$_2$ are still limited, mainly focusing on hydro/solvothermal and hydrolysis followed by calcining paths. From recent studies, the defect structure, interfacial structure, and morphology of CeO$_2$ nanomaterials are crucial in electrocatalysis. Hence, developing more facile, efficient, and low-cost methods to optimize the CeO$_2$-based materials' unique structure is desirable. More efficient synthetic methods are needed to construct CeO$_2$-based materials with controlled structures and surface physicochemical properties. Increasing the concentration of the structural defects and contact of the intimate interfaces and designing hollow nanocages and multi-shelled structures are promising ways to fabricate composite catalysts with advanced structures, better compositions, and high catalytic activity.

(3) Combinations of theoretical calculations and advanced experimental techniques in designing efficient and cost-effective ECs. Before experimentation,

the computational simulations will realize high-throughput screening and provide a reasonable guide for designing and synthesizing ECs. The theoretical calculations will help us further understand structure-activity relationships and catalytic mechanisms. In-depth insights are expected to synergistically combine theoretical studies with advanced experimental techniques using *in situ* characterization skills, such as *in situ/operando* XPS, *in situ* TEM, and *in situ* synchrotron radiation technology. These studies will serve as guides in developing and manufacturing CeO_2-based ECs.

(4) Focusing on multifunctional ECs. Most CeO_2-based nanomaterials are active in only one catalytic reaction, OER, HER, or ORR. Multifunctional ECs are more significant for practical applications, especially water splitting, fuel cells, and metal-air batteries. Designing and developing functional CeO_2-based ECs is a prospect.

Acknowledgment

We acknowledge the financial support from the National Nature Science Foundation of China (21872109) and the Fundamental Research Funds for the Central Universities (No. D5000210829). Basic Scientific Research Projects of Wenzhou City (Grant No. H20210006).

References

1 Zou, X. and Zhang, Y. (2015). *Chem. Soc. Rev.* 44: 5148–5180.
2 Hua, Y., Li, X., Chen, C., and Pang, H. (2019). *Chem. Eng. J.* 370: 37–59.
3 Seitz, L.C., Dickens, C.F., Nishio, K. et al. (2016). *Science* 353: 1011–1014.
4 Danilovic, N., Subbaraman, R., Chang, K.C. et al. (2014). *Angew. Chem. Int. Ed.* 126: 14240–14245.
5 Yan, Y., Xia, B.Y., Zhao, B., and Wang, X. (2016). *J. Mater. Chem. A* 4: 17587–17603.
6 Safizadeh, F., Ghali, E., and Houlachi, G. (2015). *Int. J. Hydrogen Energy* 40: 256–274.
7 Gong, M., Wang, D.-Y., Chen, C.-C. et al. (2016). *Nano Res.* 9: 28–46.
8 Gou, W., Zhang, M., Wu, J. et al. (2020). *MRS Bull.* 45: 555–561.
9 Lu, G., Zheng, H., Lv, J. et al. (2020). *J. Power Sources* 480: 229091.
10 Montini, T., Melchionna, M., Monai, M., and Fornasiero, P. (2016). *Chem. Rev.* 116: 5987–6041.
11 Yu, J., Cao, Q., Li, Y. et al. (2019). *ACS Catal.* 9: 1605–1611.
12 Zhang, Q., Zhang, S., Tian, Y., and Zhan, S. (2018). *ACS Sustain. Chem. Eng.* 6: 15411–15418.
13 Liu, P., Lu, Y., Ma, W. et al. (2019). *Sustain. Energy Fuels* 3: 3344–3351.
14 Xu, H., Cao, J., Shan, C. et al. (2018). *Angew. Chem. Int. Ed.* 130: 8790–8794.
15 Yang, W., Wang, X., Song, S., and Zhang, H. (2019). *Chem* 5: 1743–1774.

16 Zhu, Y., Wang, J., Chu, H. et al. (2020). *ACS Energy Lett.* 5: 1281–1291.
17 Deng, Y. and Yeo, B.S. (2017). *ACS Catal.* 7: 7873–7889.
18 Xuning, L., Wang, H.-Y., Yang, H. et al. (2018). *Small Methods* 2: 1700395.
19 Wu, T., López, N., Vegge, T., and Hansen, H.A. (2020). *J. Catal.* 388: 1–10.
20 Qi, J., Gao, L., Wei, F. et al. (2019). *ACS Appl. Mater. Interfaces* 11: 47525–47534.
21 Ma, Y., Gao, W., Zhang, Z. et al. (2018). *Surf. Sci. Rep.* 73: 1–36.
22 Rodriguez, J.A., Grinter, D.C., Liu, Z. et al. (2017). *Chem. Soc. Rev.* 46: 1824–1841.
23 Schmitt, R., Nenning, A., Kraynis, O. et al. (2020). *Chem. Soc. Rev.* 49: 554–592.
24 Wang, J., Xiao, X., Liu, Y. et al. (2019). *J. Mater. Chem. A* 7: 17675–17702.
25 Song, X.Z., Zhu, W.Y., Wang, X.F., and Tan, Z. (2021). *ChemElectroChem* 8: 1–26.
26 Singh, K.R., Nayak, V., Sarkar, T., and Singh, R.P. (2020). *RSC Adv.* 10: 27194–27214.
27 Li, Y. and Shen, W. (2014). *Chem. Soc. Rev.* 43: 1543–1574.
28 Tang, W.-X. and Gao, P.-X. (2016). *MRS Commun.* 6: 311–329.
29 Li, J., Zhang, Z., Tian, Z. et al. (2014). *J. Mater. Chem. A* 2: 16459–16466.
30 Li, J., Liu, H.-X., Gou, W. et al. (2019). *Energy Environ. Sci.* 12: 2298–2304.
31 Zhan, Y., Zhou, X., Nie, H. et al. (2019). *J. Mater. Chem. A* 7: 15599–15606.
32 Li, J., Zhou, X., Xia, Z. et al. (2015). *J. Mater. Chem. A* 3: 13066–13071.
33 Zhang, T., Wu, M.-Y., Yan, D.-Y. et al. (2018). *Nano Energy* 43: 103–109.
34 Ling, T., Yan, D.-Y., Jiao, Y. et al. (2016). *Nat. Commun.* 7: 1–8.
35 Zheng, T., Sang, W., He, Z. et al. (2017). *Nano Lett.* 17: 7968–7973.
36 Gao, W., Yan, M., Cheung, H.-Y. et al. (2017). *Nano Energy* 38: 290–296.
37 Chen, T., Zhang, R., Ye, B. et al. (2020). *Nanotechnology* 31: 125402.
38 Li, J., Xia, Z., Zhang, M. et al. (2019). *J. Mater. Chem. A* 7: 17775–17781.
39 Jadhav, H.S., Roy, A., Desalegan, B.Z., and Seo, J.G. (2020). *Sustain. Energy Fuels* 4: 312–323.
40 Gao, W., Wen, D., Ho, J.C., and Qu, Y. (2019). *Mater. Today Chem.* 12: 266–281.
41 Demir, E., Akbayrak, S., Önal, A.M., and Özkar, S. (2018). *ACS Appl. Mater. Interfaces* 10: 6299–6308.
42 Gao, T., Yang, J., Nishijima, M. et al. (2018). *J. Electrochem. Soc.* 165: F1147–F1153.
43 Gao, X., Yu, G., Zheng, L. et al. (2019). *ACS Appl. Energy Mater.* 2: 966–973.
44 Sung, M.-C., Lee, G.-H., and Kim, D.-W. (2019). *J. Alloys Compd.* 800: 450–455.
45 Sun, Z., Zhang, J., Xie, J. et al. (2018). *Inorg. Chem. Front.* 5: 3042–3045.
46 Wu, T., Vegge, T., and Hansen, H.A. (2019). *ACS Catal.* 9: 4853–4861.
47 Weng, Z., Liu, W., Yin, L.-C. et al. (2015). *Nano Lett.* 15: 7704–7710.
48 Sun, H., Tian, C., Li, Y. et al. (2020). *Adv. Sustain. Syst.* 4: 2000122.
49 Sheng, M., Weng, W., Wang, Y. et al. (2018). *J. Alloys Compd.* 743: 682–690.
50 Sun, Z., Zhang, J., Xie, J. et al. (2018). *Dalton Trans.* 47: 12667–12670.
51 Wang, X., Yang, Y., Diao, L. et al. (2018). *ACS Appl. Mater. Interfaces* 10: 35145–35153.

52 Wang, X., Sun, C., He, F. et al. (2019). *ACS Appl. Mater. Interfaces* 11: 32460–32468.
53 Zheng, Z., Li, N., Wang, C.-Q. et al. (2012). *Int. J. Hydrogen Energy* 37: 13921–13932.
54 Xiong, L., Bi, J., Wang, L., and Yang, S. (2018). *Int. J. Hydrogen Energy* 43: 20372–20381.
55 Lan, Q., Lin, Y., Li, Y., and Liu, D. (2018). *J. Mater. Sci.* 53: 12123–12131.
56 Zhang, R., Ren, X., Hao, S. et al. (2018). *J. Mater. Chem. A* 6: 1985–1990.
57 Zhang, L., Ren, X., Guo, X. et al. (2018). *Inorg. Chem.* 57: 548–552.
58 Wang, Z., Du, H., Liu, Z. et al. (2018). *Nanoscale* 10: 2213–2217.
59 Liu, M., Ji, Z., Shen, X. et al. (2018). *Eur. J. Inorg. Chem.* 2018: 3952–3959.
60 Bajdich, M., García-Mota, M., Vojvodic, A. et al. (2013). *J. Am. Chem. Soc.* 135: 13521–13530.
61 Zheng, Y.R., Gao, M.R., Gao, Q. et al. (2015). *Small* 11: 182–188.
62 Zhang, Y.-C., Liu, Y.-K., Zhang, L. et al. (2018). *Appl. Surf. Sci.* 452: 423–428.
63 Paier, J., Penschke, C., and Sauer, J. (2013). *Chem. Rev.* 113: 3949–3985.
64 Ng, J.W.D., García-Melchor, M., Bajdich, M. et al. (2016). *Nat. Energy* 1: 16053.
65 Xu, H., Wang, B., Shan, C. et al. (2018). *ACS Appl. Mater. Interfaces* 10: 6336–6345.
66 Xu, H., Shan, C., Wu, X. et al. (2020). *Energy Environ. Sci.* 13: 2949–2956.
67 Wen, Y., Wei, Z., Liu, J. et al. (2021). *J. Energy Chem.* 52: 412–420.
68 Dinari, M., Allami, H., and Momeni, M.M. (2020). *J. Electroanal. Chem.* 877: 114643.
69 Kanwal, F., Batool, A., Akbar, R. et al. (2020). *Bull. Chem. Soc. Ethiop.* 34: 353–363.
70 Song, S., Wang, K., Yan, L. et al. (2015). *Appl. Catal., B* 176–177: 233–239.
71 Demir, E., Akbayrak, S., Önal, A.M., and Özkar, S. (2019). *J. Colloid Interface Sci.* 534: 704–710.
72 He, X., Yi, X., Yin, F. et al. (2019). *J. Mater. Chem. A* 7: 6753–6765.
73 Li, T., Li, S., Liu, Q. et al. (2019). *ACS Sustain. Chem. Eng.* 7: 17950–17957.
74 Zhou, X., Shen, X., Xia, Z. et al. (2015). *ACS Appl. Mater. Interfaces* 7: 20322–20331.
75 Zhou, X., Xia, Z., Zhimin, T. et al. (2015). *J. Mater. Chem. A* 3: 8107–8114.
76 Qiu, B., Wang, C., Zhang, N. et al. (2019). *ACS Catal.* 9: 6484–6490.
77 Liu, Y., Ma, C., Zhang, Q. et al. (2019). *Adv. Mater.* 31: 1900062.
78 Feng, J.X., Ye, S.H., Xu, H. et al. (2016). *Adv. Mater.* 28: 4698–4703.
79 Yu, J., Wang, J., Long, X. et al. (2021). *Adv. Energy Mater.* 11: 2002731.
80 Xia, J., Zhao, H., Huang, B. et al. (2020). *Adv. Funct. Mater.* 30: 1908367.
81 Li, M., Pan, X., Jiang, M. et al. (2020). *Chem. Eng. J.* 395: 125160.
82 Li, X., You, S., Du, J. et al. (2019). *J. Mater. Chem. A* 7: 25853–25864.
83 Sun, Z., Cao, X., Gonzalez Martinez, I.G. et al. (2018). *Electrochem. Commun.* 93: 35–38.
84 Liu, G., Wang, M., Wu, Y. et al. (2020). *Appl. Catal., B* 260: 118199.
85 Zhao, D., Pi, Y., Shao, Q. et al. (2018). *ACS Nano* 12: 6245–6251.
86 Liu, Z., Li, N., Zhao, H. et al. (2017). *Chem. Sci.* 8: 3211–3217.

87 Wu, X., Yang, Y., Zhang, T. et al. (2019). *ACS Appl. Mater. Interfaces* 11: 39841–39847.
88 Yang, L., Liu, R., and Jiao, L. (2020). *Adv. Funct. Mater.* 30: 1909618.
89 Grimaud, A., May, K.J., Carlton, C.E. et al. (2013). *Nat. Commun.* 4: 2439.
90 Gao, W., Xia, Z., Cao, F. et al. (2018). *Adv. Funct. Mater.* 28: 1706056.
91 Yu, J., Wang, Z., Wang, J. et al. (2020). *ChemSusChem* 13: 5273–5279.
92 Sun, W., Zaman, W.Q., Ma, C. et al. (2020). *ACS Appl. Energy Mater.* 3: 4432–4440.
93 Wang, Y., Hao, S., Liu, X. et al. (2020). *ACS Appl. Mater. Interfaces* 12: 37006–37012.
94 Audichon, T., Morisset, S., Napporn, T.W. et al. (2015). *ChemElectroChem* 2: 1128–1137.
95 Xu, S., Lv, C., He, T. et al. (2019). *J. Mater. Chem. A* 7: 7526–7532.
96 Zhou, J., Zheng, H., Luan, Q. et al. (2019). *Sustain. Energy Fuels* 3: 3201–3207.
97 Huang, X., Zheng, H., Lu, G. et al. (2019). *ACS Sustain. Chem. Eng.* 7: 1169–1177.
98 Hao, J., Luo, W., Yang, W. et al. (2020). *J. Mater. Chem. A* 8: 22694–22702.
99 Galani, S.M., Mondal, A., Srivastava, D.N., and Panda, A.B. (2020). *Int. J. Hydrogen Energy* 45: 18635–18644.
100 Yu, Y., Peng, X., Ali, U. et al. (2019). *Inorg. Chem. Front.* 6: 3255–3263.
101 Zhou, X., Guo, S., Cai, Q., and Huang, S. (2019). *Nanoscale Adv.* 1: 3686–3692.
102 Shao, M., Chang, Q., Dodelet, J.-P., and Chenitz, R. (2016). *Chem. Rev.* 116: 3594–3657.
103 Zhou, X., Tian, Z., Li, J. et al. (2014). *Nanoscale* 6: 2603–2607.
104 Yang, Z., Zhou, X., Jin, Z. et al. (2014). *Adv. Mater.* 26: 3156–3161.
105 Escudero-Escribano, M., Malacrida, P., Hansen, M.H. et al. (2016). *Science* 352: 73–76.
106 Greeley, J., Stephens, I., Bandarenka, A. et al. (2009). *Nat. Chem.* 1: 552–556.
107 Velázquez-Palenzuela, A., Masini, F., Pedersen, A.F. et al. (2015). *J. Catal.* 328: 297–307.
108 Malacrida, P., Escudero-Escribano, M., Verdaguer-Casadevall, A. et al. (2014). *J. Mater. Chem. A* 2: 4234–4243.
109 Qin, J., Zhang, Y., Leng, D., and Yin, F. (2020). *Sci. Rep.* 10: 1–7.
110 Park, S.H., Choi, C.H., Koh, J.K. et al. (2013). *ACS Comb. Sci.* 15: 572–579.
111 Fugane, K., Mori, T., Yan, P. et al. (2015). *ACS Appl. Mater. Interfaces* 7: 2698–2707.
112 Chauhan, S., Mori, T., Masuda, T. et al. (2016). *ACS Appl. Mater. Interfaces* 8: 9059–9070.
113 Yousaf, A.B., Imran, M., Uwitonze, N. et al. (2017). *J. Phys. Chem. C* 121: 2069–2079.
114 Xu, F., Wang, D., Sa, B. et al. (2017). *Int. J. Hydrogen Energy* 42: 13011–13019.
115 Masuda, T., Fukumitsu, H., Fugane, K. et al. (2012). *J. Phys. Chem. C* 116: 10098–10102.
116 Li, Y., Zhang, X., Wang, S., and Sun, G.J.C. (2018). *ChemElectroChem* 5: 2442–2448.

117 Carrillo-Rodríguez, J.C., García-Mayagoitia, S., Pérez-Hernández, R. et al. (2019). *J. Power Sources* 414: 103–114.
118 Kwon, K., Lee, K.H., Jin, S.-a. et al. (2011). *Electrochem. Commun.* 13: 1067–1069.
119 Wang, W., Chen, J.-Q., Tao, Y.-R. et al. (2019). *ACS Catal.* 9: 3498–3510.
120 Hong, Q., Lu, H., and Wang, J. (2017). *ACS Sustain. Chem. Eng.* 5: 9169–9175.
121 Qian, J., Wang, T., Zhang, Z. et al. (2020). *Nano Energy* 74: 104948.
122 Choi, H., Fuller, A., Davis, J. et al. (2012). *Appl. Catal., B* 127: 336–341.
123 Bian, L., Duan, C., Wang, L. et al. (2017). *J. Mater. Chem. A* 5: 15253–15259.
124 Kolchina, L.M., Lyskov, N.V., Kuznetsov, A.N. et al. (2016). *RSC Adv.* 6: 101029–101037.
125 Lin, X., Zhou, L., Huang, T., and Yu, A. (2012). *Int. J. Electrochem. Sci.* 7: 9550–9559.
126 Hou, Y., Wang, J., Hou, C. et al. (2019). *J. Mater. Chem. A* 7: 6552–6561.
127 Li, X., Li, Z., Yang, X. et al. (2017). *J. Mater. Chem. A* 5: 3320–3329.
128 Sudarshan, K., Sharma, S., Gupta, R. et al. (2017). *Mater. Chem. Phys.* 200: 99–106.
129 Kalubarme, R.S., Jadhav, H.S., Park, C.-N. et al. (2014). *J. Mater. Chem. A* 2: 13024–13032.
130 Jiang, Y., Zou, L., Cheng, J. et al. (2018). *J. Power Sources* 400: 1–8.
131 Bae, J., Yang, H., Son, J. et al. (2016). *J. Am. Ceram. Soc.* 99: 4050–4056.
132 Hota, I., Soren, S., Mohapatra, B.D. et al. (2019). *J. Electroanal. Chem.* 851: 113480.
133 Parwaiz, S., Bhunia, K., Das, A.K. et al. (2017). *J. Phys. Chem. C* 121: 20165–20176.
134 Yang, L., Zhuang, C., Hao, L. et al. (2017). *ACS Appl. Mater. Interfaces* 9: 22518–22529.
135 Sun, S., Xue, Y., Wang, Q. et al. (2017). *Chem. Commun.* 53: 7921–7924.
136 Yu, Y., Wang, X., Gao, W. et al. (2017). *J. Mater. Chem. A* 5: 6656–6663.
137 Jiang, Y., Cheng, J., Zou, L. et al. (2016). *Electrochim. Acta* 210: 712–719.
138 Yuan, X., Ge, H., Liu, X. et al. (2016). *J. Alloys Compd.* 688: 613–618.
139 Pi, L., Jiang, R., Cai, W. et al. (2020). *ACS Appl. Mater. Interfaces* 12: 3642–3653.
140 Hou, Y., Wang, J., Liu, J. et al. (2019). *Adv. Energy Mater.* 9: 1901751.
141 Li, J.-C., Maurya, S., Kim, Y.S. et al. (2020). *ACS Catal.* 10: 2452–2458.
142 Goswami, C., Yamada, Y., Matus, E.V. et al. (2020). *Langmuir* 36: 15141–15152.
143 Wang, Y., Wang, J., Mohamed, Z. et al. (2020). *Appl. Mater. Today* 19: 100603.
144 Hota, I., Debnath, A., Muthe, K. et al. (2020). *ChemistrySelect* 5: 6608–6616.
145 Zhang, Z., Gao, D., Xue, D. et al. (2019). *Nanotechnology* 30: 395401.
146 Bai, S., Zhang, X., Yu, Y. et al. (2019). *ChemElectroChem* 6: 4040–4048.
147 Liu, K., Huang, X., Wang, H. et al. (2016). *ACS Appl. Mater. Interfaces* 8: 34422–34430.
148 Meng, G., Chang, Z., Cui, X. et al. (2020). *Chem. Eng. J.* 127913.
149 Yang, J., Wang, J., Zhu, L. et al. (2019). *Mater. Lett.* 234: 331–334.
150 Yang, Z.-D., Chang, Z.-W., Xu, J.-J. et al. (2017). *Sci. China Chem.* 60: 1540–1545.

151 Yu, Y., He, B., Liao, Y. et al. (2018). *ChemElectroChem* 5.
152 Sivanantham, A., Ganesan, P., and Shanmugam, S. (2018). *Appl. Catal., B* 237: 1148–1159.
153 Cao, C., Xie, J., Zhang, S. et al. (2017). *J. Mater. Chem. A* 5: 6747–6755.
154 Zhu, Y., Liu, S., Jin, C. et al. (2015). *J. Mater. Chem. A* 3: 13563–13567.
155 Xue, Y., Huang, H., Miao, H. et al. (2017). *J. Power Sources* 358: 50–60.
156 Bai, Y., Wu, J., Qiu, X. et al. (2007). *Appl. Catal., B* 73: 144–149.
157 Xu, C. and Shen, P.K. (2005). *J. Power Sources* 142: 27–29.
158 Wang, Y., Wang, S., and Wang, X. (2009). *Electrochem. Solid-State Lett.* 12: B73.
159 Srivastava, M., Das, A.K., Khanra, P. et al. (2013). *J. Mater. Chem. A* 1: 9792–9801.
160 Ding, Y., Huang, L., Zhang, J. et al. (2020). *J. Mater. Chem. A* 8: 7229–7234.
161 Zhang, S., Zhao, C., Liu, Y. et al. (2019). *Chem. Commun.* 55: 2952–2955.
162 Gao, D., Zhang, Y., Zhou, Z. et al. (2017). *J. Am. Chem. Soc.* 139: 5652–5655.
163 Dong, H., Zhang, L., Li, L. et al. (2019). *Small* 15: e1900289.
164 Zhang, Q., Du, J., He, A. et al. (2019). *J. Solid State Chem.* 279: 120946.
165 Chen, Q., Zhang, X., Jin, Y. et al. (2020). *Chem. Asian J.* 15: 4131–4152.
166 Xu, B., Xia, L., Zhou, F. et al. (2019). *ACS Sustain. Chem. Eng.* 7: 2889–2893.
167 Li, S.J., Bao, D., Shi, M.M. et al. (2017). *Adv. Mater.* 29: 1700001.
168 Chu, K., Cheng, Y.-h., Li, Q.-q. et al. (2020). *J. Mater. Chem. A* 8: 5865–5873.
169 Lv, C., Yan, C., Chen, G. et al. (2018). *Angew. Chem. Int. Ed.* 130: 6181–6184.
170 van der Ham, C.J.M., Koper, M.T.M., and Hetterscheid, D.G.H. (2014). *Chem. Soc. Rev.* 43: 5183–5191.
171 Aresta, M., Dibenedetto, A., and Angelini, A. (2014). *Chem. Rev.* 114: 1709–1742.
172 Jin, H., Guo, C., Liu, X. et al. (2018). *Chem. Rev.* 118: 6337–6408.
173 Wang, Y., Chen, Z., Han, P. et al. (2018). *ACS Catal.* 8: 7113–7119.
174 Zong, X., Zhang, J., Zhang, J. et al. (2020). *Electrochem. Commun.* 114: 106716.
175 Varandili, S.B., Huang, J., Oveisi, E. et al. (2019). *ACS Catal.* 9: 5035–5046.

3

Metal-Free Carbon-Based Nanomaterials: Fuel Cell Applications as Electrocatalysts

Lai-Hon Chung, Zhi-Qing Lin, and Jun He

School of Chemical Engineering and Light Industry, Guangdong University of Technology, Guangzhou, China

3.1 Introduction

In the last century, rapid anthropological development along with fast-paced technological advancement betters the quality of humans' living. However, these are acquired by excessive consumption of fossil fuel that leads to shortage of fuel, heavy air pollution, and deteriorating global warming. These issues drive scientific researchers to search for more sustainable and environmentally benign energy sources as alternatives to traditional fossil fuels. To tackle this issue, hydrogen is one of the most promising and sustainable candidates because of the following reasons: (i) it has one of the highest energy densities (120–142 MJ kg^{-1}) [1], (ii) massive production of hydrogen can be achieved by water splitting electrolyzers run by renewable energy (i.e. solar energy, wind energy, hydro-energy, etc.) [2], and (iii) energy generation by hydrogen emits no carbon dioxide and produces harmless water as product.

Fuel cell is a device that converts hydrogen into energy electrochemically. Various types of fuel cells have been applied to different aspects of daily life on the ground that they give high energy conversion efficiency and produce harmless products during generation of electricity [3–10]. There are two important reactions in fuel cell for electricity generation – *oxygen reduction reaction* (ORR) and *hydrogen oxidation reaction* (HOR). ORR is an O_2 consumption process at cathode, while HOR is a H_2 dissipation process at anode. The final energy output by fuel cell in terms of standard potential is theoretically +1.23 V vs. reversible hydrogen electrode (RHE), yet the potential is practically <+1.0 V with overpotential as the activation cost. HOR is relatively simpler than ORR because HOR involves a simple oxidation of H_2 molecule while ORR can be attained through different reaction pathways with or without intermediates. In HOR, H_2 molecules are oxidized to H^+ ions in acidic medium (Eq. (3.1)) or to H_2O in alkaline medium (Eq. (3.2)):

$$H_{2(g)} \rightarrow 2H^+_{(aq)} + 2e^- \quad E^0 = 0.00\text{ V} \tag{3.1}$$

$$H_{2(g)} + 2OH^-_{(aq)} \rightarrow 2H_2O_{(l)} + 2e^- \quad E^0 = +0.83\text{ V} \tag{3.2}$$

Functional Nanomaterials: Synthesis, Properties, and Applications, First Edition.
Edited by Wai-Yeung Wong and Qingchen Dong.
© 2022 WILEY-VCH GmbH. Published 2022 by WILEY-VCH GmbH.

On the other hand, in ORR, the reduction of O_2 molecules is also different under acidic and alkaline conditions and proceeds through different pathways. In acidic medium, O_2 molecule is reduced to H_2O directly through four-electron process (Eq. (3.3)) or to unstable H_2O_2 (Eq. (3.4)) that is further reduced to H_2O (Eq. (3.5)):

$$O_{2(g)} + 4H^+_{(aq)} + 4e^- \rightarrow 2H_2O_{(l)} \quad E^0 = +1.23\,\text{V} \tag{3.3}$$

$$O_{2(g)} + 2H^+_{(aq)} + 2e^- \rightarrow H_2O_{2(aq)} \quad E^0 = +0.67\,\text{V} \tag{3.4}$$

$$H_2O_{2(aq)} + 2H^+_{(aq)} + 2e^- \rightarrow 2H_2O_{(l)} \quad E^0 = +1.77\,\text{V} \tag{3.5}$$

In alkaline medium, O_2 molecule is reduced to OH^- ions directly in four-electron fashion (Eq. (3.6)) or to unstable HO_2^- (Eq. (3.7)) that is subsequently reduced to OH^- ions (Eq. (3.8)):

$$O_{2(g)} + 2H_2O_{(l)} + 4e^- \rightarrow 4OH^-_{(aq)} \quad E^0 = +0.40\,\text{V} \tag{3.6}$$

$$O_{2(g)} + H_2O_{(l)} + 2e^- \rightarrow HO^-_{2(aq)} + OH^-_{(aq)} \quad E^0 = -0.065\,\text{V} \tag{3.7}$$

$$HO^-_{2(aq)} + H_2O_{(l)} + 2e^- \rightarrow 3OH^-_{(aq)} \quad E^0 = +0.87\,\text{V} \tag{3.8}$$

In practice, ORR and HOR are kinetically unfavorable stemming from the fact that these two processes are multi-electron in nature and dissociation of both O_2 and H_2 requires large activation energy. To lower the kinetic barriers, reduce the overpotentials, and improve the conversion efficiencies of both ORR and HOR, electrocatalysts have been applied to fuel cells. Pt and Pt-based alloys have long been used as electrocatalysts for both HOR and ORR because of their relatively lower overpotentials and higher current density than other commercially available electrocatalysts. Even though Pt and Pt-based alloys are still benchmark electrocatalysts for HOR and ORR, Pt is not a sustainable material for electrocatalyst with the following reasons: (i) it generally shows low tolerance toward CH_3OH/CO poisoning; (ii) upon prolonged operation, Pt-based electrocatalysts deteriorate through different processes; (iii) Pt and derivatives are scarce and expensive [11–16]. These drawbacks limit their applications in the H_2 energy conversion and drive researchers to search for cheaper, more abundant, highly tolerant, and active alternatives for sustainable development of fuel cell electrocatalysts.

The performance of an electrocatalyst in fuel cell is dictated by the electrode materials and highly dependent on its surface area, electrical conductivity, catalytic activity, long-term stability, etc. The development of such electrocatalysts lie in three directions: (i) increasing the number of active catalytic sites of electrocatalysts via structural design, (ii) enhancing the intrinsic activity of active sites of the electrocatalysts by structural modulation of the catalyst surface to reinforce the binding affinity of O_2 and H_2, (iii) promoting mass transfer of O_2 and H_2 in heterogeneous phases between liquid electrolyte and solid electrocatalyst surface (i.e. improve materials' ability to absorb/adsorb O_2/H_2; increase the O_2/H_2 decomposition efficiency of electrocatalyst to facilitate the flow of O_2/H_2). In recent decades, the rapidly emerging new classes of metal-free materials (e.g. heteroatom-substituted carbon-rich materials [17], defect-incorporated carbon-rich materials [18], and covalent organic framework (COF)) shed light on the search for metal-free carbon rich electrocatalysts for fuel cell application. Some of these metal-free carbon-rich materials exhibited ORR activity comparable and even superior to that of Pt-based electrocatalyst in alkaline medium. Nevertheless, the electrocatalytic activities of such materials are still far lower than those of metal-based electrocatalysts in acidic conditions. The excellent performance in alkaline conditions and the inferior performance in acidic medium of metal-free carbon-rich electrocatalysts highlight that such materials are of great potential to promote sustainable energy conversion and deserve further investigation for improving their performance in acidic conditions. Recently, the works of metal-free carbon-rich electrocatalysts are dominated by ORR catalysis on the ground that ORR catalyzed by Pt-based electrocatalysts is generally less effective than HOR. There arise various approaches to modulate the metal-free carbon-rich materials for better electrocatalytic capability. This chapter is aimed to review, yet not exhaustively, the advances of metal-free electrocatalyst (mainly on ORR) along with classification of different types of materials. Also, their ORR performance in correlation to their structural, electronic, and chemical properties will be discussed. In this chapter, all potentials used for comparison are with respect to RHE, and all the Pt/C electrodes mentioned for comparison refer to the one with 20% weight of Pt (20 wt% Pt) unless otherwise stated. Also, for fair comparison, the onset potential of Pt/C catalyzed ORR in alkaline medium is uniformly represented as +0.96 V vs. RHE, and corresponding onset potentials by different materials in comparison with Pt/C will be standardized.

3.2 Heteroatom-Doped Carbon Nanomaterials

Heteroatom-doped carbon-based materials are renowned for large surface area, high electrical conductivity, tunable pore size and dopant concentration, various doping modes, and unique electrochemical properties. Because of these strengths, heteroatom-doped carbon materials have been applied to fuel cells, zinc–air batteries, and even multifunctional electrocatalysis [17–23]. Among different types of heteroatom-substituted materials, nitrogen(N)-doped carbon materials represent one of the mostly reported groups in the application of ORR. It is worth mentioning that N-doped carbon materials can be generated in various morphologies such as N-graphene, N-carbon nanotubes (CNTs), N-graphyne, etc. It is generally accepted that the doped nitrogen atoms cause uneven distribution of charges on neighboring carbon atoms that facilitate chemisorption and reduction of O_2 molecules by N-doped carbon materials. In general, there are three substitution modes for nitrogen atoms to take over carbon members (graphitic-N, pyrrolic-N, and pyridinic-N) (Figure 3.1). More importantly, the ORR performance of N-doped carbon materials is governed by the types of substituted N. In this section,

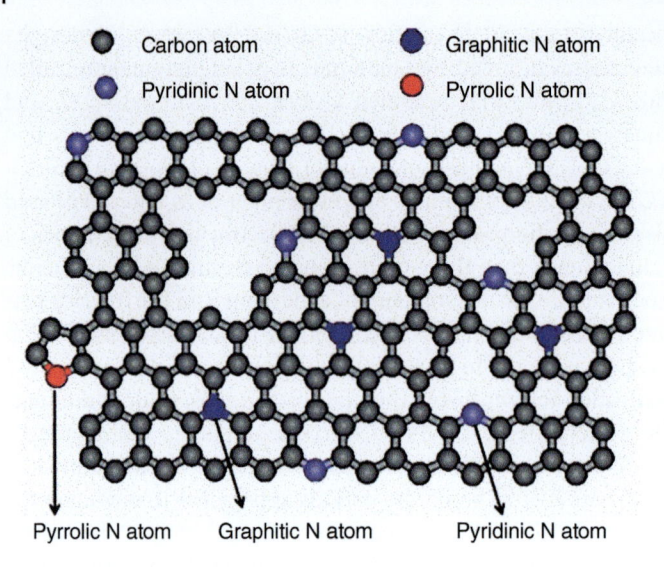

Figure 3.1 Schematic structure of nitrogen-doped graphene. Source: Reproduced with permission from Sheng et al. [24]. Copyright © 2011, American Chemical Society.

heteroatom-doped carbon materials in various morphologies will be discussed for their ORR performance.

3.2.1 Heteroatom-Doped Carbon Nanotubes

CNTs are one-dimensional (1D) cylindrical giant molecules composed of sp^2-hybridized carbon atoms resembling a rolled graphene layer. The large surface area and decent electrical conductivity make CNTs potential candidates as ORR electrocatalysts. Unfortunately, pristine CNTs exhibit poor ORR activity far inferior to that of commercially available Pt/C electrocatalyst. Recently, breakthrough has been made, and it was found that doping heteroatoms to CNTs greatly enhance the intrinsic ORR activity.

In 2009, vertically aligned nitrogen-doped carbon nanotubes (VA-NCNTs) were firstly reported by Dai and coworkers as metal-free ORR electrocatalyst in alkaline conditions [25]. These NCNTs were obtained by pyrolysis of iron(II) phthalocyanine in either presence or absence of ammonia (NH_3) atmosphere and subsequent removal of all iron content by electrochemical purification. The VA-NCNTs were found to catalyze ORR via a $4e^-$ pathway (direct reduction of O_2 into H_2O) and exhibited higher electrocatalytic activity, lower overpotential, less crossover effect (electrochemical reaction of substrates other than O_2 during electrocatalysis), and better long-term operation stability than commercially available Pt-based electrodes in alkaline conditions. The ORR half-wave potential of VA-NCNTs in alkaline medium was around +0.86 V vs. RHE that was comparable to that of commercially available Pt/C electrocatalysts and higher than that of pure CNTs (+0.66 V). Moreover, VA-NCNT electrocatalyst showed steady performance over continuous operation (identical voltammetric response before and after 100 000 cyclic voltammetric cycles), good tolerance toward crossover effect (amperometric response from ORR on the VA-NCNTs remained unchanged upon sequential addition of other fuel substrates such as H_2 gas, glucose, CH_3OH, and formaldehyde), and CO poisoning effect that represents one of the mostly encountered issues for noble metal-based ORR electrocatalysts. Theoretical calculation was employed to provide more insight into the ORR mechanism operated by VA-NCNTs. The results revealed that partially positive carbon atoms adjacent to electronegative nitrogen atoms facilitate the adsorption and reduction of O_2 molecules on carbon surface. Also, the nitrogen atom induces charge delocalization that turns the O_2 chemisorption mode from the commonly proposed end-on adsorption on pure CNT surface to a side-on adsorption on N-doped CNTs.

In 2011, Hu and coworkers reported using boron-doped carbon nanotubes (BCNTs) as metal-free electrocatalysts for ORR [26]. These nanotubes were synthesized by chemical vapor deposition (CVD) with benzene, triphenylborane (TPB), and ferrocene as precursors and catalyst. The boron content of BCNTs were in the range of 0–2.24 at% based on different TPB concentrations. When compared with pure CNTs, BCNTs showed positive shifts in both onset and peak potentials and noticeable surge of current density along with increasing boron content (e.g. peak position shifted from +0.64 V in non B-doped CNT to +0.72 V in 2.24 at% B-doped CNT in 1.0 M NaOH, also the current densities increased with higher boron content, and the maximum peak current of 2.24 at% is 8.0 mA mg^{-1}). These results indicated that ORR performance of these BCNTs was strongly dependent on the boron content. It is noteworthy that BCNTs exhibited decent tolerance toward CH_3OH crossover effect and CO poisoning. Besides, theoretical calculations were performed on B-doped armchair (5,5) single-walled CNT before and after O_2 adsorption and found that boron was extracted electrons by more electronegative neighboring carbon atoms. The partially positively charged boron centers were responsible for capturing the O_2 molecules (Figure 3.2). In contrast to the

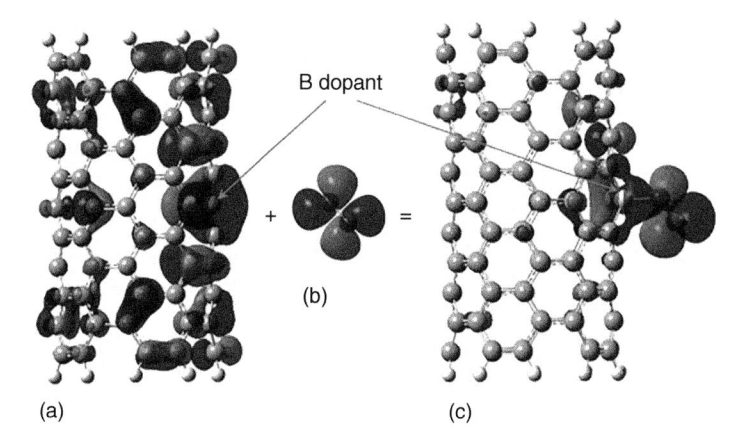

Figure 3.2 Important molecular orbitals involved in the O_2 adsorption on BCNT(5,5). (a) Spin-down HOMO-1 of BCNT(5,5). (b) LUMO of triplet O_2. (c) Spin-down HOMO-2 of O_2-BCNT(5,5). Source: Reproduced with permission from Yang et al. [26]. Copyright © 2011, Wiley-VCH.

NCNTs where O_2 molecules were adsorbed on the three carbon atoms adjacent to nitrogen atoms [25], O_2 molecules were adsorbed on the boron dopants of BCNTs directly.

Meanwhile, Dai and coworkers extended the work of heteroatom-substituted CNTs and reported using vertically aligned B and N co-doped carbon nanotubes (VA-BCN nanotubes) as efficient metal-free electrocatalysts for the ORR [27]. VA-BCN nanotubes were prepared by pyrolysis of melamine diborate as a single source of C, B, and N. In 0.1 M KOH medium, the half-wave potential of the VA-BCN nanotubes is +0.76 V that was higher than both VA-BCNT (+0.51 V) and VA-NCNT (+0.71 V). In comparison with the Pt/C electrode, VA-BCN showed an even more positive half-wave potential in linear sweep voltammetry (LSV) curves and higher diffusion current density (Figure 3.3a). Far higher carbon content (85.5%) than boron (4.2%) and nitrogen (10.3%) content in VA-BCN ensures its better electrical conductivity than other BCN-based materials. The Tafel plots showed that the kinetic current density on the VA-BCN nanotubes was around 10.13 mA cm^{-2} that was higher than those of Pt/C (8.34 mA cm^{-2}), VA-CNT (0.10 mA cm^{-2}), VA-BCNT (0.28 mA cm^{-2}), and VA-NCNT (4.24 mA cm^{-2}) (Figure 3.3b). Also, ORR on VA-CNT and VA-BCNT electrodes were found to proceed through a 2e$^-$ pathway along with generation of hydroperoxide anions while those on VA-NCNT and VA-BCN electrodes were found to go through a 4e$^-$ pathway with direct production of OH$^-$ ions as final product.

After the success in N-, B-doped and even N/B-co-doped CNTs, Hu and coworkers published a work in 2013 to investigate whether co-doping of N and B really improves ORR activity of CNTs [28]. Two types of B and N co-doped CNTs were intentionally prepared as BCNTs-BA (benzylamine as N source with dominantly B and N atoms bonded) and BCNTs-NH$_3$ (ammonia as N source with mostly B and N atoms separated). In BCNT-BA, the N contents were within 1.39–1.57 at%, and the B contents lay within 0.35–1.68 at%. On the other hand, in BCNT-NH$_3$, the N contents were within 0.58–2.19 at%, and the B contents lay in the range of 0.84–1.93 at%. Electrochemical measurements showed that the ORR performance of BCNT-BA

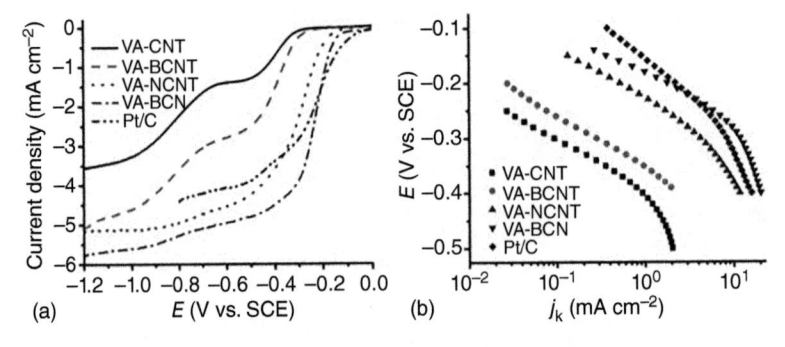

Figure 3.3 (a) LSV curves of various electrodes in O_2-saturated 0.1 M KOH electrolyte at a scan rate of 10 mV s^{-1} and a rotation rate of 1000 rpm. (b) Tafel plots derived from (a) in the low-current region. Source: Reproduced with permission from Wang et al. [27]. Copyright © 2011, Wiley-VCH.

gradually dropped until the level of the pristine CNTs along with increasing B/N ratio while the onset potentials and current densities of BCNT-NH$_3$ got better along with increasing content in both B and N. Specifically, BCNT-NH$_3$ of 2.19 at% N and 1.93 at% B exhibited the highest current density of 8.9 mA mg^{-1} and peak potential at +0.76 V in 1.0 M NaOH solution. Theoretical calculation was performed to investigate the poor activity of BCNTs-BA. Highest occupied molecular orbital (HOMO) plot showed that interaction between O$_2$ molecules and BCNT-BA surface was weak with little charge transferred to O$_2$. Hence it led to low ORR activity (Figure 3.4b). Also, the second-order orbital perturbation analysis revealed that neutralization of extra electrons on N atom by the vacant orbital from B atom took place, and this disfavored chemisorption of O$_2$ molecules by the co-doped CNTs (Figure 3.4a). Alternatively, the separation of B dopants from N dopants prevented the neutralization between electronegative N and electropositive B; therefore, N and B can synergistically boost the ORR of N/B-co-doped CNTs.

Heteroatom-substituted CNTs are not restricted to B and N dopants but also open to other elements such as sulfur (S) and silicon (Si). In 2015, Li and coworkers reported using sulfur-doped carbon nanotubes (SCNTs) as electrocatalyst for ORR in alkaline medium [29]. The SCNTs were prepared by direct annealing oxidized CNTs and benzenedithiol (BDT) under the protection of N$_2$ atmosphere. It was found that *p*SCNT-900 produced by 900 °C annealing of CNTs and BDT in mass ratio of 1 : 2 exhibited the best ORR performance over SCNTs obtained from other synthetic conditions and BDT isomers. To be more specific, *p*SCNT (5.79 mA cm^{-2}) possessed a higher diffusion-limiting current density than Pt/C counterpart (4.90 mA cm^{-2}) at +0.16 V in O$_2$-saturated 0.1 M KOH solution, and the onset potentials of *p*SCNT (+0.90 V) and Pt/C electrode are comparable (+0.96 V). Also, the high kinetic current density of *p*SCNT (34.6 mA cm^{-2} at +0.63 V) indicated that ORR catalyzed by *p*SCNT proceeded through a 4e$^-$ O$_2$ reduction pathway. It is important that *p*SCNT showed more tolerance toward CH$_3$OH crossover effect and better long-term operation stability than the Pt/C catalysts in alkaline electrolytes.

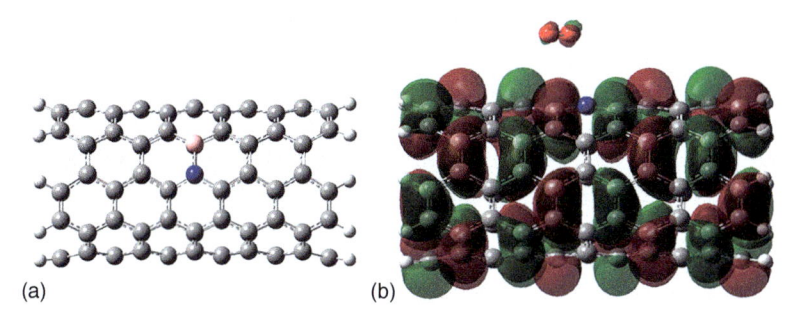

Figure 3.4 (a) Bonded B and N co-doped CNT(5,5). (b) HOMO plot of the corresponding O$_2$ adsorption configuration (isodensity value of 0.007 a.u.). N, blue; B, pink; C, black; H, grey; O, red. Source: Reproduced with permission from Zhao et al. [28]. Copyright © 2013, American Chemical Society.

In the same year, Peng and coworkers reported the preparation of Si doped and Si/N co-doped nanostructures for ORR application [30]. Si-doped nanosphere (Si-CNS) and Si/N-co-doped nanotube (SiN-CNT) were synthesized by the thermolysis of dimethylsilicone oil and 3-aminopropyl-triethoxysilane using FeMo/Al_2O_3 as catalyst via CVD, respectively. The ORR activity of SiN-CNT (+0.94 V) was higher than that of N-CNT (+0.90 V) as reflected by the onset potentials (Figure 3.5a). Furthermore, even though SiN-CNT had a lower onset potential than commercially available Pt/C electrocatalyst (+0.96 V), the current density of SiN-CNT below +0.51 V exceeds that of Pt/C electrode (Figure 3.5b). It is noteworthy that the Si-CNS showed a more positive onset potential and around 1.6–3.5 times the current density over the range of −0.04 to +0.46 V when compared with non-modified CNS (Figure 3.5c,d). Besides, both Si-CNSs and SiN-CNTs showed tolerance toward CH_3OH crossover effect and performed constantly upon continuous operation for 10 000 seconds while Pt/C were obviously inferior in CH_3OH tolerance and stability upon continuous operation (Figure 3.5e,f). Theoretical calculation was employed to provide more insight to explain how Si dopants enhance CNTs and CNSs on ORR. It was found that doping Si or co-doping both Si and N into carbon graphene sheets led to a change in net charge distribution on graphene surface and hence a change of the O_2 chemisorption mode in which O_2 molecules were originally adsorbed to pure carbon surface through end-on adsorption mode and turned to carry out chemisorption via a side-on adsorption mode after introduction of Si or Si/N dopants. Moreover, the incorporation of Si or Si/N elements into graphene narrowed the band gap that contributed to easier reduction of O_2 molecules on graphene surface. These two reasons account for how Si-doped or Si/N-co-doped carbon nanomaterials facilitate ORR.

These fruitful research works highlight that heteroatom-doped CNTs exhibit far more attractive performance than primitive CNTs in ORR application and prove CNTs as one of the potential candidates to replace well-established noble metal-based materials as electrocatalysts for ORR.

3.2.2 Heteroatom-Doped Graphenes

Graphene, as an extended two-dimensional (2D) carbon allotrope, exists as honeycomb-like hexagonal lattice [31, 32]. Since graphene is electrically conducting, it has long been investigated for its application as electrode materials. Related research burst out being a hot topic especially after the discovery of graphene preparation by "Scotch tape" method in which graphene flakes are peeled off from a chunk of graphite by tearing the graphite-attached Scotch tape off the substrate surface [31]. Within graphene layer, each sp^2-hybridized carbon atom connects to three adjacent carbon atoms through three strong bonds, and the remaining p-orbital on each carbon atom overlaps with those of other neighboring atoms constructing a band of fully filled orbitals and a band of vacant orbitals. Hence, graphene is considered as a material with zero band gap that hinders the application of graphene in different aspects such as electronics, fuel storage, electrocatalysis, etc. [33–35]. To "activate" graphene, several morphological regulation methods

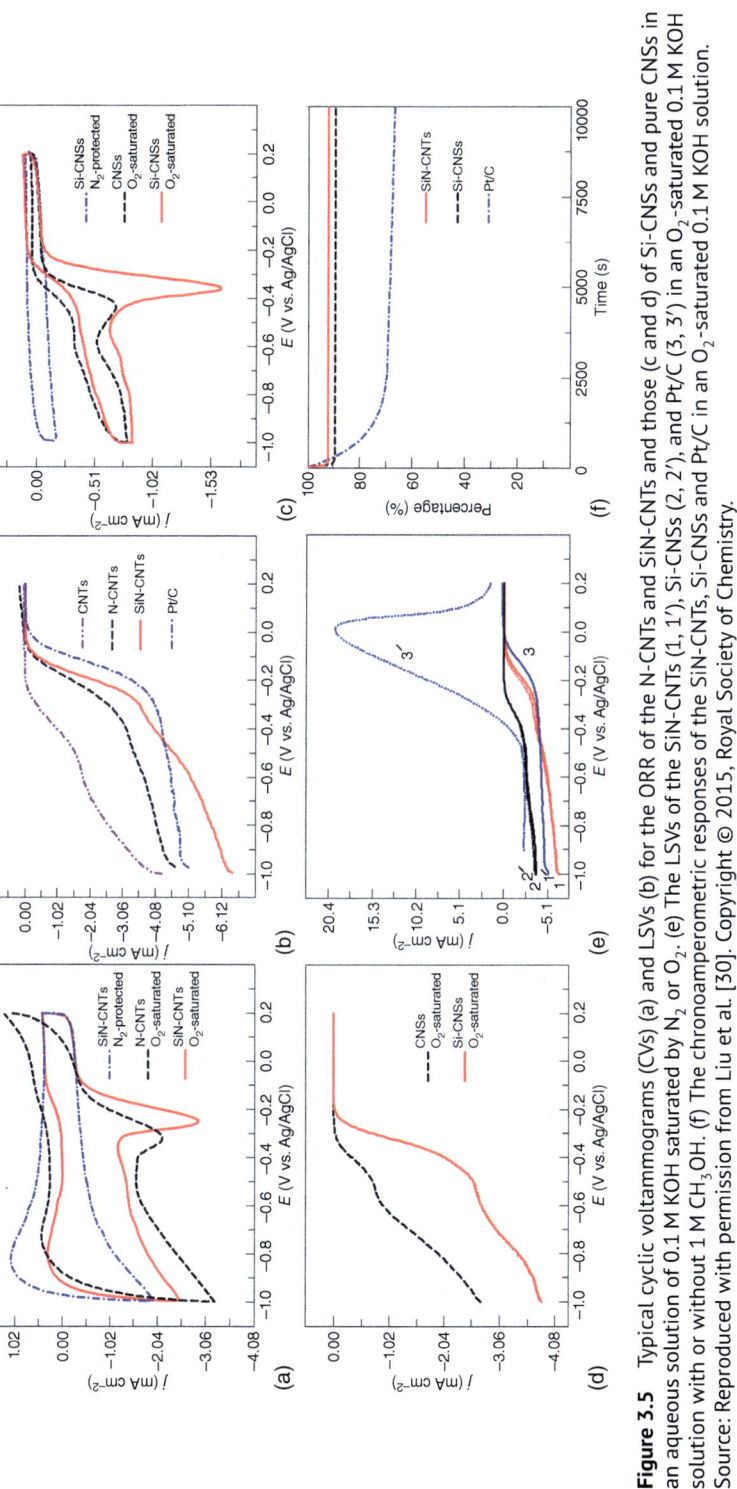

Figure 3.5 Typical cyclic voltammograms (CVs) (a) and LSVs (b) for the ORR of the N-CNTs and SiN-CNTs and those (c and d) of Si-CNSs and pure CNSs in an aqueous solution of 0.1 M KOH saturated by N_2 or O_2. (e) The LSVs of the SiN-CNTs (1, 1'), Si-CNSs (2, 2'), and Pt/C (3, 3') in an O_2-saturated 0.1 M KOH solution with or without 1 M CH_3OH. (f) The chronoamperometric responses of the SiN-CNTs, Si-CNSs and Pt/C in an O_2-saturated 0.1 M KOH solution. Source: Reproduced with permission from Liu et al. [30]. Copyright © 2015, Royal Society of Chemistry.

have been proposed to induce a band gap in graphene (i.e. chemical functionalization, establishment of multilayer graphene, introduction of guest molecules, etc.) [36–39]. Beyond morphological modulation, doping of heteroatoms also represents one of the most effective ways to modify electronic properties, surface activeness, and elemental compositions of graphene materials. Pinpointing ORR application in alkaline medium, nitrogen (N), phosphorus (P), sulfur (S), and halogen (X) atoms have been utilized as dopants for preparation of heteroatom-functionalized graphene and graphene-based derivatives that will be discussed in detail in this section.

Heteroatom-doped graphenes usually exhibit better ORR capability than pure graphene. This finding is inspiring and encouraging yet at the same time raises a question: what is the magic of heteroatoms on graphene for ORR performance? With respect to this question, in 2011, Xia attempted to give answer with the work utilizing theoretical calculation to investigate the ORR mechanism on N-doped graphene for fuel cells [40]. Through simulation, the ORR mechanism on N-doped graphene was found to proceed via a 4e$^-$ pathway (direct conversion from O_2 to H_2O) that is consistent with experimental findings [41, 42]. Also, the introduction of N-dopants to graphene results in asymmetry of spin density and atomic charge density that are possibly responsible for the high electrocatalytic activities of N-doped graphene toward ORR.

In 2010, Dai and coworkers reported N-doped graphene as an electrocatalyst for ORR, and this work represented the first example of using graphene and its derivatives as metal-free catalysts for ORR [41]. The N-doped graphene was prepared by CVD using $NH_3/CH_4/H_2/Ar$ gas mixture as reactants and nickel as catalyst. The high-resolution X-ray photoelectron spectroscopy (XPS) M 1s spectrum confirmed the existence of pyridinic (~398.3 eV) and pyrrolic (~400.5 eV) N atoms within the graphene structure. When compared with commercially available Pt-based electrodes (C2-20, 20% platinum on VulcanXC-72R; E-TEK), the reported N-doped graphene exhibited better electrocatalytic activity (e.g. the steady-state catalytic current density at N-doped graphene electrode was around 3 times higher than that of the Pt/C electrode over a broad potential range) (Figure 3.6a), long-term operation stability (e.g. no obvious decrease in current after 200 000 continuous cyclic voltammetric measurement) (Figure 3.6d), and tolerance toward crossover/poison effects (e.g. the amperometric signals from the ORR on the N-doped graphene electrode remained unchanged after the introduction of H_2, glucose, and CH_3OH (Figure 3.6b) and steady amperometric responses from the ORR on the N-doped graphene electrode after the addition of a 10% (v/v) CO in air into the electrolyte (Figure 3.6c)). This work demonstrates the potential of graphene to be an alternative class of metal-free ORR electrocatalysts after appropriate heteroatom doping.

In the next year, Xia and coworkers reported a facile and catalyst-free method for preparation of N-doped graphene by thermal annealing graphite oxide with commercially readily available melamine as N source [24]. The doping of N content in graphene layers could be achieved up to 10.1 at%, and it was found that annealing temperature played a vital role in controlling the doping efficiency. Besides, the dominant XPS peak located at 398.2 eV revealed that the major N species in graphene

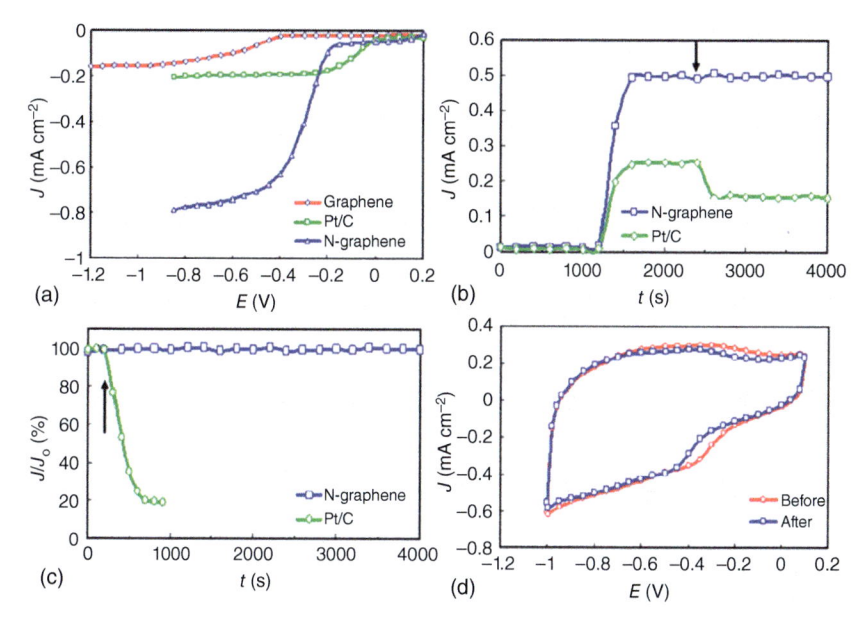

Figure 3.6 (a) Rotating ring-disk electrode (RRDE) voltammograms for the ORR in air-saturated 0.1 M KOH at the C-graphene electrode (red line), Pt/C electrode (green line), and N-graphene electrode (blue line). Electrode rotating rate: 1000 rpm. Scan rate: 0.01 V s^{-1}. Mass(graphene) = Mass(PtC) = Mass(N-graphene) 7.5 g. (b) Current density (j)–time (t) chronoamperometric responses obtained at the Pt/C (circle line) and N-graphene (square line) electrodes at −0.4 V in air-saturated 0.1 M KOH. The arrow indicates the addition of 2% (w/w) CH$_3$OH into the air-saturated electrochemical cell. (c) Current (j)–time (t) chronoamperometric response of Pt/C (circle line) and N-graphene (square line) electrodes to CO. The arrow indicates the addition of 10% (v/v) CO into air-saturated 0.1 M KOH at −0.4 V; j_0 defines the initial current. (d) Cyclic voltammograms of N-graphene electrode in air-saturated 0.1 M KOH before (circle line) and after (square line) a continuous potentiodynamic swept for 200 000 cycles at room temperature (25 °C). Scan rate: 0.1 V s^{-1}. Source: Reproduced with permission from Qu et al. [41]. Copyright © 2010, American Chemical Society.

layer was pyridinic nitrogen. The onset potential of ORR for the N-doped graphene is +0.86 V that was about 0.1 V more positive than pristine graphene and almost the same potential of VA-NCNTs reported by Dai and coworkers [25]. Noteworthy, cyclic voltammograms of N-doped graphene showed that the onset potential was insensitive to the change of N content. Also, N-doped graphene exhibited a larger reduction current than pristine graphene and was found to undergo ORR through a 4e$^-$ pathway (2e$^-$ pathway for original graphene). With these experimental findings, more positive onset potential and enhanced reduction current of N-doped graphene than pristine graphene could be attributed to a faster reaction kinetics with a higher delivered electron number per O$_2$ molecule in the case of N-doped graphene.

In 2012, Qu and coworkers reported N-doped graphene quantum dots (N-GQDs) prepared from electrochemical decomposition of graphene layers at high applied potentials with tetrabutylammonium perchlorate as the nitrogen source [43]. The resultant N-GQDs had an N/C atomic ratio of around 4.3% that was close to

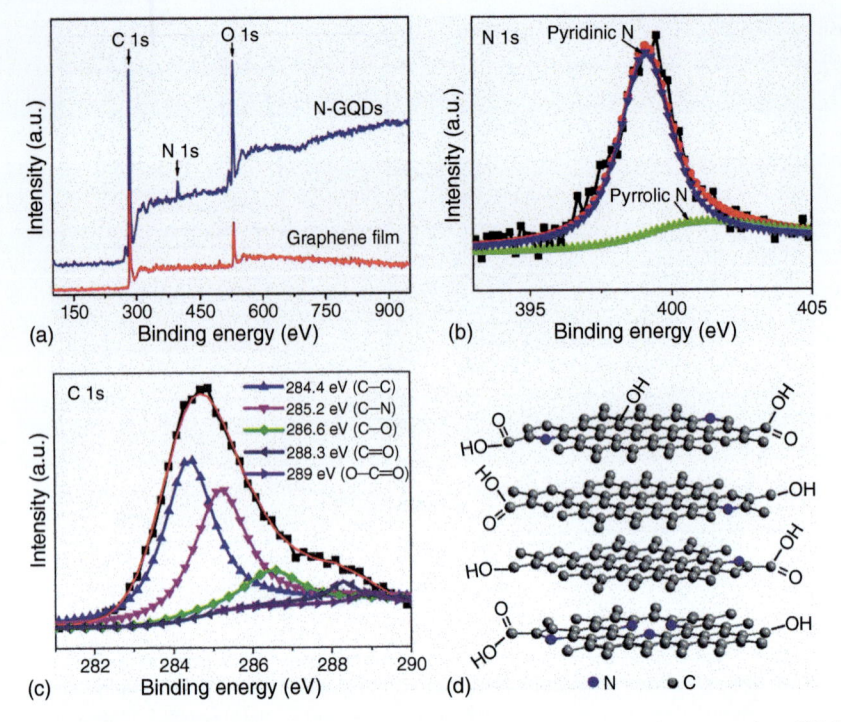

Figure 3.7 (a) XPS spectra of the original graphene film (red) and the as-produced N-GQDs (blue). (b, c) High-resolution N 1s (b) and C 1s (c) peaks of N-GQDs. (d) Possible structure of an O-rich N-GQD (not drawn to scale). Source: Reproduced with permission from Li et al. [43]. Copyright © 2012, American Chemical Society.

previously reported VA-NCNTs [25] and N-doped graphene [41]. From XPS spectra, it was found that the N-GQDs bore not only C and N but also high content of O (Figure 3.7a). Moreover, the N species in the N-GQDs existed mostly as pyridinic N (398.5 eV) (Figure 3.7b), and there are some signals corresponding to O-rich groups that are consistent with C—O (286.6 eV), C=O (288.3 eV), and O—C=O (289 eV) groups (Figure 3.7c). These O-rich groups in N-QGDs were tentatively assigned as terminal carboxylic acid and alcohol groups on each N-QGD unit (Figure 3.7d). These N-GQDs exhibited intense blue emission upon photoexcitation by a 365 nm lamp and ORR activity (i.e. onset potential = +0.80 V) close to that of commercially available Pt/C electrodes and comparable with that of N-doped graphene [41] in alkaline medium. Also, the N-GQDs were found to go through a 4e$^-$ pathway for ORR. Based on the remarkable quantum confinement and edge effect of 0D GQDs, it is reasonable to expect that GQDs doped with chemically bonded nitrogen could enormously change their electronic properties and offer more active sites for ORR. With the discovery from this work, depositing N-GQDs on graphene layer as ORR-active catalyst opens another direction toward using N-doped nanocarbon materials for ORR application.

Besides N-GQDs, the same research group also searches for other nanoscale carbon-based materials for ORR application. In 2012, they published a work about ultralight, 3D, N-doped graphene framework (NGF) that was prepared by hydrothermal treatment of graphene oxide and 5 vol% pyrrole followed by thermal annealing [44]. The as-prepared NGF possesses an ultralow density of 2.1 ± 0.3 mg cm^{-3}, comparable with that of the lightest silica aerogels (2–3 mg cm^{-3}) [45] and close to that of CNT sheets (1.5 mg cm^{-3}) [46]. It was found that NGF has an N/C atomic ratio of 4.9% and high-resolution XPS spectra reflect the existence of pyridinic (398.5 eV) and pyrrolic (401 eV) N atoms in the graphene layers. Encouragingly, NGF also exhibited decent ORR activity with onset potential of around +0.78 V and remarkably higher diffusion current density than that of Pt/C electrode over a large potential range (+0.16 to +0.56 V).

With reference to the success of N/B-co-doped CNTs [27], Dai and coworkers reported a facile approach to generate N/B-co-doped graphene by thermal annealing graphite oxide in the presence of boric acid and NH$_3$ [47]. The resultant co-doped graphene demonstrated superior electrocatalytic activity to commercially available Pt/C electrocatalyst. Three types of N/B-co-doped graphene labeled according to their corresponding ratio ($B_{38}C_{28}N_{34}H_{26}$, $B_{12}C_{77}N_{11}H_{26}$, $B_{7}C_{87}N_{6}H_{26}$) were studied with boric acid–graphene oxide mixture (B-GO) as reference material. As shown in the high-resolution XPS C1s spectra, almost total loss of peak at 286 eV and rise of signals at 284.5 eV from B-GO to N/B-co-doped graphene signified the loss of oxygen components (Figure 3.8a). The peaks at 192 and 190 eV in the high-resolution XPS B1s spectra corresponded to boron oxide in B-GO and B species in N/B-co-doped graphene, respectively (Figure 3.8b). These B signals indicated that B atoms have been successfully incorporated into the graphene lattice network. LSV curves showed that $B_{12}C_{77}N_{11}H_{26}$ gave the optimal ORR performance close to commercial Pt/C electrocatalyst (comparable onset potential, almost higher current density within the whole potential range) (Figure 3.8b) while the low ORR activity of $B_{38}C_{28}N_{34}H_{26}$ was proposed to be associated with the low C content. Theoretical calculation was performed to explain the difference in ORR activity between pure graphene and N/B-co-doped graphene. It was found that the band gap had the order of $B_{38}C_{28}N_{34}H_{26}$ (1.23 eV) > $C_{100}H_{26}$ (0.80 eV) > $B_{7}C_{87}N_{6}H_{26}$ (0.79 eV) > $B_{12}C_{77}N_{11}H_{26}$ (0.71 eV). This echoed the low conductivity proposed in $B_{38}C_{28}N_{34}H_{26}$. Although $B_{38}C_{28}N_{34}H_{26}$ was found to contain the most atoms with large charge density, B and N dopants were arranged in the form of insulating B–N cluster that hindered electron transfer over the graphene network. On the contrary, both $B_{7}C_{87}N_{6}H_{26}$ and $B_{12}C_{77}N_{11}H_{26}$ had more C atoms with relatively high spin density and charge density when compared with the pristine graphene that accounted for the better ORR performance of both $B_{7}C_{87}N_{6}H_{26}$ and $B_{12}C_{77}N_{11}H_{26}$ than pure graphene. These findings highlight that even though N- and B-doping help enhance the ORR catalytic activity of graphene, appropriate doping ratio is a key factor to activate the synergistic effect of N- and B-dopants for improving graphene ORR performance.

Beyond the heteroatom-doped graphene mentioned above, Qiao and coworkers reported mesoporous S- and N-dual-doped graphene (dual-doped NSG) as electrocatalyst for enhanced ORR performance [48]. The dual-doped NSG was prepared by thermal annealing (900 °C) of graphene oxide/colloidal silica composite in the

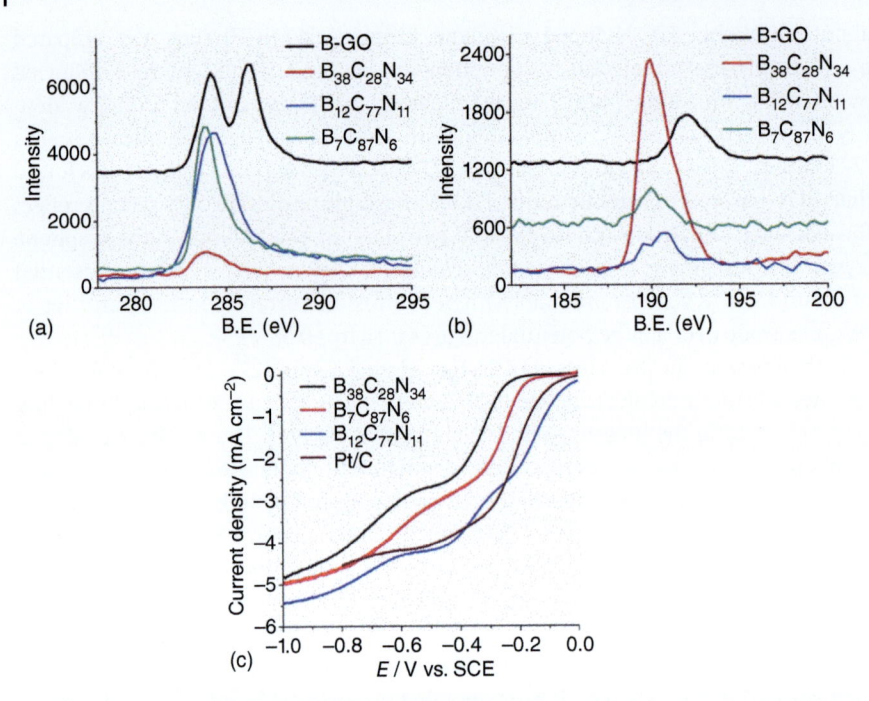

Figure 3.8 High-resolution (a) XPS C 1s and (b) XPS B 1s of BCN graphene in comparison with N-doped graphene, graphene oxide, or boric acid and graphene oxide mixtures. (c) LSV curves of ORR on BCN graphene with different compositions in O_2-saturated 0.1 M KOH solution at 10 mV s^{-1} and compared with the commercial Pt/C electrocatalyst. Source: Reproduced with permission from Wang et al. [47]. Copyright © 2012, Wiley-VCH.

presence of melamine (N source) and benzyl disulfide (S source) followed by the removal of silica nanoparticles by hydrofluoric acid. The elemental contents of N and S were found to be 4.5 and 2.0 at% that are comparable with those of singly doped analogues (N = 5.1 at% and S = 1.3 at%, respectively). Remarkably, dual-doped NSG exhibited a high ORR onset potential of +0.90 V, close to that of commercially available Pt/C electrocatalyst (+0.96 V) and far better than those of N-doped graphene, S-doped graphene, and pristine graphene (c. +0.78 V) (Figure 3.9a). Furthermore, dual-doped NSG also showed a larger ORR current than those N-doped, S-doped, and pristine graphenes (Figure 3.9a). The theoretical calculation suggested that the enhanced ORR activity of dual-doped NSG might possibly arise from the uneven distribution of electron density on C atoms neighboring to both S and N atoms (Figure 3.9b).

Besides N and S, graphene dopants extend to the use of P. In 2013, Hou and coworkers reported the first example of P-doped graphene (PG) as ORR electrocatalyst [49]. PG was obtained by thermal annealing method using graphite oxide (GO) and triphenylphosphine (TPP) as C and P sources, respectively. The resultant PG exhibited remarkable ORR catalytic activity, satisfactory tolerance to CH_3OH crossover effect, and good long-term stability. Specifically, PG gave a much higher ORR current density than pristine graphene (e.g. PG > pristine graphene of

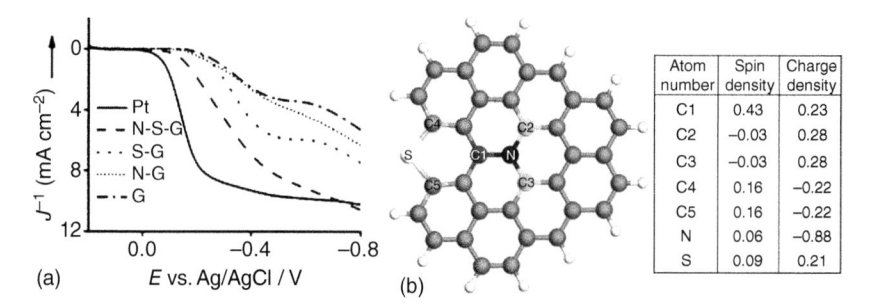

Atom number	Spin density	Charge density
C1	0.43	0.23
C2	−0.03	0.28
C3	−0.03	0.28
C4	0.16	−0.22
C5	0.16	−0.22
N	0.06	−0.88
S	0.09	0.21

Figure 3.9 (a) LSVs of different samples at 1600 rpm. (b) Spin and charge density of graphene network (gray) dual-doped by N (black) and S (white). C1 has an extremely high spin density, C2 and C3 have high positive charge density, and C4 and C5 have moderately high positive spin densities. Source: Reproduced with permission from Liang et al. [48]. Copyright © 2012, Wiley-VCH.

about 1.4 times at −0.2 V). Also, PG displayed a significantly increased ORR onset potential of around +0.90 V, much more positive than that of pristine graphene (c. +0.80 V) and close to that of commercially available Pt/C electrocatalyst (c. +0.96 V). The electron transfer number (n) of PG was calculated to be 3.0–3.8 based on the Koutecky–Levich (K–L) equation indicating that PG catalyzed ORR via combination of 2e$^-$ and 4e$^-$ pathways. Besides, PG showed high tolerance toward CH_3OH crossover effect as reflected by steady chronoamperometric responses upon introduction of CH_3OH during ORR. Furthermore, PG exhibited better durability than commercially available Pt/C electrocatalyst as shown by the prolonged operation at +0.7 V (current loss for PG was 13%, whereas that for commercially available Pt/C was 33%). Driven by the enhanced ORR activity of carbon black (CB)-modified reduced graphene oxide [50], PG was modified with CB (ratio: PG/CB = 5 : 1 w/w), and as expected PG/CB gave a more positive ORR onset potential (c. +0.92 V) and higher cathodic current density (−4.18 mA cm^{-2}) than CB (c. +0.83 V; −3.26 mA cm^{-2}) and PG (c. +0.92 V; −3.67 mA cm^{-2}). The extraordinary ORR performance of PG/CB composite was proposed to originate from higher electrical conductivity of PG/CB than pristine CB.

In 2013, Kim and coworkers reported heteroatom-doped graphene nanosheets prepared from covalent functionalization of GO precursors with various amines (for N-doped), cystamine (for S-doped), or boronic acid (for B-doped) followed by thermal annealing [51]. Distinctive peaks corresponding to N, B, and S confirmed the incorporation of N, B, and S into the resultant graphene nanosheets. Among these three types of nanosheets, N-doped analogues were investigated with the ORR catalytic capability and found to exhibit good electrocatalytic activity that was attributed to the presence of different types of N atoms (pyridinic-N, pyrrolic-N, graphitic-N, and N-oxide) on the graphene plane (Scheme 3.1). All the resultant graphene nanosheets showed better ORR catalytic performance than thermally reduced GO. In consistency with the general proposal that pyridinic-N is the most active catalytic site for ORR in N-doped graphene structure because of delocalization of the electrons from pyridinic-N unit [52], the counterpart with most pyridinic-N

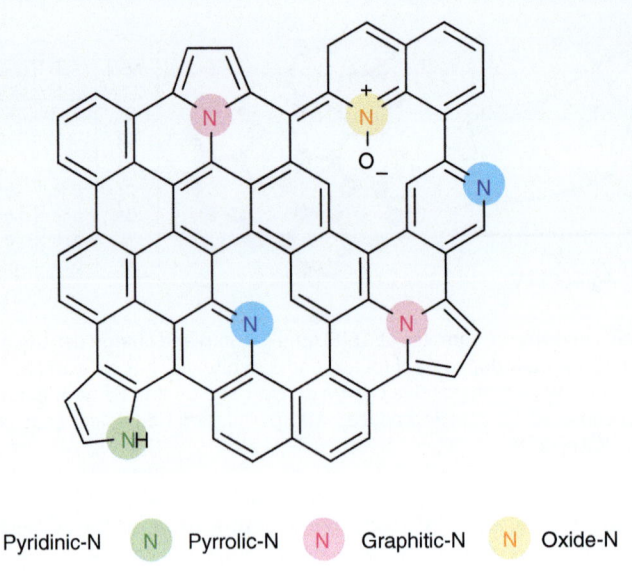

Scheme 3.1 Schematic representation of nitrogen bonding configurations in N-doped graphene nanosheets. Source: Reproduced with permission from Park et al. [51]. Copyright © 2013, Royal Society of Chemistry.

content (NRGO3) gave the best ORR catalytic performance (onset potential of c. +0.89 V; the highest limiting current density of −4.55 mA cm^{-2}) that was close to that of commercial Pt/C catalyst (onset potential of c. +0.96 V). Besides, NRGO3 was found to undergo a one-step, direct 4e$^-$ ORR pathway based on K–L equation and be more durable than commercial Pt/C (chronoamperometric response of NRGO3 decreased slowly with current retention of 73% that was higher than that of commercial Pt/C of 60%). The facile synthesis of N-, S-, and B-functionalized GO precursors introduced in this work provides alternative approaches to more accessible introduction of heteroatoms on the graphene nanosheets for more ORR-related investigation.

In the same year, Baek, Dai, and Xia collaboratively reported edge-selectively sulfurized graphene nanoplatelets (SGnP) as efficient metal-free electrocatalysts for ORR [53]. SGnP were prepared by ball milling the pristine graphite along with sulfur (S$_8$) (Scheme 3.2). It is worth mentioning that scanning electron microscopy (SEM) images showed that pristine graphite with large grain size (c. 150 mm) and plane flake morphology was dramatically altered upon ball milling. The resultant SGnP showed largely reduced grain size and flowerlike morphology. It was proposed to be stemming from graphitic C—C bond homolytic cleavages that favored generation of reactive carboradicals [54] for sulfur pick up to yield SGnP. Also as shown in high-resolution tunneling electron microscope (HR-TEM), SGnP had a highly ordered structure with honeycomb lattice in the central area and some distortion at the edge region. As reflected in TEM images, the electron densities along the edge lines seemed to be higher than those of basal area, and this probably arose from the higher electron density of sulfur moieties on edge line than carbon moieties

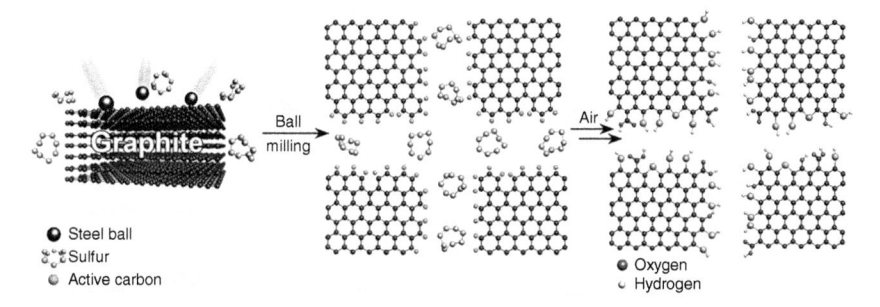

Scheme 3.2 A schematic representation of the ball-milling process. Source: Reproduced with permission from Jeon et al. [53]. Copyright © 2013, Wiley-VCH.

located at basal area. More importantly, electrocatalytic measurement illustrated that SGnP exhibited better electrocatalytic activity than pristine graphite and commercially available Pt/C electrocatalyst in terms of long-term stability and CH_3OH crossover/CO poisoning tolerance. For example, the onset potential of ORR for SGnP (+0.74 V) was much more positive than pristine graphite (+0.56 V). Besides, the limiting current density up to +0.16 V for SGnP was −4.55 mA cm^{-2} that was 2.7 times higher than that of pristine graphite (−1.67 mA cm^{-2}) and comparable with that of commercial Pt/C electrocatalyst (−4.79 mA cm^{-2}). Theoretical calculation was performed to interpret the electrocatalytic activeness of SGnP. It was found that the covalently bonded sulfur or oxidized sulfur at the edges of graphene remotely induced both charge and spin densities on the graphene where carbon members with high positive charge and spin densities served as electrocatalytic active sites. Those sulfur atoms on the edges bore increased charge density that was probably a reason for the high electrocatalytic activity of SGnP. SGnP were also oxidized to give oxidized sulfur-doped graphene nanoplatelets (SOGnP) that as expected showed a more positive onset potential. In addition, the electron transfer number of SOGnP was calculated based on K–L equation to be 3.6 that was dominantly a more efficient 4e$^-$ pathway while that of SGnP was 3.3 indicative of a combination of 2e$^-$ and 4e$^-$ pathway. These results signify that graphene edge functionalized by sulfur and oxidized sulfur can boost ORR activity of pristine graphite and opens an avenue to tune ORR activity of graphite-based materials.

Edge functionalization is not limited to sulfur and has been extended to other elements. In the same year, Jeon et al. reported edge-halogenated graphene nanoplatelets (XGnPs; X = Cl, Br, I) as efficient metal-free electrocatalysts for ORR [55]. These XGnPs were obtained by simple ball milling of graphite flake in the presence of chlorine (Cl_2), bromine (Br_2), or iodine (I_2) (Scheme 3.3). These XGnPs showed about 2–4.4 times higher current values than that of pristine graphite and 60–133% that of the commercial Pt/C electrode. More importantly, the numbers of electron transfer during ORR at +0.16 V for ClGnP, BrGnP, and IGnP were calculated to be 3.5, 3.8, and 3.9, respectively, suggesting that all XGnPs underwent ORR via an ideal 4e$^-$ process far better than pristine graphite (2.0) that probably carried out ORR via a classical 2e$^-$ pathway. The onset potentials of XGnPs positively shifted in an order of ClGnP < BrGnP < IGnP. It is noteworthy that the

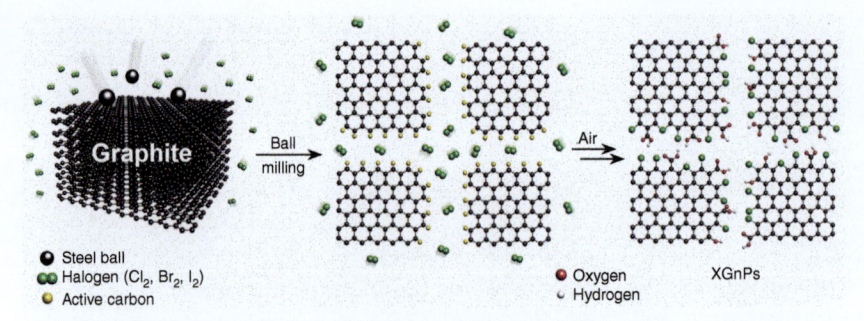

Scheme 3.3 A schematic representation for mechanochemically driven edge-halogenation reaction between the *in situ* generated active carbon species (gold balls) and reactant halogens (twin green balls). Active carbon species were generated by homolytic bond cleavages of graphitic C—C bonds and reacted with halogen molecules to produce edge-halogenated graphene nanoplatelets (XGnPs) in a sealed ball-mill capsule, and the remnant active carbon species are terminated upon subsequent exposure to air moisture. Red and gray balls stand for oxygen and hydrogen, respectively. Source: Reproduced with permission from Jeon et al. [55]. Copyright © 2013, The Author(s).

ORR activity enhancement in N-doped carbon-based materials has been proposed to stem from the fact that more electronegative atom polarizes adjacent carbon atoms in the graphitic framework and promotes the O_2 adsorption and charge transfer [25]. Interestingly, the electronegativity differences between halogens (Cl = 3.16, Br = 2.96, and I = 2.66) and carbon (2.55) show a reverse order to the ORR activities and appears to contradict the proposed doping-induced charge transfer mechanism during O_2 reduction. It was pointed out that the ease of electron donation increased in an order of I > Br > Cl because of increasing atomic size down the group of halogens. Theoretical calculation results suggested that binding strength with O_2 increased in an order of IGnP > BrGnP > ClGnP in which IGnP weakened the O_2 bond to the highest extent (Figure 3.10a–c). This work presents an alternative concept on dopant-induced ORR enhancement in graphite-based materials.

Following the success of N-doped graphene [56], S-doped graphene [57], and N/S-co-doped graphene [48] as ORR catalyst, He and coworkers reported another

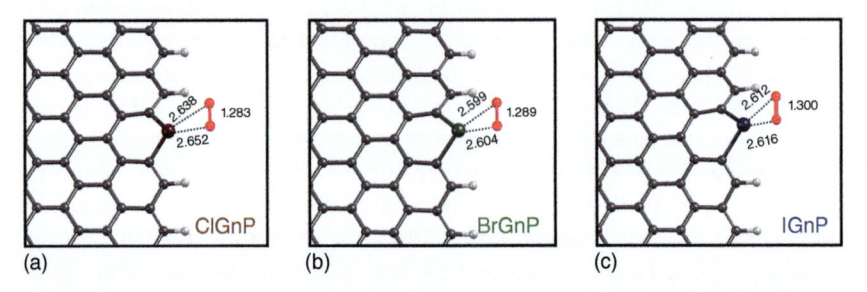

Figure 3.10 The optimized O_2 adsorption geometries onto XGnPs, in which halogen covalently linked to two sp^2 carbons. (a) ClGnP, (b) BrGnP, and (c) IGnP. Source: Reproduced with permission from Jeon et al. [55]. Copyright © 2013, The Author(s).

type of N/S-co-doped graphene (N-S-G) that possessed decent ORR performance when compared with commercial Pt/C catalyst [58]. The N-S-G catalysts were prepared by pyrolysis of poly[3-amino-5-mercapto-1,2,4-triazole] and graphite oxide composite (PAMTa-GO). It was found that the N-S-G catalysts prepared from different pyrolysis temperature gave different catalytic performance. The catalyst prepared from pyrolysis at 1000 °C (N-S-G 1000) had the highest content of graphitic-N and thiophene-S that were found to play important roles for ORR in N-doped graphene [56] and S-doped graphene [57], respectively. As expected, N-S-G 1000 gave the best ORR catalytic performance when compared with N-S-G prepared from pyrolysis at other temperatures and showed an onset potential (c. +0.91 V) comparable with that of commercially available Pt/C catalyst (c. +0.96 V). Besides, N-S-G 1000 exhibited higher limiting current density than commercial Pt/C catalyst at +0.73 V (Figure 3.11a) that was proposed to be due to the higher kinetic facileness of N-S-G 1000 for ORR [59]. Also, based on the K–L plot, electron transfer number for N-S-G 1000 during ORR was calculated to be 3.6, suggesting that the ORR process catalyzed by N-S-G 1000 proceeded dominantly via a 4e$^-$ pathway. As other reported carbon-based nanomaterials with enhanced ORR activity, N-S-G 1000

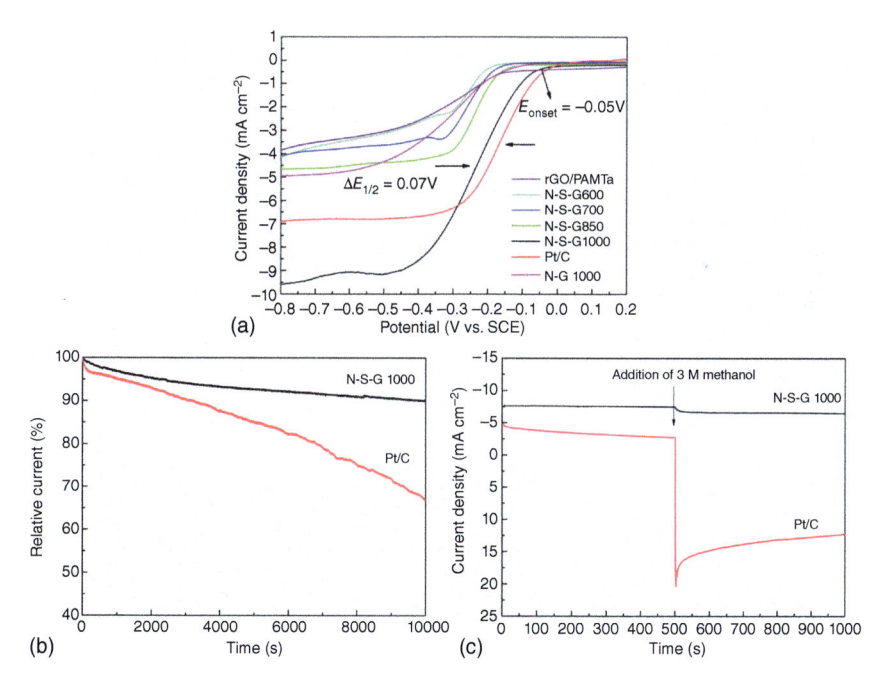

Figure 3.11 (a) LSVs of PAMTa/rGO, N-S-G 600, N-S-G 700, N-S-G 850, N-S-G 1000, N-G 1000, and Pt/C in O_2-saturated 0.1 M NaOH at a scan rate of 10 mV s^{-1} with a rotation speed of 2000 rpm. (b) Stability evaluation of N-S-G 1000 and Pt/C for 10 000 seconds in an O_2-saturated 0.1 M NaOH solution at −0.3 V vs. SCE with a rotation speed of 1600 rpm. (c) Current–time (i–t) chronoamperometric responses for ORR at −0.3 V vs. SCE in an O_2-saturated 0.1 M NaOH solution at Pt/C and N-S-G 1000-modified electrode with the addition of 3 M CH_3OH at 500 seconds, which is marked with an arrow. Source: Reproduced with permission from Li et al. [58]. Copyright © 2014, Royal Society of Chemistry.

demonstrated excellent CH_3OH tolerance and better stability than commercial Pt/C in alkaline medium. For example, N-S-G 1000 kept 89.93% of its initial current after a chronoamperometric test for 10 000 seconds while Pt/C catalyst had 33.15% current attenuation loss (Figure 3.11b). Also, N-S-G 1000 showed no change in current after the introduction of CH_3OH during chronoamperometric test while Pt/C showed a strong amperometric response at the time of CH_3OH addition (Figure 3.11c). The outstanding ORR performance of N-S-G 1000 reported in this work was ascribed to the high electroactive surface area, large number of active sites of highly catalytically active graphitic-N and thiophene-S groups, and also synergistic effect induced by appropriate ratio between graphitic-N and thiophene-S.

In the same year, Wang and coworkers further advanced the preparation of N/S-co-doped graphene (NSG) by a one-pot approach in which NSG was prepared by direct annealing GO and thiourea in argon with thiourea serving as both the N and S sources [60]. The annealing temperature was found to play an important role in N- and S-doping over graphene. For example, the content of N decreased along with increasing annealing temperature like that reported in N-doped graphene prepared by annealing GO in the presence of NH_3 [61]. On the other hand, the content of S increased with elevating annealing temperature indicating that doping of S required higher temperature. Higher annealing temperature produced more graphitic N that was beneficial to ORR performance of doped graphene. However, with the annealing temperature ranging from 600 to 1000 °C, the NSG with the best performance in this work was NSG700 (700 was the annealing temperature); this signified that the ORR performance was subjected to the interplay between content of N- and S-dopants at the most appropriate annealing temperature. When compared with undoped graphene, NSG700 showed more positive onset potential and higher current density (Figure 3.12). On the one hand, when compared with

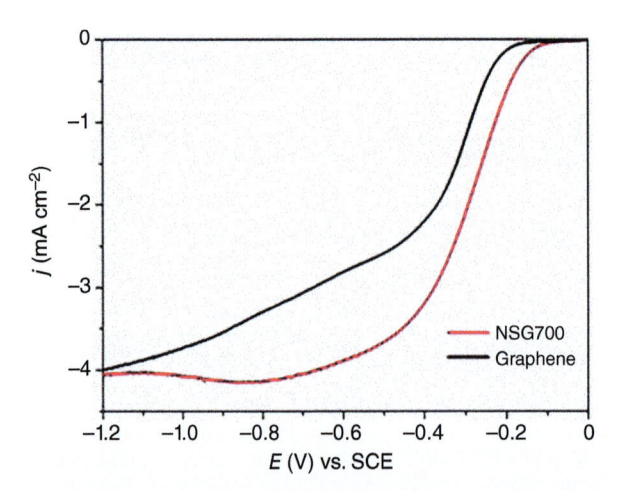

Figure 3.12 The disk current density obtained from the LSV testing on electrode materials modified RRDE in oxygen-saturated 0.1 M KOH electrolyte. Source: Reproduced with permission from Wang et al. [60]. Copyright © 2014, Royal Society of Chemistry.

commercial Pt/C electrocatalyst, NSG700 gave constant performance after 2000 cyclic voltammetric cycles while the onset potential of commercial Pt/C catalyst shifted negatively by 45 mV. This highlighted the stability of NSG700 as ORR electrocatalyst for continuous operation. With competent ORR performance, NSG reported in this work can be prepared through a rather simple way, and this in no doubt lowers the barrier of putting NSG into practice for application in fuel cell, metal–air batteries, and so forth.

Normally, heteroatom-doped graphene with enhanced ORR activity is obtained via CVD, plasma treatment, or high-temperature annealing methods [41, 62] that usually involve large amount of energy dissipation and harsh reaction conditions. Interestingly, in 2015, Wang and coworkers reported an alternative method to obtain S-doped graphene (SG) that came from cycled lithium–sulfur battery (Li–S battery) with graphene–sulfur composites as cathode materials [63]. It was found that S-doping into graphene was achieved by the continuous charge/discharging cycling of the Li–S battery. High-resolution XPS C1s spectra of the SG gave signals at 284.6 and 285.8 eV that corresponded to S-incorporated graphene and the C–O or C–S, respectively. Also, the high-resolution S2p spectra showed peaks at 164.5 and 163.5 eV that confirmed the doping of S in graphene network. Besides, higher proportion of defective carbon in SG than pristine graphene consolidated the doping of S over the graphene network. Noteworthy, S^{2-} species were detected on the SG sample probably coming from the residual polysulfides. It has been proposed that charge transfer behavior from S^{2-} to graphene might alter the electronic and O_2 adsorption properties, leading to the ORR performance enhancement [64]. SG showed more positive ORR halfway potential, higher limiting diffusion current, and larger number of electron transfer during ORR than undoped graphene (Figure 3.13a,b). Not only that, SG gave better performance than Pt/C did in terms of CH_3OH crossover (current density of SG remained constant upon introduction

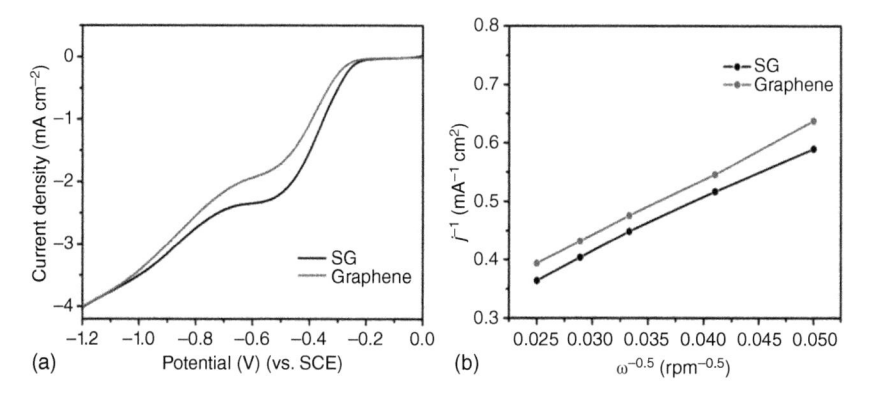

Figure 3.13 (a) LSV curves of the ORR on SG and graphene in an O_2-saturated KOH solution (0.1 M) at a scan rate of 10 mV s^{-1} and a rotation rate of 1600 rpm. (b) K–L plots for SG and graphene at −0.8 V. Source: Reproduced with permission from Ma et al. [63]. Copyright © 2015, Wiley-VCH.

of CH$_3$OH during measurement while Pt/C totally malfunctioned at the same scenario) and durability (onset potential of SG remained unchanged while that of Pt/C shifted negatively upon 2000 cycles of cyclic voltammetric scans). Although the ORR enhancement of SG compared with pristine graphene in this work is not as significant as previously reported heteroatom-functionalized graphene is, SG prepared from this methodology adds a new dimension to heteroatom-doped graphene preparation and improves the practicability of these graphene in various applications.

Graphene is an attractive class of nanomaterial because it is intrinsically an electrical conductor. Through various approaches, heteroatoms have been incorporated into graphene network, and it leads to a leap on ORR performance when compared with the pristine graphene. The works described in this section illustrate that the potential graphene shows to be a new generation of electrocatalyst for ORR.

3.2.3 Heteroatom-Doped Graphdiyne

Graphdiyne, constructed by sp^2- and sp-hybridized carbon atoms, represents a new class of 2D nanostructured carbon materials with high degree of π-conjugation, intrinsic porosity, and chemical and mechanical stability [65–69]. Graphene and graphdiyne are related since both are also single layered with infinite planar extension. On the other hand, graphdiyne is structurally different from graphene. Graphene is constituted by a network of sp^2-hybridized carbon, while graphdiyne has a benzene ring bearing six acetylenic groups as a central unit jointed repeatedly through acetylene coupling to give the final structure. Considering the four-carbon diacetylenic bridges, π orbitals confer graphdiyne ability to attract molecules through extra π interaction that is absent in graphene. This property along with intrinsic porosity makes graphdiyne a competent candidate for small-molecule activation (e.g. O$_2$). Based on the similarities and differences when compared with graphene, graphdiyne has been explored for its potential in ORR application. Taking reference to the success of other carbon-based nanomaterials, heteroatom functionalization is an interesting approach to introduce ORR catalytic activity. As this direction is still green, N atom is the only candidate used as heteroatom functionalization for enhancing ORR activity of graphdiyne so far.

In 2014, Zhang and coworkers reported the use of N-doped graphdiyne (N-GD) as a metal-free electrocatalyst for ORR application [70]. The N-GD were prepared through cross-coupling reaction of hexaethynylbenzene in the presence of pyridine on copper surface followed by annealing at different temperature under NH$_3$/Ar atmosphere (10% NH$_3$). Field-emission scanning electron microscope (FESEM) and energy-dispersive X-ray spectrum (EDS) showed homogeneous distribution of C, N, and O in the skeletons of N 550-GD (550 denoted the annealing temperature of N-GD to be 550 °C) (Figure 3.14). The elemental O was proposed to be due to incorporation of O$_2$ from air via physical adsorption. The high-resolution N 1s XPS of N 550-GD showed two peaks at 398.2 and 399.4 eV that corresponded to imine N (N substituted

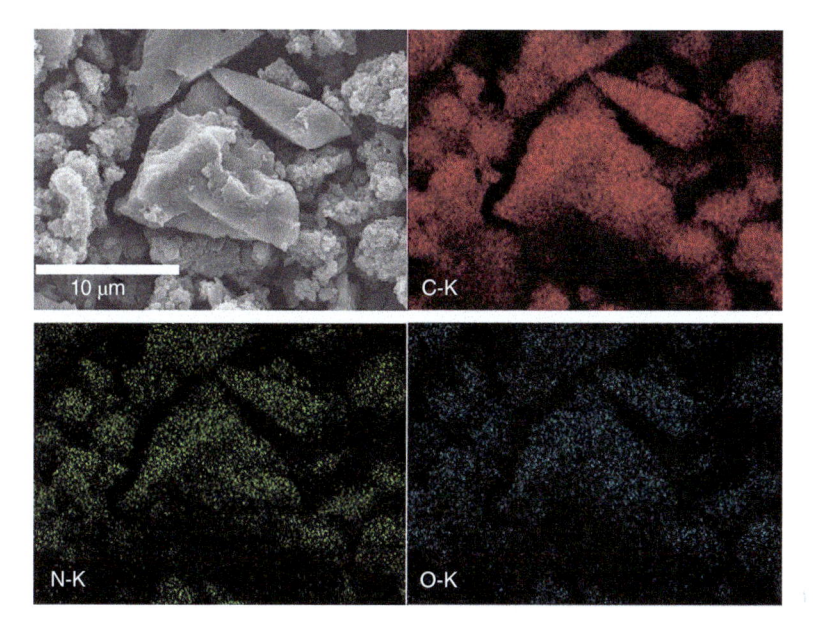

Figure 3.14 FESEM and EDS elemental mapping images of the prepared N 550-GD powder. Source: Liu et al. [70]/Reproduced with permission from Royal Society of Chemistry.

for the sp-hybridized carbon atoms on diacetylenic bridge) or pyridinic N; theoretical calculation results suggested the dominant species for N atoms in N 550-GD was imine N. It was found that N substitution at diacetylenic bridge was more favorable at temperature <550 °C while N substitution at aromatic ring was more likely to take place at temperature >550 °C (pyridinic N signals increased significantly in N 600-GD). N 550-GD exhibited an onset potential of +0.90 V vs. RHE that was more positive than GD (+0.82 V) and close to that of commercial Pt/C electrocatalyst (c. +0.96 V) (Figure 3.15). The N-GD showed better CH_3OH crossover tolerance and durability upon long-term operation. For example, N 550-GD kept steady current density while Pt/C showed significant change in current density upon introduction of CH_3OH; N 550-GD only gave a small shift of 9 mV after 6000th cycle of operation, whereas Pt/C had a shift of 42 mV upon same operation conditions.

Three years later, Huang and coworkers published a work of porous N-doped graphdiyne (N-GDY) as ORR electrocatalyst [71]. It was proposed by the authors that N-GDY with higher porosity could apparently improve the ORR activity of N-doped graphdiyne. Two types of N-doped GDY were prepared. One was N-GDY prepared from calcination of pyridine-mixed GDY at different high temperature (700–1000 °C) while another one was $N'N$-GDY prepared from calcination of N-GDY obtained at 900 °C in an NH_3 atmosphere. $N'N$-GDY was prepared since the N content in all N-GDY samples was relatively low. Also, calcination at high temperature was to introduce porosity into the GDY samples for more efficient ORR. The porosity of $N'N$-GDY had the highest Brunauer–Emmett–Teller (BET) surface area (1154 $m^2 g^{-1}$) that was far higher than all the N-GDY in this work (52–464 $m^2 g^{-1}$). Indeed, $N'N$-GDY exhibited onset potential much more positive

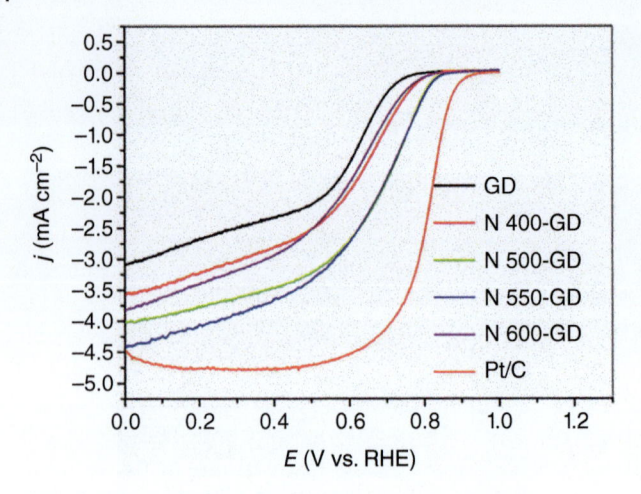

Figure 3.15 LSV curves for GD, N-doped GD, and Pt/C on an RDE in O_2-saturated 0.1 M KOH solution at a scan rate of 10 mV s^{-1}. The rotation rate is 1600 rev min^{-1}. Source: Reproduced with permission from Liu et al. [70]. Copyright © 2014, Royal Society of Chemistry.

than N-GDY did, comparable with that of Pt/C catalyst. Besides, $N'N$-GDY demonstrated better stability than commercial Pt/C catalyst ($N'N$-GDY only showed a negative shift of 19 mV at potential while Pt/C had a negative shift of 35 mV after 5000th cycle of operation) yet at the same time had a high number of transferred electrons during ORR (n is c. 3.84), indicating an almost ideal 4e$^-$ ORR pathway. Based on experimental and theoretical results, it was proposed that the high ORR catalytic activities of N-GDY and $N'N$-GDY arose from (i) enhanced electronic conjugation and hence electrical conductivity after the removal of small GDY pieces and O-containing groups during high-temperature treatment, (ii) porosity induced by high-temperature calcination, and (iii) substitution of N for C atoms as imine groups in diacetylenic bridge and pyridinic N in benzene skeletons.

Although the above two works presented the possibility of N-doped graphdiyne as promising ORR electrocatalyst, Wang and coworkers held a different perspective toward the assignment of N atoms substituted over the graphdiyne network. In 2018, a work was released by their group to present a rational and controllable approach to introduce sp-hybridized N atoms into specific sites of few-layer oxidized graphdiyne (FLGDYO) with experimentally supported mechanistic study using thermogravimetry–differential thermal gravity–mass spectrometry (TG-DTG-MS) [72]. The resultant sp-N-doped few-layer graphdiyne (sp-NFLGDY) was obtained from annealing at high temperature using FLGDYO and melamine. Based on the monitoring by TG-DTG-MS, the FLGDYO captured $NHCNH_2^+$ fragment released from melamine to form a five-member cyclic moiety on acetylenic portion that rearranged and decomposed upon the introduction of hydrogen radical to give sp-NFLGDY and $H_2C=N^+=CH_2$. The X-ray absorption near-edge structure (XANES) spectroscopy and XPS verified the existence of sp-N atoms with signals at 397.6 eV and probed the existence of graphitic-N (400.9 eV), amino-N (399.3 eV), and pyridinic N (398.3 eV). The proportion of sp-N atoms increased along with higher annealing temperature. The resultant sp-NFLGDY obtained from 900 °C (sp-NFLGDY-900) gave the best ORR performance in terms of onset potential (+0.96 V), current density (38.0 mA cm^{-2} at +0.75 V), and number of electron transfer during ORR ($n = 3.9$) with comparable onset potential and more tolerance toward CH_3OH interference when compared with Pt/C electrocatalyst. More importantly, sp-NFLGDY-900 was also tested the ORR activity in acidic medium (0.1 M $HClO_4$), and it retained higher current values (85%) than Pt/C (60%) did after the operation for 10 000 seconds. This work clarifies the formation mechanism of N-doped graphdiyne and contributes a rational approach to well-defined N-doped graphdiyne. This piece of work undoubtedly provides insight into the rational design of ORR-active high-performance N-doped graphdiyne in a controllable manner that is critically important for mass production of such material.

3.2.4 Heteroatom-Doped Porous Carbon Nanomaterials

The above-discussed materials are carbon allotropes with electronic conjugation and well-defined morphological classification (CNTs as 1D material; graphene and graphdiyne as 2D materials). Ever since the report of N-doped CNTs as ORR electrocatalyst by Dai and coworkers, studies of heteroatom-doped carbon-based nanomaterials as ORR electrocatalysts increase explosively. It is generally accepted that uneven electron density distribution induced by heteroatoms, appropriately large porosity, and good electrical conductivity contribute to the ORR activity of carbon-based materials. Based on this direction, there arise many reports of morphologically diversified heteroatom-functionalized porous carbon nanomaterials prepared via various methodologies, and these materials all possess ORR catalytic activity comparable with commercially available ORR electrocatalysts.

To make environmentally sustainable ORR catalysts more sustainable, it is favorable to obtain ORR active materials from natural precursors. Specifically, porous N-doped carbon catalysts have been obtained from cellulose [73, 74], coconut-shell based charcoal [75], and naturally available fallen ginkgo leaves [76]. Considering the cellulose derived materials, one example (N-doped carbon, NC-A) was prepared from polydopamine (PDA)-modified mixed cellulose ester filter films (MCEFs) that was further functionalized by phenol, melamine, and hexamethylenetetramine polymerization followed by KOH activation and annealing at 900 °C [73] (Scheme 3.4) while another one (pyridinic-N dominated and defect-enriched graphene-like nanocarbon, ND-GLC) was prepared from cellulose precursors, along with KOH activation and NH_3 treatments [74] (Scheme 3.5). For the natural fallen ginkgo leaves, N-doped fullerene-like carbon shell (NDCS) was obtained from pyrolysis of the fallen ginkgo leaves at 800 °C under N_2 atmosphere followed by treatment with HCl [76] (Scheme 3.6). On the other hand, N-doped activated carbon (AC-N) were prepared by annealing coconut shell-based charcoal in the presence of KOH followed by NH_3 modification. All these materials possess

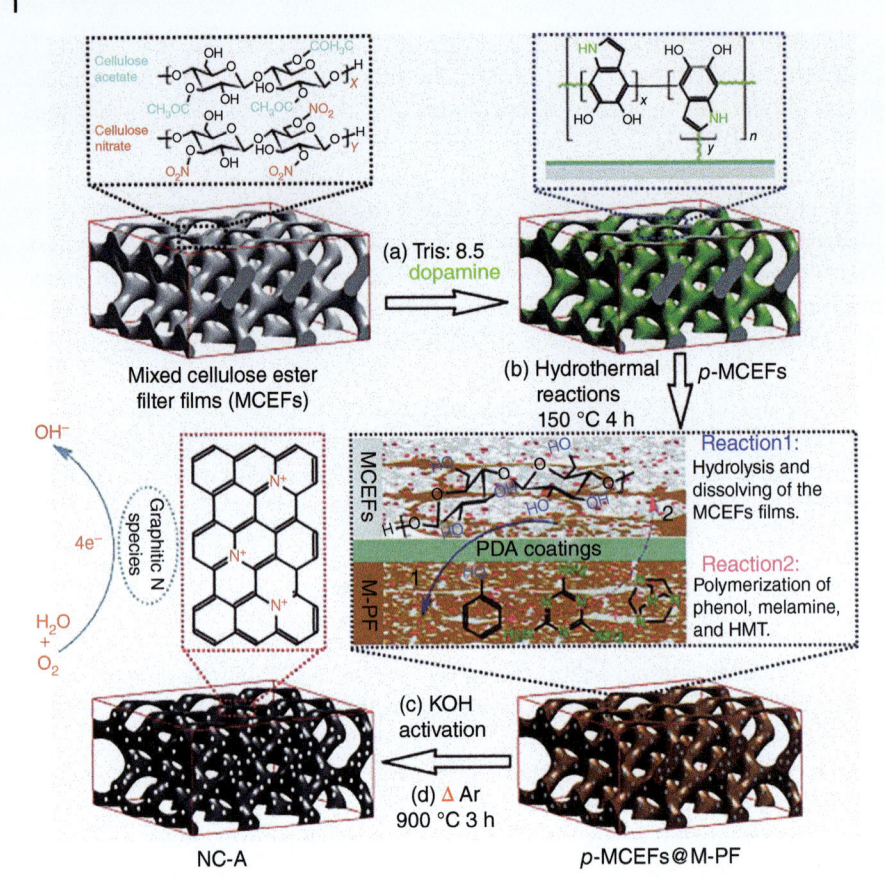

Scheme 3.4 Fabrication and formation mechanism of the 3D hierarchically porous nitrogen-doped carbon materials. Tris = tris buffer solution (pH 8.5). HMT = hexamethylenetetramine. Source: He et al. [73]. Reproduced with permission of Wiley-VCH.

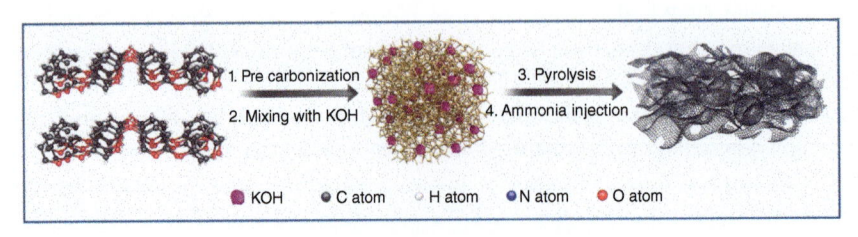

Scheme 3.5 Schematic illustration of the synthesis of ND-GLC. Source: Reproduced with permission from Zhang et al. [74]. Copyright © 2018, Elsevier Ltd.

graphitic-N, pyridinic-N, and pyrrolic-N as N-dopants in the carbon network as reflected in high-resolution XPS spectra. They had high surface area (BET surface area = 583–2441 $m^2\,g^{-1}$), proceeded ORR via a $4e^-$ pathway (number of electron transferred = 3.5–4.1), and exhibited comparable or even more positive onset and half-way potentials (specifically ND-GLC demonstrated onset and

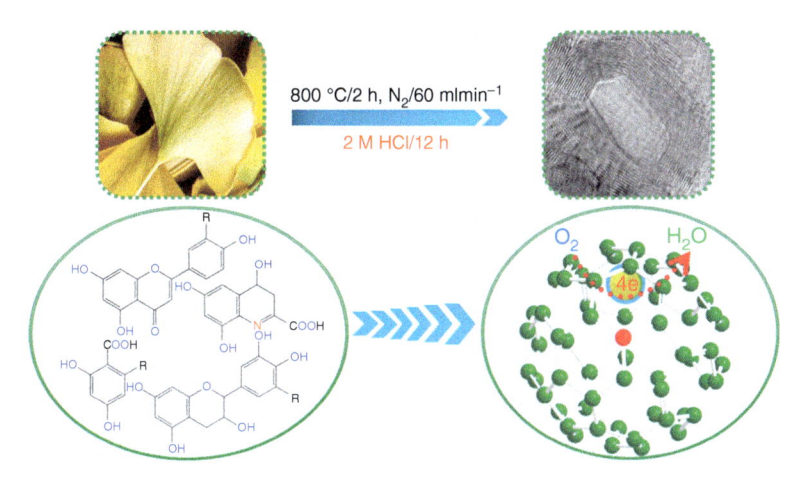

Scheme 3.6 Illustrated procedure of fabricating N-doped fullerene-like carbon shell from organic-rich fallen ginkgo leaves. Source: Reproduced with permission from Gao et al. [76]. Copyright © 2015, Elsevier Ltd.

half-way potential 23 and 19 mV positive than Pt/C catalyst) [74], comparable or higher limiting current density at designated potential (c. 4.55 mA cm^{-2}), and better CH_3OH/CO crossover tolerance and operation stability (current kept steady upon introduction of CH_3OH/CO or sufficiently long operation time, such as 10 000 seconds) when compared with commercially available Pt/C electrocatalyst.

Alternatively, porous carbon nanomaterials have been prepared by direct annealing ionic liquids or salts at high temperature. N-doped carbon foam (CF_N) was prepared from decomposition of sodium alkoxide into solid carbonized lumps that were further processed at 1000 and 1400 °C for reduction of O-content and graphitization [77]. Interestingly, a type of hierarchically porous carbon materials (HPCMs) bearing interconnected macropore channels with controlled pore sizes was synthesized using NH_4Cl and sodium or iron citrate salts through phase-transition-assisted (PTA) strategy at high temperature (600–900 °C) [78] (Scheme 3.7). As confirmed by high-resolution XPS, both CF_N and HPCMs possessed graphitic-N, pyridinic-N, and pyrrolic-N. It is noteworthy that both of these porous carbon nanomaterials with high surface area (BET surface area = 700–1141 m^2 g^{-1}) proceeded ORR via an admixture of 2e$^-$ and 4e$^-$ pathway (number of electron transferred = 3.4–3.95) and exhibited comparable or more positive onset (+0.90 to +1.00 V) and halfway (+0.80 to +0.90 V) potentials along with steady performance over long-term operation (e.g. 60 000 potential cycles) when compared with commercially available Pt/C electrocatalyst.

Apart from the abovementioned methodologies to access porous carbon nanomaterials, template-mediated synthesis represents a controllable methodology to give rise to structurally well-defined carbon-based nanomaterials. Various templates such as silica nanoparticles [79, 80], magnesium oxide (MgO) [81], polyaniline-coated melamine fiber (PANI@MF) [82], mesoporous PDA [83], metal–organic framework (MOF) [84], and COF [85]. Looking into

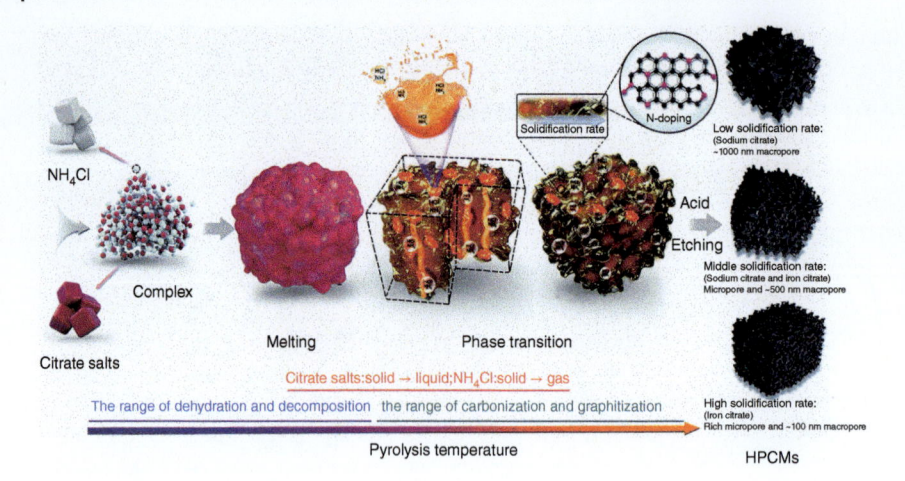

Scheme 3.7 Schematic representation of the synthesis of hierarchically porous carbon materials. Source: Reproduced with permission from Li et al. [78]. Copyright © 2018, Royal Society of Chemistry.

silica nanoparticle-templated methodology, Yang et al. reported the preparation of mesoporous N-doped carbon materials coming from carbonization of nucleobase-carrying all-organic ionic liquid through hard templating with silica nanoparticles [79]. The resultant mesoporous material exhibited the surface area as highest as 1553 m^2 g^{-1}. The best member showed onset potential approximately 35 mV more positive than the commercial Pt catalyst with number of electrons transferred during ORR to be 4.1 indicative of the 4e$^-$ process and robust CH$_3$OH crossover tolerance. Impressively, the mesoporous materials worked properly in acidic conditions and gave performance comparable with that of Pt catalyst. On the one hand, Müllen and Feng published an interesting work of N-doped mesoporous carbon nanospheres (NMCNs) prepared from ammonium persulfate triggered silica nanoparticle-templated polyaniline formation followed by carbonization and etching with HF or NaOH [80] (Scheme 3.8). The NMCNs possessed

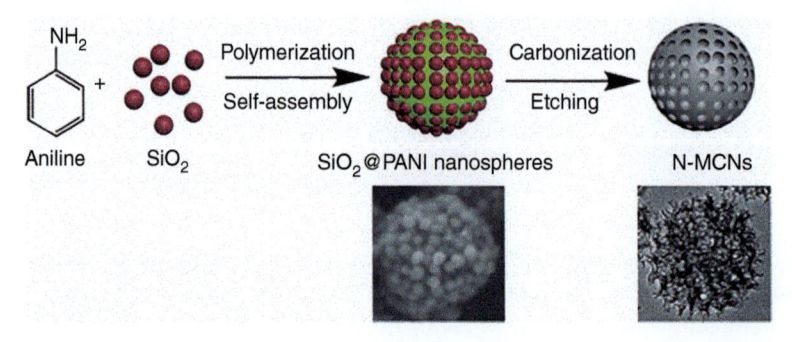

Scheme 3.8 Illustration of self-assembly of colloidal silica with PANI and the derived N-MCNs. Source: Wang et al. [80]/Reproduced with permission from WILEY-VCH.

tailorable mesopores (7–42 nm), particle sizes (90–300 nm), high surface area (785–1117 $m^2\,g^{-1}$), pore volume (1.46–2.01 $cm^3\,g^{-1}$), and N content (5.54–8.73 at%). The NMCNs performed well as ORR electrocatalyst in alkaline medium with high ORR catalytic activity (half-wave potential of +0.78 V vs. +0.85 V for commercial Pt/C), large diffusion-limited current (c. 5.5 $mA\,cm^{-2}$), high selectivity with $4e^-$ process as dominant ORR pathway (electron transfer number >3.86), satisfactory CH_3OH tolerance, and excellent long-term stability (current kept steady over operation for 30 000 seconds).

Previously, Hu and Wang developed a MgO template synthetic method to prepare carbon nanocages (CNCs) using benzene precursors [86]. This approach was utilized for preparing N-doped carbon nanocages (NCNCs) with pyridine as substitute for benzene precursors and growth temperature ranging from 600 to 900 °C [81]. The NCNCs possessed large surface areas up to 1393 $m^2\,g^{-1}$ with satisfactory onset potential (+0.86 V), electron transfer number of 3.27 signifying a blend of $2e^-$ and $4e^-$ mechanism, and superior operation stability in alkaline medium when compared with Pt/C electrocatalyst.

In 2017, Zhang and coworkers reported a one-pot strategy to attain porous N-doped carbon nanoribbon (NCNR) utilizing PANI@MF as template processed subsequently at 900 °C [82]. During the formation process, melamine first turns to be melamine fiber (MF) in the presence HNO_3; ammonium persulfate initiated polymerization of aniline on the surface of MF to yield PANI@MF that underwent carbonization at 900 °C in N_2 stream to give NCNR (Scheme 3.9). NCNRs were probed to contain pyridinic-N, pyrrolic-N, and graphitic-N as reflected by high-resolution XPS. The best NCNRs reported in this work had an onset potential of +0.99 V, a half-way potential of +0.87 V, and a diffusion-limited current density of −5.8 $mA\,cm^{-2}$ comparable with those of commercially available Pt/C catalyst. Also, the number of electron transfer during ORR was probed to be 3.9 indicative of dominantly $4e^-$ pathway. Along with superior CH_3OH crossover tolerance in alkaline medium (amperometric signals kept constant upon introduction of CH_3OH), though NCNRs showed inferior onset potential (+0.81 V) and halfway potential (+0.53 V) to those of Pt/C electrocatalyst (+0.84, +0.79 V) in acidic medium, the tolerance toward CH_3OH was the same as that in alkaline medium that is far better than that of Pt/C catalyst.

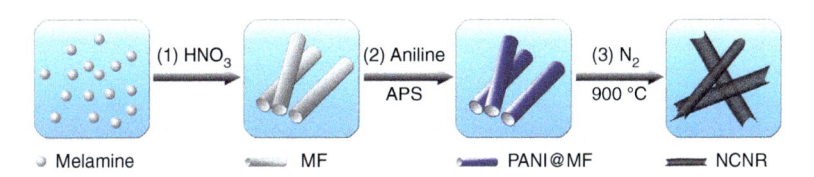

Scheme 3.9 Schematic illustration of the synthesis of NCNR using melamine fiber (MF) as the template. Source: Reproduced with permission from Huang et al. [82]. Copyright © 2017, Elsevier Ltd.

In 2016, Lou and coworkers reported mesoporous carbon nanomaterials obtained from a novel emulsion-induced interface anisotropic assembly strategy followed by calcination [83]. The process begins with the formation of emulsion system between immiscible 1,3,5-trimethylbenzene (TMB) and water interface where island-shaped mesoporous PDA assembled by block copolymer F127/TMB/PDA micelles (Scheme 3.10). Micelles proceeded the cooperative assembly to give bowl-like mesoporous PDA particles with radially oriented mesopore channels. The final mesoporous bowl-like N-doped carbon nanomaterials were obtained after the calcination of the mesoporous PDA particles. The resultant carbon particles possessed a bowl-like morphology with radially oriented structure (Figure 3.16). These particles, having a surface area of c. 619 m^2 g^{-1} based on Barrett–Joyner–Halenda (BJH) model, exhibited satisfactory ORR performance with an onset potential of +0.85 V and number of electron transfer of c. 3.5 indicative of dominantly 4e$^-$ pathway and superior stability over long-term operation when compared with Pt/C electrocatalyst.

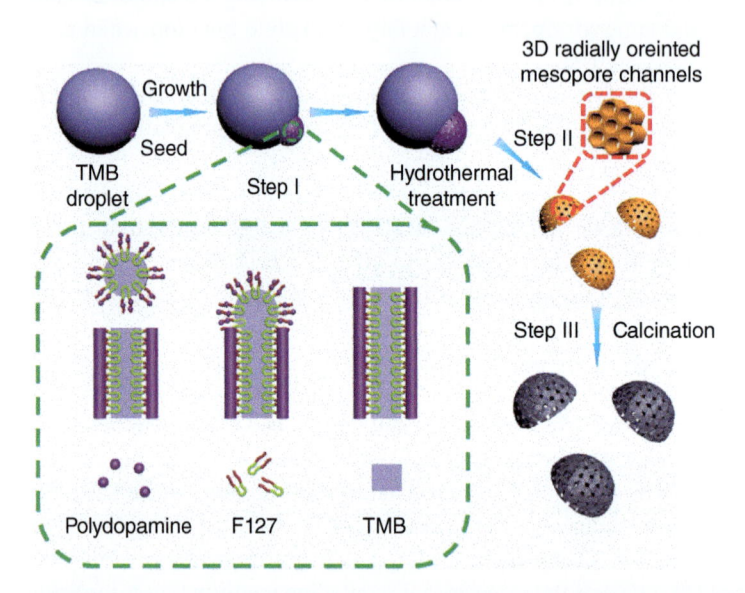

Scheme 3.10 Schematic illustration of the formation process of bowl-like mesoporous particles. Step I: formation of the block copolymer F127/TMB/polydopamine oligomer composite micelles and emulsion induced interface anisotropic assembly of asymmetric bowl-like mesostructured polydopamine particles with radially oriented large mesochannels. Step II: hydrothermal treatment of the mesostructured polydopamine nanocomposites to stabilize the structure. Step III: carbonization at 800 °C for 2 hours under nitrogen atmosphere to generate bowl-like mesoporous carbon particles with radially oriented mesochannels. Source: Reproduced with permission from Guan et al. [83]. Copyright © 2016, American Chemical Society.

Noteworthy, MOFs and COFs are novel porous materials with well-defined structure, morphology, and crystallinity [87–90]. These MOFs and COFs are rich in heteroatom content and hence favorable source for heteroatom-doped porous carbon nanomaterials. There were two reports using MOF and COF as template respectively for the formation of heteroatom-doped porous carbon-based

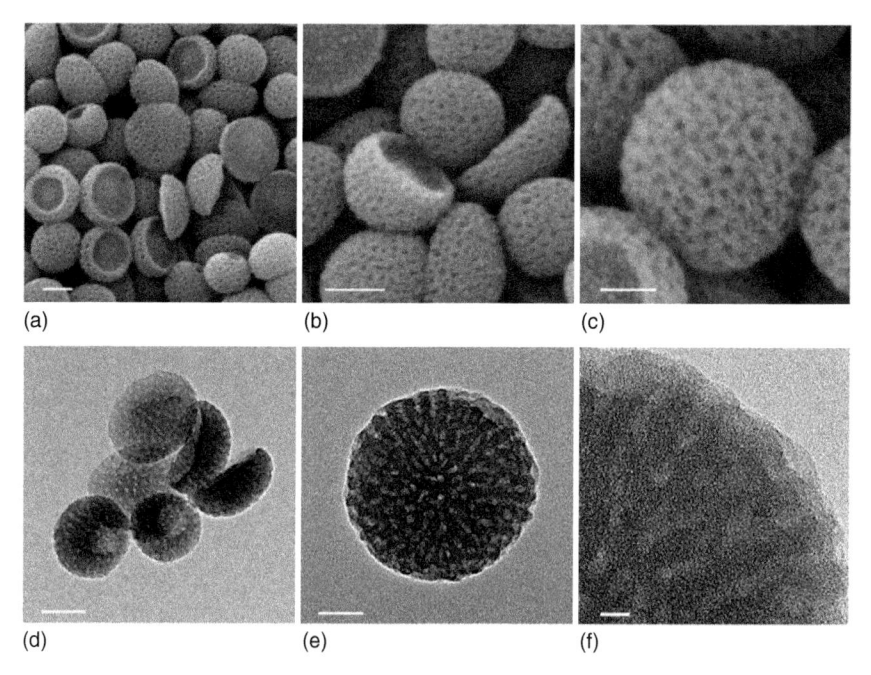

Figure 3.16 Microscopy characterizations of bowl-like mesoporous carbon particles: (a, b) FESEM images, (c) magnified FESEM image, (d) TEM image, (e) magnified TEM image, and (f) high-resolution TEM image of radially oriented mesochannels in a bowl-like carbon particle. Scale bars are 100 nm (a, b, d), 50 nm (c, e), and 10 nm (f). Source: Guan et al. [83]/ Reproduced with permission from American Chemical Society.

materials [84, 85]. Considering the MOF-derived porous materials, MOF-5 (MOF constructed by Zn–O cluster and benzenedicarboxylate) was chosen as a template along with dicyandiamide (DCDA), triarylphosphine (TArP), and dimethyl sulfoxide (DMSO) as sources of N, P, and S, respectively [84]. The porous material was prepared from first encapsulation of DCDA, TPP, and DMSO within MOF-5 in CH_3OH solution followed by carbonization at 900 °C in N_2 atmosphere. The resultant porous carbon material exhibited ORR onset potential, higher limiting current density, and superior CH_3OH crossover tolerance than commercially available Pt/C catalyst. The excellent electrocatalytic activity of the porous carbon materials was proposed to stem from the following reasons: (i) electronegative heteroatoms doped in carbon framework polarize the carbon atoms that serve as active site for O_2 adsorption and activation, (ii) heteroatoms incorporated into the carbon framework induce asymmetric spin density that effectively weaken the O—O bonding for enhanced ORR activity, (iii) incorporation of heteroatoms introduces change in porosity with increased active sites and mesopores, and (iv) synergistic effect of N, P, and S ternary-doping may promote ORR activity superior to sole elemental doping. For the porous material prepared from COF template, the N, P-co-doped porous carbon nanomaterial was obtained from encapsulation of 4,4′,4″-(1,3,5-triazine-2,4,6-triyl)trianiline (TAPT)-2,5-dihydroxyterephthalaldehyde (DHTA)-COF) with phytic acid (PA) in

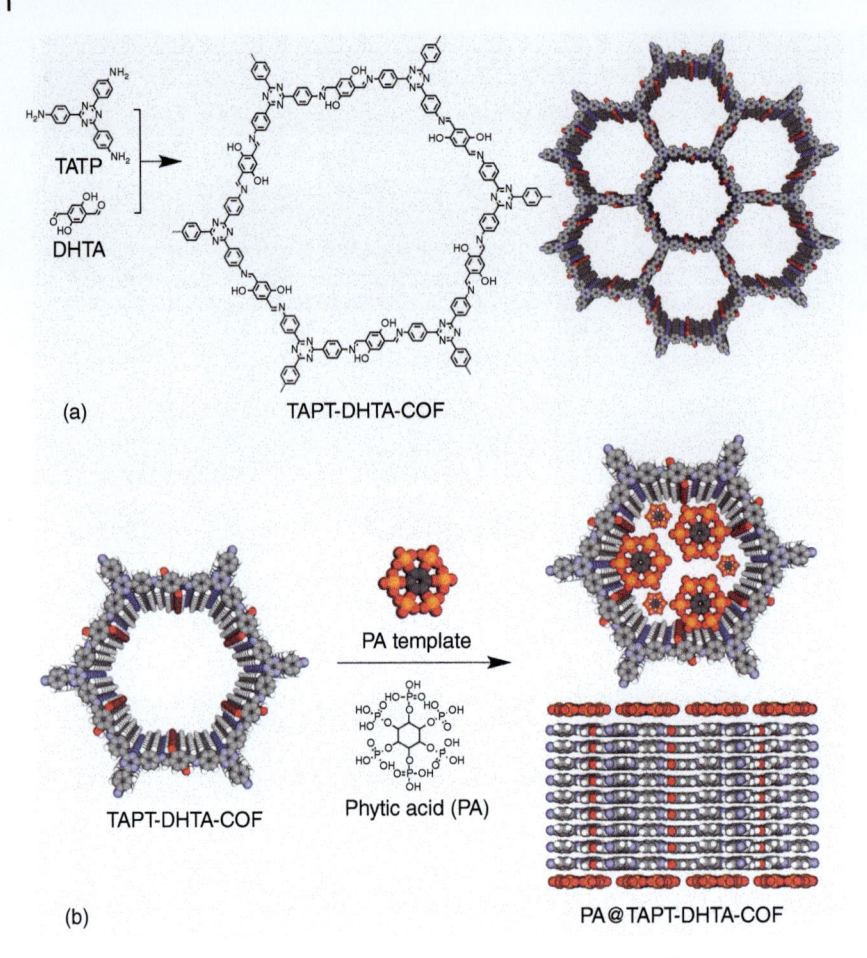

Scheme 3.11 (a) Structure of PA template and schematic for the synthesis of TAPT–DHTA–COF. (b) Template synthesis of PA@TAPT–DHTA–COF by mixing TAPT–DHTA–COF with PA. The PA molecules occupy the pores (upper for top view) and separate COFs by coating on the surface (lower for side view). Source: Reproduced with permission from Xu et al. [85]. Copyright © 2018, Wiley-VCH.

1D channels and layers first (Scheme 3.11a,b) followed by pyrolysis at 1000 °C and further NH_3 treatment [85]. The materials without and with NH_3 treatment were labeled PA@TAPT-DHTA-COF$_{1000}$ and PA@TAPT-DHTA-COF$_{1000NH3}$, respectively. Both materials possessed satisfactorily large BET surface area (495–1160 m^2 g^{-1}). The ORR onset potential, halfway potential, and limiting current density of PA@TAPT-DHTA-COF$_{1000}$ were +0.97 V, +0.78 V, and 6.5 mA cm^{-2}, while those of PA@TAPT-DHTA-COF$_{1000NH3}$ were +0.99 V, +0.88 V, and 7.2 mA cm^{-2}. These performances were comparable with or even better than the commercial Pt/C catalysts (+0.96 V, +0.83 V, and 6.0 mA cm^{-2}). The number of electrons transferred during ORR for both materials were close to 4 (3.60–3.98) indicative of their high selectivity through a 4e$^-$ ORR mechanism. These materials also showed better durability and CH_3OH tolerance when compared with the commercial Pt/C electrocatalyst.

Heteroatom-doped nanocarbon materials discussed above cover counterparts from several major morphologies. Through experimental and theoretical approaches, reasons for outstanding ORR catalytic activity exhibited by these heteroatom-doped carbonaceous materials have been proposed. However, these works failed to provide direct evidence to clarify the active sites for ORR to carry out. In 2016, Nakamura and coworkers made efforts to probe the active sites of ORR and clarify the mechanism using highly oriented pyrolytic graphite (HOPG) as model catalysts [91]. It was found that HOPG containing pyridinic-N as the only N-dopants exhibited better ORR activity than the one with graphitic-N as the unique N-dopants did. More importantly, the decrease of pyridinic-N peak along with the increase of pyridonic-N peak in XPS spectra after ORR and active temperature-programmed desorption (TPD) profile of pyridinic-N-dominant HOPG signified that the carbon adjacent to pyridinic-N served as a Lewis-base site that adsorbed O_2 molecules [92]. Since O_2 molecule adsorption was the first step of ORR, this finding evidently showed that the carbon adjacent to pyridinic-N was an active site for ORR. Supported by these experimental findings, an ORR mechanism has been proposed. Firstly, O_2 is adsorbed to the carbon adjacent to pyridinic-N to give a peroxy radical that takes up an electron and a H^+ ion to give a peroxide. The peroxide either can go through proton-coupled electron transfer to regenerate HOPG catalyst with elimination of H_2O_2 or takes up two electrons along with two protons to give H_2O and pyridonic-N bearing intermediate that undergoes proton-coupled electron transfer to regenerate the HOPG catalyst and another H_2O. Although this work only focuses on N-doped graphite, the detailed research approach delivered undoubtedly provides insight into rational design of ORR-active heteroatom-doped carbonaceous materials.

3.2.5 Heteroatom-Doped Composite Materials

Each heteroatom-doped carbon-based material discussed in Sections 3.2.1–3.2.4 has one specific morphology or one major constituent form. Materials in different morphologies possess specific advantages and disadvantages toward ORR. For example, CNTs have three-dimensional conductive network beneficiary for mass and electron transport, whereas graphene has an extensive planar network of sp^2-hybridized carbon with good electrical conductivity and large surface area. Meanwhile CNTs prefer to form bundles and graphene nanosheets easily restack together; these reduce the surface available for ORR application. Therefore, it is favorable to develop carbon-based nanomaterials as ORR electrocatalyst with the strengths from materials in different morphologies.

In 2011, Qiao and coworkers reported nanoporous graphitic carbon-nitride/carbon composite (g-C_3N_4@carbon) as an efficient ORR electrocatalyst [93]. The composite was prepared by incorporating g-C_3N_4 into the framework of a highly ordered mesoporous carbon (CMK-3) via a nanocasting method in which CMK-3 template was introduced hydrophilic groups and cyanamide as g-C_3N_4 source followed

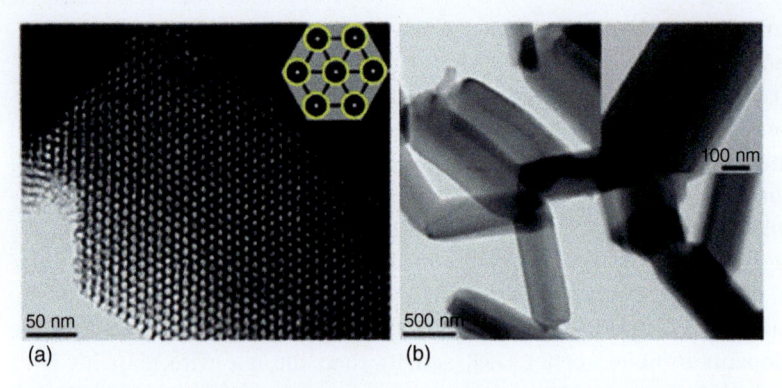

Figure 3.17 (a, b) Typical HRTEM images of ordered mesoporous g-C_3N_4@CMK-3 nanorods. Inset in panel (a) represents a schematic illustration (yellow, g-C_3N_4; black, carbon); inset in panel (b) reveals the ordered mesoporous channels. Source: Zheng et al. [93]/Reproduced with permission from American Chemical Society.

by calcination at 550 °C. The resultant composite materials arranged as ordered nanorods (Figure 3.17) and demonstrated onset potential ∼0.1 V more negative, better kinetic limiting current density (11.3 mA cm^{-2}), superior CH_3OH crossover tolerance, and similar ORR mechanism selectivity (n = c. 4) when compared with commercial Pt/C catalyst.

One year later, Huang and Yang demonstrated the successful fabrication of selenium-doped CNT/graphene composites (Se-CNT-graphene) by a facile thermal annealing process in which GO/CNTs/diphenyl diselenide (DDS) mixture was thermally annealed at 900 °C (Scheme 3.12) [94]. Se-CNT-graphene was found to have number of electron transfer of 3.95 (dominantly a 4e$^-$ pathway) and exhibit onset potential more positive and higher limiting current density than CNT-graphene along with superior CH_3OH crossover tolerance and operation durability in alkaline medium. This suggests the important role of Se-doping in composite carbon-based

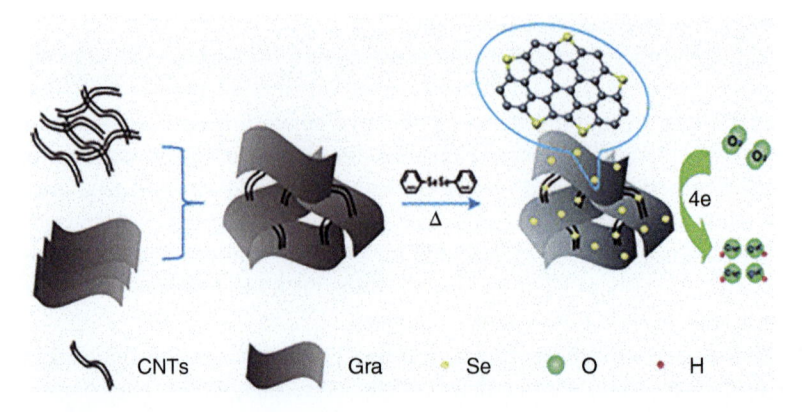

Scheme 3.12 Schematic illustration of Se-CNT–graphene preparation. Source: Reproduced with permission from Jin et al. [94]. Copyright © 2012, Royal Society of Chemistry.

materials for the enhancement of ORR catalytic activity. Noteworthy, several reasons have been proposed to account for the excellent ORR activity brought by Se-doping: (i) Based on XPS, UV–Vis, and electrochemical impedance spectroscopy (EIS) results, the increased electron transfer might arise from introduction of Se that probably restored defects at the edge of carbon network and promoted the formation of a π-conjugation system resulting in the effective electron transfer and hence enhanced ORR performance. (ii) Close electroneutrality of Se and C disfavored the previous proposition that doped electronegative atoms (e.g. N, B, P) induced electron-deficient center on adjacent carbon atoms or other heteroatoms serve as active sites for ORR [25, 26]. (iii) It was suggested that the introduction of functional groups or doping atoms induced higher strain in carbon materials [95, 96]. Introduction of Se atoms into carbon network might cause higher strain at edges, which might facilitate charge localization and associated chemisorption of O_2. (iv) EIS results revealed that the value of the interfacial charge-transfer resistance on Se-CNT-graphene was much smaller than those of single Se-CNT and Se-graphene. It is likely that CNTs in accompany with graphene synergistically improve ORR catalytic activity through formation of an interpenetration network between CNTs and graphene accelerating reactant, ion, and electron transport.

Afterward, there were two reports on composite carbon-based nanomaterials for ORR. The first one was the nanocomposite composed of N-doped reduced graphene oxide (N-rGO) and N-CNT prepared from solution assembly of CNT, rGO, and melamine followed by thermal annealing at 900 °C [97]. It was found that composite materials based on mass ratio between rGO and CNT of 1 : 5 (N-rGO-CNT-0.2) gave the best ORR performance. For example, N-rGO-CNT-0.2 exhibited a halfway potential 98 mV more positive than that of N-CNT. Besides, its current density at −0.48 V was 4.72 mA cm^{-2} that exceeded largely those of N-rGO and N-CNT. The electron transfer during ORR was calculated to be 3.9 indicating that N-rGO-CNT-0.2 underwent catalytic ORR through a 4e$^-$ pathway. The CH_3OH crossover tolerance and durability of N-rGO-CNT-0.2 were proved to be better than commercial Pt/C electrocatalysts. Another work was the preparation of hybrid nanosphere (CGHNs) from assembling CNTs and graphene through a one-step aerosol route (Figure 3.18) [98]. The CGHNs were also further N-doped by thermal treatment under NH_3 and P-doped through annealing with triphenylphosphine to give N,P-co-doped CGHNs (N,P-CGHNs) that showed E_{onset} of +0.94 V and $E_{1/2}$ of +0.82 V, slightly more positive than those of Pt/C measured under the same conditions (E_{onset} = +0.93 V and $E_{1/2}$ = +0.81 V). N,P-CGHNs gave a better diffusion-limiting current (e.g. 5.90 mA cm^{-2} at +0.2 V) than Pt/C did (5.28 mA cm^{-2}) and proceeded ORR through a 4e$^-$ pathway (number of electron transferred for ORR = 4). More importantly, N,P-CGHNs could catalyze ORR even in 0.1 M $HClO_4$ solution with high E_{onset} of c. +0.90 V that was close to that of Pt/C (E_{onset} = +0.96 V). Gratifyingly, these results show that the ORR performance exhibited by the carbon-based nanocomposite is not only limited to alkaline medium but also plausible in acidic medium.

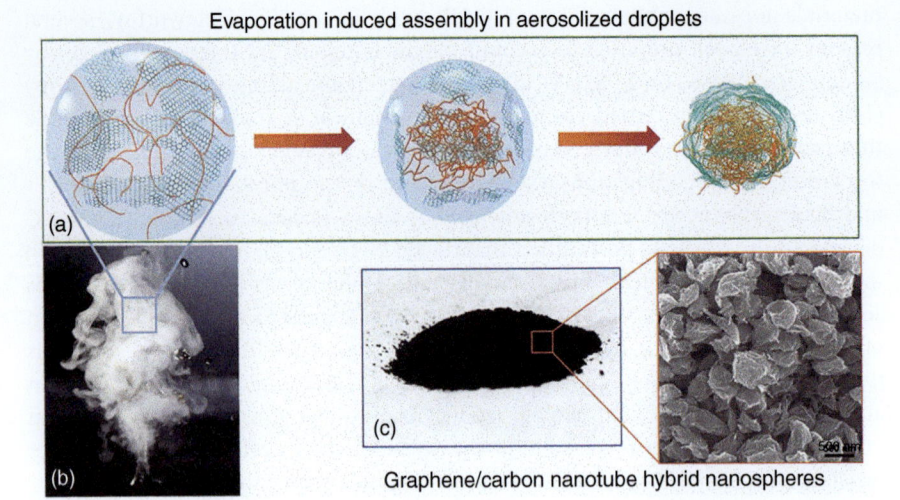

Figure 3.18 (a) Schematic illustration of the process for co-assembling carbon nanotubes and graphene into hybrid nanospheres in rapidly evaporating aerosol droplets. (b) A photograph of the ultrasonic fountain and mist generated by a high-frequency ultrasound (1.7 MHz) from an aqueous dispersion containing oxidized carbon nanotubes and graphene oxides. (c) A photograph and corresponding SEM image of the black powder collected from the filter after aerosol processing. Source: Reproduced with permission from Yang et al. [98]. Copyright © 2016, Wiley-VCH.

The works covered above reveal that carbon-based composite materials with decent ORR performance can be obtained through combination of different carbon-based nanomaterials. Apart from the type of dopants, a blend of carbon-based nanomaterials in different morphologies also contribute to the enhancement of ORR performance.

3.2.6 Better ORR Performance in Acidic Medium

Most of the materials discussed in Sections 3.2.1–3.2.5 are operable in alkaline medium while only few of them work properly in acidic medium. Considering the application of ORR catalysts in fuel cells, the operating is not only limited to alkaline but also preferred to be acidic medium in some cases such as proton-exchange membrane fuel cell (PEMFC) representing one of the most promising devices in energy conversion owing to its high energy density and environmentally benign nature [99–103]. The ORR catalysts in PEMFCs are preferably active in acidic conditions because Nafion is generally used as proton-exchange membrane in PEMFCs. Nevertheless, only Pt and derived alloys are the current benchmark catalysts regarding ORR catalysts operating efficiently in acidic medium [104–106]. The scarcity and sky-high cost of state-of-the-art Pt-based electrocatalysts stand in the way of commercializing PEMFCs in large-scale fashion [107–110]. These reasons urge the search for alternative materials (especially non-metal based) as substitutes for existing Pt-based ORR electrocatalyst in PEMFCs.

In 2013, Peng and coworkers reported a S/N-co-doped CNTs (SN-CNTs) exhibiting enhanced ORR activity in both acidic and alkaline media [111]. SN-CNTs were prepared by thermal annealing of N-CNTs, obtained from thermolysis of aniline in the presence of NH_3, with sulfur source. When compared with commercially available Pt/C, the SN-CNTs operated comparably in alkaline medium in terms of E_{onset}, $E_{1/2}$, and diffusion-limited current density. In acidic medium, though the SN-CNTs showed an E_{onset} 18 mV more negative and lower diffusion-limited current density than Pt/C, they exhibited far better CH_3OH crossover tolerance and long-term stability than Pt/C (e.g. SN-CNTs retained 89.6% current while Pt/C showed a low current retention of 49.3%). Two years later, another work with similar findings was published by Pérez-Alonso and coworkers [112]. The point to note in their work is that the CNTs were ball-milled prior to introduction of heteroatoms by means of thermal annealing. SN-CNTs prepared from CNTs ball-milled for longer time exhibited better ORR catalytic performance.

Besides CNTs, N-doped carbon nanosheets (NDCNs) with size-defined mesopores were reported as highly efficient metal-free ORR electrocatalyst [113]. The NDCNs were produced through a template synthesis (Scheme 3.13). Graphene–silica (G-silica) was first functionalized with a surfactant and sandwiched with shells comprising close-packed colloidal silica nanoparticles (NPs) to give G-silica-X (X denotes the particle size of colloidal silica NPs or the pore size of the resultant NDCN with diameter in nm scale). G-silica-X was then coated a PDA layer, thermally treated at high temperature followed by etching of silica to generate

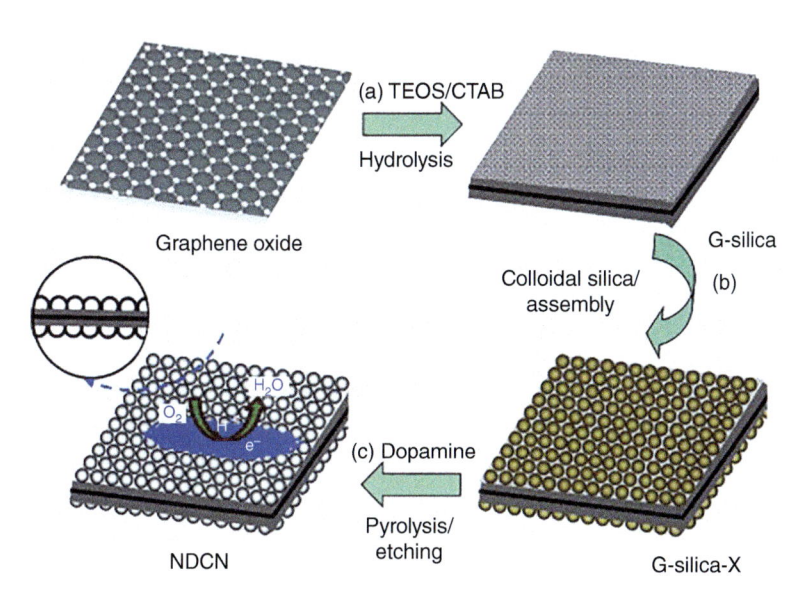

Scheme 3.13 Synthesis of NDCN-X. (a) CTAB-directed hydrolysis of TEOS on GO nanosheets (CTAB = cetyltrimethyl ammonium bromide, TEOS = tetraethylorthosilicate, and GO = graphene oxide). (b) Electrostatic assembly of G-silica and colloidal silica NPs. (c) Self-polymerization of DA, pyrolysis, and silica removal treatment (DA = dopamine). Source: Reproduced with permission from Wei et al. [113]. Copyright © 2014, Wiley-VCH.

NDCN-X with defined mesopores. Among NDCN of different pore size, NDCN-22 exhibited the best ORR performance with comparable diffusion-limiting current (5.45 vs. 5.78 mA cm^{-2}), E_{onset} of 100 mV, and $E_{1/2}$ of 130 mV more positive than Pt/C electrocatalyst in alkaline medium. In acidic medium, though NDCN-22 showed E_{onset} of 220 mV more negative than that of Pt/C, the slight negative shift in both E_{onset} and $E_{1/2}$ of about 40 mV over 4000 cycles of electrochemical operation highlighted the high durability of the NDCN catalysts.

Inspired by other reported acid-active heteroatom-functionalized carbon-based ORR catalyst, a N/S-co-doped carbon aerogel (NSCA) with high extent of defective sites was reported to exhibit outstanding ORR performance in acidic medium [114]. It was shown by XPS spectra that further annealing of as-synthesized NSCA at 1000 °C increased the portion of defective sites (Figure 3.19). NSCA showed an $E_{1/2}$ only 26 mV more negative than that of Pt/C and carried out ORR through a 4e$^-$ pathway with H_2O_2 yield below 3%. Supported by theoretical calculation, the highly efficient catalytic activity of NSCA originated from the synergistic effect of N and S atoms along with the edge pentagon defect (carbon network on edge area had pentagonal defect sites). The findings from this work shed light to the design of high-performance metal-free carbon-based ORR electrocatalysts in acidic electrolytes.

Following the previous success of acid-operable ORR catalyst, Wei and coworkers reported a N-doped catalyst active in acidic medium [115]. This catalyst was prepared by a dual template method in which polyacrylonitrile (PAN) was used as the N and C source along with ZnO and NaCl serving as mesoporous and macroporous templates. The removal of ZnO by evaporation during carbonization process and leaching during H_2SO_4 treatment and dissolution of NaCl during water washing step led to the construction of mesopores and 3D macroporous structures, respectively (Scheme 3.14). It was found that catalyst constructed from PAN/ZnO/NaCl in 1 : 1 : 10 ratio (labeled as NC-1 : 1 : 10) gave the best ORR performance when compared with other mixing ratio. For example, NC-1 : 1 : 10 had an $E_{1/2}$ only 115 mV more negative than Pt/C in 1 M HClO$_4$. It was one of

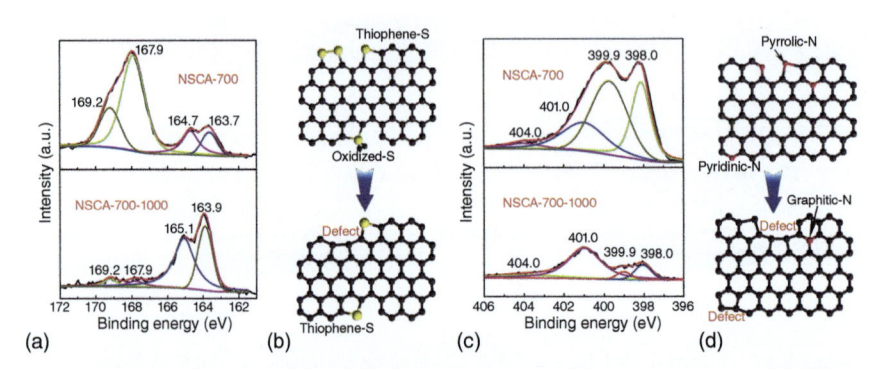

Figure 3.19 (a and c) The high-resolution XPS spectra of NSCA: (a) S 2p and (c) N 1s. (b and d) Schematic illustration of the conversion of the heteroatom dopant through annealing treatment: (b) S dopant and (d) N dopant. Source: Reproduced with permission from Li et al. [114]. Copyright © 2018, Elsevier Inc.

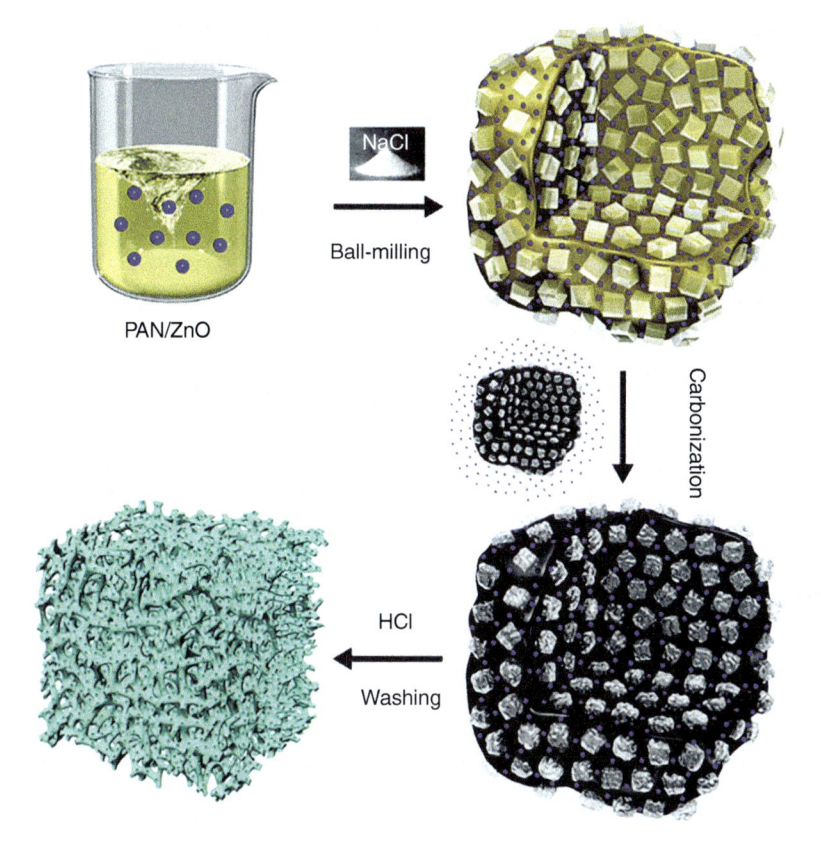

Scheme 3.14 Schematic synthesis of a NC-1 : 1 : 10 catalyst. Source: Wu et al. [115]/ Reproduced with permission from Royal Society of Chemistry.

the most active N-doped carbon-based nanomaterials operating in acidic medium. EIS spectra showed that the charge transfer resistance of NC-1 : 1 : 10 was much smaller than those of NC-1 : 0 : 0, NC-1 : 1 : 0, and NC-1 : 0 : 10 signifying that 3D hierarchical porous network structure in NC-1 : 1 : 10 allowed facile transfer of O_2 to active site for ORR. This work presents a new strategy to design metal-free ORR catalyst with satisfactory performance in acidic medium.

3.3 Undoped Carbon Nanomaterials

In Section 3.2, ORR catalysts based on heteroatom-doped carbon nanomaterials have been discussed in detail. Generally, electrocatalytic activity in heteroatom-doped carbon materials probably originates from the break of electroneutrality and electron transfer caused by doping of heteroatoms. Apart from heteroatom doping, the surface functional groups, edge area, and intrinsically/topologically defective sites in nanocarbon materials also play an important role in the electrocatalytic activity [116–120]. These findings lead to a fundamental question: are heteroatom dopants essential for metal-free carbon-based nanomaterials ORR catalysts? In other words, can undoped carbon-based nanomaterials exhibit ORR catalytic activity? Currently, dopant-free nanocarbon materials have been demonstrated to exhibit decent ORR activity that was proposed to stem from the activeness of edge sites [121]. Defects represent one of the most influential parameters to enhance the ORR catalytic activity in undoped carbon nanomaterials. It is of fundamental importance to clarify the role of defective sites of carbon materials in ORR catalysis. Therefore, along with advancement in developing heteroatom-functionalized carbon-based nanomaterials and more understanding on the ORR mechanism by such materials, researchers then turn to the origin, undoped carbon nanomaterials, and investigate how these materials catalyze ORR.

3.3.1 Edge as Defect

Several works have been carried out to show that edge areas of carbon nanomaterials exhibit higher activity than basal plane for ORR. Intriguingly, Dai and Wang published a work with direct evidence to demonstrate that the edge was more active than the basal plane using highly oriented pyrolytic graphite (HOPG) as experimental specimens [122]. Comparison based on the ORR performance was made between the air-saturated droplets deposited on edge and basal plane of graphite (Figure 3.20a–c).

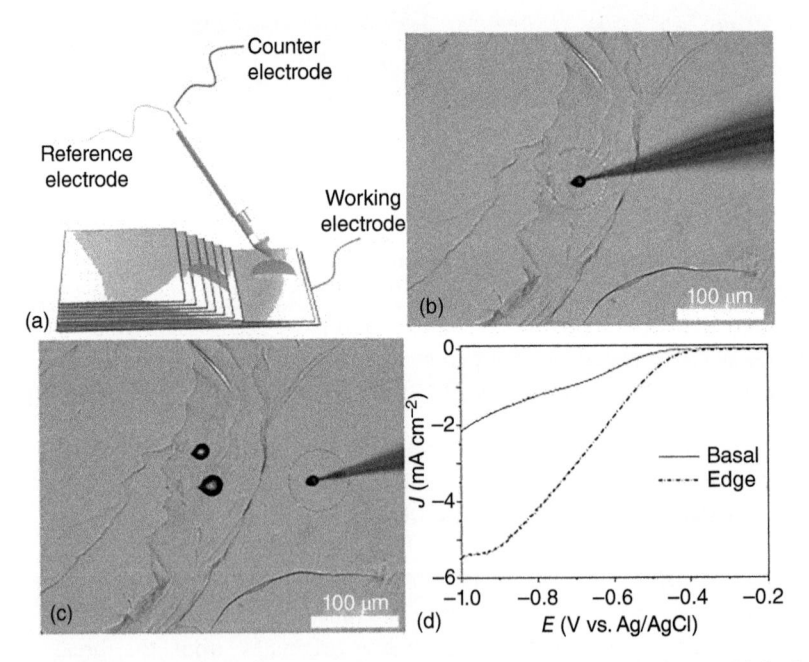

Figure 3.20 (a) Micro-apparatus for the ORR electrochemical experiment. (b) Optical photograph of the HOPG as the working electrode with the air-saturated droplet deposited on the edge of the HOPG. (c) The air-saturated droplet was deposited on the basal plane of the HOPG electrode. (d) LSV curves of the ORR tested for a droplet located either on the edge (as shown in Figure 3.20b) or on the basal plane (as shown in Figure 3.20c) of the HOPG. Source: Shen et al. [122]/Reproduced with permission from Wiley-VCH.

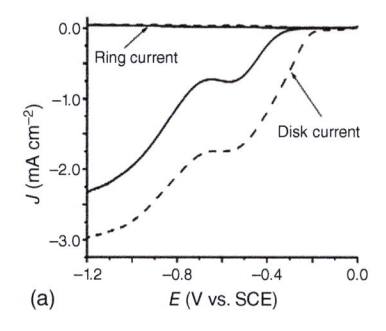

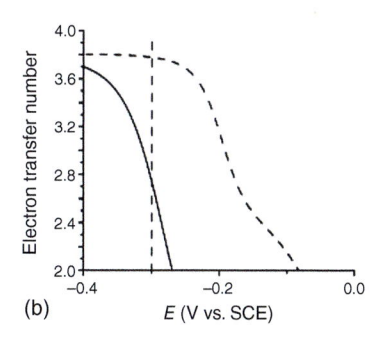

Figure 3.21 (a) RRDE experiments with pristine graphite and graphite ball-milled for 96 hours in O_2-saturated KOH electrolyte solution (0.1 M). (b) The dependence of the electron transfer number on the potential. ---, ball milled for 96 hours; —, ball milled for 0 hours. Source: Shen et al. [122]. Reproduced with permission of Wiley-VCH.

It was found that the ORR on the edges exhibited a more positive E_{onset} and a higher current density than that on the basal plane (Figure 3.20d), signifying that edges of graphite executed higher ORR activity than the basal plane. Besides, it was found that ball-milled graphite showed much better ORR performance than bulk graphite powder (Figure 3.21). These results consolidate the higher ORR activity of edges than basal planes of graphite.

Following this direction, the same group reported preparation of dopant-free and edge-rich graphene through Ar plasma etching of pristine graphene (Scheme 3.15) [121]. The intensity ratio between defect band and intact band (I_D/I_G) in Raman spectra increased from graphene to plasma-treated graphene (P-G), indicating an increase of defective sites in P-G than graphene. Furthermore, the ORR performance of P-G was better than graphene significantly in 0.1 M KOH as reflected by a more positive E_{onset} and $E_{1/2}$ of P-G (+0.91, +0.74 V) than graphene

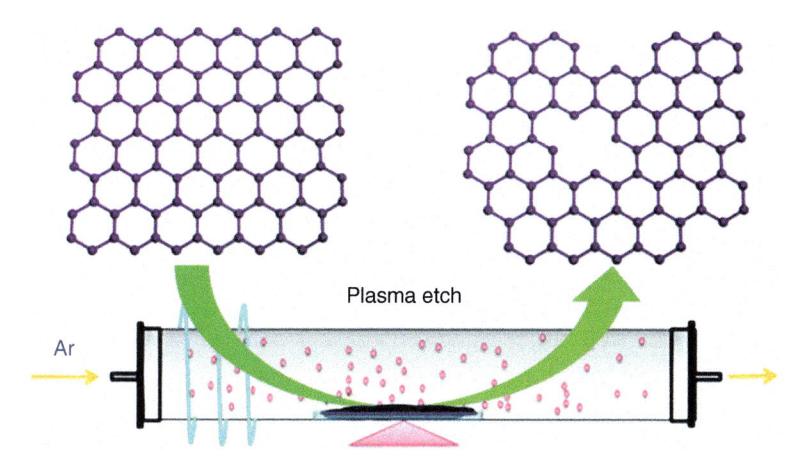

Scheme 3.15 Illustration of the preparation of the edge-rich and dopant-free graphene by Ar plasma etching. Source: Reproduced with permission from Tao et al. [121]. Copyright © 2016, Royal Society of Chemistry.

(+0.81, +0.57 V) and higher number of electron transfer of P-G ($n = 3.85$) than graphene ($n = 2.31$) during ORR. Edge-rich CNTs and graphite could be prepared from corresponding CNTs and graphite. Similar improvement on ORR performance could be observed from comparison between plasma-treated samples and primitive samples. This work highlights higher ORR activity of edge area than basal plane of carbon nanomaterials and presents a general approach to introduce defects into nanocarbon of different morphologies.

3.3.2 Intrinsic/Topological Defects

When compared with edge as defects, intrinsic/topological defects are more general because they refer to the defects present in the materials' whole framework or structure and are not limited to specific region. These defects usually can be generated by the removal of incorporated heteroatoms or metal ions from network of nanocarbon during pyrolysis. Understanding the role of intrinsic/topological defects in ORR enhancement undoubtedly synergizes the introduction of defects and heteroatom doping in nanocarbon materials to promote ORR catalysis and give insights into rational design of novel carbon-based ORR catalysts. Pinpointing this target, researchers intensively put effort to make progress recently.

In 2014, Song and Li demonstrated the introduction of topological defects into CNTs could activate CNTs as ORR catalyst [123]. The CNTs with topological defects (TCNTs) were prepared using ethanol as precursor and Al_2O_3-supported Fe–Co bimetallic catalyst via CVD method in temperature range of 650–850 °C. It was found that TCNTs gave an obvious improvement in ORR when compared with primitive CNTs in terms of E_{onset} and number of electron transfer during ORR. Density functional theory (DFT) calculation results gave the following reasons to account for the enhanced ORR performance made by TCNTs. The introduction of topological defects broke the delocalization of orbitals and promoted the formation of localized orbitals on CNTs network where the localized orbitals fostered O_2 adsorption and subsequent O_2 dissociation.

In 2015, Yao and coworkers published a work for studying the effect of vacancies on graphene network for ORR catalysis [124]. The defective graphene materials were produced through thermal treatment of porous organic framework material (PAF-40) at temperature range of 700–1000 °C. Increasing annealing temperature gave higher portion of defects in the resultant graphene (e.g. I_D/I_G increased from 0.81 to 1.02 along with increasing carbonization temperature). The graphene obtained from 1000 °C in the presence of O_2 (O_2 facilitates the removal of N atoms from graphene) gave the best performance for ORR (e.g. E_{onset} only 43 mV lower than that of Pt/C). It was proposed in this work that the unstable single-atom vacancies tended to combine and form divacancies, e.g. G585 defects (585 denotes the defect composed of pentagon–octagon–pentagon moiety) (Figure 3.22), which were stable and active sites for ORR. Nevertheless, Xia and coworkers published a work investigating the role of defects in ORR catalytic activities of graphene [125]. In this work, DFT calculation was used to scrutinize the ORR activity of point and line defects in graphene. It was found that, among the point defects (Stone–Wales defect,

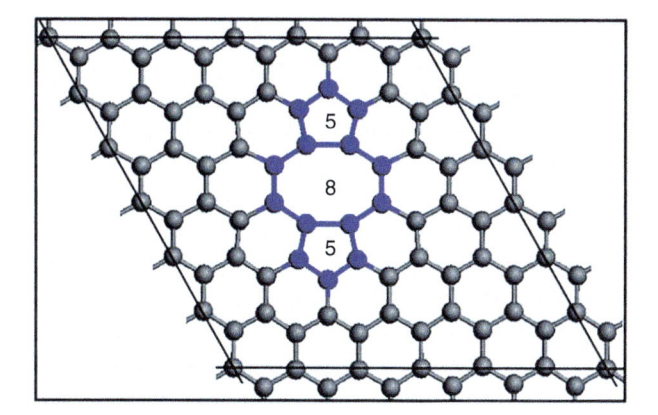

Figure 3.22 Pictorial representation of G585 defects in graphene. Source: Reproduced with permission from Zhao et al. [124]. Copyright © 2015, Royal Society of Chemistry.

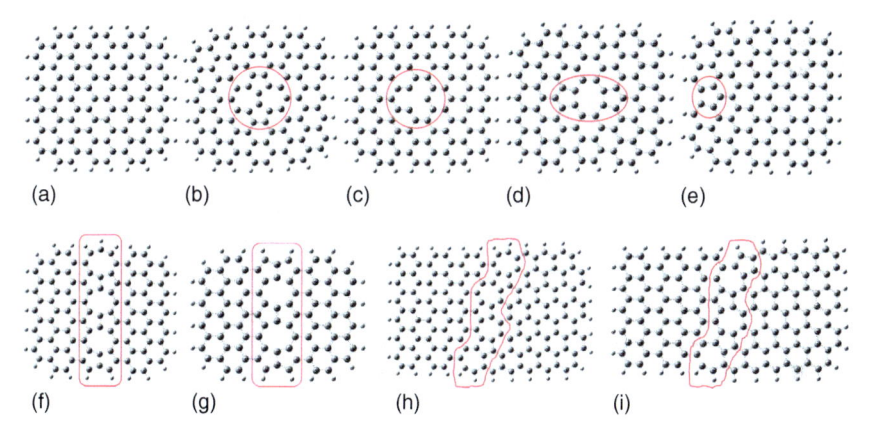

Figure 3.23 Perfect and defective graphene clusters. (a) Perfect graphene cluster, (b) Stone–Wales defect (SW), (c) single vacancy (SV), (d) double vacancies (DV), (e) edge defect with pentagon ring at zigzag edge (PZ), octagon and fused pentagon carbon rings line defect with (f) odd number of octagon rings (GLD-558-01) and (g) even number of octagon rings (GLD-558-02), and pentagon–heptagon pair line defects with (h) even number of heptagon rings (GLD-57-01) and (i) odd number of heptagon rings (GLD-57-02). The larger gray and smaller white balls denote carbon and hydrogen atoms, respectively. Source: Reproduced with permission from Zhang et al. [125]. Copyright © 2015, Royal Society of Chemistry.

single vacancy, double vacancies, one substituting pentagon ring) (Figure 3.23a–e), only the pentagon ring at the zigzag edge was catalytically active. Considering line defects (e.g. pentagon–heptagon chain, GLD-57, and pentagon–pentagon–octagon chain, GLD-558) (Figure 3.23f–i), only the structures with odd number of heptagon or octagon rings (Figure 3.23f,i) generate spin density and manage to catalyze ORR. Even though divergence exists for what defects are responsible for the enhanced ORR performance, defects in graphene network contribute to the catalytic activity of ORR.

Apart from graphene, the introduction of defects can also boost ORR performance of inactive activated carbon (AC) [126]. Defective AC (D-AC) was produced via calcination of N-doped AC at high temperature of 1050 °C. D-AC was found to exhibit comparable ORR activity with Pt/C. For example, the E_{onset}, $E_{1/2}$, and limiting current density between D-AC and Pt/C (D-AC vs. Pt/C) were +0.88 vs. +0.90 V, +0.77 vs. +0.79 V, and 4.4 vs. 5.0 mA cm^{-2}. D-AC kept 90.6% of current after operation for 20 000 seconds while Pt/C only retained 80.2% under the same conditions. The method reported in this work to activate ORR inert AC represents a facile and scalable route. Scale-up production of such materials involves cheap and earth-abundant raw materials along with a simple preparation process, highlighting the great potential of the D-AC in ORR application.

Preparation of defective carbon nanomaterials can be achieved by template-assisted formation. Specifically, ORR-active carbon spheres (CS) have been prepared through a facile solvothermal carbonization of ethanol in which ethanol was admixed with iodine and iron(III) salt followed by solvothermal treatment to give an intermediate lump that was acid-leached for removing iron ions and then calcinated to yield the resultant CS (Scheme 3.16) [127]. The most defective CS in this work possessed I_D/I_G of 2.19 signifying the defective nature of CS. The most defective CS also bore a BET surface area of 985 m^2 g^{-1} reflecting the porous nature of CS. The best candidate of CS in this work exhibited a comparable ORR performance with commercial Pt/C catalyst (in 0.1 M KOH, in E_{onset}, CS = +0.98 V, Pt/C = +0.96 V; in $E_{1/2}$, CS = +0.84 V, Pt/C = +0.86 V; for limiting current density,

Scheme 3.16 Schematic illustration of the preparation of the CSs. Source: Xiang et al. [127]/Reproduced with permission from Royal Society of Chemistry.

CS = 4.17 mA cm^{-2}, Pt/C = 5.15 mA cm^{-2}). Based on the slopes of K–L plots, the electron transfer number during ORR for the best CS was 3.9 that is close to that of Pt/C ($n = 4$) indicative of an almost ideal 4e$^-$ O$_2$ reduction pathway. The stability of CS was superior to Pt/C as revealed by the negligible change in $E_{1/2}$ upon operation of 13 500 CV cycles and negative shift in $E_{1/2}$ of 13 mV after 4000 CV cycles for Pt/C. The excellent ORR performance was ascribed to the intrinsic carbon defects and oxidized carbon over the CS.

Meanwhile, Hu and coworkers reported an ORR-active carbon nanocages (CNCs) generated by templated synthesis [128]. The CNC was produced through CVD method with growing temperature range of 700–900 °C using benzene as precursor and MgO as template followed by removal of MgO template by repeated flushing with HCl solution and H$_2$O. It was found that CNC produced at 700 °C (CNC700) possessed the highest specific surface area (1713 m^2 g^{-1}) and I_D/I_G (2.29) signifying its high concentration of defective sites. These parameters decreased along with increasing reaction temperature suggesting that crystallinity of CNC became better upon higher reaction temperature. CNC700 exhibited satisfactory ORR performance with E_{onset} = +0.88 that was attributed to the high concentration of defects within the structure. DFT calculation has been employed to provide insights into the account of the ORR activeness of CNCs. It was found that pentagon and zigzag edge defects among various modes of defects (including hole and armchair edge defects) in CNC gave a continuous downhill free energy on ORR, suggesting that the pentagon and zigzag edge defects were the ORR active sites and explained the high ORR activity of the CNCs.

When it comes to templated synthesis, MOFs, as framework with crystallinity, always serve as good template for generation of porous carbon-based nanomaterials. There were two works on ORR-active porous carbon materials derived from carbonization of MOFs. One came from the work by Yao and coworkers that reported a porous carbon material (PC-I8-950) generated via carbonization of a Zn-based MOF at 950 °C [129] while another one was contributed by the work by Wang and coworkers that reported a mesoporous carbon material (MPC-np) obtained from carbonization of Zn fumarate-based nonporous MOFs [130]. Both MOF-derived materials showed satisfactory ORR performance in 0.1 M KOH (e.g. E_{onset}: +0.91 V for both PC-I8-950 and MPC-np; number of electron transfer during ORR = 3.6–3.9; superior CH$_3$OH tolerance and long-term stability when compared with commercial Pt/C catalyst). The success of ORR-active carbon materials through carbonization of structurally well-defined MOFs highlights the possibility of modulating defective sites of ORR-active carbon-based catalyst based on the unique yet diversified structures of MOFs.

3.4 Carbon-Based Organic Framework

In Section 3.2.4, COFs have been utilized as starting materials for generation of heteroatom-doped porous carbon nanomaterials. Generally, porous organic framework like COF or metal-free carbon-based polymers can be constructed with crystallinity through various combinations of building blocks [90, 131, 132]. It was found that pyrolysis of porous metal-free carbon-based polymer produces heteroatom-doped carbon-based nanomaterials demonstrating decent ORR catalytic activity. An example of ORR catalytically active heteroatom-doped porous carbon nanomaterials prepared from pyrolysis of COF followed by NH_3 treatment has been discussed in Section 3.2.4 [85]. Interestingly, COFs themselves were demonstrated to exhibit ORR activity in several recent works that are to be discussed in this section.

In 2018, Yu and coworkers reported a carbazole-decorated covalent triazine framework (CTF-CSU1) constructed from $ZnCl_2$-catalyzed condensation of carbazoles at 400 °C (Scheme 3.17) [133]. This type of framework had rich N contents (15.33 wt% N) and large surface area (982 $m^2\,g^{-1}$) with abundant mesopores; these features are linked to the satisfactory ORR activity of CTF-CSU1. Although CTF-CSU1 was found to proceed ORR via a blended $2e^-$ and $4e^-$ pathway with E_{onset} of +0.79 V and $E_{1/2}$ of +0.57 V more negative than those of Pt/C (E_{onset} = +0.96 V and $E_{1/2}$ = +0.86 V), it exhibited a larger limited current density (5.6 mA cm^{-2}) than Pt/C (5.03 mA cm^{-2}). Additionally, it showed a more profound stability and CH_3OH tolerance than Pt/C during ORR.

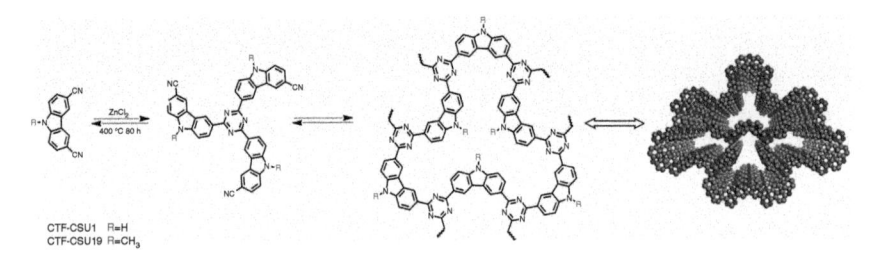

Scheme 3.17 Condensation reaction of 3,6-dicyanocarbazole to a discrete trimer and extended CTF-CSUs (inset shows highly idealized schematic representation of the CTFCSU1 (C, gray; N, blue)). Source: Reproduced with permission from Yu et al. [133]. Copyright © 2018, Elsevier Inc.

In the meantime, Maji and coworkers reported a kind of ORR catalytically active conjugated microporous polymer (TPA-BP-1 and TPA-TPE-2) arising from direct Sonogashira coupling of tris(4-bromophenyl)amine with two kinds of extended alkynes (Scheme 3.18) [134]. Both polymers demonstrated good ORR performance with E_{onset} = +0.80 and +0.82 V, progress of ORR through a $4e^-$ pathway, and excellent electrocatalytic stability (almost 100% electro-current retention after operation for 12 hours). Also, DFT calculation showed that O_2 molecule bound to the N site in a head-on fashion because of a relatively electro-deficient N center that explained the easy binding of O_2 molecules for the facile ORR.

Very recently, a COF (NDI-COF) derived from condensation between 1,4,5,8-naphthalenetetracarboxylic dianhydride (NTCDA) and 1,3,5-tris(4-aminophenyl)benzene (TAPB) via solvothermal reaction at 120 °C for 4 days has been reported to

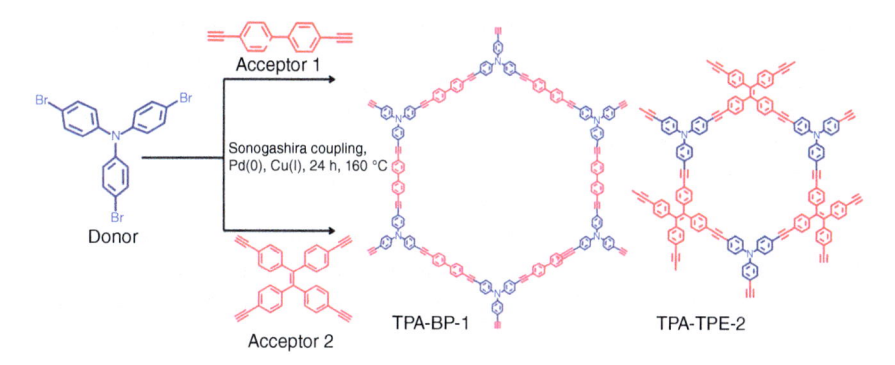

Scheme 3.18 Synthetic scheme toward fabrication of two redox active and semiconducting conjugated microporous polymers. Source: Reproduced with permission from Roy et al. [134]. Copyright © 2018, Royal Society of Chemistry.

be active in ORR catalysis (Scheme 3.19) [135]. NDI-COF was found to be porous with high BET surface area (1138 m^2 g^{-1}). Fabricated on graphite carbon electrode, it showed an E_{onset} of +0.76 V more positive than bare graphite carbon electrode by 120 mV. Also, ORR proceeded through a way of direct 4e$^-$ reduction of O$_2$ along with stability even after long period of operation (i.e. 10 000 seconds).

Scheme 3.19 Schematic representation of the synthesis of NDI-COF. Source: Reproduced with permission from Royuela et al. [135]. Copyright © 2020, Royal Society of Chemistry.

These porous organic framework materials are not merely metal-free, but also ORR catalysts themselves without further pyrolytic treatment. The development of these materials provides an alternative approach to rational design of and more facile way to porous ORR catalysts.

3.5 Application in Fuel Cells

As mentioned in the introduction, energy production through conversion of H_2 and O_2 into H_2O represents one of the most sustainable ways in recent energy technology. Fuel cell is an electrochemical device that combines H_2 and O_2 chemically to yield H_2O and energy. H_2 is oxidized at anode through HOR while O_2 is reduced at cathode via ORR. Specifically, ORR is kinetically sluggish as it involves breaking of strong O=O bond along with four coupled electrons. Therefore, electrocatalyst is essential to accelerate the ORR and improve the efficiency of energy conversion in fuel cells. Even though there are many reports of metal-free carbon-based materials as electrocatalyst for ORR, most of them are not yet applied in real case of fuel cell operation. In this section, research works about application of metal-free carbon-based nanomaterials as ORR catalyst in real scenario of fuel cells will be discussed. For alkaline and acidic fuel cells (i.e. conversion of H_2 and O_2 into H_2O along with energy production), the working principles are similar, whereas the only difference is the use of alkaline and acidic electrolytes, respectively. Until now only a few works focus on fabricating carbon-based nanomaterials as ORR catalysts in both alkaline fuel cell and acidic PEMFC. Apart from these conventional fuel cells, zinc–air (Zn–air) fuel cell emerges as a new candidate recently. In a Zn–air fuel cell, Zn is used as fuel instead of H_2 in alkaline fuel cell and PEMFC while all these fuel cells share the same oxidant – O_2. Zn–air fuel cell operates with oxidation of Zn to $Zn(OH)_2$ as anodic reaction and reduction of O_2 as cathodic reaction (Figure 3.24) [136]. The latter part is the same as other fuel cells. Most reports of metal-free carbon-based

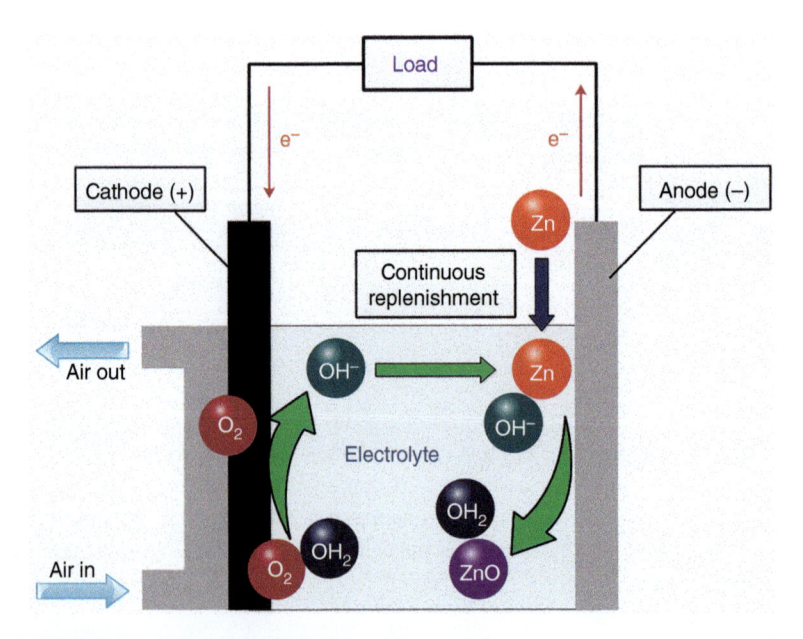

Figure 3.24 Schematic of zinc–air fuel cell.

nanomaterials as ORR catalyst in Zn–air battery that is a primary cell with electrochemical reactions during discharging are principally the same as that of Zn–air fuel cell (rechargeable). Fabrication of metal-free carbon-based nanomaterials as ORR catalyst in Zn–air battery serves as a crucial prerequisite for further application in Zn–air fuel cell. Therefore, related works will be discussed in this section.

3.5.1 Application in Alkaline Fuel Cell and PEMFC

Considering application of metal-free nanocarbon materials in alkaline fuel cell, two works are selected here for discussion. One is the CNT/heteroatom-doped carbon (CNT/HDC) core–sheath nanostructures prepared from carbonization of CNTs/ionic liquid (1-butyl-3-methylimidazolium bis(trifluoromethyl-sulfonyl)imide (BMITFSI))/tetra-ethyl orthosilicate (TEOS) monolithic gel followed by alkaline or acid etching (Scheme 3.20) [137]. The most active candidate was CNT/HDC carbonized at 1000 °C (CNT/HDC-1000) that showed E_{onset} of +0.92 V and $E_{1/2}$ of +0.82 V with high kinetic current density of 8.3 mA cm^{-2} at +0.8 V. CNT/HDC-1000 was fabricated in a membrane electrode assembly (MEA) for assessment of its catalytic capability as cathode in an alkaline fuel cell. It generated current and power densities of 368 mA cm^{-2} and 221 mW cm^{-2}, respectively, which were far better than those of CNT-based cathode and those of Pt/C-based MEA (Figure 3.25). Another related work presented dual-heteroatom co-doped graphdiyne as ORR-active electrocatalyst [138]. Dual-heteroatom co-dopants were divided into three groups – N/S,

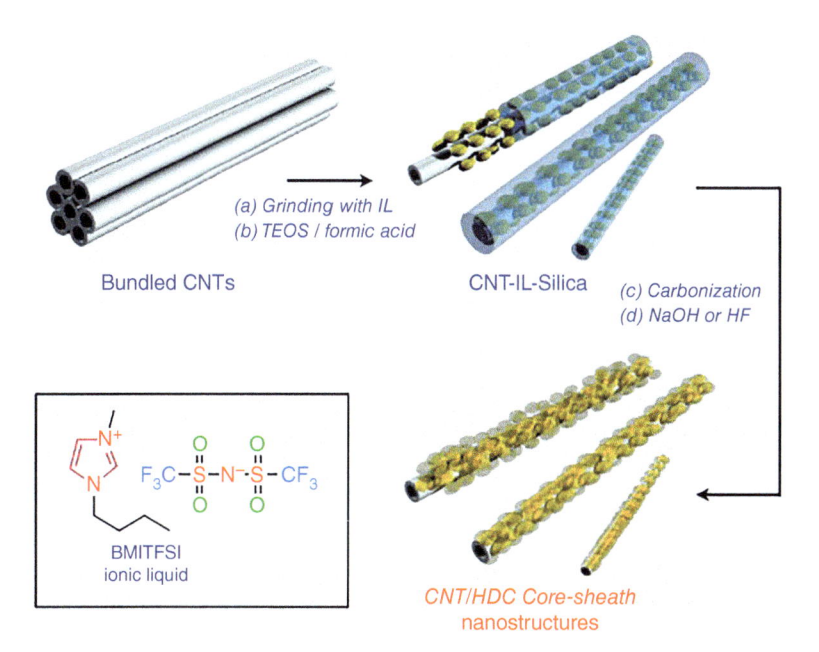

Scheme 3.20 Synthesis of the CNT/HDC core–sheath nanostructures. Source: Reproduced with permission from Sa et al. [137]. Copyright © 2014, Wiley-VCH.

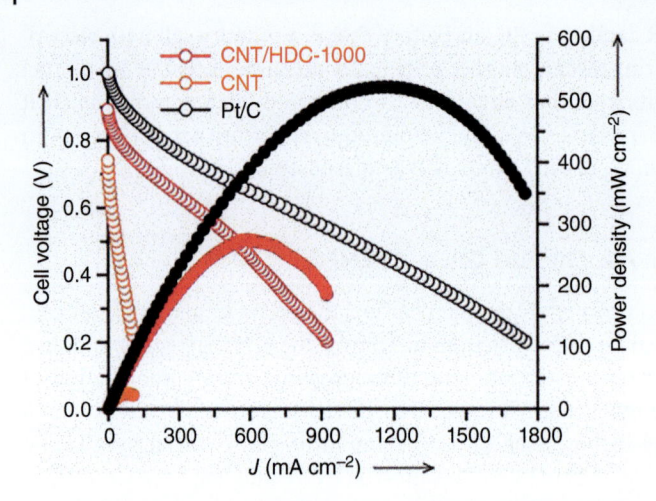

Figure 3.25 Performance of MEAs that employ CNT/HDC-1000, CNTs, or Pt/C as the cathodes in alkaline fuel cells at 50 °C; open and closed circles correspond to cell voltages and power densities, respectively. Source: Reproduced with permission from Sa et al. [137]. Copyright © 2014, Wiley-VCH.

N/B, and N/F in which N/F co-doped graphdiyne (NFGD) exhibited the best ORR performance along with E_{onset} comparable with Pt/C catalyst, high selectivity for a 4e$^-$ ORR pathway, complete CO and CH_3OH crossover tolerance, and excellent long-term stability in alkaline medium. NFGD as cathodic catalyst in a Zn–air cell showed a current density of ~103 mA cm^{-2} (at +0.8 V) and a peak power density of 86 mW cm^{-2} that were comparable with those of Pt/C fabricated electrode (~103 mA cm^{-2} at +0.8 V and 88 mW cm^{-2}).

Looking into acidic fuel cell, Dai and coworkers reported an example of previously reported VA-NCNT array [25] and a rationally designed N-doped graphene/CNT composite (N-G-CNT) as the cathode catalyst in PEMFC demonstrating high gravimetric current density [139]. Both VA-NCNT- and N-G-CNT-based MEA (current at 0.2 V = 1500–1550 A g^{-1}; peak power density = 300–320 W g^{-1}; catalyst loading 0.16–0.5 mg cm^{-2}) exhibited comparable gravimetric activities with the commonly used electrocatalyst in PEMFC (e.g. Fe/N/C: current at 0.2 V = 800–2500 A g^{-1}; peak power density = 233–400 W g^{-1}; catalyst loading of 0.9–3.9 mg cm^{-2}). It is noteworthy that N-G-CNT should be fabricated along with CB for maximal ORR catalytic performance because CB well separate N-G-CNT to allow smooth transport of O_2 through channels of the catalyst.

On the other hand, a composite of zigzag-edged graphene ribbons (GNRs) on CNTs (GNR@CNT) has been applied as cathode catalyst for ORR in acid-medium PEMFC. N-doped GNR@CNT (N-GNR@CNT) and GNR@CNT were prepared by partial unzipping of multi-walled carbon nanotubes (MWCNTs) in the presence of H_2SO_4/$KMnO_4$ mixture followed by freeze-drying and thermal treatment at 900 °C. The electrocatalysts were finally assembled into the MEA for evaluation in PEMFC. Carbon black (XC-72) was added as a spacer to separate GNR for good

PEMFC performance (Scheme 3.21) [140]. Both N-GNR@CNT and GNR@CNT exhibit comparable E_{onset} (N-GNR@CNT, +0.99 V; GNR@CNT, +0.96 V) and $E_{1/2}$ (N-GNR@CNT, +0.84 V; GNR@CNT, +0.82 V) with those of commercial Pt/C catalyst (E_{onset} = +1.03 V; $E_{1/2}$ = +0.84 V) with high selectivity toward a 4e$^-$ ORR pathway. It should be noted that GNR@CNT demonstrated the best polarization curve and the highest peak power density of 520 W g^{-1} higher than 430 W g^{-1} of N-GNR@CNT at the catalyst loading of 0.25 mg cm^{-2} while the power density of GNR@CNT declined obviously, probably due to the deteriorated O$_2$ transport in a thicker catalyst layer (Figure 3.26a,b). The stabilities of both GNR@CNT and N-GNR@CNT were better than commercially available Fe/N/C catalyst in PEMFC upon long-time operation with pure H$_2$/O$_2$ as fuel gases (Figure 3.26c), and this finding was consistent with the one in the previous report discussed above [139]. It was also clarified that the high ORR activity of the GNR@CNT originated from the zigzag carbon rather than the oxidized zigzag carbon.

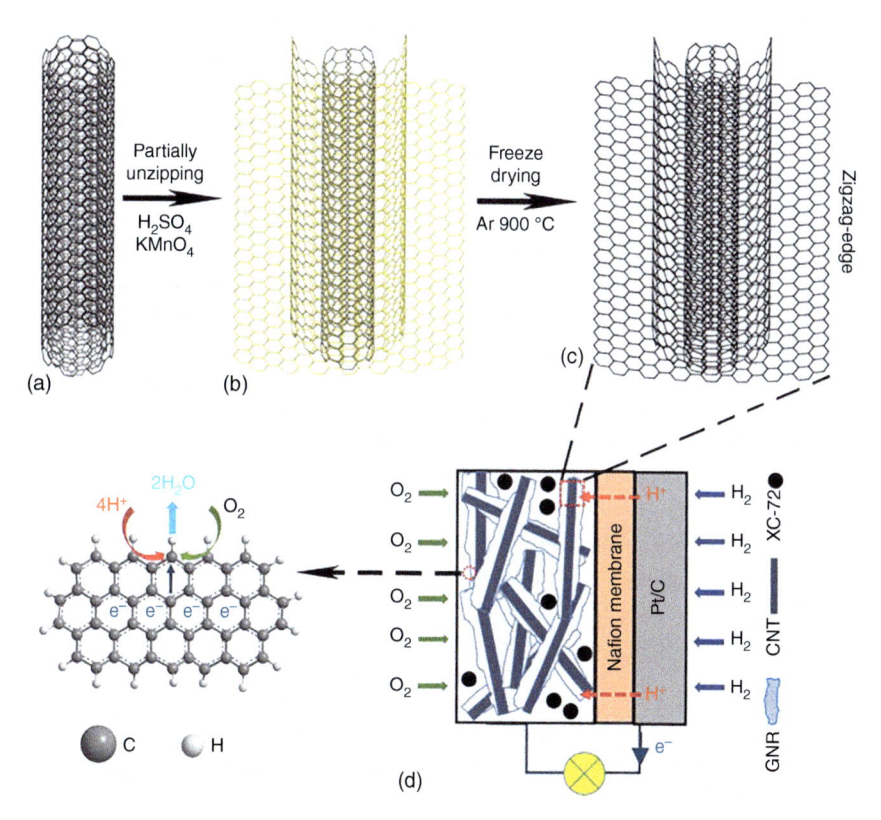

Scheme 3.21 Schematic illustration. The synthetic route of zigzag-type graphene nanoribbons on carbon nanotubes (GNR@CNT) from (a) MWCNT to (b) partially unzipped oxidized CNT and to (c) GNR@CNT. (d) The application as oxygen reduction reaction catalyst in a proton exchange membrane fuel cell (PEMFC). Carbon black XC-72 is used as spacer to prevent the stacking of active materials. Source: Xue et al. [140]/Springer Nature/CC BY 4.0.

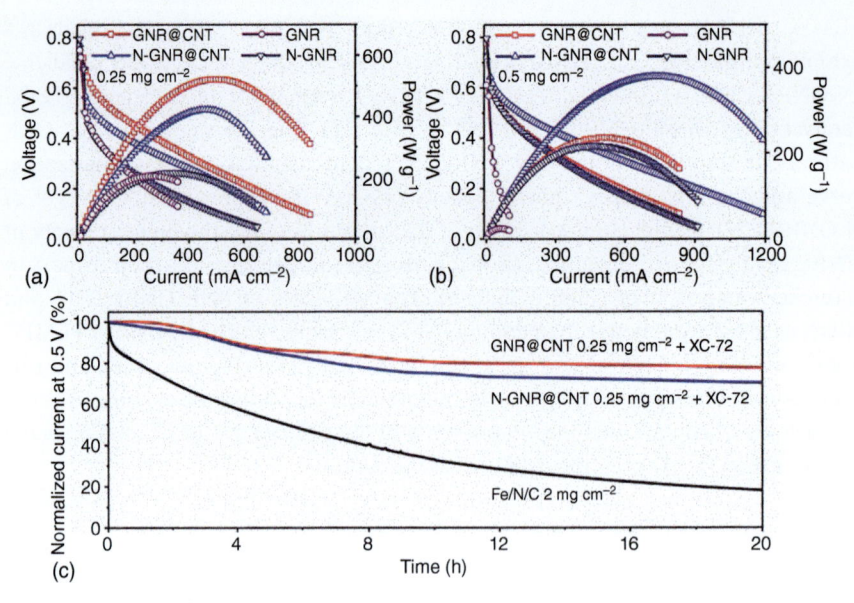

Figure 3.26 Proton exchange membrane fuel cell evaluation. Polarization and power density curves of graphene nanoribbons on carbon nanotubes (GNR@CNT), N-doped GNR@CNT (N-GNR@CNT), and N-doped graphene nanoribbons (N-GNR) as a function of the real current density with cathode catalyst loading of (a) 0.25 mg cm^{-2} and (b) 0.50 mg cm^{-2} in a proton exchange membrane fuel cell (PEMFC). (c) Stability of the indicated catalysts in PEMFC measured at 0.5 V. The absolute current densities before durability tests (at 100%) were 136, 80, and 1216 mA cm^{-2} for graphene nanoribbons on carbon nanotube (GNR@CNT), nitrogen-doped catalyst (N-GNR@CNT), and reference catalyst iron–nitrogen–carbon (Fe/N/C), respectively. Weight ratio of Nafion/catalyst/carbon black (XC-72) = 5/1/4. Cell, 80 °C; H$_2$/O$_2$, 80 °C, 100% relative humidity, 2 bar back pressure. Source: Xue et al. [140]/Springer Nature/CC BY 4.0.

3.5.2 Application in Zinc–Air Battery

Compared with conventional H$_2$/O$_2$ fuel cells, Zn-air fuel cells are easier to set up since the fuel involved is not gaseous H$_2$, but the Zn anode. On the one hand, Zn–air battery operates similarly in electrochemical point of view as Zn–air fuel cell does and only differentiates in terms of rechargeability. Therefore, successful application of metal-free carbonaceous nanomaterials as ORR cathodic catalyst in Zn–air batteries paves a way for extending these materials toward ORR catalysis in Zn–air fuel cells.

Porous nanocarbon materials have been reported as active ORR electrocatalyst and assembled into Zn–air battery for evaluating the performance in the real operation of chemical cells. Specifically, a hierarchically porous N-doped carbon (meso/micro-PoPD) prepared from silica colloid-templated polymerization of o-phenylenediamine followed by pyrolysis of the composite at 900 °C, subsequent NaOH etching, and final NH$_3$ treatment demonstrated comparable ORR performance with Pt/C at same amount of catalyst loading (e.g. $E_{1/2}$ = +0.85 V in 0.1 M KOH, loading = 0.1 mg cm^{-2}) [141]. Besides, this material was also active even in acidic medium (e.g. $E_{1/2} = +0.84$ V in 0.5 M H_2SO_4). It should be noted that the ORR catalytic performance was highly dependent on the portion of quaternary N species in the structure (meso/micro-PoPD with higher content of quaternary N gives better ORR performance than meso-PoPD that had lower content of quaternary N). Gratifyingly, meso/micro-PoPD has been fabricated in a Zn–air battery as the cathode and showed excellent performance when compared with Pt/C and meso-PoPD at different current densities. The battery ran smoothly without decline in voltage until complete consumption of the Zn anode. It was found that the battery could be regenerated by refueling Zn anode and electrolyte periodically as reflected by steady voltage intensity.

Afterward, Yu and coworkers reported an interesting work in which a N-doped carbon nanofiber aerogel derived from renewable bacterial cellulose [142] showed high density of N active sites (5.8 at%), high specific surface area (916 $m^2 g^{-1}$), and satisfactory ORR performance in alkaline medium ($E_{1/2} = +0.80$ V and high selectivity in ORR 4e^- pathway with number of electron transfer to be 3.97). When assembled into Zn–air battery as cathodic catalyst, the aerogel exhibited high voltages of 1.35 and 1.24 V at the discharge current densities of 1.0 and 10 mA cm^{-2} that were comparable with the existing Pt/C electrode. Bioavailable raw materials as precursors to generate metal-free carbonaceous materials for ORR catalyst and even fabrication as cathodic catalyst in Zn–air battery serve as a sustainable approach in fuel cell ORR catalyst development.

Recently, a 3D N/P co-doped hierarchical carbon (NPHC) was prepared from SiO_2 templated carbonization of small organic molecules (phytic acid + 4,4′-bipyridine + polytetrafluoroethylene) [143]. It is worth noting that NPHC exhibited $E_{1/2}$ of +0.81 V only 30 mV more negative than Pt/C catalyst along with a high selectivity of a 4e^- ORR pathway. Besides, NPHC was fabricated as cathode catalyst in Zn–air battery that exhibited an open-circuit voltage of 1.46 V along with power density of 145 mW cm^{-2} and a specific capacity of 706 mA h g^{-1} comparable with that of Pt/C (742 mA h g^{-1}). This facile SiO_2-templated preparation lowers the barriers in ORR catalyst development for application in fuel cells.

In Sections 3.2.2–3.2.3, graphene and graphdiynes have been discussed for their excellent ORR activity arising from good conductivity, heteroatom doping, defective morphology, etc. After the demonstration of their potential as ORR catalyst for fuel cell application, these materials have been fabricated in Zn–air battery with promising outcomes. For graphene, Dai and coworkers demonstrated N/P-doped graphene-based carbon foam as cathodic electrode with an open-circuit potential of 1.48 V, a specific capacity of 735 mA h g^{-1}, a peak power density of 55 mW cm^{-2}, and a stable operation for continuous 240 hours after replacing Zn anodes and electrolytes [144]. Also, an example of defective graphene has been fabricated as cathodic electrocatalyst in Zn–air battery that was customized to be rechargeable battery (Figure 3.27a) and could be cycled for 180 cycles with stable performance at 2 mA cm^{-2}. DFT calculation evidently showed that both N/P co-doping and edge effects on graphene were essential for the electrocatalytic activity of the graphene-based carbon foam. Meanwhile, Yao and coworkers successfully exemplified that defective graphene obtained through the removal

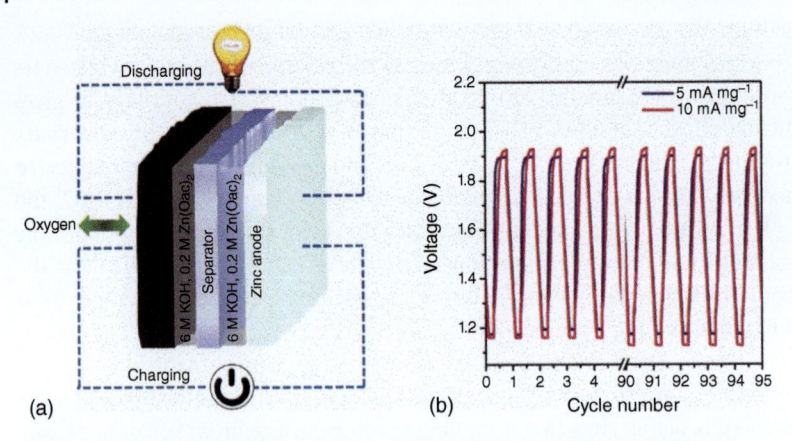

Figure 3.27 (a) Potential applications of the defective graphene as cathodic electrocatalyst in Zn–air fuel cell. (b) Performance of the battery charge–discharge cycling at 5 and 10 mA mg^{-1}, respectively. Source: Reproduced with permission from Jia et al. [145]. Copyright © 2016, Wiley-VCH.

of heteroatom dopants exhibited decent ORR activity along with good durability during charging/discharging cycle (after 90 cycles, charge potentials at current densities of both 5 and 10 mA mg^{-1} keep constant) (Figure 3.27b) and a peak power density of c. 154 mW mg^{-1} at a current density of 195 mA mg^{-1} [145] comparable with Pt/C reported previously [146].

In 2018, Huang and coworkers reported a facile preparation of pyridinic-N dominated and defect-rich graphene-like nanocarbon material (ND-GLC) from alkaline activation of cellulose and NH$_3$ [74]. The resulted ND-GLC demonstrated a better ORR performance than Pt/C with more positive E_{onset} and $E_{1/2}$ by 23 and 19 mV, respectively. Moreover, ND-GLC-based Zn–air battery gave higher power density (134 mW cm^{-2}) and discharge voltage of c. 1.3 V compared with Pt/C catalyst (110 mW cm^{-2} and c. 1.25 V) (Figure 3.28a,b). Both experimental and theoretical studies revealed that the excellent ORR catalytic activity originated from synergistic effect by both defects and pyridinic-N dopants.

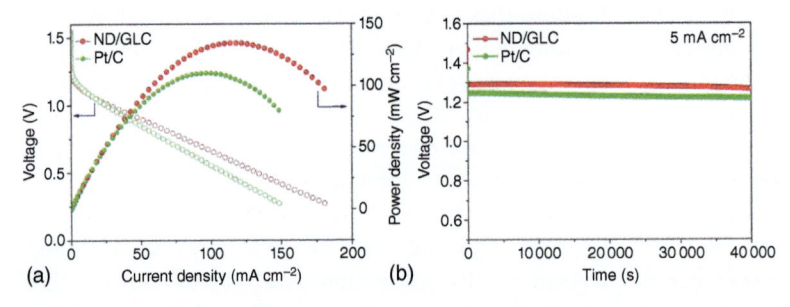

Figure 3.28 (a) Polarization curves and power density plots and (b) discharge curves of Zn–air batteries fabricated with ND-GLC and Pt/C electrode. Source: Reproduced with permission from Zhang et al. [74]. Copyright © 2018, Elsevier Ltd.

Intriguingly, Yang and coworkers reported a novel series of blend material composed of aggregated linear conjugated polymers (LCPs) on reduced graphite oxides (rGOs) (Scheme 3.22) [147]. Among three series of LCP/rGO, P-T/rGO exhibited the best catalytic performance with E_{onset} of +0.94 V and $E_{1/2}$ of +0.79 V along with dominantly a 4e$^-$ ORR pathway. P-T/rGO also gave the best performance compared with the other analogues as ORR catalyst in Zn–air battery with a maximum voltage of 1.16 V and excellent retention of voltage after 60-hour galvanostical discharge (declined only at rate of 0.077% per hour). This blending approach of LCP and rGO prevented harsh pyrolysis and heteroatom doping for fabricating ORR-active carbonaceous catalysts.

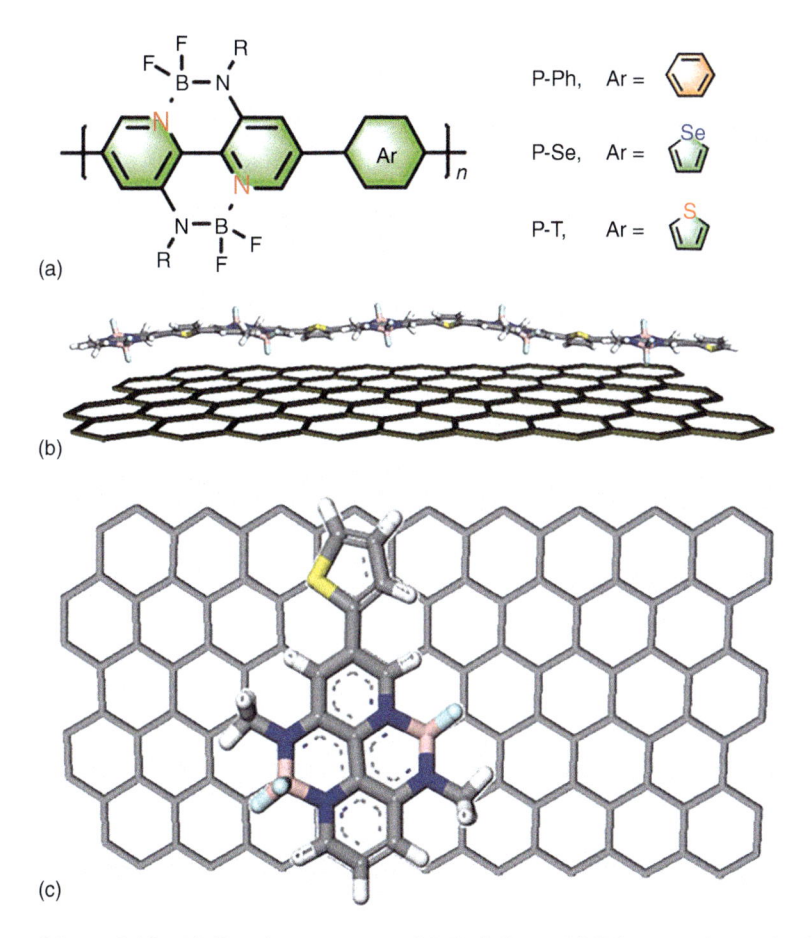

Scheme 3.22 (a) Chemical structures of P-Ph, P-Se, and P-T linear conjugated polymers. (b) Side view and (c) Top view of the molecular packing between linear conjugated polymers and rGO layers. Source: Reproduced with permission from Long et al. [147]. Copyright © 2019, Wiley-VCH.

Beyond graphene-based materials, a pyridine analogue of graphdiyne (PyN-GDY) has been reported for excellent ORR performance in both alkaline (E_{onset} of +1.00 V and $E_{1/2}$ of +0.84 V in 0.1 M KOH, better than Pt/C) and acidic media (E_{onset} of +0.81 V and $E_{1/2}$ of +0.55 V in 0.1 M HClO$_4$, worse than Pt/C) with high selectivity in ORR pathway for both media [148]. In addition, PyN-GDY-based Zn–air battery exhibited open-circuit voltage (1.51 V), power density (130 mW cm^{-2}), and specific discharging capacity (647 mA h g^{-1}) comparable with those of Pt/C-based Zn–air battery (1.50 V, 136 mW cm^{-2}, 641 mA h g^{-1}). DFT calculation results suggested that acetylenic carbon near pyridinic-N was the most active site. The findings from this work presented PyN-GDY as a close counterpart to GDY with extraordinary intrinsic ORR activity and great performance as electrocatalyst in Zn–air battery. This opens an alternative way for modulating ORR catalytic activity of metal-free carbon-based nanomaterials.

Very recently, highly crystalline metal-free thiophene-sulfur COFs (JUC-527 and JUC-528) (Scheme 3.23) were reported to exhibit superior ORR activity (in terms of E_{onset}, $E_{1/2}$, ORR kinetic, etc.) when compared with structurally similar COFs without thiophene as linkers (PDA-TAPB-COF, benzene-1,4-dicarboxaldehyde as substitute for thiophene linker) (Figure 3.29a,b) [149]. JUC-528-based Zn–air battery also gave competent performance with current density of c. 38 mA cm^{-2} at +1.0 V, power density of c. 0.164 W cm^{-2} at +0.45 V, and steady voltage of 1.12 V at

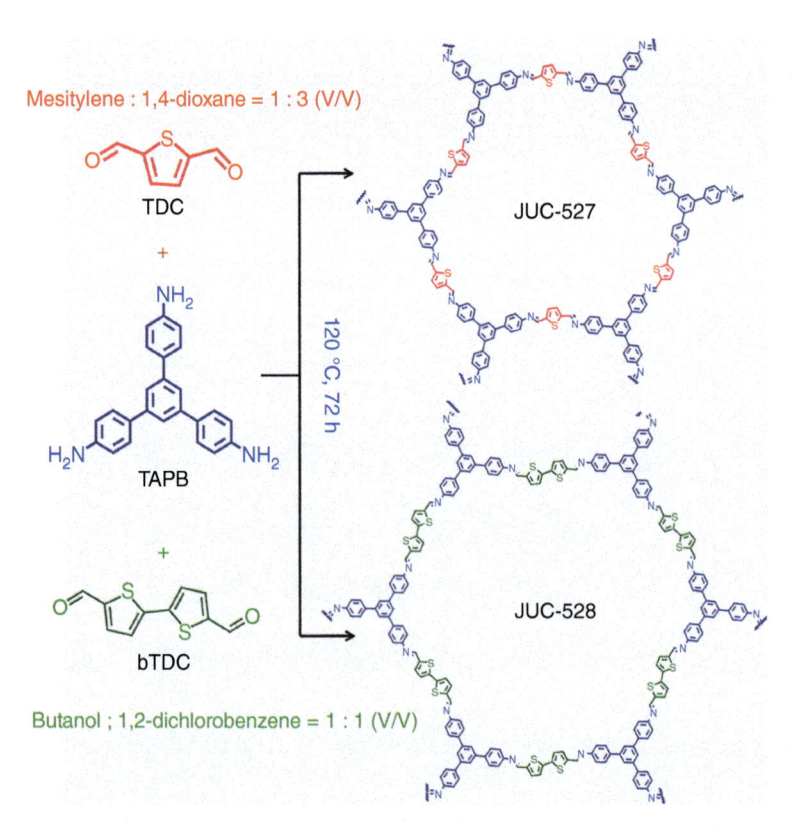

Scheme 3.23 Syntheses and structures of JUC-527 and JUC-528. Source: Reproduced with permission from Li et al. [149]. Copyright © 2020, American Chemical Society.

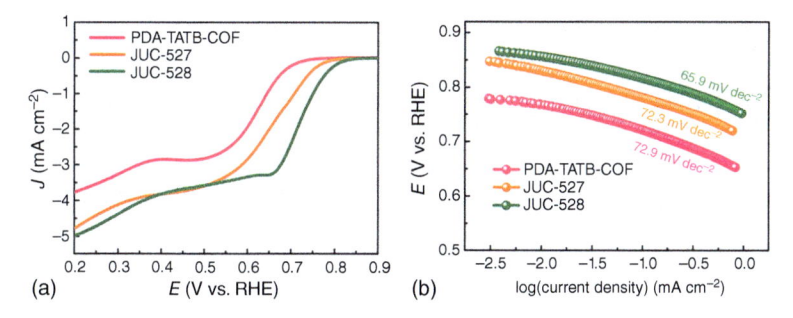

Figure 3.29 (a) LSV (at 1600 rpm) curves of PDA-TAPB-COF, JUC-527, and JUC-528 in O_2-saturated 0.1 M KOH electrolyte. (b) Comparison of Tafel plots. Source: Reproduced with permission from Li et al. [149]. Copyright © 2020, American Chemical Society.

10 mA cm^{-1} even after 60-hour operation. Additionally, DFT calculation revealed that pentacyclic thiophene linkers narrowed the band gap and hence lowered the overpotentials of ORR when compared with benzene linker. This result consistently supported the better ORR performance of both JUC-527 and JUC-528 than PDA-TAPB-COF.

Based on previous studies on the fundamentals of ORR catalytic performance by metal-free carbonaceous nanomaterials, various kinds of such materials have been manipulated as cathodic catalyst in real fuel cells and metal–air batteries. These metal-free nanocarbon materials show comparable and even better performance as cathode electrocatalyst in fuel cells and metal–air batteries. This casts light on the development of metal-free carbonaceous nanomaterials for substitution of precious metal-based ORR electrocatalyst in application of fuel cells and metal–air batteries.

3.6 Conclusion

Since the report of VA-NCNTs as active ORR catalyst in 2009, research on metal-free carbonaceous nanomaterials to substitute rare and costly Pt-based electrocatalysts has grown intensively. At the beginning, following the success of pioneering VA-NCNTs, heteroatom-doped carbon-based materials in the form of other well-known classes such as graphene and graphdiyne received the most attention. Afterward, along with the increasing understanding of fabricating heteroatom-doped nanocarbon materials, focus has been shifted to study dopant-free pure carbon nanomaterials for clarifying the role of defects in ORR catalysis. Recently, COFs have emerged as a novel and fast-growing class of candidate in ORR catalyst development. Most of these materials exhibit comparable or even superior ORR catalytic capability in alkaline medium when compared with commercial Pt/C electrocatalyst. Gratifyingly, some of currently reported materials can even execute ORR catalysis with competent performance and operational stability in acidic medium. To put these ORR catalytically active nanocarbon materials into practice, some materials have been assembled into fuel cells and Zn–air batteries as cathodic catalysts with fruitful outcomes. Even though utilizing metal-free carbon-based nanomaterials as ORR catalysts is still a green direction, intensive contribution from researchers fuels up the search for sustainable nanocarbon materials taking over existing Pt catalyst of scarcity and prohibitive cost. It is expected that cost-effective and good performing metal-free nanocarbon materials ultimately can actualize massive production of fuel cells and zinc–air batteries.

References

1 Møller, K.T., Jensen, T.R., Akiba, E., and Li, H.-W. (2017). Hydrogen - a sustainable energy carrier. *Prog. Nat. Sci.: Mater. Int.* 27: 34–40.
2 Carmo, M., Fritz, D.L., Mergel, J., and Stolten, D. (2013). A comprehensive review on PEM water electrolysis. *Int. J. Hydrogen Energy* 38: 4901–4934.
3 Gür, T.M. (2013). Critical review of carbon conversion in "carbon fuel cells". *Chem. Rev.* 113: 6179–6206.
4 Wang, W., Su, C., Wu, Y. et al. (2013). Progress in solid oxide fuel cells with nickel-nased anodes operating on methane and related fuels. *Chem. Rev.* 113: 8104–8151.
5 Liu, M., Zhang, R., and Chen, W. (2014). Graphene-supported nanoelectrocatalysts for fuel cells: synthesis, properties, and applications. *Chem. Rev.* 114: 5117–5160.
6 Nasef, M.M. (2014). Radiation-grafted membranes for polymer electrolyte fuel cells: current trends and future directions. *Chem. Rev.* 114: 12278–12329.
7 Zhang, H. and Shen, P.K. (2012). Advances in the high performance polymer electrolyte membranes for fuel cells. *Chem. Soc. Rev.* 41: 2382–2394.
8 Devanathan, R. (2008). Recent developments in proton exchange membranes for fuel cells. *Energy Environ. Sci.* 1: 101–119.
9 Hawkes, A., Staffell, I., Brett, D., and Brandon, N. (2009). Fuel cells for micro-combined heat and power generation. *Energy Environ. Sci.* 2: 729–744.
10 Chen, Z., Higgins, D., Yu, A. et al. (2011). A review on non-precious metal electrocatalysts for PEM fuel cells. *Energy Environ. Sci.* 4: 3167–3192.
11 Shao, Y., Yin, G., and Gao, Y. (2007). Understanding and approaches for the durability issues of Pt-based catalysts for PEM fuel cell. *J. Power Sources* 171: 558–566.
12 Borup, R., Meyers, J., Pivovar, B. et al. (2007). Scientific aspects of polymer electrolyte fuel cell durability and degradation. *Chem. Rev.* 107: 3904–3951.
13 de Bruijn, F.A., Dam, V.A.T., and Janssen, G.J.M. (2008). Review: durability and degradation issues of PEM fuel cell components. *Fuel Cells* 8: 3–22.
14 Sasaki, K., Naohara, H., Cai, Y. et al. (2010). Core-protected platinum monolayer shell high-stability electrocatalysts for fuel-cell cathodes. *Angew. Chem. Int. Ed.* 49: 8602–8607.
15 Chen, H., Pei, P., and Song, M. (2015). Lifetime prediction and the economic lifetime of proton exchange membrane fuel cells. *Appl. Energy* 142: 154–163.
16 Kannan, A., Kabza, A., and Scholta, J. (2015). Long term testing of start–stop cycles on high temperature PEM fuel cell stack. *J. Power Sources* 277: 312–316.

17 Zhou, M., Wang, H.-L., and Guo, S. (2016). Towards high-efficiency nano-electrocatalysts for oxygen reduction through engineering advanced carbon nanomaterials. *Chem. Soc. Rev.* 45: 1273–1307.

18 Yan, X., Jia, Y., and Yao, X. (2018). Defects on carbons for electrocatalytic oxygen reduction. *Chem. Soc. Rev.* 47: 7628–7658.

19 Dai, L., Xue, Y., Qu, L. et al. (2015). Metal-free catalysts for oxygen reduction reaction. *Chem. Rev.* 115: 4823–4892.

20 Duan, J., Chen, S., Jaroniec, M., and Qiao, S.Z. (2015). Heteroatom-doped graphene-based materials for energy-relevant electrocatalytic processes. *ACS Catal.* 5: 5207–5234.

21 Liu, X. and Dai, L. (2016). Carbon-based metal-free catalysts. *Nat. Rev. Mater.* 1: 16064.

22 Tang, C. and Zhang, Q. (2017). Nanocarbon for oxygen reduction electrocatalysis: dopants, edges, and defects. *Adv. Mater.* 29: 1604103.

23 Zhang, J., Chen, G., Müllen, K., and Feng, X. (2018). Carbon-rich nanomaterials: fascinating hydrogen and oxygen electrocatalysts. *Adv. Mater.* 30: 1800528.

24 Sheng, Z.-H., Shao, L., Chen, J.-J. et al. (2011). Catalyst-free synthesis of nitrogen-doped graphene *via* thermal annealing graphite oxide with melamine and its excellent electrocatalysis. *ACS Nano* 5: 4350–4358.

25 Gong, K., Du, F., Xia, Z. et al. (2009). Nitrogen-doped carbon nanotube arrays with high electrocatalytic activity for oxygen reduction. *Science* 323: 760–764.

26 Yang, L., Jiang, S., Zhao, Y. et al. (2011). Boron-doped carbon nanotubes as metal-free electrocatalysts for the oxygen reduction reaction. *Angew. Chem. Int. Ed.* 50: 7132–7135.

27 Wang, S., Iyyamperumal, E., Roy, A. et al. (2011). Vertically aligned BCN nanotubes as efficient metal-free electrocatalysts for the oxygen reduction reaction: a synergetic effect by co-doping with boron and nitrogen. *Angew. Chem. Int. Ed.* 50: 11756–11760.

28 Zhao, Y., Yang, L., Chen, S. et al. (2013). Can boron and nitrogen co-doping improve oxygen reduction reaction activity of carbon nanotubes? *J. Am. Chem. Soc.* 135: 1201–1204.

29 Li, W., Yang, D., Chen, H. et al. (2015). Sulfur-doped carbon nanotubes as catalysts for the oxygen reduction reaction in alkaline medium. *Electrochim. Acta* 165: 191–197.

30 Liu, Z., Fu, X., Li, M. et al. (2015). Novel silicon-doped, silicon and nitrogen-codoped carbon nanomaterials with high activity for the oxygen reduction reaction in alkaline medium. *J. Mater. Chem. A* 3: 3289–3293.

31 Novoselov, K.S., Geim, A.K., Morozov, S.V. et al. (2004). Electric field effect in atomically thin carbon films. *Science* 306: 666–669.

32 Geim, A.K. and Novoselov, K.S. (2007). The rise of graphene. *Nat. Mater.* 6: 183–191.

33 Wu, J., Pisula, W., and Müllen, K. (2007). Graphenes as potential material for electronics. *Chem. Rev.* 107: 718–747.

34 Stoller, M.D., Park, S.J., Zhu, Y. et al. (2008). Graphene-based ultracapacitors. *Nano Lett.* 8: 3498–3502.

35 Sahoo, N.G., Pan, Y., Li, L., and Chan, S.H. (2012). Graphene-based materials for energy conversion. *Adv. Mater.* 24: 4203–4210.

36 Boukhvalov, D.W. and Katsnelson, M.I. (2008). Chemical functionalization of graphene with defects. *Nano Lett.* 8: 4373–4379.

37 Güell, A.G., Ebejer, N., Snowden, M.E. et al. (2012). Structural correlations in heterogeneous electron transfer at monolayer and multilayer graphene electrodes. *J. Am. Chem. Soc.* 134: 7258–7261.

38 Chang, C.-H., Fan, X., Li, L.-J., and Kuo, J.-L. (2012). Band gap tuning of graphene by adsorption of aromatic molecules. *J. Phys. Chem. C* 116: 13788–13794.

39 Ritter, K.A. and Lyding, J.W. (2009). The influence of edge structure on the electronic properties of graphene quantum dots and nanoribbons. *Nat. Mater.* 8: 235–242.

40 Zhang, L. and Xia, Z. (2011). Mechanisms of oxygen reduction reaction on nitrogen-doped graphene for fuel cells. *J. Phys. Chem. C* 115: 11170–11176.

41 Qu, L., Liu, Y., Baek, J.-B., and Dai, L. (2010). Nitrogen-doped graphene as efficient metal-free electrocatalyst for oxygen reduction in fuel cells. *ACS Nano* 4: 1321–1326.

42 Yu, D., Zhang, Q., and Dai, L. (2010). Highly efficient metal-free growth of nitrogen-doped single-walled carbon nanotubes on plasma-etched substrates for oxygen reduction. *J. Am. Chem. Soc.* 132: 15127–15129.

43 Li, Y., Zhao, Y., Cheng, H. et al. (2012). Nitrogen-doped graphene quantum dots with oxygen-rich functional groups. *J. Am. Chem. Soc.* 134: 15–18.

44 Zhao, Y., Hu, C., Hu, Y. et al. (2012). A versatile, ultralight, nitrogen-doped graphene framework. *Angew. Chem. Int. Ed.* 51: 11371–11375.

45 Leventis, N., Sotiriou-Leventis, C., Zhang, G., and Rawashdeh, A.-M.M. (2002). Nanoengineering strong silica aerogels. *Nano Lett.* 2: 957–960.

46 Zhang, M., Fang, S., Zakhidov, A.A. et al. (2005). Strong, transparent, multifunctional, carbon nanotube sheets. *Science* 309: 1215–1219.

47 Wang, S., Zhang, L., Xia, Z. et al. (2012). BCN graphene as efficient metal-free electrocatalyst for the oxygen reduction reaction. *Angew. Chem. Int. Ed.* 51: 4209–4212.

48 Liang, J., Jiao, Y., Jaroniec, M., and Qiao, S.Z. (2012). Sulfur and nitrogen dual-doped mesoporous graphene electrocatalyst for oxygen reduction with synergistically enhanced performance. *Angew. Chem. Int. Ed.* 51: 11496–11500.

49 Zhang, C., Mahmood, N., Yin, H. et al. (2013). Synthesis of phosphorus-doped graphene and its multifunctional applications for oxygen reduction reaction and lithium ion batteries. *Adv. Mater.* 25: 4932–4937.

50 Li, Y., Li, Y., Zhu, E. et al. (2012). Stabilization of high-performance oxygen reduction reaction Pt electrocatalyst supported on reduced graphene oxide/carbon black composite. *J. Am. Chem. Soc.* 134: 12326–12329.

51 Park, M., Lee, T., and Kim, B.-S. (2013). Covalent functionalization based heteroatom doped graphene nanosheet as a metal-free electrocatalyst for oxygen reduction reaction. *Nanoscale* 5: 12255–12260.

52 Rao, C.V., Cabrera, C.R., and Ishikawa, Y. (2010). In search of the active site in nitrogen-doped carbon nanotube electrodes for the oxygen reduction reaction. *J. Phys. Chem. Lett.* 1: 2622–2627.

53 Jeon, I.-Y., Zhang, S., Zhang, L. et al. (2013). Edge-selectively sulfurized graphene nanoplatelets as efficient metal-free electrocatalysts for oxygen reduction reaction: the electron spin effect. *Adv. Mater.* 25: 6138–6145.

54 Cho, D.W., Parthasarathi, R., Pimentel, A.S. et al. (2010). Nature and kinetic analysis of carbon–carbon bond fragmentation reactions of cation radicals derived from set-oxidation of lignin model compounds. *J. Org. Chem.* 75: 6549–6562.

55 Jeon, I.-Y., Choi, H.-J., Choi, M. et al. (2013). Facile, scalable synthesis of edge-halogenated graphene nanoplatelets as efficient metal-free eletrocatalysts for oxygen reduction reaction. *Sci. Rep.* 3: 1810.

56 Lin, Z., Waller, G.H., Liu, Y. et al. (2013). Simple preparation of nanoporous few-layer nitrogen-doped graphene for use as an efficient electrocatalyst for oxygen reduction and oxygen evolution reactions. *Carbon* 53: 130–136.

57 Yang, S., Zhi, L., Tang, K. et al. (2012). Efficient synthesis of heteroatom (N or S)-doped graphene based on ultrathin graphene oxide-porous silica sheets for oxygen reduction reactions. *Adv. Funct. Mater.* 22: 3634–3640.

58 Li, Y., Li, M., Jiang, L. et al. (2014). Advanced oxygen reduction reaction catalyst based on nitrogen and sulfur co-doped graphene in alkaline medium. *Phys. Chem. Chem. Phys.* 16: 23196–23205.

59 Jiang, Z.-j., Jiang, Z., and Chen, W. (2014). The role of holes in improving the performance of nitrogen-doped holey graphene as an active electrode material for supercapacitor and oxygen reduction reaction. *J. Power Sources* 251: 55–65.

60 Wang, X., Wang, J., Wang, D. et al. (2014). One-pot synthesis of nitrogen and sulfur co-doped graphene as efficient metal-free electrocatalysts for the oxygen reduction reaction. *Chem. Commun.* 50: 4839–4842.

61 Li, X., Wang, H., Robinson, J.T. et al. (2009). Simultaneous nitrogen doping and reduction of graphene oxide. *J. Am. Chem. Soc.* 131: 15939–15944.

62 Zeng, J.-J. and Lin, Y.-J. (2014). Schottky barrier inhomogeneity for graphene/Si-nanowire arrays/n-type Si Schottky diodes. *Appl. Phys. Lett.* 104: 133506.

63 Ma, Z., Dou, S., Shen, A. et al. (2015). Sulfur-doped graphene derived from cycled lithium-sulfur batteries as a metal-free electrocatalyst for the oxygen reduction reaction. *Angew. Chem. Int. Ed.* 54: 1888–1892.

64 Wang, S., Yu, D., and Dai, L. (2011). Polyelectrolyte functionalized carbon nanotubes as efficient metal-free electrocatalysts for oxygen reduction. *J. Am. Chem. Soc.* 133: 5182–5185.

65 Li, G., Li, Y., Liu, H. et al. (2010). Architecture of graphdiyne nanoscale films. *Chem. Commun.* 46: 3256–3258.

66 Matsuoka, R., Sakamoto, R., Hoshiko, K. et al. (2017). Crystalline graphdiyne nanosheets produced at a gas/liquid or liquid/liquid interface. *J. Am. Chem. Soc.* 139: 3145–3152.

67 Li, Y., Xu, L., Liu, H., and Li, Y. (2014). Graphdiyne and graphyne: from theoretical predictions to practical construction. *Chem. Soc. Rev.* 43: 2572–2586.

68 Guo, J., Wang, Z., Shi, R. et al. (2020). Graphdiyne as a promising mid-infrared nonlinear optical material for ultrafast photonics. *Adv. Opt. Mater.* 8: 2000067.

69 Ren, Y., Dong, Y., Feng, Y., and Xu, J. (2018). Compositing two-dimensional materials with TiO_2 for photocatalysis. *Catalysts* 8: 590.

70 Liu, R., Liu, H., Li, Y. et al. (2014). Nitrogen-doped graphdiyne as a metal-free catalyst for high-performance oxygen reduction reactions. *Nanoscale* 6: 11336–11343.

71 Lv, Q., Si, W., Yang, Z. et al. (2017). Nitrogen-doped porous graphdiyne: a highly efficient metal-free electrocatalyst for oxygen reduction reaction. *ACS Appl. Mater. Interfaces* 9: 29744–29752.

72 Zhao, Y., Wan, J., Yao, H. et al. (2018). Few-layer graphdiyne doped with sp-hybridized nitrogen atoms at acetylenic sites for oxygen reduction electrocatalysis. *Nat. Chem.* 10: 924–931.

73 He, W., Jiang, C., Wang, J., and Lu, L. (2014). High-rate oxygen electroreduction over graphitic-N species exposed on 3D hierarchically porous nitrogen-doped carbons. *Angew. Chem. Int. Ed.* 53: 9503–9507.

74 Zhang, J., Sun, Y., Zhu, J. et al. (2018). Defect and pyridinic nitrogen engineering of carbon-based metal-free nanomaterial toward oxygen reduction. *Nano Energy* 52: 307–314.

75 Wang, Y., Zuo, S., and Liu, Y. (2018). Ammonia modification of high-surface-area activated carbons as metal-free electrocatalysts for oxygen reduction reaction. *Electrochim. Acta* 263: 465–473.

76 Gao, S., Wei, X., Fan, H. et al. (2015). Nitrogen-doped carbon shell structure derived from natural leaves as a potential catalyst for oxygen reduction reaction. *Nano Energy* 13: 518–526.

77 Liu, J., Cunning, B.V., Daio, T. et al. (2016). Nitrogen-doped carbon foam as a highly durable metal-free electrocatalyst for the oxygen reduction reaction in alkaline solution. *Electrochim. Acta* 220: 554–561.

78 Li, W., Ding, W., Jiang, J. et al. (2018). A phase-transition-assisted method for the rational synthesis of nitrogen-doped hierarchically porous carbon materials for the oxygen reduction reaction. *J. Mater. Chem. A* 6: 878–883.

79 Yang, W., Fellinger, T.-P., and Antonietti, M. (2011). Efficient metal-free oxygen reduction in alkaline medium on high-surface-area mesoporous nitrogen-doped carbons made from ionic liquids and nucleobases. *J. Am. Chem. Soc.* 133: 206–209.

80 Wang, G., Sun, Y., Li, D. et al. (2015). Controlled synthesis of N-doped carbon nanospheres with tailored mesopores through self-assembly of colloidal silica. *Angew. Chem. Int. Ed.* 54: 15191–15196.

81 Chen, S., Bi, J., Zhao, Y. et al. (2012). Nitrogen-doped carbon nanocages as efficient metal-free electrocatalysts for oxygen reduction reaction. *Adv. Mater.* 24: 5593–5597.

82 Huang, J., Han, J., Gao, T. et al. (2017). Metal-free nitrogen-doped carbon nanoribbons as highly efficient electrocatalysts for oxygen reduction reaction. *Carbon* 124: 34–41.

83 Guan, B.Y., Yu, L., and Lou, X.W. (2016). Formation of asymmetric bowl-like mesoporous particles *via* emulsion-induced interface anisotropic assembly. *J. Am. Chem. Soc.* 138: 11306–11311.

84 Li, J.-S., Li, S.-L., Tang, Y.-J. et al. (2014). Heteroatoms ternary-doped porous carbons derived from MOFs as metal-free electrocatalysts for oxygen reduction reaction. *Sci. Rep.* 4: 5130.

85 Xu, Q., Tang, Y., Zhang, X. et al. (2018). Template conversion of covalent organic frameworks into 2D conducting nanocarbons for catalyzing oxygen reduction reaction. *Adv. Mater.* 30: 1706330.

86 Xie, K., Qin, X., Wang, X. et al. (2012). Carbon nanocages as supercapacitor electrode materials. *Adv. Mater.* 24: 347–352.

87 Long, J.R. and Yaghi, O.M. (2009). The pervasive chemistry of metal–organic frameworks. *Chem. Soc. Rev.* 38: 1213–1214.

88 Zhou, H.C., Long, J.R., and Yaghi, O.M. (2012). Introduction to metal–organic frameworks. *Chem. Rev.* 112: 673–674.

89 Feng, X., Ding, X., and Jiang, D. (2012). Covalent organic frameworks. *Chem. Soc. Rev.* 41: 6010–6022.

90 Huang, N., Wang, P., and Jiang, D. (2016). Covalent organic frameworks: a materials platform for structural and functional designs *Nat. Rev. Mater.* 1: 16068.

91 Guo, D., Shibuya, R., Akiba, C. et al. (2016). Active sites of nitrogen-doped carbon materials for oxygen reduction reaction clarified using model catalysts. *Science* 351: 361–365.

92 Metiu, H., Chrétien, S., Hu, Z. et al. (2012). Chemistry of Lewis acid–base pairs on oxide surfaces. *J. Phys. Chem. C* 116: 10439–10450.

93 Zheng, Y., Jiao, Y., Chen, J. et al. (2011). Nanoporous graphitic-C_3N_4@carbon metal-free electrocatalysts for highly efficient oxygen reduction. *J. Am. Chem. Soc.* 133: 20116–20119.

94 Jin, Z., Nie, H., Yang, Z. et al. (2012). Metal-free selenium doped carbon nanotube/graphene networks as a synergistically improved cathode catalyst for oxygen reduction reaction. *Nanoscale* 4: 6455–6460.

95 Choi, C.H., Park, S.H., and Woo, S.I. (2011). Heteroatom doped carbons prepared by the pyrolysis of bio-derived amino acids as highly active catalysts for oxygen electro-reduction reactions. *Green Chem.* 13: 406–412.

96 Ji, L., Rao, M., Zheng, H. et al. (2011). Graphene oxide as a sulfur immobilizer in high performance lithium/sulfur cells. *J. Am. Chem. Soc.* 133: 18522–18525.

97 Zhang, Y., Jiang, W.-J., Zhang, X. et al. (2014). Engineering self-assembled N-doped graphene–carbon nanotube composites towards efficient oxygen reduction electrocatalysts. *Phys. Chem. Chem. Phys.* 16: 13605–13609.

98 Yang, J., Sun, H., Liang, H. et al. (2016). A highly efficient metal-free oxygen reduction electrocatalyst assembled from carbon nanotubes and graphene. *Adv. Mater.* 28: 4606–4613.

99 Geng, D., Chen, Y., Chen, Y. et al. (2011). High oxygen-reduction activity and durability of nitrogen-doped graphene. *Energy Environ. Sci.* 4: 760–764.

100 Liang, H.-W., Wei, W., Wu, Z.S. et al. (2013). Mesoporous metal-nitrogen-doped carbon electrocatalysts for highly efficient oxygen reduction reaction. *J. Am. Chem. Soc.* 135: 16002–16005.

101 Huang, X., Zhao, Z., Cao, L. et al. (2015). High-performance transition metal–doped Pt_3Ni octahedra for oxygen reduction reaction. *Science* 348: 1230–1234.

102 Shao, M., Chang, Q., Dodelet, J.-P., and Chenitz, R. (2016). Recent advances in electrocatalysts for oxygen reduction reaction. *Chem. Rev.* 116: 3594–3657.

103 Khan, M.A., Zhao, H., Zou, W. et al. (2018). Recent progresses in electrocatalysts for water electrolysis. *Electrochem. Energy Rev.* 1: 483–530.

104 Bu, L., Guo, S., Zhang, X. et al. (2016). Surface engineering of hierarchical platinum-cobalt nanowires for efficient electrocatalysis. *Nat. Commun.* 7: 11850.

105 He, D., Zhang, L., He, D. et al. (2016). Amorphous nickel boride membrane on a platinum-nickel alloy surface for enhanced oxygen reduction reaction. *Nat. Commun.* 7: 12362.

106 Bu, L., Zhang, N., Guo, S. et al. (2016). Biaxially strained PtPb/Pt core/shell nanoplate boosts oxygen reduction catalysis. *Science* 354: 1410–1414.

107 Wu, G., More, K.L., Johnston, C.M., and Zelenay, P. (2011). High-performance electrocatalysts for oxygen reduction derived from polyaniline, iron, and cobalt. *Science* 332: 443–447.

108 Peng, H., Liu, F., Liu, X. et al. (2014). Effect of transition metals on the structure and performance of the doped carbon catalysts derived from polyaniline and melamine for ORR application. *ACS Catal.* 4: 3797–3805.

109 Wu, G., Wang, J., Ding, W. et al. (2016). A strategy to promote the electrocatalytic activity of spinels for oxygen reduction by structure reversal. *Angew. Chem. Int. Ed.* 55: 1340–1344.

110 Li, J., Huang, W., Wang, M. et al. (2019). Low-crystalline bimetallic metal–organic framework electrocatalysts with rich active sites for oxygen evolution. *ACS Energy Lett.* 4: 285–292.

111 Shi, Q., Peng, F., Liao, S. et al. (2013). Sulfur and nitrogen co-doped carbon nanotubes for enhancing electrochemical oxygen reduction activity in acidic and alkaline media. *J. Mater. Chem. A* 1: 14853–14857.

112 Domínguez, C., Pérez-Alonso, F.J., Al-Thabaiti, S.A. et al. (2015). Effect of N and S co-doping of multiwalled carbon nanotubes for the oxygen reduction. *Electrochim. Acta* 157: 158–165.

113 Wei, W., Liang, H., Parvez, K. et al. (2014). Nitrogen-doped carbon nanosheets with size-defined mesopores as highly efficient metal-free catalyst for the oxygen reduction reaction. *Angew. Chem. Int. Ed.* 53: 1570–1574.

114 Li, D., Jia, Y., Chang, G. et al. (2018). A defect-driven metal-free electrocatalyst for oxygen reduction in acidic electrolyte. *Chem* 4: 2345–2356.

115 Wu, R., Wan, X., Deng, J. et al. (2019). NaCl protected synthesis of 3D hierarchical metal-free porous nitrogen-doped carbon catalysts for the oxygen reduction reaction in acidic electrolyte. *Chem. Commun.* 55: 9023–9026.

116 Li, M., Zhang, L., Xu, Q. et al. (2014). N-doped graphene as catalysts for oxygen reduction and oxygen evolution reactions: theoretical considerations. *J. Catal.* 314: 66–72.

117 Ma, L., Shen, X., Zhu, G. et al. (2014). FeCo nanocrystals encapsulated in N-doped carbon nanospheres/thermal reduced graphene oxide hybrids: facile synthesis, magnetic and catalytic properties. *Carbon* 77: 255–265.

118 Wang, D.-W. and Su, D. (2014). Heterogeneous nanocarbon materials for oxygen reduction reaction. *Energy Environ. Sci.* 7: 576–591.

119 Jiao, Y., Zheng, Y., Jaroniec, M., and Qiao, S.Z. (2014). Origin of the electrocatalytic oxygen reduction activity of graphene-based catalysts: a roadmap to achieve the best performance. *J. Am. Chem. Soc.* 136: 4394–4403.

120 Su, D.S., Perathoner, S., and Centi, G. (2013). Nanocarbons for the development of advanced catalysts. *Chem. Rev.* 113: 5782–5816.

121 Tao, L., Wang, Q., Dou, S. et al. (2016). Edge-rich and dopant-free graphene as a highly efficient metal-free electrocatalyst for the oxygen reduction reaction. *Chem. Commun.* 52: 2764–2767.

122 Shen, A., Zou, Y., Wang, Q. et al. (2014). Oxygen reduction reaction in a droplet on graphite: direct evidence that the edge is more active than the basal plane. *Angew. Chem. Int. Ed.* 53: 10804–10808.

123 Jiang, S., Li, Z., Wang, H. et al. (2014). Tuning nondoped carbon nanotubes to an efficient metal-free electrocatalyst for oxygen reduction reaction by localizing the orbital of the nanotubes with topological defects. *Nanoscale* 6: 14262–14269.

124 Zhao, H., Sun, C., Jin, Z. et al. (2015). Carbon for the oxygen reduction reaction: a defect mechanism. *J. Mater. Chem. A* 3: 11736–11739.

125 Zhang, L., Xu, Q., Niu, J., and Xia, Z. (2015). Role of lattice defects in catalytic activities of graphene clusters for fuel cells. *Phys. Chem. Chem. Phys.* 17: 16733–16743.

126 Yan, X., Jia, Y., Odedairo, T. et al. (2016). Activated carbon becomes active for oxygen reduction and hydrogen evolution reactions. *Chem. Commun.* 52: 8156–8159.

127 Xiang, Q., Yin, W., Liu, Y. et al. (2017). A study of defect-rich carbon spheres as a metal-free electrocatalyst for an efficient oxygen reduction reaction. *J. Mater. Chem. A* 5: 24314–24320.

128 Jiang, Y., Yang, L., Sun, T. et al. (2015). Significant contribution of intrinsic carbon defects to oxygen reduction activity. *ACS Catal.* 5: 6707–6712.

129 Zhao, X., Zou, X., Yan, X. et al. (2016). Defect-driven oxygen reduction reaction (ORR) of carbon without any element doping. *Inorg. Chem. Front.* 3: 417–421.

130 Wang, X., Li, X., Ouyang, C. et al. (2016). Nonporous MOF-derived dopant-free mesoporous carbon as an efficient metal-free electrocatalyst for the oxygen reduction reaction. *J. Mater. Chem. A* 4: 9370–9374.

131 He, H., Perman, J.A., Zhu, G., and Ma, S. (2016). Metal–organic frameworks for CO_2 chemical transformations. *Small* 12: 6309–6324.

132 Fu, S., Zhu, C., Song, J. et al. (2017). Metal–organic framework-derived non-precious metal nanocatalysts for oxygen reduction reaction. *Adv. Energy Mater.* 7: 1700363.

133 Yu, W., Gu, S., Fu, Y. et al. (2018). Carbazole-decorated covalent triazine frameworks: novel nonmetal catalysts for carbon dioxide fixation and oxygen reduction reaction. *J. Catal.* 362: 1–9.

134 Roy, S., Bandyopadhyay, A., Das, M. et al. (2018). Redox-active and semi-conducting donor-acceptor conjugated microporous polymers as metal-free ORR catalysts. *J. Mater. Chem. A* 6: 5587–5591.

135 Royuela, S., Martínez-Periñán, E., Arrieta, M.P. et al. (2020). Oxygen reduction using a metal-free naphthalene diimide-based covalent organic framework electrocatalyst. *Chem. Commun.* 56: 1267–1270.

136 Sapkota, P. and Kim, H. (2009). Zinc–air fuel cell, a potential candidate for alternative energy. *J. Ind. Eng. Chem.* 15: 445–450.

137 Sa, Y.J., Park, C., Jeong, H.Y. et al. (2014). Carbon nanotubes/heteroatom-doped carbon core-sheath nanostructures as highly active, metal-free oxygen reduction electrocatalysts for alkaline fuel cells. *Angew. Chem. Int. Ed.* 53: 4102–4106.

138 Zhang, S., Cai, Y., He, H. et al. (2016). Heteroatom doped graphdiyne as efficient metalfree electrocatalyst for oxygen reduction reaction in alkaline medium. *J. Mater. Chem. A* 4: 4738–4744.

139 Shui, J., Wang, M., Du, F., and Dai, L. (2015). N-doped carbon nanomaterials are durable catalysts for oxygen reduction reaction in acidic fuel cells. *Sci. Adv.* 1: e1400129.

140 Xue, L., Li, Y., Liu, X. et al. (2018). Zigzag carbon as efficient and stable oxygen reduction electrocatalyst for proton exchange membrane fuel cells. *Nat. Commun.* 9: 3819.

141 Liang, H.-W., Zhuang, X., Brüller, S. et al. (2014). Hierarchically porous carbons with optimized nitrogen doping as highly active electrocatalysts for oxygen reduction. *Nat. Commun.* 5: 4973.

142 Liang, H.-W., Wu, Z.-Y., Chen, L.-F. et al. (2015). Bacterial cellulose derived nitrogen-doped carbon nanofiber aerogel: an efficient metal-free oxygen reduction electrocatalyst for zinc-air battery. *Nano Energy* 11: 366–376.

143 Chen, S., Cui, M., Yin, Z. et al. (2020). Confined synthesis of N, P co-doped 3D hierarchical carbons as high efficiency oxygen reduction reaction catalysts for Zn-air battery. *ChemElectroChem* 7: 4131–4135.

144 Zhang, J., Zhao, Z., Xia, Z., and Dai, L. (2015). A metal-free bifunctional electrocatalyst for oxygen reduction and oxygen evolution reactions. *Nat. Nanotechnol.* 10: 444–452.

145 Jia, Y., Zhang, L., Du, A. et al. (2016). Defect graphene as a trifunctional catalyst for electrochemical reactions. *Adv. Mater.* 28: 9532–9538.

146 Li, Y. and Dai, H. (2014). Recent advances in zinc–air batteries. *Chem. Soc. Rev.* 43: 5257–5275.

147 Long, X., Li, D., Wang, B. et al. (2019). Heterocyclization strategy for construction of linear conjugated polymers: efficient metal-free electrocatalysts for oxygen reduction. *Angew. Chem. Int. Ed.* 58: 11369–11373.

148 Lv, Q., Wang, N., Si, W. et al. (2020). Pyridinic nitrogen exclusively doped carbon materials as efficient oxygen reduction electrocatalysts for Zn-air batteries. *Appl. Catal., B* 261: 118234.

149 Li, D., Li, C., Zhang, L. et al. (2020). Metal-free thiophene-sulfur covalent organic frameworks: precise and controllable synthesis of catalytic active sites for oxygen reduction. *J. Am. Chem. Soc.* 142: 8104–8108.

4

Rare Earth Luminescent Nanomaterials and Their Applications

Jianle Zhuang and Xuejie Zhang

South China Agricultural University, College of Materials and Energy, Key Laboratory for Biobased Materials and Energy of Ministry of Education, No. 483, Wushan Road, Tianhe District, Guangzhou 510642, P. R. China

4.1 Introduction

Rare earth luminescent materials are important among various luminescent materials. Rare earth elements, which include 15 lanthanides (from lanthanum to lutetium), scandium and yttrium, usually exist as trivalent cations. With the abundant f-orbital structure, lanthanide ions can exhibit strong fluorescence emission through 4f–4f or 4f–5d transitions. They are widely used as emitters in many luminescent materials [1–3]. However, due to the forbidden effect of the 4f transition, direct excitation of lanthanide ions is an inefficient method. Researchers usually adopt the doping technique, which incorporates a low concentration of atoms or ions to the host matrix to obtain luminescent materials with ideal properties [2]. Through doping, rare earth luminescent materials can show efficient emission in a broad range upon excitation, from ultraviolet to mid-infrared light region. Compared with the traditional lanthanide chelates, quantum dots, and organic dyes, rare earth luminescent materials have large Stokes shift, sharp emission spectrum, long life, high stability, low toxicity, and less photobleaching. Therefore, they are widely applicable in lighting, display, sensing, optoelectronic devices, solar cells, and biological medicine [1–13].

Traditionally, powdered rare earth luminescent materials were prepared by a solid reaction process, where the solid precursor reactants were directly mixed, ground, and calcinated at high temperatures (generally higher than 1000 °C). However, this process suffers from high energy consumption and impurities contamination. The final products always suffer from inhomogeneous composition and uncontrollable morphology [1, 2]. In contrast, many soft chemistry methods, such as hydro/solvothermal synthesis, coprecipitation, thermal decomposition, and sol–gel synthesis, have advantages in the preparation of rare earth luminescent materials with tunable size and morphology [2, 3]. These methods are based on the solution-phase process, where the reactants are homogeneously mixed. Thus, the composition, size, and morphology of the products can be well tuned at the nanoscale, significant for multifunctional applications.

Functional Nanomaterials: Synthesis, Properties, and Applications, First Edition.
Edited by Wai-Yeung Wong and Qingchen Dong.
© 2022 WILEY-VCH GmbH. Published 2022 by WILEY-VCH GmbH.

4 Rare Earth Luminescent Nanomaterials and Their Applications

The synthesis, characterization, properties and application of rare earth luminescent nanomaterials is introduced, including upconversion nanoparticles (UCNPs) and downconversion nanoparticles (DCNPs). The upconversion (UC) is a process where the species is excited by low-energy photons and emits high-energy photons, e.g. excited by NIR light and emits visible light. On the other hand, the downconversion (DC) is a process where the species is excited by high-energy photons and emits low-energy photons, e.g. excited by UV light and emits visible light [1, 4]. Rare earth-based UCNPs became popular in the past 20 years due to their unique properties and wide applications [1–14]. Meanwhile, rare earth-based DCNPs received less attention.

The first part will be a detailed introduction about UCNPs, from the development history, mechanism, composition, synthesis, characterization, UC emission tuning and application, while the second part will provide some examples of DCNPs.

4.2 Rare Earth Based UCNPs

4.2.1 Development of Upconversion Materials

The phenomenon of upconversion emission was discovered in the 1960s. The upconversion properties and mechanisms of doped rare earth ions were extensively investigated by researchers (represented by Auzel) from the 1970s to the 1990s [1]. With continuous development in nanotechnology, many materials have been used in new ideas and have gained new vitality. UCNPs have become the focus of research in rare earth luminescence [5]. For the first time in 2003, Haase and coworkers prepared 6–8 nm $LnPO_4$:Yb/Tm ($YbPO_4$:Er) colloidal nanoparticles and observed their upconversion emission [14]. In 2004, they first fabricated the cubic $NaYF_4$:Yb/Er colloidal nanoparticles and found the upconversion efficiency was about eight times larger than that of $LnPO_4$:Yb/Tm and $YbPO_4$:Er (Figure 4.1) [15]. This report is significant for the follow-up research on UCNPs. In the same year, Yi et al. used cubic $NaYF_4$:Yb/Er nanoparticles as biomarkers for the first time, introducing UCNPs to

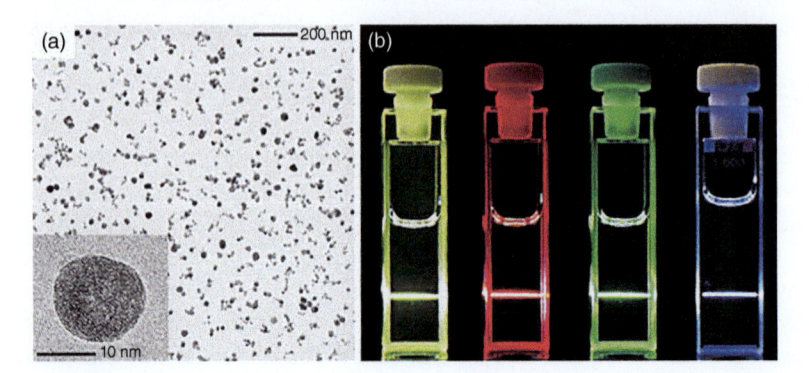

Figure 4.1 TEM image of cubic $NaYF_4$:Yb/Er UCNPs (a) and photographs of the UC emission of colloidal solution of UCNPs (b). Source: Heer et al. [15] / Reproduced with permission from Wiley-VCH.

biological applications [16]. The synthesis of hexagonal NaYF$_4$:Yb/Er nanoparticles was first reported by Li's group in 2005 [17]. In 2008, Zhang's group developed a revised method for preparing UCNPs with uniform size and regular morphology, which became the most popular synthetic route for UCNPs to date [18]. In 2010, Liu's pioneer work on simultaneous phase and size control of UCNPs through lanthanide doping was published in a top journal, Nature [19]. This is the first published paper about UCNPs, which is a milestone. Over the past decade, great success has been achieved in the controlled synthesis, upconversion emission tuning and multifunctional applications of UCNPs [5, 11, 12, 20].

4.2.2 Upconversion Mechanism

The research on upconversion mechanism focuses on the energy-level transition of rare earth ions. Upconversion mechanism involves various substrate materials and activated ions. It always develops with the emergence of new materials. At present, the main mechanisms can be summarized as follows [1, 5, 21].

4.2.2.1 Excited-State Absorption (ESA)

Excited-state absorption (ESA) is the most basic process of upconversion emission where the successive absorption of photons by a single ion results in the electron transition from the ground state to the excited state. As presented in Figure 4.2, the electron in the ground state absorbs the photon and transitions to the intermediate energy level. The electron then absorbs another photon and transitions to a higher energy level due to the long lifetime of the intermediate energy level. Upconversion emission occurs from the higher energy level. The ladder-like arrangement and near evenly-spaced energy levels of Ln^{3+} ions is required to achieve efficient ESA. Only a few rare earth ions have such energy-level structures, such as Ho^{3+}, Er^{3+}, and Tm^{3+}.

4.2.2.2 Energy Transfer Upconversion (ETU)

Energy transfer upconversion (ETU) also takes advantage of the sequential absorption of excited photons to populate the intermediate energy levels of the rare earth dopant. However, it usually takes place between two neighboring ions, namely the sensitizer (I) and the activator (II). As shown in Figure 4.2, the sensitizer is first excited from the ground state to the metastable level by absorbing a pumped photon

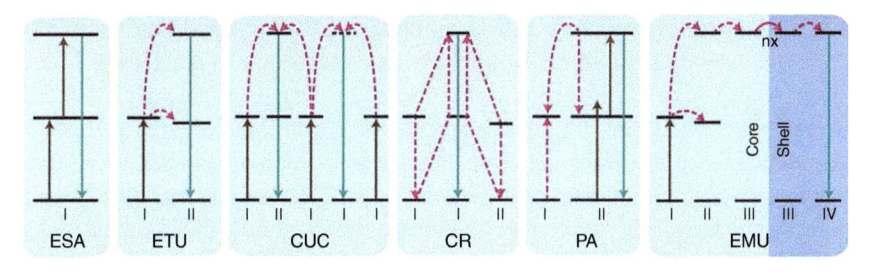

Figure 4.2 Principal UC mechanisms of lanthanide-doped materials. Source: Zheng et al. [5]. Copyright 2019, Elsevier.

and its energy transferred to the activator to make the transition to the excited state. After the sequential absorption and energy transfer of the sensitizer, the activator can transition to the higher excited state and emit high-energy photons. The Yb^{3+} ion is commonly used as a sensitizer due to its absorption cross section that matches with the wavelength of the commercially available diode laser [1]. The upconversion efficiency of an ETU process is sensitive to the distance between the sensitizer and activator, which depends on the concentrations of the doping ions.

4.2.2.3 Cooperative Upconversion (CUC)

CUC is a process involving the interaction of three ion centers. CUC generally includes cooperative luminescence and cooperative sensitization. For cooperative luminescence, the energy donor (I) and acceptor (I) are the same type of rare earth ions, e.g. Yb^{3+}–Yb^{3+} ion pair. As for cooperative sensitization, the two participants are different rare earth ions (I and II), e.g. Yb^{3+}–Pr^{3+}, Yb^{3+}–Tb^{3+}, Yb^{3+}–Eu^{3+} and ion pairs. The efficiency of the CUC process is generally (10^{-6}) lower than that of the ESA or ETU process because it involves quasi-virtual pair levels during electronic transitions. Generally, the CUC is only considered when ETU does not exist.

4.2.2.4 Cross Relaxation (CR)

The cross relaxation (CR) process can occur between the same or different types of ions. For two ions located in the excited state at the same time, one ion transfers energy to the other to make it transition to a higher excited-state level, with nonradiative relaxation of itself at the same time, as shown in Figure 4.2. The CR process is the basis of ion–ion interaction. Its efficiency is closely related to the doping concentration of the rare earth ions. CR is the main cause of the "concentration quenching mechanism," which is generally not conducive for upconversion emission. However, it can be utilized in some cases to tune the color output of UCNPs. For example, by introducing Ce^{3+} into Yb^{3+}–Ho^{3+} co-doped materials, the upconversion emission of Ho^{3+} can be tuned from green to red by using the CR between Ho^{3+} and Ce^{3+} [22].

4.2.2.5 Photon Avalanche (PA)

Photon avalanche (PA) is a process that produces upconversion above a certain threshold of excitation power. Three important factors that induce PA upconversion are weak ground state absorption, strong ESA, and efficient CR. In this process, the intermediate reservoir level is first populated by weak ground state absorption (GSA), followed by the ESA or ET from another excited ion to populate the luminescent level where upconversion occurs. After this stage, an efficient CR occurs between the excited ion and an adjacent ground state ion, resulting in two UC ions populating in the reservoir level. The feedback looping of ESA (or ET) and CR will exponentially increase the population in the reservoir and luminescent levels, producing strong upconversion emission (Figure 4.2). PA is considered to have the highest efficiency in photon upconversion. However, its disadvantages, such as weak GSA, high pump threshold, and long rise time, limit its widespread application [5].

4.2.2.6 Energy Migration-Mediated Upconversion (EMU)

The energy migration-mediated upconversion (EMU) process was first proposed by Wang and Liu in core/shell nanostructures which were never found in bulk materials [21]. To realize this novel process, four kinds of rare earth ions are needed, i.e. sensitizers (i), accumulators (ii), migrators (iii), and activators (iv). A sensitizer is used to absorb NIR photons and promote a neighboring accumulator to the excited states. A migrator, usually Gd^{3+} ion, obtains the excitation energy from the accumulator, followed by the random energy hopping throughout the core/shell interface. Finally, the activator captures the migrating energy to achieve upconversion emission (Figure 4.2). The elaborate arrangement of the four kinds of rare earth ions in the core/shell structure is essential for generating an efficient EMU process. It is the development of nanomaterials that leads to the discovery of this novel upconversion mechanism. This is an important extension of traditional upconversion processes (ESA, ETU, CU, CR, and PA), broadening the applications of UCNPs [5].

4.2.3 Composition of UCNPs

UCNPs usually comprise an inorganic host matrix and dopants (such as rare earth ions). The dopants provide the luminescence center, and the host provides the appropriate crystal field. The distance and spatial arrangement of ions to ions are important for upconversion emission. Under NIR excitation, many rare earth-doped host materials can emit visible light, but an efficient upconversion emission is determined by the host, dopant, and doping concentration [3, 5, 6, 10, 23].

4.2.3.1 Host

Choosing the upconversion host material determines the distance, relative spatial position, coordination number of doped ions, and type of surrounding anions. It also directly affects the performance of the upconversion material [3]. Therefore, it is essential to choose a suitable host material to obtain the desired high efficient and controllable luminescence properties. Host materials with low phonon energy, such as fluorides (c. $350\,cm^{-1}$), are optimal candidates for photon upconversion. They minimize the nonradiative energy loss and maximize the radiative emission during the upconversion process [3]. Among the fluoride hosts, $NaYF_4$ is known to be one of the most efficient host materials [3, 5]. Small nonradiative energy losses are also observed in heavy halogenides, but these materials suffer from low chemical stability. Metal oxides have high chemical stability, but their phonon energy is always higher than $500\,cm^{-1}$. Besides the phonon energy, the crystal structure of the host material also has a significant effect on upconversion efficiency. Generally, host materials with low symmetry have more efficient upconversion emission than those with high symmetry. For example, the upconversion efficiency of a hexagonal $NaYF_4$:Yb,Er is about 10 times higher than that of the cubic counterpart [24].

4.2.3.2 Activator

Activators (emitter) are the source of upconversion emission in UCNPs. If the energy gaps between three or more subsequent energy levels are similar in one ion, sequential excitation of the ion to a highly excited state is possible with a single monochromatic light source. Each absorption step requires the same photon energy. Rare earth dopant ions (i.e. Er^{3+}, Tm^{3+}, and Ho^{3+}) with ladder-like energy levels are usually selected as activators to generate upconversion emission under CW excitation. These energy levels can absorb multiple photons and reduce the nonradiative energy transfer between different excited energy levels, promoting the upconversion process. A low concentration (<2 mol%) of the activator is adopted to minimize the energy loss resulting from CR. After discovering the EMU process, efficient activators have expanded to Ce^{3+}, Sm^{3+}, Eu^{3+}, Gd^{3+}, Tb^{3+}, and Dy^{3+} ions in core/shell UCNPs. Other metal ions including Mn^{2+}, Cu^{2+} and Pb^{2+} have also been proven to be efficient activators for upconversion luminescence, broadening the range of upconversion luminous centers [5].

4.2.3.3 Sensitizer

The sensitizer transfers absorbed energy efficiently to the activator for the emission of high-energy light. The absorption cross section of rare earth luminescent ions in the NIR region is small, which leads to the relatively low upconversion efficiency of single ion-doped nanoparticles. Sensitizers and activators are usually co-doped in a typical upconversion nanoparticle to increase the NIR absorption, improving the upconversion efficiency. Yb^{3+} ion is used as a sensitizer due to its broad absorption cross section at 980 nm and large excitation spectrum. Furthermore, Yb^{3+} ion shows single excited state ($^2F_{5/2}$), which matches perfectly with the f–f transitions of many activators, such as Ho^{3+}, Er^{3+}, Tm^{3+}, etc. [1] Several research about co-doped Yb^{3+}–Er^{3+}, Yb^{3+}–Tm^{3+} and Yb^{3+}–Ho^{3+} in UCNPs have been reported. Nd^{3+} is another important sensitizer with efficient absorption at about 800 nm [25]. Compared with Yb^{3+}-based UCNPs pumped by 980 nm laser, Nd^{3+}-sensitized UCNPs can achieve photon upconversion at biocompatible excitation wavelength, which significantly reduces the overheating problem. This property makes Nd^{3+}-sensitized UCNPs ideal candidates for biological applications.

4.2.4 Synthesis of UCNPs

Many chemical methods, including thermal decomposition, hydro/solvothermal, coprecipitation, sol-gel and microwave-assisted synthesis, have been used to synthesize UCNPs (Table 4.1) [2, 5]. Various UCNPs with controllable component, size, and morphology have been fabricated (Figure 4.3). Some synthetic methods are introduced in this chapter.

4.2.4.1 Thermal Decomposition

Thermal decomposition is a nanomaterial synthesis by thermally induced chemical decomposition. It dissolves metallic organic compounds as precursors in organic solvents with high boiling points and synthesizes nanoparticles at high temperature. Oleamine (OM), oleic acid (OA), 1-octadecene (OD), and other organic solvents are often used as raw materials [2, 35, 36]. High-temperature reaction is provided, and long chain alkyl groups are used to cover the crystal surface to prevent the

Table 4.1 Typical synthetic methods to lanthanide-doped upconversion nanocrystals.

Synthetic methods	Host	Advantages	Disadvantages
Thermal decomposition	LaF_3, Na(Y, Gd, Yb, Lu)F_4, Li(Y, Lu)F_4, (La, Y, Tb, Ho, Er, Tm, Yb, Lu)$_2O_3$, Ba(Y, Gd, Lu)F_5, CaF_2, $BaTiO_3$, YOF, etc.	High-quality, monodisperse nanocrystals	Rigorous and harsh synthesis condition Hydrophobic resultants. Toxic by-products
Hydro(solvo) thermal synthesis	Na(Y, Gd, Yb, Lu)F_4, $LiYF_4$, $La_2(MoO_4)_3$, (La, Y)$_2O_3$, (La, Y, Gd, Lu)F_3, Ba(Y, Gd)F_5, BaY_2F_5, Ba_2GdF_7, (Ca, Sr)F_2, Sr_2ScF_7, (Y, Yb)PO_4, $Y_3Al_5O_{12}$, MnF_2, $KMnF_3$, $GdVO_4$, $PbTiO_3$, $KNbO_3$, Gd_2O_2S, etc.	Formation of highly crystalline phases at much lower temperature Excellent control over particle size and shape	Specialized autoclaves required. Impossibility of observing the crystal as it grows
Coprecipitation	Na(La, Y, Gd, Yb, Lu, Sc, Tb)F_4, (La, Y, Gd, Lu)F_3, Li(Y, Gd, Yb, Lu)F_4, (Ca, Sr)F_2, $BaYF_5$, GdOF, (La, Y, Yb, Lu)PO_4, ZrO_2, Y_2O_2S, ZnO, $Gd_4O_3F_6$, $CaMoO_4$, $Y_3Al_5O_{12}$, Y_4MoO_9, $Y_2(MoO_4)_3$, $Gd_2(WO_4)_3$, $Gd_3Ga_5O_{12}$, $BaTiO_3$, Y_2SiO_5, etc.	Simple and rapid preparation Fast growth rate Easy control of particle size and composition	Low product yield Particle aggregation Post-heat treatment typically need
Sol–gel synthesis	(La, Y)F_3, CaF_2, LaOF, (Y, Gd)$_2O_3$, TiO_2, ZnO, Ta_2O_5, GeO_2, Al_2O_3, $BaTiO_3$, $LaPO_4$, (Gd, Lu)$_3Ga_5O_{12}$, YVO_4, ZrO_2, Y_2SiO_5, $Y_2Ti_2O_7$, $BaGd_2(MoO_4)_4$, $CaGd_2(WO_4)_4$, etc.	Simple and cost-effective accessories Versatile Better homogeneity Less energy consumption	Cost of precursors High-temperature post-heat treatment required
Ionic liquid-based synthesis	Na(Y,Gd,Yb)F_4, (La, Y)F_3, CaF_2, $BaMgF_4$, etc	Easy processing Low solvent toxicity	Ratively expensive Poorly biodegradable
Microwave-assisted synthesis	(La, Gd)F_3, Na(Y, Gd)F_4, CaF_2, $BaYF_5$, $LaPO_4$, (Ca, Sr,Ba)MoO_4, Ca(La, Gd)$_2$(MoO4)$_4$, $NaLa(MoO4)_2$, $NaY(WO4)_2$, $SrY_2(MoO4)_4$, etc	Uniform heating. Rapid synthesis Higher product yields	Specialized microwave reactor required

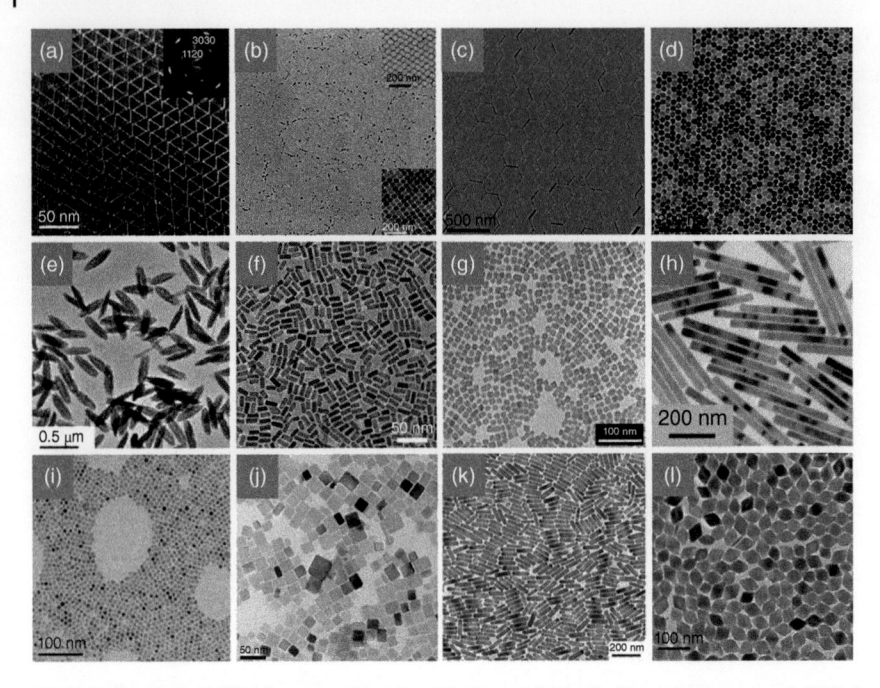

Figure 4.3 Typical TEM images of lanthanide-doped (a) LaF$_3$. Source: Zhang et al. [26] / Reproduced with permission from Wiley-VCH. (b–d) NaYF$_4$ nanoparticles synthesized by thermal decomposition method. Source: Ye et al. [27] / Reproduced with permission from National Academy of Sciences of the United States of America. (e) YF$_3$ Source: Wang et al. [28] Reproduced with permission from American Chemical Society. (f) NaGdF$_4$ Source: Ren et al. [29] / Reproduced with permission from American Chemical Society. and (g, h) NaYF$_4$ nanocrystals synthesized by hydro/solvothermal strategy. Source: Refs. [19, 30] / Reproduced with permission from WILEY-VCH, Reproduced with permission from American Chemical Society. (i) CaF$_2$ [31] / Reproduced with permission from WILEY-VCH. (j) KMnF$_3$ [32] / Reproduced with permission from WILEY-VCH. (k) KYb$_2$F$_7$ [33] / Reproduced with permission from Springer Nature. and (l) LiYF$_4$ nanoparticles [34] prepared by coprecipitation route / Reproduced with permission from Royal Society of Chemistry.

aggregation of nanoparticles. For metal-organic compounds, especially the reaction precursors of fluoride-based nanoparticles doped with rare earth, trifluoroacetate is usually selected and put in OA, OD/OM according to a stoichiometric ratio. It is heated at a temperature above 300 °C and accompanied by protective gas to prepare the nanocrystals [37]. Yan's team was the first to synthesize monodispersed fluoride nanoparticles using thermal decomposition (Figure 4.3a) [26]. Boyer et al. synthesized cubic and hexagonal lanthanum-doped NaYF$_4$ nanoparticles by thermal decomposition of rare earth salts of trifluoroacetic acid in OD, OA/OM solvents and used them for upconversion luminescence [35, 38]. This method has also been widely used in the synthesis of various high-quality fluoride nanocrystals, such as NaGdF$_4$ [39], NaLaF$_4$ [40], LiYF$_4$ [41], etc. Similar to other synthetic methods, the temperature, time, and ratio of reactants of thermal decomposition affect the morphology of the products. Researchers have done studies to synthesize nanoparticles with controllable morphologies [42, 43]. Murray and coworkers made significant progress by synthesizing homogeneous $NaYF_4$ nanospheres, nanorods, and nanoplates via thermal decomposition by controlling the reaction time and the ratio of reactants (Figure 4.3b–d) [27]. High-quality nanocrystals with narrow size distribution and strong upconversion emission can be obtained by thermal decomposition. However, the gas environment of the reaction is sensitive, the price of the reaction precursors is expensive, and the by-products produced by the reaction are highly toxic. Therefore, the use of thermal decomposition to synthesize upconversion luminescent nanoparticles also needs to overcome potential constraints [19, 36, 44].

4.2.4.2 Hydro/Solvothermal Synthesis

Hydrothermal synthesis is a method of mixing nanoparticles in water or solvent. It refers to the growing of crystalline materials from aqueous solution (or other solvents) in a closed reaction vessel under high-temperature and high-pressure reaction conditions. These reaction conditions accelerate the solubility of solid-phase reactants and promote nanoparticle formation [2, 45]. The hydrothermal method has been widely and effectively used to synthesize monodispersed nanoparticles with tunable structures and properties. This method uses special sealed containers for its reaction. The crystal phase, size, and morphology of the synthesized nanoparticles can be manipulated by controlling the reaction temperature, reaction time, reactant concentration, pH value of the reaction solution or adding different surfactants [46–50]. Based on the hydrothermal method, Li proposed a "liquid–solid-solution (LSS)" reaction strategy for the synthesis of various monodispersed nanocrystals, such as semiconductors, mono-metallic particles, and dielectric materials [50]. With rare earth elements of chloride, nitrate, and acetic acid salts as precursors and HF, NH_4F, NaF, and NH_4HF_2 as fluorine sources, the researchers synthesized high-quality fluoride nanoparticles (Figure 4.3e–h), such as $NaYF_4$, $NaGdF_4$, YF_3, LaF_3, $BaGdF_5$, and CaF_2 [28–30]. In addition to fluoride-based nanoparticles, lanthanum-oxide nanoparticles with controllable particle size and morphology, such as Er_2O_3, $YbPO_4$, $LuPO_4$ and $GdVO_4$, were also successfully prepared by hydrothermal method [51, 52].

4.2.4.3 Coprecipitation

The coprecipitation method is a simple nanocrystal synthesis. The reaction conditions, equipment, and experimental steps are simple. Lanthanum-doped nanocrystals can be synthesized by stoichiometry. van Veggel and colleagues first synthesized Ln^{3+}-doped LaF_3 nanocrystals using this method [52]. This approach generally add end-capping ligands in solvents, such as polyethyleneimine (PEI), ethylenediamine tetraacetic acid ester (EDTA), polyvinylpyrrolidone (PVP), etc. It is used as a surfactant in the process and regulation of nanoparticle growth, and to improve solubility and surface functionalization. The lanthanide-doped materials such as LaF_3, CaF_2, $NaYF_4$, $NaScF_4$, $KMnF_3$, $LiYF_4$, KYb_2F_7, SrF_2, etc. have been synthesized via coprecipitation method (Figure 4.3i–l) [31–34, 53]. The synthesis of $YbPO_4$ and $LuPO_4$ nanocrystals doped with Ln ions also used the same method. This method has become popular for the synthesis of UCNPs with controllable size and morphology.

4.2.4.4 Sol–Gel Synthesis

In sol–gel preparation, metal chelates (metal alcohols, etc.) and some inorganic salts are used as precursors. The precursors are dissolved in solvents (water or organic solvents) to produce a uniform solution. The solute then hydrolyzes or alcoholyzes in the solution, and the products are aggregated into monomer sol particles of about 1 nm by condensation polymerization process. Sol particles further aggregate and grow to form gels, which are processed by drying and sintering to obtain the required materials [54]. In polycondensation gelation, the size of the colloidal particles and the degree of cross-linking between colloidal molecules can be affected by adjusting the pH value and reaction temperature of the solution, controlling the morphology and properties of the final product. The physical and chemical properties of the products can be regulated by controlling the temperature and time during dry calcination. In 2002, Patra et al. developed a sol–gel method to synthesize Er^{3+}-doped ZrO_2 nanoparticles [55]. Researchers have also synthesized $BaTiO_3$, YVO_4, $Lu_3Ga_5O_{12}$, TiO_2 doped with lanthanide upconversion nanoparticles using this method [56–58]. Sol–gel method can be used to prepare single-component mixtures with high purity, uniform particle size distribution and high chemical activity at low temperature. It can also prepare products that are difficult to prepare by traditional methods. The experimental equipment requirements are low and the safety of the experimental process can be guaranteed. It has a wide range of applications in optics, electronics, thermal sensing, catalysts, and biological medicine.

4.2.4.5 Microwave-Assisted Synthesis

Microwave-assisted method is widely used in material synthesis. It is an environmentally friendly chemical method. This method makes use of the field strength of the electromagnetic field to generate the dielectric polarization of the microscopic particles inside the sample. It forms the polarization time consistent with the microwave frequency and the dielectric heating that provides the required energy for the reaction of the material. Wang et al. synthesized monodispersed $NaYF_4$ upconversion nanocrystals using microwave-assisted methods [59]. They investigated the factors controlling the size and shape of the MYF_4: Yb, Er (M = Na, Li, Ba) UCNPs, and reported the effects of the reaction time and reactant concentration on the products [60]. Microwave-assisted heating method has the advantages of fast heating, convenience, high efficiency, and safety. The sample has high purity and uniform particle size distribution, which is important in the preparation of functional materials such as nanomaterials and phosphors.

4.2.5 Characterization of UCNPs

Comprehensive elucidation of the properties of UCNPs is important to explain their unique photophysical properties, understand the mechanism of energy transfer, and construct high-efficiency UCNPs for various applications. There is better understanding of UCNPs with the development of testing methods. In 2015, Liu et al. gave a comprehensive overview (Figure 4.4) of the instrumentation techniques commonly utilized for the characterization of UCNPs [61]. Their overview summarized the characterization of UCNPs as follows.

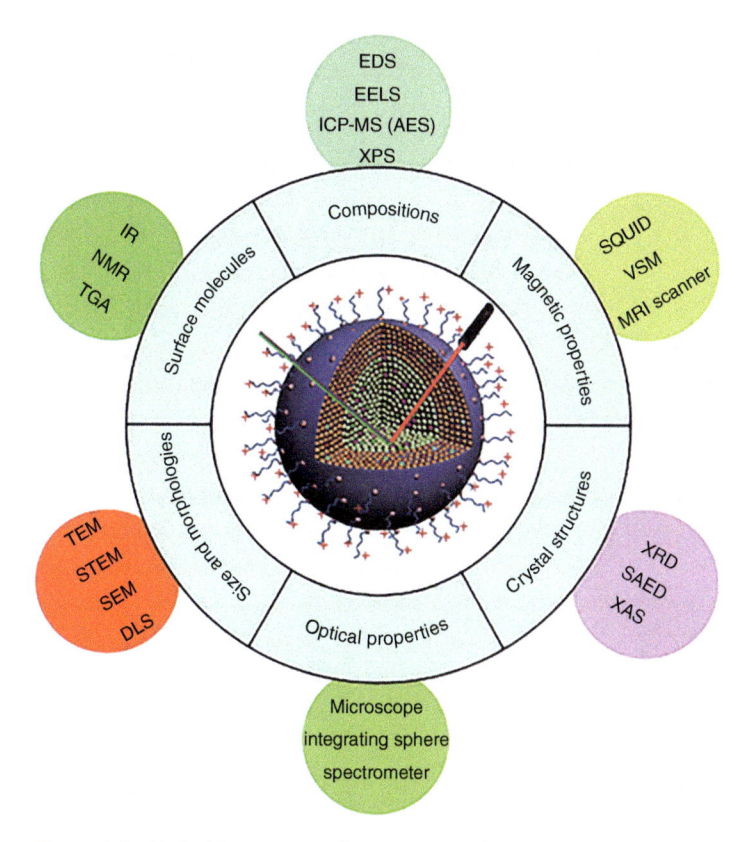

Figure 4.4 Typical instrumentation tools used for the characterization of UCNPs. Source: Liu et al. [61]. Copyright 2015, Royal Society of Chemistry.

4.2.5.1 Identification of Crystal Structures

Crystal structure is an important parameter that affects the upconversion emission of UCNPs. Common crystal structure identification techniques of UCNPs include X-ray diffraction (XRD), selected area electron diffraction (SAED), and X-ray absorption spectroscopy (XAS). The host material is the factor that determines the upconversion performance. For example, the hexagonal $NaYF_4$ has stronger upconversion emission than cubic $NaYF_4$. XRD is the basic test for crystal structure. XRD can be used to quick identify the phase and crystallinity of UCNPs with reference to a well-established database. Strong diffraction peaks can be observed from high-quality UCNPs with a hexagonal or cubic phase. XRD can also monitor the phase transformation of UCNPs. For example, the phase transformation from cubic to hexagonal in $NaYF_4$ can be clearly observed by XRD measurements upon doping with increased Gd^{3+} concentrations (Figure 4.5) [19]. Compared with XRD that measures materials in macroscopic scale, SAED is a technology that identifies the microstructure of materials. SAED can be used to directly examine the phase transition process of a single nanoparticle, e.g. from the Y_2O_3 to $NaYF_4$ [62].

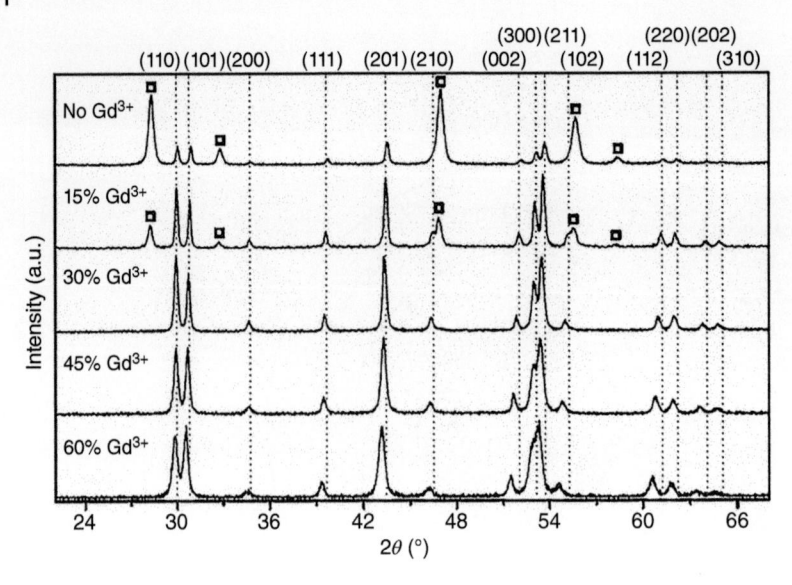

Figure 4.5 X-ray powder diffraction patterns of the NaYF$_4$:Yb/Er (18/2 mol%) nanocrystals obtained after heating for 2 hours at 200 °C in the presence of 0, 15, 30, 45, and 60 mol% Gd^{3+} dopant ions, respectively. Diffraction peaks corresponding to cubic NaYF$_4$ are marked with square boxes. A gradual decrease in diffraction peak intensities for cubic phase is observed as a function of increased Gd^{3+} dopant content. Source: Wang et al. [19].

4.2.5.2 Determination of Size and Morphology

Size and morphology are important factors that affect the luminescent properties and applications of UCNPs. For example, UCNPs with sizes smaller than 10 nm are more useful in bio-application because they are easily excreted from organisms. The methods for determining size and morphology of UCNPs are electron microscopy, including scanning electron microscopy (SEM), transmission electron microscopy (TEM), and scanning transmission electron microscope (STEM). SEM is a convenient test method for high-resolution imaging of UCNPs surfaces due to its depth of field and 3D rendering feature. SEM has a unique advantage in observing the surface characteristics of small-sized upconversion crystals such as elongated micro-octadecahedra, micro-octadecahedra, nanoslices, nanodiscs, short nanorods, and long nanorods (Figure 4.6) [63]. TEM can be used as a tool to reveal the size and morphology of UCNPs synthesized under different conditions. With the help of TEM characterization, the growth mechanism of UCNPs can be elucidated. For example, van Veggel and coworkers demonstrated the epitaxial layer-by-layer growth on NaYF$_4$ UCNPs through Ostwald ripening [64]. TEM analysis of the three aliquots confirms the complete dissolution and deposition of the α-NaYF$_4$ nanocrystals on the NaYF$_4$:Yb^{3+}/Er^{3+} core nanocrystals (Figure 4.7). Due to the contrast differences in TEM images, different components in composite materials such as NaYF$_4$:Yb/Tm-Au, NaYF$_4$:Yb/Er@SiO$_2$, NaYF$_4$:Yb/Er-CdSe, and hierarchical UCNPs can be clearly differentiated. Under the high-resolution imaging mode, the crystallographic structures of UCNPs, including crystalline domains and defects, can be examined at an atomic level. For those UCNPs with similar

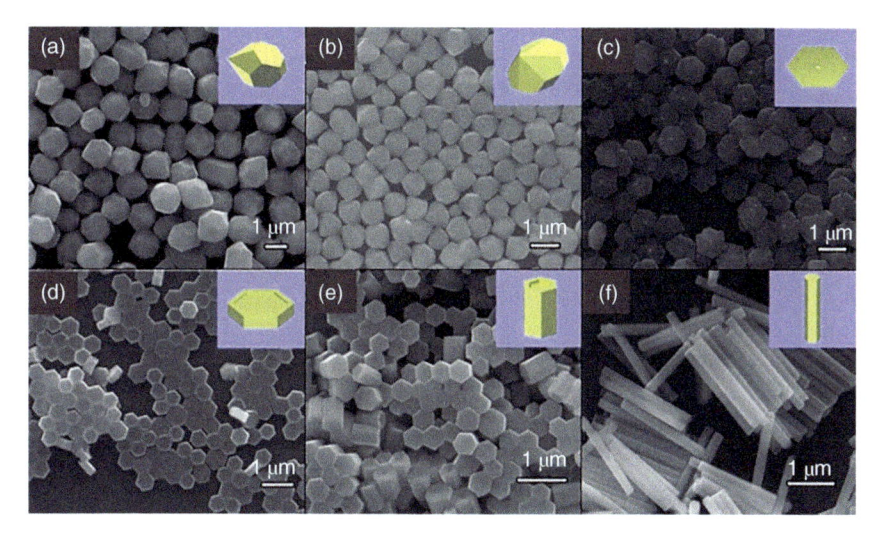

Figure 4.6 SEM images of different morphologies of NaYF$_4$:Yb,Er: (a) elongated micro-octadecahedra, (b) micro-octadecahedra, (c) nanoslices, (d) nanodiscs, (e) short nanorods, and (f) long nanorods. Source: Fu et al. [63] / Reproduced with permission from Wiley-VCH.

core and shell structure but different rare earth elements, such as NaYF$_4$@NaGdF$_4$, NaGdF$_4$:Yb/Ca/Er@NaYF$_4$, and NaGdF$_4$:Nd@NaYF$_4$@NaGdF$_4$:Nd/Yb/Er, STEM in a high-angle annular dark-field (HAADF) imaging mode is a powerful method to distinguish the different layers of the core/shell structure [5].

4.2.5.3 Characterization of Surface Moieties

Understanding the surface moieties of the UCNPs is important for their synthesis, luminescent properties, and functional applications. Moieties on the surface of UCNPs can be categorized into two classes: inorganic and organic species. Some instruments are often used to characterize the surface structure of UCNPs, such as FTIR, nuclear magnetic resonance (NMR), and thermogravimetric analysis (TGA). For example, Oleic acid has been proven to be a valuable ligand in the controlled synthesis of UCNPs. By using oleic acid as ligand in the synthesis of UCNPs, the products show FTIR absorption peaks at 2927 and 2857 cm^{-1} (stretching vibrations of –CH$_2$), 1560 and 1464 cm^{-1} (stretching vibrations of –COO$^-$), and 1705 cm^{-1} (stretching vibration of C=O). Surface coating with macromolecules like polyacrylic acid (PAA), PVP, and PEI on UCNPs can be identified by FTIR measurement. FTIR spectroscopy can also be used to identify inorganic coatings made of silicon oxide and carbon based on their characteristic absorptions (Si–O–Si: stretching vibration at 1100 cm^{-1}; O–H: stretching vibration at 3320 cm^{-1}; C–OH: stretching vibration at 1000–1300 cm^{-1}). In some circumstances, IR analysis can be used to monitor the revolution of the reaction solution. Wu and coworkers demonstrated using FTIR spectra that the amidation reaction between oleic acid and octadecylamine occurred before the nucleation of UCNPs, and the amidation product on the surface affected the luminescent performance of NaYF$_4$:Yb/Er UCNPs [65]. NMR technology has

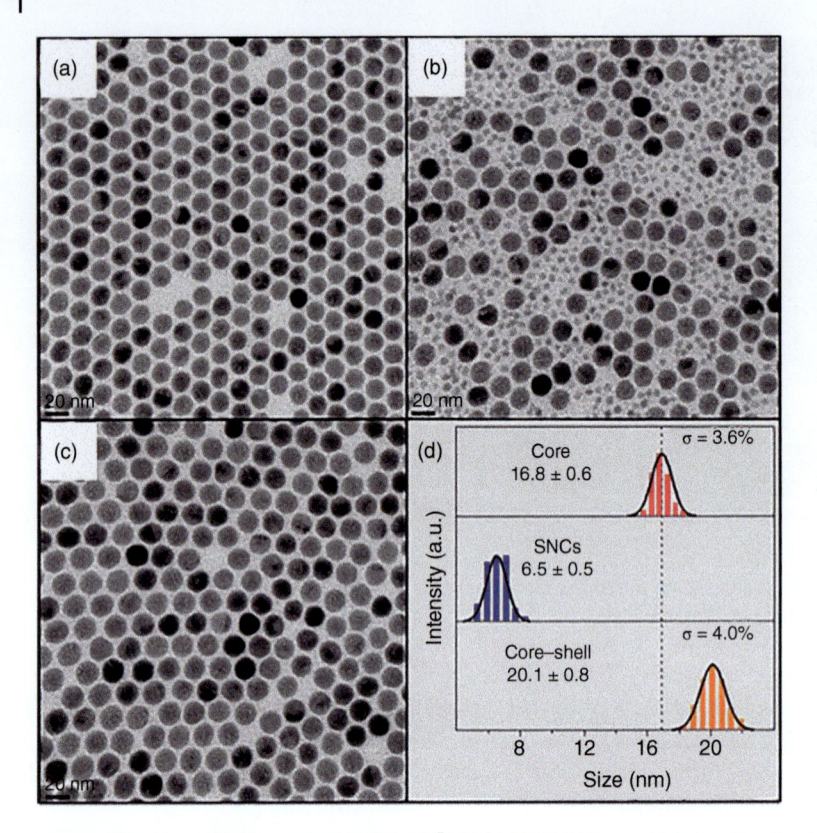

Figure 4.7 (a–c) TEM images of NaYF$_4$:Yb^{3+}/Er^{3+} (15%/2%) core NCs ($t = 0$), after injection of sacrificial α-NaYF4 NCs, and after selffocusing NaYF$_4$:Yb^{3+}/Er^{3+} (15%/2%) core/NaYF$_4$ shell NCs, respectively, and (d) size distribution of the NCs. Source: Jhonson et al. [64]/ Reproduced with permission from American Chemical Society.

become another powerful tool for characterizing organic compounds on inorganic nanoparticles. For example, the presence of the oleic acid ligand on the surface of the NaYF$_4$ nanocrystals was confirmed by the ^{1}H NMR [35]. The broaden line and shifted signals in ^{1}H NMR are two representative characteristics of the interaction between ligands and UCNPs, indicating the ligands were anchored on the surface of UCNPs [66]. NMR can also monitor the interaction between ligands to address the "cooperative effect" of carboxylic acid/amine on the size, shape, and color output of UCNPs [65].

4.2.5.4 Composition Determination

Determining the material composition helps in understanding the properties of materials. For UCNPs, it is important to determine the type and concentration of doped rare earth ions (sensitizer or activator) as it impacts luminescent properties. The common methods of composition determination include energy dispersive spectrometer (EDS), electron energy loss spectroscopy (EELS), XPS, ICP-MS, and inductively coupled plasma-atomic emission spectrometry (ICP-AES). EDS is the most used and convenient method. Its probe is always equipped in SEM or TEM.

EDS in line-scanning mode is useful for characterizing core/shell UCNPs. Veggel and coworkers [67] reported that the EDS line scan across a single NaYF$_4$/NaGdF$_4$ particle show that Y is located in the core of the particle and Gd is in the shell, verifying its core/shell structure. The 2D scanning of EDS can fully estimate the distribution of the rare earth elements in the nanostructure, helping to characterize the heterostructure of UCNPs and clarify their crystal growth mechanism [68, 69]. Like EDS, EELS in 1D or 2D scanning mode can also provide solid evidence for the formation of core/shell UCNPs [21, 70]. XPS is ideal for detecting subtle changes on the surface of UCNPs due to its high sensitivity. For example, the replacement of Y^{3+} by a small amount of Gd^{3+} in NaYF$_4$ through cation exchange can be confirmed by XPS measurement [71]. The EDS, EELS, and XPS can only be used for micro-domain element analysis, while ICP-MS and ICP-AES can determine the average elemental content of many UCNPs, which have been used in many researches [61].

4.2.5.5 Measurement of Optical Properties

Optical characteristic is the most fascinating property of UCNPs. Its unique anti-Stokes luminescence allows UCNPs to have broad application prospects. Therefore, the measurement of optical properties is important. The 980 nm laser is used as an excitation source for UCNPs because it matches well with the absorption of the sensitizer Yb^{3+} ions. The excitation source can be expanded to 800 nm when Nd^{3+} ions are used as sensitizer. The measurement of the upconversion emission spectra can be achieved by a commercial spectrometer coupled with an external laser source. The emission spectrum of the well-known NaYF$_4$:Yb/Er UCNPs shows strong emission around 540 and 654 nm. The absolute quantum yield (QY) of UCNPs is the conversion efficiency of NIR photons into visible and ultraviolet light, requiring integrating spheres and extended spectrometer. The absolute quantum yield of UCNPs is generally low, ranging from 0.0022% to 1.2% for AREF$_4$ UCNPs without core/shell structure and increased up to 7.6% in core/shell structure. It has been reported in 2018 that the quantum yield of Eu, Sm, Mn-doped CaS UCNPs was up to nearly 60% [72]. However, it is difficult to compare the quantum yield from different literatures. The quantum yield is dependent on experimental conditions, such as the laser power of the excitation and the photon detector. The study of the emission dynamics of UCNPs requires a spectrometer and a pulsed laser. Upconversion lifetime, which is related to the size, phase, structure, and doping concentration of UCNPs, can be obtained from the decay curve. For example, Jin and coworkers found that the lifetime of NaYF:Yb/Er UCNPs reduced with the decrease of particle size from 45 to 6 nm [73]. Zhao and coworkers demonstrated that the lifetime of Er^{3+} ion at the $^4S_{3/2}$ state increased from 287 to 762 μm with the increase of the shell thickness [74]. Furthermore, researchers can investigate the imaging of UCNPs *in vivo* or *in vitro* by using a confocal microscope. For small animal imaging, a detailed discussion of the customized instrumental setup was proposed by Li's group [75].

4.2.5.6 Evaluation of Magnetic Properties

UCNPs have useful optical and magnetic properties, which can be used as probes for multimode imaging. The magnetic properties of UCNPs mainly come from the unpaired 4f electrons in rare earth ions, such as Gd^{3+}, Nd^{3+}, Dy^{3+}, and Ho^{3+}. These properties are characterized using commercial instruments such as the Superconducting Quantum Interference Device (SQUID) Magnetometer or the Vibrating Sample Magnetometer (VSM). The magnetization and magnetic mass susceptibility of UCNPs are sensitive to the amount of magnetic rare earth ions in the host. In 2009, Prasad and coworkers reported the synthesis and biological application of water-soluble $NaYF_4$:RE/Gd (RE = Er, Yb or Eu) UCNPs, combining both optical and magnetic resonance (MR) imaging modalities [76]. The ability of UCNPs as magnetic contrast agents in magnetic resonance is usually examined on commercially available MR scanner like 3T Siemens Magnetom Trio [77]. UCNPs such as rare earth-doped $NaGdF_4$ and $Gd^{3+}/Yb^{3+}/Er^{3+}$ co-doped $NaYF_4$ have been extensively investigated as MRI probes [77–79].

4.2.6 Tuning of Upconversion Emission

The UC emission of UCNPs has been tuned from ultraviolet to infrared through various strategies, which is important in realizing the application of UCNPs in biomarking, medical diagnosis, biosensing, and bioanalysis. However, it is a challenge to develop a general protocol to precisely tune the color output of UCNPs over a wide spectral range and maintain the high efficiency of UC emission. The following describes various strategies for manipulating UC emission of UCNPs [2, 6].

4.2.6.1 Tuning UC Emission Changing by the Chemical Composition and Varying Dopant Concentration

Lanthanide ions have rich energy-level structures. Different combinations of lanthanide ions can cover a spectrum, from ultraviolet (UV) to near-infrared (NIR). The UC emission tuning of UCNPs can be achieved by adjusting the different doping combinations and concentrations of the lanthanide ions. Tm^{3+}, Er^{3+}, and Ho^{3+} are the most common doping activators in UCNPs, which can produce blue, green, or red emission after being sensitized by Yb^{3+}.

Haase and coworkers used Yb^{3+}/Er^{3+} and Yb^{3+}/Tm^{3+} co-doped $NaYF_4$ nanoparticles to produce strong yellow and blue emission, which was the first report of high-efficiency multicolor UC emission of lanthanide-doped UCNPs [15]. Wang and Liu reported a general method of tuning the UC emission of $NaYF_4$ nanoparticles from visible light to NIR light by controlling the doping concentration and different combinations of activators (Figure 4.8) [80]. The luminescent color output of YF_3 nanoparticles doped with different lanthanide-doped ions and their combinations could be tuned by manipulating the concentration of the sensitizer Yb^{3+} [81]. In addition, the simultaneous doping of Er^{3+} for green and red emissions and Tm^{3+} for blue emission could produce white UC emission in UCNPs. With the wide emission

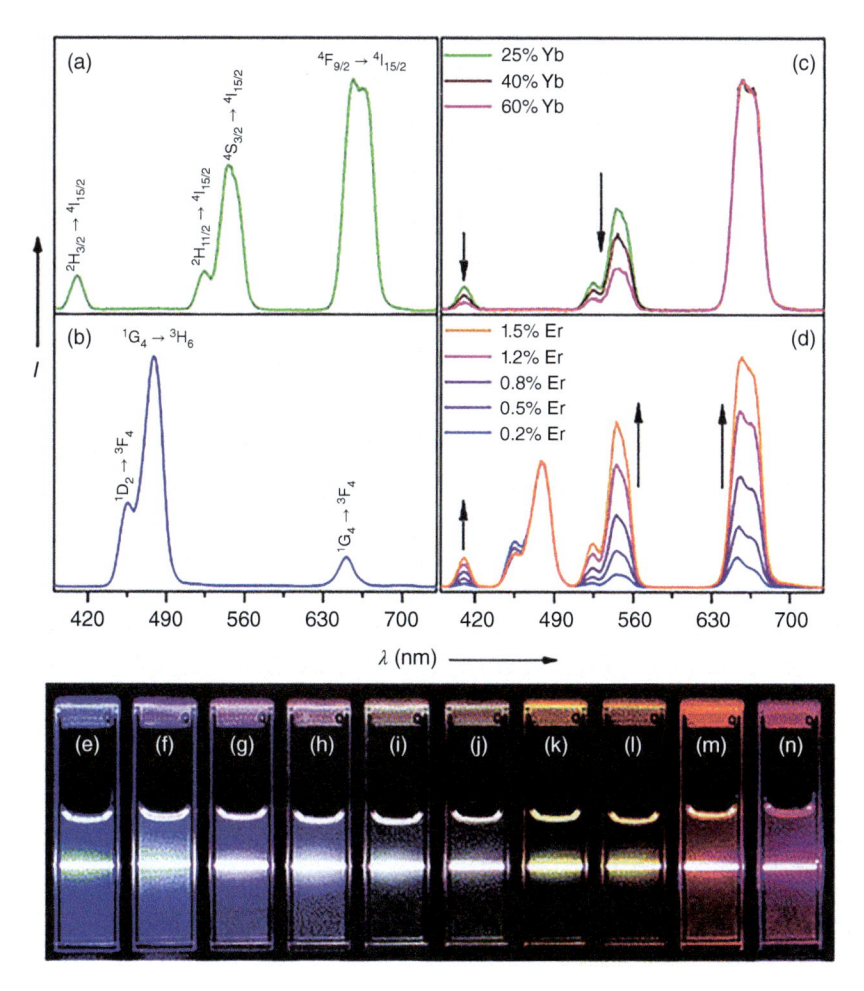

Figure 4.8 Upconversion emission spectra of (a) NaYF$_4$:Yb/Er (18/2 mol%), (b) NaYF$_4$:Yb/Tm (20/0.2 mol%), (c) NaYF$_4$:Yb/Er (25–60/2 mol%), and (d) NaYF$_4$:Yb/Tm/Er (20/0.2/0.2–1.5 mol%) particles in ethanol solutions (10 mM). Compiled luminescent photos showing corresponding colloidal solutions of (e) NaYF$_4$:Yb/Tm (20/0.2 mol%), (f–j) NaYF$_4$:Yb/Tm/Er (20/0.2/0.2–1.5 mol%), and (k–n) NaYF$_4$:Yb/Er (18–60/2 mol%). The samples were excited at 980 nm with a 600 mW diode laser. Source: Wang and Liu [80] / Reproduced with permission from American Chemical Society.

wavelength range of lanthanide dopants, specific lanthanide dopants or their combinations can produce a wide range of UC emission spectra in the UV to NIR region, which is beneficial for multiple markings and coding.

4.2.6.2 Tuning UC Emission by Host Matrix Screening

The host matrix also plays a vital role in determining the luminescence characteristics of the UCNPs. The excitation energy of the doped ions can interact with the host material through lattice vibration.

The change of the matrix will lead to the change of the phonon energy, which results to different upconversion energy transfer pathways. For example, the UC emission of ZrO_2:Er^{3+} in the monoclinic phase was higher than that of the tetragonal phase. Even with similar Er^{3+} energy levels, it was found that the UC emission intensity of Er^{3+}-doped Lu_2O_3 was higher than of Er^{3+}-doped Y_2O_3. It could be attributed to the better mixing of 4f and 5d orbitals in the monoclinic structure rather than other structures [82]. In contrast to oxide materials, fluoride matrix materials have low phonon energy and are often used to achieve high-efficiency UC emission. For example, Zeng et al. reported Mn^{2+} and Er^{3+} co-doped $KMnF_3$ UCNPs [83] where they observed that the strong interaction between different lanthanide ions in the $KMnF_3$ host NPs resulted in a decrease in the excitation energy of the $^4S_{3/2}$ energy level of Er^{3+}. Due to the low phonon energy of the host and the energy transfer from Mn^{2+} to Er^{3+} ions, this strong interaction led to an increase in the population at the $^4F_{9/2}$ energy level of Er^{3+}. This design enhanced UC emission.

4.2.6.3 Tuning UC Emission by Interparticle Energy Transfer or Antenna Effect

Sarkar et al. have reported the first observation of UC emission through interparticle energy transfer (IPET) [84]. The colloidal mixture containing 5 nm $BaLuF_5$:Tm^{3+} and $BaLuF_5$:Yb^{3+} nanoparticles showed strong blue UC emission when excited at 980 nm. In contrast, the solutions containing only $BaLuF_5$:Tm^{3+} or $BaLuF_5$:Yb^{3+} nanoparticles did not emit light, which showed the energy transfer from Yb^{3+}-doped nanoparticles to Tm^{3+}-doped nanoparticles. The result proved important as a series of color outputs can be produced using similar mechanisms.

Some limitations of UCNPs in applications is their extremely low absorption cross section and narrow absorption band in the NIR region. Zou et al. developed the first sensitization system where the IR-806 dye acted as an antenna to efficiently absorb NIR photons at 800 nm. The excitation energy was transferred to the $^2F_{5/2}$ energy level of Yb^{3+} ions through a resonance energy transfer process (Figure 4.9a) [85]. The emission of dye-sensitized β-$NaYF_4$:Yb/Er nanoparticles was significantly increased by about 3300 times than that of non-sensitized nanoparticles due to the increase of optical cross section and absorption bandwidth. Wisser et al. reported that dye sensitization improved the UC quantum yield of UCNPs with different sizes [86]. The smallest particle (10.9 nm in diameter) was enhanced up to 10 times (Figure 4.9b). Enhancing UC performance by sensitizing emission with fluorescent dyes emphasizes the importance of enhancing the radiation rate in lanthanide-doped UCNPs, which can introduce UCNP composites to new structures. Although there was no report of multicolor UC emission in these works, the antenna mechanism can be extended to other lanthanide ions or combinations to produce efficient multicolor emission.

4.2.6.4 Tuning UC Emission Through Energy Migration

Efficient UC emission is generally limited to the lanthanide activators (Er^{3+}, Tm^{3+}, and Ho^{3+} ions) when excited at 980 nm. However, Liu and coworkers reported that lanthanide activators (Eu^{3+}, Tb^{3+}, Dy^{3+}, and Sm^{3+}) without

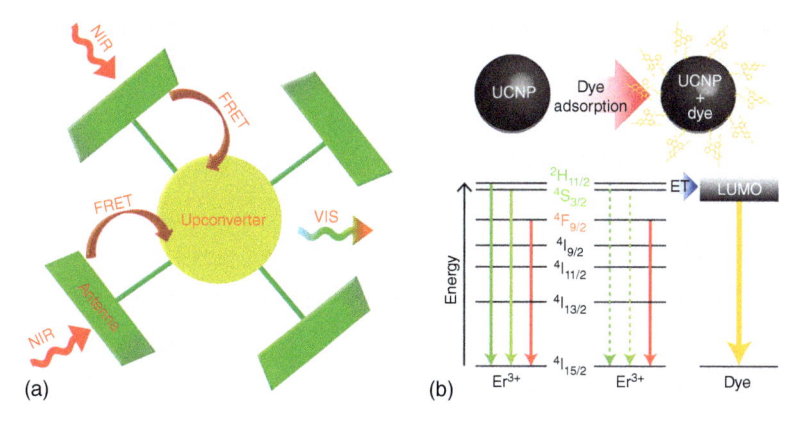

Figure 4.9 (a) Principal concept of the dye-sensitized nanoparticle. Source: Zou et al. [85]. Copyright 2012, Springer Nature. (b) Schematic summarizing the operating concept. UCNPs are decorated with dyes. The LUMO of this dye is aligned with the $^2H_{11/2}$ and $^4S_{3/2}$ states in Er^{3+}, enabling efficient energy transfer to the dye molecules. Source: Wisser et al. [86]. Copyright 2018, American Chemical Society.

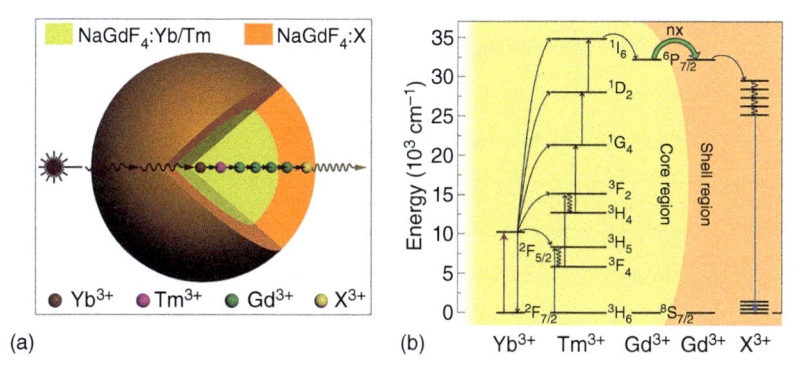

Figure 4.10 Tuning upconversion through energy migration in core–shell nanoparticles. (a) Schematic design of a lanthanide-doped NaGdF$_4$@NaGdF$_4$ core–shell nanoparticle for EMU (X: activator ion). (b) Proposed energy transfer mechanisms in the core–shell nanoparticle. Source: Wang et al. [21]. Copyright 2011, Springer Nature.

long-lived intermediate energy states could also perform efficient UC emission through an energy migration-mediated UC process (Figure 4.10) [21]. NaGdF$_4$:Yb^{3+},Tm^{3+}@NaGdF$_4$:X^{3+}(X = Eu, Tb, Dy or Sm) core/shell nanoparticles were prepared with lanthanide ions doped into the core/shell UCNPs. The sensitizer ion (Yb^{3+}) first transferred its excitation energy to the accumulator ion (Tm^{3+}) and excited it to a high-lying excited state. The resulting energy was transferred from the high-lying excited state of Tm^{3+} to the migrator ion (Gd^{3+}), followed by the energy migration through the core/shell interface via Gd^{3+} sublattice. Finally, the migration energy in the Gd^{3+} sublattice was captured by the activated ions (X^{3+}), resulting in UC emission.

4.2.6.5 Tuning UC Emission Using Cross-Relaxation Processes

The cross-relaxation (CR) process is a typical energy transfer where one ion transfers part of the excitation energy to another ion. The CR process is the result of ion–ion interaction, and its efficiency depends on the concentration of the dopants. The CR process can be utilized to modulate the color output in UCNPs by enhancing the emission of one excited energy level while suppressing the emission of another excited energy level.

Chen et al. reported that a high concentration Ce^{3+} (>10 mol%) was doped into $NaYF_4$:Yb^{3+}/Ho^{3+} UCNPs through the CR process between Ho^{3+} and Ce^{3+}, realizing the tuning of UC luminescence from green to red [87]. Capobianco and coworkers utilized CR between two Er^{3+} ions to reduce green UC emission while increasing red UC emission [88]. They also utilized the CR between Tm^{3+} and Dy^{3+} to suppress the high-energy UC emission of Tm^{3+} in the Yb^{3+}/Tm^{3+}/Dy^{3+} co-doped $LiYF_4$ colloidal UCNPs [89]. Other transition metal ions such as Mn^{2+} and Fe^{3+} are also used to manipulate UC emission.

4.2.6.6 Tuning UC Emission Using Core/Shell Structures

The use of core/shell structure with two or more lanthanide activators incorporated into the core and different shells is a strategy to regulate the multicolor luminescence of UCNPs. The core/shell structure spatially isolates the lanthanide activators and eliminates harmful CR between them. It also suppresses the surface-related quenching mechanism in core nanoparticles to yield a tunable emission range with high efficiency. At the same time, the inert shell layer grown on the luminescent core can suppress the nonradiative attenuation caused by surface defects, improving the luminescence efficiency of UCNPs [6].

$NaYF_4$:Yb^{3+}/Tm^{3+}@$NaYF_4$:Yb^{3+}/Er^{3+} UCNPs with Tm^{3+} and Er^{3+} doped separately in core and shell can be synthesized. Tunable UC emission from visible to NIR were achieved by adjusting the combination of Yb^{3+}, Tm^{3+}, and Er^{3+} at defined concentrations. A class of $NaYbF_4$:Nd^{3+}@$Na(Yb^{3+}, Gd^{3+})F_4$:X (Er^{3+}, Ho^{3+}, Tm^{3+})@$NaGdF_4$ core/shell/shell UCNPs for simultaneous tuning of excitation and emission can also be fabricated [70]. Zhang and coworkers [90] prepared a class of $NaGdF_4$:Yb^{3+}/Er^{3+}@$NaYF_4$:Yb^{3+}@$NaGdF_4$:Yb^{3+}/Nd^{3+}@$NaYF_4$@$NaGdF_4$:Yb^{3+}/Tm^{3+}@$NaYF_4$ UCNPs, which displayed blue and green/red emissions at 980 and 796 nm excitation, respectively (Figure 4.11).

The distribution of dopants in the core/shell can also be utilized to tune the emission color. For example, Li et al. altered the doping position of the dopants (core doping and shell doping) by adjusting the type and order of addition of the shell precursor [91]. As the doping site gradually shifts from the inner core to the outer shell, the intensity ratio of red-to-green emission decreased significantly.

4.2.6.7 Tuning UC Emission Using Size- and Shape-Induced Surface Effects

The size- and shape-related optical properties of UCNPs have been studied. The relative intensity of red and green gradually increased as the particle size decreased in Y_2O_3:1%Er^{3+}, 4%Yb^{3+} nanoparticles. The red-to-green luminescence intensity ratio were reported to increase as the particle size of $NaYF_4$:Yb/Er

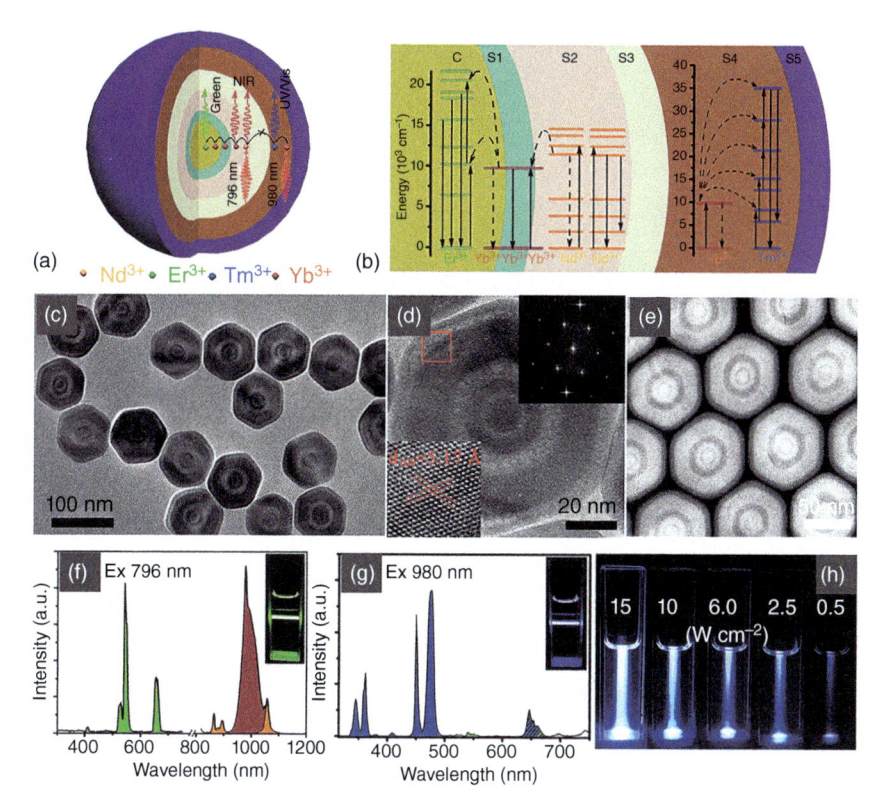

Figure 4.11 The structure (a) and energy transfer mechanisms (b), TEM (c), HRTEM and the corresponding FFT (d), HAADF-STEM (e) of the orthogonal excitations-emissions UCNPs (f–h). Source: Li et al. [90] / Reproduced with permission from John Wiley & Sons, Inc.

UCNPs decreased. It was attributed to the size-dependent surface quenching, which affected $^4I_{11/2} \to {}^4I_{13/2}$ and $^4S_{3/2} \to {}^4F_{9/2}$ relaxations of the Er^{3+}. Mai et al. also reported the size-dependent UC emission of $NaYF_4:Yb^{3+}/Er^{3+}$ nanoparticles dispersed in solution [92]. It was observed that as the size of the UCNPs decreases, the surface-induced (defects, ligands, solvents) effect becomes more prominent, producing more efficient multiphonon-assisted nonradiative relaxations to modify the relative population among close-lying excited states or to deactivate UC photoluminescence. Wang et al. investigated the UC emission of $NaGdF_4:Yb^{3+}/Tm^{3+}$ nanoparticles with different sizes (15, 20, and 30 nm) (Figure 4.12a,b) [93]. Although the particle size varied, the emission spectra were the same when coated with the inert shell $NaGdF_4$. This result provided the first evidence supporting surface-related effects and illustrated the size-dependent behavior of UC photoluminescence.

Li and Zhang and Ye et al. reported that $NaYF_4:Yb^{3+}/Er^{3+}/Tm^{3+}$ nanoplates, nanospheres, and nanoellipses or $NaYF_4:Yb^{3+}, Er^{3+}/Ho^{3+}$ nanorods, nanoplates, and nanoprisms exhibited different UC emission (Figure 4.12c–e) [18, 27]. However, detailed explanations were not provided in their works. They speculated that the different colors may be caused by the shape-related surface mechanism that

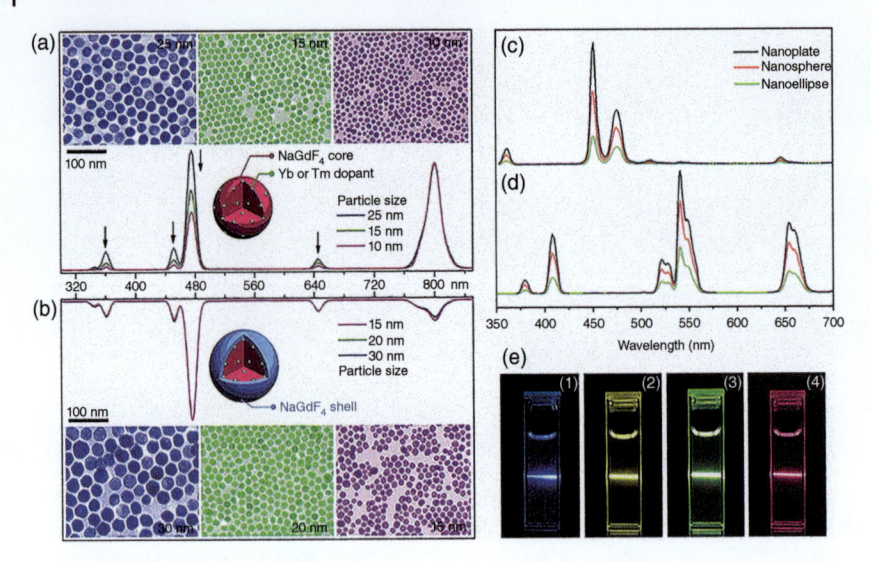

Figure 4.12 (a, b) UC emission of NaGdF$_4$:Yb^{3+}/Tm^{3+} nanoparticles with different sizes. Copyright 2010, John Wiley & Sons, Inc. UC emission of different NaYF$_4$:25%Yb, 0.3%Tm (c) and NaYF$_4$:18%Yb, 2%Er (d) and photographs of the nanospheres in hexane under excitation of 980 nm NIR laser. Source: [18, 93] / Reproduced with permission from IOP Publishing, Ltd., Reproduced with permission from John Wiley & Sons, Inc.

produces a different phonon effect on the lanthanide ions in UCNPs. In addition, the multicolor output of NaYF$_4$:Yb^{3+},Er^{3+} nanoparticles can also be affected by crystallinity, which could be controlled by changing the reaction temperature and time.

4.2.6.8 Tuning UC Emission Using FRET or RET

Fluorescence resonance energy transfer (FRET) is a mechanism that describes energy transfer from a donor to an acceptor through nonradiative dipole–dipole coupling. When using UCNPs as energy donors and organic dyes or quantum dots as acceptors, the nonradiative energy can be transferred through multipolar interactions. This process can change the energy distribution in some UCNPs energy levels, resulting in a large degree of freedom in the UC emission wavelength. The effect of this interaction depends on the distance between the two particles (normally less than 10 nm). It was reported that multicolor UC emission was obtained from core/shell NaYF$_4$:Yb^{3+},Er^{3+}/Tm^{3+} UCNPs coated with silica [94]. It was produced by FRET from UCNPs to organic dyes or quantum dots encapsulated in a silica shell. Xu et al. also achieved the UC emission tuning of Er^{3+} and Tm^{3+} in the composites of NaYF$_4$:Yb^{3+}, Er^{3+}/Tm^{3+}, and carbon dots via a similar strategy (Figure 4.13) [95]. Gold nanoparticles were also widely used as energy acceptors for FRET.

Radiative energy transfer (RET) is a process where acceptors absorb the light emitted by the donors. The difference between RET and FRET is that the energy transfer process of RET is radiative. Compared with nonradiative FRET, RET

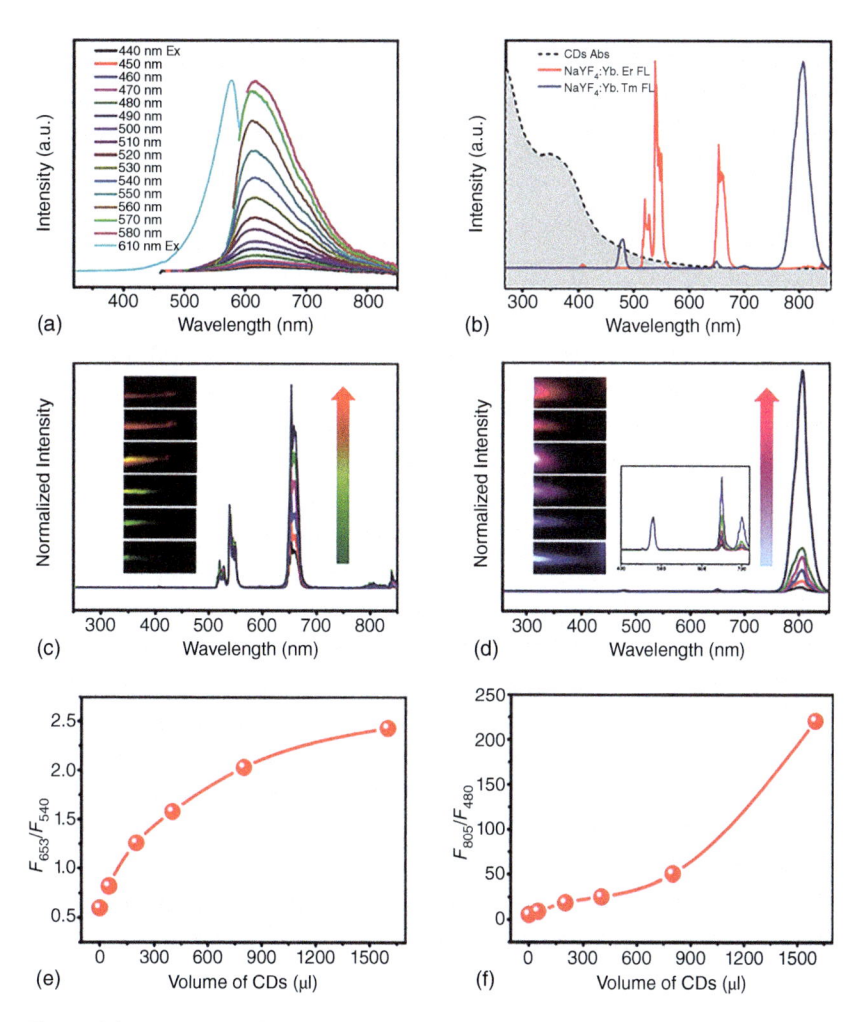

Figure 4.13 (a) PL and PLE spectra of the CD solution. (b) UV–vis absorption spectrum of CDs and UC emission spectra of NaYF$_4$:Yb,Er and NaYF$_4$:Yb,Tm under 980 nm excitation. Normalized UC emission spectra of different NaYF$_4$:Yb,Er@CDs (c) and NaYF$_4$:Yb,Tm@CDs (d) under 980 nm laser excitation. Insets are optical photos of solutions containing the different NaYF$_4$:Yb,Er@CDs and NaYF$_4$:Yb,Tm@CDs composites under 980 nm laser, respectively. The arrow indicates the increased CD amounts in the composites. Plots of the intensity ratio of red/green (e) and NIR/blue (f) vs. concentration of CDs. Source: Xu et al. [95] / Reproduced with permission from Royal Society of Chemistry.

breaks the distance limitation of energy transmission efficiency. Zheng et al. mixed LiYbF$_4$:Tm^{3+}@LiYF$_4$ and CsPbX$_3$ perovskite quantum dots in cyclohexane solution. LiYbF$_4$:Tm^{3+}@LiYF$_4$ with strong UV and blue emission was used to sensitize CsPbX$_3$ perovskite quantum dots with strong absorption in the UV and visible regions (Figure 4.14) [96]. Due to the limitations of the combination method, the distance between them was far, and RET would occur instead of FRET. The full-spectrum UC photoluminescence was achieved. This strategy can also be

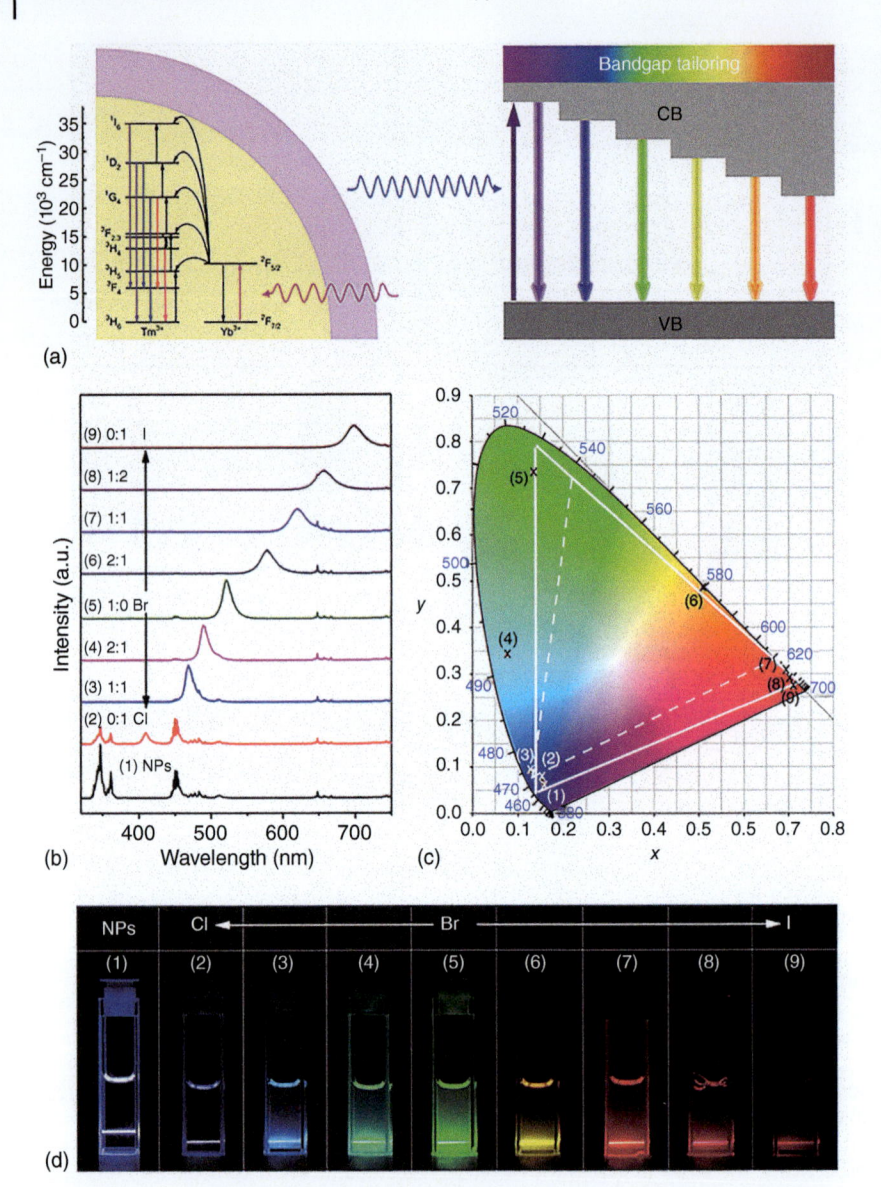

Figure 4.14 (a) Simplified energy-level scheme of LiYbF$_4$:0.5%Tm^{3+}@LiYF$_4$ core/shell NPs indicating major upconversion processes, and schematic illustration of full-color upconversion tuning in CsPbX$_3$ PeQDs through sensitization by the NPs. (b) UC emission spectra for LiYbF$_4$:0.5%Tm^{3+}@LiYF$_4$ core/shell NPs and the NP-sensitized CsPbX$_3$ PeQDs with varying halide compositions under 980 nm excitation. (c) Corresponding color gamut of the emission colors from the samples. (d) Photographs of samples under 980 nm illumination. Source: (a–c) Zheng et al. [96]/Reproduced with permission from Springer Nature.

adopted to load fluorescent and quenching molecules on UCNPs functionalized with amphiphilic polymers. Due to the RET process from UCNPs to organic dyes under NIR excitation, the formed supramolecular UCNPs-dye complex showed up to five different UC visible emission and were used for *in vivo* multicolor imaging.

4.2.6.9 Tuning Upconversion Emission Through External Stimulus

The strategies for regulating the luminescence of UCNPs require strict modification of reaction conditions or material compositions. For UCNPs with a fixed composition, it is feasible to tune the color of UC emission through external stimulus. For example, Deng et al. reported a non-steady-state UC technology to achieve full-color tuning [22]. In their research, the emission color of $Yb^{3+}/Ho^{3+}/Ce^{3+}$ co-doped core/shell nanoparticles could be dynamically tuned by manipulating the pulse excitation width of a 980 nm pulsed laser (Figure 4.15a). The emission color of the core/shell/shell UCNPs can also be tuned under different excitation power density, manipulating the photon transfer pathways. (Figure 4.15b) [97]. Chen et al. prepared $NaYbF_4:Ho^{3+}$ nanoparticles that are sensitive to the excitation power. They found that by increasing the excitation power density, the UCNPs showed the red–green emission intensity ratio from 0.37 to 5.19 [98].

Other methods such as altering the excitation temperature, polarization anisotropic modulation, electric field, and magnetic field manipulation, applying of mechanical stress, and altering the pH environment can achieve remote dynamic modulation of UC emission.

4.2.7 Applications of UCNPs

With the development of well-controlled UCNPs and their advantageous optical properties, researches on Ln-doped upconversion nanoparticles have made progress, allowing UCNPs in many applications [5, 10, 99].

4.2.7.1 Bioimaging

Compared with traditional optical probes, UCNPs have useful optical properties, such as high sensitivity, resistance to photobleaching, and light scintillation, beneficial to biological imaging and therapeutic applications. UCNPs are excited by NIR radiation and transparent to biological tissues. It minimizes background fluorescence and provides a good signal-to-noise ratio. It also allows deep tissue penetration while avoiding phototoxic effects on biological specimens.

Many studies proved the feasibility of UCNPs for *in vitro* bioimaging. Spatial and temporal distributions of HeLa, breast carcinoma and glioblastoma cells, colon cancer cells, myoblasts, and ovarian cancer cells have been demonstrated [68, 100–106]. Chatterjee et al. developed an upconversion fluorophore consisting of $NaYF_4$:Yb, Er nanoparticles and PEI coating. They reported on the imaging tissues in small mammals using UCNPs [104]. They claimed that the fluorescence emission from PEI/$NaYF_4$:Yb, Er fluorophores could be detected from a depth of 10 mm after the fluorescent particles were injected into tissues deep in the body of a Wistar rat.

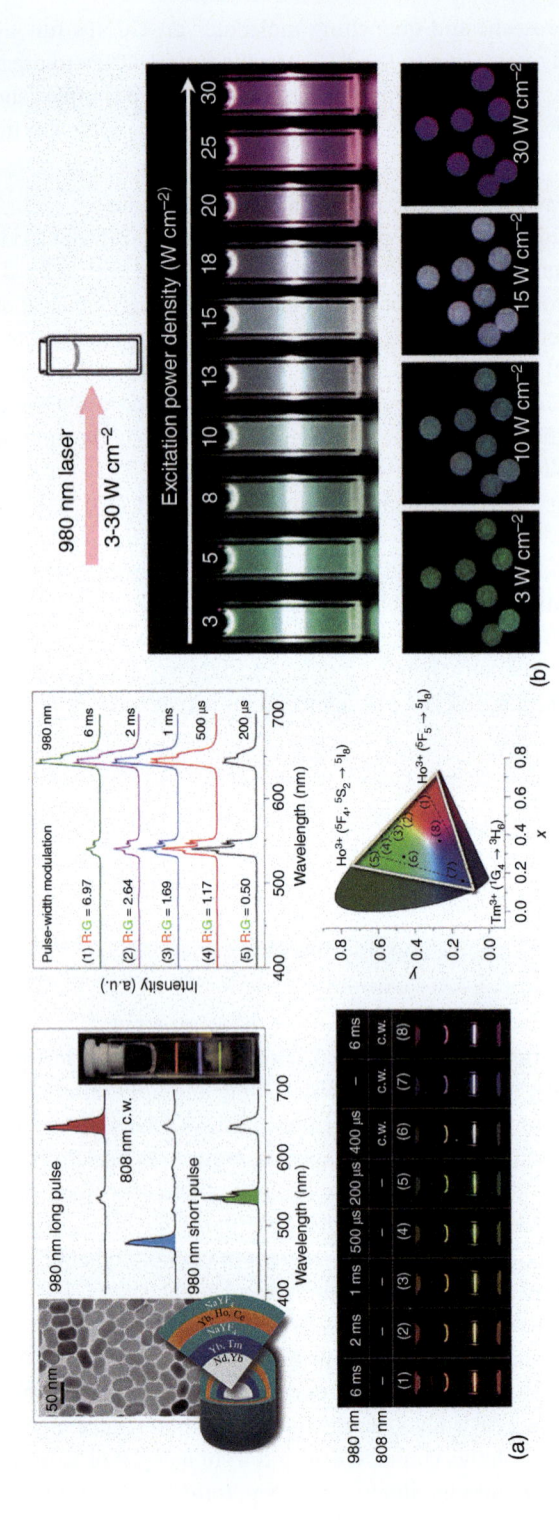

Figure 4.15 (a) Temporal multicolor tuning in NaYF$_4$-based core–shell nanocrystals. Copyright 2015, Springer Nature. (b) the emission tuning of UCNPs under different excitation power density. Source: Refs. [22, 97] / Reproduced with permission from Springer Nature, Reproduced with permission from John Wiley & Sons, Inc.

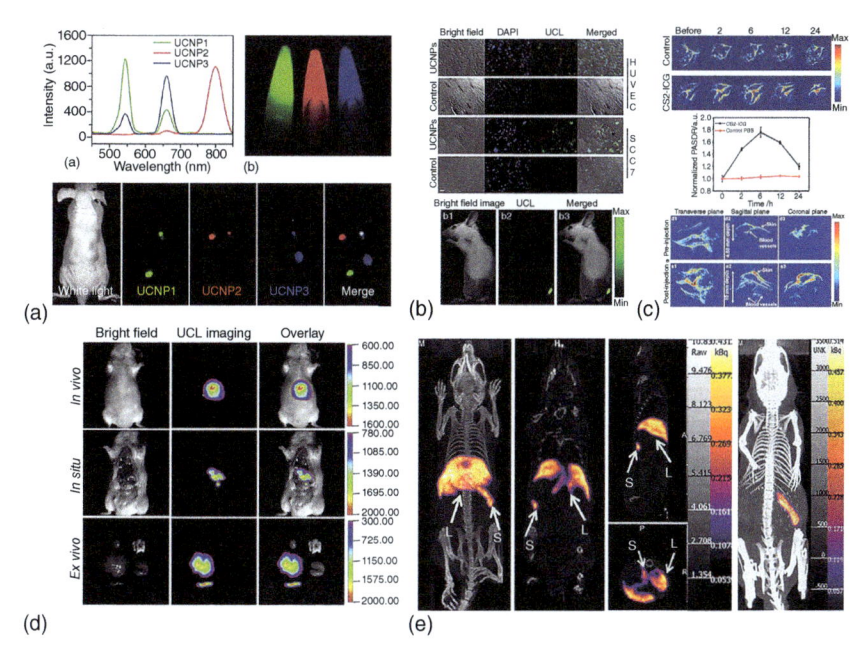

Figure 4.16 (a) Multicolor UCL imaging of three UCNPs solutions and a nude mouse subcutaneously injected with different UCNP solutions. Source: Cheng et al. [109] / Reproduced with permission from Springer Nature. (b) UCL imaging of HUVECs, SCC7 cells and mouse after incubation with UCNPs. (c) PAI of tumor blood vessels in SCC7 tumor-bearing mice with intravenous injection of PBS (up) and nanoformulation (bottom). Source: Liu et al. [110]/Reproduced with permission from John Wiley & Sons, Inc. (d) *In vivo*, *in situ* and *ex vivo* imaging of the Kunming mouse tail vein injection of the Sm-UCNPs (e) *In vivo* SPECT images after intravenous injection of Sm-UCNPs. Source: Yang et al. [111]/Reproduced with permission from John Wiley & Sons, Inc.

Multicolor imaging in biological systems can be achieved by synthesizing emitted UCNPs at different wavelengths [66, 107, 108]. As shown in Figure 4.16a, Cheng et al. synthesized three types of UCNPs with polyethylene glycol (PEG) functionalized by varying the rare earth-doping concentrations for blue, green, and red fluorescence emission [109]. The three types of UCNPs can be differentiated by observing their emission light upon excitation by a 980 nm laser. The UCNPs can be used for multicolor *in vivo* imaging, labeling, and tracking and applied in multiplexed *in vivo* lymph node mapping by subcutaneous injection.

Single imaging modalities are flawed. Recently, there were efforts to develop multimodal bioimaging methods that combine various imaging modalities into a single nanosystem, offering synergistic advantages [112]. UCNPs can enable multimodal bioimaging as they can be tailored and engineered as functional probes for magnetism, X-ray attenuation, radioactivity, etc. [113].

He et al. developed a new type of multifunctional cancer diagnosis and treatment platform, which is composed of β-NaGdF$_4$:Yb/Er@β-NaGdF$_4$:Yb@β-NaNdF$_4$:Yb@MS-Au$_{25}$-PEG for simultaneous PT/PA/UCL/MR/computed tomography (CT) bioimaging [114]. In this system, the Au$_{25}$ shell exhibits considerable photothermal and photodynamic effects, while the Gd^{3+} and Yb^{3+} ions in UCNPs have MR and CT imaging functions. In addition to multimode imaging, the platform can also inhibit tumor growth *in vivo* under 808 nm light excitation due to photothermal and photodynamic effects. It is a promising tool for realizing image-guided cancer treatments. Liu et al. designed $NaYF_4$:Yb/Er@$NaYF_4$:Yb@$NaNdF_4$:Yb@$NaYF_4$@$NaGdF_4$ UCNPs, which can provide photoacoustic (PA), fluorescence, and MR imaging functions (shown in Figure 4.16b,c) [110]. Positron emission tomography (PET) technique requires a radioactive isotope that emits positrons, which emit gamma rays for imaging when positrons encounter electrons. Li synthesized ^{18}F-labeled UCNPs through an inorganic reaction between rare earth ions and F-ions. Studies have shown that ^{18}F-labeled UCNP can be used for PET imaging and lymphatic monitoring [115]. Li and coworkers also introduced a small amount of radioactive $^{153}Sm^{3+}$ ions to prepare a UC emission and single photon emission computed tomography (SPECT) dual-modal biological imaging system composed of NaLuF4:^{153}Sm, Yb, and Tm (Figure 4.16d,e) [111].

4.2.7.2 Therapy

Photodynamic therapy (PDT) is a type of light therapy for treating cancer cells. When photosensitizers are exposed to specific wavelengths of light, they generate reactive oxygen species (ROS) and cause oxidative damage to surrounding cells. NIR light has an advantage over UV and visible light in penetrating human tissues, making UCNPs an ideal candidate for PDT applications in deep tissue cancer therapy (Figure 4.17a) [116].

Hematoporphyrin and silica phthalocyanine dihydroxide molecularly loaded $NaGdF_4$:Yb,Er@CaF_2 core/shell UCNPs exhibited excellent PDT efficiency against HeLa cancer cell damage under NIR irradiation [118]. Qian et al. constructed a dual-function $NaYF_4$:Yb,Er mesoporous SiO_2 core/shell UCNPs doped with ZnPc photosensitizer, which can be used for fluorescence imaging and PDT processing [119]. After excitation at 980 nm, the red emission of $NaYF_4$:Yb, Er UCNPs was transferred to the ZnPc photosensitizer and triggered the PDT process. The large surface area and porous structure of the mesoporous SiO_2 coated shells allow for the storage of more ZnPc photosensitizer in the shell and facilitate the release of generated ROS. Meanwhile, the wrapping also avoids direct contact between $NaYF_4$:Yb,Er nanoparticles and the environment, improving the stability of UCNPs. Idris et al. developed a method to activate dual photosensitizers using UCNPs. They applied it in enhanced PDT *in vivo* therapy [120]. Two photosensitizers, merocyanine 540 (MC540) and zinc (II) phthalocyanine (ZnPc), were loaded into the $NaYF_4$:Yb, Er@mesoporous SiO_2 composite and were simultaneously activated by the red and green emission from the NaYF4:Yb, Er UCNPs upon 980 nm laser excitation. *In vivo* studies have also shown that tumor growth in mice injected with UNCPs was inhibited after 980 nm NIR laser irradiation, which is promising for future non-invasive deep cancer treatment.

Chemotherapy is a common and effective method for treating cancer. However, it has its limitations, such as poor targeting efficiency, drug dependence, and drug resistance. The combination of luminescence and drug delivery has therapeutic potential. It enhances drug targeting by carrying the drug and delivering it to specific sites through luminescence tracking. It also allows detecting of drug delivery efficiency. Recently, UNCPs have been investigated as drug carriers because of their unique physicochemical properties. Their size, UC emission, and surface chemistry can be tailored to suit *in vitro* and *in vivo* drug delivery applications.

There are three strategies in designing a UCNPs-based drug delivery system: (i) UCNPs capped with polymer, (ii) UCNPs decorated with mesoporous SiO_2, and (iii) UCNPs encapsulated by hollow mesoporous-coated spheres. UCNPs were first functionalized with an amphiphilic polymer grafted with PEG. The anticancer drug adriamycin (DOX) was loaded by physisorption through hydrophobic interactions [117]. It was found that increasing the rate of drug dissociation in a weakly acidic environment to control the release of DOX allows for effective drug release in tumor cells (Figure 4.17b). Zhang et al. developed another multifunctional nanocomposites composed of UCNPs and thermo/pH-sensitive polymer (P(NIPAm-*co*-MAA))-gated mesoporous silica shell. The UCNPs were utilized as optical nanoprobes and the (P(NIPAm-*co*-MAA)) brushes were used as the "valve" to control the diffusion of embedded drugs through the porous structure of the silica

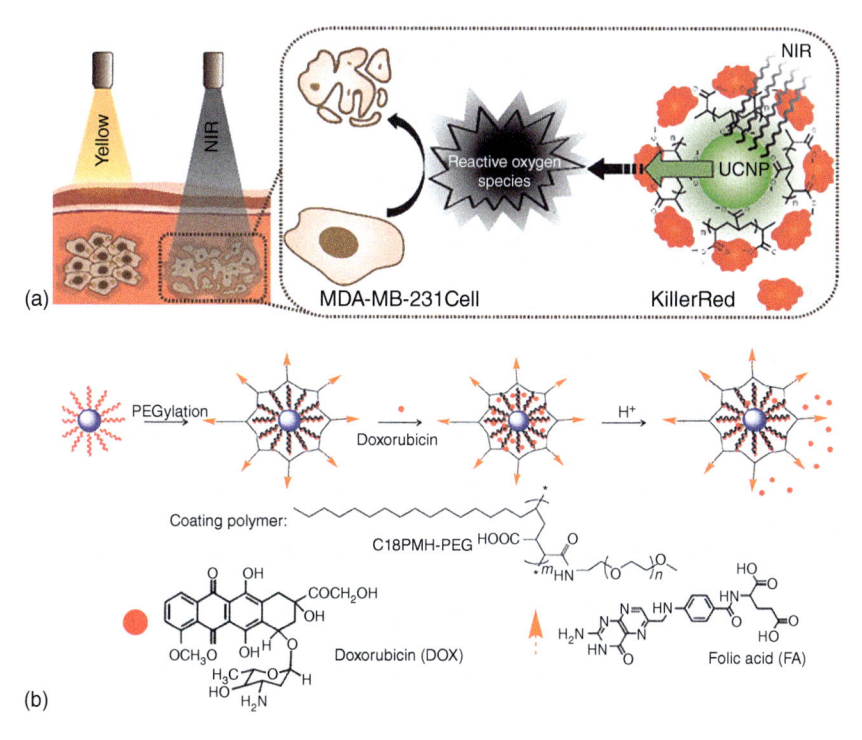

Figure 4.17 (a) A schematic illustration of KillerRed-UCNPs assembly and their application in the centimeter-deep PDT. Source: Liang et al. [116]. Copyright 2017, Elsevier. (b) Schematic illustration of the UCNPs-based drug delivery system. DOX molecules were physically adsorbed into the oleic acid layer on the nanoparticle surface by hydrophobic and released from UCNPs triggered by decreasing pH. Source: Reproduced with permission from Wang et al. [117]. Copyright 2011, Elsevier.

shell [121]. Due to its excellent thermo/pH sensitivity, the synthesized drug delivery system showed low levels of anticancer drug release at low temperature/high pH and increased drug release at high-temperature/low pH, which is essential for temperature/pH modulation of "on-off" drug release.

Radiotherapy (RT) is a noninvasive clinical treatment that uses radiation to damage cancer cells and slow the growth or spread of tumors in the body. At present, radiation therapy has been used in about 50% of cancer treatments. It suffers from two main challenges: (i) the inability to accurately identify the location of the tumor; and (ii) the radio-resistance of the cancer cells due to the complex microenvironment in the tumor area. It is known that elements with high atomic number (Z) can strongly absorb ionizing radiation. Therefore, if high-Z substances are delivered to tumor cells, they can selectively increase the radiation dose to the tumor compared to the healthy cells and result in a higher dose to the infected area. Shi and coworkers reported a universal mesoporous silica nanothermal material based on UCNPs, which can be used for radiotherapy and photodynamic therapy in hematoporphyrin (HP) sensitizer, and chemically in docetaxel (Dtxl) therapy and radiation therapy [122]. In this nanothermodynamic system, Gd-doped UCNPs are used as a contrast agent in magnetic/UC luminescence imaging. HP and Dtxl are used for X-ray irradiation and NIR excitation for coordinated chemistry/radiation/photodynamic therapy. With the synergy of the three treatment methods, it is proven that the developed nano-treatment system can completely prevent DNA repair and permanently destroy cancer cells.

Many studies have focused on UCNPs-based therapeutic systems for various therapeutic applications. Most of them can respond to various exogenous (light, temperature, magnetic fields, etc.) or endogenous stimuli (pH, enzyme concentration, etc.), selectively targeting diseased cells without affecting normal cells. However, this strategy requires the treatment platform to recognize certain tumor biomarkers and respond with specific physicochemical reactions, such as protonation, surface chemical changes, and encapsulant decomposition. These reactions make the treatment complicated and restrict targeting efficiency. In this case, the development of simpler, more effective UCNPs-based nanocarriers is a better option for cancer treatment [99].

4.2.7.3 Optogenetics

Optogenetics is a revolutionary biological technology that can precisely control the activity and function of neurons through light. Current optogenetics strategy uses visible light (470/530 nm) to stimulate light-responsive Channelrhodopsin-2 (ChR2) and halhohodopsin (NpHR) molecules. However, the visible light in deep tissues that activates the target neurons is a technical challenge for optogenetics research [5]. Compared with visible light, NIR light has a greater depth of tissue penetration. By using the light conversion effect of UCNPs, a new type of UC-mediated optogenetic system has been developed.

Hososhima first demonstrated the concept of UC-mediated NIR optogenetics in 2015 [123]. In their research, $NaYF_4$:Sc/Yb/Er(Tm) UCNPs were mixed with collagen to form a neuronal culture substrate. When a specially designed substrate

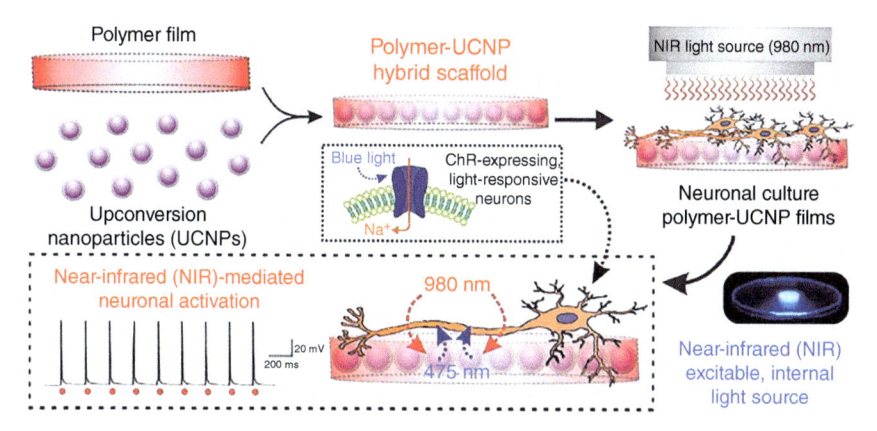

Figure 4.18 Rational design of polymer–UCNPs hybrid scaffolds for optogenetic neuronal stimulation. The UCNPs served as NIR to blue transducer thus facilitating optogenetic activation of channelrhodopsin (ChR)-expressing neurons upon NIR irradiation. Source: Shah et al. [124]/Reproduced with permission from Royal Society of Chemistry.

is irradiated with a 975 nm laser, the UC emission of NaYF$_4$:Sc/Yb/Er(Tm) particles will activate the neurons to induce neuronal responses. The photocurrent was observed in the corresponding expression cells. Shah et al. used the hybrid scaffolds containing NaYF$_4$:Yb/Tm@NaYF$_4$ UCNPs and poly(lactic-co-glycolic acid) as substrate film and investigated the optogenetic control of neuronal activity upon NIR radiation as shown in Figure 4.18 [124]. Although the high excitation power density required in their experiment may cause overheating problems and erroneous photocurrent signals, the research introduced an avenue for controlling neuronal activity in deep tissue regions. Shi and coworkers also have several studies on UC-mediated optogenetics. They first verified the feasibility of UC-mediated optogenetic modulation of targeted neural circuits in living rodents [125]. Subsequently, they developed an all-optical method for flexible wireless neural activity manipulation in freely moving animals, including activation and inhibition [126, 127]. Their research shows that UNCPs can be used for transcranial and deep brain stimulation in awake, freely moving rodents, which is essential for advanced neurological applications. Chen et al. developed an effective method to stimulate labeled neurons in deep regions of the brain [128]. UCNPs were injected into the ventral tegmental area (VTA) of the mouse brain. With NIR irradiation, the green or blue luminescence emitted by the UCNPs could stimulate or inhibit neuronal activity. Using this approach, researchers can stimulate the release of dopamine by activating neurons in the brain, inhibit neuronal activity in the hippocampus, and trigger memory recall [129]. This novel strategy provides a solution for clinical treatment of brain disorders such as Parkinson's disease and epilepsy.

4.2.7.4 Sensing and Detection

FRET has been used as a valuable tool for understanding different biological and physicochemical processes such as time distribution of biomolecules, molecular interactions, protein–protein interactions, etc. Several FRET systems are used in combination with UCNPs. This is the UCNPs-FRET platform which has been developed and applied to sense or detect various analytes, such as pH, temperature, gas, metal ions, DNA, etc.

An interesting example of an UC-based detection platform is the pH sensor, where changes in pH can be measured by proportional changes in the UC emission in different bands [130, 131]. By using this approach, Schäferling and colleagues performed intracellular pH sensing [132]. In their study, the pH indicator (pHrodo) bound to the nanoparticle surface was sensitized by energy transfer from UCNPs. Changes in the pH were detected by comparing the emission intensity of UCNPs (550 nm) and pH-sensitive pHrodo dye (590 nm). These probes provide a suitable solution for detecting changes in pH *in vitro* and *in vivo* models with high biological background fluorescence.

The UC-based nanothermometer can measure local temperature through the ratio change in different emission bands [133–137]. Capobianco and coworkers used an optical method based on UCNPs to determine the temperature of a single battery. The UC emission intensity ratio between the $^2H_{11/2} \rightarrow \,^4I_{15/2}$ (525 nm) and $^4S_{3/2} \rightarrow \,^4I_{15/2}$ (545 nm) transitions of Er ions is used to sense temperature changes (Figure 4.19a) [138]. Incubating the UCNPs with HeLa cells, this fluorescence nanothermometer can measure the internal temperature of living cells from 25 to 45 °C, useful for understanding its pathology.

Biosensors based on UC also have emerging functions that can detect specific chemical substances in living systems [141, 142]. Zhu designed a non-coding RNA sensing nanoplatform by performing DNA hybridization between DNA-modified UCNPs and gold nanoparticles [140]. In their research, it is possible to recover the quenching of the UC emission via FRET by designing the signal DNA sequence that clamps the DNA. In addition, an exonuclease III (Exo III)-assisted cycle amplification strategy was introduced to improve sensitivity (Figure 4.19c). These characteristics make the UCNPs-based sensing platform useful for various ncRNA detection and clinical diagnosis.

Achatz et al. developed a nanoplatform that can sense oxygen with NIR light (Figure 4.19b) [139]. NaYF$_4$:Yb/Tm UCNPs act as nanolamps and emit short-wave emission at 455 and 475 nm under 980 nm laser to excite the oxygen probe with light [Ir(CS)$_2$(acac)]. The absorbance is maximum at 468 nm. The green emission of the Ir(III) complex at 568 nm is quenched by oxygen due to excitation at 470 nm, allowing quantitative detection of O$_2$.

The UNCPs is also used for direct imaging and sensing of metal ions. Wei et al. combined NaYF$_4$:Yb,Er,Tm@NaGdF$_4$ UCNPs to synthesize an Fe^{3+}-responsive Nile Red Derivative (NRD) probe that showed high selectivity and sensitivity to Fe^{3+} in water and living cells [143]. The addition of Fe^{3+} changed the structure of NRD and increased the absorption of NRD, terminating the emission of UCNPs through the FRET process. Peng et al. developed a Zn^{2+} nanosensor by assembling UCNPs with chromophores. They reported that UC emission could be effectively quenched by the chromophore through the FRET process and recovered again by the addition of Zn^{2+}, allowing quantitative detection of Zn^{2+} [144]. This nanoprobe also

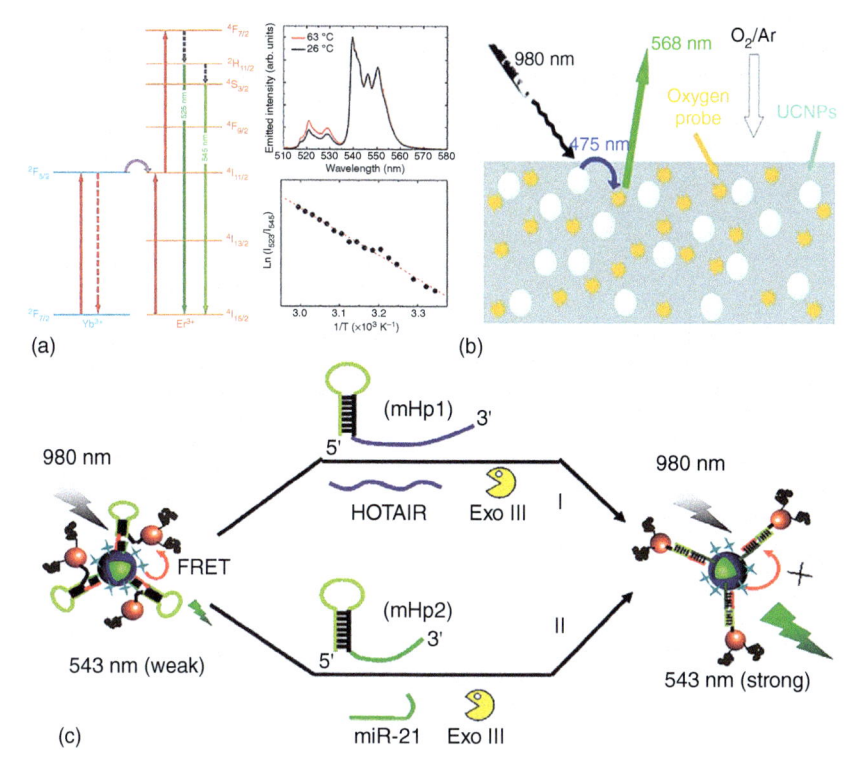

Figure 4.19 (a) The intensity ratio of green upconversion emission bands from $^2H_{11/2} \rightarrow\ ^4I_{15/2}$ and $^4S_{3/2} \rightarrow\ ^4I_{15/2}$ transitions of Er^{3+} ions is employed to detecting temperature through optical method. Source: Vetrone et al. [138]. Copyright 2010, American Chemical Society. (b) Schematic of the inner filter-effect-based UCNP O_2 sensor. Source: Achatz et al. [139]. Copyright 2011, John Wiley & Sons, Inc. (c) Schematic illustration of the established sensing nanoplatform for detecting (I) antisense RNA segment and (II) miRNA. Source: Zhang et al. [140]. Copyright 2018, John Wiley & Sons, Inc.

demonstrated effective detection of Zn^{2+} in brain slices from mice with Alzheimer's disease and zebrafish.

4.2.7.5 Photocatalysis

Photocatalysis is a chemical reaction between organic substances and free radicals. The free radicals are generated by photocatalysts with UV–vis light. It is an important clean and sustainable solar energy application. However, it is limited by its inability to collect the complete solar spectrum including NIR light. UCNPs can be introduced for their ability to convert low-energy (especially NIR) light into high-energy (UV and/or visible) light. Photocatalytic systems with UCNPs as spectral converters can operate according to general principles but with improved performance. The first visible photocatalyst based on a combination of Er_2O_3 UCNPs and TiO_2 was developed in 2005 [145]. TiO_2-coated YF_3:Yb/Tm nanoparticles were prepared and used as photocatalysts to convert NIR light into UV light to further stimulate the potential of NIR light for improved photocatalytic reactions, as shown in Figure 4.20a [146].

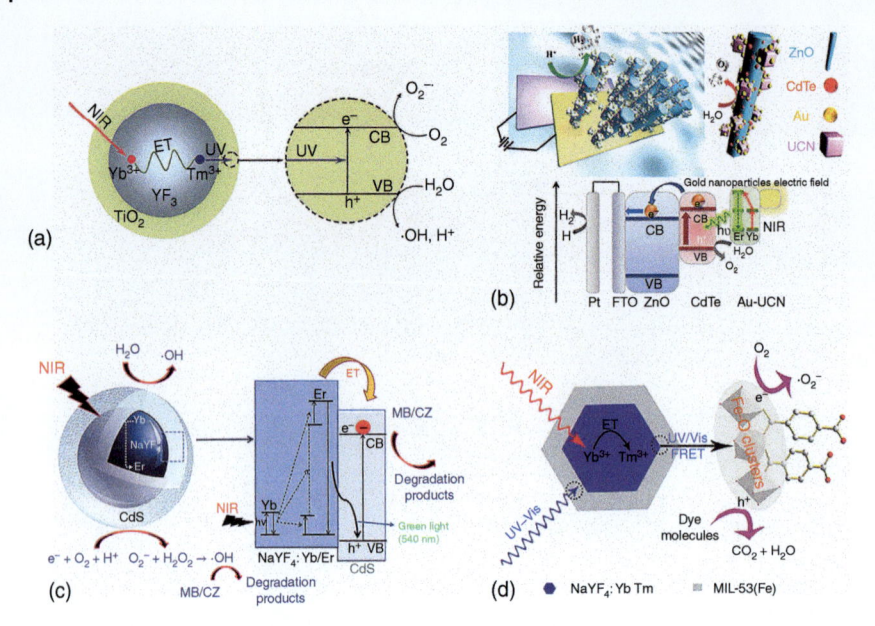

Figure 4.20 (a) Mechanism of near-infrared-activated photocatalysis. (ET: energy transfer). Source: Qin et al. [146]. Copyright 2010, Royal Society of Chemistry. (b) Sketch of the Au-UCNs–CdTe–ZnO photoelectrode and the mechanism of energy conversion from NIR to chemical fuel. Source: Chen et al. [147]. Copyright 2013, Royal Society of Chemistry. (c) Schematic representation of possible mechanism for the photocatalytic activity of NaYF$_4$:Yb/Er@CdS hybrid nanocomposites in the degradation of pollutant under NIR irradiation. Source: Balaji et al. [148]. Copyright 2017, Elsevier. (d) Schematic illustration of the photocatalytic mechanism of UCNP@MIL-53(Fe) NP. Source: Li et al. [149]. Copyright 2017, American Chemical Society.

Figure 4.20b shows that ZnO nanorod arrays decorated with CdTe quantum dots and equipartition excitations enhanced with NaYF$_4$:Yb/Er UCNPs were developed for NIR-driven photo-electrochemical water splitting [147]. The optical band gap of a semiconductor matches the radiated light energy, triggering the photocatalytic process. Highly reactive radicals can oxidize pollutants and break them down into CO_2 and H_2O.

The combination of conventional semiconductor photocatalysts with UCNPs has become popular for its optical properties. The composites prepared by combining luminescent ions such as Eu^{3+}, Ho^{3+}, Tm^{3+}, Er^{3+}, and Nd^{3+} with semiconductor photocatalysts absorb low-energy photons to generate high-energy light, activating the photocatalysts. Krishnan developed a simplistic approach for the synthesis of a novel NIR-driven core/shell NaYF$_4$:Yb/Er@ CdS photocatalyst [148]. Figure 4.20c shows that the core UCNPs are in hexagonal phase and the CdS shell is in cubic phase. The synthesized photocatalysts were subjected to photocatalytic environmental remediation by degrading the colored (methyl orange) and colorless (carbendazim) pollutants under NIR irradiation. Significant UC energy is transferred from the Er^{3+} ions in the nucleus to the CdS shell. Overall, the

NaYF$_4$:Yb/Er@CdS core/shell structure has demonstrated the use of NIR light from the solar spectrum for photocatalytic applications.

Metal-organic framework (MOF) is another material that can be used in photocatalytic applications. Due to its structural properties and tunability, many MOF-based materials have been available for photocatalysis. Li et al. proposed a facile method to integrate UCNPs with iron (Fe)-based MOFs to prepare NIR-responsive photocatalysts [149]. The UCNPs absorb NIR light and emit UV visible light. The MOF is activated by the excited UCNPs and produces photogenerated electrons and holes, shown in Figure 4.20d. When electrons migrate to the particle surface, they will react with the surrounding oxygen molecules to produce O^{2-} radicals that can be used as oxidants for dye degradation.

4.2.7.6 UCNPs-Mediated Molecular Switches

Conventional molecular switches use UV and visible light to induce photochemical reactions between two isomers. This process may lead to deleterious effects associated with low storage density for optical storage and low penetration depth for photorelease and photodynamic therapy. UCNPs offer a possible solution to these problems. By using UCNPs as antennas, UC-based photoswitching systems have shown to reversibly modulate the properties and structures of light-switchable molecules such as spiropyrans, diarylethenes, and azobenzenes [150–153].

Branda and coworkers reported structural conversion of photo-responsive dithienylethene (DTE) molecules by UC-mediated techniques [154]. In their study, a mixture of UCNPs and DTE molecules was cast into acrylate films. Under 980 nm laser irradiation, UV emission of NaYF$_4$:Yb/Tm and green emission of NaYF$_4$:Yb/Er UCNPs were produced, triggering the open- and closed-loop reactions of the DTE molecules shown in Figure 4.21a. They also reported a bidirectional reversible optical switch relying on a nanoparticle where NaYF$_4$:Yb/Tm@NaYF$_4$:Yb/Er@NaYF$_4$ UCNPs are used as energy donors under NIR radiation [155]. Figure 4.21b shows that the predominant UV emission induced by Tm can drive the closed-loop reaction of DTE molecules. On the contrary, low power density leads to open-loop reactions of organic molecules due to the green emission from Er. Li demonstrated the reversible tuning of self-organized helical superstructures by doping chiral switch molecules and UCNPs into the liquid crystal body. The power density-dependent reversible conversion process was observed under irradiation at 980 nm, accompanied by a change in the reflected wavelength of the photonic superstructure [156].

Photoisomerization is another type of photoinitiated process where certain molecules (e.g. azobenzene) undergo trans → cis isomerization when irradiated with light of a specific wavelength. Wu et al. doped UCNPs nanophosphors into a cross-linked liquid crystal polymer (CLCP) film containing azotolane. After exposure to NIR light at 980 nm, the composite film produced rapid bending due to trans → cis photoisomerization of the azopentane unit and orientation change of the liquid crystal. Once the laser is turned off, the bent film can still be fully restored to its initial state. NIR light is a more suitable stimulation source for light-driven organic actuators in biological systems applications. Thus, NIR

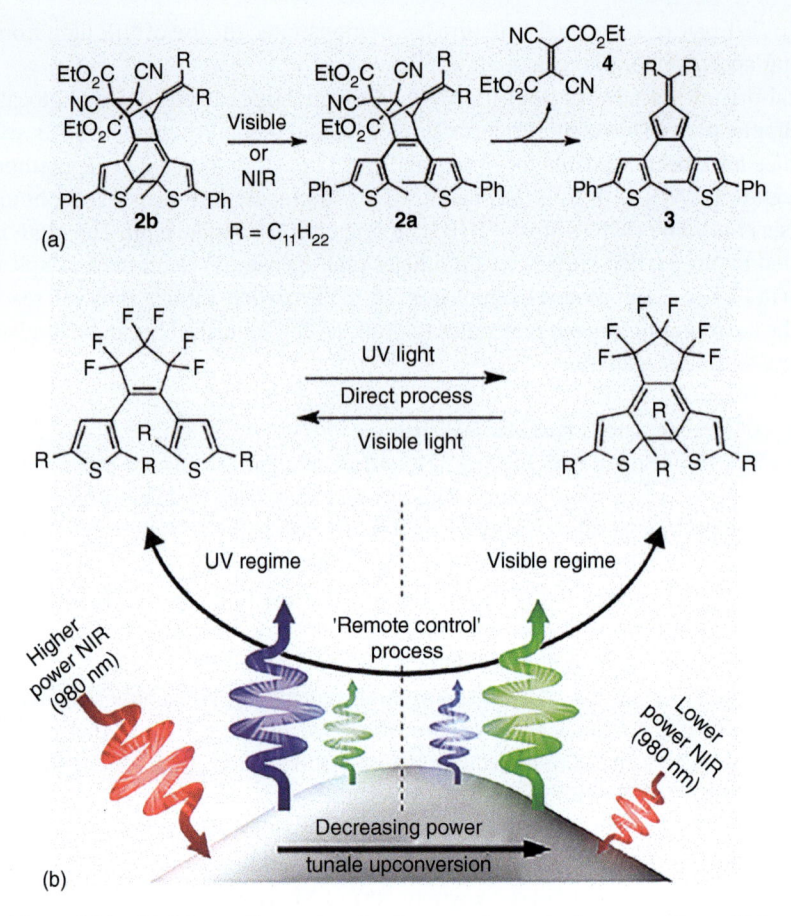

Figure 4.21 (a) Scheme illustrating the ring-opening and release reactions of bicyclic compound 2b as it is irradiated with visible or NIR light. Source: Carling et al. [154]. Copyright 2009, American Chemical Society. (b) The "direct" photoreactions of the DTE derivatives used in this study are triggered by UV light (for ring-closing) and visible light (for ring-opening). These reactions can also be triggered in a "remote control" process using the UV light generated under high excitation power densities and the visible light generated under low excitation power densities when the core–shell–shell UCNPs (ErTm and TmEr) absorb near-infrared light (980 nm). The sizes of the colored arrows represent the relative amount of each type of light excited or emitted during the multiphoton process. Source: Boyer et al. [155]. Copyright 2010, American Chemical Society.

light-induced deformable UCNP-CLCP systems have great potential for various biological applications, such as artificial muscle-like actuators and all-optical switches [157].

4.2.7.7 Other Technological Applications

UCNPs can also be applied in anti-counterfeiting, fingerprint detection, lighting, transparent displays, and solar cells [10, 158–163].

Ye et al. reported a new strategy for fabricating UC emission patterns based on UCNP-doped liquid crystal networks. The identification of luminescence patterns is

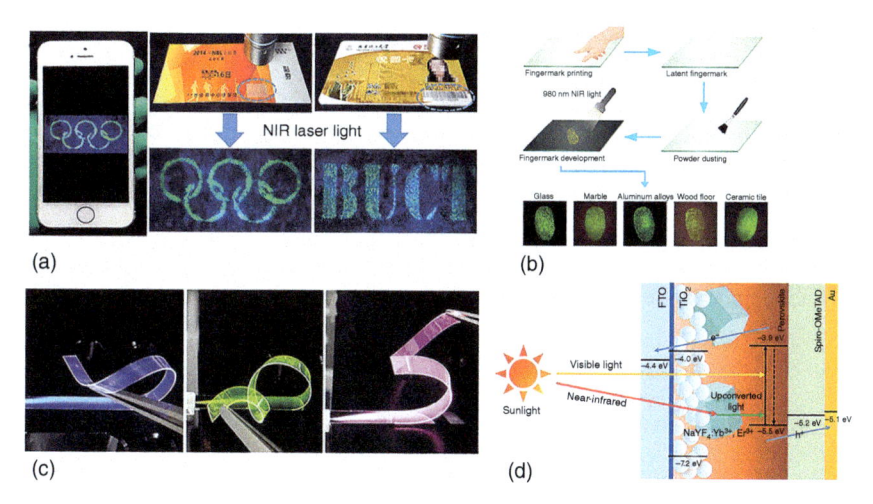

Figure 4.22 (a) Pattern image identification with a portable apparatus consisting of a cell phone and a ×20 objective. Source: Ye et al. [164]/Reproduced with permission from John Wiley & Sons, Inc. (b) Illustration of the development of latent fingermarks using a UCNP powder dusting process. A fingermark was printed on a substrate such as glass, marble, alumina alloys, wood floor, or ceramic tiles. Source: Wang et al. [165]/Reproduced with permission from Springer Nature. (c) Photographs showing the luminescence from severely bended blue-, green-, and red-emitting UCNP-incorporated polymer waveguides, respectively. Source: Park et al. [162]/Reproduced with permission from Royal Society of Chemistry. (d) Schematic diagrams of the energy transfer process in perovskite solar cells using an upconverting mesoporous layer. Source: Roh et al. [166]. Copyright 2016, American Chemical Society.

not limited to optical microscopy. It can also be detected by portable devices (e.g. cell phone). The identification process of portable devices is simpler and more straightforward compared to microscope identification (Figure 4.22a) [164]. UCNPs with human-safe NIR photoactivity have been shown to stain fingerprints on different substrate surfaces [165]. UCNPs were used as fluorescent markers for fingerprint identification with high sensitivity, high efficiency, low background, and low toxicity on a wide range of substrate surfaces, such as glass, marble, aluminum alloy sheets, wood floors, and ceramic tile (Figure 4.22b). This work demonstrates that UCNPs are universal fluorescent markers that can detect fingerprints on the surface of almost any material, making it useful for practical applications in forensic science.

A flexible transparent display based on a core/shell structure incorporated with a polymer waveguide of UCNPs was demonstrated by Park et al. [162]. UCNPs with green and blue UC emissions were synthesized. After doping with UCNPs, bisphenol A ethoxylated diacrylate was used as the core material of the polymer waveguide. The improved strip-shaped polymer waveguide was transparent and flexible. The polymer waveguides show blue, green, and red luminescence when irradiated with NIR laser. In addition, fabrication of pattern-based polymer waveguide-based displays was performed by a reactive ion etching process. They display bright blue, green, and red characters under NIR irradiation. Figure 4.22c shows photographs of luminescence from severely bended blue-, green-, and red-emitting UCNP-incorporated polymer waveguides.

Roh et al. reported the application of UCNPs in perovskite solar cells for sunlight NIR harvesting [166]. UCNPs were synthesized and doped into the perovskite solar cells as upconversion mesoporous layers. The UCNPs-based perovskite solar cell exhibited a large power conversion efficiency compared to the TiO_2 nanoparticle-based perovskite solar cell. This work demonstrates that UCNPs can expand the absorption range of the solar cell via UC process, which leads to an increase in photocurrent. The energy transfer process is schematically illustrated in Figure 4.22d.

4.3 Rare Earth Based DCNPs

4.3.1 $Y_3Al_5O_{12}$:RE (RE = Ce^{3+}, Tb^{3+})

YAG nanoparticles with good control over size (<50 nm) and shape has various emerging applications including tagging, imaging, scintillation, solid-state illumination, and micro-LED applications. However, controlling size and shape is still in its infancy. Optical properties are not yet at the same level as other factors in the nanoscale. There are currently three main methods to synthesize nano-YAG: coprecipitation method, sol–gel method, and solvothermal method.

4.3.1.1 Coprecipitation Approach

In a coprecipitation synthesis, Al- and Y-based hydroxides are formed first, then calcined at high temperatures (≥800 °C), which are lower than the temperatures needed for a solid-state reaction (≥1500 °C). The method aims to achieve a homogeneous mixture of the elements before calcination. Less diffusion is needed during crystallization, and possible at lower temperatures. A homogeneous mixture of the elements is important, otherwise phase impurities will form next to the desired YAG phase. To obtain individual nanoscale (size <50 nm) YAG particles, it is important to acquire the small and individual precursor particles and keep their size during the calcination treatment. A method to keep the particles small during calcination is to use a protective matrix that keeps the precursor particles separated to prevent sintering, but does not interfere with the crystallization process. This matrix can also be removed after the calcination treatment without deteriorating the YAG nanoparticles (Figure 4.23).

Song et al. used K_2SO_4 as matrix material, which has a higher melting point (1067 °C) and is soluble in water [167]. The YAG precursor particles were mixed with an excess of K_2SO_4 (5 : 1 molar ratio) via a microemulsion method, ensuring a good coating of the precursor particles. Using an excess of K_2SO_4 is also important, as the commonly observed large sintered structures resulted in little salt. The authors obtained the smallest crystalline YAG:Ce particles (a particle size of ~5 nm) via precipitation method after a heat treatment for 2 hours at 1000 °C.

By comparing the two emission spectra, the peak emission of nano-YAG (565 nm) is significantly redshifted compared with that of micro-YAG (554 nm). Comparing

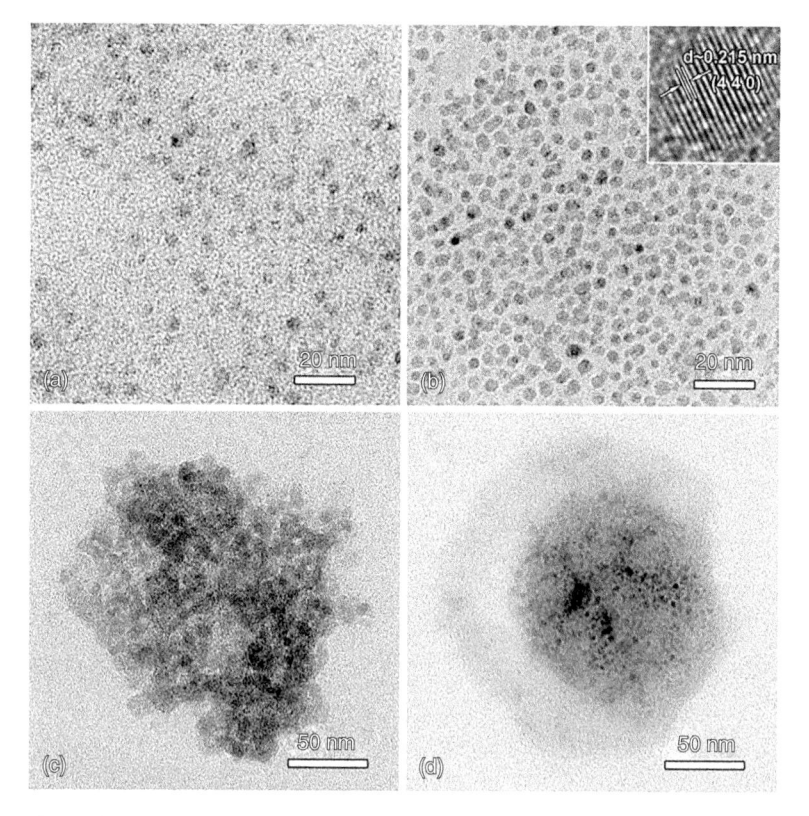

Figure 4.23 TEM images of (a) YAG precursor, (b) YAG obtained after heating at 1000 °C for 4 hours, (c) YAG precursor encapsulated by K_2SO_4 salt. (d) TEM image obtained after continuous irradiation with high-energy electron beam. Source: Song et al. [167].

the excitation spectra shown in Figure 4.24, the excitation peak intensity of the nano-YAG at 335 nm is significantly lower than that of bulk YAG. This condition indicates that the transition probability between 4f and 5d ($^2B_{1g}$) in the nano-YAG is weakened, and the transition probability between 4f and 5d ($^2A_{1g}$) is increased. The full width at half maximum (FWHM) of the emission spectrum of the nano-YAG is also larger than that of bulk YAG.

4.3.1.2 Sol–Gel Method

The sol–gel-based synthesis for nanoscale YAG particles has some similarities to coprecipitation. There is a first wet chemistry synthesis step where Y and Al precursor salts are mixed in a solvent to form a solution. However, a chelating agent is added to ensure the formation of a gel when the solution is heated and dried. The nature of the sol–gel reaction, where a homogeneous solution turns into a gel, can allow mixing of the elements at the atomic scale. Sol–gel reactions should be performed at low pH (<5) to prevent formation of hydroxide precipitates. After drying the gel, a combustion reaction takes place, where the chelating species acts as fuel. The combustion reaction product needs further calcination at temperatures ≥700 °C

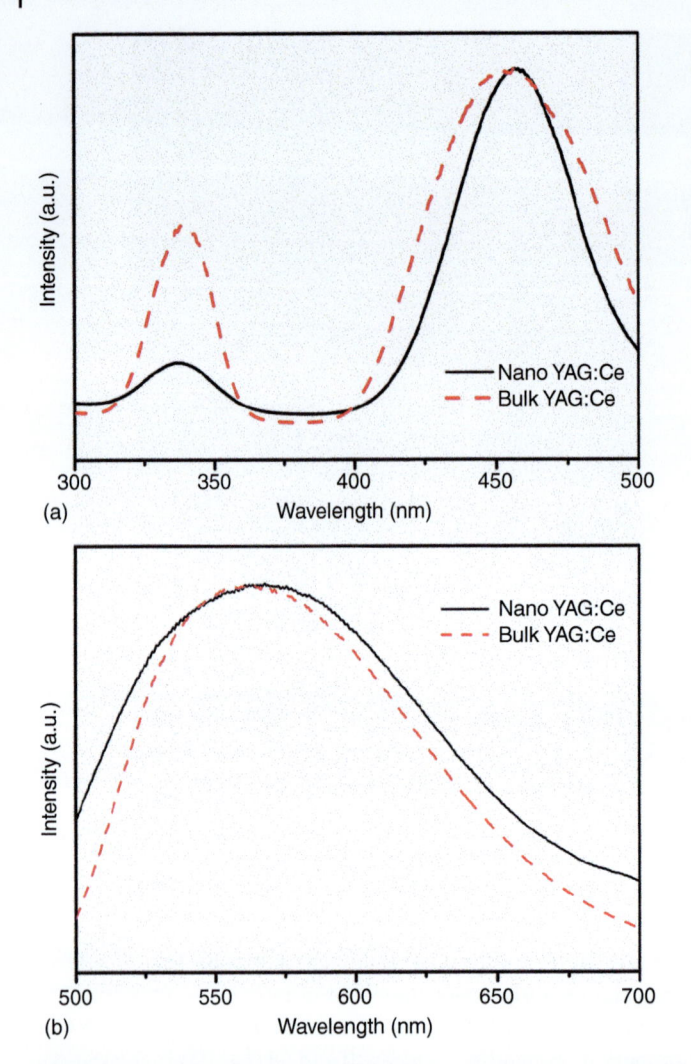

Figure 4.24 (a) Excitation and (b) emission spectra of nano-YAG:0.06 Ce and bulk YAG:0.06 Ce. Source: Song et al. [167].

to obtain phase pure YAG. This last calcination step is needed for crystallization. However, it leads to the aggregation of the YAG nanoparticles.

Calcination time and temperature are important in the final YAG particle size and morphology. The effect of the calcination temperature on particle sizes was studied in detail by Zhang et al. [168]. They prepared YAG:Tb particles using nitrate salts and citric acid as fuel, followed by calcination for 2 hours at temperatures ranging from 800 to 1600 °C. An increase in size was observed (Figure 4.25a–e) when the calcination temperature is increased from 800 °C (50 nm sintered particles) to 1000 °C (100 nm sintered particles) to 1600 °C (~500 nm).

Aboulaich et al. also reported a method to synthesize highly crystalline YAG:Tb nanophosphor inside the pores of a mesoporous silica monolith (MSM) [169]. This

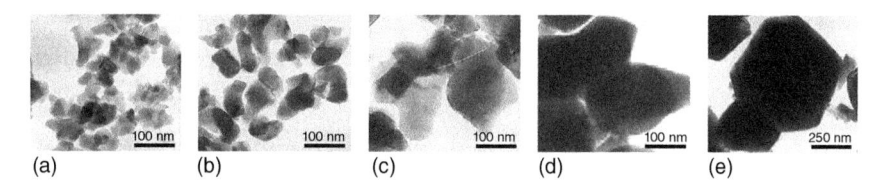

Figure 4.25 Transmission electron microscopy images of nano-YAG:Tb particles made via a sol–gel method, with different calcination parameters. (a–e) YAG:Tb particles calcined at increasing temperatures: (a) 800, (b) 1000, (c) 1200, (d) 1400, (e) 1600 °C. Source: Zhang et al. [168].

porous material was permeated by a solution of YCl_3, $TbCl_3$, and Al-isopropoxide in isopropanol and calcined for 3 hours at 1100 °C. The silica is dissolved overnight in warm 1 M NaOH, yielding final YAG:Tb with particle sizes of ∼30 nm that appeared slightly sintered. The MSM–YAG:Tb composite exhibited a strong green fluorescence emission with the characteristic main emission band of Tb^{3+} located at 548 nm.

4.3.1.3 Solvothermal Method

Solvothermal reactions are another approach to obtain nanoscale YAG particles. These reactions happen at temperatures above the normal boiling point of the solvent used in a closed vessel at high pressure. Hydrothermal or glycothermal synthesis methods are used for YAG synthesis, which are classes of solvothermal reactions where the solvents used are water or glycerols. The advantage of this synthesis over sol–gel or coprecipitation is that crystalline material can be produced in a one-step reaction. Important reaction parameters are the temperature and pressure during the reaction and the chemical nature of the precursor materials and solvents.

Aboulaich et al. reported a rapid solvothermal synthesis (t_h = 5 minutes) leading to well-crystallized and highly monodispersed YAG:Ce^{3+} NPs of about 30 nm diameter by heating mixed yttrium, aluminum, and cerium carbonate precursors in an autoclave at 300 °C using a 1,4-butylene glycol–water mixture as solvent (Figure 4.26a) [170].

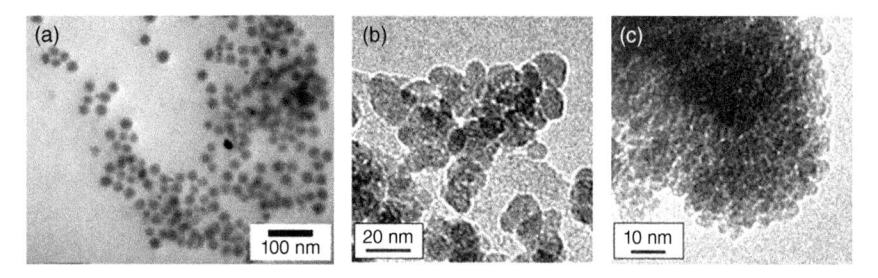

Figure 4.26 (a) TEM images of YAG:Ce^{3+}(1%) NPs prepared using the solvothermal synthesis route only 5 minutes heating to 300 °C. Source: Aboulaich et al. [170]. (b) After heating for 2 hours at 300 °C in 1,4 butanediol, 10 nm YAG particles form. (c) Adding citric acid as surfactant leads so smaller but severely agglomerated YAG particles. Source: Asakura et al. [171].

Agglomeration or aggregation of the YAG nanoparticles during the reaction is a major problem when small individual particles are targeted. During the early stages of the reaction, aggregation results in rapid growth of the nanoparticles above 50 nm. Asakura et al. added citric acid as surfactant to their reaction with aluminum isopropoxide, acetates, and 1,4 butanediol at 300 °C for 2 hours. The carboxyl groups of citric acid coordinate metallic ions stronger than the hydroxyl groups of 1,4-butanediol, and suppress particle growth, leading to YAG particle sizes of 4 nm compared to 10 nm obtained in a synthesis without citric acid (Figure 4.26b,c). However, the 4 nm particles show severe agglomeration. It is ascribed to the bridging function of citric acid, arising from its three carboxylic groups.

In applications, absorption strength and quantum yield are the main material characteristics to optimize. These two characteristics are strongly interconnected, as both benefit from high crystallinity (i.e. reduced crystal defects), homogeneous Ce distribution, and Ce in the right oxidation state (i.e. Ce(III)). Ce doping concentration is a compromising factor, since high doping concentration will increase the effective absorption of the YAG:Ce material. However, concentration quenching and self-absorption can occur when the doping concentration is too high. In addition, excessive Ce^{3+} concentration can lead to distortion of the YAG crystal lattice, since its ionic radius (r_{Ce3+} = 1.143 Å in an eightfold coordination) is much larger than that of Y^{3+} (r_{Y3+} = 1.019 Å in an eightfold coordination). The reported optimal doping concentrations for nano-YAG:Ce vary from 0.4% to 4% Ce. The reported QYs for nano-YAG:Ce (up to 50%) are lower than the near-unity QYs reported for bulk YAG:Ce, which is due to the following reasons: (i) formation of Ce(IV) during the synthesis; (ii) lower crystallinity (i.e. more defects) due to low synthesis or calcination temperatures; (iii) an inhomogeneous Ce distribution leading to concentration quenching; and (iv) surface defects which arise from the large surface to volume ratio of nanoparticles.

4.3.2 $SrAl_2O_4:Eu^{2+}, Dy^{3+}$

The rare earth-doped strontium aluminate luminescent material is currently the most researched and commercialized luminescent material. It has advantages including high luminous intensity, long persistent luminescence time, and environmental protection. With Eu^{2+} and Dy^{3+} co-doped, it can emit intense green phosphorescence. The application of the strontium aluminate luminescent material has been involved in daily life. Its applications are mainly in low-brightness emergency lighting, luminous indicators, etc. It is necessary to make it into DCNPs to realize its applications in biological imaging, anti-counterfeiting and encryption. The specific methods are as follows.

4.3.2.1 Hydrothermal Method

Xu et al. used $Sr(NO_3)_2$, $Al(NO_3)_3 \cdot 9H_2O$, $CO(NH_2)_2$ as raw materials, Eu and Dy as activating ions, and hydrothermally reacted the mixture at 160 °C for 24 hours [172]. The product was heated in a weak reducing atmosphere at 1300 °C for 2 hours

Figure 4.27 SEM images of hydrothermal sheet products under different magnifications (a, b). Source: Xu et al. [172].

to obtain a nano-flaky strontium aluminate long persistence luminescent material with a thickness of about 20 nm and a length of about 100 nm. Its morphology is shown in Figure 4.27. The results also indicate that the slow hydrolysis of urea is beneficial to the formation of sheet-like structures.

4.3.2.2 Sol–Gel Method

Hwang et al. [173] reported a 3 mol% of europium-doped strontium aluminate ($SrAl_2O_4:Eu^{2+}$) coatings on silicon substrates were prepared by electrostatic spray deposition method using a salted sol–gel-derived solution as a starting material. The deposited films at 100 °C for 5 hours were heated at 1100 °C for 2 hours under a reducing ambient atmosphere of 95% N_2 and 5% H_2. The obtained Eu-doped $SrAl_2O_4$ phosphor film has a grain size of around 60 nm. The excitation peak is between 300 and 500 nm, and the emission peak has a broadband at 512 nm.

4.3.2.3 Microwave Method

In 2014, Elsagh et al. produced $SrAl_2O_4:Eu^{2+},Dy^{3+}$ nanophosphors by microwave synthesis method [174]. The microwave output power was 720 W, the frequency was 2.45 GHz, and the reaction time was 9 minutes. The particle size of the product was about 60 nm. In 2015, Shan et al. synthesized $SrAl_2O_4: Eu^{2+}, Dy^{3+}$ phosphor via coprecipitation method assisted by microwave irradiation in a weak reductive atmosphere of active carbon, which has a flaky structure of 500 nm width and 70 nm thickness. Its morphology and structure are shown in Figure 4.28a,b [175].

4.3.2.4 Electrospinning

Cheng et al. reported strontium aluminate luminescent fiber in 2010 using electrospinning approach [176]. Ethanol and deionized water were mixed as a solvent. After dissolving aluminum nitrate, strontium aluminate, and rare earth ions, 4.84 g PVP was added to prepare a spinning solution. Strontium aluminate nanofibers were prepared under 30 kV conditions. After a heating rate of 2 °C min^{-1} to 1200 °C for heat treatment under reduction, the final product was obtained. Figure 4.29 shows the SEM images of the electrospun product before and after heat treatment. The diameter of the spinning fiber is about 300 nm. After the heat treatment, the fiber shrinks to

Figure 4.28 SEM images of the samples prepared with different precipitating agent: (a) ammonium carbonate and (b) ammonia solution. Source: Shan et al. [175].

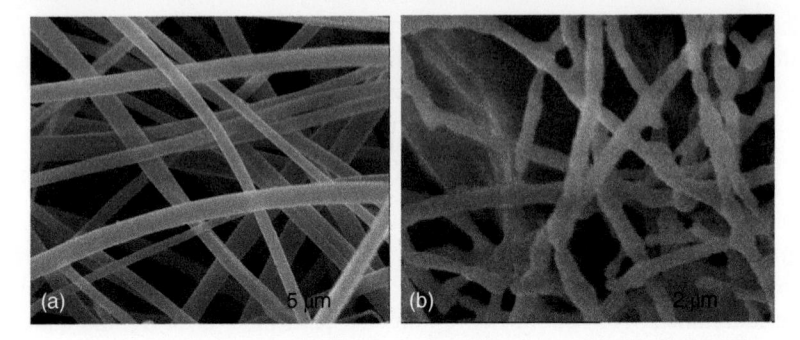

Figure 4.29 SEM images of strontium aluminate fiber before (a) and after (b) heat treatment. Source: Cheng et al. [176].

a certain extent, but the morphology of the fiber remains unchanged. The excitation peak is at 346–375 nm, and the emission peak is at 509 nm.

4.3.3 Y_2O_3:Eu^{3+}

Yttrium oxide (Y_2O_3) is an important advanced engineering material that has been studied in many applications such as transparent ceramics, sensors, catalysts, magnets, and so on. It has advantageous properties such as good thermal, mechanical, and chemical stability, high corrosion resistance, and low toxicity. It also has a large band gap about 5.8 eV, small phonon energy about 380 cm^{-1}, high dielectric constant, and is optically isotropic. Y_2O_3 can be a host matrix for rare earth (Eu^{3+}, Dy^{3+}, Sm^{3+}, etc)-doped phosphors that hold prospective application in solid-state lasers (SSL), cathode ray tubes (CRT), field emission displays (FEDs), and biological imaging. The performance of nanomaterials depend strongly on their morphology and size distribution, which plays an important role in practical applications. The morphology and size controllable synthesis of Y_2O_3 and RE-doped Y_2O_3 has become prominent in recent years. Several approaches, including hydrothermal method, combustion method, spray pyrolysis method, precipitation method, and microwave-assisted synthesis methods, were employed

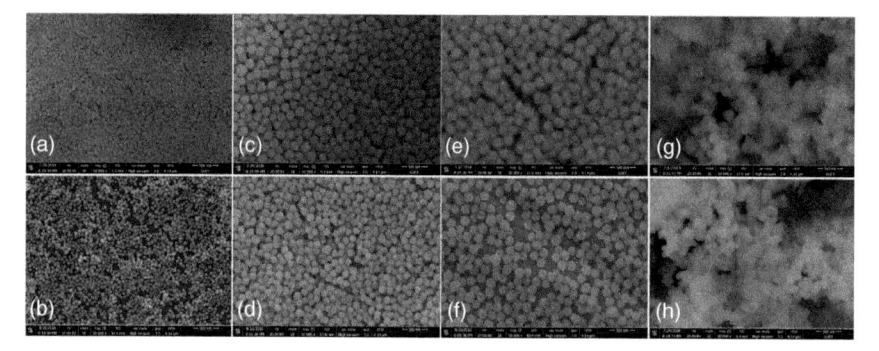

Figure 4.30 FE-SEM of the as-prepared precursor (upside) and calcined Y_2O_3 (downside) with different Y^{3+} concentration, 0.05 M (a, b); 0.35 M (c, d); 0.50 M (e, f); 0.65 M (g, h). Source: Zhao et al. [177].

to achieve Y_2O_3 and RE-doped Y_2O_3 with various morphologies, such as nanowires or nanorods, nanotubes, hollow microspheres, nanosheets, and monodispersed nanosphere.

Monodispersed nanospherical phosphor, including RE-doped Y_2O_3 with narrow size distribution, has potential in practical and technological applications due to its high packing density and low light scattering. Zhao et al. reported a modified homogeneous urea precipitation process to achieve monodispersed spherical Y_2O_3 powders using PVP (PVP-K30) as a surface active agent [177]. The average particle sizes of the as-obtained Y_2O_3 are about 65, 105, 140, and 180 nm, while the Y^{3+} concentration increase from 0.05–0.15, 0.35, and 0.50 M, respectively (Figure 4.30). The Eu-doped Y_2O_3 phosphor displayed a strong red emission peak at 613 nm ascribed to $^5D_0 \rightarrow {}^7F_2$ forced electric dipole transition of Eu^{3+} under 394 nm excitation. With the monodispersed spherical shape, narrow size distribution and high yield, they can be useful in practical and technological applications in bioimaging, solid-state lighting, display devices, and functional ceramics.

Dhanaraj et al. used a gel-polymer pyrolysis method to prepare cubic Y_2O_3:Eu red luminescent nanomaterials with an average particle size of 10–20 nm [178]. Compared with industrially produced bulk Y_2O_3:Eu phosphors, it has higher quantum yield and stronger fluorescence emission. Zhai et al. also used metal nitrate and EDTA as raw materials to synthesize Y_2O_3:Eu fluorescent nanomaterials at a lower temperature via sol–gel method [179].

Li prepared Y_2O_3 nanopowders by oxalic acid precipitation at room temperature [180]. In a micro reactor, non-ionic surfactant and oxalic acid solution were added to YCl_3 solutions. The solution is stirred until white yttrium oxalate precipitates. The resulting precipitate was washed and dried several times. It was gradually heated to 800 °C and calcined for 10 minutes to obtain Y_2O_3 powder, which has an average particle size of less than 30 nm. The results indicate that the purity, particle size, and dispersibility of the Y_2O_3 nanopowders were depended on the concentration of the YCl_3 solution, the mass fraction of the oxalic acid solution, and the concentration of the non-ionic surfactant.

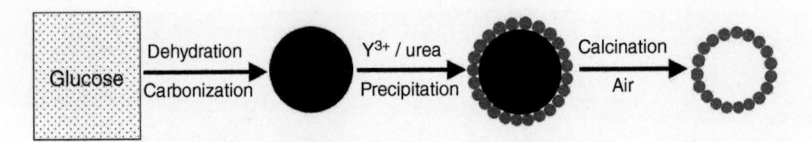

Figure 4.31 Schematic illustration of the formation on process of Y_2O_3:Eu hollow microspheres. Source: Li et al. [180].

Jia et al. prepared Y_2O_3:Eu hollow microsphere nanomaterials using the colloidal carbon sphere as the hard template through the heat treatment process by precipitation method [181]. The preparation process is shown in Figure 4.31. The Y_2O_3:Eu fluorescent nanomaterial obtained by this method exhibits strong red light emission under the excitation of UV light with lifetime of 1.76 ms. The precipitation method is very sensitive to reaction conditions, and it is difficult to prepare uniformly dispersed nanomaterials.

Ye et al. used a glycine-nitrate solution combustion method to prepare Y_2O_3:Eu nanophosphor [182]. The particle size of the prepared material is related to the combustion temperature. Its optical properties also depend on the particle size. The particle size of the product can be controlled through changing the combustion temperature and the ratio of glycine to nitrate. Compared with Y_2O_3:Eu obtained by other preparation methods, the Y_2O_3:Eu nanomaterials prepared by combustion method have higher quenching concentration.

4.3.4 $LnVO_4$:Ln^{3+} (Ln = La, Gd, Y; Ln^{3+} = Eu^{3+}, Dy^{3+}, Sm^{3+})

Rare earth vanadate can achieve excellent UC and DC emission and multicolor light adjustment through doping. It has advantageous chemical and optical properties and is widely used in light-emitting devices, medicine, phosphors, and other applications. Its simple synthetic route and easy-control particle size make it an ideal rare earth compound host material. Among them, $LaVO_4$, $GdVO_4$, and YVO_4 have been studied as high-quality rare earth vanadate host materials.

Li and coworker reported the synthesis of lanthanide-doped $LaVO_4$ nanocrystals using a self-assembly method [183]. As shown in Figure 4.32, Ln^{3+}-$LaVO_4$ can self-assemble into an ordered array on the crystal surface through the molecular interaction of oleic acid molecules. The morphology and size can be controlled during synthesis. These nanocrystals show well-grown crystal cross-sections and high morphology uniformity. They are soluble in non-polar solvents. They can also be attached to the crystal surface of nanocarbides through long alkyl chains and self-assemble a high-order colloidal structure in organic solvents. Ln^{3+}-$LaVO_4$ nanocrystals can be arranged into uniform films by coating technology, and exhibit typical doped-ion luminescence.

Liu et al. has synthesized YVO_4:Ln^{3+} (Ln = Eu, Sm, and Dy) nanoparticles via a fast and facile microwave method using NH_4VO_3 as vanadium source [184]. The results indicate that the morphology and size of YVO_4 products can be modified by varying the pH value of the solution or using different solvents (Figure 4.33). The emission spectra of the YVO_4:Eu^{3+} products synthesized with different solvents are similar

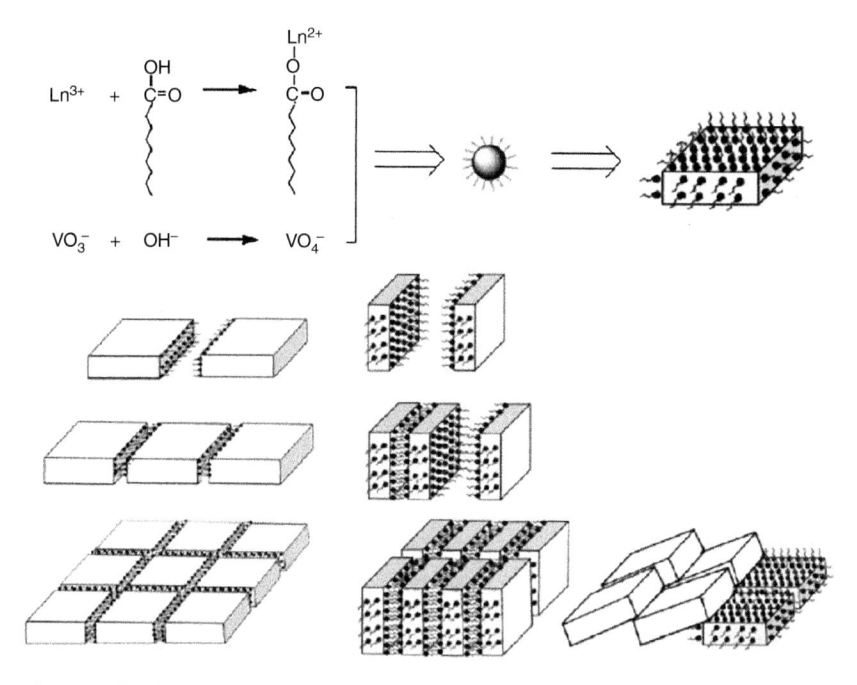

Figure 4.32 Schematic diagram showing the formation and self-assembly process of LaVO$_4$ nanocrystals. Source: Ye et al. [182].

in shape, but different in intensity. It indicates that the luminescent properties are closely correlated with the morphologies of the materials. When the excitation wavelength is fixed at around 280 nm, no emission from the VO$_4^{3-}$ groups is observed, indicating that the nonradiative energy transfer process from the VO$_4^{3-}$ groups to the Eu^{3+}, Dy^{3+}, or Sm^{3+} ions is efficient.

As an excellent host material, yttrium vanadate can be doped with different rare earth ions to achieve multi-emission. As shown in Figure 4.34, Liu and coworkers synthesized (Y, P)VO$_4$ doped with rare earth ions in an aqueous solution containing PVP. By choosing a suitable host-activator system and controlling the doping concentration, the color of the emitted light can be tuned under single-wavelength excitation [185].

4.3.5 LaPO$_4$:Ce^{3+},Tb^{3+}

Despite the single-doping model, activators codoping with sensitizer ions, which have higher absorption coefficients, could lead to more efficient luminescence emission by taking advantage of the energy transfer from sensitizers to activators. Codoping of Ce^{3+} ions improves the luminescence intensity of Ln^{3+} (Ln = Eu, Tb, Dy, Sm) ions, since the emission band of Ce^{3+} ions matches with several f–f absorption bands of Ln^{3+} ions. The sensitization effect of Ce^{3+} is especially efficient in Ce^{3+}–Tb^{3+} couple. The reports proved that the emission intensity of Tb^{3+} ions is enhanced by the codoping of Ce^{3+} ions.

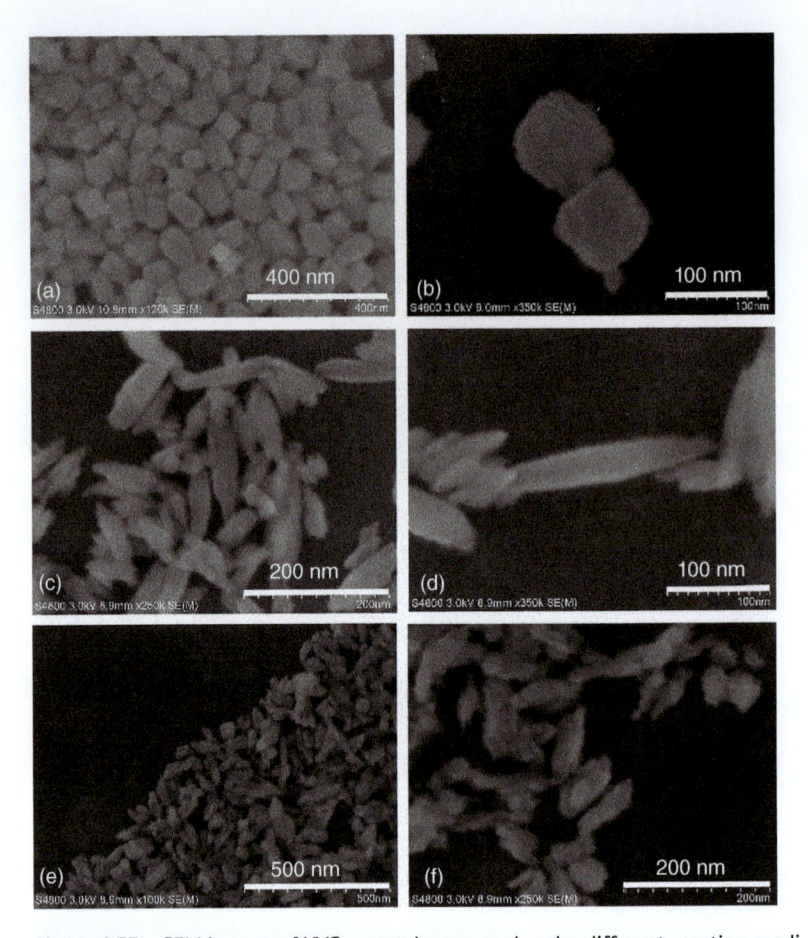

Figure 4.33 SEM images of YVO$_4$ crystals prepared under different reaction mediums: (a, b) ethylene glycol–water, (c, d) ethanol–water and (e, f) isopropanol–water. Source: Liu et al. [184].

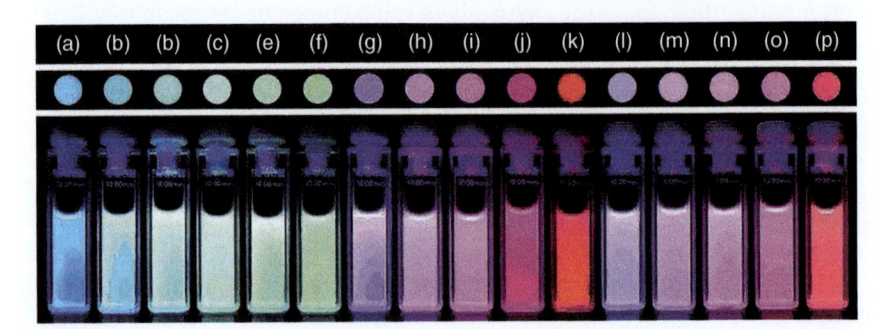

Figure 4.34 Fluorescence photos of (Ln,P)Y-VO$_4$ nanoparticles in glass slides and aqueous solutions. Source: Wang et al. [185].

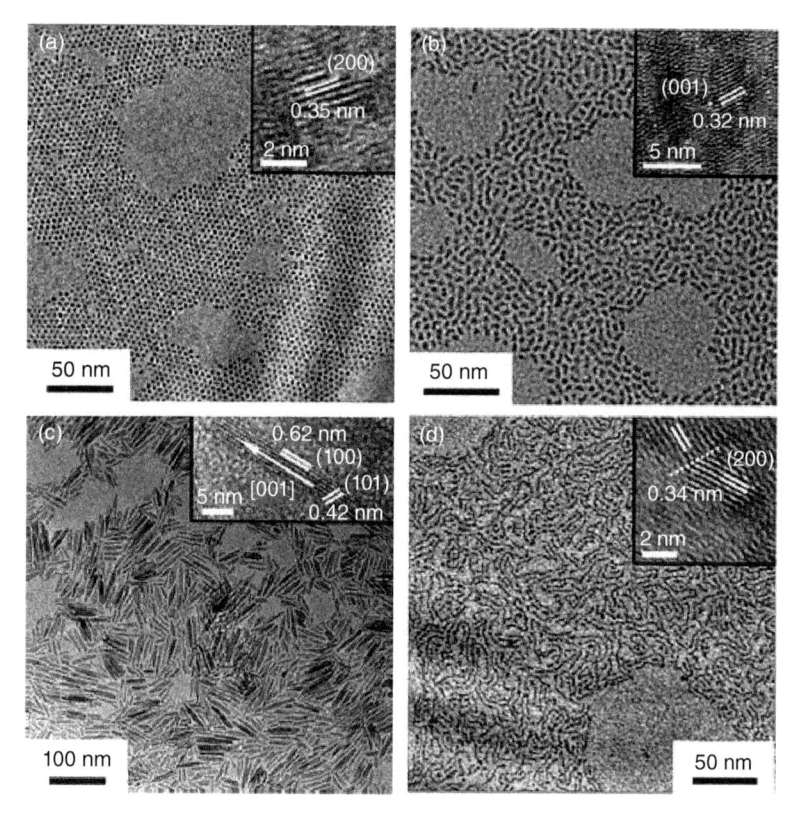

Figure 4.35 TEM and HRTEM images (inset) of the LaPO$_4$:Ce,Tb. (a) Nanopolyhedra. (b) Quasi-nanorods. (c) Nanorods. (d) Wormlike nanowires (highlighted: the twined (200) plane). Source: Mai et al. [186].

Mai et al. reported LaPO$_4$:Ce,Tb nanocrystals with diverse shapes (Figure 4.35), synthesized at a temperature range of 180–260 °C in long alkyl-chain solvents (organic acids and amines) via limited anion-exchange reaction (LAER). The resulting nanocrystals show intensive green under UV excitation with quantum yields higher than 40%. The luminescence spectra of monodispersed La$_{0.4}$Ce$_{0.45}$Tb$_{0.15}$PO$_4$ nanocrystals with different morphologies are similar and showed no significant difference from that of the bulk material. On the UV excitation of the cerium absorption band, quantum yields of 50%, 47%, 45%, and 50% are obtained from the terbium emission of LaPO$_4$:Ce,Tb polyhedra, quasirods, rods, and wormlike wire dispersion, respectively, in comparison with that of an ethanol (spectroscopic grade) solution of rhodamine 6G (Lambda Physics, laser grade) under identical optical density at the same wavelength.

4.3.6 Applications

4.3.6.1 Biological Imaging

Escudero et al. synthesized near-ultraviolet and visible excitable Eu- and Bi-doped NPs based on rare earth vanadates (REVO$_4$, RE = Y, Gd) via a facile route from

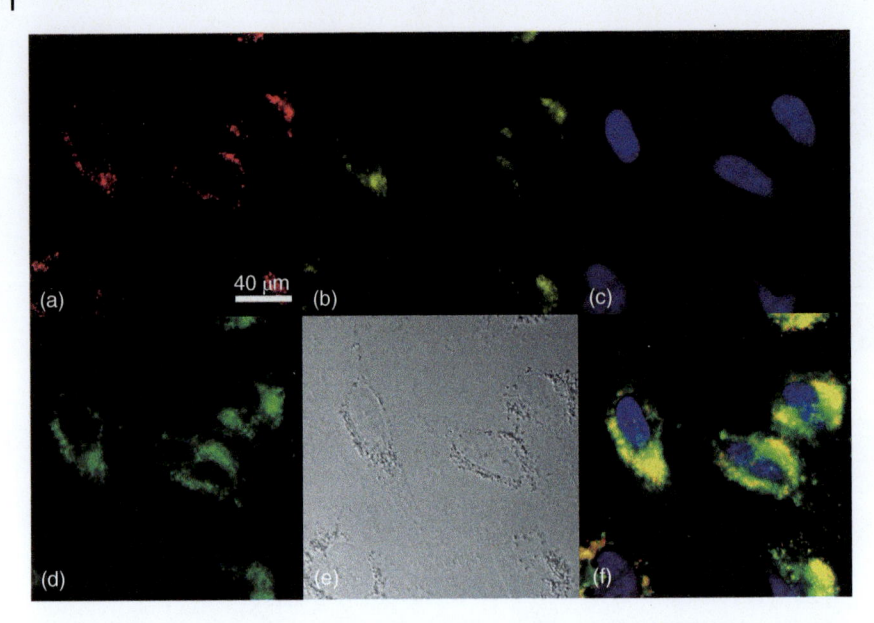

Figure 4.36 Fluorescence images of HeLa cells incubated with Bi20YPAA2 nanoparticles for 24 hours. (a) Red channel, nanoparticles; (b) yellow channel, lysosomes; (c) blue channel, cell nuclei; (d) green channel, cell membranes; (e) transmission image; and (f) merged all channels. Source: Escudero et al. [187].

appropriate RE precursors, europium and bismuth nitrate, and sodium orthovanadate, by homogeneous precipitation in an ethylene glycol/water mixture at 120 °C. The particle size could be tuned by modifying the amount of the added PAA. These Eu-Bi-doped YVO_4 nanoparticles with low cell toxicity and good morphology can be used in biological imaging of human cervical cancer cells (He La). The broad absorption band of the NPs with a maximum at 342 nm permits their excitation with near-ultraviolet light in a fluorescence microscope. On the other hand, the shift of the absorption band of the Bi-doped vanadate matrices toward longer wavelengths makes their excitation possible with a visible radiation by using a laser as the excitation source in a confocal laser scanning microscope (cLSM). As shown in Figure 4.36, a strong red luminescence in the interior of the cells could be observed after irradiation at (340 ± 24) nm and 24 hours incubation. This fluorescence was co-localized with the yellow produced by LAMP1, a marker for the membrane of lysosomes, indicating that the NPs are located in the cell's lysosomes after being internalized.

4.3.6.2 Tumor Treatment

Lin and coworkers deposited noble metal Au on $NdVO_4$ to form a metal/semiconductor hybrid nanostructure to improve Vis/NIR light absorption [188]. $NdVO_4$/Au heterojunction nanocrystals (NC) were synthesized with $NdVO_4$ nanorods (NRs) and plasmonic gold nanoparticles (NPs), and PVP was introduced to enhance stability and biocompatibility. The photothermal conversion rate was improved to 32.15%. The ability to generate ROS was also increase. $NdVO_4$/Au can be effectively internalized through endocytosis, resulting in phototoxicity

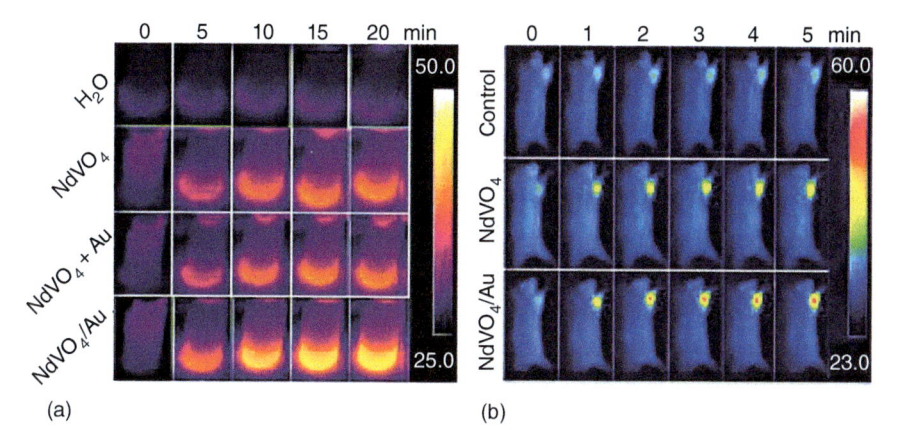

Figure 4.37 (a) The thermal images of H_2O, $NdVO_4$, $NdVO_4$ + Au and $NdVO_4$/Au (300 μg ml^{-1}, 1 ml) exposed to 808 nm laser irradiation (1.3 W cm^{-2}) for different times (0, 5, 10, 15, and 20 minutes). (b) The thermal images of U14-tumor-bearing mice with intravenous injection of normal saline, $NdVO_4$ and $NdVO_4$/Au (20 mg kg^{-1}, 100 μl) for 12 hours exposed to 808 nm laser light irradiation (0.5 W cm^{-2}) for different times (0, 1, 2, 3, 4, and 5 minutes). Source: Chang et al. [188].

to He La cells. As shown in Figure 4.37, *in vivo* experiments further prove that $NdVO_4$/Au can be used as an effective NIR light-triggered anticancer agent and has excellent anti-tumor effects. Based on the photothermal conversion performance and thermal expansion effect under near-infrared radiation, $NdVO_4$/Au provides a photothermal (PT) and PA dual-modal imaging platform for precise cancer diagnosis and treatment.

4.3.6.3 Fluorescent Ink

Chen et al. synthesized $(Y,Gd)VO_4:Bi^{3+},Eu^{3+}$ fluorescent inks and studied the photoluminescence phenomenon by exciting them with different wavelengths of excitation light at different temperatures [189]. By adjusting the concentration of Bi^{3+} and Eu^{3+}, the wide emission color from green to yellow to orange can be adjusted. The temperature-dependent luminescence and wavelength-selective excitation of $(Y,Gd)VO_4:Bi^{3+}, Eu^{3+}$ are observed, which provided different encryption methods for anti-counterfeiting. Figure 4.38a shows the anti-counterfeiting pattern of the Chinese name of the Hefei University of Technology prepared from $(Y_{0.955}Bi_{0.04}Eu_{0.005})VO_4$ fluorescent ink. When heated and excited at different wavelengths, different emission colors with different intensities are observed. Patterns can be designed and manufactured using different phosphors to enhance the encryption ability by overcoming a single color. Figure 4.38b shows the anti-counterfeiting logos of the famous ceramic sanitary product brands "HCG" and "TOTO," which alternately used the $(Y_{0.955}Bi_{0.04}Eu_{0.005})VO_4$ and $(Y_{0.945}Bi_{0.04}Eu_{0.015})VO_4$ inks. By combining artistic design and fluorescence spectra, sophisticated encryption can be achieved as every spectrum has its own fingerprint.

Kumar et al. used a hydrothermal method to prepare multicolor luminescent lanthanide-doped Y_2O_3 nanorods [190]. $Y_2O_3:Eu^{3+}$, $Y_2O_3:Tb^{3+}$, and $Y_2O_3:Ce^{3+}$

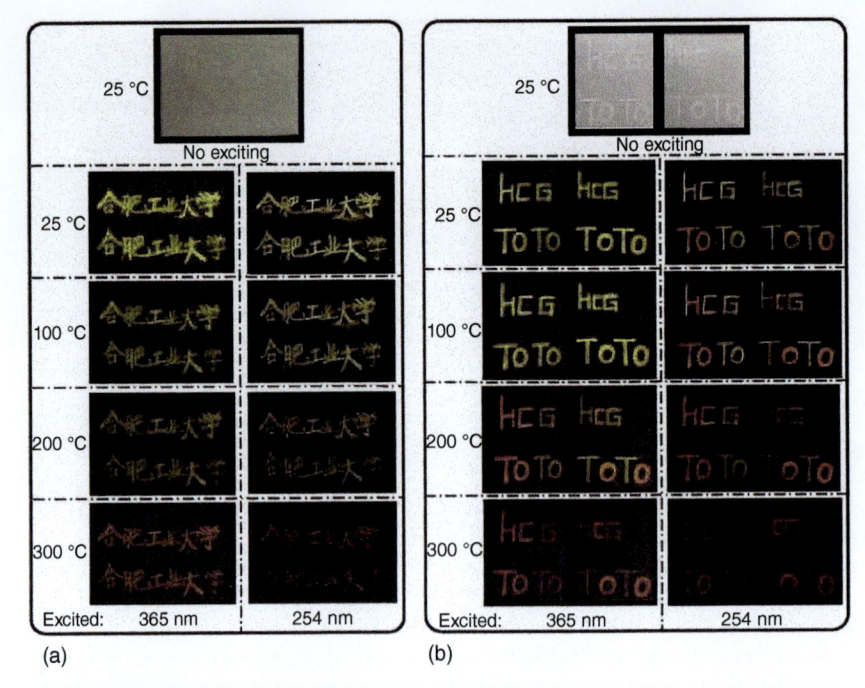

Figure 4.38 (a) Anti-counterfeiting patterns consisting of (a) the letter symbols of the Chinese name of Hefei University of Technology written from $(Y_{0.955}Bi_{0.04}Eu_{0.005})VO_4$ ink and (b) the letter symbols of "HCG" and "TOTO" written by using $(Y_{0.955}Bi_{0.04}Eu_{0.005})VO_4$ and $(Y_{0.945}Bi_{0.04}Eu_{0.015})VO_4$ inks alternately and their luminescence pictures under 365 and 254 nm excitation at 25, 100, 200, and 300 °C, respectively. Source: Chen et al. [189].

nanorods emit strong green (541 nm), bright blue (438 nm), and strong red (611 nm) light when excited at 254, 305, and 381 nm, respectively. Figure 4.39a shows the red, green, and blue light-emitting QR codes printed on black paper under ultraviolet light. Under normal circumstances, the printed QR code will not display any color. These QR codes are unreadable under normal light. However, these codes are displayed in red, green, and blue when under ultraviolet light. Therefore, the QR code can only be read under ultraviolet light.

The research group also printed the CSIR-NPL logo on black paper using red, green, and blue inks. The printed logo of CSIR-NPL shows strong red, green, and blue on black paper under ultraviolet light, as shown in Figure 4.39b–d.

4.4 Summary and Outlook

This chapter summarized the developments in the synthesis, properties, and applications of rare earth luminescent nanomaterials including UCNPs and DCNPs. Significant progress has been made in the synthesis of UCNPs and DCNPs with controllable size, shape, morphology, and phase by various synthetic methods over the last two decades. The successful synthesis of high-quality UCNPs and

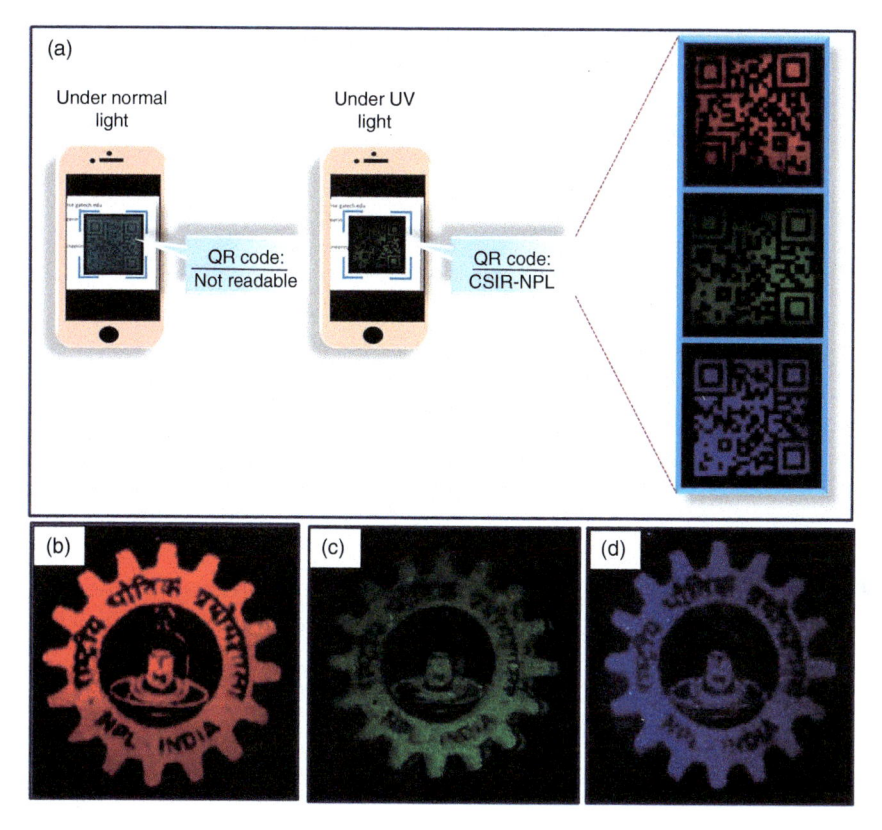

Figure 4.39 (a) Schematic diagram of a smartphone with a QR code scanning application, and red, green and blue QR codes under ultraviolet light. (b–d) Optical image of red, green and blue printed logo under ultraviolet light. Source: Kumar et al. [190].

DCNPs has also expanded the range of applications from lighting, display, sensing, bioimaging, to therapy. UCNPs have become popular especially in biomedicine. The excitation of NIR irradiation is considered more effective than UV or visible light in biological tissue. One main obstacle to the introduction of UCNPs in clinical applications is the long-term biological effects on the human body. Thus, it is necessary to study the long-term degradability of UCNPs in biological systems.

For DCNPs, more research is needed to improve their luminescent properties, which are not as good as their bulk counterpart. They are more useful in conventional applications.

References

1 Auzel, F. (2004). Upconversion and anti-stokes processes with f and d ions in solids. *Chem. Rev.* 104 (1): 139–173.
2 Gai, S., Li, C., Yang, P. et al. (2014). Recent progress in rare earth micro/nanocrystals: soft chemical synthesis, luminescent properties, and biomedical applications. *Chem. Rev.* 114 (4): 2343–2389.

3 Haase, M. and Schäfer, H. (2011). Upconverting nanoparticles. *Angew. Chem. Int. Ed.* 50 (26): 5808–5829.

4 Loo, J.F.-C., Chien, Y.-H., Yin, F. et al. (2019). Upconversion and downconversion nanoparticles for biophotonics and nanomedicine. *Coord. Chem. Rev.* 400: 213042.

5 Zheng, K., Loh, K.Y., Wang, Y. et al. (2019). Recent advances in upconversion nanocrystals: expanding the kaleidoscopic toolbox for emerging applications. *Nano Today* 29: 100797.

6 Chen, G., Qiu, H., Prasad, P.N. et al. (2014). Upconversion nanoparticles: design, nanochemistry, and applications in theranostics. *Chem. Rev.* 114 (10): 5161–5214.

7 Zhou, J., Liu, Q., Feng, W. et al. (2014). Upconversion luminescent materials: advances and applications. *Chem. Rev.* 115 (1): 395–465.

8 Zheng, W., Huang, P., Tu, D. et al. (2015). Lanthanide-doped upconversion nano-bioprobes: electronic structures, optical properties, and biodetection. *Chem. Soc. Rev.* 44 (6): 1379–1415.

9 Gu, B. and Zhang, Q. (2018). Recent advances on functionalized upconversion nanoparticles for detection of small molecules and ions in biosystems. *Adv. Sci.* 5 (3): 1700609.

10 Lingeshwar Reddy, K., Balaji, R., Kumar, A. et al. (2018). Lanthanide doped near infrared active upconversion nanophosphors: fundamental concepts, synthesis strategies, and technological applications. *Small* 14 (37): 1801304.

11 Chen, B. and Wang, F. (2019). Combating concentration quenching in upconversion nanoparticles. *Acc. Chem. Res.* 53 (2): 358–367.

12 Fan, Y., Liu, L., and Zhang, F. (2019). Exploiting lanthanide-doped upconversion nanoparticles with core/shell structures. *Nano Today* 25: 68–84.

13 Dong, H., Sun, L.-D., and Yan, C.-H. (2020). Upconversion emission studies of single particles. *Nano Today* 35: 100956.

14 Heer, S., Lehmann, O., Haase, M. et al. (2003). Blue, green, and red upconversion emission from lanthanide-doped $LuPO_4$ and $YbPO_4$ nanocrystals in a transparent colloidal solution. *Angew. Chem. Int. Ed.* 42 (27): 3179–3182.

15 Heer, S., Kompe, K., Gudel, H.U. et al. (2004). Highly efficient multicolour upconversion emission in transparent colloids of lanthanide-doped $NaYF_4$ nanocrystals. *Adv. Mater.* 15 (23–24): 2102–2105.

16 Yi, G.S., Lu, H.C., Zhao, S.Y. et al. (2004). Synthesis, characterization, and biological application of size-controlled nanocrystalline $NaYF_4$:Yb,Er infrared-to-visible up-conversion phosphors. *Nano Lett.* 4 (11): 2191–2196.

17 Zeng, J.H., Su, J., Li, Z.H. et al. (2005). Synthesis and upconversion luminescence of hexagonal-phase $NaYF_4$:Yb, Er^{3+}, phosphors of controlled size and morphology. *Adv. Mater.* 17 (17): 2119–2123.

18 Li, Z. and Zhang, Y. (2008). An efficient and user-friendly method for the synthesis of hexagonal-phase $NaYF_4$:Yb, Er/Tm nanocrystals with controllable shape and upconversion fluorescence. *Nanotechnology* 19 (34): 345606.

19 Wang, F., Han, Y., Lim, C.S. et al. (2010). Simultaneous phase and size control of upconversion nanocrystals through lanthanide doping. *Nature* 463 (7284): 1061–1065.

20 Li, Z., Liang, T., Wang, Q. et al. (2019). Strategies for constructing upconversion luminescence nanoprobes to improve signal contrast. *Small* 16 (1): 1905084.

21 Wang, F., Deng, R., Wang, J. et al. (2011). Tuning upconversion through energy migration in core–shell nanoparticles. *Nat. Mater.* 10 (12): 968–973.

22 Deng, R., Qin, F., Chen, R. et al. (2015). Temporal full-colour tuning through non-steady-state upconversion. *Nat. Nanotechnol.* 10 (3): 237–242.

23 Nadort, A., Zhao, J., and Goldys, E.M. (2016). Lanthanide upconversion luminescence at the nanoscale: fundamentals and optical properties. *Nanoscale* 8 (27): 13099–13130.

24 Kramer, K.W., Biner, D., Frei, G. et al. (2004). Hexagonal sodium yttrium fluoride based green and blue emitting upconversion phosphors. *Chem. Mater.* 16 (7): 1244–1251.

25 Xie, X., Gao, N., Deng, R. et al. (2013). Mechanistic investigation of photon upconversion in Nd^{3+}-sensitized core–shell nanoparticles. *J. Am. Chem. Soc.* 135 (34): 12608–12611.

26 Zhang, Y.-W., Sun, X., Si, R. et al. (2005). Single-crystalline and monodisperse LaF_3 triangular nanoplates from a single-source precursor. *J. Am. Chem. Soc.* 127 (10): 3260–3261.

27 Ye, X., Collins, J.E., Kang, Y. et al. (2010). Morphologically controlled synthesis of colloidal upconversion nanophosphors and their shape-directed self-assembly. *Proc. Natl. Acad. Sci. U.S.A.* 107 (52): 22430–22435.

28 Wang, X., Zhuang, J., Peng, Q. et al. (2006). Hydrothermal synthesis of rare-earth fluoride nanocrystals. *Inorg. Chem.* 45 (17): 6661–6665.

29 Ren, J., Jia, G., Guo, Y. et al. (2016). Unraveling morphology and phase control of $NaLnF_4$ upconverting nanocrystals. *J. Phys. Chem. C* 120 (2): 1342–1351.

30 Wang, L. and Li, Y. (2007). Controlled synthesis and luminescence of lanthanide doped $NaYF_4$ nanocrystals. *Chem. Mater.* 19 (4): 727–734.

31 Zheng, W., Zhou, S., Chen, Z. et al. (2013). Sub-10 nm lanthanide-doped CaF_2 nanoprobes for time-resolved luminescent biodetection. *Angew. Chem. Int. Ed.* 52 (26): 6671–6676.

32 Wang, J., Wang, F., Wang, C. et al. (2011). Single-band upconversion emission in lanthanide-doped $KMnF_3$ nanocrystals. *Angew. Chem. Int. Ed.* 50 (44): 10369–10372.

33 Wang, J., Deng, R., MacDonald, M.A. et al. (2014). Enhancing multiphoton upconversion through energy clustering at sublattice level. *Nat. Mater.* 13 (2): 157–162.

34 Zhang, Y., Huang, P., Wang, D. et al. (2018). Near-infrared-triggered antibacterial and antifungal photodynamic therapy based on lanthanide-doped upconversion nanoparticles. *Nanoscale* 10 (33): 15485–15495.

35 Boyer, J.C., Vetrone, F., Cuccia, L.A. et al. (2006). Synthesis of colloidal upconverting $NaYF_4$ nanocrystals doped with Er^{3+}, Yb^{3+} and Tm^{3+}, Yb^{3+} via thermal decomposition of lanthanide trifluoroacetate precursors. *J. Am. Chem. Soc.* 128 (23): 7444–7445.

36 Mahalingam, V., Naccache, R., Vetrone, F. et al. (2009). Sensitized Ce^{3+} and Gd^{3+} ultraviolet emissions by TM^{3+} in Colloidal $LiYF_4$ nanocrystals. *Chem. Eur. J.* 15 (38): 9660–9663.

37 Chen, D., Yu, Y., Huang, F. et al. (2012). Lanthanide dopant-induced formation of uniform sub-10 nm active-core/active-shell nanocrystals with near-infrared to near-infrared dual-modal luminescence. *J. Mater. Chem.* 22 (6): 2632–2640.

38 Boyer, J.C., Cuccia, L.A., and Capobianco, J.A. (2007). Synthesis of colloidal upconverting $NaYF_4$: Er^{3+}/Yb^{3+} and Tm^{3+}/Yb^{3+} monodisperse nanocrystals. *Nano Lett.* 7 (3): 847–852.

39 Naccache, R., Vetrone, F., Mahalingam, V. et al. (2009). Controlled synthesis and water dispersibility of hexagonal phase $NaGdF_4$:$Ho3^+/Yb^{3+}$ nanoparticles. *Chem. Mater.* 21 (4): 717–723.

40 Yi, G.S., Lee, W.B., and Chow, G.M. (2007). Synthesis of $LiYF_4$, $BaYF_5$, and $NaLaF_4$ optical nanocrystals. *J. Nanosci. Nanotechnol.* 7 (8): 2790–2794.

41 Mahalingam, V., Vetrone, F., Naccache, R. et al. (2009). Colloidal Tm^{3+}/Yb^{3+}-doped $LiYF_4$ nanocrystals: multiple luminescence spanning the UV to NIR regions via low-energy excitation. *Adv. Mater.* 21 (40): 4025–4028.

42 Liu, X., Zhang, X., Tian, G. et al. (2014). A simple and efficient synthetic route for preparation of $NaYF_4$ upconversion nanoparticles by thermo-decomposition of rare-earth oleates. *CrystEngComm* 16 (25): 5650–5661.

43 Na, H., Woo, K., Lim, K. et al. (2013). Rational morphology control of β-$NaYF_4$:Yb,Er/Tm upconversion nanophosphors using a ligand, an additive, and lanthanide doping. *Nanoscale* 5 (10): 4242–4251.

44 Yi, G.S. and Chow, G.M. (2006). Synthesis of hexagonal-phase $NaYF_4$:Yb,Er and $NaYF_4$:Yb,Tm nanocrystals with efficient up-conversion fluorescence. *Adv. Funct. Mater.* 16 (18): 2324–2329.

45 Feng, S. and Xu, R. (2001). New materials in hydrothermal synthesis. *Acc. Chem. Res.* 34 (3): 239–247.

46 Zhang, F., Li, J., Shan, J. et al. (2009). Shape, size, and phase-controlled rare-earth fluoride nanocrystals with optical up-conversion properties. *Chem. Eur. J.* 15 (41): 11010–11019.

47 Yan, Z.G. and Yan, C.H. (2008). Controlled synthesis of rare earth nanostructures. *J. Mater. Chem.* 18 (42): 5046–5059.

48 Guo, H., Li, Z.Q., Qian, H.S. et al. (2010). Seed-mediated synthesis of $NaYF_4$:Yb, Er/$NaGdF_4$ nanocrystals with improved upconversion fluorescence and MR relaxivity. *Nanotechnology* 21 (12): 125602.

49 Niu, W., Wu, S., Zhang, S. et al. (2011). Multicolor output and shape controlled synthesis of lanthanide-ion doped fluorides upconversion nanoparticles. *Dalton Trans.* 40 (13): 3305–3314.

50 Wang, X., Zhuang, J., Peng, Q. et al. (2005). A general strategy for nanocrystal synthesis. *Nature* 437 (7055): 121–124.

51 Nguyen, T.-D., Dinh, C.-T., and Do, T.-O. (2010). Shape- and size-controlled synthesis of monoclinic ErOOH and cubic Er_2O_3 from micro- to nanostructures and their upconversion luminescence. *ACS Nano* 4 (4): 2263–2273.

52 Stouwdam, J.W. and van Veggel, F.C.J.M. (2002). Near-infrared emission of redispersible Er^{3+}, Nd^{3+}, and Ho^{3+} doped LaF_3 nanoparticles. *Nano Lett.* 2 (7): 733–737.

53 Ai, Y., Tu, D., Zheng, W. et al. (2013). Lanthanide-doped $NaScF_4$ nanoprobes: crystal structure, optical spectroscopy and biodetection. *Nanoscale* 5 (14): 6430–6438.

54 Lin, J., Yu, M., Lin, C. et al. (2007). Multiform oxide optical materials via the versatile pechini-type sol–gel process: synthesis and characterics. *J. Phys. Chem. C* 111 (16): 5835–5845.

55 Patra, A., Friend, C.S., Kapoor, R. et al. (2002). Upconversion in Er^{3+}:ZrO_2 nanocrystals. *J. Phys. Chem. B* 106 (8): 1909–1912.

56 Patra, A., Friend, C.S., Kapoor, R. et al. (2003). Fluorescence upconversion properties of Er^{3+}-doped TiO_2 and $BaTiO_3$ nanocrystallites. *Chem. Mater.* 15 (19): 3650–3655.

57 Venkatramu, V., Falcomer, D., Speghini, A. et al. (2008). Synthesis and luminescence properties of Er^{3+}-doped $Lu_3Ga_5O_{12}$ nanocrystals. *J. Lumin.* 128 (5): 811–813.

58 Kuisheng, Y., Fang, Z., Rina, W. et al. (2006). Upconversion luminescent properties of YVO_4:Yb^{3+}, Er^{3+} nano-powder by sol–gel method. *J. Rare Earths* 24 (1, Suppl. 1): 162–166.

59 Wang, H.-Q. and Nann, T. (2009). Monodisperse upconverting nanocrystals by microwave-assisted synthesis. *ACS Nano* 3 (11): 3804–3808.

60 Wang, H.-Q., Tilley, R.D., and Nann, T. (2010). Size and shape evolution of upconverting nanoparticles using microwave assisted synthesis. *CrystEngComm* 12 (7): 1993–1996.

61 Liu, X., Deng, R., Zhang, Y. et al. (2015). Probing the nature of upconversion nanocrystals: instrumentation matters. *Chem. Soc. Rev.* 44 (6): 1479–1508.

62 Zhuang, J., Wang, J., Yang, X. et al. (2009). Tunable thickness and photoluminescence of bipyramidal hexagonal beta-$NaYF_4$ microdisks. *Chem. Mater.* 21 (1): 160–168.

63 Fu, J., Fu, X., Wang, C. et al. (2013). Controlled growth and up-conversion improvement of sodium yttrium fluoride crystals. *Eur. J. Inorg. Chem.* 2013 (8): 1269–1274.

64 Johnson, N.J.J., Korinek, A., Dong, C. et al. (2012). Self-focusing by ostwald ripening: a strategy for layer-by-layer epitaxial growth on upconverting nanocrystals. *J. Am. Chem. Soc.* 134 (27): 11068–11071.

65 Niu, W.B., Wu, S.L., and Zhang, S.F. (2011). Utilizing the amidation reaction to address the "cooperative effect" of carboxylic acid/amine on the size, shape, and multicolor output of fluoride upconversion nanoparticles. *J. Mater. Chem.* 21 (29): 10894–10902.

66 Niu, W.B., Wu, S.L., and Zhang, S.F. (2010). A facile and general approach for the multicolor tuning of lanthanide-ion doped NaYF$_4$ upconversion nanoparticles within a fixed composition. *J. Mater. Chem.* 20 (41): 9113–9117.

67 Abel, K.A., Boyer, J.C., Andrei, C.M. et al. (2011). Analysis of the shell thickness distribution on NaYF$_4$/NaGdF$_4$ core/shell nanocrystals by EELS and EDS. *J. Phys. Chem. Lett.* 2 (3): 185–189.

68 Cheng, L., Yang, K., Li, Y.G. et al. (2011). Facile preparation of multifunctional upconversion nanoprobes for multimodal imaging and dual-targeted photothermal therapy. *Angew. Chem. Int. Ed.* 50 (32): 7385–7390.

69 Liu, X., Li, X., Qin, X. et al. (2017). Hedgehog-like upconversion crystals: controlled growth and molecular sensing at single-particle level. *Adv. Mater.* 29 (37): 1702315.

70 Wen, H.L., Zhu, H., Chen, X. et al. (2013). Upconverting near-infrared light through energy management in core-shell-shell nanoparticles. *Angew. Chem. Int. Ed.* 52 (50): 13419–13423.

71 Liu, Q., Sun, Y., Li, C.G. et al. (2011). F-18-Labeled magnetic-upconversion nanophosphors via rare-earth cation-assisted ligand assembly. *ACS Nano* 5 (4): 3146–3157.

72 Wang, J., He, N., Zhu, Y. et al. (2018). Highly-luminescent Eu,Sm,Mn-doped CaS up/down conversion nano-particles: application to ultra-sensitive latent fingerprint detection and *in vivo* bioimaging. *Chem. Commun.* 54 (6): 591–594.

73 Zhao, J.B., Lu, Z.D., Yin, Y.D. et al. (2013). Upconversion luminescence with tunable lifetime in NaYF$_4$:Yb,Er nanocrystals: role of nanocrystal size. *Nanoscale* 5 (3): 944–952.

74 Zhang, F., Che, R., Li, X. et al. (2012). Direct Imaging the upconversion nanocrystal core/shell structure at the subnanometer level: shell thickness dependence in upconverting optical properties. *Nano Lett.* 12 (6): 2852–2858.

75 Xiong, L.Q., Chen, Z.G., Tian, Q.W. et al. (2009). High contrast upconversion luminescence targeted imaging *in vivo* using peptide-labeled nanophosphors. *Anal. Chem.* 81 (21): 8687–8694.

76 Kumar, R., Nyk, M., Ohulchanskyy, T.Y. et al. (2009). Combined optical and MR bioimaging using rare earth ion doped NaYF$_4$ nanocrystals. *Adv. Funct. Mater.* 19 (6): 853–859.

77 Zhou, J., Sun, Y., Du, X.X. et al. (2010). Dual-modality *in vivo* imaging using rare-earth nanocrystals with near-infrared to near-infrared (NIR-to-NIR) upconversion luminescence and magnetic resonance properties. *Biomaterials* 31 (12): 3287–3295.

78 Ren, G.Z., Zeng, S.J., and Hao, J.H. (2011). Tunable multicolor upconversion emissions and paramagnetic property of monodispersed bifunctional lanthanide-doped NaGdF$_4$ nanorods. *J. Phys. Chem. C* 115 (41): 20141–20147.

79 Zhou, J., Yu, M.X., Sun, Y. et al. (2011). Fluorine-18-labeled Gd^{3+}/Yb^{3+}/Er^{3+} co-doped NaYF$_4$ nanophosphors for multimodality PET/MR/UCL imaging. *Biomaterials* 32 (4): 1148–1156.

80 Wang, F. and Liu, X. (2008). Upconversion multicolor fine-tuning: Visible to near-infrared emission from lanthanide-doped NaYF$_4$ nanoparticles. *J. Am. Chem. Soc.* 130 (17): 5642–5643.

81 Chen, G., Qiu, H., Fan, R. et al. (2012). Lanthanide-doped ultrasmall yttrium fluoride nanoparticles with enhanced multicolor upconversion photoluminescence. *J. Mater. Chem.* 22 (38): 20190.

82 Vetrone, F., Boyer, J.C., Capobianco, J.A. et al. (2002). NIR to visible upconversion in nanocrystalline and bulk Lu$_2$O$_3$:Er^{3+}. *J. Phys. Chem. B* 106 (22): 5622–5628.

83 Zeng, J.H., Xie, T., Li, Z.H. et al. (2007). Monodispersed nanocrystalline fluoroperovskite up-conversion phosphors. *Cryst. Growth Des.* 7 (12): 2774–2777.

84 Sarkar, S., Meesaragandla, B., Hazra, C. et al. (2013). Sub-5 nm Ln^{3+}-doped BaLuF$_5$ nanocrystals: a platform to realize upconversion via interparticle energy transfer (IPET). *Adv. Mater.* 25 (6): 856–860.

85 Zou, W., Visser, C., Maduro, J.A. et al. (2012). Broadband dye-sensitized upconversion of near-infrared light. *Nat. Photonics* 6 (8): 560–564.

86 Wisser, M.D., Fischer, S., Siefe, C. et al. (2018). Improving quantum yield of upconverting nanoparticles in aqueous media via emission sensitization. *Nano Lett.* 18 (4): 2689–2695.

87 Chen, G., Liu, H., Somesfalean, G. et al. (2009). Upconversion emission tuning from green to red in Yb^{3+}/Ho^{3+}-codoped NaYF$_4$ nanocrystals by tridoping with Ce^{3+} ions. *Nanotechnology* 20 (38): 385704.

88 Vetrone, F., Boyer, J.C., Capobianco, J.A. et al. (2003). Concentration-dependent near-infrared to visible upconversion in nanocrystalline and bulk Y$_2$O$_3$:Er^{3+}. *Chem. Mater.* 15 (14): 2737–2743.

89 Mahalingam, V., Naccache, R., Vetrone, F. et al. (2011). Preferential suppression of high-energy upconverted emissions of Tm^{3+} by Dy^{3+} ions in Tm^{3+}/Dy^{3+}/Yb^{3+}-doped LiYF$_4$ colloidal nanocrystals. *Chem. Commun.* 47 (12): 3481.

90 Li, X., Guo, Z., Zhao, T. et al. (2016). Filtration shell mediated power density independent orthogonal excitations-emissions upconversion luminescence. *Angew. Chem. Int. Ed.* 55 (7): 2464–2469.

91 Li, X., Shen, D., Yang, J. et al. (2012). Successive layer-by-layer strategy for multi-shell epitaxial growth: shell thickness and doping position dependence in upconverting optical properties. *Chem. Mater.* 25 (1): 106–112.

92 Mai, H.-X., Zhang, Y.-W., Sun, L.-D. et al. (2007). Highly efficient multicolor up-conversion emissions and their mechanisms of monodisperse NaYF$_4$:Yb,Er core and core/shell-structured nanocrystals. *J. Phys. Chem. C* 111 (37): 13721–13729.

93 Wang, F., Wang, J., and Liu, X. (2010). Direct evidence of a surface quenching effect on size-dependent luminescence of upconversion nanoparticles. *Angew. Chem. Int. Ed.* 49 (41): 7456–7460.

94 Li, Z., Zhang, Y., and Jiang, S. (2008). Multicolor core/shell-structured upconversion fluorescent nanoparticles. *Adv. Mater.* 20 (24): 4765–4769.

95 Xu, X., Zhang, X., Hu, C. et al. (2019). Construction of NaYF$_4$:Yb,Er(Tm)@CDs composites for enhancing red and NIR upconversion emission. *J. Mater. Chem. C* 7 (21): 6231–6235.

96 Zheng, W., Huang, P., Gong, Z. et al. (2018). Near-infrared-triggered photon upconversion tuning in all-inorganic cesium lead halide perovskite quantum dots. *Nat. Commun.* 9 (1): 3462.

97 Zhang, C., Yang, L., Zhao, J. et al. (2015). White-light emission from an integrated upconversion nanostructure: toward multicolor displays modulated by laser power. *Angew. Chem. Int. Ed.* 54 (39): 11531–11535.

98 Chen, B., Liu, Y., Xiao, Y. et al. (2016). Amplifying excitation-power sensitivity of photon upconversion in a NaYbF$_4$:Ho nanostructure for direct visualization of electromagnetic hotspots. *J. Phys. Chem. Lett.* 7 (23): 4916–4921.

99 Zhu, X., Zhang, J., Liu, J. et al. (2019). Recent progress of rare-earth doped upconversion nanoparticles: synthesis, optimization, and applications. *Adv. Sci.* 6 (22): 1901358.

100 Jin, J.F., Gu, Y.J., Man, C.W.Y. et al. (2011). Polymer-coated NaYF$_4$:Yb^{3+}, Er^{3+} upconversion nanoparticles for charge-dependent cellular imaging. *ACS Nano* 5 (10): 7838–7847.

101 Dong, N.N., Pedroni, M., Piccinelli, F. et al. (2011). NIR-to-NIR two-photon excited CaF$_2$:Tm^{3+},Yb^{3+} nanoparticles: multifunctional nanoprobes for highly penetrating fluorescence bio-imaging. *ACS Nano* 5 (11): 8665–8671.

102 Yang, Y.M., Shao, Q., Deng, R.R. et al. (2012). *In vitro* and *in vivo* uncaging and bioluminescence imaging by using photocaged upconversion nanoparticles. *Angew. Chem. Int. Ed.* 51 (13): 3125–3129.

103 Xing, H.Y., Bu, W.B., Zhang, S.J. et al. (2012). Multifunctional nanoprobes for upconversion fluorescence, MR and CT trimodal imaging. *Biomaterials* 33 (4): 1079–1089.

104 Chatterjee, D.K., Rufalhah, A.J., and Zhang, Y. (2008). Upconversion fluorescence imaging of cells and small animals using lanthanide doped nanocrystals. *Biomaterials* 29 (7): 937–943.

105 Jalil, R.A. and Zhang, Y. (2008). Biocompatibility of silica coated NaYF$_4$ upconversion fluorescent nanocrystals. *Biomaterials* 29 (30): 4122–4128.

106 Boyer, J.C., Manseau, M.P., Murray, J.I. et al. (2010). Surface modification of upconverting NaYF$_4$ nanoparticles with PEG-phosphate ligands for NIR (800 nm) biolabeling within the biological window. *Langmuir* 26 (2): 1157–1164.

107 Kobayashi, H., Kosaka, N., Ogawa, M. et al. (2009). *In vivo* multiple color lymphatic imaging using upconverting nanocrystals. *J. Mater. Chem.* 19 (36): 6481–6484.

108 Ehlert, O., Thomann, R., Darbandi, M. et al. (2008). A four-color colloidal multiplexing nanoparticle system. *ACS Nano* 2 (1): 120–124.

109 Cheng, L.A., Yang, K., Zhang, S.A. et al. (2010). Highly-sensitive multiplexed *in vivo* imaging using PEGylated upconversion nanoparticles. *Nano Res.* 3 (10): 722–732.

110 Liu, Y., Kang, N., Lv, J. et al. (2016). Deep photoacoustic/luminescence/magnetic resonance multimodal imaging in living subjects using high-efficiency upconversion nanocomposites. *Adv. Mater.* 28 (30): 6411–6419.

111 Yang, Y., Sun, Y., Cao, T.Y. et al. (2013). Hydrothermal synthesis of NaLuF$_4$:Sm-153,Yb,Tm nanoparticles and their application in dual-modality upconversion luminescence and SPECT bioimaging. *Biomaterials* 34 (3): 774–783.

112 Jennings, L.E. and Long, N.J. (2009). 'Two is better than one'—probes for dual-modality molecular imaging. *Chem. Commun.* (24): 3511–3524.

113 Liu, Q., Feng, W., and Li, F.Y. (2014). Water-soluble lanthanide upconversion nanophosphors: synthesis and bioimaging applications *in vivo*. *Coord. Chem. Rev.* 273: 100–110.

114 He, F., Yang, G.X., Yang, P.P. et al. (2015). A new single 808 nm NIR light-induced imaging-guided multifunctional cancer therapy platform. *Adv. Funct. Mater.* 25 (25): 3966–3976.

115 Sun, Y., Yu, M.X., Liang, S. et al. (2011). Fluorine-18 labeled rare-earth nanoparticles for positron emission tomography (PET) imaging of sentinel lymph node. *Biomaterials* 32 (11): 2999–3007.

116 Liang, L.U., Lu, Y.Q., Zhang, R. et al. (2017). Deep-penetrating photodynamic therapy with KillerRed mediated by upconversion nanoparticles. *Acta Biomater.* 51: 461–470.

117 Wang, C., Cheng, L.A., and Liu, Z.A. (2011). Drug delivery with upconversion nanoparticles for multi-functional targeted cancer cell imaging and therapy. *Biomaterials* 32 (4): 1110–1120.

118 Qiao, X.F., Zhou, J.C., Xiao, J.W. et al. (2012). Triple-functional core-shell structured upconversion luminescent nanoparticles covalently grafted with photosensitizer for luminescent, magnetic resonance imaging and photodynamic therapy *in vitro*. *Nanoscale* 4 (15): 4611–4623.

119 Qian, H.S., Guo, H.C., Ho, P.C.L. et al. (2009). Mesoporous-silica-coated up-conversion fluorescent nanoparticles for photodynamic therapy. *Small* 5 (20): 2285–2290.

120 Idris, N.M., Gnanasammandhan, M.K., Zhang, J. et al. (2012). In vivo photodynamic therapy using upconversion nanoparticles as remote-controlled nanotransducers. *Nat. Med.* 18 (10): 1580–U190.

121 Zhang, X., Yang, P.P., Dai, Y.L. et al. (2013). Multifunctional up-converting nanocomposites with smart polymer brushes gated mesopores for cell imaging and thermo/pH dual-responsive drug controlled release. *Adv. Funct. Mater.* 23 (33): 4067–4078.

122 Fan, W.P., Shen, B., Bu, W.B. et al. (2014). A smart upconversion-based mesoporous silica nanotheranostic system for synergetic chemo-/radio-/photodynamic therapy and simultaneous MR/UCL imaging. *Biomaterials* 35 (32): 8992–9002.

123 Hososhima, S., Yuasa, H., Ishizuka, T. et al. (2015). Near-infrared (NIR) up-conversion optogenetics. *Sci. Rep.-UK* 5: 16533.

124 Shah, S., Liu, J.J., Pasquale, N. et al. (2015). Hybrid upconversion nanomaterials for optogenetic neuronal control. *Nanoscale* 7 (40): 16571–16577.

125 Lin, X.D., Wang, Y., Chen, X. et al. (2017). Multiplexed optogenetic stimulation of neurons with spectrum-selective upconversion nanoparticles. *Adv. Healthc. Mater.* 6 (17): 1700446.

126 Wang, Y., Lin, X.D., Chen, X. et al. (2017). Tetherless near-infrared control of brain activity in behaving animals using fully implantable upconversion microdevices. *Biomaterials* 142: 136–148.

127 Lin, X.D., Chen, X., Zhang, W.C. et al. (2018). Core-shell-shell upconversion nanoparticles with enhanced emission for wireless optogenetic inhibition. *Nano Lett.* 18 (2): 948–956.

128 Chen, S., Weitemier, A.Z., Zeng, X. et al. (2018). Near-infrared deep brain stimulation via upconversion nanoparticle-mediated optogenetics. *Science* 359 (6376): 679–683.

129 Feliu, N., Neher, E., and Parak, W.J. (2018). Toward an optically controlled brain. *Science* 359 (6376): 632–634.

130 Du, S.R., Hernandez-Gil, J., Dong, H. et al. (2017). Design and validation of a new ratiometric intracellular pH imaging probe using lanthanide-doped upconverting nanoparticles. *Dalton Trans.* 46 (40): 13957–13965.

131 Liu, X., Zhang, S.Q., Wei, X. et al. (2018). A novel "modularized" optical sensor for pH monitoring in biological matrixes. *Biosens. Bioelectron.* 109: 150–155.

132 Nareoja, T., Deguchi, T., Christ, S. et al. (2017). Ratiometric sensing and imaging of intracellular pH using polyethylenimine-coated photon upconversion nanoprobes. *Anal. Chem.* 89 (3): 1501–1508.

133 Zheng, K.Z., Liu, Z.Y., Lv, C.J. et al. (2013). Temperature sensor based on the UV upconversion luminescence of Gd^{3+} in Yb^{3+}-Tm^{3+}-Gd^{3+} codoped $NaLuF_4$ microcrystals. *J. Mater. Chem. C* 1 (35): 5502–5507.

134 Zhang, J., Ji, B.W., Chen, G.B. et al. (2018). Upconversion luminescence and discussion of sensitivity improvement for optical temperature sensing application. *Inorg. Chem.* 57 (9): 5038–5047.

135 Fischer, L.H., Harms, G.S., and Wolfbeis, O.S. (2011). Upconverting nanoparticles for nanoscale thermometry. *Angew. Chem. Int. Ed.* 50 (20): 4546–4551.

136 Liu, G.F., Sun, Z., Fu, Z.L. et al. (2017). Temperature sensing and bio-imaging applications based on polyethylenimine/CaF_2 nanoparticles with upconversion fluorescence. *Talanta* 169: 181–188.

137 Sedlmeier, A., Achatz, D.E., Fischer, L.H. et al. (2012). Photon upconverting nanoparticles for luminescent sensing of temperature. *Nanoscale* 4 (22): 7090–7096.

138 Vetrone, F., Naccache, R., Zamarron, A. et al. (2010). Temperature sensing using fluorescent nanothermometers. *ACS Nano* 4 (6): 3254–3258.

139 Achatz, D.E., Meier, R.J., Fischer, L.H. et al. (2011). Luminescent sensing of oxygen using a quenchable probe and upconverting nanoparticles. *Angew. Chem. Int. Ed.* 50 (1): 260–263.

140 Zhang, K.Y., Yang, L., Lu, F. et al. (2018). A universal upconversion sensing platform for the sensitive detection of tumour-related ncRNA through an exo III-assisted cycling amplification strategy. *Small* 14 (10): 1703858.

141 Ge, X.Q., Sun, L.N., Ma, B.B. et al. (2015). Simultaneous realization of Hg^{2+} sensing, magnetic resonance imaging and upconversion luminescence *in vitro* and *in vivo* bioimaging based on hollow mesoporous silica coated UCNPs and ruthenium complex. *Nanoscale* 7 (33): 13877–13887.

142 Zhu, H., Lu, F., Wu, X.C. et al. (2015). An upconversion fluorescent resonant energy transfer biosensor for hepatitis B virus (HBV) DNA hybridization detection. *Analyst* 140 (22): 7622–7628.

143 Wei, R., Wei, Z., Sun, L. et al. (2016). Nile red derivative-modified nanostructure for upconversion luminescence sensing and intracellular detection of Fe(3+) and MR imaging. *ACS Appl. Mater. Interfaces* 8 (1): 400–410.

144 Peng, J., Xu, W., Teoh, C.L. et al. (2015). High-efficiency *in vitro* and *in vivo* detection of Zn^{2+} by dye-assembled upconversion nanoparticles. *J. Am. Chem. Soc.* 137 (6): 2336–2342.

145 Wang, J., Wen, F.Y., Zhang, Z.H. et al. (2005). Degradation of dyestuff wastewater using visible light in the presence of a novel nano TiO_2 catalyst doped with upconversion luminescence agent. *J. Environ. Sci.* 17 (5): 727–730.

146 Qin, W.P., Zhang, D.S., Zhao, D. et al. (2010). Near-infrared photocatalysis based on YF_3:Yb^{3+},Tm^{3+}/TiO_2 core/shell nanoparticles. *Chem. Commun.* 46 (13): 2304–2306.

147 Chen, C.K., Chen, H.M., Chen, C.J. et al. (2013). Plasmon-enhanced near-infrared-active materials in photoelectrochemical water splitting. *Chem. Commun.* 49 (72): 7917–7919.

148 Balaji, R., Kumar, S., Reddy, K.L. et al. (2017). Near-infrared driven photocatalytic performance of lanthanide-doped $NaYF_4$@CdS core-shell nanostructures with enhanced upconversion properties. *J. Alloys Compd.* 724: 481–491.

149 Li, M.H., Zheng, Z.J., Zheng, Y.Q. et al. (2017). Controlled growth of metal-organic framework on upconversion nanocrystals for NIR-enhanced photocatalysis. *ACS Appl. Mater. Interfaces* 9 (3): 2899–2905.

150 Zhou, Z.G., Hu, H., Yang, H. et al. (2008). Up-conversion luminescent switch based on photochromic diarylethene and rare-earth nanophosphors. *Chem. Commun.* (39): 4786–4788.

151 Liu, J.A., Bu, W.B., Pan, L.M. et al. (2013). NIR-Triggered anticancer drug delivery by upconverting nanoparticles with integrated azobenzene-modified mesoporous silica. *Angew. Chem. Int. Ed.* 52 (16): 4375–4379.

152 Zhou, L., Chen, Z.W., Dong, K. et al. (2014). DNA-Mediated construction of hollow upconversion nanoparticles for protein harvesting and near-infrared light triggered release. *Adv. Mater.* 26 (15): 2424–2430.

153 Yan, B., Boyer, J.C., Habault, D. et al. (2012). Near infrared light triggered release of biomacromolecules from hydrogels loaded with upconversion nanoparticles. *J. Am. Chem. Soc.* 134 (40): 16558–16561.

154 Carling, C.J., Boyer, J.C., and Branda, N.R. (2009). Remote-control photoswitching using NIR light. *J. Am. Chem. Soc.* 131 (31): 10838–10839.

155 Boyer, J.C., Carling, C.J., Gates, B.D. et al. (2010). Two-way photoswitching using one type of near-infrared light, upconverting nanoparticles, and changing only the light intensity. *J. Am. Chem. Soc.* 132 (44): 15766–15772.

156 Wang, L., Dong, H., Li, Y.N. et al. (2014). Reversible near-infrared light directed reflection in a self-organized helical superstructure loaded with upconversion nanoparticles. *J. Am. Chem. Soc.* 136 (12): 4480–4483.

157 Wu, W., Yao, L.M., Yang, T.S. et al. (2011). NIR-Light-induced deformation of cross-linked liquid-crystal polymers using upconversion nanophosphors. *J. Am. Chem. Soc.* 133 (40): 15810–15813.

158 Kim, W.J., Nyk, M., and Prasad, P.N. (2009). Color-coded multilayer photopatterned microstructures using lanthanide (III) ion co-doped $NaYF_4$ nanoparticles with upconversion luminescence for possible applications in security. *Nanotechnology* 20 (18).

159 Ma, R.L., Bullock, E., Maynard, P. et al. (2011). Fingermark detection on non-porous and semi-porous surfaces using $NaYF_4$:Er,Yb up-converter particles. *Forensic Sci. Int.* 207 (1–3): 145–149.

160 Ma, R.L., Shimmon, R., McDonagh, A. et al. (2012). Fingermark detection on non-porous and semi-porous surfaces using YVO_4:Er,Yb luminescent upconverting particles. *Forensic Sci. Int.* 217 (1–3): E23–E26.

161 Suzuki, S., Teshima, K., Wakabayashi, T. et al. (2011). Novel fabrication of NIR-vis upconversion $NaYF_4$:Ln (Ln = Yb, Er, Tm) crystal layers by a flux coating method. *J. Mater. Chem.* 21 (36): 13847.

162 Park, B.J., Hong, A.R., Park, S. et al. (2017). Flexible transparent displays based on core/shell upconversion nanophosphor-incorporated polymer waveguides. *Sci. Rep.-UK* 7: 45659.

163 de Wild, J., Meijerink, A., Rath, J.K. et al. (2011). Upconverter solar cells: materials and applications. *Energy Environ. Sci.* 4 (12): 4835–4848.

164 Ye, S.M., Teng, Y.X., Juan, A. et al. (2017). Modulated visible light upconversion for luminescence patterns in liquid crystal polymer networks loaded with upconverting nanoparticles. *Adv. Opt. Mater.* 5 (4): 1600956.

165 Wang, M., Li, M., Yang, M.Y. et al. (2015). NIR-induced highly sensitive detection of latent fingermarks by $NaYF_4$:Yb, Er upconversion nanoparticles in a dry powder state. *Nano Res.* 8 (6): 1800–1810.

166 Roh, J., Yu, H., and Jang, J. (2016). Hexagonal beta-$NaYF_4$:Yb^{3+},Er^{3+} nanoprism-incorporated upconverting layer in perovskite solar cells for near-infrared sunlight harvesting. *ACS Appl. Mater. Interfaces* 8 (31): 19847–19852.

167 Song, L., Dong, Y., Shao, Q. et al. (2018). Preparation of $Y_3Al_5O_{12}$:Ce nanophosphors using salt microemulsion method and their luminescent properties. *J. Mater. Sci.* 53 (21): 15196–15203.

168 Zhang, J.-J., Ning, J.-w., Liu, X.-J. et al. (2003). Synthesis of ultrafine YAG:Tb phosphor by nitrate–citrate sol–gel combustion process. *Mater. Res. Bull.* 38 (7): 1249–1256.

169 Aboulaich, A., Caperaa, N., El Hamzaoui, H. et al. (2015). *In situ* synthesis of a highly crystalline Tb-doped YAG nanophosphor using the mesopores of silica monoliths as a template. *J. Mater. Chem. C* 3 (19): 5041–5049.

170 Aboulaich, A., Deschamps, J., Deloncle, R. et al. (2012). Rapid synthesis of Ce^{3+}-doped YAG nanoparticles by a solvothermal method using metal carbonates as precursors. *New J. Chem.* 36 (12): 2493–2500.

171 R. Asakura; T. Isobe; K. Kurokawa; T. Takagi; H. Aizawa; M. Ohkubo (2007). Effects of citric acid additive on photoluminescence properties of YAG:Ce^{3+} nanoparticles synthesized by glycothermal reaction. 127(2), 416–422. doi:https://doi.org/10.1016/j.jlumin.2007.02.046.

172 Xu, Y.-F., Ma, D.-K., Guan, M.-L. et al. (2010). Controlled synthesis of single-crystal SrAl$_2$O$_4$:Eu^{2+},Dy^{3+} nanosheets with long-lasting phosphorescence. *J. Alloys Compd.* 502 (1): 38–42.

173 Hwang, K.-S., Kang, B.-A., Kim, S.-D. et al. (2011). Cost-effective electrostatic-sprayed SrAl$_2$O$_4$:Eu^{2+} phosphor coatings by using salted sol–gel derived solution. *Bull. Mater. Sci.* 34 (5): 1059–1062.

174 Elsagh, M., Rajabi, M., and Amini, E. (2014). Characterization of SrAl$_2$O$_4$:Eu^{2+}, Dy^{3+} phosphor nano-powders produced by microwave synthesis route. *J. Mater. Sci. - Mater. Electron.* 25 (4): 1612–1619.

175 Shan, W., Wu, L., Tao, N. et al. (2015). Optimization method for green SrAl$_2$O$_4$:Eu^{2+},Dy^{3+} phosphors synthesized via co-precipitation route assisted by microwave irradiation using orthogonal experimental design. *Ceram. Int.* 41 (10): 15034–15040.

176 Cheng, Y., Zhao, Y., Zhang, Y. et al. (2010). Preparation of SrAl$_2$O$_4$:Eu^{2+}, Dy^{3+} fibers by electrospinning combined with sol–gel process. *J. Colloid Interface Sci.* 344 (2): 321–326.

177 Zhao, Y., Zhang, X., Li, X. et al. (2020). Monodispersed spherical Y$_2$O$_3$ and Y$_2$O$_3$:Eu^{3+} particles synthesized from modified homogeneous urea precipitation process. *J. Alloys Compd.* 829: 154562.

178 Dhanaraj, J., Jagannathan, R., Kutty, T.R.N., and Lu, C.-H. (2001). Photoluminescence characteristics of Y$_2$O$_3$:Eu^{3+} nanophosphors prepared using sol–gel thermolysis. *J. Phys. Chem. B* 105 (45): 11098–11105.

179 Zhai, Y.Q., Yao, Z.H., Ding, S.W. et al. (2003). Synthesis and characterization of Y$_2$O$_3$:Eu nanopowder via EDTA complexing sol–gel process. *Mater. Lett.* 57 (19): 2901–2906.

180 Li, L. (2005). Synthesis of Y$_2$O$_3$ nano-powder from yttrium oxalate under ambient temperature. *J. Rare Earths* 23 (3): 358–361.

181 Jia, G., Yang, M., Song, Y. et al. (2009). General and facile method to prepare uniform Y$_2$O$_3$:Eu hollow microspheres. *Cryst. Growth Des.* 9 (1): 301–307.

182 Ye, T., Guiwen, Z., Weiping, Z., and Shangda, X. (1997). Combustion synthesis and photoluminescence of nanocrystalline Y$_2$O$_3$:EU phosphors. *Mater. Res. Bull.* 32 (5): 501–506.

183 Liu, J. and Li, Y.D. (2007). Synthesis and self-assembly of luminescent Ln^{3+}-doped LaVO$_4$ uniform nanocrystals. *Adv. Mater.* 19 (8): 1118–1122.

184 Liu, Y., Xiong, H., Zhang, N. et al. (2015). Microwave synthesis and luminescent properties of YVO$_4$:Ln^{3+} (Ln = Eu, Dy and Sm) phosphors with different morphologies. *J. Alloys Compd.* 653: 126–134.

185 Wang, F., Xue, X., and Liu, X. (2008). Multicolor tuning of (Ln, P)-doped YVO_4 nanoparticles by single-wavelength excitation. *Angew. Chem.* 120 (5): 920–923.

186 Mai, Hao-Xin; Zhang, Ya-Wen; Sun, Ling-Dong; Yan, Chun-Hua (2007). Orderly aligned and highly luminescent monodisperse rare-earth orthophosphate nanocrystals synthesized by a limited anion-exchange reaction. *Chemistry of Materials*, 19(18), 4514–4522. doi:https://doi.org/10.1021/cm0710731

187 Escudero, Alberto; Carrillo-Carrión, Carolina; Zyuzin, Mikhail Valeryevich; Ashraf, Sumaira; et al. (2016). Synthesis and functionalization of monodisperse near-ultraviolet and visible excitable multifunctional Eu^{3+}, Bi^{3+}:$REVO_4$ nanophosphors for bioimaging and biosensing applications. *Nanoscale*, doi:https://doi.org/10.1039/C6NR03369E.

188 Chang, M., Wang, M., Shu, M. et al. (2019). Enhanced photoconversion performance of $NdVO_4$/Au nanocrystals for photothermal/photoacoustic imaging guided and near infrared light-triggered anticancer phototherapy. *Acta Biomater.* 99: 295–306.

189 Chen, L., Zhang, Y., Luo, A. et al. (2012). The temperature-sensitive luminescence of $(Y,Gd)VO_4$:Bi^{3+},Eu^{3+} and its application for stealth anti-counterfeiting. *Phys. Status Solidi RRL* 6 (7): 321–323.

190 Kumar, P., Nagpal, K., and Gupta, B.K. (2017). Unclonable security codes designed from multicolor luminescent lanthanide-doped Y_2O_3 nanorods for anticounterfeiting. *ACS Appl. Mater. Interfaces* 9 (16): 14301–14308.

5

Metal Complex Nanosheets: Preparation, Property, and Application

Ryota Sakamoto

Department of Chemistry, Graduate School of Science, Tohoku University, and Division for the Establishment of Frontier Sciences of Organization for Advanced Studies at Tohoku University, 6-3 Aramaki-Aza-Aoba, Aoba-ku, Sendai, Miyagi 980-8578, Japan

5.1 Introduction

Two-dimensional (2D) materials have gained attention in various research fields as a promising nanomaterial useful in many applications. The research has been led by inorganic nanosheets, such as graphene [1], metal oxide nanosheets [2], transition metal dichalcogenides (TMDCs) [3], and few-atoms-thick layers of transition metal carbides, nitrides, or carbonitrides (MXenes) [4]. These inorganic nanomaterials have unique physical and chemical properties based on their 2D structures. For example, graphene possesses large carrier mobilities, large absorptivities per layer against visible to infrared light, great thermal conductivities, excellent mechanical strength, long spin-diffusion lengths, and large surface areas [1]. These characteristics underlie the anticipation from researchers and industries.

Prospects for inorganic nanosheets led researchers to also consider molecule-based nanosheets [5–7], which are tethered from molecular components, including organic monomers. The concept of molecule-based nanosheets (or 2D polymers) was already proposed in the 1930s [8]. However, the synthesis and identification of molecule-based nanosheets were attained only recently. There are two advantages for using molecule-based nanosheets over their inorganic counterpart: (i) variations of molecule-based nanosheets from combinations of organic monomers, and (ii) intrinsic porous structures in stacked forms that are beneficial in mass transfer and molecular storage.

Metal complexes feature photo-, opto-, and magneto-properties that are absent in pure organic molecules. The characteristic physical properties of metal complexes may be incorporated into the nanosheet. Metal complexes also serve as important components for molecular superstructures such as metal–organic frameworks (MOFs) [9, 10] with coordination bonds featuring unique high-symmetry orientations, often formed in a self-assembly fashion. Thus, metal complexes are functional motifs for molecule-based nanosheets. There might be an increase in the number of publications for molecule-based nanosheets in recent years.

Functional Nanomaterials: Synthesis, Properties, and Applications, First Edition.
Edited by Wai-Yeung Wong and Qingchen Dong.
© 2022 WILEY-VCH GmbH. Published 2022 by WILEY-VCH GmbH.

This section focuses on metal complex nanosheets. Section 5.2 focuses on the synthesis and characterization of new nanomaterials, while Section 5.3 describes their unique properties and functions. Finally, an outlook on metal complex nanosheets is discussed in Section 5.4.

5.2 Preparation of Metal Complex Nanosheets

5.2.1 Vacuum Phase Fabrication

Seitsonen, Barth, and coworkers fabricated a metal complex nanosheet network in the vacuum phase [11, 12]. Terephthalic acid (TPA) and metallic Fe were co-deposited on a Cu(100) flat substrate in an ultra-high vacuum (UHV) chamber by organic molecular beam epitaxy (OMBE) and electron beam heating, respectively. The Cu(100) substrate was heated at 400 or 450 K, allowing the adsorbates to move on the surface and providing a driving force to form coordination bonds. This method led to polymorphism. It also featured a two-dimensional network (Figure 5.1a), visualized by scanning tunneling microscopy (STM). Figure 5.1b–d shows a density functional theory (DFT)-optimized structure of the two-dimensional network. Four TPA ligands coordinated two Fe nuclei. Two of them featured an η^2-coordination mode, while the rest adopted a μ-coordination mode. The bite angles were 117.3° and 124.5° for the η^2- and μ-TPA ligands, respectively. There was a slight distortion in the TPA ligands, where they sat above the plane defined by the Fe atoms. The Fe–Fe spacing is 4.4 Å. A simulated STM image (Figure 5.1b) was consistent with the experimental (Figure 5.1a). The DFT calculation also suggested strong hybridization between the Fe center and Cu substrate, a magnetic moment of $3.4\,\mu_B$, and magnetic coupling between the two Fe nuclei. The 2D network structures produced by the vacuum phase deposition method are unstable under ambient conditions. However, they may feature significant coordination structures and physical properties.

5.2.2 Mechanical Exfoliation

The preparation of graphene was attained by mechanical exfoliation of highly ordered pyrolytic graphite (HOPG) using plastic tape [13]. However, this method was rarely applied for metal complex nanosheets. The first report was contributed by Clemente-Léon and Coronado [14]. The authors fabricated bulk crystals of [Fe(acac$_2$-trien)][MnCr(Br$_2$-An)$_3$]·(CH$_3$CN)$_2$ (Figure 5.2a–d). [Fe(acac$_2$-trien)][MnCr(Br$_2$-An)$_3$]·(CH$_3$CN)$_2$ featured a layered structure, where the anionic [MnCr(Br$_2$-An)$_3$] part comprises a hexagonal network with the cationic [Fe(acac$_2$-trien)] part embedded in the pore (Figure 5.2a,b). Then, the bulk crystal was exfoliated using plastic tape and transferred onto an oxide-passivated silicon substrate. Figure 5.2e shows a representative atomic force microscopy (AFM) image of the metal complex nanosheet with a thickness of up to 2 nm (Figure 5.2f). The authors ascribed the thickness to the bilayer nanosheet.

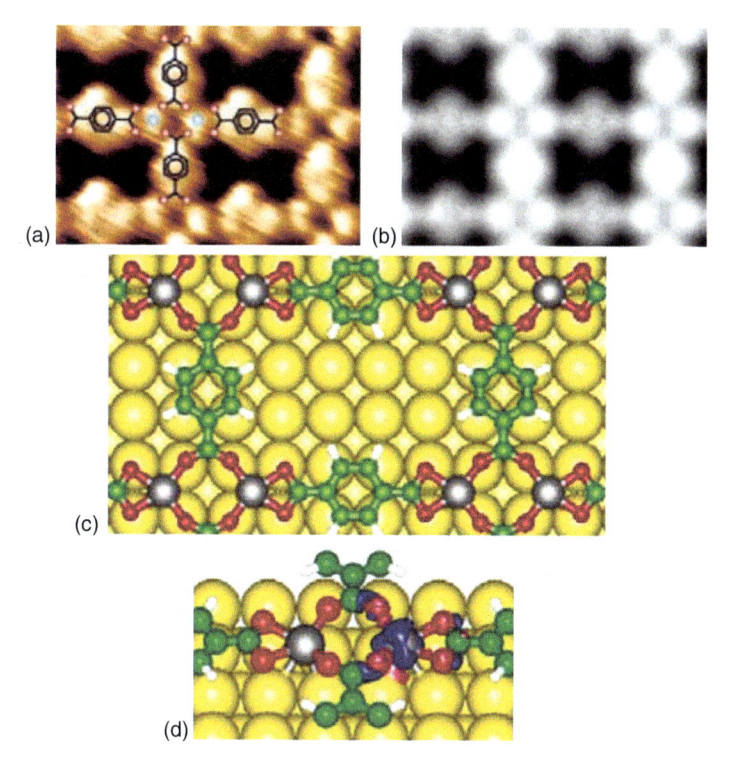

Figure 5.1 Fully reticulated nanoporous Fe-diterephthalate grid assembled on a Cu(100) substrate. (a) Constant current mode STM image showing a (6 × 4) unit cell (image size 40 Å × 30 Å). The arrangement of the TPA backbone indicates that a molecule is engaged in either two bidentate or four unidentate carboxylate bonds to the Fe−Fe centers (two blue spheres). (b) STM image simulation showing the contours of the constant local density of states at the sample Fermi level derived from the DFT model of the optimized structural arrangement in the model in (c). (d) Perspective view of the Fe−Fe unit. The Fe charge rearrangement contour levels indicated on the right, drawn with respect to a removed Fe atom, are $\pm 0.004\,e^-/\text{Å}^3$. Purple indicates increased electron density, and blue indicates decreased electron density. Carboxylate moieties and Fe centers lie in almost the same plane ($\Delta z \approx 0.2$ Å), with the Fe centers displaced from the fourfold hollow sites, resulting in a lateral Fe−Fe spacing of 4.4 Å. Source: Seitsonen et al. [12], Figures 01 & 02 [p. 5634 & 5635]/with permission from American Chemical Society.

5.2.3 Liquid-Phase Exfoliation

Bulk materials featuring 2D-layered motifs may be subjected to liquid-phase exfoliation, producing thinner nanosheets. A weakly held layered structure can be separated by ultrasonication with solvent. The domain size does not exceed the area of the crystal facet, and crystal defects and mosaicity reduce the domain size. An example of liquid-phase exfoliation was reported by Zamora and coworkers [15]. Figure 5.3a shows the crystal structure of a 2D MOF synthesized by a solvothermal process using $CuBr_2$, isonicotinic acid, KOH, and KBr in water, with $Cu_2Br(\text{isonicotinate})_2$. The 2D laminar frameworks stacked with each other without strong chemical bonding, facilitating liquid-phase exfoliation in water (10^{-10} mg ml^{-1}). AFM found nanosheet

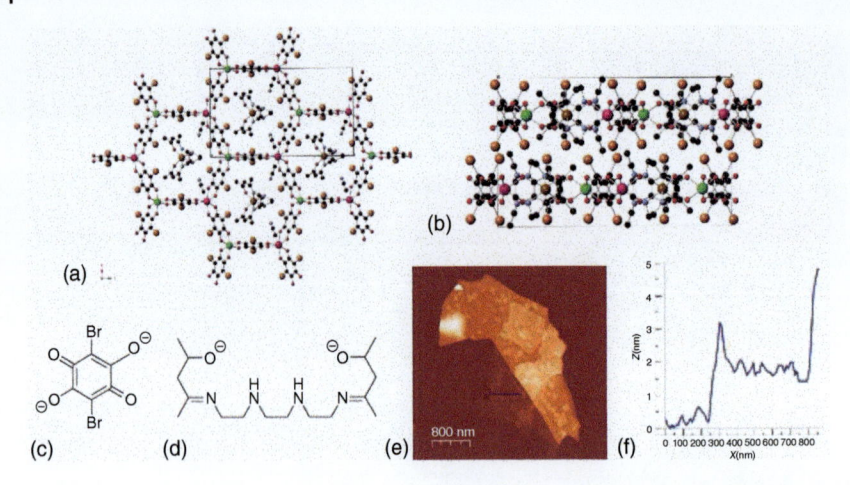

Figure 5.2 (a, b) Projections of [Fe(acac$_2$-trien)][MnCr(Br$_2$-An)$_3$]·(CH$_3$CN)$_2$ in the *ab* and *bc* planes (Fe [brown], Cr [green], Mn [pink] C [black], N [blue], O [red], Br [orange]). Hydrogen atoms are omitted for clarity. (c, d) Chemical structures of Br$_2$-An and acac$_2$-trien. (e, f) AFM images and height profile of the [Fe(acac$_2$-trien)][MnCr(Br$_2$-An)$_3$] nanosheet on a SiO$_2$/Si substrate. Source: Abhervé et al. [14], Figures 01&02&06 and Schemes 01&02 [p. 4666&4667&4669]/with permission from The Royal Society of Chemistry/CC BY 3.0.

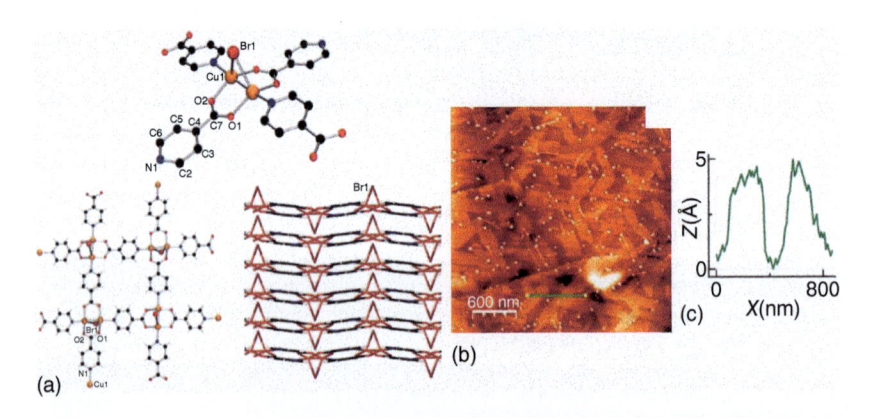

Figure 5.3 (a) Crystal-phase structure of the MOF studied herein: detail view of the coordination sphere of the dinuclear copper component (top); top-view of the single-layer framework (left), and side-view of the laminated layers (right). Source: Hermosa et al. [15]/Royal Society of Chemistry/CC BY 3.0. (b) AFM topography image of the single-layer MOF. Source: Amo-Ochoa et al. [15], Figures 01&02 [p. 3263&3264]/with permission from the Royal Society of Chemistry. (c) Cross-section analysis across the green line in (b).

domains with a lateral size of ∼600 nm (Figure 5.3b; along the longer side) and a thickness of 5 ± 0.15 Å (Figure 5.3c), consistent with the single-layer nature of the exfoliated MOF.

A series of chemically-assisted exfoliations of MOFs was also proposed. Jiang and Zhou applied the method to a layered MOF comprising [tetrakis(4-carboxyphenyl) porphyrinato]palladium and Zn dinuclear paddlewheel unit (Figure 5.4a) [16].

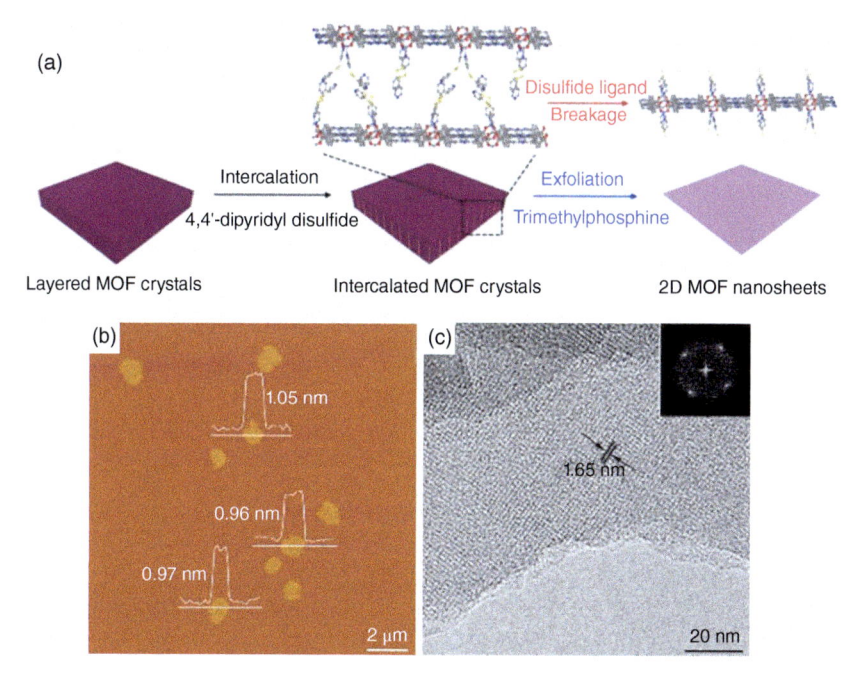

Figure 5.4 (a) Schematic illustration of the overall process developed to produce a MOF nanosheet via an intercalation and chemical exfoliation approach. (b) AFM image of the exfoliated MOF nanosheet with corresponding height profiles. (c) High-resolution TEM image of the exfoliated multilayer MOF nanosheet. The corresponding FFT pattern is shown in the inset. Source: Ding et al. [16], Figures 01&04 [p. 9136&9138]/with permission from American Chemical Society.

Upon addition of 4,4′-dipyridyl disulfide (DPDS) in a suspension of MOF crystals in N,N-diethylformamide and ethanol, DPDS coordinates with the Zn paddlewheel, expanding the out-of-plane lattice parameter from 19.604 to 45.237 Å. The intercalated MOF was then suspended in ethanol containing triethylphosphine under N_2. Triethylphosphine induced DPDS reduction, leading to the spontaneous exfoliation of the layered MOF. Figure 5.4b,c shows AFM and transmission electron microscopy (TEM) images, featuring a 1 nm thick single layer and 1.65 nm interplanar distance of the (100) plane of the intercalated MOF.

5.2.4 Liquid/Liquid Interfacial Synthesis

Sakamoto, Wong, and Nishihara created a metal complex nanosheet featuring the bis(dipyrrinato)zinc(II) complex motif (Figure 5.5a) using liquid/liquid interfacial synthesis [17]. The synthetic procedure corresponded to a dichloromethane solution of the dipyrrin ligand covered with aqueous $Zn(OAc)_2$ under ambient conditions (Figure 5.5b). Spontaneous complexation between the dipyrrin ligand and Zn^{2+} ions occurred at the liquid/liquid interface, forming a multilayer nanosheet (Figure 5.5c). The multilayer nanosheet featured a sheet morphology on the centimeter scale (Figure 5.5d). Optical microscopy (OM), scanning electron microscopy (SEM), and

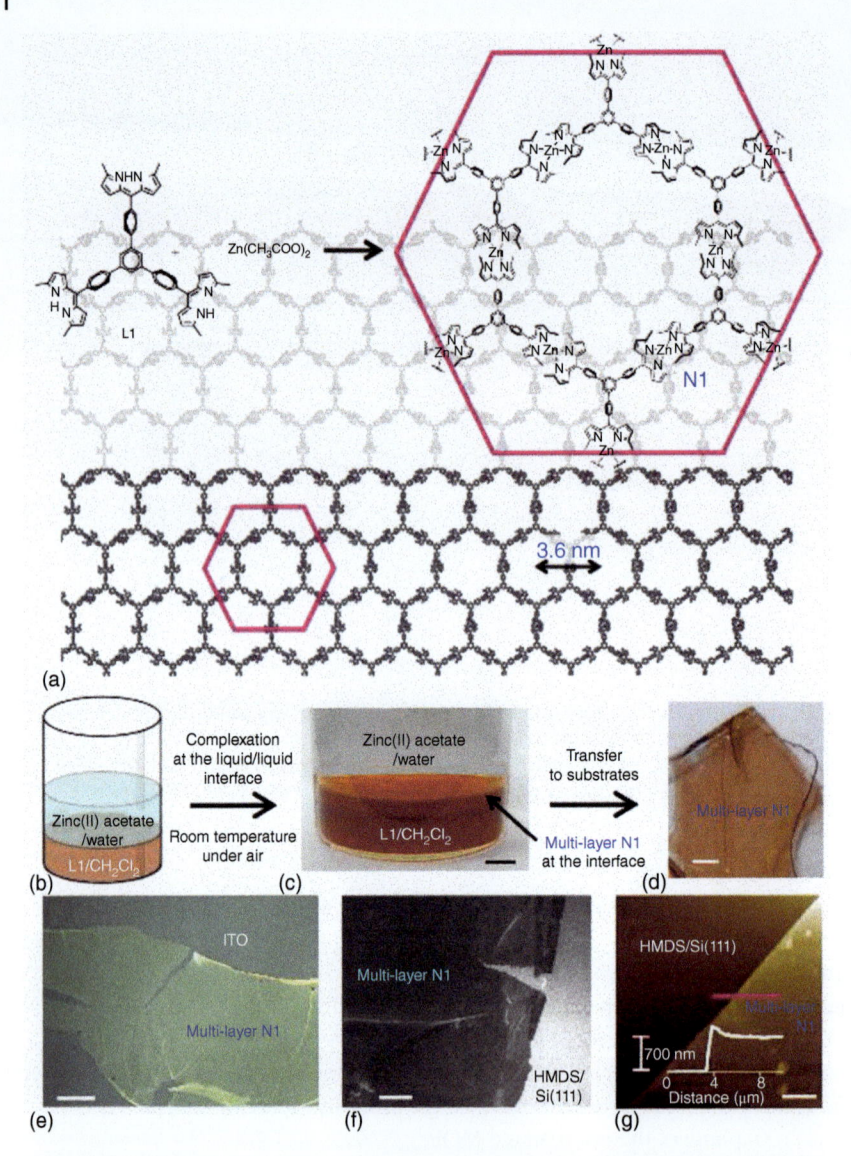

Figure 5.5 (a) Chemical structures of the three-way dipyrrin ligand molecule and the bis(dipyrrinato)zinc(II) complex nanosheet. (b) Schematic illustration of the liquid/liquid interfacial synthesis. Source: (a, b) Sakamoto et al. [17]/Springer Nature/CC BY 4.0. (c) Photograph of the liquid/liquid interfacial reaction holding a multilayer nanosheet at the interface. (d) Photograph of the multilayer nanosheet transferred onto an ITO substrate. (e) Optical microscopic image. (f) FE-SEM image on a HMDS/Si(111) substrate. (g) AFM height image and its cross-section analysis along the magenta line. Source: (c–g) Sakamoto et al. [17], Figures 01&02 [p. 2&3]/with permission from Macmillan Publishers Limited/CC BY 4.0.

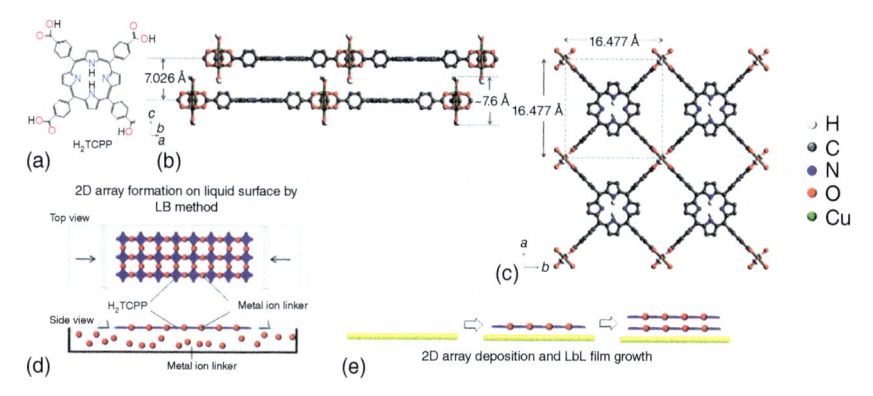

Figure 5.6 (a) Molecular structures of H$_2$TCPP. (b, c) Structure of NAFS-2. (d) Gas/liquid interfacial synthesis of NAFS-2. (e) Layer-by-layer growth of NAFS-2. Source: Makiura et al. [19], Figures 01&02 [p. 5641]. Copyright 2011 American Chemical Society.

AFM also visualized the sheet morphology on smaller scales (Figure 5.5e–g). The dipyrrin ligand concentration could control the nanosheet thickness ranging from 6 to 800 nm (corresponding to 5–670 layers). The crystallinity of the nanosheet was verified by selected area electron diffraction in TEM (TEM/SAED).

5.2.5 Gas/Liquid Interfacial Synthesis

A prototype of the gas/liquid interfacial synthesis for metal complex nanosheets was reported by Varaksa, Magnera, and Michl [18]. 1,3,5-tris[10-(3-ethylthiopropyl)dimethylsilyl-1,10-dicarba-*closo*-decaboran-1-yl]benzene formed a network structure on Hg in a Langmuir–Blodgett trough upon application of electrochemical potential. The authors estimated the structure of the coordination network using Austin Model 1 (AM1) semi-empirical calculation, where the triangular sulfide molecules were connected through S—Hg^{2+}—S or S—Hg$_2^{2+}$—S bonds. The 2D coordination network structure was not identified and isolated. Nevertheless, the authors pioneered the research in metal complex nanosheets.

Makiura et al. demonstrated that the air/liquid interfacial protocol was applied to fabricate single-layer MOFs [19]. They employed 5,10,15,20-tetrakis(4-carboxyphenyl)porphyrin (H$_2$TCPP) (Figure 5.6a) as a quasi-fourfold symmetry carboxylate ligand molecule to create a 2D MOF nanosheet (NAFS-2; Figure 5.6b,c). An aqueous CuCl$_2$ solution was used as the subphase of a Langmuir–Blodgett trough, where Cu^{2+} serves as a metal linker. To the subphase, a solution of H$_2$TCPP in a mixture of toluene and ethanol was spread, forming the 2D network at the air/liquid interface upon compression (Figure 5.6d). The resultant nanosheet on the interface was transferred onto a flat substrate by Langmuir–Schäfer technique.

Moreover, the iteration of the series of processes led to stacked NAFS-2 with the desired thickness (Figure 5.6e). The authors confirmed the crystallinity of the stacked sample by synchrotron grazing incidence X-ray diffraction (GIXRD). The average domain size was estimated to be 20 nm from GIXRD and AFM.

Sakamoto, Schlüter, and coworkers created a molecule-based nanosheet featuring the bis(terpyridine)metal complex motif (Figure 5.7a) [20]. Before forming the nanosheet between a terpyridine ligand with sixfold symmetry (Figure 5.7a) and Fe ions at an air/water interface, the authors employed Brewster angle microscopy (BAM), AFM, and Langmuir isotherm measurements, confirming that the terpyridine ligand molecules on a pure water subphase form a dense and reversible monolayer that adopts a horizontal conformation. The addition of $Fe(NH_4)_2(SO_4)_2$ to the subphase completed the nanosheet formation. During the polymerization, the monolayer was pressurized moderately ($2\,m\,Nm^{-1}$). The resultant nanosheet was identified by AFM, OM, and TEM (Figure 5.7b–d). The AFM height image quantified the monolayer thickness to be 1.4 nm, with overestimation from the theoretical value of 0.8 nm. This overestimation is often seen in nanosheets. The OM picture showed a large domain (>500 × 500 μm), and the bright-field TEM image of the nanosheet was free-standing to traverse a grid filled with holes.

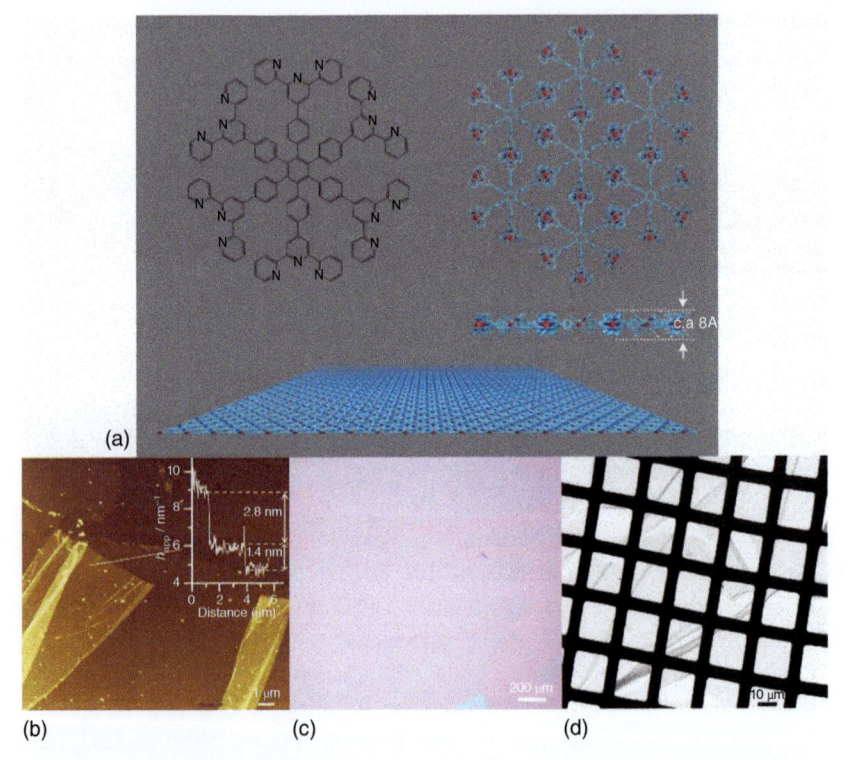

Figure 5.7 (a) Chemical structure of the terpyridine ligand molecule and schematic illustrations for the ideal two-dimensional network obtained from the ligand through complexation with Fe^{2+} ions (Fe, red; C, turquoise; N, blue). (b) Tapping-mode AFM image with a height profile measured along the white line in (a). (c) Optical microscope image after vertical transfer onto 300 nm SiO_2/Si. (d) TEM image after horizontal transfer from the top onto a Cu grid with 20 μm × 20 μm holes. Source: Bauer et al. [20], Figures 01&04 [p. 7880&7882]/with permission from Wiley-VCH Verlag GmbH & Co. KGaA.

5.3 Properties of Metal Complex Nanosheets

5.3.1 Electroproperties

Several metal complex nanosheets were employed as active materials for lithium-ion batteries (LIBs). Their porous and layered structures were advantageous. An example was reported by Sakaushi, Nishihara, and coworkers [21], where a conductive bis(diimino)nickel framework (NiDI) was employed (Figure 5.8a). These metal complex nanosheets feature electroconductivity [22]. The neutral form of a mononuclear model complex undergoes reversible redox reactions in oxidation and reduction directions (Figure 5.8b). The authors anticipated that the redox characteristics could enhance LIB capacity. Figure 5.8c presented a close-up TEM image of NiDI with a hexagonal periodicity consistent with the expected 2D lattice. Cyclic voltammetry found anodic and cathodic peaks at 3.73 and 3.56 V vs. Li^+/Li, respectively. These results were assigned to the conversion between the neutral and oxidized states (Figure 5.8d). On the other hand, a cathodic peak at 3.21 V coupled with a broad anodic peak ascribed the conversion between the neutral and reduced state. The capacity with a negative cut-off potential of 2.0 V showed more than 100 mA h g^{-1} in both positive and negative scans, while it decreased to c. 40 mA h g^{-1} with a potential window of 3.0–4.5 V. The performance of NiDI as a LIB cathode was further investigated using a coin cell (Figure 5.8e). Two sets of plateau regions were found in the potential range of 2.0–4.3 V, stemming from the insertion and desertion of Li^+ or PF_6^-. The specific capacity reached 155 mA h g^{-1} at 10 mA g^{-1}, corresponding to a specific energy density of 434 W h kg^{-1}. NiDI has one of the highest specific capacities among MOF-based cathode materials, comparable to $LiCoO_2$ and $LiFePO_4$. The cell also featured durability up to 300 cycles at 250 mA g^{-1}, and the coulombic efficiency was >99% (Figure 5.8f).

Nishihara and coworkers demonstrated that the bis(terpyridine)metal complex nanosheet (Figure 5.7a) served as an active layer for a quasi-solidified electronic device (Figure 5.9) [23]. The bis(terpyridine)metal complex motif underwent rapid, robust, and reversible redox reactions on the metal center, which underlay the application. The authors employed the liquid/liquid interfacial reaction to obtain stacked nanosheets with thicknesses of 180 and 120 nm for the Fe and Co centers, respectively. The Fe nanosheet was deposited on an indium tin oxide (ITO) electrode and was subjected to electrochemical polarization in an electrolyte solution (Figure 5.9a). The original valence of the Fe center was +2, making the nanosheet deep purple with a metal-to-ligand charge transfer (MLCT) transition. When the Fe nanosheet was oxidized, the MLCT band disappeared, changing the color to pale yellow. The electrochromic response was quick and durable. The faradaic current for the redox reaction decayed in 0.35 seconds and observed negligible color fading after 800 cycles. The Fe nanosheet on the ITO electrode was then incorporated into a solidified device with nanosheet-ITO/gel electrolyte/ITO (Figure 5.9b). By applying a voltage to the Fe nanosheet (+3.0 V vs. the counter electrode), a color change to pale yellow was observed (Figure 5.9c).

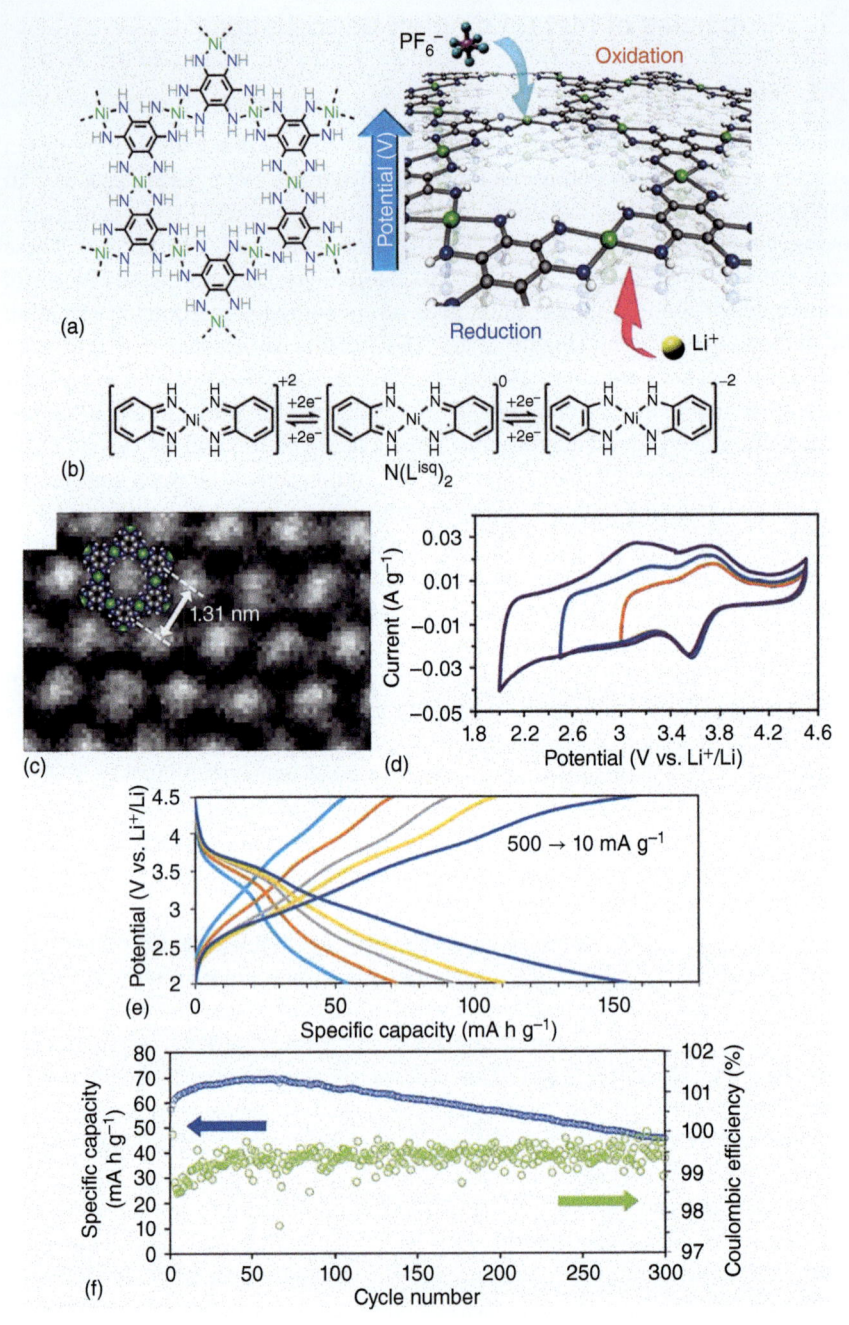

Figure 5.8 (a) Chemical structure and redox reactions associated with counter ion uptake of NiDI. (b) Redox reactions in a mononuclear model complex. (c) High-magnification TEM image of a crystalline domain of NiDI (Ni: green, N: blue, C: gray, H: white). (d) Cyclic voltammograms of NiDI at 0.1 mV s^{-1} in various potential windows. (e) Charge–discharge curves at 10 to 500 mA g^{-1}. (f) Cycling test and coulombic efficiency for up to 300 cycles at a current density of 250 mA g^{-1}. Source: (a, b, d–f) Reproduced with permission from Wada et al. [21]. Copyright 2018 Wiley-VCH Verlag GmbH & Co. KGaA. (c) Wada et al. [21], Figures 01&02&03 [p. 8886&8887&8888]/with permission from Wiley-VCH Verlag GmbH & Co. KGaA.

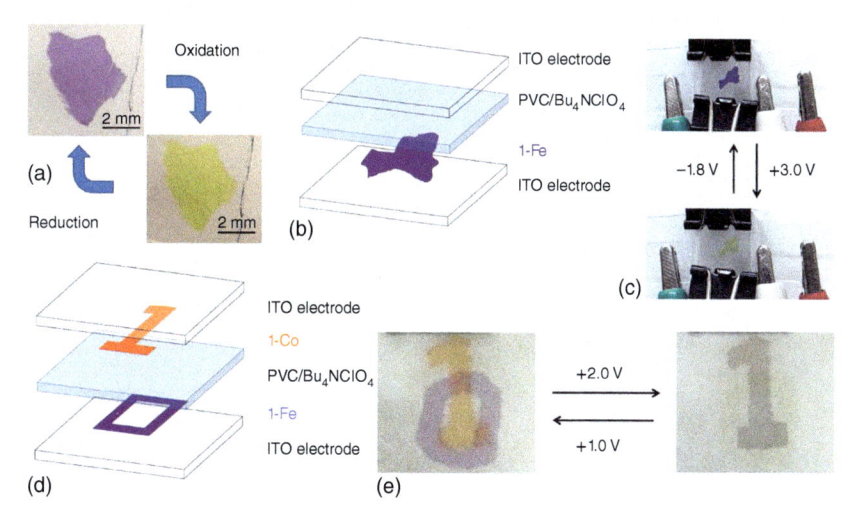

Figure 5.9 (a) Color change in the Fe-terpyridine nanosheet upon the reversible Fe^{3+}/Fe^{2+} redox reaction on an ITO electrode. (b, c) Structure of a solidified electrochromic device and its operation. (d, e) Structure of a solidified dual electrochromic device and its operation. Source: Takada et al. [23], Figures 04&07 [p. 4684&4686]/with permission from American Chemical Society.

In contrast, a negative voltage (−1.8 V) application restored the original deep purple color. The color changes were similar to those in the solution assigned to the redox reaction on the Fe^{3+}/Fe^{2+} center. Therefore, the authors proposed a dual electrochromic device (Figure 5.9d,e). An ITO electrode on one side was modified with the Fe nanosheet prepared in the shape of "0" and on the other side with the Co nanosheet engraved in the shape of "1." The original color of the Co nanosheet was orange with the Co^{2+} center. By applying a voltage of +2.0 V to the Fe nanosheet against the Co nanosheet, the colors of "0" and "1" were altered to pale-yellow and purple. The Fe^{2+} and Co^{2+} centers underwent oxidation (to Fe^{3+}) and reduction (to Co^+). On the other hand, a voltage of +1.0 V turned the device into its original state. The series of electrochromic behavior may be utilized in electronic paper.

5.3.2 Photoproperties

In Section 5.3.1, the Fe- and Co-terpyridine nanosheets underwent distinctive electrochromic behavior. Sakamoto, Nishihara, and coworkers demonstrated that the introduction of Zn^{2+} instead of Fe^{2+} and Co^{2+} changed the functionality of the terpyridine nanosheet [24]. The metal center swap led to the loss of redox ability, but simultaneously, the acquisition of luminescent property. The Zn-terpyridine nanosheet with a thickness of 65 nm was colorless and transparent, but emitted blue fluorescence at 480 nm upon excitation with UV light (Figure 5.10a,b). This emission is ascribed to a ligand-centered $\pi-\pi^*$ transition. The hexagonal framework of the Zn-terpyridine nanosheet is cationic, which is accompanied by BF_4^- as a counter anion. The cationicity allowed the Zn-terpyridine nanosheet to

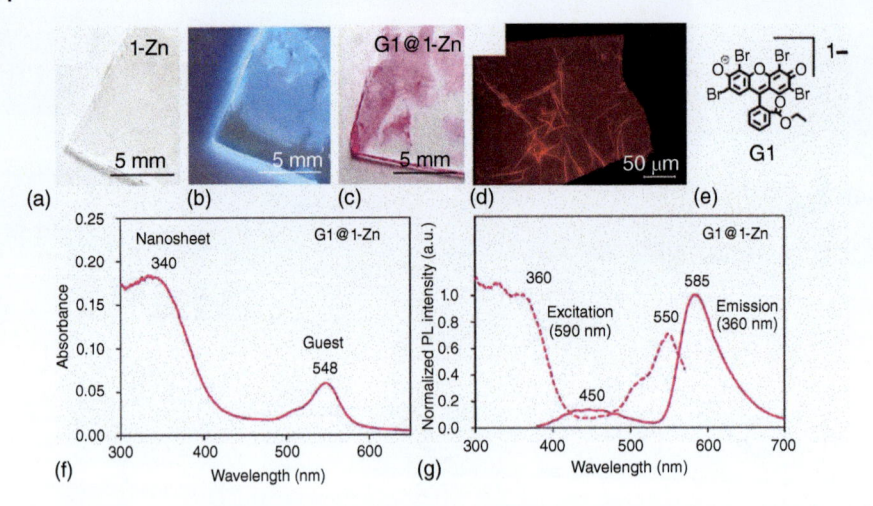

Figure 5.10 (a, b) Photos of the Zn-terpyridine nanosheet under ambient and 365 nm light. (c) Photo of G1@Zn-terpyridine nanosheet under ambient light. (d) Fluorescent microscopic image for G1@Zn-terpyridine nanosheet under illumination with UV light. (e) Molecular structure of G1. (f, g) Absorption, emission, and excitation spectra for G1@Zn-terpyridine nanosheet. Images and spectra were recorded on quartz. Source: Tsukamoto et al. [24], Figures 02&05 [p. 5360&5363]/with permission from American Chemical Society.

uptake an anionic dye molecule (G1) through a solution-phase anion exchange reaction (Figure 5.10c,e,f). Upon excitation of the G1-incorporated nanosheet with UV light, red luminescence from G1 overwhelmed the original blue from the nanosheet (Figure 5.10d,g). The authors concluded from the luminescence quantum yield measurements that quasi-quantitative energy transfer took place from the nanosheet to G1.

Sakamoto, Wong, and Nishihara pursued the application of a porphyrin-conjugated bis(dipyrrinato)zinc complex nanosheet as an active layer for photoelectric conversion (Figure 5.11a) [25]. The authors designed a porphyrin-hybridized dipyrrin ligand, where the porphyrin core provided fourfold symmetry, allowing them to fabricate a 2D grid framework in the nanosheet. The porphyrin core was also expected to enhance the photoelectric conversion ability, because it is an established dye molecule. Indeed, the chessboard nanosheet exhibited performance better than the bis(dipyrrinato)zinc complex nanosheet shown in Figure 5.5a without the porphyrin core (Figure 5.11b,c). The porphyrin-conjugated nanosheet featured a maximum quantum efficiency of 2.02%, more than twice that of the plain nanosheet (0.86%). Thus, the bis(dipyrrinato)zinc(II) complex nanosheet served as the first photofunctional bottom-up coordination nanosheet. Furthermore, the photoresponsive wavelength of the porphyrin-conjugated nanosheet was expanded to cover the whole visible region (400–650 nm), which was wider than that of the plain nanosheet (450–550 nm). This result was caused by the absorption bands of the porphyrin (Soret band: 400–450 nm; Q band: 550–650 nm) and dipyrrin ($^1\pi$–π^*: 450–550 nm) units, which complemented each other (Figure 5.11c).

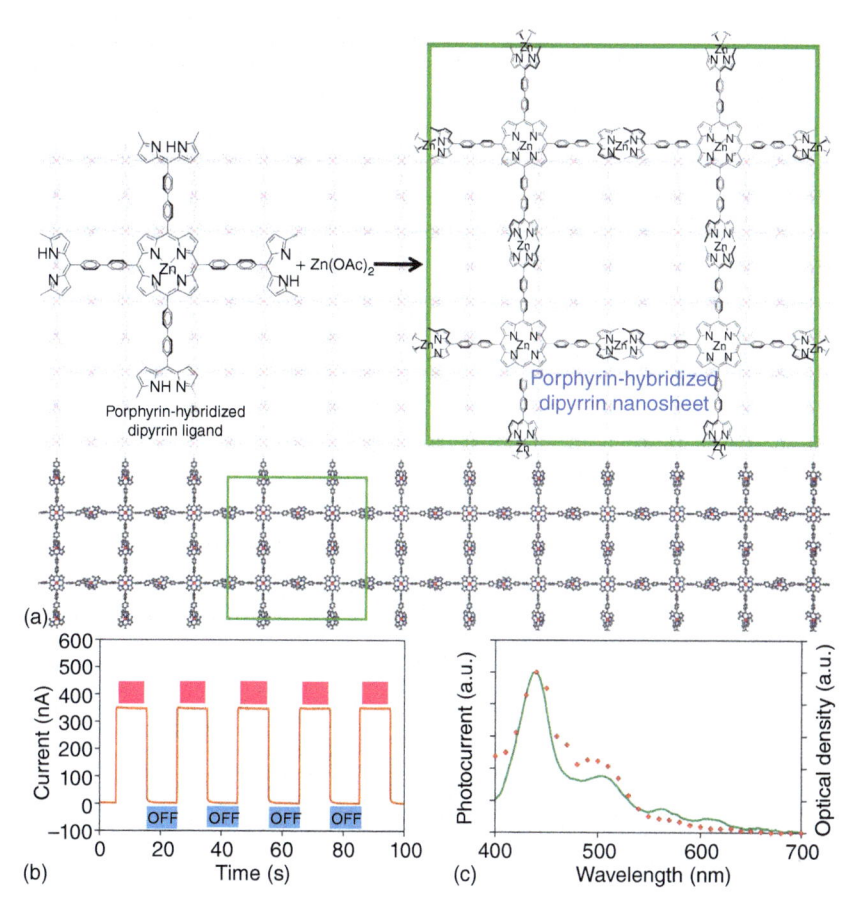

Figure 5.11 (a) Chemical structure of the porphyrin-hybridized bis(dipyrrinato)zinc(II) complex nanosheet. (b) Typical anodic current response upon irradiation of a working electrode (SnO$_2$ substrate modified with the nanosheet) with intermittent 440 nm light in the acetonitrile electrolyte solution. (c) Action spectrum for the photocurrent generation (orange dots) and absorption spectrum of the nanosheet (solid green line). Source: Sakamoto et al. [25], Figures 01&07 [p. 3527&3529]/with permission from Wiley-VCH Verlag GmbH & Co. KGaA.

5.3.3 Magnetoproperties

Magnetic order in atomically thin layers of metal complex nanosheets was observed and reported by Espallargas, Coronado, and coworkers [26]. The authors synthesized MUV-1-Cl with a formula of Fe(bimCl)$_2$ (HbimCl = 5-chlorobenzimidazole) (Figure 5.12a,b). The iron centers adopted a distorted tetrahedral coordination sphere, which was bridged by bimCl.

Bulk crystalline MUV-1-Cl underwent antiferromagnetic coupling among the high-spin Fe(II) centers. Its magnetic parameters were $\theta = -80.6$ K, $T_N = 20$ K, and $J = -22.9$ cm^{-1} for Curie–Weiss temperature, Néel temperature, and the exchange parameter, respectively (Figure 5.12c,d). Bulk MUV-1-Cl was exfoliable

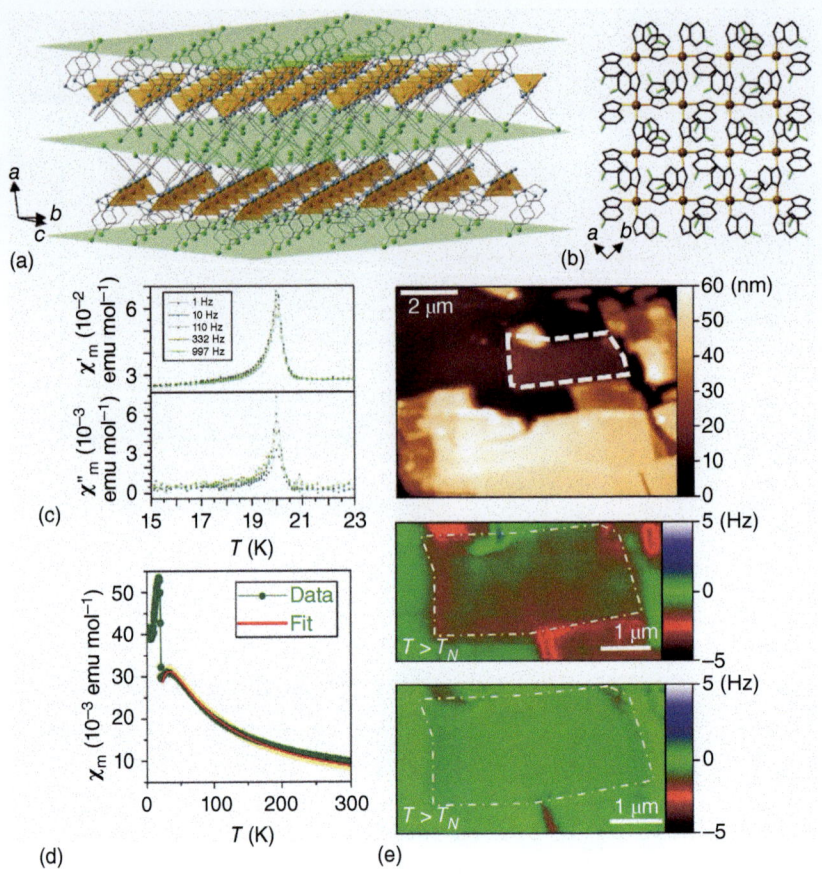

Figure 5.12 (a) Layered structure of MUV-1-Cl showing the Cl atoms located at the surface of the layers, represented as green planes. Iron centers are shown in orange (polyhedral representation), nitrogen atoms in blue, carbon atoms in black, and chlorine atoms in green. Hydrogen atoms are omitted for clarity. (b) Structure of a single layer of MUV-1-Cl viewed along the c axis (a–b plane). (c) Temperature dependence of the in-phase (top) and out-of-phase (bottom) dynamic a.c. susceptibility of MUV-1-Cl measured at different frequencies. (d) Thermal dependence of magnetic susceptibility at 2–300 K. The data are fitted (red line) following a Lines expansion for a quadratic-layer Heisenberg antiferromagnet with $S = 2$. The yellow line represents the prediction bands with a confidence interval of 95%. (e) LT-MFM measurements of a 5.7-nm-thick flake of MUV-1-Cl. Top: general topography image of the selected region with the flake highlighted by a dashed white line; middle and bottom: MFM images of the flake (highlighted by dot-dashed white lines) showing the difference in frequency shift below (middle) and above (bottom) T_N. Note the change in color contrast below T_N, which indicates an attractive tip-sample interaction in the ordered state that disappears at temperatures above T_N. Source: López-Cabrelles et al. [26], Figures 02&03 [p. 1003&1004]/with permission from Springer Nature Publishing AG.

into single- or few-layer nanosheets by mechanical method using plastic tape. A 5.7-nm thick MUV-1-Cl was deposited on a silicon substrate and subjected to low-temperature magnetic force microscopy (LT-MFM; Figure 5.12e). The MUV-1-Cl nanosheet domain had attractive tip-sample interaction below T_N (red region) while canceled above T_N. The series of LT-MFM results were associated with the antiferromagnetism observed in the bulk form.

5.4 Outlook on Metal Complex Nanosheets

There is potential for metal complexes as functional nanomaterials. However, further exploration of the basic and material science of these complexes is needed. The definition of nanosheets is still under discussion. Are nanosheets exclusively single-layered? Are multilayer nanosheets acceptable? How thick may they be? Most applications utilize metal complex nanosheets as bulk forms, or unidentified stacking patterns or layers. In contrast, they are most useful in applications in single-, or few-layer forms.

Recent innovation and improvement in analytical methods, such as AFM, STM, and TEM have accelerated the science of nanosheets. However, the quality (i.e. crystallinity, degrees of disordered, or defect sites) of single- or few-layer metal complex nanosheets has been seldom quantified. Practical and rapid analytic methods evaluating metal complex nanosheet qualities are still insufficient.

Recent efforts have demonstrated the functionality of metal complex nanosheets, which could lead to practical applications in the future. However, metal complex nanosheets have lagged behind inorganic nanosheets.

Another challenge is fabricating heterostructures, such as stacked layers comprising different nanosheets and lateral heterojunctions. These elaborated structures will enhance the functionality of metal complex nanosheets, as demonstrated in inorganic nanosheets [27].

References

1 Allen, M.J., Tung, V.C., and Kaner, R.B. (2010). Honeycomb carbon: a review of graphene. *Chem. Rev.* 110: 132–145.
2 Wang, L. and Sasaki, T. (2014). Titanium oxide nanosheets: graphene analogues with versatile functionalities. *Chem. Rev.* 114: 9455–9486.
3 Manzeli, S., Ovchinnikov, D., Pasquier, D. et al. (2017). 2D transition metal dichalcogenides. *Nat. Rev. Mater.* 2: 17033.
4 Fu, Z., Wang, N., Legut, D. et al. (2019). Rational design of flexible two-dimensional MXenes with multiple functionalities. *Chem. Rev.* 119: 11980–12031.
5 Sakamoto, R., Takada, K., Pal, T. et al. (2017). Coordination nanosheets (CONASHs): strategies, structures and functions. *Chem. Commun.* 53: 5781–5801.
6 Feng, X. and Schlüter, A.D. (2018). *Angew. Chem. Int. Ed.* 57: 13478–13763.
7 Rodríguez-San-Miguel, D., Montoro, C., and Zamora, F. (2020). Covalent organic framework nanosheets: preparation, properties and applications. *Chem. Soc. Rev.* 49: 2291–2302.

8 Gee, G. and Rideal, E.K. (1935). Reactions in monolayers of drying oils. I. The oxidation of the maleic anhydride compound of β-elaeostearin. *Proc. R. Soc. London Ser. A* 153: 116–128.

9 Krause, S., Hosono, N., and Kitagawa, S. (2020). Chemistry of soft porous crystals: structural dynamics and gas adsorption properties. *Angew. Chem. Int. Ed.* 59: 15325–15341.

10 Furukawa, H., Cordova, K.E., O'Keeffe, M., and Yaghi, O.M. (2013). The chemistry and applications of metal–organic frameworks. *Science* 341: 1230444.

11 Lingenfelder, M.A., Spillmann, H., Dmitriev, A. et al. (2004). Towards surface-supported supramolecular architectures: tailored coordination assembly of 1,4-benzenedicarboxylate and Fe on Cu(100). *Chem. Eur. J.* 10: 1913–1919.

12 Seitsonen, A.P., Lingenfelder, M., Spillmann, H. et al. (2006). Density functional theory analysis of carboxylate-bridged diiron units in two-dimensional metal–organic grids. *J. Am. Chem. Soc.* 128: 5634–5635.

13 Novoselov, K.S., Geim, A.K., Morozov, S.V. et al. (2004). Electric field effect in atomically thin carbon films. *Science* 306: 666–669.

14 Abhervé, A., Mañas-Valero, S., Clemente-León, M., and Coronado, E. (2015). Graphene related magnetic materials: micromechanical exfoliation of 2D layered magnets based on bimetallic anilate complexes with inserted [Fe^{III}(acac$_2$-trien)]$^+$ and [Fe^{III}(sal$_2$-trien)]$^+$ molecules. *Chem. Sci.* 6: 4665–4673.

15 Amo-Ochoa, P., Welte, L., González-Prieto, R. et al. (2010). Single layers of a multifunctional laminar Cu(I,II) coordination polymer. *Chem. Commun.* 46: 3262–3264.

16 Ding, Y., Chen, Y.-P., Zhang, X. et al. (2017). Controlled intercalation and chemical exfoliation of layered metal–organic frameworks using a chemically labile intercalating agent. *J. Am. Chem. Soc.* 27: 9136–9139.

17 Sakamoto, R., Hoshiko, K., Liu, Q. et al. (2015). A photofunctional bottom-up bis(dipyrrinato)zinc(II) complex nanosheet. *Nat. Commun.* 6: 6713.

18 Varaksa, N., Pospíšil, L., Magnera, T.F., and Michl, J. (2002). Self-assembly of a metal-ion-bound monolayer of trigonal connectors on mercury: an electrochemical Langmuir trough. *Proc. Natl. Acad. Sci. U.S.A.* 99: 5012–5017.

19 Makiura, R., Motoyama, S., Umemura, Y. et al. (2011). Highly crystalline nanofilm by layering of porphyrin metal–organic framework sheets. *J. Am. Chem. Soc.* 133: 5640–5643.

20 Bauer, T., Zheng, Z., Renn, A. et al. (2011). Synthesis of free-standing, monolayered organometallic sheets at the air/water interface. *Angew. Chem. Int. Ed.* 50: 7879–7884.

21 Wada, K., Sakaushi, K., Sasaki, S., and Nishihara, H. (2018). Multielectron-transfer-based rechargeable energy storage of two-dimensional coordination frameworks with non-innocent ligands. *Angew. Chem. Int. Ed.* 57: 8886–8890.

22 Kambe, T., Sakamoto, R., Hoshiko, K. et al. (2013). π-Conjugated nickel bis(dithiolene) complex nanosheet. *J. Am. Chem. Soc.* 135: 2462–2465.

23 Takada, K., Sakamoto, R., Yi, S.-T. et al. (2015). Electrochromic bis(terpyridine)metal complex nanosheets. *J. Am. Chem. Soc.* 137: 4681–4689.

24 Tsukamoto, T., Takada, K., Sakamoto, R. et al. (2017). Coordination nanosheets based on terpyridine–zinc(II) complexes: as photoactive host materials. *J. Am. Chem. Soc.* 139: 5359–5366.

25 Sakamoto, R., Yagi, T., Hoshiko, K. et al. (2017). Photofunctionality in porphyrin-hybridized bis(dipyrrinato)zinc(II) complex micro- and nanosheets. *Angew. Chem. Int. Ed.* 56: 3526–3530.

26 López-Cabrelles, J., Mañas-Valero, S., Vitórica-Yrezábal, I.J. et al. (2018). Isoreticular two-dimensional magnetic coordination polymers prepared through pre-synthetic ligand functionalization. *Nat. Chem.* 10: 1001–1007.

27 Gong, Y., Lin, J., Wang, X. et al. (2014). Vertical and in-plane heterostructures from WS_2/MoS_2 monolayers. *Nat. Mater.* 13: 1135–1142.

6

Synthesis, Properties, and Applications of Metal Halide Perovskite-Based Nanomaterials

Mei-Li Sun, Cai-Xiang Zhao, Jun-Feng Shu, and Xiong Yin

Beijing University of Chemical Technology, State Key Laboratory of Chemical Resource Engineering, College of Chemistry, Beijing 100029, China

6.1 Introduction

6.1.1 Crystal Structure and Phase of Metal Halide Perovskites

Perovskite, a class of materials with the same crystal structure as calcium titanate ($CaTiO_3$), was first discovered by Gustav Rose in 1839 [1] and later named by a Russian mineralogist, Count Lev A. Perovskiy. Metal halide perovskites (MHP) represent a crystal structure with the chemical formula ABX_3, in which A and B are cations with different ionic radii, and X is a halide anion that can be bonded with both cations. The ideal perovskite structure is a cubic close packing structure belonging to the space group *Pm3m*. The crystal structure of MHP is shown in Figure 6.1a, in which B ion is located at the center of the cubic cell and formed an octahedral structure of [BX_6] with X ion, while A is located in the interstitial space between the octahedral structure to balance the charge. A usually refers to univalent cation with a larger radius, such as MA^+ ($CH_3NH_3^+$), FA^+ ($HC(NH_2)_2^+$), Cs^+, etc.; B is commonly a group IV divalent cation, such as Pb^{2+}, Sn^{2+}, Ge^{2+}, etc.

The crystal structure of MHP will change with the size of A ion and the change of the interaction force between A and the [BX_6] octahedral structure [6]. In the hybrid halide perovskite, only a few A-site cations are constructed. The common organic cations are shown in Figure 6.1b [3, 7]. They must be small enough to fit into the $(BX_3)^-$ cage and be single-charged to fit the prerequisite for perovskite formation. In order to judge the stability of the crystal structure of MHP, Goldschmidt proposed the tolerance factor (τ, $\tau = (R_A + R_X)/\sqrt{2}(R_B + R_X)$) and the octahedron factor (μ, $\mu = R_B/R_X$), where R_A, R_B, and R_X represent the radii of the corresponding ions [8]. Gregor Kieslich et al. calculated τ for over 2500 amine-metal-anion permutations of the periodic table, as shown in Figure 6.1c [4]. For MHP, when $\tau < 0.8$, the crystal structure has a positive phase; when $0.8 < \tau < 1.0$ and $0.44 < \mu < 0.90$, the crystal structure has a cubic phase; and when $\tau > 1.0$, the perovskite material has a hexagonal crystal structure. The shape, size, and charge distribution of A ion

Functional Nanomaterials: Synthesis, Properties, and Applications, First Edition.
Edited by Wai-Yeung Wong and Qingchen Dong.
© 2022 WILEY-VCH GmbH. Published 2022 by WILEY-VCH GmbH.

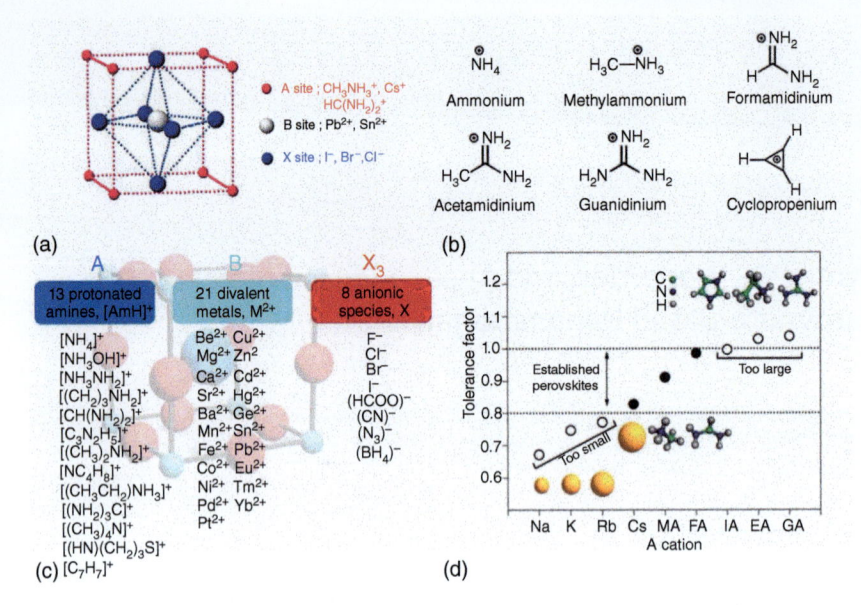

Figure 6.1 (a) The crystal structure of metal halide perovskite. Source: Reprinted with permission from Ref. [2]. Copyright 2019 American Chemical Society. (b) Common organic A-site cations in PbI_3^--based perovskite. Source: Reprinted with permission from Ref. [3]. Copyright 2015 American Chemical Society. (c) Amine-metal-anions used to calculate the tolerance factor. Source: Reprinted with permission from Ref. [4]. Copyright 2015 the Royal Society of Chemistry. (d) Tolerance factor in $APbI_3$ with different cation radius. Source: Reprinted with permission from Ref. [5]. Copyright 2016 the American Association for the Advancement of Science.

are the key factors for the structural stability of MHP [9, 10]. Michael Saliba et al. reported the tolerance factor of A cations with different ionic radii, as shown in Figure 6.1d, indicating that not all univalent cations can participate in perovskite structure formation [5]. τ is not the only factor determining the crystal phase and stability of MHP. There are other non-geometric factors, such as ionic stability, valence bond, etc.

Another trademark feature of MHP is successive phase transitions originating from the double bridging halide ion (B–X–B) [11, 12]. When external conditions (pressure or temperature, etc.) change, the ideal perovskite structure, where all B–X–B angles are 180°, will be distorted, resulting in phase transitions during which the perovskite will change from the cubic phase (α-phase) with high symmetry to the tetravalent (β-phase) or rhombic (γ-phase) phase with low symmetry [13]. As shown in Figure 6.2a, the crystal phase of $CsPbI_3$ exhibited significant temperature dependence. The phase transition can be achieved by tilting the BX_6 octahedron and displacement changes at A/B or X position, as shown in Figure 6.2c. Although the B–X–B angle change from 180° to 150°, below which the crystal phase will change or become amorphous, it does not affect the overall three-dimensional structure [14]. In addition to temperature or pressure, interactions between the cations and the perovskite frame by forming hydrogen bonds (N—H…X) that

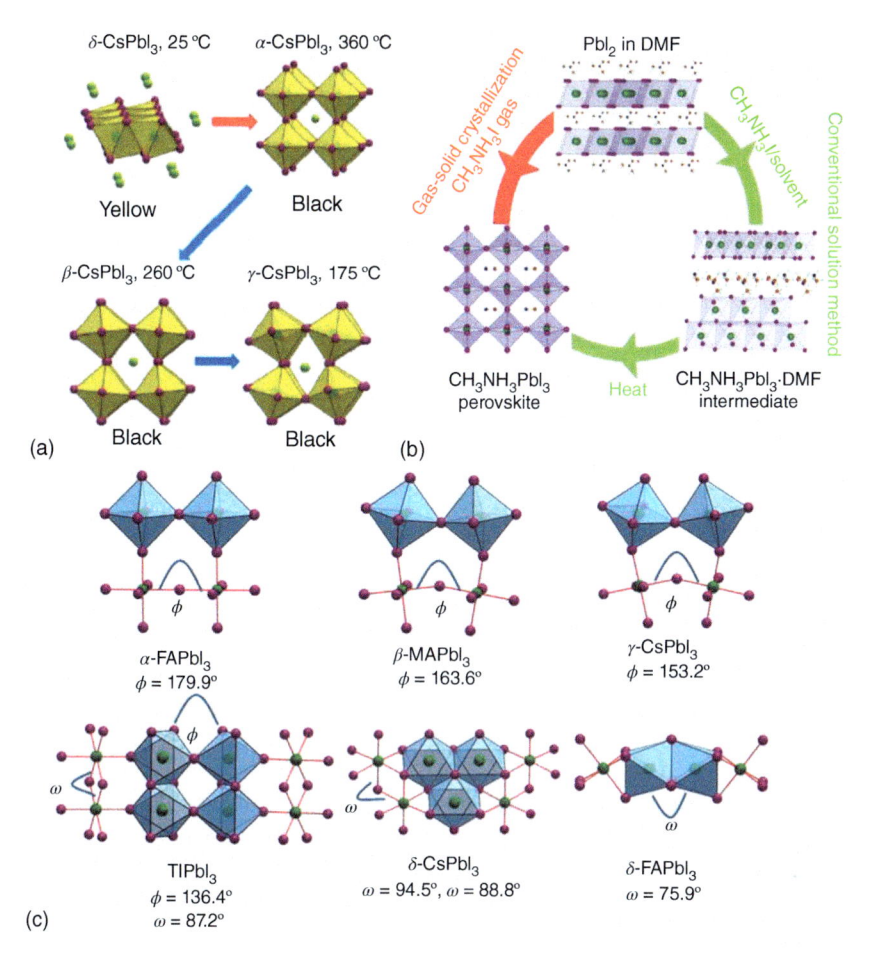

Figure 6.2 (a) The crystal phase of CsPbI$_3$ in different temperature. Source: Reprinted with permission from Ref. [14]. Copyright 2015 American Chemical Society. (b) The scheme of solvated CH$_3$NH$_3$PbI$_3$·DMF intermediate phase formation via conventional solution process. Source: Reprinted with permission from Ref. [15]. Copyright 2014 American Chemical Society. (c) The crystal structure and B–X–B angles of APbI$_3$ perovskite with different A-site cations. Source: Reprinted with permission from Ref. [14]. Copyright 2015 American Chemical Society.

produce the CH$_3$NH$_3$PbI$_3$·X solvates with solvent (H$_2$O, DMF, DMSO, etc.) can also lead to phase transitions. The instability of CH$_3$NH$_3$PbI$_3$ is closely related to this [15], as shown in Figure 6.2b. This phase transition can be avoided by using poor hydrogen bonding or keeping the temperature above the phase transition temperature during the growth of perovskite crystals.

6.1.2 Classification of Metal Halide Perovskite-Based Nanomaterials

MHP materials can be divided into organic–inorganic hybrid perovskite and all-inorganic perovskite based on the difference of the A-site cation.

6.1.2.1 Organic–Inorganic Hybrid Perovskite Materials

According to the tolerance factor theory, when the B-site element is Pb and the X-site element is halogen, the ion radius of the A-site element can be calculated. Based on halide perovskite, Gregor Kieslich et al. studied more than 2500 prospective organic cations and inorganic metals that are likely to be A ions. They found that only more than 700 ions could meet the requirements of the tolerance factor for perovskite formation. Among these ions, more than 600 ions only exist in theory, which has not been synthesized or discovered by researchers [4]. Recently, the most studied A-site ions that meet the prerequisites for perovskite formation are methylammonium (MA, $CH_3NH_3^+$), formamidinium (FA, $CH(NH_2)_2^+$), ethylammonium (EA^+, $C_2H_5NH_3^+$), guanidinium (($NH_2)_3C^+$), Rb^+, Cs^+, and so on. Their ionic radiuses are shown in Table 6.1.

MAPbI$_3$ $MAPbI_3$ with a bandgap of 1.55 eV is the most studied perovskite material in the early stages for its photoelectric properties [16, 17]. At room temperature, $MAPbI_3$ is a tetragonal phase. When the temperature rises to 57 °C, the phase transitions from tetragonal to cubic [18]. Polycrystalline $MAPbI_3$ films have high mobility of charge carriers (1–70 $cm^2\,V^{-1}\,s^{-1}$) [19, 20]. The MA cation in the octahedral cage can be redirected in a very short time (~14 ps). It indicates that the interaction between MA cation and PbX_6 octahedron is very weak, which is the main reason why $MAPbI_3$ is easy to decompose under illumination or at high temperatures (>150 °C). Although $MAPbI_3$ has excellent optoelectronic performance, its instability is a limitation in various applications.

FAPbI$_3$ The FA^+ ion was the first cation used to replace MA^+ in perovskite. The ionic radius of FA is slightly larger than that of MA (253 pm vs. 217 pm), which results in a slight expansion of the crystal structure, thus reducing the bandgap of perovskite. The bandgap is shown in Figure 6.3a. Based on the bandgap value of 1.48 eV, the absorption edge of $FAPbI_3$ was extended to 800 nm. The absorption coefficient is comparable to that of $MAPbI_3$, as shown in Figure 6.3b [21, 22]. In

Table 6.1 The ionic radius of ion used in ABX_3 perovskite.

Ion	Effective ionic radius (pm)
Methylammonium (MA, $CH_3NH_3^+$)	217
Formamidinium (FA, $CH(NH_2)_2^+$)	253
Dimethylammonium (DMA^+, $(CH_3)_2NH_2^+$)	272
Guanidinium (GA+, $(NH_2)_3C^+$)	278
Rb^+	172
Cs^+	188
Pb^{2+}	119
Sn^{2+}	69

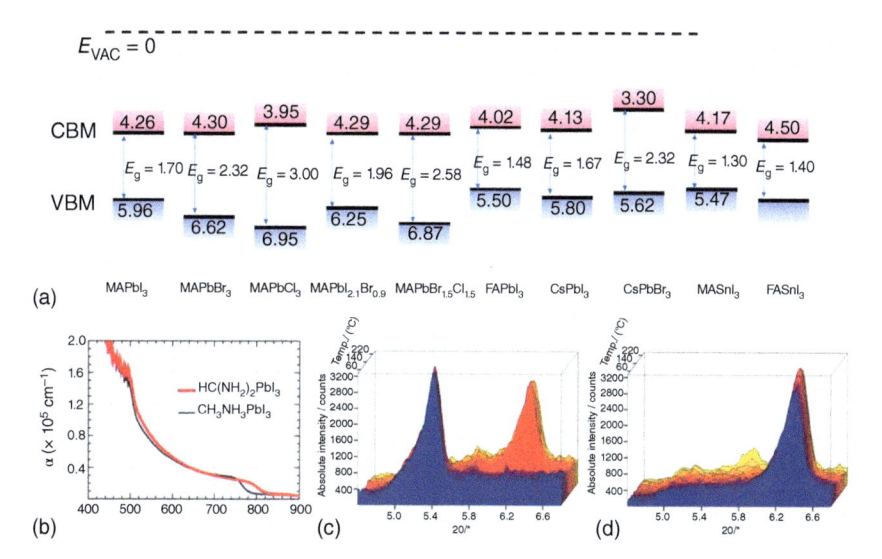

Figure 6.3 (a) The bandgap of ABX_3 perovskite with different components. (b) Absorption coefficient (α) of $HC(NH_2)_2PbI_3$ ($FAPbI_3$) and $CH_3NH_3PbI_3$ ($MAPbI_3$) estimated from diffuse transmittance and reflectance of 250 nm-thick perovskite films. Source: Reprinted with permission from Ref. [15]. Copyright 2014 WILEY-VCH Verlag GmbH & Co. KGaA, Weinheim. Temperature-dependent X-ray diffraction measurements of non-stabilized $FAPbI_3$ with phase transition (c) and stabilized $FAPbI_3$ (d) without phase transition. Source: Reprinted with permission from Ref. [15]. Copyright 2014 American Chemical Society.

addition, the carrier lifetime and diffusion length of $FAPbI_3$ are longer than that of $MAPbI_3$. Although bandgap reduction is conducive to the absorption of light by perovskite materials, it also leads to the phase instability of FA-based perovskite. $FAPbI_3$ crystallizes into a photo-inactive hexagonal phase with a bandgap of 2.43 eV at room temperature [23], which can be converted to a photoactive cubic phase only when annealed above 150 °C. With a high potential barrier, the cubic phase $FAPbI_3$ is stable under inert conditions. However, phase conversion is significantly accelerated when exposed to moisture. Water vapor will cause surface defects by attacking the surface of $FAPbI_3$ crystals, reducing the formation energy of the non-perovskite phase [24]. Therefore, the low-phase instability of $FAPbI_3$ perovskite needs to be addressed.

Mixed A-Site Cation Perovskite Although $MAPbI_3$ and $FAPbI_3$ show excellent photoelectric properties, the volatile nature of $MAPbI_3$ and the environmental phase instability of $FAPbI_3$ are undeniable. Researchers are currently developing perovskites with mixed A-site cation, such as $FA/MAPbX_3$, $FA/CsPbX_3$, $MA/CsPbX_3$, $MA/FA/CsPbX_3$, and $MA/FA/Cs/RbPbX_3$. Andreas Binek et al. found that introducing a small amount of MA^+ could stabilize cubic phase $FAPbI_3$ at room temperature [25]. They found that the phase transition from cubic to hexagonal did not occur within temperatures around 25–125 °C by temperature-dependent X-ray diffraction (Figure 6.3c,d). In addition, the introduction of MA^+ significantly increased the PL lifetime of the perovskite, improving the photovoltaic performance. Xiaojia Zheng et al. studied the intrinsic instability mechanism of α-FAPbI$_3$ at room temperature [26]. The results showed that the (111) plane of FAPbI$_3$ has anisotropic-strained lattice, which was conducive to the transformation of α-phase to β-phase with the environmental conditions accelerating the process. The introduction of MA$^+$ relaxed the strain in the lattice, stabilizing the α-phase of FAPbI$_3$. The corresponding mechanism is shown in Figure 6.4a. Although the introduction of MA$^+$ will enhance the phase stability of perovskite, the volatility of MA$^+$ will reduce the thermal and photostability of perovskite. The initial activation energy of thermal degradation of FAPbI$_3$ was 115 ± 3 kJ mol^{-1}, higher than that of MAPbI$_3$ with 93 ± 8 kJ mol^{-1} [29].

The photo and moisture instability of FAPbI$_3$ can be overcome by introducing Cs$^+$ instead of MA$^+$, forming binary perovskite FA/CsPbX$_3$. Organic cations (FA$^+$ or MA$^+$) have a weak interaction with iodine in perovskite, while Cs$^+$ has a strong chemical bond with iodine in perovskite lattice. It improves the photo and moisture stability of perovskite. Nam-Gyu Park and coworkers introduced 10% Cs$^+$ into FAPbI$_3$ to partially replace FA$^+$. They found that the introduction of Cs$^+$ obtained cubic phase FAPbI$_3$ without any treatment, which was confirmed by the absorbance at 630 nm and X-ray diffraction (Figure 6.4b,c) [30]. They believed that the enhanced phase stability was due to the contraction of the cubic octahedron volume caused by the introduction of Cs$^+$, thus strengthening the interaction between FA$^+$ and iodide. Compared with the FA/MA system, the introduction of Cs$^+$ enhances the phase stability of perovskite and improves the stability under moisture and illumination conditions [31, 32]. As shown in Figure 6.4e, FAPbI$_3$ degraded rapidly when exposed to humid conditions for 18 days, while FA/CsPbI$_3$ films remained stable. In addition to stability, the crystallinity of perovskite was also enhanced [27]. Subsequently, the researchers introduced Cs$^+$ into the FA/MA system to form ternary-cation perovskite, Cs/FA/MAPbX$_3$ [33, 34], which further improved the thermal stability of the perovskite material and reduced its sensitivity to processing conditions. When only MA$^+$ was present, the black phase of FA-based perovskite was affected by the temperature at the beginning of the crystallization process. When Cs$^+$ and MA$^+$ were added simultaneously, the thermal stability of the perovskite improved. Saliba et al. studied the relationship between thermal stability, film formation, and processing conditions by UV–visible absorption spectra [28]. As shown in Figure 6.4g3 and g4, when Cs$^+$ was added, the black phase formed without annealing. However, when the perovskite was prepared at 18 °C without Cs$^+$, the black phase of perovskite cannot be formed even if annealed at 100 °C for 1 hour. Only when the perovskite was prepared at 25 °C was the black phase obtained by annealing. From Figure 6.4f, the thermal stability of Cs-containing perovskite was enhanced compared to the perovskite without Cs$^+$.

In addition to inorganic cation Cs$^+$, other alkali metal ions are also important in the crystal structure and intrinsic properties of hybrid perovskites [35]. Rb$^+$, a small and oxidation-stable ion, can be embedded into an organic cation forming a perovskite with excellent properties. According to Figure 6.1d, the tolerance factor of RbPbI$_3$ is close to 1, and the ion radius of Rb$^+$ is only slightly smaller than Cs$^+$ (172 pm vs. 188 pm). However, even when annealed at 460 °C, RbPbI$_3$

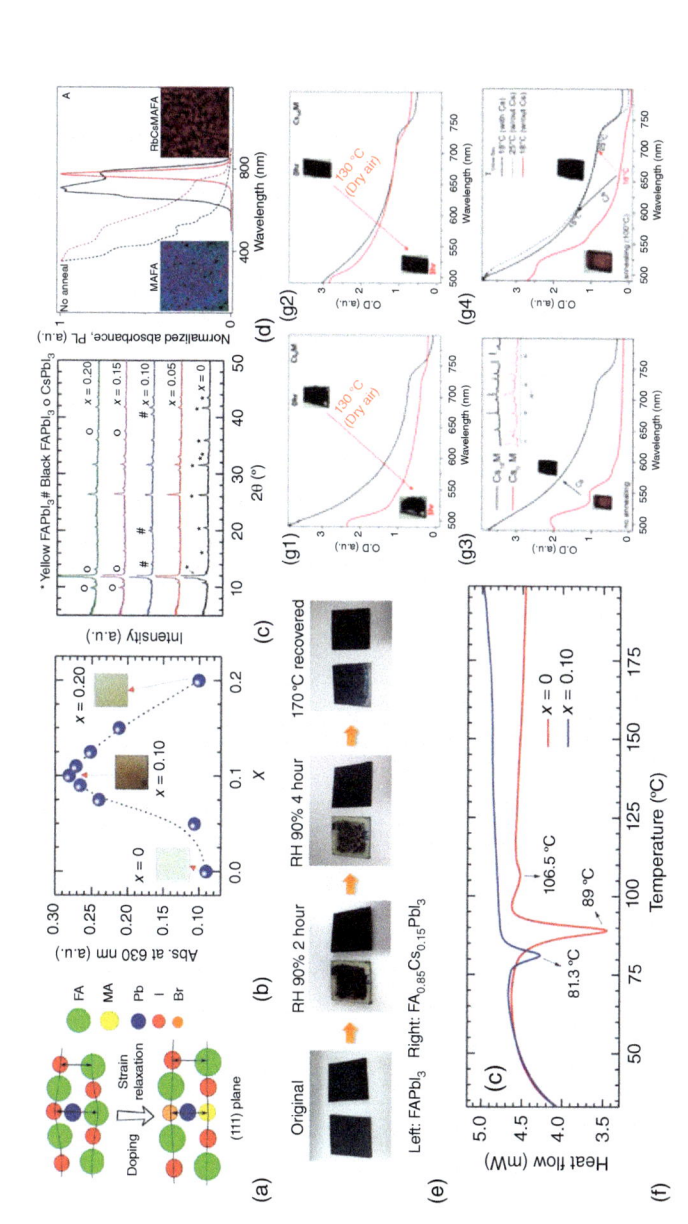

Figure 6.4 (a) Schematic representation of strain relaxation during MABr alloying. Source: Reprinted with permission from Ref. [22]. Copyright 2016 American Chemical Society. (b) Absorbance at 630 nm and (c) X-ray diffraction of the $FA_{1-x}Cs_xPbI_3$ films coated on glass. Source: Reprinted with permission from Ref. [25]. Copyright 2015 WILEY-VCH Verlag GmbH & Co. KGaA, Weinheim. (d) UV–vis (dashed lines) and PL (solid lines) of unannealed MAFA (black) and RbCsMAFA (red) films. The inset images show fluorescence microscopy measurements of MAFA and RbCsMAFA films. Source: Reprinted with permission from Ref. [26]. Copyright 2016 The American Association for the Advancement of Science. (e) Photos of $FAPbI_3$ and $FA_{0.85}Cs_{0.15}PbI_3$ thin films under various high-humidity conditions. Source: Reprinted with permission from Ref. [5]. Copyright 2016 American Chemical Society. (f) Differential scanning calorimetry (DSC) of in $FA_{1-x}Cs_xPbI_3$ (x = 0 and 0.10) perovskite. Source: Reprinted with permission from Ref. [27]. Copyright 2015 WILEY-VCH Verlag GmbH & Co. KGaA, Weinheim. (g) UV–visible adsorption of Cs_0M (g1) and $Cs_{10}M$ (g2) films with the corresponding images annealed at 130 °C for 3 hours in dry air; Adsorption spectra of as-fabrication films without annealing process containing Cs^+ and without Cs^+ at room temperature (g3) and 18 °C (g4). Source: Reprinted with permission from Ref. [28]. Copyright 2016 the Royal Society of Chemistry.

was still yellow phase rather than black phase [5], which explains why it cannot be used as a light absorption layer. Although RbPbI$_3$ alone is not feasible as an absorbent material, adding Rb$^+$ to perovskite as an additive can improve the properties of perovskite. Michael Saliba et al. embedded Rb$^+$ into four mixed-A-site perovskite, Rb/FA, Rb/Cs/FA, Rb/MA/FA, and Rb/Cs/MA/FA [5]. It was found that the RbCsMAFA-based perovskite exhibited the best performance. The effects of the addition of Rb$^+$ on the starting conditions of the perovskite crystallization process were studied by UV–vis spectroscopy, photoluminescence spectroscopy, and fluorescence microscopy maps (Figure 6.4d). It can be seen from the fluorescence microscopy maps for the unannealed films that the addition of Rb$^+$ caused perovskite to crystallize in a more homogeneous starting condition, while the MAFA films crystallized under inhomogeneous starting conditions. Other alkali metal ions also play an important role in the crystal structure. The intrinsic properties of hybrid perovskites and different alkali metal ions play different roles, as shown in Figure 6.5a,b. Different alkali metals have different positions in perovskite. Rb$^+$ and K$^+$ tend to be distributed at the grain boundary, while other ions are distributed in the crystal lattice of perovskite [35].

In addition to inorganic ions, organic cations, such as guanidinium (GA$^+$) and dimethylammonium (DMA$^+$), are often introduced into perovskite. According to Table 6.1, the ionic radii of GA$^+$ and DMA$^+$ are larger than that of MA$^+$ and FA$^+$, leading to the tolerance factor of pure GAPbI$_3$ and DMAPbI$_3$ being larger than 1 (GAPbI$_3$ is 1.04 [38], and DMAPbI$_3$ is 1.03 [37]). Nicholas De Marco et al. found that introducing GA$^+$ to MAPbI$_3$ could significantly enhance the carrier lifetimes and obtain the devices with high open-circuit voltages. They believed that the hydrogen bonding ability of GA$^+$ enhanced the grain size and continuity, which passivated the uncoordinated iodide species between the adjacent grains and enhanced the carrier lifetimes. Alexander D. Jodlowski et al. also demonstrated that the introduction of GA$^+$ increased the number of hydrogen bonding and reduced the distance between H and I, which stabilized the perovskite structure. As shown in Figure 6.5c, the addition of GA$^+$ enhanced the number of hydrogen bonds in MAPbI$_3$ perovskites from 1–2H bonds to 6H bonds [36]. Similarly, Hao Chen et al. used DMA$^+$ to partially replace MA$^+$ in MAPbI$_3$, which inhibited defect state density and carrier recombination [37]. At the same time, the presence of secondary amines improved the hydrophobicity of perovskite, as shown in Figure 6.5d.

6.1.2.2 All-Inorganic Perovskite Materials

All-inorganic perovskite materials have high thermal stability due to the absence of volatile organic cation [39, 40]. Types of all-inorganic MHP are limited. Only the alkali metal Cs at A meets the tolerance factor requirements if the B element is Pb. In contrast to organic cations, Cs$^+$ ions form a strong chemical bond with the perovskite lattice rather than a weak hydrogen bond. This is related to the activation energy of thermal degradation of perovskite. The activation energy of thermal degradation of CsPbI$_3$ is 650 ± 90 kJ mol^{-1}, which is higher than that of MAPbI$_3$ (93 ± 8 kJ mol^{-1}) and FAPbI$_3$ (115 ± 3 kJ mol^{-1}) [29]. Therefore, CsPbX$_3$ has higher thermal stability than the organic–inorganic hybrid perovskite [41, 42].

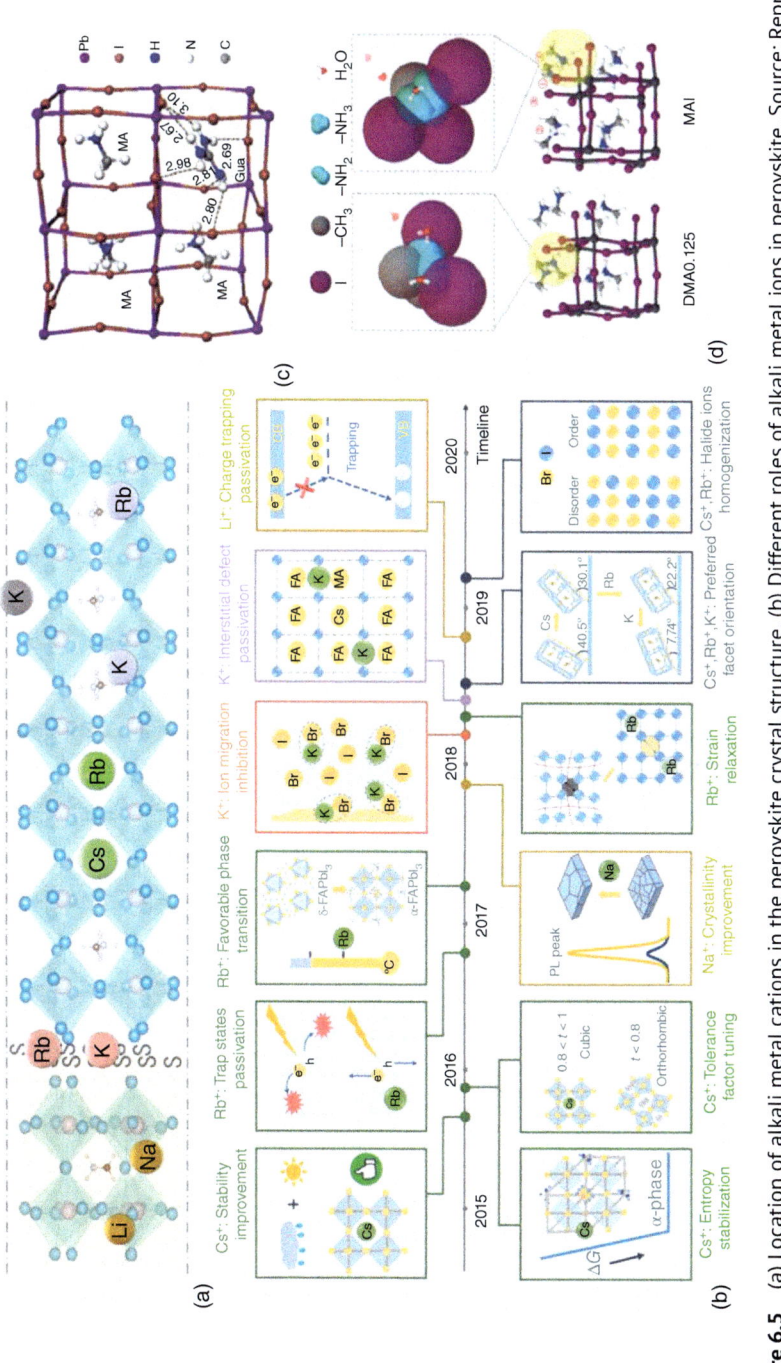

Figure 6.5 (a) Location of alkali metal cations in the perovskite crystal structure. (b) Different roles of alkali metal ions in perovskite. Source: Reprinted with permission from Ref. [35]. Copyright 2021 the Royal Society of Chemistry. (c) Optimized simulated structure of the unit cell of MA$_{0.75}$GA$_{0.25}$PbI$_3$ including the six H-bond distances with respect to I atoms. Source: Reprinted with permission from Ref. [36]. Copyright 2017 Springer Nature. (d) Water adsorption structure and adsorption model for DMA$_{0.125}$ and DMA molecule (left) and MA1 and MA molecule (right). Source: Reprinted with permission from Ref. [37]. Copyright 2019 WILEY-VCH Verlag GmbH & Co. KGaA, Weinheim.

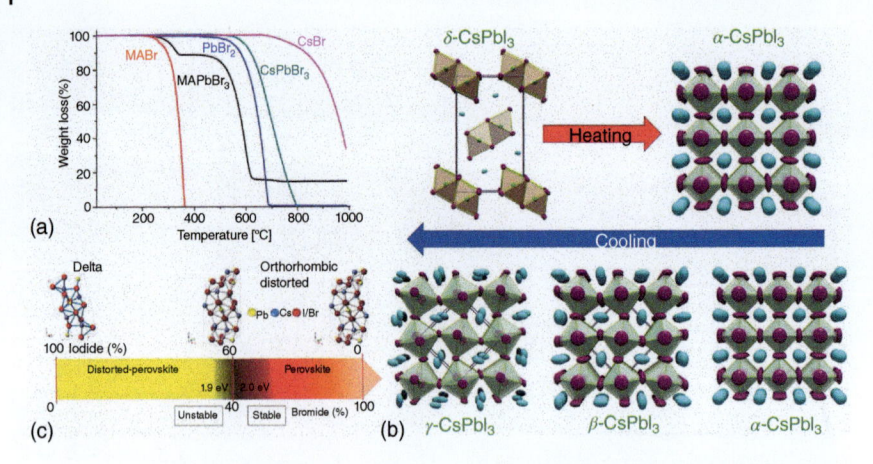

Figure 6.6 (a) Thermogravimetric analyses of MABr, MAPbBr$_3$, PbBr$_2$, CsPbBr$_3$, and CsBr. Source: Reprinted with permission from Ref. [44]. Copyright 2016 American Chemical Society. (b) The corresponding phase structures of CsPbI$_3$ at the different temperatures. Source: Reprinted with permission from. Copyright 2018 American Chemical Society. (c) Illustration of the crystal structure for perovskite as a function of the iodine/bromine ratio. Source: Reprinted with permission from Ref. [45]. Copyright 2018 WILEY-VCH Verlag GmbH & Co. KGaA, Weinheim.

For example, CsPbBr$_3$ films exhibited good thermal stability, which was annealed at 250 °C as prepared by Michael Kulbak et al. [43]. According to the TGA analysis, the decomposition temperature of CsPbBr$_3$ was 580 °C (Figure 6.6a), which was higher than that of MAPbI$_3$ (220 °C). However, the decomposition product of CsPbBr$_3$ was PbBr$_2$ instead of CsBr, which indicated that CsBr was more stable than PbBr$_2$ [44].

As shown in Figure 6.6b, there are four crystal structures of CsPbX$_3$: the cubic structure (α-CsPbX$_3$), the tetragonal structure (β-CsPbX$_3$), the orthorhombic structure (γ-CsPbX$_3$), and the non-perovskite structure (δ-CsPbX$_3$) [46]. At room temperature, the stable phase of CsPbX$_3$ is not a photoactive cubic phase. For example, CsPbI$_3$ is the non-perovskite yellow phase, while CsPbBr$_3$ is the orthorhombic phase, related to the small ionic radius of Cs$^+$ [47]. However, the photoactive cubic phase (α-phase) is a requirement for practical applications. The researchers found that the phase stability of CsPbX$_3$ could be improved by additive engineering. Adding a small amount of Br$^-$ to CsPbI$_3$ can improve phase stability [48, 49]. Sandy Sanchez et al. found that adding at least 40% Br$^-$ to CsPbBr$_3$ could maintain the phase stability of CsPbI$_3$ at room temperature, as shown in Figure 6.6c [45]. But at the same time, the addition of Br$^-$ will increase the bandgap of CsPbI$_3$, which is unfavorable to the absorption performance of the perovskite. In addition to Br$^-$, other ions, such as DMA$^+$, EDA$^+$, Bi^{3+}, Sb^{3+}, Sr^{2+}, and SCN$^-$, can also improve the phase stability of CsPbX$_3$ at room temperature [50, 51].

6.1.2.3 Lead-Free Perovskite Materials and Low-Lead Perovskite Material

Although lead-based perovskite materials exhibit excellent photoelectric properties, their toxicity is a major obstacle to their commercialization. The contamination of lead ions to soil and water sources negatively impacts human, animal, and plant survival [52, 53]. Lead toxicity can interfere with various body functions and may cause irreversible health effects [54]. According to the United States Environmental Protection Agency (U.S. EPA), the maximum permissible amount of Pb is 15 and 0.15 µg l^{-1} in water and air, respectively [55]. They are much lower than the amount of Pb estimated in a cell with the thickness of a perovskite layer. Thus, low-cost and high-efficiency lead-free perovskite has become popular. According to the requirements of ionic size and Goldschmidt tolerance factor, many cations are predicted to replace lead in Pb-based perovskite. The elements proposed to replace lead in halide perovskite materials include group IV elements (Sn^{2+}, Ge^{2+}), alkali metal ions (Be^{2+}, Mg^{2+}, Ca^{2+}), transition metal ions (V^{2+}, Mn^{2+}, Co^{2+}, Ni^{2+}, Zn^{2+}), and lanthanides metal ions (Eu^{2+}, Tm^{2+}), as shown in Figure 6.1c. However, considering the stability and photovoltaic performance of a perovskite structure, the current research focuses on lead-free perovskite (Sn-based and Ge-based perovskite) and some low-lead perovskite materials. Sn-based perovskite has been explored as the first lead-free perovskite material. This is due to the similar valence electron configuration between Sn and Pb ions. Thus, Sn-based perovskites have gained popularity in photovoltaics [56, 57]. The ionic radius of Sn^{2+} (69 pm) is smaller than that of Pb^{2+} (119 pm), which leads to large differences in the band structure and bandgaps. As shown in Figure 6.3a, the bandgaps of Sn-based perovskite are also smaller than those of Pb-based perovskites, which could be favorable for light-absorbing materials [58].

Sn-based perovskites possess high electron conductivity, long diffusion length, and superior electron mobility [59]. However, Sn-based perovskite materials are volatile under environmental conditions compared with their lead-based counterparts. The Sn-based perovskite devices showed low performance [60]. The high defect densities of Sn^{2+}-based perovskites and easy oxidization of Sn^{2+} to Sn^{4+} lead to the collapse of the perovskite structure. Further studies revealed that introducing additives, such as SnI_2, SnF_2, and $SnCl_2$ into the Sn^{2+} precursor could compensate for the loss of Sn^{2+} [61, 62]. Song et al. reported that the excess SnI_2 dispersed uniformly into the perovskite films and function as a compensator and a suppressor of Sn^{2+} vacancies, reducing p-type conductivity [63]. Among the three additives mentioned, SnF_2 was the most popular. Mathews and coworkers found that introducing SnF_2 into $CsSnI_3$ reduced the background carrier density, enhancing the photovoltaic performance with a high photocurrent density of up to 22 mA cm^{-2} [64]. However, excess SnF_2 induced phase separation on the surface of the perovskite film. Seok and coworkers reported the homogeneous dispersion of SnF_2 via the formation of the SnF_2–pyrazine complex [65]. They found that the pyrazine restricted the phase separation through interaction with SnF_2, and the interaction reduced the Sn vacancies effectively. As shown in Figure 6.7a,b, the SEM and XRD results confirmed that the addition of pyrazine significantly improved the surface morphology of perovskite and prevented the oxidation of Sn^{2+}. Besides pyrazine, the hypophosphorous acid (H_3PO_2) was also used in Sn-perovskites as a reducing agent to combine with SnF_2 [66].

The metal ions containing the external ns^2 electronic structures with low ionization energy can enhance the light absorption efficiency and carrier diffusion length of ABX_3 [6]. The Ge-based perovskites ($AGeX_3$) possess similar transport and optical

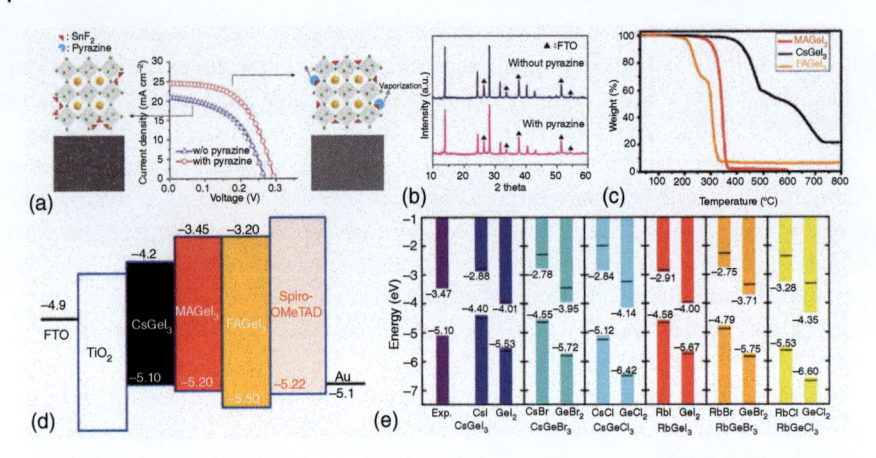

Figure 6.7 (a) SEM images, J–V curves and (b) XRD of FASnI$_3$ perovskite films and devices with and without pyrazine. Source: Reprinted with permission from Ref. [66]. Copyright 2016 American Chemical Society. (c) Thermogravimetric analysis (TGA) thermogram of Germanium perovskites and (d) schematic energy level diagram of (d) Ge-based perovskite with different A-site ions. Source: Reprinted with permission from Ref. [67]. Copyright 2015 the Royal Society of Chemistry. (e) Schematic energy level diagrams of Ge-based perovskite with different halide ions. Source: Reprinted with permission from Ref. [68]. Copyright 2019 American Chemical Society.

properties compared with Pb or Sn-based perovskites. This is due to Ge^{2+} having 4s^2 electrons and an ionic radius similar to Pb^{2+}. Krishnamoorthy et al. reported the synthesis of three Ge-based halide perovskite AGeI$_3$ (A = Cs, MA, FA) materials [67]. These compounds are stable up to 150 °C, as shown by TGA in Figure 6.7c. The corresponding bandgap was related to the A-site cation size (Figure 6.7d). Moreover, the bandgap was also dependent on the halide ions. For example, the bandgap of AGeX$_3$ (A = Cs and Rb) increased with the decrease in the size of halide ions (Figure 6.7e) [68]. Sun et al. investigated the structural and electronic properties of MAGeX$_3$ using density functional theory methods [69]. MAGeI$_3$ exhibited an analogous bandgap, substantial stability, remarkable optical properties, and significant hole and electron conductive behavior compared with MAPbI$_3$. However, the efficiency of solar cells with Ge-based perovskite as the absorbent layer was only 0.2% [67] due to the poor stability of Ge-based perovskite under the air atmosphere.

Although advances in lead-free perovskite materials have been achieved, the performance of related materials-based devices is still low compared with Pb-based perovskite materials. Reducing lead content is another approach to eliminate the toxicity of Pb. Thus, Pb–Sn mixed perovskites have been studied extensively [70, 71]. The introduction of Sn reduced the bandgap of Pb-based perovskites, which will be suitable as a bottom cell configuration in tandem architecture. Hao et al. reported that the energy bandgaps of the mixed Pb/Sn compounds did not follow the linear trend (the Vegard's law) within the range of 1.35–1.55 eV (Figure 6.8b). The resultant compound had a narrow bandgap of 1.3 eV, and its light absorption extended into the near-infrared range (~1050 nm). The absorption properties were confirmed by IPCE measurement, shown in Figure 6.8a [72]. Subsequently, Zhao et al. found that

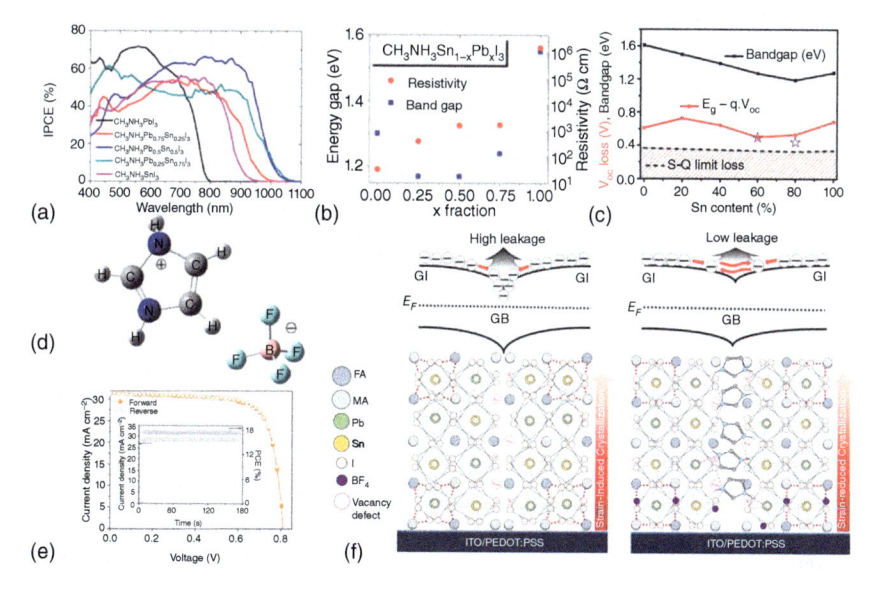

Figure 6.8 (a) IPCE spectra of the devices based on $CH_3NH_3Sn_{1-x}Pb_xI_3$ ($x = 0, 0.25, 0.5, 0.75$, and 1) perovskites. (b) Dependence of resistivity and optical bandgap of the $CH_3NH_3Sn_{1-x}Pb_xI_3$ on the x fraction. Source: Reprinted with permission from Ref. [72]. Copyright 2014 American Chemical Society. (c) The V_{oc} loss compared to the bandgap (E_g-qV_{oc}) related to the Sn content. Source: Reprinted with permission from Ref. [73]. Copyright 2016 WILEY-VCH Verlag GmbH & Co. KGaA, Weinheim. (d) Chemical structure of the $IMBF_4$ additive. (e) J–V curves under AM 1.5 G illumination for the best-performing PSCs with $IMBF_4$ additive under forward and reverse bias. The inset shows the steady-state J_{sc} and corresponding stabilized PCE at MPP. (f) Schematic illustrations of possible mechanisms underlying the synergistic effects of IM cation and BF_4 anion on the mixed Pb–Sn perovskite films. Source: Reprinted with permission from Ref. [74]. Copyright 2020 Wiley-VCH GmbH.

the high open-circuit voltage approaching the prediction of the Shockley–Queisser model was obtained using the lowest bandgap perovskite $CH_3NH_3(Pb_xSn_{1-x})I_3$ with 60% Sn in the perovskite layer, as shown in Figure 6.8c. The high open-circuit voltage originated from the high intrinsic charge-carrier mobility in low-bandgap semiconductors [73]. The various preparation approaches of perovskite films can also lead to differences in the bandgap values of perovskite [70, 73].

The methods of tuning the morphologies and crystallinities of Pb-based perovskites are also suitable for improving the performance and stability of Pb-Sn-based perovskites [75, 76]. In Sn-Pb perovskites, the Sn^{2+} ions are also oxidized to Sn^{4+}, which must be resolved. Various additives have been used to mitigate the tin oxidation process and enhance the device's performance. They include antioxidant additives (e.g. SnF_2, $SnCl_2$, $SnBr_2$), halide anions (e.g. Br^-, Cl^-), metallic Sn, thiocyanate-based additives (e.g. MASCN, GuaSCN, $Pb(SCN)_2$), transition metal ions (Cd^{2+}), and organic acid additives. In addition, ionic liquid-type additives composed of organic cations and inorganic anions can play complementary roles for mixed Pb-Sn perovskites. Kim et al. introduced the ionic additives imidazolium tetrafluoroborate (IMBF$_4$) (Figure 6.8d) into Pb–Sn-based perovskite layers, achieving a power conversion efficiencies (PCE) above 19% with remarkable operational stability (Figure 6.8e) [77]. The chemical interaction between IM cation and charged uncoordinated Pb^{2+} or Sn^{2+} could effectively passivate the defects. The BF$_4^-$ anions intercalated into the perovskite lattice for lattice expansion, relaxing the lattice strain. The possible mechanism diagram is shown in Figure 6.8f.

6.2 Properties of Metal Halide Perovskite Materials

6.2.1 Tunable Bandgap

MHP materials possess intense and wide absorption from the visible to the near-infrared region within the solar spectrum [74]. The bandgap is the basic property of light-absorbing material, related to the maximum value of the theoretical conversion power efficiency of solar cells. A suitable bandgap is an important parameter for the applications of perovskites in various fields, such as light-emitting diodes [78], perovskite solar cells [39, 40], photodetectors [79, 80] and sensors [81], and other photoelectric devices [16]. The changes in the bandgap are related to the composition selection of metals, halogens, and A-site cations [82, 83]. There are three main strategies for tuning the bandgap: (i) changing the halogen ions, (ii) changing the A-site cations, and (iii) changing the B-site elements.

Changing the halogen ions. The halogen anions could influence the level of the valence band. The observed bandgaps changed upon halide substitution. The influences are ascribed to the differences in electronic states of the anions. For example, the valence band composition of Cl, Br, and I is 3p, 4p, and 5p, respectively. The corresponding electron binding energy (lower ionization potential) decreases accordingly from 3p to 5p. As shown in Figure 6.9a [84], the valence band energy varied up to 0.6 eV by changing the chloride to iodide in the methylammonium-based perovskites.

Changing the B-site element. Substitution of the B-site element can alter the conduction band and electronic properties. Different B-site ions indicate that perovskites have different M—X—M bond angles, which impact the bandgap. In AMI$_3$ (M = Ge^{2+}, Sn^{2+}, Pb^{2+}), the M–I–M angle in the MI$_6$ octahedra is 166.27(8)°, 159.61(5)°, and 155.19(6)° for Ge^{2+}, Sn^{2+}, and Pb^{2+}, respectively. The bandgap of AMI$_3$ with different M ions followed the order of AGeI$_3$ < ASnI$_3$ < APbI$_3$ (Figure 6.9d) [53]. Tin-based perovskites were used in photovoltaic devices due to their ideal bandgaps. Ogomi et al. reported on Sn/Pb halide-based perovskites [86]. The electronic absorption edge varied with x in the CH$_3$NH$_3$Sn$_x$Pb$_{(1-x)}$I$_3$. The edge shifted from 1000 to 1200 nm when x increased from 0.3 to 1.0 (Figure 6.9b), with the corresponding bandgap shown in Figure 6.9c.

Changing the A-site cations. The size of the A-site cation can affect the symmetry of the BX$_6^{4-}$ octahedral network, leading to the differences of tolerance factor and modifying the bandgap. As shown in Table 6.1, the symmetry of the three perovskites follows the order of FAPbI$_3$ > MAPbI$_3$ > CsPbI$_3$. The bridging

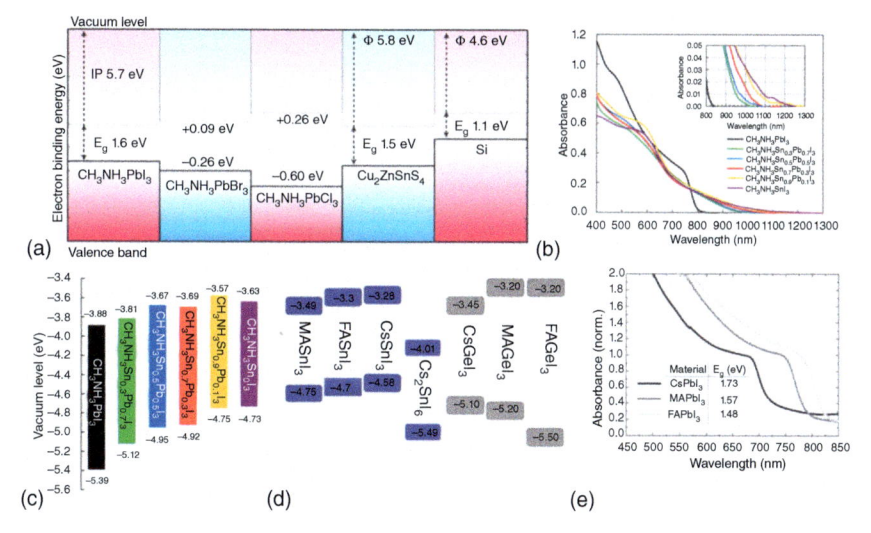

Figure 6.9 (a) Calculated natural band offsets of $CH_3NH_3PbI_3$ and related materials based on density functional calculations (with quasi-particle corrections). Source: Reprinted with permission from Ref. [84]. Copyright 2015 American Chemical Society. (b) Electronic absorption spectra of $CH_3NH_3Sn_xPb_{1-x}I_3$ perovskites coated on porous TiO_2. (c) Energy diagram for $CH_3NH_3Sn_xPb_{1-x}I_3$ perovskites. Source: Reprinted with permission from Ref. [85]. Copyright 2014 American Chemical Society. (d) Energy Diagram of perovskites with different B-site elements. Source: Reprinted with permission from Ref. [53]. Copyright 2015 Elsevier. (e) UV–vis spectra for the $APbI_3$ perovskites, where A is either caesium (Cs), methylammonium (MA) or formamidinium (FA). Source: Reprinted with permission from Ref. [85]. Copyright 2014 American Chemical Society.

angles of Pb–I–Pb gradually deviated from the ideal linear conformation when perovskites changed from $FAPbI_3$ to $CsPbI_3$. The orbital overlap of Pb–I was reduced. The resultant bandgap increased in this order: $FAPbI_3$ (1.48 eV) < $MAPbI_3$ (1.57 eV) < $CsPbI_3$ (1.73 eV) (Figure 6.9e) [85].

Meanwhile, the absorption edge red-shifted with increasing radius of the A ions. When the mixture of FA^+, MA^+, and Cs^+ was used in the A-site, the optical absorption onset of $MA_xFA_{1-x}PbI_3$ also red-shifted compared to that of $MAPbI_3$. Thus, solar light harvesting property [87] and stability, charge separation, and carrier transport properties were tuned [88].

6.2.2 High Absorption Coefficient

The MHP materials possess a higher light absorption coefficient than the traditional photovoltaic materials. For example, the absorption coefficient of $CH_3NH_3PbI_3$ at 500 nm is up to 1.3×10^5 cm^{-1}. The absorption cutoff wavelength is about 800 nm, which means that its absorption range almost covers the entire UV–visible wavelength [89, 90]. The high absorption coefficient of $CH_3NH_3PbI_3$ indicates that the film thickness required for perovskite materials is thin under the same illumination. It is beneficial to the collection efficiency of the photo-generated carriers, reducing the carrier recombination.

6.2.3 Excellent Charge Transport Performance

MHP materials have high carrier mobility, such as 1–30 cm^2 V^{-1} s^{-1} in polycrystalline thin films and up to 200 cm^2 V^{-1} s^{-1} in single-crystal thin films, indicating perovskite materials present long carrier diffusion length [91, 92]. Moreover, the effective masses of electron and hole are equal in MHP materials, resulting in the characteristics of the bipolar transmission. These unique properties will benefit photovoltaic applications [93, 94]. Long carrier diffusion length can reduce the carrier recombination and enhance the charge collection properties. In MAPbI$_3$, the reported minimum diffusion distance of the carrier is about 100–300 nm. In single-crystal MAPbI$_3$, the diffusion length of the carrier exceeds 175 μm [95].

6.2.4 Photoluminescence Properties

Metal halide perovskite materials MHP, as direct bandgap semiconductors, could exhibit photoluminescence (PL) properties at room temperature. The principle of photoluminescence is as follows: When the MHP absorbs the energy larger than its bandgap energy, the electrons in the valence band will transfer to the conduction band, forming an excited state. The excited state is unstable. When the electrons return to the ground state, the photons will be emitted (Figure 6.10a). The photoluminescence effect is generally observed in halogen perovskites. The photoluminescence intensity is very high, especially for the low-dimensional perovskite materials [99–101]. The high photoluminescence intensity is due to the limiting effect in the low-dimensional perovskite materials. The limiting effect makes its exciton binding energy large and improves the efficiency of photoluminescence. Exciton binding energy refers to the interaction between excited electrons in the conduction band and holes in the valence band during the optical transition. The stronger the binding force is, the larger the probability of electron-hole recombination. Three-dimensional perovskite presents a small exciton binding energy of only 9–60 meV. Thus, its photoluminescence efficiency is low [102, 103]. The exciton binding energy of low-dimensional perovskite can reach more than 200 meV, increasing recombination probability. The photoluminescence quantum yield (PLQY) of low-dimensional perovskite nanocrystals can reach more than 90% [104, 105]. PLQY is an important parameter to evaluate the photoluminescence performance of materials, and its formula is as follows:

$$\text{PLQY} = \frac{\text{Number of emitted photons}}{\text{Number of absorbed photons}}$$

Where the number of absorbed photons refers to the number of photons of incident light absorbed by perovskite materials, and the number of emitted photons refers to the number of photons emitted by perovskite after absorbing incident photons. In addition, 2D perovskites will form a multi-quantum well structure after absorbing photons due to different layers of perovskites. The perovskites with a few layers will transfer energy to the perovskites with a high number of layers. The perovskites with a narrow bandgap will emit light [106]. Quan et al. summarized the relationship between the number of layers of 2D perovskite and exciton binding energy

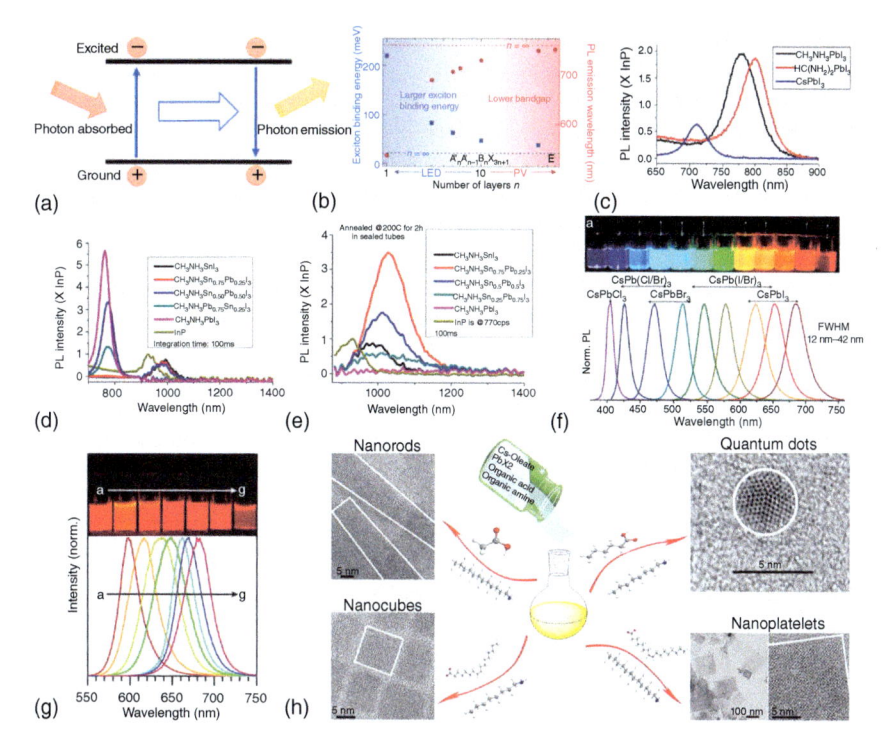

Figure 6.10 (a) The scheme for the physical process of photoluminescence. (b) Exciton binding energy and PL emission wavelength as a function of number of layers. Source: Reprinted with permission from Ref. [96]. Copyright 2018 WILEY-VCH Verlag GmbH & Co. KGaA, Weinheim. (c) PL spectra of $APbI_3$ obtained from the open tube reaction, PL properties of $CH_3NH_3Sn_{1-x}Pb_xI_3$ solid solution obtained from (d) mildly grinding the precursors (interface reaction) and (e) by annealing the same specimen at 200 °C for 2 hours in a sealed tube. Source: Reprinted with permission from Ref. [59]. Copyright 2013 American Chemical Society. (f) The photographs of colloidal solutions of $CsPbX_3$ in toluene under UV lamp and corresponding PL spectra. Source: Reprinted with permission from Ref. [97]. Copyright 2015 American Chemical Society. (g) Normalized photoluminescence spectra and photographs under UV illumination of the QDs with different sizes. Source: Reprinted with permission from Ref. [98]. Copyright 2016 The American Association for the Advancement of Science. (h) Schematic illustrating the formation process for different $CsPbX_3$ (X = Cl, Br, I) nanocrystals mediated by organic acid and amine ligands at room temperature. Source: Reprinted with permission from Ref. [98]. Copyright 2016 American Chemical Society.

(Figure 6.10b) [96]. According to the mechanism, the excellent photoluminescence performance of 2D perovskite is expected.

The PL properties of perovskite materials are related to particle size and environmental temperature. The PL emission can be facile at cryogenic temperatures. It can be observed in thin film and nanocrystal samples, compared with bulk samples. In bulk samples, the PL properties are also dependent on the synthesis methods [59]. Constantinos C. Stoumpos et al. reported the synthesis of $CH_3NH_3Sn_{1-x}PbxI_3$ solids using solid-state chemistry and chemical solution approaches [59]. The materials prepared from the chemical solution method were not luminescent. However, the

242 | *6 Synthesis, Properties, and Applications of Metal Halide Perovskite-Based Nanomaterials*

samples from solid-state chemistry methods exhibited strong PL (Figure 6.10c–e). The large difference in PL emission is ascribed to the differences in defects and impurities during the various synthetic processes.

The PL emission spectra of MHP can be adjusted by changing the composition, size, and morphology of the perovskites [97]. For example, the PL emission of $CsPbX_3$ nanocrystals shows a strong dependence on size and composition (Figure 6.10f) [97]. The PL properties dependent on the size/morphology of perovskite have been studied. In Figure 6.10g, with the decrease of particle size, the PL emission peak of $CsPbBr_3$ perovskite quantum dots (PQDs) moved to the short wavelength region. A similar phenomenon was also observed in $CsPbI_3$ PQDs [41]. The composition of perovskite showed a larger impact than that in the size of perovskite nanocrystals. The composition changed the exciton Bohr diameter of perovskites. In addition, the morphology also influences the TRPL, as shown in Figure 6.10h. Sun et al. investigated the $CsPbBr_3$ perovskites with different morphologies, including spherical QDs, nanocrystals, nanorods (NRs), and NPs [98]. The resultant photoluminescence decay lifetime varied from a few nanoseconds to tens or hundreds of nanoseconds for the different morphologies. The changes in decay lifetime could be attributed to the different surface states of perovskites with different morphologies.

6.3 Synthesis of Metal Halide Perovskite-Based Nanomaterials

The synthesis methods for MHP are as follows: hot injection method, ligand-assisted reprecipitation method, one-step spin-coating method, two-step spin-coating method, vapor-assisted solution method [107, 108], painting method, ultrasonic spraying technology, and blade coating method [109]. The approaches are summarized in Figure 6.11.

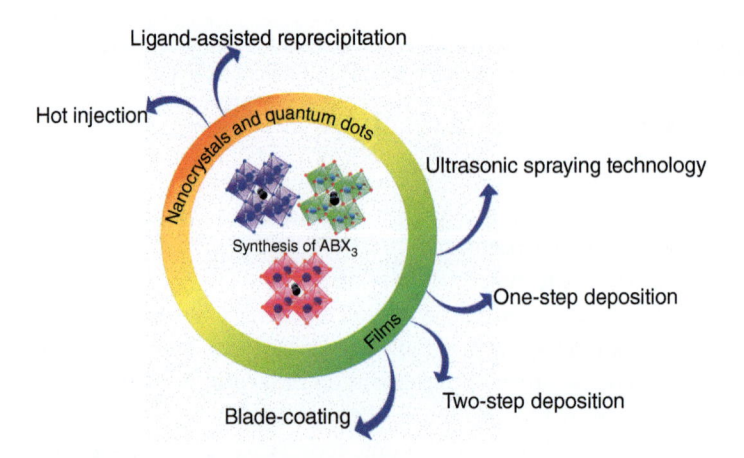

Figure 6.11 The synthesis approaches of metal halide perovskite-based nanomaterials.

6.3.1 Hot Injection Method

Hot injection method is the most common approach to synthesize MHP nanocrystals. The method requires the long-chain organic ligands to adjust the size and morphology of perovskite nanocrystals. The synthesis process is shown in Figure 6.12a [112]. First, the metal salts, i.e. PbX_2 (X = Cl, Br, I), were dissolved in octadecene at a certain temperature (80–200 °C). The organic ligands, such as oleylamine and oleic acid, were also added to improve dissolution. Then, the cation precursors were injected into the solution for pyrolysis to form the perovskite nanocrystallines [113]. When the reaction finished, the reacted solution was cooled in an ice bath, and the perovskite nanocrystals were collected by centrifugation.

Organic ligands are essential for synthesizing MHP nanocrystals. They consist of acid ligands (oleic acid (OA), diisooctylphosphinic acid (DIOP), etc.) and alkaline ligands (oleylamine (OAm), dioctylamine (DOAm), etc.). The mixture of OA and OAm is popular. The molecular structures of the organic ligands are shown in Figure 6.12b [110]. The length of the ligand chain showed an important effect on the size and morphology of perovskite nanocrystals.

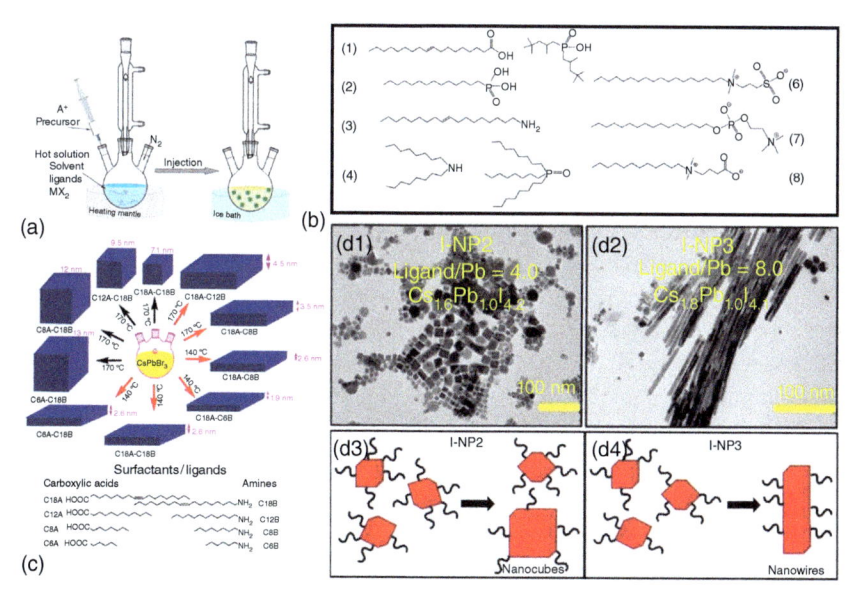

Figure 6.12 (a) Synthesis scheme of perovskite nanocrystals using the hot injection method. Source: Reprinted with permission from Ref. [14]. Copyright 2019 American Chemical Society. (b) The molecular structure of different organic ligands. Source: Reprinted with permission from Ref. [110]. Copyright 2020 American Chemical Society. (c) The effect of the length of ligand chain for the size and morphology of perovskite nanocrystals. Source: Reprinted with permission from Ref. [111]. Copyright 2016 American Chemical Society. (d) TEM images of different ligand-to-lead molar ratio and the observed elemental composition obtained by EDX measurements and corresponding reaction pathway. Source: Reprinted with permission from Ref. [110]. Copyright 2020 WILEY-VCH Verlag GmbH & Co. KGaA, Weinheim.

Pan et al. summarized the influence of different ligand chain lengths on the morphology and size of $CsPbBr_3$ nanocrystals. As shown in Figure 6.12c, besides the effect of temperature, the length of the alkyl chain had an independent correlation with the size and morphology of $CsPbBr_3$. The ratio of ligand-to-metal ion also determined the morphology of perovskite nanocrystals. Grisorio et al. reported the influence of the ligand-to-metal ratio on the morphology of iodine-based perovskite nanocrystals. The ratios easily changed in the hot injection process [113]. When the ligand-to-metal ratio increased from 4.0 to 8.0, the morphology of $CsPbI_3$ nanocrystals changed from nanocubes to nanowires (Figure 6.12d1, d2). Nanowires can be formed from the attachment of nanocubes in the presence of high ligand-to-metal ratios (Figure 6.12d3–d4). Although the hot injection method is popular, it requires a high reaction temperature and a strict inert gas atmosphere. The synthesis process is unstable and expensive, which is not suitable for large-scale productions.

6.3.2 Ligand-Assisted Reprecipitation Method

Ligand-assisted reprecipitation (LARP) method can be performed at room temperature. This method is simple and fast. All the precursors are first dissolved in their good solvents (e.g. DMF), and then added to the bad solvents (e.g. toluene) drop by drop, as shown in Figure 6.13a. The solubility of perovskite in the good solvent and the bad solvent are different. When the solvents are mixed, the perovskite nanocrystals will begin to crystallize. During crystal precipitation, some ligands are added to assist crystallization [116]. The method is heat-free and conducted at room temperature. Thus, it is suitable for the synthesis of all-inorganic halide perovskites and inorganic–organic halide perovskites. However, the crystallization process at low temperatures results in more structural defects and trap states in perovskite materials. He Huang et al. found that improving the crystallization temperature enhanced the quantum yield. Emission peaks of PQDs were also tuned (Figure 6.13b). When the temperature increased from 0 to 60 °C, the quantum yield of $CH_3NH_3PbBr_3$ quantum dots increased from 74% to 93%.

The enhanced quantum yield was due to the good crystallinity and surface passivation at high temperatures [114]. The solvents could influence the defects of perovskites in the LARP method. Zhang et al. investigated the interaction between the precursor and the good solvent in the crystallization process of $MAPbI_3$ QDs. The precursors could form intermediates in some coordinated solvents (e.g. DMSO, DMF, THF). Meanwhile, the intrinsic PbI_2 could be maintained in some non-coordinated solvents (e.g. GBL, ACN) [115] (Figure 6.13c). The formation of the intermediate led to the formation of some defects in $MAPbI_3$ QDs. The resultant defects concluded in their fast degradation.

6.3.3 Solution Deposition Methods

The solution deposition method has been used to prepare MHP-based films. Solution deposition methods could be divided into the one-step method and the two-step method.

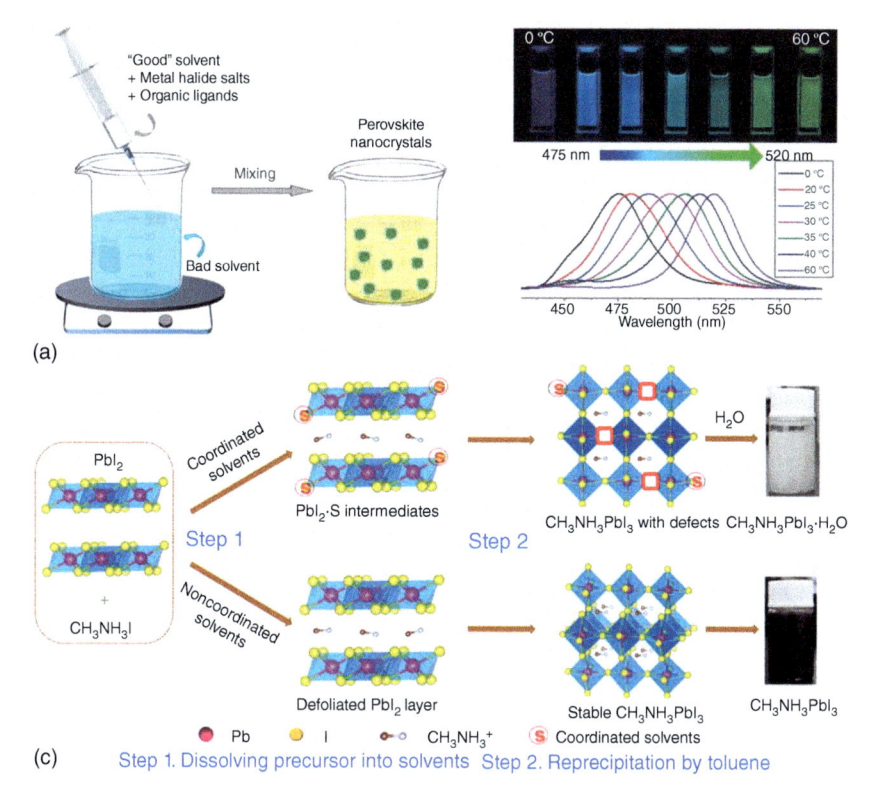

Figure 6.13 (a) Synthesis process of perovskite nanocrystals by the ligand-assisted reprecipitation method. Source: Reprinted with permission from Ref. [14]. Copyright 2019 American Chemical Society. (b) Photograph and PL spectra of colloidal solutions of MAPbI$_3$ QDs in different temperature. Source: Reprinted with permission from Ref. [114]. Copyright 2015 WILEY-VCH Verlag GmbH & Co. KGaA, Weinheim. (c) Schematic illustrations of the transformation process from a precursor (CH$_3$NH$_3$I and PbI$_2$) solution to CH$_3$NH$_3$PbI$_3$ perovskite in coordinated solvents (top) and non-coordinated solvents (bottom). Source: Reprinted with permission from Ref. [115]. Copyright 2017 American Chemical Society.

6.3.3.1 One-Step Method

The formation of perovskite films in the one-step method involves the nucleation, growth, and crystallization processes. The first step of the one-step method is dissolving all the reactants in a polar solvent (such as dimethylformamide (DMF), dimethyl sulfoxide (DMSO), or γ-butyrolactone (GBL)) to obtain the perovskite precursor solution. The precursor solution is spin-coated on a substrate to form a thin film. Finally, the thin film is heated at a certain temperature to enhance the crystallization (Figure 6.14a). The quality of the perovskite film was determined by the concentration of the precursor solution, spinning speed, annealing temperature, and the external environment (humidity or specific solvent atmosphere, etc.) during its annealing process. In the one-step method, the crystal formation of perovskite and the removal of residual solvent coincide during the spinning, coating, and annealing processes. Thus, it is easy to form pinholes in the perovskite film and uneven film covering (Figure 6.14b) [121]. The evaporation of the solvent with a high boiling point, i.e. DMF, is difficult. The existence of the solvent prevents the uniform growth of grains in the film. Therefore, the one-step anti-solvent method has been developed. The anti-solvent method refers to adding the anti-solvent into the perovskite precursor during the spinning process. The anti-solvent cannot dissolve the perovskite, but it can be miscible with residual solvent. The added anti-solvent can remove the residual solvent and make the precursor solution reach the supersaturated state to promote perovskite formation. In 2014, Spiccia and coworkers reported a one-step fast crystallization-deposition (FDC) method. The approach resulted in flat, highly uniform $CH_3NH_3PbI_3$ thin films. It involved the spin-coating of a DMF solution of $CH_3NH_3PbI_3$ on a substrate, followed by exposure of the wet film to a second solvent, such as chlorobenzene (CBZ), to induce crystallization, as shown in Figure 6.14c [122]. Miyasaka and coworker investigated the effect of three different anti-solvents, chlorobenzene (CB), ethyl acetate (EA), and toluene (T), on the quality of Cs/FA/MA perovskite film. The perovskite quality

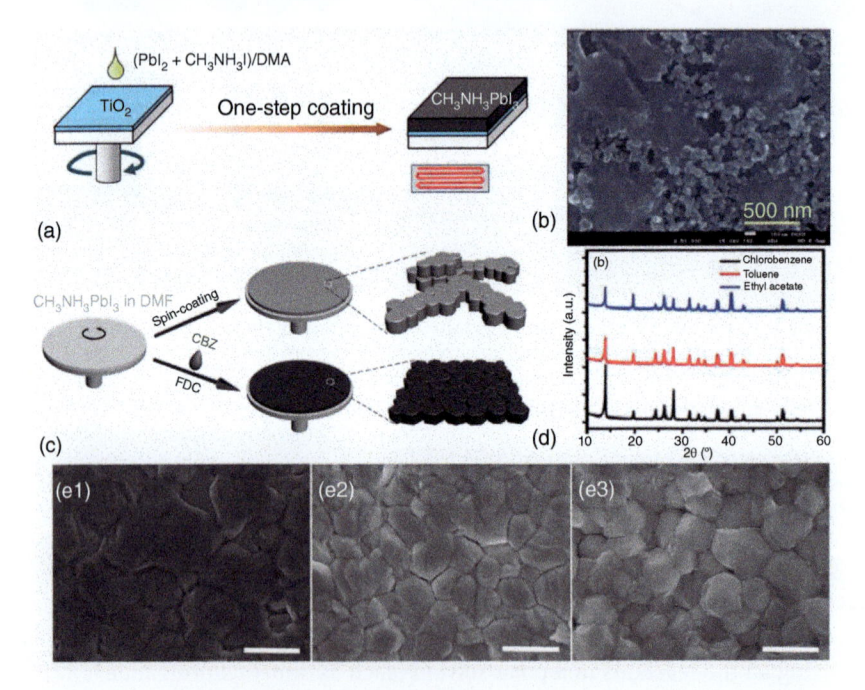

Figure 6.14 (a) One-step spin-coating procedures for the formation of $CH_3NH_3PbI_3$. Source: Reprinted with permission from Ref. [117]. Copyright 2014 Creative Commons Attribution (CC BY). (b) SEM images of $CH_3NH_3PbI_3$ on the mesoporous TiO_2 layer using the one-step deposition method. Source: Reprinted with permission from Ref. [118]. Copyright 2013 WILEY-VCH Verlag GmbH & Co. KGaA, Weinheim. (c) Schematic illustration of the FDC process and conventional spin-coating process for fabricating perovskite films. Source: Reprinted with permission from Ref. [119]. Copyright 2014 WILEY-VCH Verlag GmbH & Co. KGaA, Weinheim. (d) XRD pattern of perovskite absorber layer deposited using various dripping solvents during spin-coating. Top SEM images of perovskite films with different anti-solvents (e1) chlorobenzene, (e2) ethyl acetate, and (e3) toluene dripping. Scale bar is 1 μm. Source: Reprinted with permission from Ref. [120]. Copyright 2017 WILEY-VCH Verlag GmbH & Co. KGaA, Weinheim.

and crystallization process were dependent on the solvent dripping recipe [120]. The peak intensity of (110) plane (≈13.8°) in CB dripping was strong, compared with EA and T dripping samples. It indicated good crystallinity in CB dripping (Figure 6.14d). Furthermore, a large grain size of up to 1.5 μm was also observed for CB-based samples (Figure 6.14e1).

6.3.3.2 Two-Step Method

The two-step method involves two main processes: (i) preparing the lead halide film and (ii) forming the perovskite layer via the reaction of lead halide and halogen cation. Nowadays, high-efficiency perovskite solar cells have been prepared using the two-step method. In 2013, Grätzel and coworkers developed a sequential deposition method to prepare the perovskite within the porous metal oxide film [118]. The findings introduced new fabrication methods of perovskite-based photovoltaic devices. The PbI_2 solution was first added to the mesoporous titanium dioxide film and then immersed into the CH_3NH_3I solution. Within several minutes, the PbI_2 changed to the $CH_3NH_3PbI_3$ perovskite. This method has been used to fabricate bulk-junction perovskite solar cells. Yang and coworkers demonstrated the vapor-assisted solution process (VASP) to fabricate perovskite thin films and related PV devices with planar heterojunction structures [119]. As shown in Figure 6.15a, the key step of the method was the growth of film via *in situ* reaction between the as-deposited film of PbI_2 and CH_3NH_3I vapor. The as-obtained perovskite film exhibited full surface coverage, uniform grain structure, large grain size (up to micrometers), and 100% precursor transformation, confirmed by AFM and TEM in Figures 6.15b and 6.10c.

Nam-Gyu Park and coworkers reported a two-step spin-coating method to prepare perovskite films. The schematic diagram is shown in Figure 6.15d [117]. They found that the size of the $MAPbI_3$ cuboids was dependent on the concentration of CH_3NH_3I and the exposure time of PbI_2 to the CH_3NH_3I solution before spin-coating. As confirmed by the SEM image in Figure 6.15e, the size of the $MAPbI_3$ cuboids increased with decreasing CH_3NH_3I concentration. Its average size was determined to be ~720 nm for 0.038 M, ~360 nm for 0.044 M, ~190 nm for 0.050 M, ~130 nm for 0.057 M, and ~90 nm for 0.063 M. The two-step method was also applied to prepare the high-quality perovskite films for several carbon-based perovskite solar cells [124]. Different from the two-step deposition process mentioned previously, the solvent for CH_3NH_3I (MAI) solution at the second step changed from isopropanol (IPA) to a mixed solvent of IPA/cyclohexane. The mixed solvent accelerated the conversion of PbI_2 to $CH_3NH_3PbI_3$ and suppressed the Ostwald ripening process. The resultant high-quality perovskite layer possessed pure phase, even surface, and compact capping top layer. The perovskite layer also resulted in a compact contact at the perovskite/carbon interface, as shown in Figure 6.15g.

In the two-step method, the quality of PbI_2 films determined the growth mechanism and quality of perovskite films. Thus, constructing PbX_2 nanostructures is an effective strategy to convert PbX_2 into perovskites. Choy and coworkers incorporated a small number of additives into the PbI_2 precursor solution to form self-assembled porous PbI_2, which facilitated perovskite conversion without any

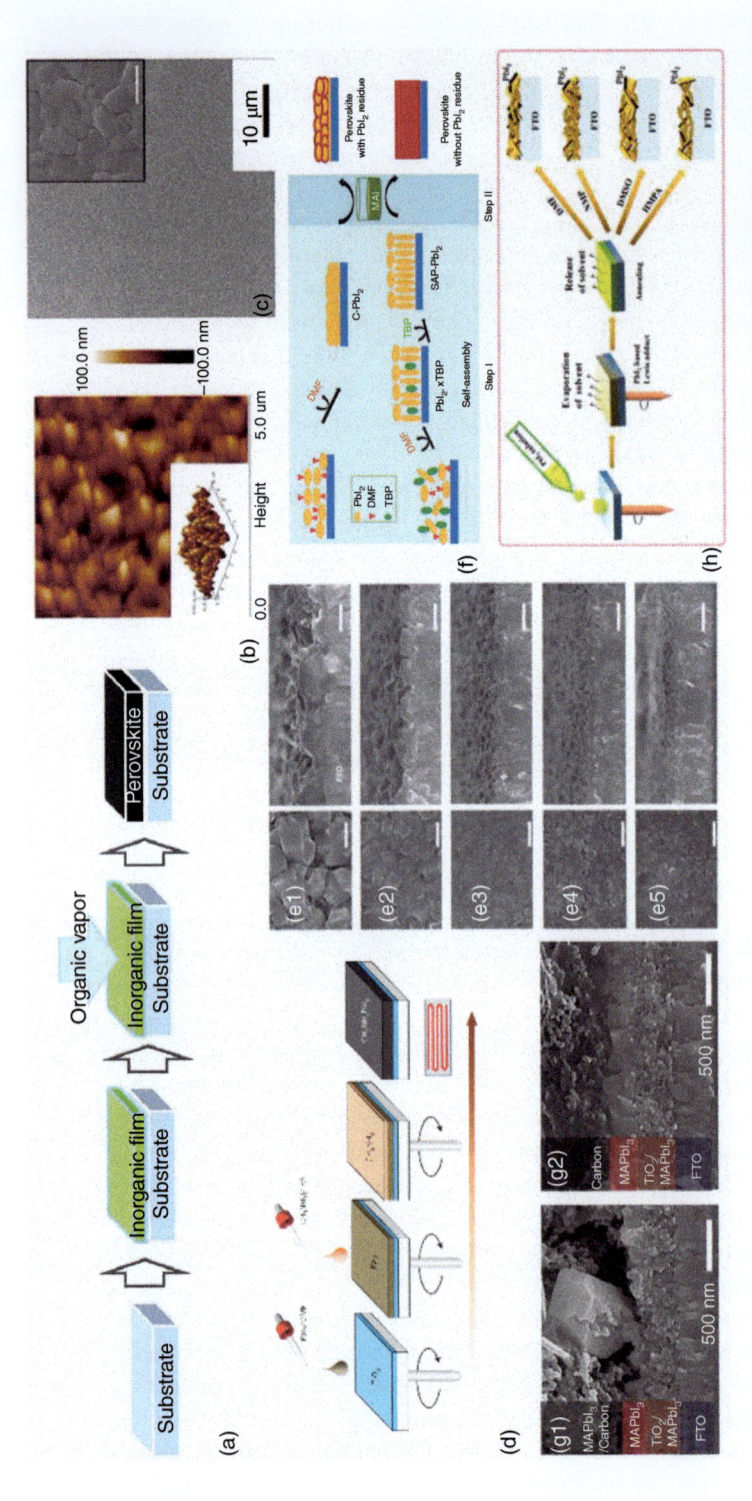

Figure 6.15 (a) Schematic illustration of perovskite film formation through vapor-assisted solution process. (b) Tapping-mode AFM height images and (c) top-view SEM images of perovskite film on the FTO/c-TiO$_2$ substrate, obtained by reacting PbI$_2$ film with CH$_3$NH$_3$I vapor at 150 °C for 2 hours in N$_2$ atmosphere. Source: Reprinted with permission from Ref. [119]. Copyright 2014 American Chemical Society. (d) Two-step spin-coating procedure for CH$_3$NH$_3$PbI$_3$ cuboids. (e) Surface cross-sectional SEM images of MAPbI$_3$ cuboid and its size dependence on the CH$_3$NH$_3$I concentration. Source: Reprinted with permission from Ref. [117]. Copyright 2014 Springer Nature. (f) Schematics of the processes of fabricating PbI$_2$ and perovskite films. Source: Reprinted with permission from Ref. [123]. Copyright 2015 WILEY-VCH Verlag GmbH & Co. KGaA, Weinheim. (g) Cross-sectional SEM images of the perovskite layers prepared from the IPA (g1) and IPA/CYHEX solvents (g2) at the reaction time of 4 hours. Source: Reprinted with permission from Ref. [124]. Copyright 2016 WILEY-VCH Verlag GmbH & Co. KGaA, Weinheim. (h) Schematic of the fabrication process of the mesoporous PbI$_2$ film from solutions with different strong Lewis bases. Source: Reprinted with permission from Ref. [125]. Copyright 2017 the Royal Society of Chemistry.

PbI$_2$ residue [126]. As shown in Figure 6.15f, when 4-*tert*-butylpyridine (TBP) was added into the PbI$_2$-DMF precursor, PbI$_2$ coordinated with the TBP molecules, a nitrogen-donor ligand to form coordination complexes (PbI$_2$·xTBP) after volatilization of DMF at room temperature. PbI$_2$·xTBP complexes will be decomposed to PbI$_2$ and TBP via annealing at 70 °C for several minutes. The sites where TBP resided became small holes. The size and amount of the holes can be controlled by changing the concentration of TBP. Cao et al. introduced some strong Lewis bases into PbI$_2$/DMF solutions in the first step, which controlled the morphologies of PbI$_2$ films with a mesoporous structure [125]. The mesoporous structure in PbI$_2$ films provided space for volume expansion during the reaction between PbI$_2$ and CH$_3$NH$_3$I and channels for CH$_3$NH$_3$I solution to diffuse into the PbI$_2$ films. The special structure eliminated the residual PbI$_2$ and helped the dissolution and recrystallization processes.

Consequently, the smooth perovskite films without any residual PbI$_2$ were obtained. The diagram of preparing mesoporous PbI$_2$ films with different Lewis base solutions is shown in Figure 6.15h. Various ligands, e.g. H$_2$O [127] and MACl [123], are used to induce nanostructure formation.

6.3.3.3 Other Solution-Processing Methods

The blade-coating process, which exhibits rapid film growth and less material consumption, is a promising method for fabricating scalable perovskite films. Deng et al. reported the compact, smooth, and pure phase of mixed cations (FA and MA) perovskite films at grain scale using this method [128]. The process of the blade coating method is shown in Figure 6.16a. PbI$_2$, MAI, and FAI were first dissolved in DMF at room temperature. After that, the solution was poured onto the preheated (100–145 °C) substrate placed on a hot plate. Finally, the solution was swiped on the substrate by the blade coating process. The perovskite films were dried in a few seconds. The changes in the initial formation stage (e.g. *in situ* temperature, precursor stoichiometry) resulted in high-quality perovskite films with a tunable composition. A rotating magnetic field technique (RMF) was proposed to manipulate the crystallization of perovskite thin films during the blade coating process to achieve controllable film growth with high quality [131]. The perovskite films are prepared using home-made equipment containing an RMF. The fabrication procedure and crystallization process of films are illustrated in Figure 6.16b. Compared with the traditional preparation method, the perovskite films manipulated by RMF showed improved uniformity and good coverage (Figure 6.16c). This approach could be achieved by physically manipulating the crystallization of perovskites and low-cost fabrication of high-performance perovskite devices.

Slot-die coating is a versatile deposition technique, which is also applied as a solution-processing method for the large-scale production of perovskite films. In this method, all injected solutions can be transferred to the film. Moreover, the film thickness is tuned by controlling the solution speed [130]. For a slot-die coating process, a coating head is placed close to and across a substrate. The ink is pumped into the head by a syringe. The ink is forced out of a narrow slit along the length of the coating head, as shown in Figure 6.16d [130]. Perovskite solar cells have also

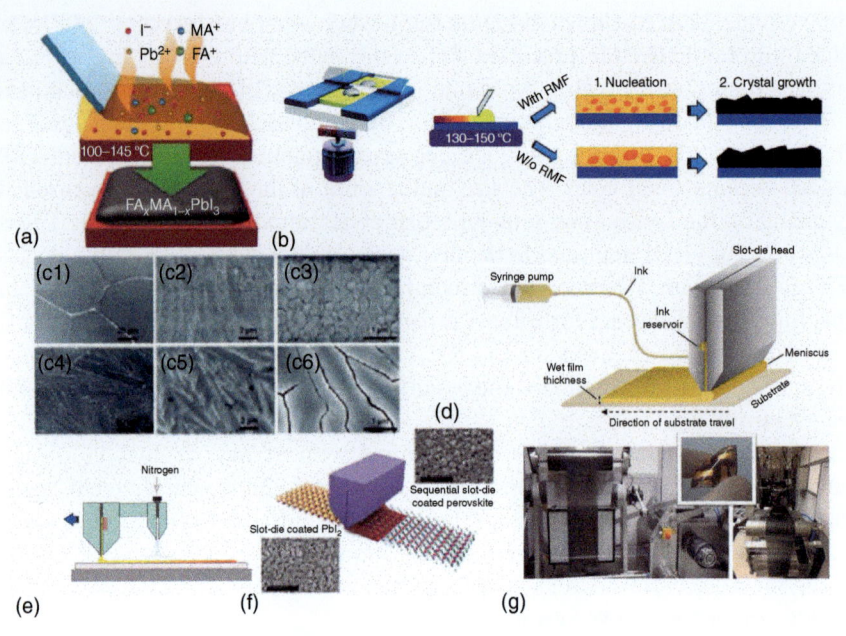

Figure 6.16 (a) Illustration of doctor-blade coating of mixed cation perovskite film from precursor solution heated on a hot plate. Source: Reprinted with permission from Ref. [128]. Copyright 2016 WILEY-VCH Verlag GmbH & Co. KGaA, Weinheim. (b) A schematic representation of the home-made equipment for the RMF-induced doctor-blading process and the deposition, nucleation and grain growth mechanisms of perovskite thin films. (c) SEM images of the (c1–c3) RMF-induced perovskite layer, (c4–c6) conventionally produced perovskite layer. Source: Reprinted with permission from Ref. [129]. Copyright 2018 the Royal Society of Chemistry. (d) Schematic of a slot-die coating process, showing the delivery of ink to the head from a syringe pump and formation of an ink wet film between the coating head lips and the substrate. Source: Reprinted with permission from Ref. [130]. Copyright 2020 Elsevier. (e) Schematic illustration of slot-die coating with a gas-quenching process for the fabrication of pinhole-free PbI_2 layer and (f) sequential slot-die coating of MAI. Source: Reprinted with permission from Ref. [129]. Copyright 2015 WILEY-VCH Verlag GmbH & Co. KGaA, Weinheim. (g) Photograph images of the roll-to-roll coated perovskite layer ($Cs_{0.15}FA_{0.85}PbI_{2.85}Br_{0.15}$) and an example of the manufactured flexible devices. Source: Reprinted with permission from Ref. [130]. Copyright 2020 WILEY-VCH Verlag GmbH & Co. KGaA, Weinheim.

been fabricated via slot-die coating by Hwang et al. except for the metal electrode [132]. The schematic illustration of slot-die coating with a gas-quenching process to fabricate the pinhole-free PbI_2 layer is shown in Figure 6.16e. The gas-quenching treatment was used to dry the PbI_2 film. The SEM images of perovskite are shown in Figure 6.16f. Large-scale perovskite films are prepared using this approach. For example, perovskite solar modules with active areas of 10.1 and 100 cm^2 showed PCE of 10.4% and 4.3%, respectively. Furthermore, the slot-die coating could be applied on flexible glass and plastic substrates. Galagan et al. prepared perovskite solar cells by roll-to-roll slot-die coating on flexible substrates with a width of 30 cm and a web speed of 3–5 m min^{-1} [129]. A photograph of the perovskite is shown in Figure 6.16g. The resultant stabilized PCE of the resultant photovoltaic device was up to 13.5%.

6.4 Application of Metal Halide Perovskite-Based Nanomaterials

MHP-based nanomaterials have been used as light-harvesting materials in low-cost, high-efficiency photovoltaic devices due to their excellent optical and electrical properties and solution fabrication process [133, 134]. MHP-based photovoltaic cells with PCEs >25% [43] and light-emitting diodes (LEDs) with external quantum efficiencies >20% [135] have been developed in the last 12 years, leading to many potential applications. Photodetectors and lasers based on these nanoparticles have also been realized.

6.4.1 Perovskite Solar Cells

Perovskite solar cells (PSCs) are the most promising photovoltaic devices due to their solution-prepared process and high PCE. The exploration of MHP-based nanomaterials as light absorbers in photovoltaics was first reported by Miyasaka's group in 2009 [136]. Starting with a PCE of 9.7% with pure MAPbI$_3$ in 2012 [137], a certified PCE of 25.5% has been achieved in 2021 [34] using a mixed cation and mixed anion composition. Organic–inorganic hybrid halide PSCs were developed in the past decade and used in commercial applications. MHP films with large grain sizes and excellent surface morphology have high-performance PSCs. However, in the solution deposition procedures, interstitial and antisite defects related to trap-states originated on the surfaces of perovskite films [138].

Therefore, various strategies, e.g. precursor solvent engineering [139], anti-solvent washing [140], composition designing [141], and post-growth treatment [133], have been used to prepare high-quality perovskite films. Since the bandgaps of perovskite materials can be adjusted from 1.2 to 3.0 eV by component engineering, the perovskites with different compositions and bandgaps have become popular in various PV applications [142], such as serving as a sub-cell (low-bandgap perovskites and/or wide-bandgap perovskites) in tandem solar cells [143], colorful displays [144] and building-integrated photovoltaics (wide-bandgap CsPbI$_{3-x}$Br$_x$) [145] etc. (Figure 6.17a). Lewis acids have unique passivation capabilities for defects with lone electron pairs like free iodide ions and lead iodine antiside defects; Lewis bases can strongly bond electron-deficient defects such as Pb^{2+} interstitials. The most prominent feature of Lewis bases is the presence of atoms containing lone electron pairs, especially nitrogen [150], sulfur [151], oxygen [152], and phosphorus atoms [153]. Small molecules with functional groups, such as C=O, P=O, and –NH$_2$, improved film quality. They can passivate undercoordinated lead due to their strong bonding with it [154].

Yang and coworkers introduced different configurations of small molecules with the same functional groups, such as theophylline, caffeine, and theobromine, into the MA-based MHP photovoltaics [146]. When N—H and C=O in the molecule were in the optimal configuration, the hydrogen bond formation between N—H and I was conducive to the combination of the main C=O and the antisite Pb defect to maximize the surface-defect binding (Figure 6.17b). The stable efficiency

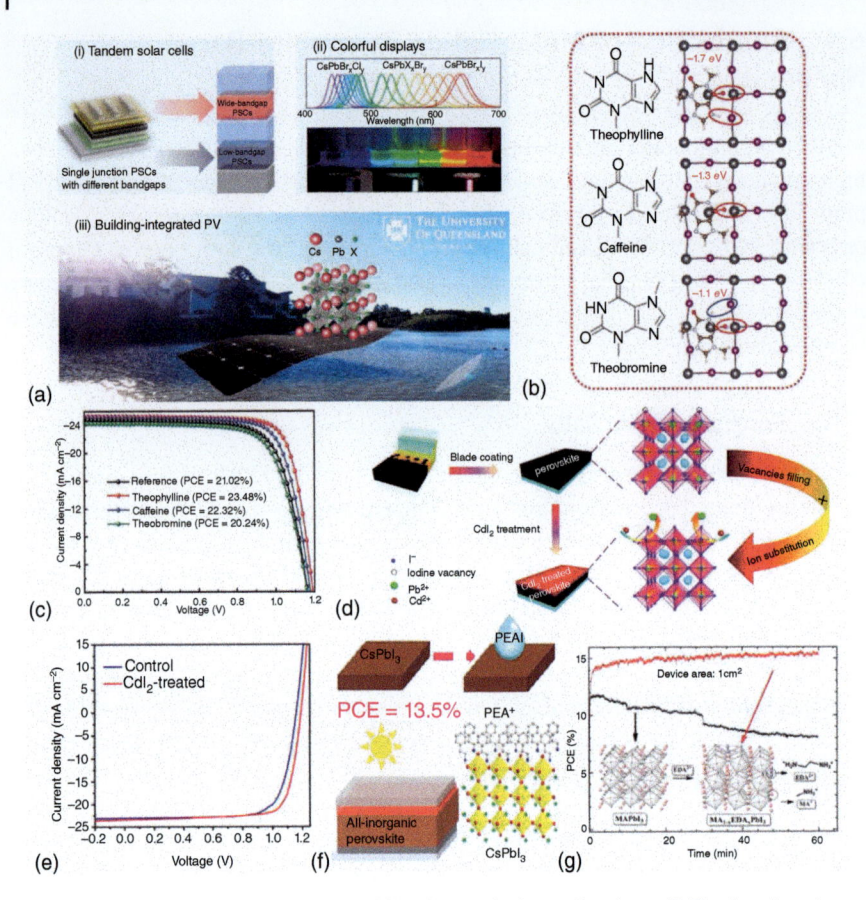

Figure 6.17 (a) The potential perovskite photovoltaic applications: (i) The low-bandgap PSCs and wide-bandgap PSCs serve as the sub-cell in tandem solar cells, (ii) Colorful displays based on inorganic perovskites with tunable halide combination, (iii) Building-integrated photovoltaics. Source: Reprinted with permission from Ref. [142]. Copyright 2021 Elsevier. (b) Theoretical model for the passivation of Pb_I defects in perovskites. (c) J–V curves of PSCs with reverse and forward scanning. Source: Reprinted with permission from Ref. [146]. Copyright 2019 The American Association for the Advancement of Science. (d) A schematic illustration of a blade-coated perovskite film with CdI_2 surface treatment. (e) J–V curves of PSCs with or without CdI_2 treatment. Source: Reprinted with permission from Ref. [147]. Copyright 2020 American Chemical Society. (f) Schematic illustration of organic cation surface termination using PEAI and procedures for the deposition of PEAI-$CsPbI_3$ perovskite. Source: Reprinted with permission from Ref. [148]. Copyright 2018 Elsevier. (g) Photovoltaic measurements with an aperture of 1.04 cm². The inside image shows the structure of $MAPbI_3$ upon modification with diammonium cation. Source: Reprinted with permission from Ref. [149]. Copyright 2017 WILEY-VCH Verlag GmbH & Co. KGaA, Weinheim.

of the resultant PSCs reached 22.6% (Figure 6.17c). The rubidium and potassium cations were incorporated into the cesium methyl ammonium and formamidine cationic standard perovskite film to improve the film quality of efficient PSCs. The coexistence of Rb^+ and K^+ cations can reduce the recombination within

PSCs and improve photovoltaic performance [2, 155]. Both Rb^+ and Cs^+ cations have been incorporated into perovskites for constructing efficient and stable PSCs [156, 157]. This approach has opened a new avenue to reduce surface point defects and compensate for the surface halide deficiencies to prepare efficient and stable PSCs.

The surface passivation strategy is the most essential and useful tool to reduce nonradiative recombination loss in PSCs. The desirable V_{oc} improvement and V_{loss} suppression in PSCs are achieved via deliberate control over the surface chemistry of perovskite films [142]. Modifying the defective surface of $Rb_{0.025}Cs_{0.025}FA_{0.70}MA_{0.25}PbI_3$ perovskite with cadmium iodide (CdI_2) reduced surface halide deficiency. It suppressed the vacancies formation energy by stabilizing the iodine ions via the formation of strong Cd—I ionic bonds, ameliorating the interfacial charge recombination loss [147]. The optimized CdI_2-treated device showcased a high PCE of 21.9% and a high V_{oc} of 1.20 V for blade-coated PSCs (Figure 6.17d,e). Moreover, composition engineering with organic ions has been adopted, resulting in stable 2D perovskite structures. For example, the phenylethylammonium iodide (PEAI) was introduced into the perovskite precursor solution to induce reduced-dimensional perovskite formation [158]. The resulting quasi-2D perovskite inhibited the phase transition, reduced the defect state density of the film, and increased the device PCE to 12.4%. The introduction of PEAI formed an organic cationic PEA^+ terminal on the surface and realized defect passivation, stabilizing $CSPbI_3$ PSCs (Figure 6.17f) [159]. Apart from PEAI, PTABr was also applied as a bifunctional additive to achieve defect passivation and Br doping [160]. The PTABr-treated $CsPbI_3$-based perovskite device promoted the stability of the perovskite phase, and the PSCs reached a PCE of 17.06%. The organic PTA cation can passivate the surface and improve the water resistance of perovskite materials.

Degradation of perovskite materials can be caused by humidity, heat, oxygen, and light [148, 161]. Water has been considered the main factor of perovskite instability [162, 163]. The hydrophobic barrier has been applied to form a completed film coverage without pinholes to protect the perovskite from water penetration. However, the deposition of a hydrophobic layer on hydrophilic perovskite due to adverse surface interactions can form a thick film and pinholes [164]. The thick insulating barriers on the perovskite film can also cause large series of resistance. Thus, the hydrophobic layers should be combined with the additional charge transfer networks [165, 166]. For example, polymer intermixing in the perovskite bulk phase has achieved the high stability of devices by improving the morphology of the perovskite film [167]. Another efficient strategy to improve water resistance is mixing 2D structures with 3D structures [2]. Incorporating ethylenediammonium iodide (EDAI) into the 3D $MAPbI_3$ structure improved the efficiency but not the stability. $MAPbI_3$-EDAI devices retained about 75% of their initial performance after 72 hours of continuous operation under illumination, while the $MAPbI_3$ devices lost 90% of their initial performance in just 15 hours of operation, as shown in Figure 6.17g [149].

6.4.2 Perovskite Light-Emitting Diode

MHPs have gained popularity for applications in light-emitting diodes (LEDs) due to their excellent optical and electronic properties, such as high charge-carrier mobility, saturated emission colors, easy color tunability, and balanced electron/hole transportation. Through defect passivation, grain size control, and balanced carrier injection, EQE for bromine-based green light (luminescence wavelength ~520 nm) and near-infrared (luminescence wavelength ~750 nm) perovskite light-emitting diode (PeLED) has reached more than 20% [168, 169]. In 2014, Lee and coworkers pioneered the development of a PeLED that could replace organic light-emitting diodes (OLEDs) and quantum dot light-emitting diodes (QDLEs) [170, 171]. PeLEDs demonstrated room-temperature electroluminescence with narrow emission, which opened new opportunities in research (Figure 6.18a) [172, 175]. The advancement of PeLEDs is partially due to the extensive material design and processing experience gained from PSCs [176]. However, several challenges are associated with the stability of PeLEDs, such as the imbalanced charge injection and the rapid degradation of electroluminescence during operation [177, 178]. Due to intrinsic

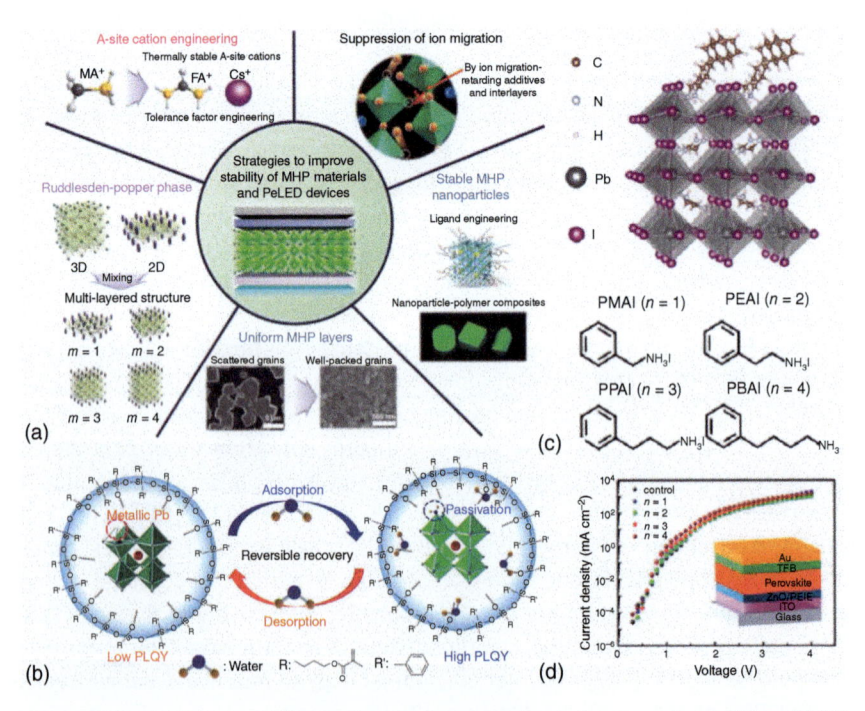

Figure 6.18 (a) Schematic illustration for the strategies to improve the stability of MHP materials and PeLEDs. Source: Reprinted with permission from Ref. [172]. Copyright 2018 WILEY-VCH Verlag GmbH & Co. KGaA, Weinheim. (b) Schematic illustration of reversible defect healing in crosslinked perovskite NP (CPN) by H_2O molecules. Source: Reprinted with permission from Ref. [173]. Copyright 2020 WILEY-VCH Verlag GmbH & Co. KGaA, Weinheim. (c) Schematic diagram of molecular passivated perovskite surface and structures of the passivation molecules. (d) J–V curves for the control and treated devices.
Source: Reprinted with permission from Ref. [174]. Copyright 2021 Springer Nature.

defects, ion migration, poor temperature tolerance, and film morphology, most PeLEDs degrade quickly, within several minutes to several hours of operation/after turn-off, even under a low current density of less than 10 mA cm^{-2}.

The degradation mechanism of perovskite nanoparticles (PeNPs) varies with composition. The A-site cation is a key parameter to control the degradation mechanism. For example, organic A-site cation, e.g. MA$^+$ and FA$^+$, can be decomposed into volatile products such as methylamine, methyl halide, or formamidine [179, 180]. Mixing A-site cations can cause changes in the lattice constants, stabilizing the perovskite crystal structure and enhancing its stability. Several approaches have been used to increase the stability of MHPs, such as the formation of an inorganic shell (e.g. Al_2O_3 or SiO_2), heterojunction structures, and nanoparticles/polymer composites [181, 182]. The ionic bonding of perovskite decomposes during silica coating. Thus, the silica coating source and coating conditions should be selected carefully [183]. Creating an ideal core–shell structure by forming a mono-dispersed crystalline shell on the surface of individual particles was also reported [184, 185]. The shell can improve the chemical stability of perovskites by preventing oxygen and water from penetrating the core. Cubic $CsPbBr_3$ at amorphous $CsPbBr_x$ shell perovskite QDs have been obtained using hot injection method and centrifugation process [186]. The amorphous shell is a protective strategy for the perovskite $CsPbBr_3$ core to enhance its ability. It also contributed to the formation of excitons, which improved PLQY. In addition, the crosslinking of ligands can also increase the chemical stability of MHPs. Lee and coworkers reported a simple but effective materials design approach to achieve long stability of crosslinked $MAPbBr_3$ NPs (Figure 6.18b) in various environments (air, water, chemicals, high temperature [85 °C] with high relative humidity [85% RH]) by employing methacrylate-functionalized matrix [173]. Unsaturated hydrocarbons in the acid and base ligands of NPs are chemically crosslinked with a methacrylate-functionalized matrix, preventing the decomposition of the perovskite crystals.

Defect deteriorates optical properties and stability of MHPs. Additional passivation strategies are required to achieve high-efficiency PeLEDs. The defects in solution-processed PeLEDs can cause severe nonradiative recombination and low stability. Xiao and coworkers have designed a double passivation molecular additive 4-fluorophenylmethylammonium-trifluoroacetate (FPMATFA) by acid-base neutralization reaction, simultaneously passivating lead and halide defects in the perovskite layer [187]. The FPMA cations can coordinate with halide dangling bonds at the perovskite crystal surface and passivate halide defects, while the TFA anions can bond strongly with lead and passivate lead defects. The dual passivation effect of FPMATFA has also been proved by density functional theory (DFT) calculations. As a result, they boosted the EQE of mixed halide-based PeLEDs to 20.9% with electroluminescence (EL) peak at 694 nm.

Guo et al. used phenylalkylammonium iodide with different alkyl chain lengths to treat the surface of perovskite films (Figure 6.18c) [188]. The phenylalkylammonium groups can stabilize the perovskite lattice by suppressing iodide ion migration (Figure 6.18d). The stabilization effect was enhanced with increasing chain length due to the stronger bonding of the passivating molecule to the perovskite surface and the increased steric hindrance for the diffusion of iodide (I$^-$) on the perovskite surface. With the optimized passivation molecule, phenylpropylammonium iodide (PPAI) achieved a record T_{50} lifetime of 130 hours under 100 mA cm^{-2}, together with an EQE of 17.5% and a record radiance of 1282.8 W sr^{-1} m^{-2} on glass substrates. Along with developing 3D perovskite emitters, perovskite films made of mixtures of 2D/quasi-2D/3D phases have become popular for high-efficiency PeLEDs [174, 189, 190]. These perovskites demonstrated much-enhanced confinement of charge carriers and effective radiative recombination within the lowest-energy perovskite phase [191, 192]. Kim et al. obtained 3D/2D hybrid perovskite by adding a small amount of neutral benzylamine to methylammonium lead bromide, inducing a proton transfer from methylammonium to benzylamine [193]. Benzylammonium in the perovskite lattice suppresses the formation of deep-trap states and ion migration, enhancing both operational stability and luminous efficiency based on its retardation effect in reorientation.

6.4.3 Sensing

The success and popularity of MHPs in PSCs and PeLEDs have also accelerated the development of MHPs in sensing. Although MHPs show potential, their application is limited by their poor stability caused by changing external conditions, e.g. varied temperatures, humid air, polar solvents, and electron-accepting/donating gases [194, 195]. These problems reflect that MHPs are sensitive to external changes. On the other hand, these factors provide advantages for highly sensitive sensing [196, 197]. MHPs-based nanomaterials have been applied in many analytes for sensing by detecting changes in various signals, such as phosphorescence, fluorescence, and I–V fluctuations. Various applications in sensing include temperature sensing, humidity sensing, ion sensing, and visual UV-light sensing [198, 199].

Many perovskite nanomaterial-based devices have demonstrated super sensing in solid and solution states to various chemical and biological species. Perovskite nanomaterials can detect small molecules, such as O_2, NO_2, CO_2, etc. [197]. Optical sensing utilizes ABX$_3$ type perovskites, while electrochemical sensing involves ABO$_3$ or ABO$_4$-type perovskites. Gas sensing involves both ABX$_3$ and ABO$_3$ type perovskites. The research focuses on two main aspects: (i) the exploitation of the high fluorescence of perovskites for optical sensing and (ii) the utilization of the redox ability of perovskites for electrochemical sensing [200].

Gas sensing. Recently, numerous researches have been focused on the development of gas sensors using perovskites. Perovskites exhibited useful properties for gas sensing, such as tunable bandgap, large adsorption coefficient, and long carrier lifetimes. Llobet and coworkers reported that the anions and cations in perovskite structure (ABX$_3$) significantly affect gas sensing performance [201]. In particular, gas sensors employing graphene decorated with different perovskite configurations were successfully employed to detect volatile organic compounds

(VOCs) vapors at ppm levels [202, 203]. The devices developed achieve highly reproducible, reversible, sensitive, and ultrafast detection of VOCs at room temperature.

Optical sensing. The high fluorescence of MHPs confers high sensitivity when they are used as optical sensors. The interaction of MHPs and metal ions cause fluorescence quenching or fluorescence enhancement mechanism. For instance, a hybrid $CH_3NH_3PbBr_3$ perovskite quantum dot (PQD) was used as a fluorescent nanosensor for rapid visual determination of ultra-trace mercury ions (Hg^{2+}) [204]. Hg^{2+} suppressed the PL peak at 520 nm for $CH_3NH_3PbBr_3$ QDs, and a blue shift in their optical properties was observed with an increase in the amounts of Hg^{2+}. The phenomena could be attributed to the surface ion-exchange reaction. Good selectivity, sensitivity, and low detection limit of 0.124 nM in the range of 0–100 nM were achieved. Various interfering metal ions, such as Cd^{2+}, Pb^{2+}, Na^+, K^+, Zn^{2+}, Ba^{2+}, Mn^{2+}, Cu^{2+}, Mg^{2+}, Ca^{2+}, and Ag^+ did not influence the fluorescence intensity of perovskite QDs.

Humidity sensing. Humidity measurement is crucial for various applications, such as weather forecasts, agriculture, and medical instruments [205]. A good humidity sensor displays high sensitivity, short response and recovery times, excellent linearity, and good stability and reliability properties. The active materials of humidity sensors based on resistance or capacitance measurement can be divided into metal oxide [206], perovskite [207], and organic polymers [208, 209]. MHPs are sensitive to moisture exposure with increasing relative humidity (RH) via photoluminescence intensity or resistance, resulting in an ultrasensitive performance [210, 211].

However, the moisture sensitivity of lead halide perovskite has rarely been developed into an applicable humidity sensor due to its intrinsic instability and toxicity. Zhan and coworkers demonstrated the application of lead-free halide perovskite as a humidity sensor [209]. The humidity sensor was fabricated using a stable $Cs_2BiAgBr_6$ thin film, which exhibited a superfast response time (1.78 seconds) and recovery time (0.45 seconds) in relative humidity (RH) ranging from 5% to 75%. At room temperature, MHPs ($CsPbBr_3$ and $CsPb_2Br_5$) combined with diverse ceramics (Al_2O_3, TiO_2, and $BaTiO_3$) achieved excellent sensing performance and stability.

Temperature sensing. MHPs-based nanomaterials of $CH_3NH_3PbX_3$ (X = Cl, Br, I) exist in the cubic, tetragonal, or orthorhombic phases. The related transitions among them can be realized by varying the treatment temperature. For example, $CH_3NH_3PbI_3$ underwent a cubic-to-tetragonal phase transition at 330 K and a subsequent tetragonal-orthorhombic transition at 160 K [18]. This property makes them suitable as fluorescent temperature sensors [212].

6.4.4 Other Devices

A photodetector is a device that converts light signals into electrical signals and acts like a human eye. Its important parameters include spectral response range, sensitivity, response time, EQE, and stability. Recently, MHPs are becoming ideal materials for high-performance photodetectors due to their low trap density, high quantum efficiency, small exciton binding energy, long carrier diffusion length, and long carrier life [213, 214]. MHP-based photodetectors can be divided into three types: photoconductors, photodiodes, and phototransistors, according to the configuration of the devices [215]. In 2014, solution-processed $MAPbX_3$ was first used as a photoconductor for a broadband wavelength from 310 to 780 nm. Although the rise time and decay time were about $0.2\,s^{-1}$, the EQE reached $1.19\% \times 103\%$ at 365 nm [216]. Several theoretical and experimental studies have recently proposed that the migration of ions is the origin of the hysteretic behavior in MHPs [217, 218]. Strategies such as interface engineering or solvent engineering have been proposed to prevent performance degradation in MHP-based optoelectronic devices due to the migration of ions [219, 220]. Jang and coworkers reported reliable $MAPbI_3$ broadband photodetectors with a buffer-layer-free simple metal/semiconductor/metal (MSM) lateral structure and a high on/off ratio ($I_{on}/I_{off} = 10^4$ under 0.05 sun conditions) [221]. The J–V characteristics of photodetectors exhibit rate-dependent hysteresis, especially in the dark, due to the migration of defect ions in $MAPbI_3$ films. Tang's group reported highly sensitive and fast-response UV photodetectors Pb-free double perovskite single crystals ($Cs_2AgInCl_6$), showcasing their advantages for optoelectronic applications [222]. In addition, photodetectors play an important role in signal processing, communication, and biological imaging. This application is promising because the heavy lead-containing perovskite materials have high absorption coefficients for X-ray radiation. Recently, using lead in X-ray imaging has been accepted in diagnostics equipment. The range adjustment of the photodetection response depends on the optical absorption characteristics of the material. Currently, the MHPs photodetectors can cover ultraviolet, visible, infrared, X-ray, and γ-rays. Narrow-band photoelectric detection continuously tuned from blue to red was achieved by the growth of $MAPbCl_3$, $MAPbBr_3$, and $MAPbI_3$ single-halide perovskite and $MAPbBr_{3-x}Cl_x$ and $MAPbI_{3-x}Br_x$ mixed-halide perovskite crystals [222].

A laser is a device that converts input light or electrical energy into light. Because of the high luminescence uniformity, a laser is used in industry, medical treatment, information, scientific research, and other fields [223–225]. MHPs have application prospects in lasers due to their low-cost, adjustable luminescence wavelength, stable emission spectrum, and solution process ability. The lasing from an MHPs polycrystalline film was reported in 2014 [226]. To overcome the instability of perovskite materials, Qu and coworkers constructed plasmonic nanolasers, where perovskite QDs were embedded in dual-mesoporous silica [227]. Organic–inorganic hybrid silica shells of the nanocomposites protected the perovskite QDs from water and endowed the nanocomposites with superhydrophobicity. Using a distributed-feedback cavity with a high-quality factor and applying triplet management strategies, Adachi and coworkers achieved stable green quasi-2D perovskite lasers under continuous-wave (CW) optical pumping in the air at room temperature [228]. It is expected that the findings will pave the way to future current-injection perovskite lasers.

References

1 Chakhmouradian, A.R. and Woodward, P.M. (2014). Celebrating 175 years of perovskite research: a tribute to Roger H. Mitchell. *Phys. Chem. Miner.* 41 (6): 387–391.

2 Jena, A.K., Kulkarni, A., and Miyasaka, T. (2019). Halide perovskite photovoltaics: background, status, and future prospects. *Chem. Rev.* 119 (5): 3036–3103.

3 Stoumpos, C.C., Frazer, L., Clark, D.J. et al. (2015). Hybrid germanium iodide perovskite semiconductors: active lone pairs, structural distortions, direct and indirect energy gaps, and strong nonlinear optical properties. *J. Am. Chem. Soc.* 137 (21): 6804–6819.

4 Kieslich, G., Sun, S., and Cheetham, A.K. (2015). An extended tolerance factor approach for organic–inorganic perovskites. *Chem. Sci.* 6 (6): 3430–3433.

5 Saliba, M., Matsui, T., Domanski, K. et al. (2016). Incorporation of rubidium cations into perovskite solar cells improves photovoltaic performance. *Science* 354 (6309): 206–209.

6 Kasel, T.W., Murray, A.T., and Hendon, C.H. (2018). Cyclopropenium $(C_3H_3)^+$ as an aromatic alternative A-site cation for hybrid halide perovskite architectures. *J. Phys. Chem. C* 122 (4): 2041–2045.

7 Zong, Y., Zhou, Y., Ju, M. et al. (2016). Thin-film transformation of NH_4PbI_3 to $CH_3NH_3PbI_3$ perovskite: a methylamine-induced conversion–healing process. *Angew. Chem. Int. Ed.* 55 (47): 14723–14727.

8 Goldschmidt, V.M. (1926). Die Gesetze der Krystallochemie. *Naturwissenschaften* 14 (21): 477–485.

9 Kim, M., Kim, G.-H., Lee, T.K. et al. (2019). Methylammonium chloride induces intermediate phase stabilization for efficient perovskite solar cells. *Joule* 3 (9): 2179–2192.

10 Meggiolaro, D., Mosconi, E., and De Angelis, F. (2018). Modeling the interaction of molecular iodine with $MAPbI_3$: a probe of lead-halide perovskites defect chemistry. *ACS Energy Lett.* 3 (2): 447–451.

11 Li, J., Munir, R., Fan, Y. et al. (2018). Phase transition control for high-performance blade-coated perovskite solar cells. *Joule* 2 (7): 1313–1330.

12 Woodward, P. (1997). Octahedral tilting in perovskites. I. Geometrical considerations. *Acta Crystallogr., Sect. B* 53 (1): 32–43.

13 Sutton, R.J., Filip, M.R., Haghighirad, A.A. et al. (2018). Cubic or orthorhombic? Revealing the crystal structure of metastable black-phase $CsPbI_3$ by theory and experiment. *ACS Energy Lett.* 3 (8): 1787–1794.

14 Stoumpos, C.C. and Kanatzidis, M.G. (2019). The renaissance of halide perovskites and their evolution as emerging semiconductors. *Acc. Chem. Res.* 48 (10): 2791–2802.

15 Hao, F., Stoumpos, C.C., Liu, Z. et al. (2014). Controllable perovskite crystallization at a gas–solid interface for hole conductor-free solar cells with steady power conversion efficiency over 10%. *J. Am. Chem. Soc.* 136 (46): 16411–16419.

16 She, X.-J., Chen, C., Divitini, G. et al. (2020). A solvent-based surface cleaning and passivation technique for suppressing ionic defects in high-mobility perovskite field-effect transistors. *Nat. Electron.* 3 (11): 694–703.

17 Xiong, Y., Xu, L., Wu, P. et al. (2019). Bismuth doping–induced stable seebeck effect based on MAPbI$_3$ polycrystalline thin films. *Adv. Funct. Mater.* 29 (16): 1900615.

18 Baikie, T., Fang, Y., Kadro, J.M. et al. (2013). Synthesis and crystal chemistry of the hybrid perovskite (CH$_3$NH$_3$)PbI$_3$ for solid-state sensitised solar cell applications. *J. Mater. Chem. A* 1 (18): 5628–5641.

19 Miyata, A., Mitioglu, A., Plochocka, P. et al. (2015). Direct measurement of the exciton binding energy and effective masses for charge carriers in organic–inorganic tri-halide perovskites. *Nat. Phys.* 11 (7): 582–587.

20 Herz, L.M. (2017). Charge-carrier mobilities in metal halide perovskites: fundamental mechanisms and limits. *ACS Energy Lett.* 2 (7): 1539–1548.

21 Xue, J., Lee, J.-W., Dai, Z. et al. (2018). Surface ligand management for stable FAPbI$_3$ perovskite quantum dot solar cells. *Joule* 2 (9): 1866–1878.

22 Lee, J.-W., Seol, D.-J., Cho, A.-N. et al. (2016). High-efficiency perovskite solar cells based on the black polymorph of HC(NH$_2$)$_2$PbI$_3$. *Adv. Mater.* 26 (29): 4991–4998.

23 Ma, F., Li, J., Li, W. et al. (2017). Stable α/δ phase junction of formamidinium lead iodide perovskites for enhanced near-infrared emission. *Chem. Sci.* 8 (1): 800–805.

24 Lin, J., Lai, M., Dou, L. et al. (2018). Thermochromic halide perovskite solar cells. *Nat. Mater.* 17 (3): 261–267.

25 Binek, A., Hanusch, F.C., Docampo, P. et al. (2015). Stabilization of the trigonal high-temperature phase of formamidinium lead iodide. *J. Phys. Chem. Lett.* 6 (7): 1249–1253.

26 Zheng, X., Wu, C., Jha, S.K. et al. (2016). Improved phase stability of formamidinium lead triiodide perovskite by strain relaxation. *ACS Energy Lett.* 1 (5): 1014–1020.

27 Li, Z., Yang, M., Park, J.-S. et al. (2015). Stabilizing perovskite structures by tuning tolerance factor: formation of formamidinium and cesium lead iodide solid-state alloys. *Chem. Mater.* 28 (1): 284–292.

28 Saliba, M., Matsui, T., Seo, J.-Y. et al. (2016). Cesium-containing triple cation perovskite solar cells: improved stability, reproducibility and high efficiency. *Energy Environ. Sci.* 9 (6): 1989–1997.

29 Juarez-Perez, E.J., Ono, L.K., and Qi, Y. (2019). Thermal degradation of formamidinium based lead halide perovskites into sym-triazine and hydrogen cyanide observed by coupled thermogravimetry-mass spectrometry analysis. *J. Mater. Chem. A* 7 (28): 16912–16919.

30 Lee, J.-W., Kim, D.-H., Kim, H.-S. et al. (2015). Formamidinium and cesium hybridization for photo- and moisture-stable perovskite solar cell. *Adv. Energy Mater.* 5 (20): 1501310.

31 Qiao, L., Sun, X., and Long, R. (2019). Mixed Cs and FA cations slow electron–hole recombination in FAPbI$_3$ perovskites by time-domain ab initio study: lattice contraction versus octahedral tilting. *J. Phys. Chem. Lett.* 10 (3): 672–678.

32 Zhao, Y., Tan, H., Yuan, H. et al. (2018). Perovskite seeding growth of formamidinium-lead-iodide-based perovskites for efficient and stable solar cells. *Nat. Commun.* 9 (1): 1607.

33 Yi, X., Zhang, Z., Chang, A. et al. (2019). Incorporating CsF into the PbI_2 film for stable mixed cation-halide perovskite solar cells. *Adv. Energy Mater.* 9 (40): 1901726.

34 Kubicki, D.J., Prochowicz, D., Hofstetter, A. et al. (2021). Phase segregation in Cs-, Rb- and K-doped mixed-cation $(MA)_x(FA)_{1-x}PbI_3$ hybrid perovskites from solid-state NMR. *J. Am. Chem. Soc.* 139 (40): 14173–14180.

35 Kausar, A., Sattar, A., Xu, C. et al. (2021). Advent of alkali metal doping: a roadmap for the evolution of perovskite solar cells. *Chem. Soc. Rev.* https://doi.org/10.1039/D0CS01316A.

36 Jodlowski, A.D., Roldán-Carmona, C., Grancini, G. et al. (2017). Large guanidinium cation mixed with methylammonium in lead iodide perovskites for 19% efficient solar cells. *Nat. Energy* 2 (12): 972–979.

37 Chen, H., Wei, Q., Saidaminov, M.I. et al. (2019). Efficient and stable inverted perovskite solar cells incorporating secondary amines. *Adv. Mater.* 31 (46): 1903559.

38 Kieslich, G., Sun, S., and Cheetham, A.K. (2014). Solid-state principles applied to organic–inorganic perovskites: new tricks for an old dog. *Chem. Sci.* 5 (12): 4712–4715.

39 Liu, C., Li, W., Li, H. et al. (2019). Structurally reconstructed $CsPbI_2Br$ perovskite for highly stable and square-centimeter all-inorganic perovskite solar cells. *Adv. Energy Mater.* 9 (7): 1803572.

40 Wang, G., Lei, M., Liu, J. et al. (2020). Improving the stability and optoelectronic properties of all inorganic less-Pb perovskites by B-site doping for high-performance inorganic perovskite solar cells. *Solar RRL* 4 (12): 2000528.

41 Swarnkar, A., Marshall, A.R., Sanehira, E.M. et al. (2016). Quantum dot–induced phase stabilization of α-$CsPbI_3$ perovskite for high-efficiency photovoltaics. *Science* 354 (6308): 92–95.

42 Gao, Y., Wu, Y., Lu, H. et al. (2019). $CsPbBr_3$ perovskite nanoparticles as additive for environmentally stable perovskite solar cells with 20.46% efficiency. *Nano Energy* 59: 517–526.

43 Kulbak, M., Cahen, D., and Hodes, G. (2015). How important is the organic part of lead halide perovskite photovoltaic cells? Efficient $CsPbBr_3$ cells. *J. Phys. Chem. Lett.* 6 (13): 2452–2456.

44 Kulbak, M., Gupta, S., Kedem, N. et al. (2016). Cesium enhances long-term stability of lead bromide perovskite-based solar cells. *J. Phys. Chem. Lett.* 7 (1): 167–172.

45 Sanchez, S., Christoph, N., Grobety, B. et al. (2018). Efficient and stable inorganic perovskite solar cells manufactured by pulsed flash infrared annealing. *Adv. Energy Mater.* 8 (30): 1802060.

46 Marronnier, A., Roma, G., Boyer-Richard, S. et al. (2018). Anharmonicity and disorder in the black phases of cesium lead iodide used for stable inorganic perovskite solar cells. *ACS Nano* 12 (4): 3477–3486.

47 Eperon, G.E., Paternò, G.M., Sutton, R.J. et al. (2015). Inorganic caesium lead iodide perovskite solar cells. *J. Mater. Chem. A* 3 (39): 19688–19695.

48 Hoke, E.T., Slotcavage, D.J., Dohner, E.R. et al. (2015). Reversible photo-induced trap formation in mixed-halide hybrid perovskites for photovoltaics. *Chem. Sci.* 6 (1): 613–617.

49 Zhou, W., Zhao, Y., Zhou, X. et al. (2017). Light-independent ionic transport in inorganic perovskite and ultrastable Cs-based perovskite solar cells. *J. Phys. Chem. Lett.* 8 (17): 4122–4128.

50 Kajal, S., Kim, J., Shin, Y.S. et al. (2020). Unfolding the influence of metal doping on properties of $CsPbI_3$ perovskite. *Small Methods* 4 (9): 2000296.

51 Bian, H., Wang, H., Li, Z. et al. (2020). Unveiling the effects of hydrolysis-derived $DMAI/DMAPbI_x$ intermediate compound on the performance of $CsPbI_3$ solar cells. *Adv. Sci.* 7 (9): 1902868.

52 Hailegnaw, B., Kirmayer, S., Edri, E. et al. (2015). Rain on methylammonium lead iodide based perovskites: possible environmental effects of perovskite solar cells. *J. Phys. Chem. Lett.* 6 (9): 1543–1547.

53 Wang, R., Wang, J., Tan, S. et al. (2015). Opportunities and challenges of lead-free perovskite optoelectronic devices. *Trends Chem.* 1 (4): 368–379.

54 Babayigit, A., Ethirajan, A., Muller, M. et al. (2016). Toxicity of organometal halide perovskite solar cells. *Nat. Mater.* 15 (3): 247–251.

55 Turkevych, I., Kazaoui, S., Ito, E. et al. (2017). Photovoltaic rudorffites: lead-free silver bismuth halides alternative to hybrid lead halide perovskites. *ChemSusChem* 10 (19): 3754–3759.

56 Dong, J., Shao, S., Kahmann, S. et al. (2020). Crystal formation: mechanism of crystal formation in Ruddlesden–Popper Sn-based perovskites. *Adv. Funct. Mater.* 30 (24): 2070154.

57 Yang, Z., Zhong, M., Liang, Y. et al. (2019). SnO_2-C_{60} pyrrolidine tris-acid (CPTA) as the electron transport layer for highly efficient and stable planar Sn-based perovskite solar cells. *Adv. Funct. Mater.* 29 (42): 1903621.

58 Yang, Z., Rajagopal, A., Chueh, C.-C. et al. (2016). Stable low-bandgap Pb–Sn binary perovskites for tandem solar cells. *Adv. Mater.* 28 (40): 8990–8997.

59 Stoumpos, C.C., Malliakas, C.D., and Kanatzidis, M.G. (2013). Semiconducting tin and lead iodide perovskites with organic cations: phase transitions, high mobilities, and near-infrared photoluminescent properties. *Inorg. Chem.* 52 (15): 9019–9038.

60 Saidaminov, M.I., Spanopoulos, I., Abed, J. et al. (2020). Conventional solvent oxidizes Sn(II) in perovskite inks. *ACS Energy Lett.* 5 (4): 1153–1155.

61 Marshall, K.P., Walker, M., Walton, R.I. et al. (2016). Enhanced stability and efficiency in hole-transport-layer-free $CsSnI_3$ perovskite photovoltaics. *Nat. Energy* 1 (12): 16178.

References

62 Koh, T.M., Krishnamoorthy, T., Yantara, N. et al. (2015). Formamidinium tin-based perovskite with low Eg for photovoltaic applications. *J. Mater. Chem. A* 3 (29): 14996–15000.

63 Song, T.-B., Yokoyama, T., Aramaki, S. et al. (2017). Performance enhancement of lead-free tin-based perovskite solar cells with reducing atmosphere-assisted dispersible additive. *ACS Energy Lett.* 2 (4): 897–903.

64 Kumar, M.H., Dharani, S., Leong, W.L. et al. (2014). Lead-free halide perovskite solar cells with high photocurrents realized through vacancy modulation. *Adv. Mater.* 26 (41): 7122–7127.

65 Lee, S.J., Shin, S.S., Kim, Y.C. et al. (2016). Fabrication of efficient formamidinium tin iodide perovskite solar cells through SnF_2–pyrazine complex. *J. Am. Chem. Soc.* 138 (12): 3974–3977.

66 Li, W., Li, J., Li, J. et al. (2016). Addictive-assisted construction of all-inorganic $CsSnIBr_2$ mesoscopic perovskite solar cells with superior thermal stability up to 473 K. *J. Mater. Chem. A* 4 (43): 17104–17110.

67 Krishnamoorthy, T., Ding, H., Yan, C. et al. (2015). Lead-free germanium iodide perovskite materials for photovoltaic applications. *J. Mater. Chem. A* 3 (47): 23829–23832.

68 Jong, U.-G., Yu, C.-J., Kye, Y.-H. et al. (2019). First-principles study on structural, electronic, and optical properties of inorganic Ge-based halide perovskites. *Inorg. Chem.* 58 (7): 4134–4140.

69 Sun, P.-P., Li, Q.-S., Yang, L.-N. et al. (2016). Theoretical insights into a potential lead-free hybrid perovskite: substituting Pb^{2+} with Ge^{2+}. *Nanoscale* 8 (3): 1503–1512.

70 Zhu, Z., Li, N., Zhao, D. et al. (2019). Improved efficiency and stability of Pb/Sn binary perovskite solar cells fabricated by galvanic displacement reaction. *Adv. Energy Mater.* 9 (7): 1802774.

71 Chen, Q., Wu, J., Matondo, J.T. et al. (2020). Optimization of bulk defects in Sn/Pb mixed perovskite solar cells through synergistic effect of potassium thiocyanate. *Solar RRL* 4 (12): 2000584.

72 Hao, F., Stoumpos, C.C., Chang, R.P.H. et al. (2014). Anomalous band gap behavior in mixed Sn and Pb perovskites enables broadening of absorption spectrum in solar cells. *J. Am. Chem. Soc.* 136 (22): 8094–8099.

73 Zhao, B., Abdi-Jalebi, M., Tabachnyk, M. et al. (2016). High open-circuit voltages in tin-rich low-bandgap perovskite-based planar heterojunction photovoltaics. *Adv. Mater.* 29 (2): 1604744.

74 Adjogri, S.J. and Meyer, E.L. (2020). A review on lead-free hybrid halide perovskites as light absorbers for photovoltaic applications based on their structural, optical, and morphological properties. *Molecules* 25 (21): 5039–5079.

75 Li, L., Zhang, F., Hao, Y. et al. (2017). High efficiency planar Sn–Pb binary perovskite solar cells: controlled growth of large grains via a one-step solution fabrication process. *J. Mater. Chem. C* 5 (9): 2360–2367.

76 Zhu, H.L., Xiao, J., Mao, J. et al. (2017). Controllable crystallization of $CH_3NH_3Sn_{0.25}Pb_{0.75}I_3$ perovskites for hysteresis-free solar cells with efficiency reaching 15.2%. *Adv. Funct. Mater.* 27 (11): 1605469.

77 Kim, H., Lee, J.W., Han, G.R. et al. (2021). Synergistic effects of cation and anion in an ionic imidazolium tetrafluoroborate additive for improving the efficiency and stability of half-mixed Pb-Sn perovskite solar cells. *Adv. Funct. Mater.* 31: 2008801.

78 Service, R.F. (2019). Perovskite LEDs begin to shine. *Science* 364 (6444): 918–918.

79 Dong, R., Fang, Y., Chae, J. et al. (2015). High-gain and low-driving-voltage photodetectors based on organolead triiodide perovskites. *Adv. Mater.* 27 (11): 1912–1918.

80 Gu, H., Chen, S.-C., and Zheng, Q. (2021). Emerging perovskite materials with different nanostructures for photodetectors. *Adv. Opt. Mater.* 9: 2001637.

81 Zhao, Z., Li, Y., Du, Y. et al. (2020). Preparation and testing of anisotropic $MAPbI_3$ perovskite photoelectric sensors. *ACS Appl. Mater. Interfaces* 12 (39): 44248–44255.

82 Masada, S., Yamada, T., Tahara, H. et al. (2020). Effect of A-site cation on photoluminescence spectra of single lead bromide perovskite nanocrystals. *Nano Lett.* 20 (5): 4022–4028.

83 Si, H., Zhang, Z., Liao, Q. et al. (2020). A-Site management for highly crystalline perovskites. *Adv. Mater.* 32 (4): 1904702.

84 Walsh, A. (2015). Principles of chemical bonding and band gap engineering in hybrid organic-inorganic halide perovskites. *J. Phys. Chem. C Nanomater. Interfaces* 119 (11): 5755–5760.

85 Eperon, G.E., Stranks, S.D., Menelaou, C. et al. (2014). Formamidinium lead trihalide: a broadly tunable perovskite for efficient planar heterojunction solar cells. *Energy Environ. Sci.* 7 (3): 982–988.

86 Ogomi, Y., Morita, A., Tsukamoto, S. et al. (2014). $CH_3NH_3Sn_xPb_{1-x}I_3$ perovskite solar cells covering up to 1060 nm. *J. Phys. Chem. Lett.* 5 (6): 1004–1011.

87 Jeon, N.J., Noh, J.H., Yang, W.S. et al. (2015). Compositional engineering of perovskite materials for high-performance solar cells. *Nature* 517 (7535): 476–480.

88 Bhunia, H., Chatterjee, S., and Pal, A.J. (2018). Band edges of hybrid halide perovskites under the influence of mixed-cation approach: a scanning tunneling spectroscopic insight. *ACS Appl. Energy Mater.* 1 (8): 4351–4358.

89 De Wolf, S., Holovsky, J., Moon, S.-J. et al. (2014). Organometallic halide perovskites: sharp optical absorption edge and its relation to photovoltaic performance. *J. Phys. Chem. Lett.* 5 (6): 1035–1039.

90 Wu, Z., Bai, S., Xiang, J. et al. (2014). Efficient planar heterojunction perovskite solar cells employing graphene oxide as hole conductor. *Nanoscale* 6 (18): 10505–10510.

91 Lim, J., Hörantner, M.T., Sakai, N. et al. (2019). Elucidating the long-range charge carrier mobility in metal halide perovskite thin films. *Energy Environ. Sci.* 12 (1): 169–176.

92 Biewald, A., Giesbrecht, N., Bein, T. et al. (2019). Temperature-dependent ambipolar charge carrier mobility in large-crystal hybrid halide perovskite thin films. *ACS Appl. Mater. Interfaces* 11 (23): 20838–20844.

93 Ziffer, M.E., Mohammed, J.C., and Ginger, D.S. (2016). Electroabsorption spectroscopy measurements of the exciton binding energy, electron–hole reduced effective mass, and band gap in the perovskite $CH_3NH_3PbI_3$. *ACS Photonics* 3 (6): 1060–1068.

94 Jiao, Y., Ma, F., Wang, H. et al. (2017). Strain mediated bandgap reduction, light spectrum broadening, and carrier mobility enhancement of methylammonium lead/tin iodide perovskites. *Part. Part. Syst. Char.* 34 (4): 1600288.

95 Mehdizadeh-Rad, H. and Singh, J. (2019). Influence of urbach energy, temperature, and longitudinal position in the active layer on carrier diffusion length in perovskite solar cells. *ChemPhysChem* 20 (20): 2712–2717.

96 Quan, L.N., Garcia De Arquer, F.P., Sabatini, R.P. et al. (2018). Perovskites for light emission. *Adv. Mater.* 30 (45): e1801996.

97 Protesescu, L., Yakunin, S., Bodnarchuk, M.I. et al. (2015). Nanocrystals of cesium lead halide perovskites ($CsPbX_3$, X = Cl, Br, and I): novel optoelectronic materials showing bright emission with wide color gamut. *Nano Lett.* 15 (6): 3692–3696.

98 Sun, S., Yuan, D., Xu, Y. et al. (2016). Ligand-mediated synthesis of shape-controlled cesium lead halide perovskite nanocrystals via reprecipitation process at room temperature. *ACS Nano* 10 (3): 3648–3657.

99 Pang, P., Jin, G., Liang, C. et al. (2020). Rearranging low-dimensional phase distribution of quasi-2D perovskites for efficient sky-blue perovskite light-emitting diodes. *ACS Nano* 14 (9): 11420–11430.

100 Cheng, T., Tumen-Ulzii, G., Klotz, D. et al. (2020). Ion migration-induced degradation and efficiency roll-off in quasi-2D perovskite light-emitting diodes. *ACS Appl. Mater. Interfaces* 12 (29): 33004–33013.

101 Zhao, L., Roh, K., Kacmoli, S. et al. (2020). Thermal management enables bright and stable perovskite light-emitting diodes. *Adv. Mater.* 32 (25): 2000752.

102 Savenije, T.J., Ponseca, C.S., Kunneman, L. et al. (2014). Thermally activated exciton dissociation and recombination control the carrier dynamics in organometal halide perovskite. *J. Phys. Chem. Lett.* 5 (13): 2189–2194.

103 Wu, K., Bera, A., Ma, C. et al. (2014). Temperature-dependent excitonic photoluminescence of hybrid organometal halide perovskite films. *Phys. Chem. Chem. Phys.* 16 (41): 22476–22481.

104 Mikhnenko, O.V., Blom, P.W.M., and Nguyen, T.-Q. (2015). Exciton diffusion in organic semiconductors. *Energy Environ. Sci.* 8 (7): 1867–1888.

105 Pinto, R.M., Gouveia, W., Maçôas, E.M.S. et al. (2015). Impact of molecular organization on exciton diffusion in photosensitive single-crystal halogenated perylenediimides charge transfer interfaces. *ACS Appl. Mater. Interfaces* 7 (50): 27720–27729.

106 Zhang, S., Yi, C., Wang, N. et al. (2017). Efficient red perovskite light-emitting diodes based on solution-processed multiple quantum wells. *Adv. Mater.* 29 (22): 1606600.

107 Li, J., Liu, X., Xu, J. et al. (2019). Fabrication of sulfur-incorporated bismuth-based perovskite solar cells via a vapor-assisted solution process. *Solar RRL* 3 (9): 1900218.

108 Li, M.-H., Yeh, H.-H., Chiang, Y.-H. et al. (2018). Highly efficient 2D/3D hybrid perovskite solar cells via low-pressure vapor-assisted solution process. *Adv. Mater.* 30 (30): 1801401.

109 Lee, K.-M., Chan, S.-H., Hou, M.-Y. et al. (2021). Enhanced efficiency and stability of quasi-2D/3D perovskite solar cells by thermal assisted blade coating method. *Chem. Eng. J.* 405: 126992.

110 Zhang, Y., Siegler, T.D., Thomas, C.J. et al. (2020). A "tips and tricks" practical guide to the synthesis of metal halide perovskite nanocrystals. *Chem. Mater.* 32 (13): 5410–5423.

111 Pan, A., He, B., Fan, X. et al. (2016). Insight into the ligand-mediated synthesis of colloidal $CsPbBr_3$ perovskite nanocrystals: the role of organic acid, base, and cesium precursors. *ACS Nano* 10: 7943–7954.

112 Shamsi, J., Urban, A.S., Imran, M. et al. (2019). Metal halide perovskite nanocrystals: synthesis, post-synthesis modifications, and their optical properties. *Chem. Rev.* 119 (5): 3296–3348.

113 Grisorio, R., Fanizza, E., Striccoli, M. et al. (2020). Shape tailoring of iodine-based cesium lead halide perovskite nanocrystals in hot-injection methods. *ChemNanoMat* 6 (3): 356–361.

114 Huang, H., Susha, A.S., Kershaw, S.V. et al. (2015). Control of emission color of high quantum yield $CH_3NH_3PbBr_3$ perovskite quantum dots by precipitation temperature. *Adv. Sci.* 2 (9): 1500194.

115 Zhang, F., Huang, S., Wang, P. et al. (2017). Colloidal synthesis of air-stable $CH_3NH_3PbI_3$ quantum dots by gaining chemical insight into the solvent effects. *Chem. Mater.* 29 (8): 3793–3799.

116 Wang, K.-H., Wu, L., Li, L. et al. (2016). Large-scale synthesis of highly luminescent perovskite-related $CsPb_2Br_5$ nanoplatelets and their fast anion exchange. *Angew. Chem. Int. Ed.* 55 (29): 8328–8332.

117 Im, J.H., Jang, I.H., Pellet, N. et al. (2014). Growth of $CH_3NH_3PbI_3$ cuboids with controlled size for high-efficiency perovskite solar cells. *Nat. Nanotechnol.* 9 (11): 927–932.

118 Burschka, J., Pellet, N., Moon, S.-J. et al. (2013). Sequential deposition as a route to high-performance perovskite-sensitized solar cells. *Nature* 499 (7458): 316–319.

119 Chen, Q., Zhou, H., Hong, Z. et al. (2014). Planar heterojunction perovskite solar cells via vapor-assisted solution process. *J. Am. Chem. Soc.* 136 (2): 622–625.

120 Singh, T. and Miyasaka, T. (2017). Stabilizing the efficiency beyond 20% with a mixed cation perovskite solar cell fabricated in ambient air under controlled humidity. *Adv. Energy Mater.* 8 (3): 1700677.

121 Eperon, G.E., Burlakov, V.M., Docampo, P. et al. (2014). Morphological control for high performance, solution-processed planar heterojunction perovskite solar cells. *Adv. Funct. Mater.* 24 (1): 151–157.

122 Xiao, M., Huang, F., Huang, W. et al. (2014). A fast deposition-crystallization procedure for highly efficient lead iodide perovskite thin-film solar cells. *Angew. Chem. Int. Ed.* 53 (37): 9898–9903.

123 Zhao, Y. and Zhu, K. (2015). Three-step sequential solution deposition of PbI_2-free $CH_3NH_3PbI_3$ perovskite. *J. Mater. Chem. A* 3 (17): 9086–9091.

124 Chen, H., Wei, Z., He, H. et al. (2016). Solvent engineering boosts the efficiency of paintable carbon-based perovskite solar cells to beyond 14%. *Adv. Energy Mater.* 6 (8): 1502087.

125 Cao, X., Zhi, L., Li, Y. et al. (2017). Control of the morphology of PbI_2 films for efficient perovskite solar cells by strong Lewis base additives. *J. Mater. Chem. C* 5 (30): 7458–7464.

126 Zhang, H., Mao, J., He, H. et al. (2015). A smooth $CH_3NH_3PbI_3$ film via a new approach for forming the PbI_2 nanostructure together with strategically high CH_3NH_3I concentration for high efficient planar-heterojunction solar cells. *Adv. Energy Mater.* 5 (23): 1501354.

127 Chiang, C.H., Nazeeruddin, M.K., Gratzel, M. et al. (2017). The synergistic effect of H_2O and DMF towards stable and 20% efficiency inverted perovskite solar cells. *Energy Environ. Sci.* 10 (3): 808–817.

128 Deng, Y., Dong, Q., Bi, C. et al. (2016). Air-stable, efficient mixed-cation perovskite solar cells with Cu electrode by scalable fabrication of active layer. *Adv. Energy Mater.* 6 (11): 1600372.

129 Galagan, Y., Di Giacomo, F., Gorter, H. et al. (2015). Roll-to-roll slot die coated perovskite for efficient flexible solar cells. *Adv. Energy Mater.* 8 (32): 1801935.

130 Patidar, R., Burkitt, D., Hooper, K. et al. (2020). Slot-die coating of perovskite solar cells: an overview. *Mater. Today Commun.* 22: 100808.

131 Lin, Y., Ye, X., Wu, Z. et al. (2018). Manipulation of the crystallization of perovskite films induced by a rotating magnetic field during blade coating in air. *J. Mater. Chem. A* 6 (9): 3986–3995.

132 Hwang, K., Jung, Y.-S., Heo, Y.-J. et al. (2015). Toward large scale roll-to-roll production of fully printed perovskite solar cells. *Adv. Mater.* 27 (7): 1241–1247.

133 Ye, H.-Y., Tang, Y.-Y., Li, P.-F. et al. (2018). Metal-free three-dimensional perovskite ferroelectrics. *Science* 361 (6398): 151–155.

134 Liao, W.-Q., Tang, Y.-Y., Li, P.-F. et al. (2017). Large piezoelectric effect in a lead-free molecular ferroelectric thin film. *J. Am. Chem. Soc.* 139 (49): 18071–18077.

135 Du, P., Gao, L., and Tang, J. (2020). Focus on performance of perovskite light-emitting diodes. *Front. Optoelectron.* 13 (3): 235–245.

136 Kojima, A., Teshima, K., Shirai, Y. et al. (2009). Organometal halide perovskites as visible-light sensitizers for photovoltaic cells. *J. Am. Chem. Soc.* 131 (17): 6050–6051.

137 Lee, M.M., Teuscher, J., Miyasaka, T. et al. (2012). Efficient hybrid solar cells based on meso-superstructured organometal halide perovskites. *Science* 338 (6107): 643–647.

138 Zou, Y., Wang, H.-Y., Qin, Y. et al. (2019). Reduced defects of $MAPbI_3$ thin films treated by FAI for high-performance planar perovskite solar cells. *Adv. Funct. Mater.* 29 (7): 1805810.

139 Luo, D., Zhao, L., Wu, J. et al. (2017). Dual-source precursor approach for highly efficient inverted planar heterojunction perovskite solar cells. *Adv. Mater.* 29 (19): 1604758.

140 Noel, N.K., Habisreutinger, S.N., Wenger, B. et al. (2017). A low viscosity, low boiling point, clean solvent system for the rapid crystallisation of highly specular perovskite films. *Energy Environ. Sci.* 10 (1): 145–152.

141 Min, H., Kim, M., Lee, S.-U. et al. (2019). Efficient, stable solar cells by using inherent bandgap of alpha-phase formamidinium lead iodide. *Science* 366 (6466): 749–753.

142 Zhang, C., Lu, Y.-N., Wu, W.-Q. et al. (2021). Recent progress of minimal voltage losses for high-performance perovskite photovoltaics. *Nano Energy* 81: 105634.

143 Mahesh, S., Ball, J.M., Oliver, R.D.J. et al. (2020). Revealing the origin of voltage loss in mixed-halide perovskite solar cells. *Energy Environ. Sci.* 13 (1): 258–267.

144 Yoon, Y.J., Lee, K.T., Lee, T.K. et al. (2018). Reversible, full-color luminescence by post-treatment of perovskite nanocrystals. *Joule* 2 (10): 2105–2116.

145 Guo, Z., Jena, A.K., Takei, I. et al. (2020). VOC Over 1.4 V for amorphous tin-oxide-based dopant-free $CsPbI_2Br$ perovskite solar cells. *J. Am. Chem. Soc.* 142 (21): 9725–9734.

146 Wang, R., Xue, J., Wang, K.-L. et al. (2019). Constructive molecular configurations for surface-defect passivation of perovskite photovoltaics. *Science* 366 (6472): 1509–1513.

147 Wu, W.-Q., Rudd, P.N., Ni, Z. et al. (2020). Reducing surface halide deficiency for efficient and stable iodide-based perovskite solar cells. *J. Am. Chem. Soc.* 142 (8): 3989–3996.

148 Deng, W., Liang, X., Kubiak, P.S. et al. (2018). Molecular interlayers in hybrid perovskite solar cells. *Adv. Energy Mater.* 8 (1): 1701544.

149 Lu, J., Jiang, L., Li, W. et al. (2017). Diammonium and monoammonium mixed-organic-cation perovskites for high performance solar cells with improved stability. *Adv. Energy Mater.* 7 (18): 1700444.

150 Zhang, Y., Grancini, G., Fei, Z. et al. (2019). Auto-passivation of crystal defects in hybrid imidazolium/methylammonium lead iodide films by fumigation with methylamine affords high efficiency perovskite solar cells. *Nano Energy* 58: 105–111.

151 You, S., Wang, H., Bi, S. et al. (2018). A biopolymer heparin sodium interlayer anchoring TiO_2 and $MAPbI_3$ enhances trap passivation and device stability in perovskite solar cells. *Adv. Mater.* 30 (22): 1706924.

152 Wang, B., Wu, F., Bi, S. et al. (2019). A polyaspartic acid sodium interfacial layer enhances surface trap passivation in perovskite solar cells. *J. Mater. Chem. A* 7 (41): 23895–23903.

153 Qin, P.-L., Yang, G., Ren, Z.-W. et al. (2018). Stable and efficient organo-metal halide hybrid perovskite solar cells via π-conjugated lewis base polymer induced trap passivation and charge extraction. *Adv. Mater.* 30 (12): 1706126.

154 Gao, F., Zhao, Y., Zhang, X. et al. (2020). Recent progresses on defect passivation toward efficient perovskite solar cells. *Adv. Energy Mater.* 10 (13): 1902650.

155 Dae-Yong, S., Seul-Gi, K., Ja-Young, S. et al. (2018). Universal approach toward hysteresis-free perovskite solar cell via defect engineering. *J. Am. Chem. Soc.* 140 (4): 1358–1364.

156 Zhao, W., Yao, Z., Yu, F. et al. (2018). Alkali metal doping for improved $CH_3NH_3PbI_3$ perovskite solar cells. *Adv. Sci.* 5 (2): 1700131.

157 Travis, W., Glover, E.N.K., Bronstein, H. et al. (2016). On the application of the tolerance factor to inorganic and hybrid halide perovskites: a revised system. *Chem. Sci.* 7 (7): 4548–4556.

158 Jiang, Y.Z., Yuan, J., Ni, Y.X. et al. (2018). Reduced-dimensional alpha-$CsPbX_3$ perovskites for efficient and stable photovoltaics. *Joule* 2 (7): 1356–1368.

159 Wang, Y., Zhang, T.Y., Kan, M. et al. (2018). Efficient alpha-$CsPbI_3$ photovoltaics with surface terminated organic cations. *Joule* 2 (10): 2065–2075.

160 Wang, Y., Zhang, T.Y., Kan, M. et al. (2018). Bifunctional stabilization of all-inorganic alpha-$CsPbI_3$ perovskite for 17% efficiency photovoltaics. *J. Am. Chem. Soc.* 140 (39): 12345–12348.

161 Christians, J.A., Miranda Herrera, P.A., and Kamat, P.V. (2015). Transformation of the excited state and photovoltaic efficiency of $CH_3NH_3PbI_3$ perovskite upon controlled exposure to humidified air. *J. Am. Chem. Soc.* 137 (4): 1530–1538.

162 Draguta, S., Christians, J.A., Morozov, Y.V. et al. (2018). A quantitative and spatially resolved analysis of the performance-bottleneck in high efficiency, planar hybrid perovskite solar cells. *Energy Environ. Sci.* 11 (4): 960–969.

163 Chen, C.-Y., Lin, H.-Y., Chiang, K.-M. et al. (2017). All-vacuum-deposited stoichiometrically balanced inorganic cesium lead halide perovskite solar cells with stabilized efficiency exceeding 11%. *Adv. Mater.* 29 (12): 1605290.

164 Kumar, G.R., Savariraj, A.D., Karthick, S.N. et al. (2016). Phase transition kinetics and surface binding states of methylammonium lead iodide perovskite. *Phys. Chem. Chem. Phys.* 18 (10): 7284–7292.

165 Habisreutinger, S.N., Leijtens, T., Eperon, G.E. et al. (2014). Carbon nanotube/polymer composites as a highly stable hole collection layer in perovskite solar cells. *Nano Lett.* 14 (10): 5561–5568.

166 Zhang, Y., Elawad, M., Yu, Z. et al. (2016). Enhanced performance of perovskite solar cells with P3HT hole-transporting materials via molecular p-type doping. *RSC Adv.* 6 (110): 108888–108895.

167 Jiang, J., Wang, Q., Jin, Z. et al. (2018). Polymer doping for high-efficiency perovskite solar cells with improved moisture stability. *Adv. Energy Mater.* 8 (3): 1701757.

168 Xu, W., Hu, Q., Bai, S. et al. (2019). Rational molecular passivation for high-performance perovskite light-emitting diodes. *Nat. Photonics* 13 (6): 418–424.

169 Kim, Y.-H., Kim, S., Kakekhani, A. et al. (2021). Comprehensive defect suppression in perovskite nanocrystals for high-efficiency light-emitting diodes. *Nat. Photonics* 15 (2): 148–155.

170 Lim, K.-G., Kim, H.-B., Jeong, J. et al. (2014). Boosting the power conversion efficiency of perovskite solar cells using self-organized polymeric hole extraction layers with high work function. *Adv. Mater.* 26 (37): 6461–6466.

171 Tan, Z.K., Moghaddam, R.S., Lai, M.L. et al. (2014). Bright light-emitting diodes based on organometal halide perovskite. *Nat. Nanotechnol.* 9 (9): 687–692.

172 Cho, H., Kim, Y.-H., Wolf, C. et al. (2018). Improving the stability of metal halide perovskite materials and light-emitting diodes. *Adv. Mater.* 30 (42): 1704587.

173 Jang, J., Kim, Y.-H., Park, S. et al. (2020). Extremely stable luminescent crosslinked perovskite nanoparticles under harsh environments over 1.5 years. *Adv. Mater.* 33 (3): 2005255.

174 Pengyun, L., Ning, H., Wei, W. et al. (2021). High-quality Ruddlesden–Popper perovskite film formation for high-performance perovskite solar cells. *Adv. Mater.* 33 (10): 2002582.

175 Quan, L.N., Rand, B.P., Friend, R.H. et al. (2019). Perovskites for next-generation optical sources. *Chem. Rev.* 119 (12): 7444–7477.

176 Kim, H.-B., Yoon, Y.J., Jeong, J. et al. (2017). Peroptronic devices: perovskite-based light-emitting solar cells. *Energy Environ. Sci.* 10 (9): 1950–1957.

177 Shen, Y., Cheng, L.-P., Li, Y.-Q. et al. (2019). High-efficiency perovskite light-emitting diodes with synergetic outcoupling enhancement. *Adv. Mater.* 31 (24): 1901517.

178 Lu, M., Zhang, Y., Wang, S. et al. (2019). Metal halide perovskite light-emitting devices: promising technology for next-generation displays. *Adv. Funct. Mater.* 29 (30): 1902008.

179 Kim, Y.H., Lee, G.H., Kim, Y.T. et al. (2017). High efficiency perovskite light-emitting diodes of ligand-engineered colloidal formamidinium lead bromide nanoparticles. *Nano Energy* 38: 51–58.

180 Cho, H., Wolf, C., Kim, J.S. et al. (2017). High-efficiency solution-processed inorganic metal halide perovskite light-emitting diodes. *Adv. Mater.* 29 (31): 1700579.

181 Park, D.H., Han, J.S., Kim, W. et al. (2018). Facile synthesis of thermally stable $CsPbBr_3$ perovskite quantum dot inorganic SiO_2 composites and their application to white light-emitting diodes with wide color gamut. *Dyes Pigm.* 149: 246–252.

182 Hu, H.C., Wu, L.Z., Tan, Y.S. et al. (2018). Interfacial synthesis of highly stable $CsPbX_3$/oxide Janus nanoparticles. *J. Am. Chem. Soc.* 140 (1): 406–412.

183 Lee, H., Park, J., Kim, S. et al. (2020). Perovskite emitters as a platform material for down-conversion applications. *Adv. Mater. Technol.* 5 (10): 2000091.

184 Bhaumik, S., Veldhuis, S.A., Ng, Y.F. et al. (2016). Highly stable, luminescent core-shell type methylammonium-octylammonium lead bromide layered perovskite nanoparticles. *Chem. Commun.* 52 (44): 7118–7121.

185 Jia, C., Li, H., Meng, X.W. et al. (2018). $CsPbX_3/Cs_4PbX_6$ core/shell perovskite nanocrystals. *Chem. Commun.* 54 (49): 6300–6303.

186 Wang, S., Bi, C., Yuan, J. et al. (2018). Original core–shell structure of cubic $CsPbBr_3$@amorphous $CsPbBr_x$ perovskite quantum dots with a high blue photoluminescence quantum yield of over 80%. *ACS Energy Lett.* 3 (1): 245–251.

187 Fang, Z., Chen, W., Shi, Y. et al. (2020). Dual passivation of perovskite defects for light-emitting diodes with external quantum efficiency exceeding 20%. *Adv. Funct. Mater.* 30 (12): 1909754.

188 Guo, Y., Apergi, S., Li, N. et al. (2021). Phenylalkylammonium passivation enables perovskite light emitting diodes with record high-radiance operational lifetime: the chain length matters. *Nat. Commun.* 12 (1): 644.

189 Vashishtha, P., Bishnoi, S., Li, C.H.A. et al. (2020). Recent advancements in near-infrared perovskite light-emitting diodes. *ACS Appl. Electron. Mater.* 2 (11): 3470–3490.

190 Lingmei, K., Xiaoyu, Z., Yunguo, L. et al. (2021). Smoothing the energy transfer pathway in quasi-2D perovskite films using methanesulfonate leads to highly efficient light-emitting devices. *Nat. Commun.* 12 (1): 1246.

191 Jiang, Y., Wei, J., and Yuan, M. (2021). Energy-funneling process in quasi-2D perovskite light-emitting diodes. *J. Phys. Chem. Lett.* 12 (10): 2593–2606.

192 Jiang, Y., Qin, C., Cui, M. et al. (2019). Spectra stable blue perovskite light-emitting diodes. *Nat. Commun.* 10 (1): 1868.

193 Kim, H., Kim, J.S., Heo, J.-M. et al. (2020). Proton-transfer-induced 3D/2D hybrid perovskites suppress ion migration and reduce luminance overshoot. *Nat. Commun.* 11 (1): 3378.

194 Jeong, M., Choi, I.W., Go, E.M. et al. (2020). Stable perovskite solar cells with efficiency exceeding 24.8% and 0.3-V voltage loss. *Science* 369 (6511): 1615–1620.

195 Huang, W., Sadhu, S., and Ptasinska, S. (2017). Heat- and gas-induced transformation in $CH_3NH_3PbI_3$ perovskites and its effect on the efficiency of solar cells. *Chem. Mater.* 29 (19): 8478–8485.

196 Boyd, C.C., Cheacharoen, R., Leijtens, T. et al. (2019). Understanding degradation mechanisms and improving stability of perovskite photovoltaics. *Chem. Rev.* 119 (5): 3418–3451.

197 Shellaiah, M. and Sun, K.W. (2020). Review on sensing applications of perovskite nanomaterials. *Chemosensors* 8 (3): 55.

198 Ding, N., Zhou, D., Pan, G. et al. (2019). Europium-doped lead-free $Cs_3Bi_2Br_9$ perovskite quantum dots and ultrasensitive Cu^{2+} detection. *ACS Sustainable Chem. Eng.* 7 (9): 8397–8404.

199 Wu, H., Zhang, W., Wu, J. et al. (2019). A visual solar UV sensor based on paraffin-perovskite quantum dot composite film. *ACS Appl. Mater. Interfaces* 11 (18): 16713–16719.

200 George, K.J., Halali, V.V., Sanjayan, C.G. et al. (2020). Perovskite nanomaterials as optical and electrochemical sensors. *Inorg. Chem. Front.* 7 (14): 2702–2725.

201 Casanova-Chafer, J., Garcia-Aboal, R., Atienzar, P. et al. (2020). The role of anions and cations in the gas sensing mechanisms of graphene decorated with lead halide perovskite nanocrystals. *Chem. Commun.* 56 (63): 8956–8959.

202 Chen, H.J., Zhang, M., Bo, R.H. et al. (2018). Superior self-powered room-temperature chemical sensing with light-activated inorganic halides perovskites. *Small* 14 (7): 7.

203 Berger, G., Frangville, P., and Meyer, F. (2020). Halogen bonding for molecular recognition: new developments in materials and biological sciences. *Chem. Commun.* 56 (37): 4970–4981.

204 Lu, L.-Q., Tan, T., Tian, X.-K. et al. (2017). Visual and sensitive fluorescent sensing for ultratrace mercury ions by perovskite quantum dots. *Anal. Chim. Acta* 986: 109–114.

205 Farahani, H., Wagiran, R., and Hamidon, M.N. (2014). Humidity sensors principle, mechanism, and fabrication technologies: a comprehensive review. *Sensors* 14 (5): 7881–7939.

206 Chou, K.-S., Lee, C.-H., and Liu, B.-T. (2016). Effect of microstructure of ZnO nanorod film on humidity sensing. *J. Am. Ceram. Soc.* 99 (2): 531–535.

207 Duan, Z., Xu, M., Li, T. et al. (2018). Super-fast response humidity sensor based on $La_{0.7}Sr_{0.3}MnO_3$ nanocrystals prepared by PVP-assisted sol-gel method. *Sens. Actuators, B* 258: 527–534.

208 Fei, T., Dai, J., Jiang, K. et al. (2016). Stable cross-linked amphiphilic polymers from a one-pot reaction for application in humidity sensors. *Sens. Actuators, B* 227: 649–654.

209 Weng, Z., Qin, J., Umar, A.A. et al. (2019). Lead-free $Cs_2BiAgBr_6$ double perovskite-based humidity sensor with superfast recovery time. *Adv. Funct. Mater.* 29 (24): 1902234.

210 Ren, K., Huang, L., Yue, S. et al. (2017). Turning a disadvantage into an advantage: synthesizing high-quality organometallic halide perovskite nanosheet arrays for humidity sensors. *J. Mater. Chem. C* 5 (10): 2504–2508.

211 Xu, W., Li, F., Cai, Z. et al. (2016). An ultrasensitive and reversible fluorescence sensor of humidity using perovskite $CH_3NH_3PbBr_3$. *J. Mater. Chem. C* 4 (41): 9651–9655.

212 Niu, Y., Zhang, F., Bai, Z. et al. (2015). Aggregation-induced emission features of organometal halide perovskites and their fluorescence probe applications. *Adv. Opt. Mater.* 3 (1): 112–119.

213 Wu, W., Han, X., Li, J. et al. (2021). Ultrathin and conformable lead halide perovskite photodetector arrays for potential application in retina-like vision sensing. *Adv. Mater.* 33 (9): 2006006.

214 Lee, W., Lee, J., Yun, H. et al. (2017). High-resolution spin-on-patterning of perovskite thin films for a multiplexed image sensor array. *Adv. Mater.* 29 (40): 1702902.

215 Liu, K., Jiang, Y., Jiang, Y. et al. (2019). Chemical formation and multiple applications of organic–inorganic hybrid perovskite materials. *J. Am. Chem. Soc.* 141 (4): 1406–1414.

216 Hu, X., Zhang, X., Liang, L. et al. (2014). High-performance flexible broadband photodetector based on organolead halide perovskite. *Adv. Funct. Mater.* 24 (46): 7373–7380.

217 Liu, L., Huang, S., Lu, Y. et al. (2018). Grain-boundary "patches" by in situ conversion to enhance perovskite solar cells stability. *Adv. Mater.* 30 (29): 1800544.

218 Chen, B., Rudd, P.N., Yang, S. et al. (2019). Imperfections and their passivation in halide perovskite solar cells. *Chem. Soc. Rev.* 48 (14): 3842–3867.

219 Sutanto, A.A., Szostak, R., Drigo, N. et al. (2020). In situ analysis reveals the role of 2D perovskite in preventing thermal-induced degradation in 2D/3D perovskite interfaces. *Nano Lett.* 20 (5): 3992–3998.

220 Shan, X., Wang, S., Meng, G. et al. (2019). Interface engineering of electron transport layer/light absorption layer of perovskite solar cells. *Prog. Chem.* 31 (5): 714–722.

221 Kwon, K.C., Hong, K., Van Le, Q. et al. (2016). Inhibition of ion migration for reliable operation of organolead halide perovskite-based metal/semiconductor/metal broadband photodetectors. *Adv. Funct. Mater.* 26 (23): 4213–4222.

222 Luo, J., Li, S., Wu, H. et al. (2018). $Cs_2AgInCl_6$ double perovskite single crystals: parity forbidden transitions and their application for sensitive and fast UV photodetectors. *ACS Photon.* 5 (2): 398–405.

223 You, P., Li, G., Tang, G. et al. (2020). Ultrafast laser-annealing of perovskite films for efficient perovskite solar cells. *Energy Environ. Sci.* 13 (4): 1187–1196.

224 Kim, H., Lim, K.-G., and Lee, T.-W. (2016). Planar heterojunction organometal halide perovskite solar cells: roles of interfacial layers. *Energy Environ. Sci.* 9 (1): 12–30.

225 Dong, H., Zhang, C., Liu, X. et al. (2020). Materials chemistry and engineering in metal halide perovskite lasers. *Chem. Soc. Rev.* 49 (3): 951–982.

226 Xing, G., Mathews, N., Lim, S.S. et al. (2014). Low-temperature solution-processed wavelength-tunable perovskites for lasing. *Nat. Mater.* 13 (5): 476–480.

227 Chen, Y., Yu, M., Ye, S. et al. (2018). All-inorganic $CsPbBr_3$ perovskite quantum dots embedded in dual-mesoporous silica with moisture resistance for two-photon-pumped plasmonic nanoLasers. *Nanoscale* 10 (14): 6704–6711.

228 Qin, C., Sandanayaka, A.S.D., Zhao, C. et al. (2020). Stable room-temperature continuous-wave lasing in quasi-2D perovskite films. *Nature* 585 (7823): 53–57.

7

Progress in Piezo-Phototronic Effect on 2D Nanomaterial-Based Heterostructure Photodetectors

Yuqian Zhao[1], Ran Ding[1], Feng Guo[1], Zehan Wu[1,2], and Jianhua Hao[1,2]

[1] The Hong Kong Polytechnic University, Department of Applied Physics, 11 Yuk Choi Rd, Hung Hom, Hong Kong, China
[2] Shenzhen Research Institute, The Hong Kong Polytechnic University, 18 Yuexing 1st Rd, Shenzhen 518057, China

7.1 Introduction

Since the demonstration of graphene isolated from graphite in 2004 [1], numerous two-dimensional (2D) materials have been widely exploited [2], including hexagonal boron nitride [3, 4], transition metal dichalcogenides (TMDs) [5, 6], and black phosphorus (BP) [7, 8]. Distinct from their bulk counterparts, these atomically thin-layered materials exhibit suitable band gap, chemical stability, and diverse exceptional properties owing to the quantum confinement effect [9]. In particular, the discovery of piezoelectricity in 2D materials has attracted much attention for the development of next-generation multifunctional devices [10, 11]. Some certain layered materials are centrosymmetric in the form of bulks while becoming non-centrosymmetric as their thickness decreases down to atomic level. This loss of centrosymmetry gives a guarantee of piezoelectricity within the 2D materials [12]. For example, atomically thin MoS_2 was firstly discovered with piezoelectric properties from the experimental observation where it was used for the fabrication of nanogenerators. Inspired by this pioneering work, the study of piezoelectricity in 2D materials provides a unique platform for exploring the coupling effects between the piezoelectricity and other prominent material features [13, 14].

Piezotronic is raised by coupling the promising semiconducting properties with piezoelectricity. Under external applied strain, piezopotential is created by the induced polarization charges within piezoelectric semiconducting materials [15]. The piezotronic effect will handle with the piezopotential acting as a "gate" voltage to modulate charge-carrier transport across the device contact interface. After that, the term of piezo-phototronic was first proposed by Wang in 2008, and then this new effect was applied onto a photodetector based on monolayer MoS_2 in 2016 [16, 17]. Accompanied with the optoelectronic excitation processes, piezo-phototronics is employed for controlling the separation and transport of photogenerated carriers, targeting the improvement of optoelectronic devices [18].

Functional Nanomaterials: Synthesis, Properties, and Applications, First Edition.
Edited by Wai-Yeung Wong and Qingchen Dong.
© 2022 WILEY-VCH GmbH. Published 2022 by WILEY-VCH GmbH.

The concept of piezo-phototronics opens up a new horizon for implementing the active flexible optoelectronics, which enables a direct control of generated electronic signals under mechanical strains [19, 20].

In general, 2D materials compose of atomically thin, layered crystalline solids via weak van de Waals (vdW) interactions, which enable to be separated by scotch tape or liquid-phase exfoliation. The simple strategy provides opportunities to construct vdW heterostructures by stacking dangling-bond-free 2D materials with the materials in other dimensions in either vertical or lateral form without concerning lattice mismatch [21, 22]. Due to the unique interlayer coupling and semiconducting properties, 2D heterostructures can be utilized as an example to qualitatively illustrate the piezo-phototronic effects on electro-optical processes [23–25]. To date, the piezo-phototronic effect has been widely exploited in various photodetectors based on vdW heterostructures, which received significant attention for future optoelectronic devices with improved performance, as summarized in Figure 7.1 [26–29].

In this chapter, we have reviewed recently developed photodetectors based on 2D heterostructures by piezo-phototronic effect. Upon introducing external straining, the optoelectronic processes can be modulated by the piezopotential across the local contacts, including metal–semiconductor Schottky contact or P–N junction. Some novel fundamental phenomena and extraordinary device performance are outlined. It is promising that the piezo-phototronic effect may greatly contribute to the development of high-performance photodetectors based on heterojunction devices. The parameters for evaluating the performance of photodetectors are presented in Table 7.1 [30].

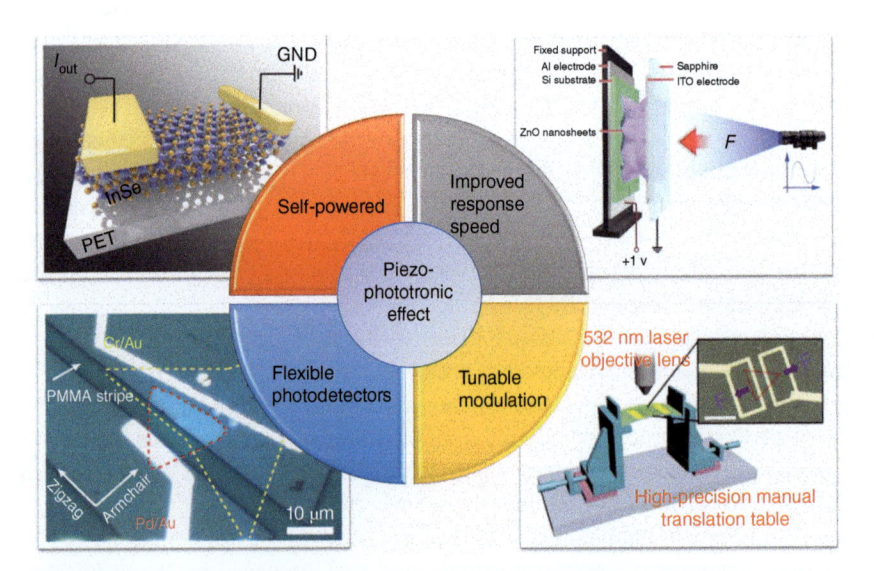

Figure 7.1 Summary of advantages and properties of 2D piezoelectric materials for photodetections. Source: Dai et al. [26, 27]. Reproduced with permission of American Chemical Society, 2019; Wiley-VCH, 2019; Wiley-VCH, 2018 and The Author(s), 2020; Lin et al. [28], Figure 02 [p. 003]/with permission from John Wiley & Sons, Inc.; Li et al. [29], Figure 16 [p. 106]/Springer Nature/CC BY 4.0.

Table 7.1 Parameters used for photodetector performance evaluation.

Parameters	Expression	Unit	Definition
Responsivity	$R_i = \dfrac{I_{Ph}}{P_{in}}$	A W^{-1}	I_{Ph} is the photocurrent, and P_{in} is the optical power
Response time	τ_r/τ_f	s	The τ is defined as consumption time of 10–90% of rising or falling
Noise equivalent power	$NEP = \dfrac{in}{R_i}$	W Hz$^{-1/2}$	in is the noise current of device
Specific detectivity	$D^* = \dfrac{\sqrt{A}}{in} R_i$	cm Hz$^{1/2}$ W^{-1} (Jones)	A is the effective area (unit of cm^{-2}) of device
External quantum efficiency (EQE)	$EQE = R_i \dfrac{h\nu}{e}$	—	$h\nu$ is the energy of one single photon, and e is unit charge

Source: Based on Han et al. [30].

7.2 Piezo-Phototronic Effect on the Junctions

7.2.1 Fundamental Physics of Piezo-Phototronics

The piezoelectricity is regarded as a phenomenon that the certain materials can generate electric charges in response to the mechanical strain. According to the electromechanical effects, it can be classified into two aspects including the direct piezoelectric effect and the converse piezoelectric effect. Direct piezoelectric effect is a process of generating electricity under applied strain, whereas the converse piezoelectric effect refers to the strain generation upon an external electric field [31]. When a piezoelectric material in non-centrosymmetric structure is applied with mechanical strain, the positive charge center will move away from the negative charge center, resulting in an electric dipole moment within the piezoelectric lattice. The existence of the dipole moments may induce polarization charges on the material surface, which contribute to the emerging of piezopotential [32].

Since the discovery of piezoelectricity in ZnO nanowires by Wang et al., a new term piezotronics is coined by coupling piezoelectric effect with semiconducting feature [33, 34]. In general, the piezotronic effect could be observed in a heterojunction system, which is usually constructed by one piezoelectric semiconductor in contact with the other material that could be a metal, a semiconductor, or even an electrolyte [35, 36]. When tensile or compressive strain is applied to the heterojunction, piezoelectric polarization charges generated at the contact junction will induce piezopotential, which is capable of modifying the interfacial band structure and carrier transporting processes of the heterojunction system [15]. For example, the induced positive piezopotential will reduce the local Schottky barrier height (SBH) at the contact, while the negative piezopotential may increase the SBH [37]. Thus, the carrier transporting property becomes a tunable dynamic based on externally applied strain [38].

The coupling of piezoelectricity, semiconductor behavior, and photonic excitation gives rise to a new effect named as piezo-phototronic effect, which also relies on the piezopotential created by the piezoelectric materials. Upon light illumination, the local piezoelectric polarization charges will significantly affect the optoelectronic processes of photogenerated excitons, including the charge-carrier separation, transport, and recombination [15]. Based on this piezo-phototronic effect, the performance of optoelectronic devices, such as photodetectors, light-emitting diodes, and solar cells, can be improved by using piezopotential as a gate voltage [19, 39]. The piezotronic and piezo-phototronic effects have been widely exploited for implementing novel flexible electronics, which enable a direct control of electronic signals by mechanical actions.

7.2.2 Piezo-Phototronic Effect on P–N Junction

The P–N junction is one of the common structures used in semiconductor optoelectronics, which is usually constructed by a p-type semiconductor connecting to the other n-type semiconductor at the interface [40]. There is a concentration gradient of donors and acceptors in the p-type and n-type semiconducting materials, and the diffusion of charges across the concentration gradient is known as the diffusion current. The ionized donors and acceptors near the junction boundary generate an electric field across the interface, where the n-side is at higher electric potential than the p-side region. The potential difference exerts a force and shifts the positive and negative charge carriers toward opposite directions, and the resultant current is called the drift current. The direction of drift current that arose from the drifting of charge carriers is identical, directing from p-side to n-side. However, the diffusion current due to the diffusion of charge carriers is in the opposite direction to drift current, pointing from n-side to p-side. The total current flowing through the junction is the sum of drift (I_{Drift}) and diffusion current ($I_{\text{Diff.}}$). At equilibrium, the net current equals to zero ($I(_{\text{Net}}) = I(_{\text{Diff.}}) - |I(_{\text{Drift}})| = 0$, when $V_{\text{equilibrium}} = 0$), indicating that the holes and electrons in the drift and diffusion components cancel out of each other, resulting in the zero net currents for each type of carrier. A depletion layer is then formed where holes and electrons are moved away, leaving only ionized donors and acceptors in this region. The potential difference across the depletion region at equilibrium is known as the built-in potential (V_{Bi}).

When an external bias is applied with the positive terminal connected to the p-side and the negative terminal to the n-side, the potential barrier is reduced and current can flows from the p-side to the n-side, known as the forward bias and forward current. When an external bias is applied in the opposite direction, the potential difference across the P–N junction is further enhanced with only a very limited amount of current that can pass through, known as the reverse bias and reverse current. The asymmetric current flow characteristic of the P–N junctions makes them very suitable for the construction of modern electronic devices. Under forward bias condition, the electric potential decreases at the p-side with respect to the n-side, hence lowering the potential difference across the P–N junction, and the probability of carrier diffusion across the junction increases, leading to an increase in the net current [41]. In contrast, the potential barrier becomes higher under reverse bias condition as the potential at the p-side is increased relative to that of the n-side. An enlarged potential barrier will occur at the junction resulting in negligible diffusion of both electrons and holes. The net current is constituted by the diffusion of minority carriers or the relatively small drift current from the collection of carriers generated in the transition region.

Regarding the strain-free condition, the P–N junction is at an equilibrium state with equal and opposite charges at each side, and there is no shifting in the center positions of the positive and negative charges in the piezoelectric material. Under external strain, the induced piezo-polarization charges modify the built-in potential at the junction and modulate the band structure of the P–N junction [15], in which the effective Fermi levels in the n-side and p-side shift toward opposite directions.

Photodetector is a type of optoelectronic device composed mainly with effective light-absorbing layer(s) and is able to covert the incident light photons into current. Theoretically, each photon received by the device from an optical source can be transferred into a free electron instantly. Therefore, the output current should be proportional to the power intensity of incident light. Nevertheless, in practical conditions, not every photon can effectively excite free electron in the photodetector. This is due to the non-efficient photon absorption and carrier collection of the practical material used during experiments [42]. Also, an external voltage source is usually required to assist the movement of the photoexcited electrons and holes to opposite directions [43].

In a typical P–N junction, the space charges caused by ionized donors and acceptors are available near the depletion zone, and the built-in electric potential enhances the separation of photogenerated carriers under illumination. When light illuminates on the P–N junction, electron–hole pairs are generated. Under the built-in electric field, photogenerated carriers travel through the depletion layer, thus inducing photocurrent [44, 45]. Under the application of external strain, the introduction of piezoelectric charges results in the deformation of depletion layer, which affects the built-in potential for carrier transport at the interface, thus modulating the output current of the P–N junction photodetector.

Taking a device with a P–N junction formed with n-type piezoelectric material as an example, when a compressive strain is applied to the device made of the piezoelectric material, negative piezoelectric charges are produced near the n-side of the depletion zone. The depletion region expands and shifts to the n-side. The charge injection leads to an increase in net current, and the separation of photogenerated carriers is consequently enhanced due to the suppression of the recombination of electrons and holes induced by the upward bending of the edges of the bands, as displayed in Figure 7.2a. Conversely, when a tensile strain is applied, positive charges are induced, and electrons are attracted to the interface. The depletion zone expands and shifts to the p-side that hinders the separation of photogenerated carriers, and therefore the performance of the photodetector device is deteriorated. A dip is generated by the positive piezoelectric charges at the band edge of the n-side, where

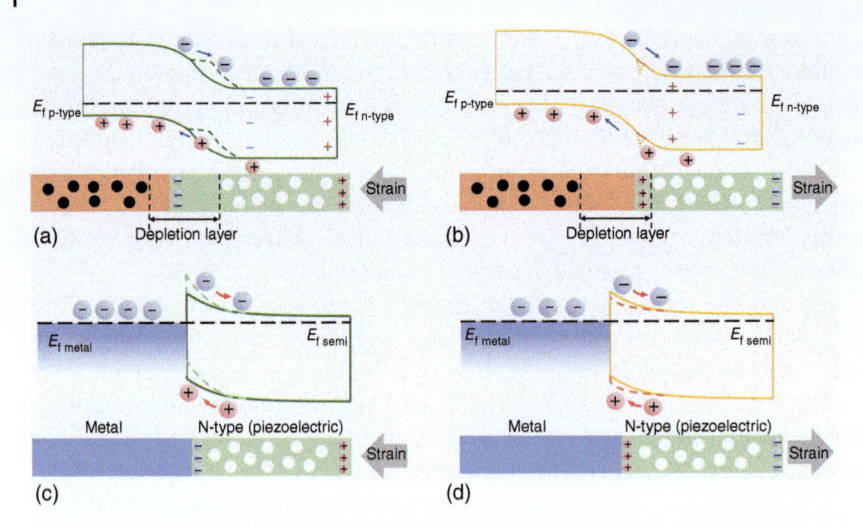

Figure 7.2 Schematic energy band diagrams. P–N junction constructed with p-type and n-type semiconductors (a) under compressive strain condition and (b) under tensile strain condition. Schottky contact between metal and semiconductor, (c) under compressive strain condition, and (d) under tensile strain condition. The solid and dashed lines symbolize without and with external strain applied, respectively.

photogenerated carriers are trapped, impeding the separation of electrons and holes and leading to the decrease in photocurrent (Figure 7.2b) [30].

Among all 2D materials, molybdenum disulfide (MoS_2), which is a typical TMDs material, is widely used as the semiconducting layer in electronic and optoelectronic applications as it possesses remarkable optoelectronic properties and stability [5, 46]. The MoS_2 is extensively explored as a typical n-type semiconductor. Besides, the odd number atomic layers of MoS_2 are reported to be piezoelectric [47, 48]. A flexible vdW heterojunction diode is constructed by stacking monolayer n-type MoS_2 and few-layer p-type WSe_2 and demonstrated with improved device performance with piezo-phototronic effect [28].

Figure 7.3a shows the optical image of the MoS_2/WSe_2 P–N heterojunction. When a negative voltage is applied, only small current is allowed to flow through the junction. On the contrary, when a positive voltage is applied, a much larger current can pass through the device, as displayed in Figure 7.3b. The built-in potential in the P–N junction can stimulate the photogenerated excitons to dissociate into free carriers spontaneously, driving electrons and holes to MoS_2 and WeS_2 sides, respectively. Diffusion and transportation of these free carriers lead to an increase in the photocurrent. Figure 7.3c demonstrates the band diagram of the MoS_2/WSe_2 P–N heterojunction under strain-free condition. Because of the crystal orientation of the triangular MoS_2, positive piezo-polarization charges are induced when compressive strain is applied. Comparatively, the band slope of the WSe_2 becomes steeper, while the band slope of MoS_2 becomes gentler when a compressive stain is applied as presented in Figure 7.3d. The positive polarization charges function like a forward bias on the P–N junction that enhance the performance of the device

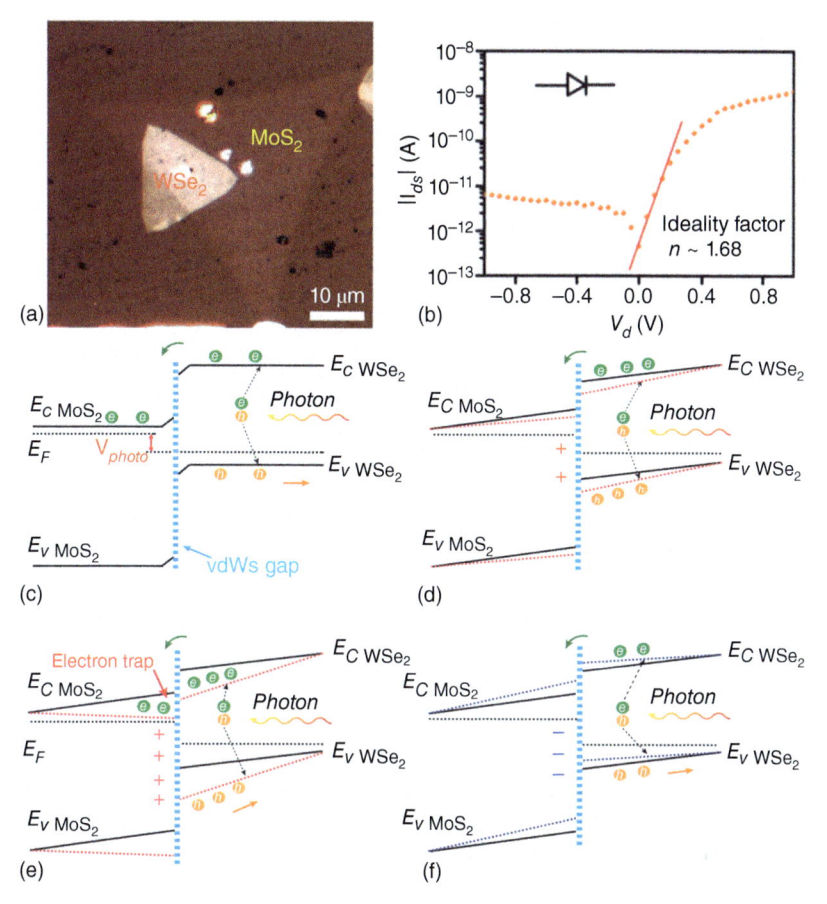

Figure 7.3 (a) Optical image of MoS$_2$/WSe$_2$ P–N heterojunction transferred onto PET substrate. Source: Lin et al. [28], Figure 02 [p. 003]/with permission from John Wiley & Sons, Inc. (b) I–V curve of MoS$_2$/WSe$_2$ P–N heterojunction diode, and schematic diagram of working mechanism (c) under strain-free and zero bias condition (d) with modulate positive piezocharges, (e) with large-density positive piezocharges, and (f) with negative piezoelectric charges. Source: Reproduced with permission from Lin et al. [28].

as they can provide an additional driving force for the separation of photoinduced excitons. When the strength of the applied compressive strain is further increased to the device as shown in Figure 7.3e, the local trapping effect causes the photogenerated electrons to be confined within the junction. The possibility of recombination between the interlayer electrons and holes hence increases, resulting in a depression in the rate of collection of the carriers and a drop in the photocurrent. When a tensile strain is applied, holes from the WSe$_2$ are attracted by negative polarization charges induced in the MoS$_2$ that reduces the built-in potential and driving force for exciton dissociation. Also, the photogenerated electrons in the WSe$_2$ are repelled by these negative polarization charges, and their injections into the MoS$_2$ are inhibited as presented in Figure 7.3f [25].

7.2.3 Piezo-Phototronic Effect on Metal–Semiconductor Junction

Another conventional design for the piezotronic heterostructure consists of two back-to-back Schottky contacts between metal electrodes and piezoelectric semiconductor [40]. When a Schottky contact is formed, the Fermi level of the semiconductor aligns to the energy level of the work function of the metal. After the redistribution of charges at the junction region, the metal–semiconductor (M–S) system reaches an equilibrium state [49, 50]. Theoretically, the mismatch in the work function and conduction band of the metal and semiconducting material is known as SBH [51, 52]. The SBH at the metal–semiconductor junction can be used to determine the movement of charge carriers at the interface. However, the effect of "Fermi level pinning" due to the presence of incomplete covalent bonds on the semiconductor surface influences the SBH in practical situations. The surface states cause additional energy states within the band gap of the semiconductor. When the semiconductor is in contact with metal, the energy states trap and absorb charges from metal and modify the Fermi energy level until these states are completely filled. Eventually, the actual SBH can be greatly depending on the density of surface states and be irrelevant of the work function of the metal. The existence of surface states can bring challenges in designing semiconductor-based device structures. For instance, the density of incomplete covalent bonds on silicon surface is around 10^{15} cm^{-2}, which is 3 magnitudes larger than the donor density in a 0.1 μm depletion layer of a material with a doping density of 10^{17} cm^{-3}.

In a Schottky contact constructed with an n-type semiconductor, the electrons travel from the semiconductor side to the metal side until an equilibrium is achieved. When a forward bias is applied to the Schottky barrier, the Fermi energy of the metal shifts to a lower position to that of the semiconductor, and the contact voltage is reduced. This leads to an increase in the forward current as more electrons from semiconductor side are able to travel through the depletion layer and reach the metal. The magnitude of current increases rapidly with increasing forward biasing voltage. In contrast, if a reverse bias is applied, the flow of electrons from semiconductor to metal is suppressed owing to the rise in the barrier height [41].

When a photosensitive Schottky contact device is exposed to a light source, the incoming photons generate electron–hole pairs in the semiconductor. The electrons generated in the conduction band tend to move away from the M–S interface, while the excited holes tend to move from semiconductor side toward the interface. The movements of the generated electrons and holes are mainly driven by the electric potential caused by the Schottky contact between semiconductor and metal. The magnitude of the SBH affects the efficiency of the separation and collection of electron–hole pairs and thus influences the output photocurrent. Therefore, the performance of the photodetectors can be coordinated as the barrier height can be modulated by piezo-phototronic effect.

For example, when compressive strain is applied on an n-type piezoelectric semiconductor–metal Schottky contact, the piezo-semiconductor is compressed, and negative polarization charges are induced. The negative piezo-polarization charges repel electrons and attract holes at the depletion zone, leading to a higher

SBH (as shown in Figure 7.2c). The magnitude of the photocurrent is raised since the separation and collection of electron–hole pairs are improved. Conversely, when a tensile strain is applied, positive polarization charges are induced. As presented in Figure 7.2d, electrons are attracted to the interface that cause a lower SBH. Through a specific selection of the piezoelectric constant of the piezo-semiconductor and careful control of the magnitude of external strain, modulation on the local SBH becomes feasible with the induced piezoelectric potential. Furthermore, with the different application directions of strain (e.g. from compressive to tensile), the property of the Schottky contact is also variable.

2D layered piezoelectric semiconductors show exclusive optoelectronic properties compared with conventional semiconducting materials. When an external strain is applied, piezopotential is generated. The photocurrent generated by the separation and collection of electrons and holes in the Schottky junction is tunable by applying an external piezoelectric field [53]. Therefore, the performance of the photodetectors can be enhanced by the piezo-phototronic effect [30].

Application of 2D materials related to piezoelectric is usually restricted by the conditions as most of the piezoresponse only exhibits with few odd number atomic layers and particularly monolayer. As a non-centrosymmetric material, 2D indium selenide (InSe) holds strong second harmonic generation (SHG) effect that facilitates broken inversion symmetry in both odd and even atomic layers [54]. In addition to intense SHG effect, InSe also possesses high mobility and prominent photoelectric properties [55, 56]. Along with a great improvement in the performance, an asymmetric Schottky junction-based γ-InSe nanosheet photodetector was fabricated and demonstrated the piezopotential with strain [26]. The γ-InSe nanosheets were first mechanically exfoliated from commercial bulk material subsequently transferred onto polyethylene terephthalate (PET) substrate, and gold electrodes were fabricated on it using standard photolithography procedure. A polydimethylsiloxane (PDMS) thin film with a thickness of 200 μm was introduced to prevent slippage of the device during stretching. Optical image and schematic diagram of the photodetector are presented in Figure 7.4a,b, respectively. The device showed a response under the irradiation of 0.368 mW cm^{-2} illuminate intensity at 650 nm wavelength light source and detected an open-circuit voltage of 40 mV along with a short-circuit current (I_{sc}) of 0.5 nA. Figure 7.4c depicts that the photocurrent ($I_{ph} = I_{light} - I_{dark}$) switched 3 orders of magnitude instantly with high reproducibility and reliability, demonstrating the extraordinary performance of this asymmetric Schottky junction self-powered photodetector (SPPD). The carrier conversion efficiency was further determined by the piezo-current response. The positive current output was induced when the strain was applied in an armchair direction, whereas the negative current output was detected when the strain was released (Figure 7.4d). The result indicates the mechanical deformation of the device manipulated the electric polarization, and the tensile strain induced a positive piezopotential that improved the built-in electric field created by the asymmetric Schottky junction. Great enhancement in responsivity and detectivity was observed under a 0.31% tensile strain under light irradiation ranging from 400 to 800 nm (Figure 7.4e). Under the illumination of 400 nm light, the resistivity of the

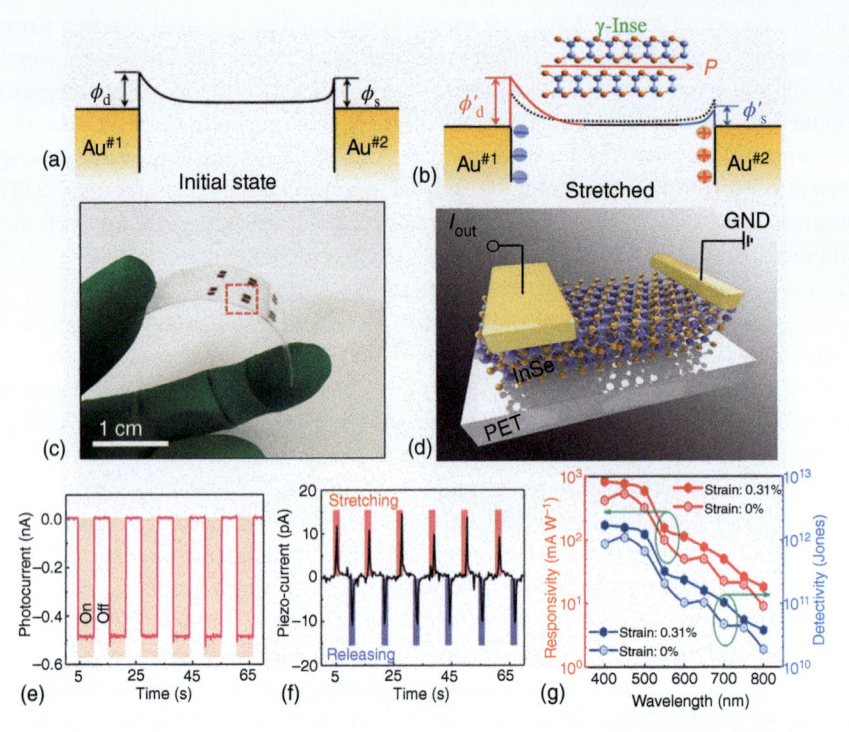

Figure 7.4 Multilayer γ-InSe self-powered photodetector. (a) Optical image of the device in PET substrate. (b) Schematic diagram of the SPPD with asymmetric contact structure. (c) Photoresponse under 650 nm illumination without voltage bias. (d) Piezoresponse under periodic tensile strain. (e) Responsivity and detectivity under different strain condition. (f) Band diagram of device under initial state. (g) Band diagram of device under tensile strain. Source: Dai et al. [26], Figure 03 [p. 7294]/with permission from American Chemical Society.

device raised from 416 to 824 mA W^{-1}, and detectivity increased from 8.7×10^{11} to 1.7×10^{12} J. A schematic energy band diagram demonstrates the effect of piezoelectric polarization on the Schottky barriers at the two metal–semiconductor junctions (Figure 7.4f,g). When tensile strain applies, SBH at source contact (ϕ_s) decreases, which indicates the improvement of photocurrent due to stretching [26]. Under the influence of piezoresistivity and piezoelectricity on γ-InSe SPPD, the piezo-charges produced by the external built-in electrical field enhanced the asymmetry of two back-to-back Schottky junctions and thus improved the photocurrent of SPPD.

7.3 Piezo-Phototronic Effect on the Performance of P–N Junction Photodetectors

The design and fabrication of low-dimensional heterostructure show great potential in enhancing the performance of photodetectors by piezo-phototronic effect. It has been reported that the performance of photodetector based on Schottky contact between 1D ZnO nanowire and silver paste was largely enhanced with applying external strain. Under 4.1 pW corresponding light power illumination, the responsivity was increased by 530% under −0.36% compressive strain. Besides, the responsivity of device can reach up to sub-µW cm^{-2} by utilizing the piezo-phototronic effect [57]. However, in 2D materials, there are few experimental works reported performance of Schottky contact photodetectors modulated by piezo-phototronic effect comparing with P–N junction heterostructure photodetectors constructed with 2D semiconductors. Therefore, P–N junction photodetectors combined with different materials will be introduced and discussed in more detail.

7.3.1 Photodetector Based on 2D Homojunction

Homojunction is a semiconductor interface formed between two blocks of the same material, which possess equal intrinsic band gap but differential Fermi levels. Benefiting from the equivalent lattice structure and approximate energy levels, continuous and smooth band bending can be generated in homojunctions, and this is conducive to the carrier separation and charge transfer across the interface. In conventional on-chip technology, homojunctions are fabricated through differentiated doping via ion implantation, surface modification, or regulatory homoepitaxy, so as to create portions with different impurity levels (donors or acceptors) within the same semiconductor. Regarding atomically thin layered materials, the shortened surface-to-bulk distance enables an acceleration in the photogenerated carrier separation and transfer with low probability of bulk recombination. Hence the surface (or interlayer) modification might become a highly effective approach for fabricating 2D material-based homojunctions. Recent attempt on such application is made by Wang et al., where a monolayer-MoS$_2$-based P–N homojunction photodiode was constructed via surface chemical-doping procedures. It is worth noting that MoS$_2$ is recognized to possess respectable stability and mechanical strength even in atomically thin level and outstanding current-switching and photoresponse properties [47, 48, 58]. Therefore, constructing MoS$_2$-based homojunctions would be an effective approach in the development of 2D material-based functional devices with high reliability, high speed, high efficiency, and high integration. Figure 7.5a exhibits the general view of a working piezo-photonic device based on monolayer-MoS$_2$ P–N homojunction. By modulating the built-in electric field and the depletion region of the junction under external strain, the photoresponse performance of MoS$_2$ can be remarkably enhanced (Figure 7.5b). When an external tensile strain is applied on the homojunction, the energy band of pristine MoS$_2$ side is lowered due to the positive piezopotential generated by polarization charges, while the band in the doped side is raised because of the negative piezopotential, which leads to higher barrier height and wider depletion zone within the homojunction. Thus, carrier separation and charge transfer across the junction interface will benefit from the adjustment of band structure, resulting in an advanced photoresponse performance. Such improvement was more significant at lower light illumination intensity, where an advanced photoresponsivity of 1200 A W^{-1} and a high detectivity up to 1.72×10^{12} Jones can be obtained under ∼0.01 mW cm^{-2}

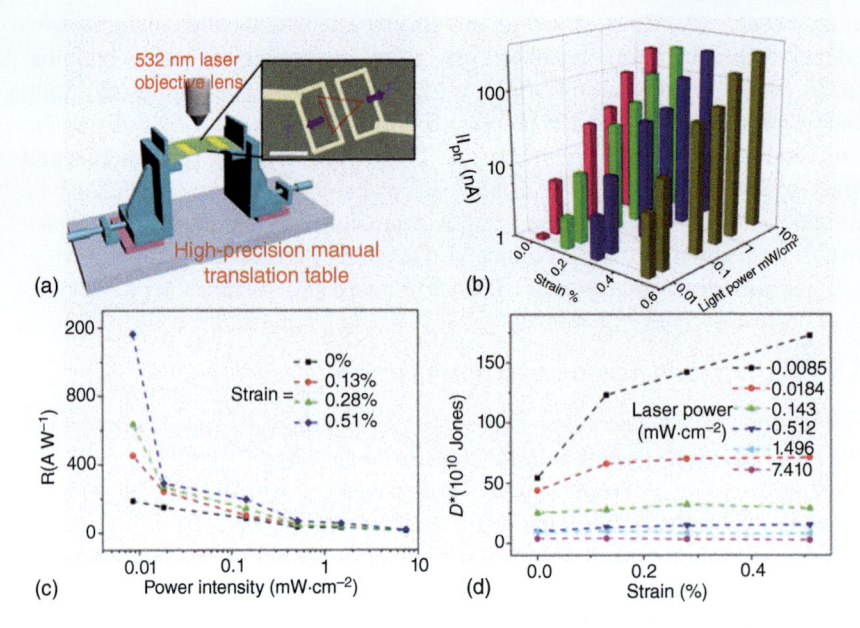

Figure 7.5 MoS$_2$ P–N junction photodiode. (a) Self-made apparatus for loading strain. Source: Li et al. [29], Figure 16 [p. 106]/Springer Nature/CC BY 4.0. (b) Photocurrent at −10 V bias response to different strains and light illumination intensities. (c) Power intensity dependence of responsivity under various strain conditions. (d) Strain dependence of detectivity under various light illumination intensities. Source: Zhang et al. [48]/IOP Publishing.

light illumination with ∼0.5% tensile strain (Figure 7.5c,d). This indicates that the constructed in-plane homojunction is substantially conducive to the photo-generated carrier separation and charge transfer across the interface. Although the screening effect of free charges balances out the piezo-polarization charges and makes such structure showing less significant improvement under high light illumination intensity, its respectable performance under low light illumination conditions makes it possible to become a photodetector with high sensitivity or a strain sensor with low power-consumption.

7.3.2 Photodetectors Based on 1D–2D Heterostructure

Composite semiconductor structures constructed from multiple layers of dissimilar materials, typically with unequal band gaps and different Fermi levels, are commonly described as heterostructures. Similar to homojunctions, heterojunctions provide advantages in engineering the electronic energy bands and carrier transportation behavior, thus facilitating the device performance such as photoresponse, strain sensing, and so on. In addition to conventional 3D heterostructures, the mix-dimensional (MD) heterojunctions, especially the vdW heterojunctions based on atomically thin 2D materials, have been increasingly accepted as the promising candidate for future piezo- and optoelectronic applications. Benefiting from the bond-free vdW integration, vdW heterojunctions provide new opportunities to break through the obstacles caused by lattice mismatch and interfacial impurity in conventional 3D heterostructures [25]. Therefore, vdW heterostructures allow unprecedented flexibility in combining materials with different chemical compositions and crystal structures together at the atomic level and further create new research directions on fundamental exploration and functional design in the electrical structures.

In the rapidly prospering 2D family, TMD is widely accepted to be a promising platform for constructing vdW heterostructures due to their considerable electrical properties and reliable stability. For example, efficient photodetectors constructed from 1D cesium lead bromide ($CsPbBr_3$) nanowires and 2D tungsten disulfide (WS_2) platform were demonstrated by Xu et al. in 2019 [59]. Through pyridinium chlorochromate (PCC)/PDMS-assisted dry-transfer approach under microscope, the chemical vapor deposition (CVD)-grown $CsPbBr_3$ nanowires and the mechanically exfoliated WS_2 flakes are stacked in controlled alignment. The obtained MD vdW heterostructure showed an advanced optoelectric performance with significantly reduced dark-current and enhanced photoresponse. Also, the on/off ratio was improved by 5 orders of magnitude (from $\sim 10^5$ to $\sim 10^{10}$), and the detectivity reached a remarkably high level of 1.36×10^{14} Jones (Figure 7.6a,b). Moreover, such heterostructure showed tunable properties under external strain, as indicated by the light-exited current and the photoresponsivity (Figure 7.6c,d). This is because

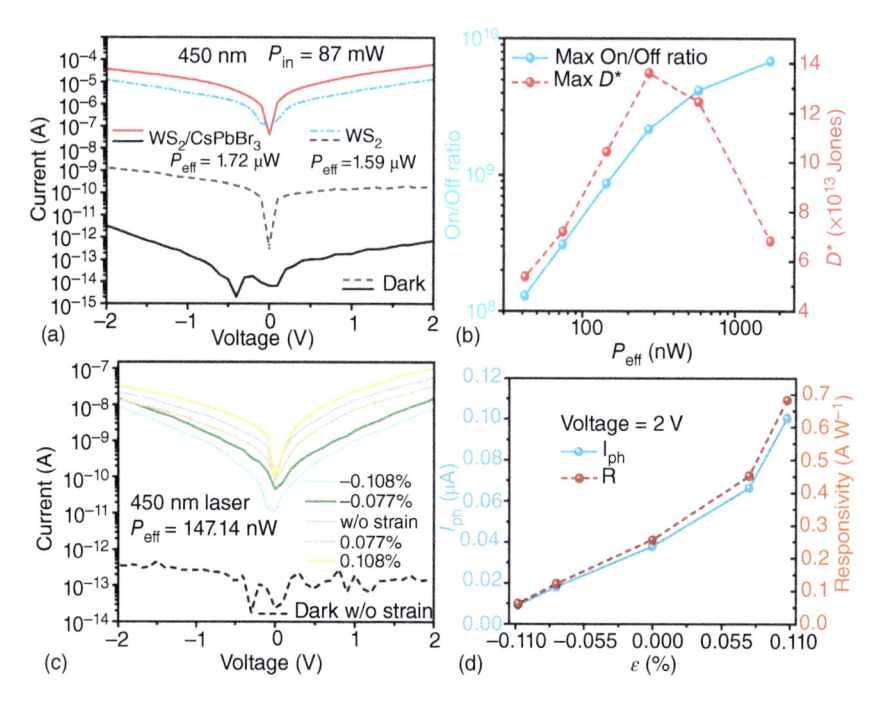

Figure 7.6 (a) $I-V$ curve of WS_2 and $WS_2/CsPbBr_3$ heterostructure photodetector. (b) On/off ratio and detectivity of the measured photodetectors. (c) $I-V$ curve of $WS_2/CsPbBr_3$ heterostructure detector under different strain conditions. (d) Stain dependence of I_{ph} and responsivity. Source: Xu et al. [59]. Reproduced with permission of Elsevier Ltd., 2019.

CsPbBr$_3$ possess a low-symmetry orthorhombic structure, which would form into distorted lattice under external strain. Thereout, piezo-polarization charges will be generated and further result in the formation of piezopotential [60]. In particular, when tensile strain is applied, it generates more negative polarization charge, so the energy bands bend reversely, and the barrier for carrier transferring across the junction is weakened. This would result in a remarkable enhancement on the carrier mobility across the junction, and higher photoresponsivity can be obtained. On the contrary, the applied compressive strain might induce more positive polarization charge, leading to a higher barrier for carrier transferring across the junction. Notably, the concentration of photoexcited carriers generated under weak excitation light here is far from sufficient to overcome the blocking effect of the barrier, so this structure can act as a bipolar strain sensor with high sensitivity and low power consumption.

Another example of piezo-photonic application constructing from TMD-based MD vdW heterostructures was demonstrated by Du et al. in 2019, where 1D ZnO nanowires and 2D tungsten diselenide (WSe$_2$) platform were combined to form P–N junctions via transfer strategy (Figure 7.7a) [61]. The study of ZnO nanowire has been an interesting research direction because it reveals remarkable piezoelectric property and great electron mobility in nano scale [62], and the coordination with 2D material platform will further broaden its application field and value. As shown in Figure 7.7b, photovoltaic effect exists obviously in the obtained WSe$_2$/ZnO P–N junction, and accordingly a self-powered mode is expectable to be realized in this structure. The photoexcited carriers can be further modified by the piezo polarization effect contributing from the strain-excited ZnO. As a result, a significantly enhanced photocurrent is obtained under external tensile strain (Figure 7.7c,d), showing the great potential of this structure in piezo-photonic applications.

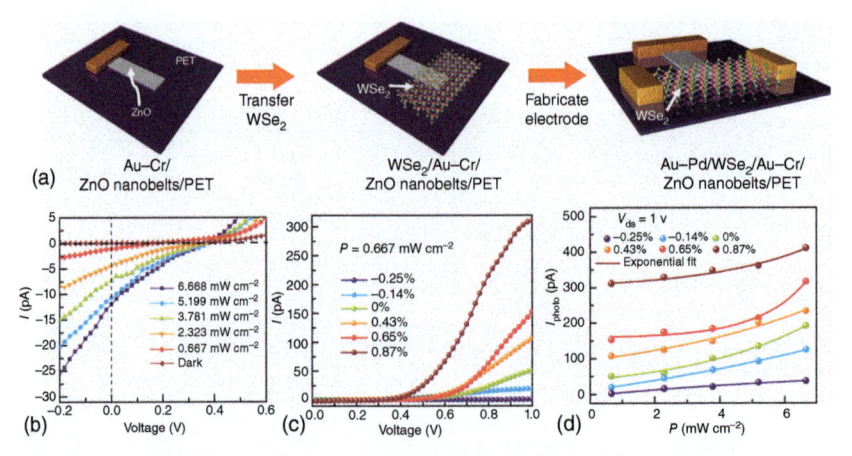

Figure 7.7 (a) Schematic diagram of fabrication process of flexible WSe$_2$/ZnO heterostructure detector. (b) I–V curve WSe$_2$/ZnO heterostructure detector under different light illumination condition. (c) I–V curve WSe$_2$/ZnO heterostructure detector under different strain condition. (d) Stain dependence of I$_{ph}$ and light illumination density. Source: Du et al. [61]. Reproduced with permission of Elsevier Ltd., 2019.

Due to their great mechanical and piezoelectric properties, n-type cadmium sulphide (CdS) nanowire [63] and p-type WSe$_2$ flakes [64] were chosen as the materials for constructing vdW P–N heterojunction photodetector. The CdS nanowires were prepared through physical vapor deposition (PVD) with a three-temperature-zone furnace [63]. Also, WSe$_2$ flakes were obtained by mechanical exfoliation. Then, the materials were transferred precisely on PET substrate to form heterostructure photodetector. Figure 7.8a shows the current responds to applied strain without light illumination. Compared with strain-free condition, applying 0.67% tensile strain leads to a decrease in current from 0.32 to 0.14 nA at 2 V bias. In contrast, when −0.73% compressive strain is applied, current increases from 0.32 to 1.75 nA under 2 V bias. As presented in Figure 7.8b, photocurrent raises with increasing compressive stain under different illumination intensity, which suggests that the performance of vdW heterostructure photodetector is enhanced. When −0.73% compressive strain is applied, photocurrent increases from 0.32 to 0.65 nA under 19.6 μW cm^{-2} illumination density. Under this condition, photoresponsivity can reach up to 33.4 A W^{-1} as depicted in Figure 7.8c. Electron transportation is affected remarkably by the band offset at valence and conduction band. In this case, the barrier height of electrons is much smaller than that of holes, resulting in that the diffusion of electrons contributes more to the current under forward bias condition. The electron concentration in WSe$_2$ flakes is much lower than that in CdS nanowires; thus, the built-in potential mostly falls in CdS nanowires at the heterojunction [65]. When mechanical strain is applied to the device, piezoelectric polarization charges are produced in CdS nanowires, which lead to carrier redistribution near interface of heterostructure [66]. Under compressive strain condition, positive polarization charges are generated to induce downward bending in energy bands of both WSe$_2$ and CdS, which result in a lower barrier height for electron transportation at heterostructure interface, and an enhanced photocurrent is obtained. Conversely, when tensile strain is applied, the reduced photocurrent can be attributed to upward bended energy bands of WSe$_2$ and CdS caused by the generated negative polarization charges of piezoelectric CdS [67].

7.3.3 Photodetectors Based on 2D–2D Heterostructure

Heterostructures based on 2D materials allow better development in the next-generation flexible electronic devices and provide possibilities for high-performance

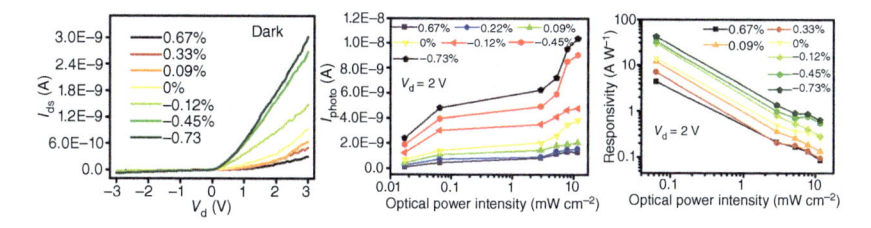

Figure 7.8 I–V curve of WSe$_2$/CdS heterostructure detector (a) with increasing tensile strain under dark condition. (b) Stain dependence of I_{ph} and light illumination density. (c) Strain dependence of responsivity under various power density. Source: Lin et al. [67]. Reproduced with permission of The Royal Society of Chemistry, 2018.

vdW structure photodetectors with selectable properties of material such as fast photoresponse and strong interaction with light. MoS_2 is a promising 2D material candidates for optoelectronic applications due to its high mobility of carriers and wide absorption spectrum [68, 69]. Moreover, the optical and electrical properties of MoS_2 are tunable by changing the number of layers and modulating defects, which is beneficial for improving the absorption efficiency of material [70, 71].

Graphene (Gr) has attracted numerous interests due to its distinguished Fermi velocity and strong optical absorption in the infrared radiation (IR) spectra since its first successful exfoliation from graphite [72, 73]. Recently, a large-area and flexible Gr/MoS_2 broadband photodetector was reported with effective modulation and separation of photogenerated carriers at junction, where the two-step hydrothermal method through processing solution was used for manufacture of Gr/MoS_2 structure on cellulose paper [74]. Hydrothermal synthesis is an effective method to fabricate materials directly on various substrates. With the advantages such as increased reactivity of reactants and decreased energy consumption during reaction, it has been widely used in fabrication of multiple morphologies by manipulating hydrothermal parameters (e.g. pH, temperature, etc.). Besides, crystallinity and size of the grown material can also be tuned by manipulating interface reaction [75]. When a tensile strain was applied to the device, a trilayer MoS_2 nanosheet is separated from the other MoS_2 nanosheets, which resulted in a decline in current due to the raised the potential barrier. In addition, the induced negative piezopotential and reduced conduction band energy in the MoS_2 led to an increase in the height of barrier and a decrease in the current [76]. On the other hand, separation of individual Gr flakes also took place. Similar to the particular MoS_2 nanosheets, isolation of individual flakes resulted in an increase in potential barrier and decrease in current. Although the current decreased with applying tensile strain, the responsivity increased. Also, responsivity under visible light was higher for visible light than IR illumination, since the absorption was dominated MoS_2 and Gr was mainly in charge of transportation. Electrons traveled from MoS_2 to Gr to align the Fermi energy levels. Thus, potential barrier and electric field built in near the interface. Under tensile strain, both potential barrier height and effective electric field increased. Therefore, illumination-generated electrons and holes were separated more effectively contributed to an increase in photocurrent [77].

Interestingly, a CuO/MoS_2-based heterostructure flexible photodetector was fabricated by the process displayed in Figure 7.9a [78, 79]. When the applied tensile strain increased from 0% to 0.65% under dark condition, current increased from 0.039 to 0.12 nA with 10 V bias as shown in Figure 7.9b. With an increase of tensile strain from 0% to 0.65%, the photocurrent rose to a maximum of 108 nA under 1656 mW cm^{-2} illumination power density. Attributing to the piezo-phototronic effect, the photocurrent was improved by 27 times in contrast to strain-free condition as presented in Figure 7.9c. Figure 7.9d indicates the relationship between detectivity (D^*) and strain. The detectivity reached up to a maximum of 3.7×10^8 J under 0.65% tensile strain with a relatively low illumination power of 2.66 mW cm^{-2}. When laser illuminates on the flexible photodetector, electron and hole pairs are generated and separated by a built-in field that contributes

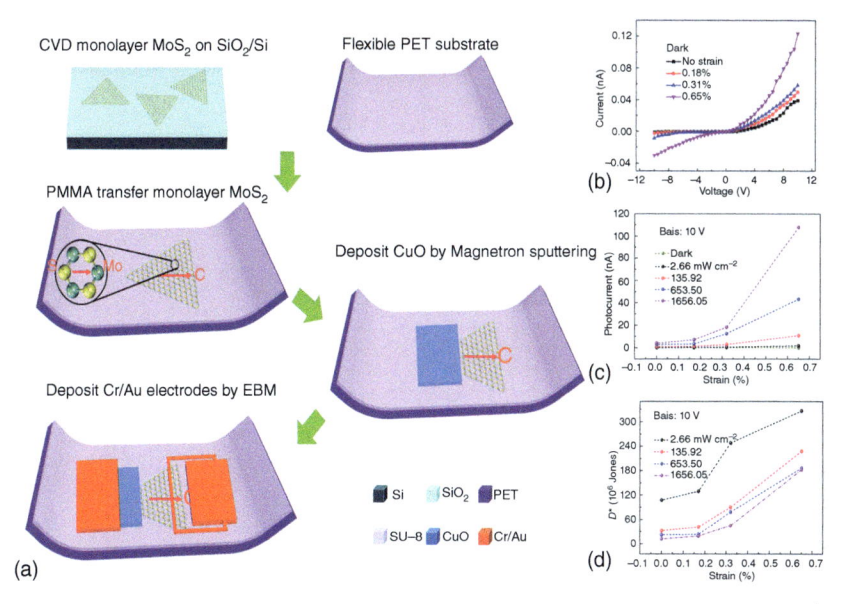

Figure 7.9 (a) Fabrication process of CuO/MoS$_2$ heterostructure photodetector. (b) I–V curve with increasing tensile strain under dark condition. (c) Strain dependence of photocurrent under various illumination power density. (d) Strain dependence of detectivity (D^*) under various power density. Source: Zhang et al. [78]. Reproduced with permission of The Royal Society of Chemistry, 2017.

to an external photocurrent. Single-layer MoS$_2$ deformed mechanically when tensile strain was applied to the flexible photodetector. Due to the piezoelectric property of single-layer MoS$_2$, positive piezo-polarization charges induced at interface lead to the formation of positive piezopotential under tensile strain. Bending of valence band and conduction band of MoS$_2$ near junction leads to depletion zone broadening; thus photogenerated carriers (electrons and holes) can be separated more efficiently. Consequently, recombination between electrons and holes was depressed, and carrier transportation was enhanced. It resulted in higher photocurrent and enhanced performance of the p-CuO/n-MoS$_2$ heterostructure photodetector [78].

Beneficial to the lack of structural centrosymmetry, many monolayer TMDs possess the intrinsic piezoelectricity and high flexibility to collaborate with external mechanical stimuli. To date, most of the reported fabrication of 2D vdW P–N junction was realized by electrostatic gating on hard substrates, and the study in flexible device is not enough. Recently, a flexible vdW MoS$_2$/WSe$_2$ heterojunction photodiode was constructed and demonstrated the dominating effect of the performance modulation in a heterostructure [25]. A high-quality monolayer n-MoS$_2$ and multilayer p-WSe$_2$ was first grown using CVD method and transferred onto the PET substrate for constructing a heterostructure. The metal electrode of Cr/Au (15/50 nm) was then fabricated to provide a robust connection to the device by electron beam lithography (EBL) and electron beam deposition method, as presented in Figure 7.10a. Significant photodiode response was depicted in the J–V

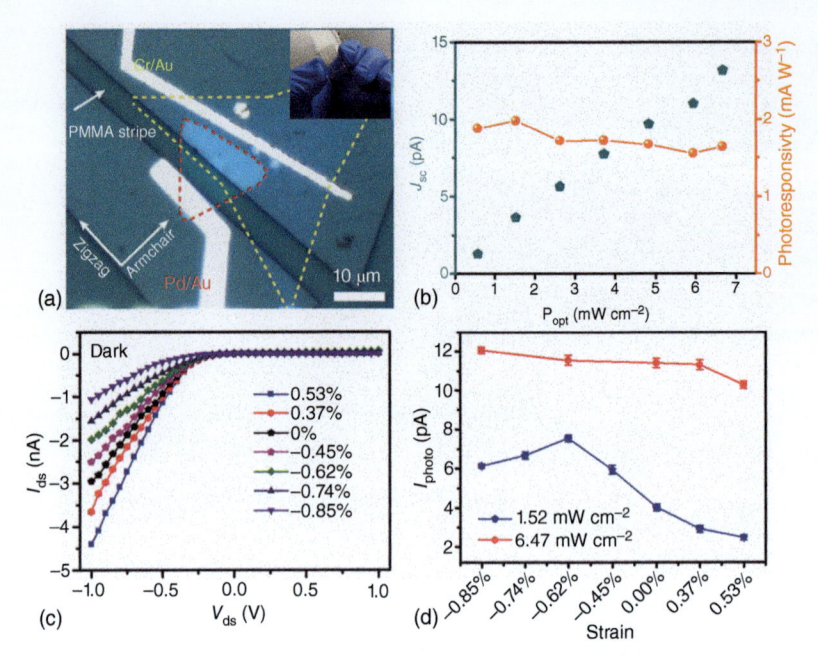

Figure 7.10 (a) Optical image of MoS$_2$/WSe$_2$ P–N heterojunction photodiode. Source: Lin et al. [28], Figure 02 [p. 003]/with permission from John Wiley & Sons, Inc. (b) Light illumination intensity dependence of J_{sc} and photoresponsivity of MoS$_2$/WSe$_2$ P–N heterojunction photodiode under strain-free condition. (c) I–V curve MoS$_2$/WSe$_2$ P–N heterostructure photodiode under different strain condition without light illumination. (d) Strain dependence of I_{photo} of MoS$_2$/WSe$_2$ P–N heterojunction photodiode under two different light illumination conditions. Source: Lin et al. [28]. Reproduced with permission of Wiley-VCH, 2018.

characteristics curve in Figure 7.10b. The curve illustrates the short-circuit current (J_{sc}) increased linearly with the incident light from 0.57 to 6.67 mW cm^{-2} and reached the highest value of intensity-independent photoresponsivity of 1.8 mA W^{-1} with a negligible dark current at 0 V bias. The strain-gated vdW junction property is demonstrated in Figure 7.10c. The current in forward bias increased with the compressive strain, whereas the current decreased with tensile strain. The transport change in the device is possibly attributed to two effects: the piezotronics effect and the piezoresistive effect. The former is due to the strain-induced piezo-polarization charges, which turn the interface barrier height and the carrier transport properties by acting as a "gate" voltage. The latter is due to a difference in electrical resistivity from the change in the band structure and the density of states of carriers, which result from mechanical strain. Figure 7.10d shows the photoresponse modulation contributed from the effects and the strain-induced light absorption change under the impact of mechanical strain on the device. Under low-intensity illumination of 1.52 mW cm^{-2} of using 532 nm light, the photocurrent decreased with the tensile strain. When the compressive strain was applied on the device, the photocurrent first rose and then fell. The device reached a responsivity of 3.4 mA W^{-1} and the maximum increase in photocurrent of 86% under a compressive strain of −0.62%.

Under higher illumination density of 6.47 mW cm^{-2}, in contrast to the low-intensity illumination, the photocurrent was slightly increased to 6.1%, and no turning point appears in the curve under entire range of strain. By combining the performance of the device in various light intensities, the results indicate that the influence of the external mechanical strain on the device is highly dependent on the optical power of the incident light. Moreover, the results imply that variation in electrical transportation and photoresponsivity arising from external strain is mainly due to the piezo-phototronic effect. The conclusion holds a good agreement in the theoretical calculation since the change in absorption spectra of MoS$_2$ and WeS$_2$ under strain condition is small. Thus, it is not capable of causing a large difference in photocurrent density if there is no contribution from the piezo-phototronic effect. Notably, the thickness of the TMDs and the illumination direction are also affected by the domination effect in the absorption. For example, WSe$_2$ dominated the absorption due to its greater thickness in the case when both materials resulted in a photoexcited electron–hole pair under the illumination of visible light.

Recently, n-type III-VI compound α-In$_2$Se$_3$ is discovered with high electrical mobility, great optical sensitivity, and piezoelectric properties, which promises a broad wavelength range photodetection prospect. Therefore, a flexible α-In$_2$Se$_3$/WSe$_2$ vdW P–N heterojunction photodetector has been purposed aiming for better optoelectronic behaviors by the piezo-phototronic effect. Due to the non-centrosymmetric structure of multilayer α-In$_2$Se$_3$, strain-induced piezopotential modulates the band slope near the junction interface, which improves the transfer characteristics and separation efficiency of photogenerated electron–hole pairs by piezo-phototronic effect [80].

7.3.4 Photodetectors Based on 3D–2D Heterostructure

Taking advantages to the distinct characteristics of bulk ZnO wurtzite crystal in optical, electronic, and piezoelectric aspects, 2D ZnO nanosheet is widely investigated in catalysis and electronics. A heterostructure photodetector constructed by p-Si and V-doped ferroelectric-ZnO nanosheets was proposed to investigate the performance of the 2D ZnO [81, 82]. For increasing the photosensing area, the Si wafer was first treated with wet chemical etching method to etch micropyramids with a bottom length of 3–10 μm on it. A ZnO seed layer was then deposited through spin coating, and a high-dense V-doped ZnO nanosheets were grown vertically and uniformly by using hydrothermal method. The schematic structure diagram of the device is shown in Figure 7.11a. Figure 7.11b indicates that dark current increased from 4.9 to 5.7 μA when strain increased from 0% to −0.02%. Moreover, the photocurrent raised from 28.5 to 44.7 μA in strain-free condition and 0.02% compressive strain, illustrating the strong influence of the piezo-phototronic effect. Figure 7.11c shows the photocurrent response to different strain and light illumination conditions. Under 442 nm light irradiation, the photocurrent was enhanced with increasing strain. For instance, photocurrent increased from 50.1 to 96.2 μA under strain-free and 0.02% compressive strain conditions at 10 mW cm^{-2}

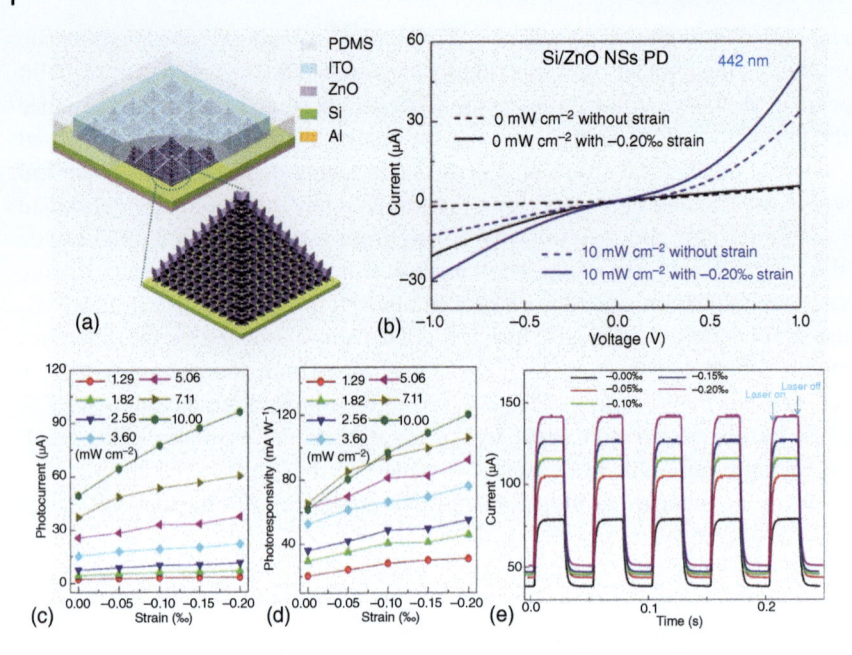

Figure 7.11 V-doped ferroelectric-ZnO nanosheets and p-Si heterostructure photodetector. (a) Schematic diagram of the device structure. (b) I–V curve with and without −0.20‰ compressive strain under dark and 442 nm light illumination with 1 V bias is applied. (c) Strain dependence of photocurrent under various illumination intensities. (d) Strain dependence of responsivity under different light illumination intensities. (e) Temporal photocurrent response to time under different compressive strain conditions with 10 mW cm^{-2} 442 nm light illumination and +1 V bias. Source: Dai et al. [27]/John Wiley & Sons/CC BY 4.0.

illumination intensity. As presented in Figure 7.11d, photoresponsivity also rose with applying compressive strain. The photoresistivity under strain-free condition was 61.6 mA W^{-1} and reached 120.3 mA W^{-1} with −0.02% strain applied. Figure 7.11e indicates that output current was improved when compressive strain (0.02%) applied to the device with +1 V bias, under an irradiation intensity of 10 mW cm^{-2} at 442 nm light. Generally, the photo-switching performance of the device is stable and highly repeatable, and the maximum output increases with compressive strain. Under the 0.02% compressive strain condition, the rise time and the fall time, which are known as the time interval for output current increase from 10% to 90% and release from 90% to 10%, were greatly improved. The rise time and fall time were, respectively, shortened from 3.34 to 3.07 ms and from 3.45 to 3.22 ms and enhanced the performance of the photodetector. The enhancement is attributed to the electron displacement polarization improved from the polarity of electric dipoles in ZnO under compressive strain condition, and the energy bands in both Si and ZnO sides are bending upward, leading to the enhancement of photodetector performance [27].

7.4 Conclusion and Future Perspectives

2D materials are featured in an ultrathin planar structure and the remarkable physical and chemical properties in contrast to bulk materials, and they are regarded as the research focus in the following decades. Advantageous to their excellent ductility and high carrier mobility, 2D materials are promising for fabricating the optoelectronic devices down to atomic thicknesses. Piezoelectric, piezotronic, and piezo-phototronic effects in 2D materials are the focus of the recent research and development as they can facilitate the enhancement in the electrical transfer behavior of carriers and optoelectronic performance. Additionally, modulation in the band structure with strain-induced polarization in piezoelectric materials is effective to enhance the effects. On the one hand, heterostructures constructed with various materials in different energy band structures can achieve outstanding electronic and optoelectronic characteristics, which differs from that of the single crystal. On the other hand, heterojunctions or homostructures formed with various n-type or p-type 2D semiconductors usually demonstrate a low dark current with a fast photoresponse, which is particularly important in high-performance photodetectors. Such features make 2D materials hold great potential to compete with the conventional bulk materials. The photoresponse of 2D material-based photodetectors and their heterostructures promoted by the piezoelectric effect is summarized in Table 7.2.

Efforts have been invested in the piezo-phototronic devices and demonstrated their merits. Still, further work and investigation are essential and crucial for this research field. Particularly, the low production efficiency in material preparation and fabrication is highly restricting the construction of 2D material-based device. Up to now, only one part of the 2D materials can be prepared by well-developed fabrication approaches and produced in a large area in wafer scale. Also, 2D materials are possessing diverse optical and electrical properties, and thus choosing the most suitable candidate for specific applications is a difficulty in device design. Moreover, the introduction of internal defects is highly uncontrollable. The accuracy of doping is a major challenge as it greatly affects the properties of 2D materials. Although some 2D materials can be doped through chemical methods, they cannot be treated compatible with conventional semiconducting fabrication technology. Compared with bulk materials, the surface conditions of 2D materials become significant during the fabrication process as they only have an atomically thin thickness. Moreover, though vdW heterostructure-based photodetectors and the advanced piezo-photonic effect have attracted great interest in recent years, the coordination of interfacial charge behaviors and the piezo-photonic performance still need further theoretical investigation and application development. For instance, the interface excitons in vdW heterojunctions, analogous to the spatially indirect excitons in III–V quantum wells [83], show reduction on the hole-electron wave function overlap along the normal axis of the interface. This would suppress the exciton-oscillator strength and the exchange interaction, resulting in ascending

Table 7.2 Piezo-phototronic effect enhanced performance of heterostructure photodetectors based on 2D materials.

Materials	Thickness	Method	Strain[a] (%)	Wavelength (nm)	Illumination intensity (W cm^{-2})	R (AW^{-1})	Enhancement[a] (%)	References
MoS_2	Monolayer	Mechanical exfoliation	−0.38	442	3.4×10^{-6}	2.3×10^4	178	[55]
α-In_2Se_3	12–15 nm	Mechanical exfoliation	−0.15	405	0.45×10^{-3}	~99	200	[87]
γ-InSe	~30 nm	Mechanical exfoliation	0.62	650	0.368×10^{-3}	198.2×10^{-3}	696	[26]
MoS_2 P–N homojunction	Monolayer	CVD followed by chemical doping	0.51	532	8.5×10^{-6}	1162	619	[48]
MoS_2/WSe_2	Monolayer/~10 nm	CVD/CVD	−0.62	532	1.52×10^{-3}	3.4×10^{-3}	86	[28]
WSe_2/CdS	~7 nm/—	Mechanical exfoliation/physical vapor deposition	−0.73	680	16.9×10^{-6}	33.4	110	[67]
MoS_2/CuO	~0.7 nm/—	CVD followed by magnetron sputtering	0.65	532	1656×10^{-3}	—	2700	[78]
WS_2/$CsPbBr_3$	—	Mechanical exfoliation/PDMS template-confined antisolvent crystallization	0.108	450	$P_{\text{effective}} = 147\,\text{nW}$, where $P_{\text{eff}} = P_{\text{incident}} \times$ (area of device/area of incident laser)	0.68	150	[59]
V-doped ZnO nanosheet	15–20 nm	Hydrothermal method	−0.02	442	10×10^{-3}	120.3×10^{-3}	195	[27]
ZnO/WSe_2	50 nm/5 nm	CVD/CVD	0.78	532	0.667×10^{-3}	394	236	[61]
MoS_2/Gr cellulose paper	Trilayer/—	Hydrothermal growth method	2	Visible light/IR	$8.1 \times 10^{-3}/83 \times 10^{-3}$	~3.15×10^{-3}/~3.1×10^{-3}	~750/~850	[77]

a) The "enhancement" column indicates the increased percentage of photoresponsivity when the strain is applied. The positive and negative signs represent for uniaxial tensile and compressive strain, respectively.

population and spin life of the exciton oscillators for advanced piezo-photonic modulation. In addition, the 2D TMD-based vdW heterojunctions possess more individual characteristics in the band structure and exciton behaviors, such as the unique internal degrees of freedom and dark momentum behavior under quantum constraints [84] and the inheritance of valley physics in twisted heterostructures [85], which provides abundant dimensions in excitonic optoelectric modulation. Besides, for atomic scale materials and interface, the band energies and structure are of significant dependence with the crystal size because of the quantum size effects [86]. Therefore, novel investigations on band alignment engineering and 2D piezo-photonic behaviors constrained at the vdW interface should be expected in the future. Furthermore, the piezoelectric-enhanced performance of the photodetectors is always restricted by the effective range of the applied strain up to a maximum value at around 1–2%. The possible improvement in the performance by applying a large strain for commercialization is restrained, as most of the reported 2D material-based photodetectors are very sensitive to light and applied external strain. Overall, there are still many challenges needed to be tackled before the realization of 2D material-based devices into practical applications, and there is a large research opportunity in this uncharted territory.

Acknowledgments

This work was supported by the Research Grant Council (RGC) of Hong Kong (RGC GRF No. PolyU 153023/18P), National Natural Science Foundation of China (No. 51972279), and RGC CRF No. C7036-17W.

References

1 Novoselov, K.S., Geim, A.K., Morozov, S.V. et al. (2004). Electric field effect in atomically thin carbon films. *Science* 306: 666.
2 Ding, R., Liu, C.K., Wu, Z. et al. (2020). A general wet transferring approach for diffusion-facilitated space-confined grown perovskite single-crystalline optoelectronic thin films. *Nano Lett.* 20: 2747.
3 Michel, K.H. and Verberck, B. (2009). Theory of elastic and piezoelectric effects in two-dimensional hexagonal boron nitride. *Phys. Rev. B* 80: 224301.
4 Sharma, V., Kagdada, H.L., Jha, P.K. et al. (2020). Thermal transport properties of boron nitride based materials: a review. *Renewable Sustainable Energy Rev.* 120: 109622.
5 Wang, Q.H., Kalantar-Zadeh, K., Kis, A. et al. (2012). Electronics and optoelectronics of two-dimensional transition metal dichalcogenides. *Nat. Nanotechnol.* 7: 699.
6 Zhang, X., Qiao, X.F., Shi, W. et al. (2015). Phonon and Raman scattering of two-dimensional transition metal dichalcogenides from monolayer, multilayer to bulk material. *Chem. Soc. Rev.* 44: 2757.

7 Luo, Z., Maassen, J., Deng, Y. et al. (2015). Anisotropic in-plane thermal conductivity observed in few-layer black phosphorus. *Nat. Commun.* 6: 1.

8 Yang, Z. and Hao, J. (2018). Recent Progress in black-phosphorus-based heterostructures for device applications. *Small Methods* 2: 1700296.

9 Tan, C., Cao, X., Wu, X.J. et al. (2017). Recent advances in ultrathin two-dimensional nanomaterials. *Chem. Rev.* 117: 6225.

10 Gao, L. (2017). Flexible device applications of 2D semiconductors. *Small* 13: 1603994.

11 Yang, Z. and Hao, J. (2019). Recent progress in 2D layered III–VI semiconductors and their heterostructures for optoelectronic device applications. *Adv. Mater. Technol.* 4: 1900108.

12 Peng, Y., Que, M., Tao, J. et al. (2018). Progress in piezotronic and piezo-phototronic effect of 2D materials. *2D Mater* 5: 042003.

13 Wu, W. and Wang, Z.L. (2016). Piezotronics and piezo-phototronics for adaptive electronics and optoelectronics. *Nature Rev. Mater.* 1: 16031.

14 Zhang, Y., Jie, W., Chen, P. et al. (2018). Ferroelectric and piezoelectric effects on the optical process in advanced materials and devices. *Adv. Mater.* 30: 1707007.

15 Wang, Z.L. (2010). Piezopotential gated nanowire devices: piezotronics and piezo-phototronics. *Nano Today* 5: 540.

16 Wu, W. and Wang, Z. (2016). Piezotronics and piezo-phototronics for adaptive electronics and optoelectronics. *Nat. Rev. Mater.* 1: 1.

17 Wu, W., Wang, L., Yu, R. et al. (2016). Piezophototronic effect in single-atomic-layer MoS_2 for strain-gated flexible optoelectronics. *Adv. Mater.* 28: 8463.

18 Hao, J. and Xu, C.N. (2018). Piezophotonics: from fundamentals and materials to applications. *MRS Bull.* 43: 965.

19 Que, M., Zhou, R., Wang, X. et al. (2016). Progress in piezo-phototronic effect modulated photovoltaics. *J. Phys. Condens. Matter* 28: 433001.

20 Bao, R., Hu, Y., Yang, Q. et al. (2018). Piezo-phototronic effect on optoelectronic nanodevices. *MRS Bull.* 43: 952.

21 Wang, H., Liu, F., Fu, W. et al. (2014). Two-dimensional heterostructures: fabrication, characterization, and application. *Nanoscale* 6: 12250.

22 Pomerantseva, E. and Gogotsi, Y. (2017). Two-dimensional heterostructures for energy storage. *Nat. Energy* 2: 1.

23 Ding, R., Lyu, Y., Wu, Z. et al. (2021). Effective piezo-phototronic enhancement of flexible photodetectors based on 2D hybrid perovskite ferroelectric single-crystalline thin-films. *Adv. Mater.* 33: 2101263.

24 Liu, Y., Zhang, S., He, J. et al. (2019). Recent progress in the fabrication, properties, and devices of heterostructures based on 2D materials. *Nano-Micro Lett.* 11: 13.

25 Jariwala, D., Marks, T.J., Hersam, M.J. et al. (2017). Mixed-dimensional van der Waals heterostructures. *Nat. Mater.* 16: 170.

26 Dai, M., Chen, H., Wang, F. et al. (2019). Robust piezo-phototronic effect in multilayer γ-InSe for high-performance self-powered flexible photodetectors. *ACS Nano* 13: 7291.

27 Dai, Y., Wu, C., Wu, Z. et al. (2019). Ferroelectricity-enhanced piezo-phototronic effect in 2D V-doped ZnO nanosheets. *Adv. Sci.* 6: 1900314.

28 Lin, P., Zhu, L., Li, D. et al. (2018). Piezo-phototronic effect for enhanced flexible MoS_2/WSe_2 van der Waals photodiodes. *Adv. Funct. Mater.* 28: 1802849.

29 Li, F., Shen, T., Wang, C. et al. (2020). Recent advances in strain-induced piezoelectric and piezoresistive effect-engineered 2D semiconductors for adaptive electronics and optoelectronics. *Nano-Micro Lett.* 12: 1.

30 Han, X., Chen, M., Pan, C. et al. (2016). Progress in piezo-phototronic effect enhanced photodetectors. *J. Mater. Chem. C* 4: 11341.

31 Damjanovic, D. (2006). Hysteresis in piezoelectric and ferroelectric materials. *Sci. Hysteresis* 3 (337).

32 Wang, Z.L. (2012). Progress in piezotronics and piezo-phototronics. *Adv. Mater.* 24: 4632.

33 Song, J., Zhou, J., Wang, Z.L. et al. (2006). Piezoelectric and semiconducting coupled power generating process of a single ZnO belt/wire. A technology for harvesting electricity from the environment. *Nano Lett.* 6 (1656).

34 Hu, Y., Chang, Y., Fei, P. et al. (2010). Designing the electric transport characteristics of ZnO micro/nanowire devices by coupling piezoelectric and photoexcitation effects. *ACS Nano* 4: 1234.

35 Liu, C., Peng, M., Yu, A. et al. (2016). Interface engineering on p-CuI/n-ZnO heterojunction for enhancing piezoelectric and piezo-phototronic performance. *Nano Energy* 26: 417.

36 Yuan, S., Io, W.F., Mao, J. et al. (2020). Enhanced piezoelectric response of layered In_2Se_3/MoS_2 nanosheet-based van der Waals heterostructures. *ACS Appl. Nano Mater.* 3: 11979.

37 Wang, Z.L. (2010). Piezotronic and piezophototronic effects. *J. Phys. Chem. Lett.* 1: 1388.

38 Zhang, Y., Liu, Y., Wang, Z.L. et al. (2011). Fundamental theory of piezotronics. *Adv. Mater.* 23: 3004.

39 Que, M.-L., Wang, X.D., Peng, Y.Y. et al. (2017). Flexible electrically pumped random lasing from ZnO nanowires based on metal–insulator–semiconductor structure. *Chin. Phys. B* 26: 67301.

40 Pan, C., Zhai, J., Wang, Z.L. et al. (2019). Piezotronics and piezo-phototronics of third generation semiconductor nanowires. *Chem. Rev.* 119: 9303.

41 Streetman, B.G. and Banerjee, A.K. (1995). *Solid State Electronic Devices*, vol. 4. Englewood Cliffs, NJ: Prentice Hall.

42 Hui, R. (2019). *Introduction to Fiber-Optic Communications*. Academic Press.

43 Tian, W., Sun, H., Chen, L. et al. (2019). Low-dimensional nanomaterial/Si heterostructure-based photodetectors. *InfoMat* 1: 140.

44 Chen, Q., Khan, M.A., Sun, C.J. et al. (1995). Visible-blind ultraviolet photodetectors based on GaN pn junctions. *Electron. Lett* 31: 1781.

45 Monroy, E., Muñoz, E., Sánchez, F.J. et al. (1998). High-performance GaN pn junction photodetectors for solar ultraviolet applications. *Semicond. Sci. Technol.* 13: 1042.

46 Mak, K.F., Lee, C., Hone, J. et al. (2010). Atomically thin MoS$_2$: a new direct-gap semiconductor. *Phys. Rev. Lett.* 105: 136805.

47 Wu, W., Wang, L., Li, Y. et al. (2014). Piezoelectricity of single-atomic-layer MoS$_2$ for energy conversion and piezotronics. *Nature* 514: 470.

48 Zhang, K., Zhai, J., and Wang, Z.L. (2018). A monolayer MoS$_2$ pn homogenous photodiode with enhanced photoresponse by piezo-phototronic effect. *2D Mater* 5: 35038.

49 Brillson, L.J. and Lu, Y. (2011). ZnO Schottky barriers and Ohmic contacts. *J. Appl. Phys.* 109: 8.

50 Rhoderick, E.H. (1982). Metal-semiconductor contacts. *IEE Proc. I-Solid-State Electron Devices* 129: 1.

51 Sze, S.M., Li, Y., and Ng, K.K. (2006). *Physics of Semiconductor Devices*. Wiley.

52 Padovani, F.A. and Stratton, R. (1966). Field and thermionic-field emission in Schottky barriers. *Solid State Electron.* 9: 695.

53 Liao, Z.M., Liu, K.J., Zhang, J.M. et al. (2007). Effect of surface states on electron transport in individual ZnO nanowires. *Phys. Lett. A* 367: 207.

54 Leisgang, N., Roch, J.G., Froehlicher, G. et al. (2018). Optical second harmonic generation in encapsulated single-layer InSe. *AIP Adv.* 8: 105120.

55 Yang, Z., Jie, W., Mak, C.H. et al. (2017). Wafer-scale synthesis of high-quality semiconducting two-dimensional layered InSe with broadband photoresponse. *ACS Nano* 11: 4225.

56 Savchyn, V.P. and Kytsai, V.B. (2000). Photoelectric properties of heterostructures based on thermo-oxidated GaSe and InSe crystals. *Thin Solid Films* 361: 123.

57 Yang, Q., Guo, X., Wang, W. et al. (2010). Enhancing sensitivity of a single ZnO micro-/nanowire photodetector by piezo-phototronic effect. *ACS Nano* 4: 6285.

58 Ebnonnasir, A., Narayanan, B., Kodambaka, S. et al. (2014). Tunable MoS$_2$ bandgap in MoS$_2$-graphene heterostructures. *Appl. Phys. Lett.* 105: 31603.

59 Xu, Q., Yang, Z., Peng, D. et al. (2019). WS$_2$/CsPbBr$_3$ van der Waals heterostructure planar photodetectors with ultrahigh on/off ratio and piezo-phototronic effect-induced strain-gated characteristics. *Nano Energy* 65: 104001.

60 Yang, Z., Lu, J., Zhuge, M. et al. (2019). Controllable growth of aligned monocrystalline CsPbBr$_3$ microwire arrays for piezoelectric-induced dynamic modulation of single-mode lasing. *Adv. Mater.* 31: 1900647.

61 Du, J., Liao, Q., Hong, M. et al. (2019). Piezotronic effect on interfacial charge modulation in mixed-dimensional Van der Waals heterostructure for ultrasensitive flexible photodetectors. *Nano Energy* 58: 85.

62 Zhang, Y., Yang, Y., Gu, Y. et al. (2015). Performance and service behavior in 1-D nanostructured energy conversion devices. *Nano Energy* 14: 30.

63 Lin, Y.-F., Song, J., Ding, Y. et al. (2008). Piezoelectric nanogenerator using CdS nanowires. *Appl. Phys. Lett.* 92: 22105.

64 Lee, J., Park, J.Y., Cho, E.B. et al. (2017). Reliable piezoelectricity in bilayer WSe$_2$ for piezoelectric nanogenerators. *Adv. Mater.* 29: 1606667.

65 Zhai, T., Bando, Y., Golberg, D. et al. (2010). One-dimensional CdS nanostructures: synthesis, properties, and applications. *Nanoscale* 2, 168.

66 Zhang, Y., Zhai, J., Wang, Z.L. et al. (2017). Piezo-phototronic matrix via a nanowire array. *Small* 13: 1702377.

67 Lin, P., Zhu, L., Li, D. et al. (2018). Tunable WSe_2–CdS mixed-dimensional van der Waals heterojunction with a piezo-phototronic effect for an enhanced flexible photodetector. *Nanoscale* 10: 14472.

68 Xie, Y., Zhang, B., Wang, S. et al. (2017). Ultrabroadband MoS_2 photodetector with spectral response from 445 to 2717 nm. *Adv. Mater.* 29: 1605972.

69 Bao, W., Cai, X., Kim, D. et al. (2013). High mobility ambipolar MoS_2 field-effect transistors: substrate and dielectric effects. *Appl. Phys. Lett.* 102: 42104.

70 Newaz, A.K.M., Prasai, D., Ziegler, J.I. et al. (2013). Electrical control of optical properties of monolayer MoS_2. *Solid State Commun.* 155: 49.

71 Ye, M., Winslow, D., Zhang, D. et al. (2015). Recent advancement on the optical properties of two-dimensional molybdenum disulfide (MoS_2) thin films. *Photonics* 2: 288.

72 Wang, R., Ren, X.G., Yan, Z. et al. (2019). Graphene based functional devices: a short review. *Front. Phys.* 14: 13603.

73 Rodrigo, D., Limaj, O., Janner, D. et al. (2015). Mid-infrared plasmonic biosensing with graphene. *Science* 349: 165.

74 Zhang, W., Chuu, C.P., Huang, C.H. et al. (2014). Ultrahigh-gain photodetectors based on atomically thin graphene-MoS_2 heterostructures. *Sci. Rep.* 4: 3826.

75 Feng, S. and Xu, R. (2001). New materials in hydrothermal synthesis. *Acc. Chem. Res.* 34: 239.

76 Zhu, H., Wang, Y., Xiao, J. et al. (2015). Observation of piezoelectricity in free-standing monolayer MoS_2. *Nat. Nanotechnol.* 10: 151.

77 Sahatiya, P. and Badhulika, S. (2017). Strain-modulation-assisted enhanced broadband photodetector based on large-area, flexible, few-layered Gr/MoS_2 on cellulose paper. *Nanotechnology* 28: 455204.

78 Zhang, K., Peng, M., Wu, W. et al. (2017). A flexible p-CuO/n-MoS_2 heterojunction photodetector with enhanced photoresponse by the piezo-phototronic effect. *Mater. Horiz.* 4: 274.

79 Salvatore, G.A., Münzenrieder, N., Barraud, C. et al. (2013). Fabrication and transfer of flexible few-layers MoS_2 thin film transistors to any arbitrary substrate. *ACS Nano* 7: 8809.

80 Zhao, Y.Q., Guo, F., Ding, R. et al. (2021). Piezo-Phototronic Effect in 2D α-In_2Se_3/WSe_2 van der Waals heterostructure for photodetector with enhanced photoresponse. *Adv. Opt. Mater.* 9: 2100864.

81 Wang, Y.W., Zhang, L.D., Wang, G.Z. et al. (2002). Catalytic growth of semiconducting zinc oxide nanowires and their photoluminescence properties. *J. Cryst. Growth* 234: 171.

82 Tang, Q., Li, Y., Zhou, Z. et al. (2010). Tuning electronic and magnetic properties of wurtzite ZnO nanosheets by surface hydrogenation. *ACS Appl. Mater. Interfaces* 2: 2442.

83 Rivera, P., Seyler, K.L., Yu, H. et al. (2016). Valley-polarized exciton dynamics in a 2D semiconductor heterostructure. *Science* 351: 688.

84 Yu, H., Wang, Y., Tong, Q. et al. (2015). Anomalous light cones and valley optical selection rules of interlayer excitons in twisted heterobilayers. *Phys. Rev. Lett.* 115: 187002.

85 Fogler, M.M., Butov, L.V., Nocoselov, K.S. et al. (2014). High-temperature superfluidity with indirect excitons in van der Waals heterostructures. *Nat. Commun.* 5: 1.

86 Ivanov, S.A., Piryatinski, A., Nanda, J. et al. (2007). Type-II core/shell CdS/ZnSe nanocrystals: synthesis, electronic structures, and spectroscopic properties. *J. Am. Chem. Soc.* 129: 11708.

87 Hou, P., Lv, Y., Chen, Y. et al. (2019). In-plane strain-modulated photoresponsivity of the α-In_2Se_3-based flexible transistor. *ACS Appl. Electron. Mater.* 2: 140.

8

Synthesis and Properties of Conducting Polymer Nanomaterials

Ziyan Zhang[1], Tianyu Sun[1], Mingda Shao[1], and Ying Zhu[1,2]

[1] Beihang University, Key Laboratory of Bio-Inspired Smart Interfacial Science and Technology of Ministry of Education, School of Chemistry, 37 Xueyuan Road, Haidian District, Beijing 100191, China
[2] Beihang University, Beijing Advanced Innovation Center for Biomedical Engineering, 37 Xueyuan Road, Haidian District, Beijing 100191, China

8.1 Introduction

Since the discovery of the first conducting polymer, polyacetylene (PA), in 1977, conducting polymers (CPs) as a novel generation of organic materials have attracted much attention due to their wide range of industrial applications and economic viability [1–3]. The Nobel Prize in Chemistry 2000 was awarded jointly to Alan J. Heeger, Alan G. MacDiarmid, and Hideki Shirakawa for the discovery and development of electronically CPs [3–5]. CPs are polymers with highly π-conjugated polymeric chains, which helps in the migration of electrons throughout its polymeric chain. They have excellent electrical and optical properties similar to those of metals and inorganic semiconductors but also exhibit the mechanical flexibility and processability.

Associated with conventional polymers [2], which were also named as conjugated polymers or "synthetic metals." The most extensively studied CPs are PA, polypyrrole (PPy), polythiophenes (PTh), poly(*p*-phenylene) (PPP), poly(*p*-phenylenevinylene) (PPV) polyaniline (PANI), and derivatives thereof (Figure 8.1a) [6]. Generally, they could be divided into two principal classes: namely, polymers with degenerate ground state and polymers with nondegenerate. The prototype of degenerate polymers is transpolyacetylene, which has two identical structures in the ground state. On the other hand, most conjugated polymers such as PPy and PANI belong to nondegenerate polymers [7].

Conductivity is an important property for CPs, which is usually determined by the density and mobility of charge carriers. Charge carrier in a CP is different from either free electron in a metal or electron/hole in an inorganic semiconductor. For CPs, solitons [8], polarons and bipolarons [9] are proposed to transfer charges in π-conjugated polymers (Figure 8.1b). Usually, solitons are served as the charge carrier for degenerated CPs (e.g. PA), whereas polarons or bipolarons are used as charge carrier in nondegenerated CPs (e.g. PPy and PANI) [7, 10]. The doping process

Functional Nanomaterials: Synthesis, Properties, and Applications, First Edition.
Edited by Wai-Yeung Wong and Qingchen Dong.
© 2022 WILEY-VCH GmbH. Published 2022 by WILEY-VCH GmbH.

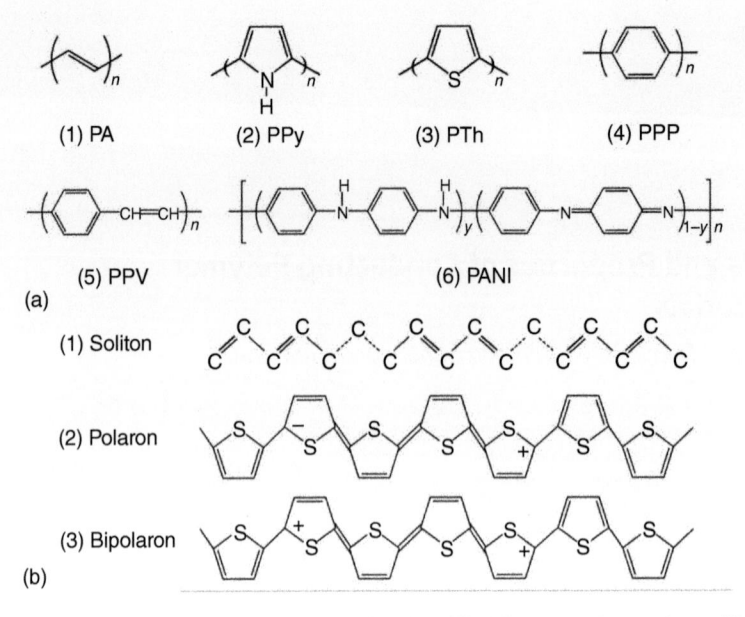

Figure 8.1 (a) Molecular structure of typical CPs: (1) trans-polyacetylene, (2) polythiophenes, (3) poly(p-phenylene), (4) polypyrrole, (5) poly(p-phenylenevinylene), and (6) poly(2,5-thienylenevinylene). (b) Schematic structure of charge carriers in CPs: (1) solitons in polyacetylene, (2) a positive polaron in polythiophenes, and (3) a positive bipolaron in polythiophenes.

facilitates the formation of radical cations/anions (polarons) or dications/dianions (bipolarons) in the backbone, while counterions from the solution enter into the polymeric material to counterbalance the charge [6]. Conductivity of the CPs is affected by chain structure that includes π-conjugated structure and length, crystalline and substituted grounds, and bounded fashion to the polymeric chain [10]. In addition, dopant structure and doping degree, morphology, and diameter of CP nanostructures also have significant impact on the overall properties of CPs [6]. The conductivity of CPs could be tuned in a wide range from 10^{-10} to 10^4 S cm^{-1} by the utilization of doping principles, endowing the conductive polymers acting as insulators, semiconductors, or conductors [6]. In addition, doping and dedoping of CPs is a reversible process that provides them with attractive electronic properties, mechanical properties, magnetic properties, wettability, optical properties, microwave absorbing property, etc., enabling a number of applications (Figure 8.2) such as photocatalysis [11], solar cell [12], E-skin [17], supercapacitor [13], drug delivery [14], biosensor [15], organic battery [18], electromagnetic interference (EMI) shield [16], etc. In addition, CPs can also be rationally designed and fabricated to nanostructures, thus providing new exciting features including tunable conductivities, flexibility, and mixed conductive mechanism that lowers the interfacial impedance between electrodes and electrolytes [19]. This chapter provides an overview of the recent advances in the synthesis and properties of CPs.

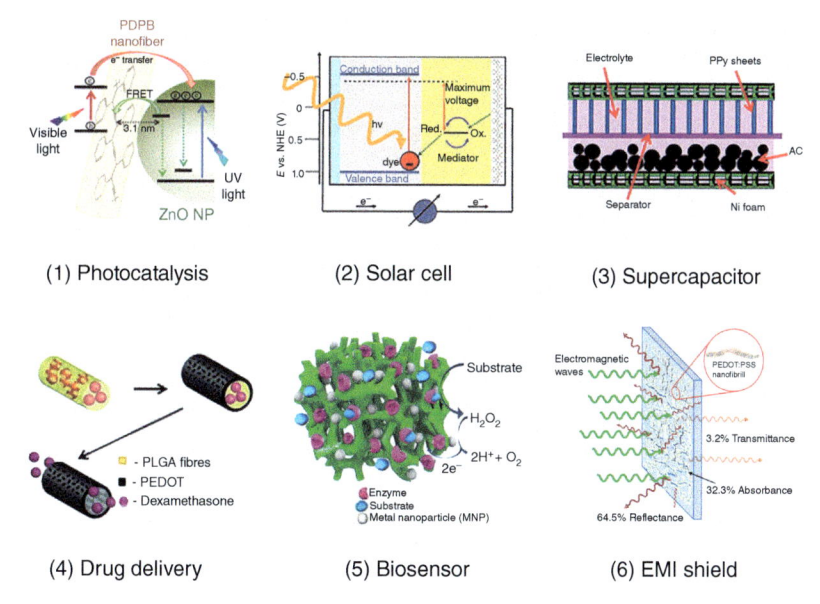

Figure 8.2 Several examples of applications of CPs. (1) Photocatalysis. Source: Sardar et al. [11]/Springer Nature/CC BY 4.0. (2) Solar cell. Source: Ahmad et al. [12]. Reproduced with permission of Royal Society of Chemistry. (3) Supercapacitor. Source: Yang et al. [13]. Reproduced with permission of RSC Pub. (4) Drug delivery. Source: Abidian et al. [14]. Reproduced with permission of Wiley. (5) Biosensor. Source: Li et al. [15]. Reproduced with permission of American Chemical Society. (6) EMI shield. Source: Hosseini et al. [16]. Reproduced with permission of American Chemical Society.

8.2 Synthesis and Properties

Conventionally, CPs could be achieved by electrochemical or chemical methods, mainly through oxidative polymerization, which allow one to obtain fine control of polymeric structures. Compared with bulk polymers, CP nanostructures displayed superior performances for applications that are associated with the nanoscale size giving a superior electrical conductivity, high surface area, high carrier mobility, improved electrochemical activity, good mechanical properties, and so on [19–21]. The template-synthesizing routes of CPs include hard template and soft template methods. The former replicates existing nanostructure by physical or chemical interactions [22, 23], and the latter relies on molecular self-assembly to form nanostructures [24, 25]. While the template-based synthesis is the most commonly used for producing morphology and size tunable nanostructures, template-free synthetic strategies such as interfacial polymerization (self-assembly), electrospinning, and radiolysis are considered to be simple, straightforward, and cost-effective techniques for the synthesis of CP micro/nanostructures without the need of a template or posttreatment for template removal [20, 26–28]. In the part, we will summary and discusses the synthesis and properties of CPs. Some selected works are discussed, particularly with regard to designing and synthesizing fine nanostructures of CPs with high performance in energy storage devices, sensors, flexible electronics, bio-scaffolds, and so on.

8.2.1 Chemical Synthesis and Properties

CPs came out formally in the 1970s, when Hideki Shirakawa prepared a PA film with higher conductivity for the first time [29]. He put the polyacetylene semiconductor film prepared by the Ziegler–Natta polymerization method in the vapor of halogen to prepare conductive PA, whose observed maximum conductivity was 38 Ω^{-1} cm^{-1}, laying the foundation for chemical synthesis of polymers [29]. With the development of CPs, the chemical synthesis method has become the most basic method for preparation of CPs, which is also called chemical oxidation method. In general, polymer monomers are used as starting materials to add oxidants, such as ammonium persulfate (APS) (($NH_4)_2S_2O_8$), iron nitrate ($Fe(NO_3)_3$), iron chloride ($FeCl_3$), etc., to precipitate the polymer. Some polymer monomers need to provide excessive protons when they are oxidized and react under acidic conditions (pH < 3) [30], for instance, PANI. The specific reaction environment is related to the polymerization mechanism. There are two main methods for synthesizing CP nanomaterials, namely, template method and template-free method [31, 32], which be used to synthesize zero-dimensional (0D) [33], one-dimensional (1D) [34], two-dimensional (2D) [35], and three-dimensional (3D) [36], CP nanomaterials (Figure 8.3a–d).

The template method of polymerization proposed by Martin and De Vito [37] is an effective technique to synthesize arrays of aligned polymer micro/nanotubes (NTs) and wires with controllable lengths and diameter. Microporous polycarbonate filters were used as the template membranes. The desired template membrane was immersed into an aqueous solution of pyrrole at 0 °C and an equal volume of an aqueous solution that was in $FeCl_3$ and *p*-toluene sulfonic acid also at 0 °C. This resulted in oxidative polymerization of the pyrrole monomer within the pores and on the faces of the template membrane; a polymerization time of one hour was used. After polymerization, PPy films that coated both faces of the membrane were removed by polishing with a methanol-wetted laboratory tissue. The PPy NTs with the diameter of 50 and 200 nm can be obtained, whose diameters depended on the pore size of templates. So far, many porous materials including anodic aluminum oxide (AAO) templates, mesoporous silica–carbon templates, and particle track-etched membranes (PTM) have been used as templates for the fabrication of nanofibers and tubes. Park et al. [38] demonstrated the use of AAO template to synthesize poly(3,4-ethylenedioxythiophene) (PEDOT) NTs modified with silver nanoparticles (Ag NPs) in one-pot method. First, $Fe(NO_3)_3$ and $AgNO_3$ are absorbed on the AAO template, and the vapor deposition polymerization (VDP) of monomer 3,4-ethylenedioxythiophene (EDOT) is used to obtain PEDOT. The silver ions on the surface are reduced to silver nanoparticles, and the template is removed with hydrochloric acid solution. PEDOT NTs modified with Ag NPs show rapid response and recovery in ammonia gas detection, which is due to the synergistic effects of silver nanoparticles, such as increased surface area and enhanced conductivity. Furthermore, the small size and uniform distribution of

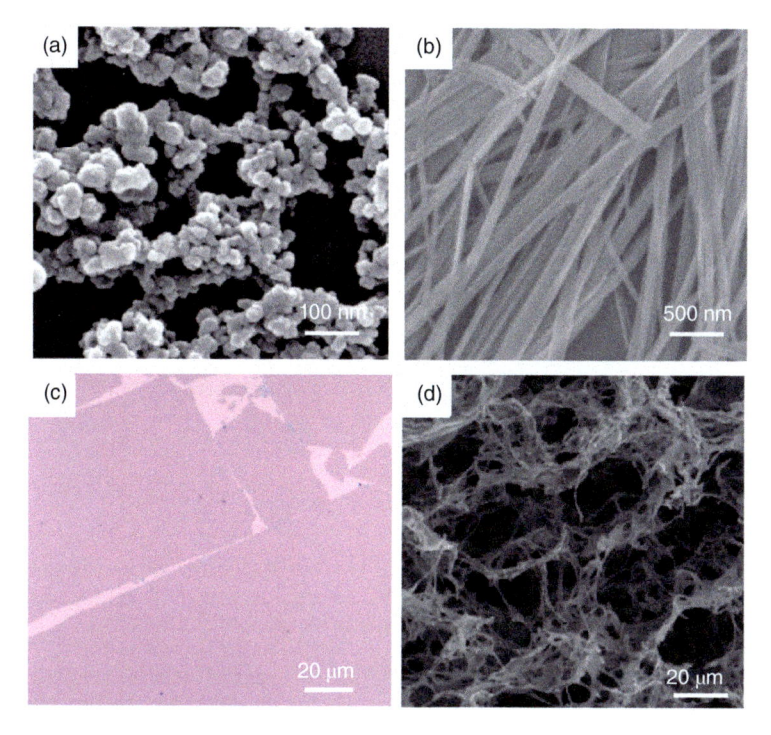

Figure 8.3 (a) SEM image of zero-dimensional PEDOT/LS(sulfonated lignin) nanoparticles. Source: Gan et al. [33]/Springer Nature/CC BY 4.0. (b) SEM image of polyoxovanadate (POV)/polymer sulfonate (PSS) nanofibers. Source: Zhang et al. [34]/with permission from Elsevier. (c) Optical image of quasi-two-dimensional PANI. Source: Zhang et al. [35]/Springer Nature/CC BY 4.0. (d) SEM image of three-dimensional PEDOT aerogels. Source: Chen et al. [36]/with permission from Elsevier.

the Ag NPs allowed the Ag NP/PEDOT NTs to be reversible, to be reproducible, and to return to the original sensitivity. The conductivity of pristine PEDOT NTs is about of 26.4 S cm^{-1}, while conductivities of Ag NPs/PEDOT NTs with 5%, 10%, 30%, and 40% (w/w) of AgNO$_3$ were measure to be about 54.9, 75.3, 291, and 302 S cm^{-1}, respectively. Interestingly, Ag NP/PEDOT NTs can be used as sensor for detecting NH$_3$ gas; the real-time response to NH$_3$ of PEDOT NTs with 30% (w/w) Ag NPs is within two seconds, while that of pristine PEDOT NTs is three seconds (Figure 8.4a,b). Moreover, the recovery time of PEDOT NTs with 30% (w/w) Ag NPs is seven seconds, decreased by 80% compared with pristine PEDOT NTs. It is well known that the removal of the template often leads to the destruction of the nanostructure and affects its conductivity. Wan et al. [39] employed the octahedral cuprous oxide (Cu$_2$O) as a template and phosphoric acid and APS as dopants and oxidants for the preparation of hollow octahedral PANI micro/nanostructures (Figure 8.4c). Compared with the conventional template method reported, the Cu$_2$O template does not need to be removed after polymerization, because it reacts with APS during the polymerization to form a soluble Cu^{2+} salt. The conductivities and morphologies of PANI is adjusted by changing the ratio of [Cu$_2$O]/[An]. The

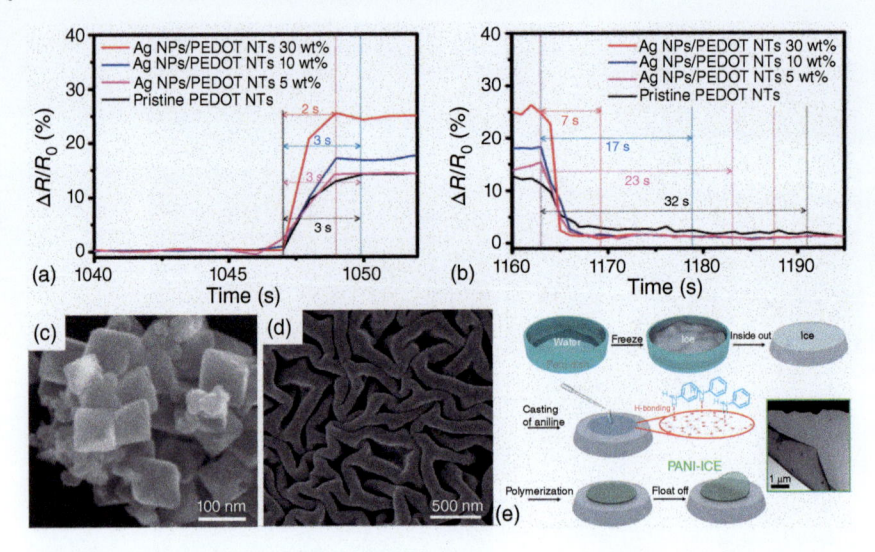

Figure 8.4 (a,b) The response and recovery times of Ag NPs/PEDOT NTs with increasing concentration of Ag NPs in PEDOT NTs at 50 ppm NH_3. Source: Park et al. [38]. Reproduced with permission of Royal Society of Chemistry. (c) Typical SEM images of octahedral-structured PANI synthesized at [An]/[H_3PO_4] = 2 : 1, [Cu_2O]/[An] = 1 : 2, and $T = 0-5\,°C$. Source: Zhang et al. [39]/with permission from Wiley. (d) SEM image of the aniline/CA salt, as viewed from above. Source: Zhu et al. [40]/with permission from Wiley. (e) The synthetic procedures for 2D PANI nanosheets on ice surfaces. Source: Choi et al. [41]/with permission from Wiley.

conductivity of PANI is about $1.28 \times 10^{-2}\,S\,cm^{-1}$ at the [Cu_2O]/[An] ratio of 1 : 10. Moreover, Zhu et al. [40] fabricated PANI nanostructures with brain-like convolutions (140–170 nm in diameter) (Figure 8.4d) by a gas/solid reaction using chlorine gas as the oxidant and using an aniline/citric acid (CA) salt grown in organic solvent as the template. In the work, the aniline/CA salts could act as template and reactants, which were prepared in an organic solution by an acid/base reaction in organic solvents, including nonpolar dichloromethane and polar ethanol and tetrahydrofuran (THF). The method provided is a universal and efficient approach to prepare brain-like nanostructures of PANI, which is completely different from the traditional chemical polymerization that utilizes an acidic aqueous solution. The surface resistance of the brain-like nanostructures was measured to be about $4.9\,M\Omega\,cm^{-2}$, thus suggesting the conducting state of the PANI. Park and coworkers [41] synthesized 2D PANI nanosheets using ice as a removable hard template by chemical oxidant polymerization. The work is that aniline solution in $1\,mol\,l^{-1}$ HCl was cast on surface of the ice frozen in a petri dish and then $(NH_4)_2S_2O_8$ solution in HCl was immediately dropwise on the ice with the aniline deposit to create chemical oxidation of aniline at $0\,°C$. (Figure 8.4e depicts the synthetic procedures for the 2D PANI nanosheets on the ice surface.) It was worth noting that aniline prevents the ice from being melted by HCl because aniline can limit proton transfer from HCl to the ice. The unique advantage of ice template method ensures the integrity and continuity of the film. As a result, PANI nanosheets showed a high current flows of 5.5 mA at 1 V and a high electrical conductivity of 35 S cm^{-1}, which was ascribed to the vertical growth of PANI that has the long-range-ordered edge-on p-stacking of the quinoid ring. The ice-template method is easy not only to transfer PANI nanosheet to various types of substrates but also to achieve any patterns of PANI sheet using predesigned masks, which facilitate the CP application in electronic devices.

In addition, the organic surfactants can be used as soft templates for synthesis of CP nanostructures that rely on surfactant micelles [42]. In water or polar solvents, these surfactants can self-assemble into supramolecular aggregates called micelles. By changing temperature, pressure, and surfactant concentration, micelles undergo a phase transition to a lyotropic nematic liquid crystal phase, which are envisaged to serve as the templates during the polymerization process. After removing the micelles, a porous mesostructure material is formed. Compared with the hard template method, this method is relatively simple and low in cost. Commonly used cationic surfactants are octyltrimethylammonium bromide (OTAB), decyltrimethylammonium bromide (DTAB), and cetyltrimethylammonium bromide (CTAB). For instance, Guo et al. [43] used sodium dodecyl sulfate (SDS) as soft template to control the formation of PANI at the hydrochloric acid solution. Containing low-concentration SDS and hydrochloric acid can produce a rectangular submicron structure of PANI. The authors found that the morphologies of self-assembled PANI depend on the pH of the solution. By changing the polymerization conditions, such as pH and surfactant concentration, different PANI nanostructures, such as particles, nanofibers, nanosheets, and rectangular NTs, can be produced. The pressed pellet conductivity was about 7.3×10^{-5} S cm^{-1} at room temperature. Zhang et al. [44] synthesized the wire-, ribbon-, and sphere-like nanostructures of PPy by chemical oxidation polymerization in the presence of various surfactants (anionic, cationic, or nonionic surfactant) with oxidizing agents APS or FeCl$_3$. The surfactants and oxidizing agents used in this study have played a key role in tailoring the nanostructures of PPy during the polymerization. It is inferred that the lamellar structures of a mesophase are formed by self-assembly between the cations of a long-chain cationic surfactant CTAB or DTAB and the anions of APS oxidizing agent. These layered mesostructures are presumed to act as templates for the formation of wire- and ribbon-like PPy nanostructures. The conductivity of PPy nanostructures was measured to be about 7.3×10^{-3} S cm^{-1} at room temperature. Yuan et al. [45] reported a novel hollow nanospheres of PANI with mesoporous, brain-like, convex-fold shell structures via a micelle-mediated phase transfer method, using perfluorooctanoic acid (PFOA)/aniline as a soft template. Figure 8.5a give the fabrication process of PANI hollow spheres with mesoporous convex-fold surface. The SEM and TEM images demonstrated that these spheres are hollow structure constituted with interconnected ravines similar to the brain cortex with width of 10–20 nm (Figure 8.5b,c). These self-assembled hollow spheres possess high specific surface areas (835.7 m^2 g^{-1}) and can be narrowly dispersed with uniform morphologies by adjusting PFOA/aniline molar ratio. Notably, the PANI spheres exhibited superhydrophobicity and high oleophobicity simultaneously, with contact angles (CAs) of $165 \pm 0.9°$, $134 \pm 0.8°$, $131 \pm 0.9°$,

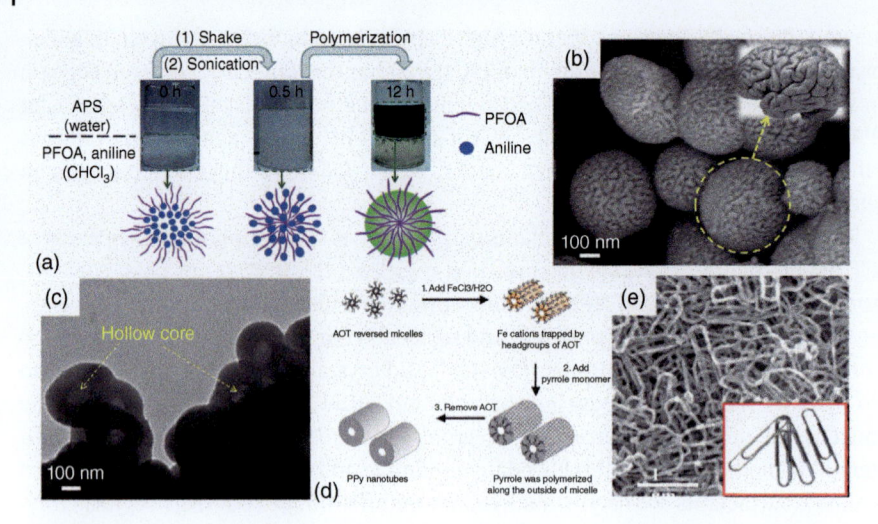

Figure 8.5 (a) Photos of polymerization processes via the micelle-mediated phase transfer method. (b) The magnification of the brain-like surface of PANI spheres. Inset: photograph of human brain structures. (c) TEM image of the hollow PANI spheres. Source: Yuan et al. [45]/with permission from Royal Society of Chemistry. (d) Schematic diagram of PPy nanotube fabrication using reverse microemulsion polymerization. Source: Jang and Yoon [46]. Reproduced with permission of Elsevier. (e) SEM images of PPy nanoclips (inset: digital picture of paper clips). Source: Liu et al. [47]/with permission from American Chemical Society.

and $125 \pm 0.7°$ for water, glycerin, ethylene glycol, and corn oil, respectively. Such complex nanostructures were difficult prepared by hard template. Park et al. [48] developed a technique for the anisotropic growth control of PANI nanostructures, specifically nanospheres, nanorods, and nanofibers by employing a polymeric stabilizer, poly(N-vinylpyrrolidone) (PVP). Aniline was first dissolved in aqueous HCl solution, and then PVP was added to the solution. Then APS acted as oxidant was added to mixture for triggering a chemical oxidation polymerization of aniline. The polymerization rate became slower in the presence of the stabilizer, since PVP can sterically restrict the directional fiber growth of PANI chain in aqueous solution. The electron transfer capability of the nanostructures was found to increase in the order of nanospheres < nanorods < nanofibers, for example, the specific capacitors based on the nanostructures were calculated from discharge curves to be $71\,F\,g^{-1}$ for nanospheres, $133\,F\,g^{-1}$ for nanorods, and $192\,F\,g^{-1}$ for nanofibers, respectively, showing a morphology-dependent capacitance behavior. The authors proposed that the oxidation level of the PANI nanostructures was the crucial factors for determining the capacitor performance.

The aggregation of surfactant molecules in the reverse microemulsion is due to the solvation of polar groups and hydrogen bonding, which act as a nanoreactor to promote the growth of nanomaterials. Jang et al. [46] fabricated PPy NTs by chemical oxidation polymerization in the presence of sodium bis(2-ethylhexyl) sulfosuccinate (sodium bis(2-ethylhexyl) sulphosuccinate [AOT]) as a reverse micelle and $FeCl_3$ as oxidant. The anionic polar head group of AOT is extracted from the aqueous solution into the reverse micelle phase. The concentrated Fe^{3+} ions in the reverse micelles are terminated by the anionic polar head group of AOT, so they will maintain the rod shape of AOT (Figure 8.5d). Then, the pyrrole monomer is added to the solvent, and the pyrrole monomer is polymerized along the outside of the rod-shaped micelle through Fe^{3+} ions. Compared with PPy tubes manufactured by self-assembly methods, the diameter of PPy is smaller, and the aspect ratio is higher, which facilitates the charge transfer in the material, leading to the increase of electrical conductivity. However, due to the poor stability of the soft template, it is difficult to accurately control the morphology of nanomaterials, so this method needs to be further optimized. Zhang et al. [47] developed a new universal oxidation template assembly (OTA) one-step synthesis method, in which CTAB is dissolved in hydrochloric acid and then added APS to produce a white precipitate of CTAB/APS complex, which is used as an oxidation template for polymerization. After the reaction is complete, pyrrole monomer is added, and a black precipitate of PPY is formed. Using CTAB/APS complex as templates, two-dimensional clip-like nanostructure PPy was also fabricated (Figure 8.5e). The conductance values of this material are in the range of $2–5\,S\,cm^{-1}$. The nanoclip PPy can be used as precursors for microwave-initiated nanocarbonization, a surprisingly large improvement in thermal stability after microwave heating. The advantage of this OTA synthesis method is that it can control the two-dimensional morphology of the material and allow large-scale production of nanomaterials.

Compared with the template method, the template-free method is simpler and inexpensive, because it omits cumbersome steps such as template synthesis and subsequent processing. By controlling the synthesis conditions, such as temperature, monomer concentration, and dopant concentration, high-performance CP nanomaterials can be synthesized. Zhu et al. [49] fabricated 3D hollow spheres of PANI self-assembled from 1D nanofibers with a conductivity of $1.11 \times 10^{-1}\,S\,cm^{-1}$ by template-free method using salicylic acid (SA) and $FeCl_3$ as the dopant and oxidant, respectively. The morphology and diameter are strongly affected by the molar ratio of aniline/SA/$FeCl_3$ and the concentration of SA and $FeCl_3$. It was found that two reactions took place at the same time. One is the reaction of aniline with SA and $FeCl_3$ to form conductive PANI, and the other is the reaction of SA with $FeCl_3$ to produce a color complex. Thereby, micelles composed of AN/SA served as a soft template for nanofiber formation and the color complex as the soft template for sphere formation coexisted, allowing the cooperation of the two soft templates to form 3D hollow sphere constructed of 1D nanofibers via a self-assembly process. Moreover, Zhu et al. [50] also explored the influence of the polymerization environment on the polymer. For the first time, under 80% high relative humidity, using APS as the oxidant and dopant, the template-free method was used to synthesize the self-assembled 3D rose-shaped PANI microstructure. When the relative humidity increases from 25% to 80%, not only the morphology of micro/nanostructured PANI undergoes a change from one-dimensional nanofibers to two-dimensional nanosheets to 3D rose-like microstructures (Figure 8.6a). It is proposed that the synergy of oriented water molecules on the water vapor–water interface and the difference in hydrogen bond energy between the interface and the body caused by the relatively high humidity.

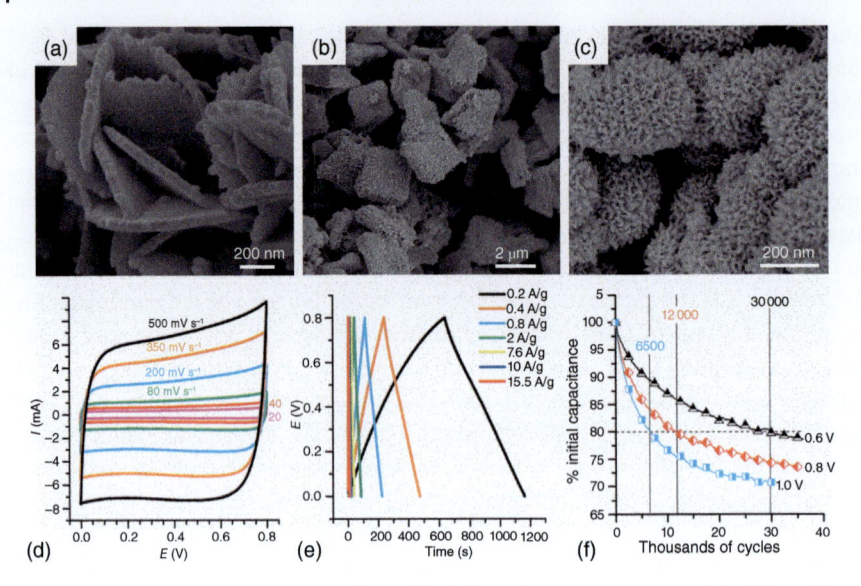

Figure 8.6 (a) SEM image of the 3D-boxlike PANI microstructures assembled from 1D nanofibers. Source: Zhu et al. [51]/with permission from Wiley. (b) Typical SEM image of template-free-synthesized, urchin-like PANI–PFOSA hollow spheres. Source: Zhu et al. [52], Figure 01 [p. 3420]/with permission from ELSEVIER. (c) 3D roselike microstructures of PANI, which are self-assembled from 2D nanosheets consisted of 1D nanofibers. Source: [50]), Figure 01 [p. 2861]/with permission from John Wiley & Sons, Inc. (d) Cyclic voltammograms for a two-electrode EC carried out at scan rates from 20 to 500 mV s^{-1} with a maximum charging voltage of 0.8 V. (e) Galvanostatic charge–discharge curves from 0 to 0.8 V at specific current densities ranging between 0.2 and 15.5 A g^{-1}. (f) Cycling stability of devices with various maximum charging voltages carried out by galvanostatic charge–discharge at 5 A g^{-1}. Source: (d–f) Santino et al. [53]. Reproduced with permission of Royal Society of Chemistry.

Zhu et al. [54] also reported on conductive and superhydrophobic rambutan-like hollow spheres of PANI prepared by a self-assembly method in the presence of perfluorooctane sulfonic acid (PFOSA), which served as dopant, soft template, and induced superhydrophobicity at the same time. Room-temperature conductivity of the hollow spheres was measured to be about 9.6×10^{-1} S cm^{-1}. Also, the water CA on the conductive hollow spheres was as high as 164.5°, revealing the superhydrophobic nature of the material. The large specific area and superhydrophobic properties of PANI hollow spheres might find potential in various fields of applications. Moreover, the authors used the template-free method to prepare conductive polymers with various nanostructures, such as cubes (Figure 8.6b) [51], sea urchins (Figure 8.6c) [55], and so on. D'Arcy and coworkers [53] reported a nanofibrillar PPy nanobrushes by low-temperature evaporative vapor-phase chemical polymerization without the use of hard templates. The microdroplets of oxidant solution serve as templates for nanofiber nucleation during polymerization. A low-temperature modified vapor-phase synthesis allows for the deposition of high aspect ratio and conductive and capacitive one-dimensional nanostructures of PPy onto three-dimensional fibrous substrates. Deposited on conductive substrates, the PPy nanofibers as capacitor exhibit a three-electrode specific capacitance of 144.7 F g^{-1} over a 1.2 V window in 1 mol l^{-1} lithium perchlorate with good reversibility. (Symmetric PPy supercapacitors with a 1 M lithium perchlorate electrolyte show a rectangular cyclic voltammogram at scan rates from 20 to 500 mV s^{-1} when charged to 0.8 V (Figure 8.6d).) A rectangular curve at high scan rates indicates a high rate of charging, allowing the device to be used for power-intensive applications. Galvanostatic charge–discharge involves applying a constant current to a device until it reaches a set maximum charging voltage and then reversing the current to fully discharged voltage of 0 V. Figure 8.6e shows nearly a symmetric triangular curve at a low specific current of 0.2 A g^{-1} and low charge-transfer resistance during charging and discharging. Triangular GCD curves with very little ohmic drop are seen even at high current densities, indicating excellent rate performance. PPy devices exhibit a high cycling stability, reaching 200 000 cycles while retaining 70% of their initial capacitance when charged to 0.6 V (Figure 8.6f).

Recently, Kaner and coworker [56] developed interfacial polymerization method for synthesis of CP nanomaterials without the need for any templates of functional dopants, in which monomers and the oxidant were dissolved in two mutually incompatible solvents, respectively, and carry out an chemical oxidation polymerization at the interface of the two phases. Figure 8.7a–e showed the color change of the two-phase interface during interfacial polymerization of aniline in a water/chloroform system. It can be found that the polymerization of aniline is performed in the organic layer, migrating into the water phase and finally filling the entire water layer. They have successfully synthesized PANI nanofibers with diameters of 30 and 120 nm obtained by using hydrochloric acid and perchloric acid. The electrical conductivity of PANI nanofibers made with 1.0 M HCl is about 0.5 S cm^{-1}, comparable with that of conventional PANI powders. Matsui and coworkers [57] also reported the synthesis of a single crystal PEDOT nanoneedles by interfacial polymerization (Figure 8.7f). The aqueous/organic interface was the organic phase of EDOT dispersed in dichloromethane solvent and aqueous phase of ferric chloride dissolved in deionized (DI) water. Herein, ferric chloride was used as an oxidant in the precipitation polymerization of thiophenes. The PEDOT nanoneedles obtained from this synthesis are about 50 nm long and 15 nm wide. Note that the tunneling current of PEDOT nanoneedles abruptly jumped to a saturated amplitude (tunneling current >10 nA) at applying the bias voltage of 3 V, while that of reached to insulating state (tunneling current <0.04 nA). The abrupt switching behavior with a response time of milliseconds near −3 and 3 V might due to the formation of single-crystalline domains of PEDOT nanoneedles. In addition, Manohar and coworkers who developed the seed polymerization method use the nanostructured seed templates to synthesize rapidly bulk quantities of PANI nanofibers without the need for conventional templates, surfactants, polymers, or organic solvents [58]. They fabricated fibrillar morphology of PANI by employing PANI fibers with an average diameter in the range 20–60 nm (Figure 8.7g) as seeds but obtained PANI nanospheres by using nanosphere seeds of PANI. The conductivities of PANI nanofibers were in the range 2–10 S cm^{-1}, similar to conventional the emeraldine PANI particles. However, a significant difference in the capacitance values for

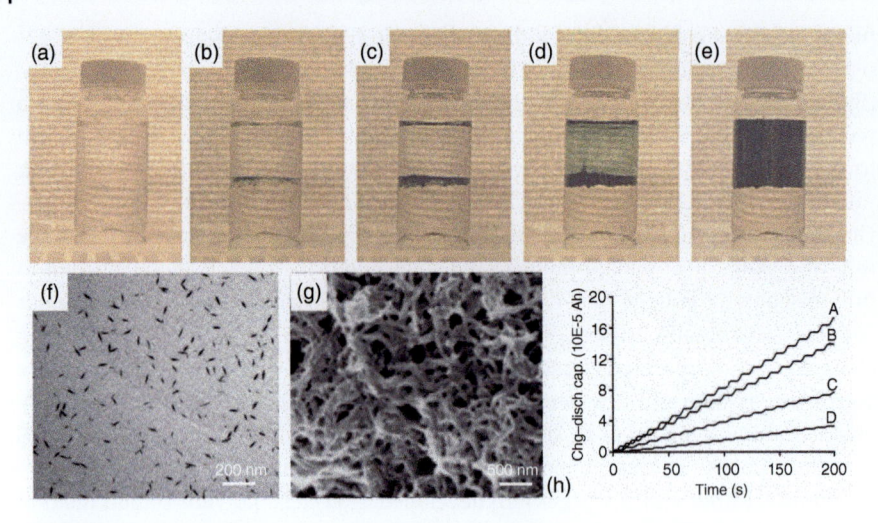

Figure 8.7 (a–e) Snapshots showing interfacial polymerization of aniline in a water/chloroform system, from a to e, in which the reaction times are 0, 1.5, 2.5, 4, and 10 minutes, respectively. The top layer is an aqueous solution of 1.0 M perchloric acid and ammonium peroxydisulfate; the bottom layer is aniline dissolved in the organic solvent chloroform. Source: (a) Huang et al. [56]/with permission from American Chemical Society. (f) TEM image showing PEDOT nanoneedles. Source: Su et al. [57]/with permission from Wiley. (g) SEM images of emeraldine PANI nanofibers synthesized by seeding the reaction using the following. (h) Charge–discharge capacity plot of emeraldine, HCl powder in the range 0.4–0.5 V (vs. SCE) in aqueous 1.0 M camphorsulfonic acid electrolyte. Charge (curve A), discharge (curve B) cycles for nanofibers and charge (curve C), and discharge (curve D) cycles for conventional (nonfibrillar) polyaniline. Source: Zhang et al. [58]/with permission from American Chemical Society.

PANI nanofibers, for instance, a capacitance value of PANI nanofibers synthesized using seed template was about 122 F g^{-1}, better than that of conventional PANI nanofibers (33 F g^{-1}) (Figure 8.7h). The authors proposed that CP nanostructures with multiple length scales can be triggered by small amounts of added nanoscale templates.

In summary, CPs with different nanostructures can be synthesized by chemical polymerization discussed above. The preparation methods and conidiations will determine the size, morphology, and properties of final CPs. Therefore, preparation approach should be carefully selected, according to the requirement of size and morphology of final CPs, the physical and chemical properties, and the initial raw materials, which will earn a broader application prospect in engineering applications.

8.2.2 Electrochemical Synthesis and Properties

In addition to chemical synthesis, electrochemical polymerization is also a typical method for preparing CPs (such as PANI, PPy, PTh, and PPP) [59], which make the CP monomer polymerize on the electrode surface to form conductive polymer deposits with controllable morphologies and properties, without the need to use binders or other conductive additives. During electrochemical polymerization, organic compounds undergo electron transfer at the electrodes to form free radical ions, which can initiate chain polymerization, or monomers are polymerized through self-coupling [60]. The IBM research group first synthesized PPy and PANI by electrochemical methods in 1979 [61]. Using methyl cyanide/tetraethylammonium tetrafluoroborate (MeCN/Et$_4$NBF$_4$) and pyrrole as the electrolyte, a PPy film is synthesized at a constant current on the platinum electrode surface in a two-electrode cell. Generally, electrochemical preparation methods can also be distinguished according to the presence or absence of templates. Template method is divided into hard template and soft template. The use of nano-pore template on the electrode surface is the most direct method for electropolymerization of porous conductive nanomaterials [62]. Wang and coworkers [63] employed AAO as a template to grow PPy nanowires by electrochemical polymerization and finally dissolved the AAO template in NaOH solution to obtain the desired nanostructure (Figure 8.8a–e). The specific step was that the AAO template was fixed on the Ti substrate with epoxy resin, and the PPy was grown in the pores of the AAO template under electric drive to form PPy nanowires, whose sizes were well matched. PPy nanowires deposited on Ti substrate were designed a new triboelectric nanogenerator (TENG), as shown in Figure 8.1b. In this case, PPy on Ti was used as triboelectrode, and polyvinylidene fluoride (PVDF) was chosen as the counter triboelectrode due to its excellent processability and flexibility, which showed opposite triboelectric polarities. When they are pressed into contact, the PPy tend to lose electrons to bear positive charges, while the PVDF would be negatively charged through accepting electrons. The PPy-based TENG shows high output performance with a maximum short-circuit current density of 23.4 mA m^{-2} and output voltage of 351 V, which can light 372 commercial red LEDs. Luo et al. [64] prepared 3D nanoporous conducting PEDOT decorated with gold nanoparticles (AuNPs/PEDOT) by electrochemical polymerization of PEDOT in the presence of the multilayer PS nanospheres as templates, followed by electrodepositing AuNPs (Figure 8.8f–h). AuNPs were electrodeposited onto the porous PEDOT surface to increase the interface conductivity and provide more active sites to load identification elements, which make AuNPs/PEDOT as a promising biosensor for the detection of miRNA24. In detail, the electroactive surface area of the 3D porous biosensor is 2.4 times that of the planar one, and the corresponding sensitivity is prominently increased by 2.07 times. Under optimal conditions, the biosensor exhibited good selectivity and stability and high sensitivity to detect miRNA24 with a low detection limit down to 3.8 fM.

Gyurcsanyi et al. [65] used a track-etched polycarbonate membrane (PCM) filter as a sacrificial microreactor to fabricate surface-imprinted polymer sulfonate (PSS)-doped PEDOT microrods on the surface of a gold electrode by electrochemical method. The key step of the method is to adsorb the protein onto the PCM surface by simple physical adsorption due to a native hydrophobic property of membrane. After the removal of the PCM template, the PEDOT microrods confined to the surface of the gold electrode possess the complementary imprint of the target protein on their surface (Figure 8.9a–d). Moreover, Dong and coworkers [66] prepared DNA–PANI complex nanowires by electropolymerization of aniline using

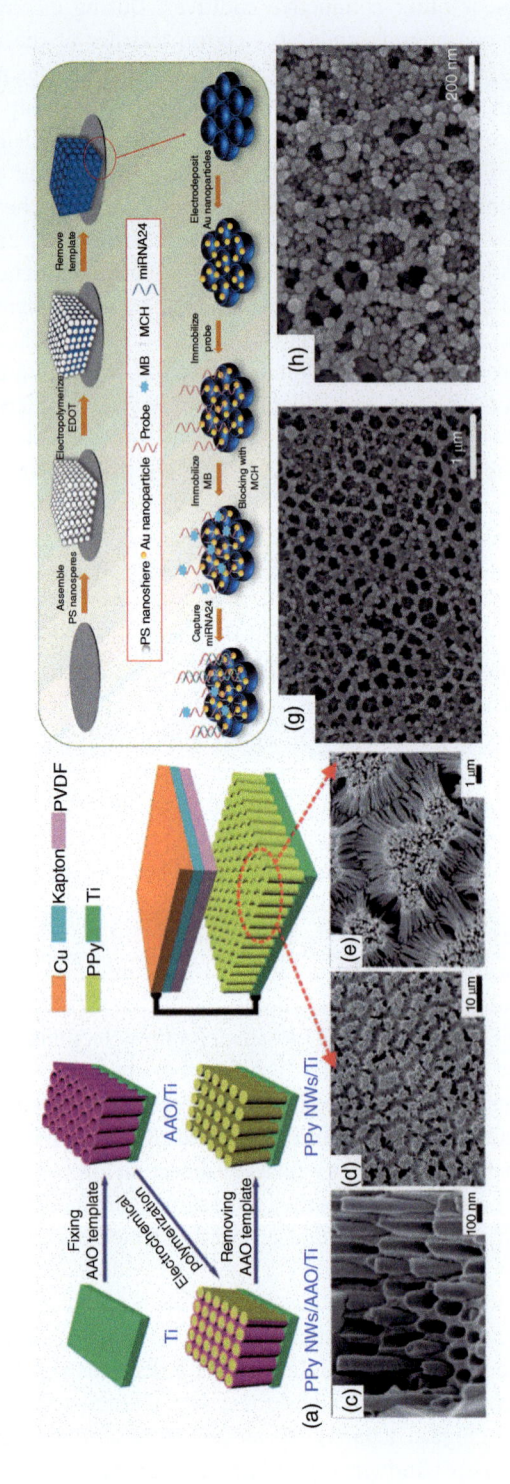

Figure 8.8 (a) Schematic depiction of PPy nanowire preparation by electrochemical polymerization with AAO as the template. (b) Schematic illustration of the designed PPy nanowire-based TENG. (c–e) FESEM images of PPy nanowires before and after dissolving the AAO template. Source: Cui et al. [63]/Royal Society of Chemistry/CC BY 4.0. (f) Construction process of the electrochemical biosensor based on 3D nanoporous AuNP/PEDOT film [64]. (g–h) SEM images of the porous AuNP/PEDOT film at different magnification. Source: (g) Ma et al. [64]/with permission from Elsevier.

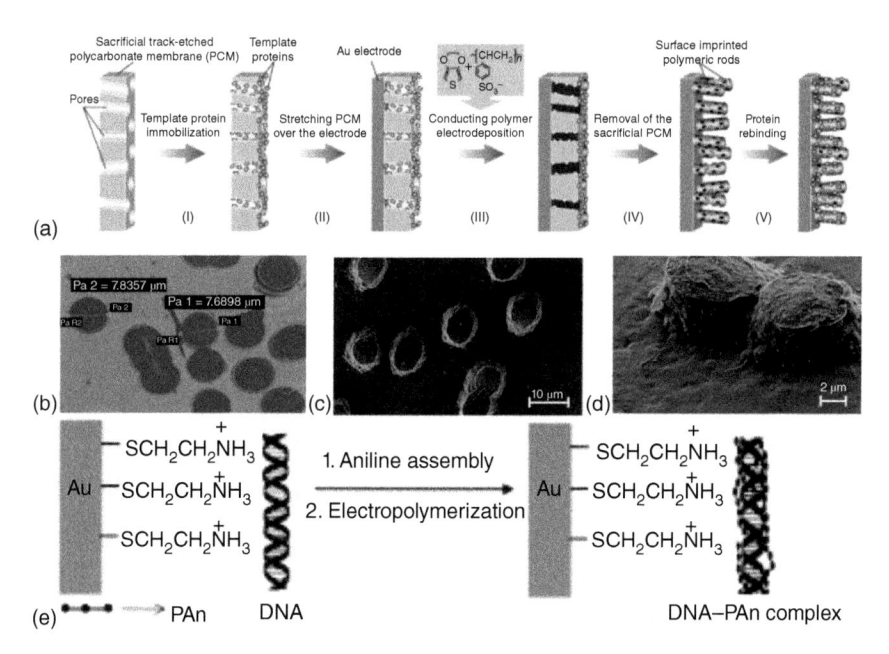

Figure 8.9 (a) Schematics of the surface-imprinting strategy for fabrication of molecularly imprinted polymers for protein assays. (b) SEM images of PEDOT/PSS rods confined to the micropores. (c,d) Freestanding microrods on the surface of the electrode after the removal of the PCM. Source: (b) Menaker et al. [65]/with permission from Wiley. (e) The formation of DNA-PANI complex on Au surface. Source: Shao et al. [66]. Reproduced with permission of Wiley.

immobilized DNA as a biological template (Figure 8.9e). The main steps include the gold electrode coated with 2-aminoethanethiol that was immersed in DNA solution, followed by the DNA-templated assembly and electropolymerization of protonated aniline. This work provides noncovalent method for building DNA–PANI complex. Bobacka and coworkers [67] prepared a molecularly imprinted PEDOT in the presence of 2,2′-methylenebis(2-methoxy-4-methylphenol) as template by co-electropolymerization of EDOT and 3-acetic acid thiophene, which can be employed detect the concentrations of lignin markers dissolved in pure solvents ranging from 1×10^{-6} to 1×10^{-2} mol l^{-1}.

Fradin et al. [68] fabricated the vertically aligned NTs using reverse micelles stabilized by monomer and electrolytes as a soft template by electropolymerization of naphtho[2,3-*b*]thieno[3,4-*e*][1,4]dioxine (NaphDOT) in dichloromethane and chloroform containing varying amounts of water (Figure 8.10). With an increasing amount of water in a solvent, the number of micelles increases but not their size, leading to a surface with more densely packed nanostructures. The authors also demonstrated that the micelles observed in organic solvent are stabilized by monomers and electrolytes playing the role of a surfactant for the formation of nanotubular structures, whose size correlated with the inner diameter of NTs that formed by subsequent electropolymerization.

The physical adsorption of the protein onto the PCM surface in the first step of the synthesis was demonstrated using a fluorescent derivative of the avidin.

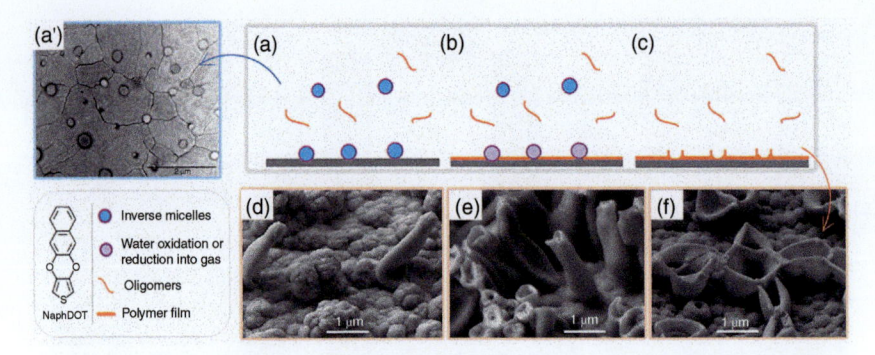

Figure 8.10 (a) Reverse micelles are adsorbed on the substrate. (a′) TEM observations of monomer solutions in $CHCl_3$ before and after polymerization. (b) Micelles act as a soft template for polymer deposition, and water is oxidized or reduced inside the micelles. (c) Gas bubbles are released and polymer growth continues. (d–f) SEM observations of NaphDOT cyclic voltammetry scan depositions in CH_2Cl_2 with various water contents. Source: Fradin et al. [68], [p. 001]/with permission from Wiley.

In recent years, the template-free method has been also used to prepare CP nanostructures. Darmanin and coworkers [69] fabricated nanoporous structures using a template-less electropolymerization of thieno [3,4-b]thiophene monomers with polar carbamate linkers and various substituents in organic solvent (CH_2Cl_2) and without the aid of surfactant. Importantly, the authors found that water plays a vital role for the formation of nanoporous structures. When polymerization of thieno-Ph occurred, the formation of nanotubular structures was observed at a low H_2O content ($CH_2Cl_2 + H_2O$ (35%)). When the H_2O content increased ($CH_2Cl_2 + H_2O$ (65%)), an increase in the number of nanotubular structures was found. However, the formation of nanoporous membranes was obtained especially with 100% of CH_2Cl_2. Moreover, the polar linkers incorporated onto thieno[3,4-b]thiophene derivatives can adjust the surface structures, for example, the thieno[3,4-b]thiophene monomers with alkyl chains (thieno-C2 and thieno-C4) was polymerized to obtain spherical structures while thieno[3,4-b]thiophene monomers with aromatic groups was synthesized to get nanotubular structures. Furthermore, the surface hydrophobicity and water adhesion were adjusted by changing polar carbamate linkers and various substituents, which may provide potential applications of water harvesting systems, separations membranes, and optoelectronic devices, as well as for sensors. In addition, Darmanin et al. [70] also produced densely packed vertically aligned NTs by using thieno[3,4-b]thiophene monomer combined with pyrene substituent in $CH_2Cl_2 + H_2O$. In this way, densely packed, open NTs obtained allowed trapping of a high amount of air and thus increased the water CA and could become even more hydrophobic by modifying the deposition method or the electrolyte. Yu et al. [71] fabricated the functionalized PEDOT and poly(3,4-propylenedioxythiophene) (PProDOT) nanofibers, nanodots, nanonetwork, and nano/microsize tubes by using a template-free electropolymerization method on indium–tin oxide substrates. The authors demonstrated the influence of temperature and functional groups on the formation of PEDOT and PProDOT nanostructures (Figure 8.11a). When nonpolar functional groups such as alkyl chains were attached, the polymers of ProDOT and EDOT tend to the formation nanofibers and nanoporous structures in general. With EDOT and ProDOT containing polar functional groups, such as hydroxyl, carboxylic acid, and triethylene glycol, the polymers formed nanodots at 25 °C, while they formed tubular structures at 0 °C. They demonstrated that the impedances of functionalized PEDOT films were lower than Au when the frequency was lower than 100 Hz mainly due to an increase in effective surface area of PEDOT electrodes. Liao et al. [72] tracked the self-assembly process of PPy NT arrays during template-free electrochemical polymerization using *in situ* electrochemical atomic force microscopy (EC-AFM). The observed self-assembly process is as follows, first forming nano-protrusions, then forming nanoneedle structures, and finally forming a one-dimensional nanostructure array (Figure 8.11b,c). This work is helpful for people to understand the principle and process of template-less electropolymerization of CPs.

Recently, bipolar electropolymerization was considered as powerful method for obtaining hybrid materials of CPs and conductive materials, in which bipolar electrodes (BPEs) are available as wireless electrodes that undergo anodic and cathodic reactions simultaneously. It is easy to achieve site-selective anisotropic modification of BPEs with CPs by the wireless electropolymerization of aromatic monomers (Figure 8.12a–c) [73]. In addition, alternating current (AC) bipolar electropolymerization was exploited to induce CP fiber formation from the ends of BPEs that germinate parallel to the direction of the electric field (Figure 8.12c). Janus-type modification of BPEs was achieved by simultaneous bipolar electropolymerization of monomer and electroreduction of metal ions. For example, Watanabe et al. [74] used gold wire as electrode to carry out AC bipolar electropolymerization, in which PEDOT fibers were grown on the both ends of the electrode (Figure 8.12d,e). Furthermore, PEDOT films and fibers were also obtained bipolar electropolymerization with different applied frequencies. Moreover, Loget et al. [75] generated Janus-like carbon tubes modified at its ends with metal and CPs by BPE position, which can allow the controlled and localized functionalization of tubes. In the work, the authors not only deposit a material selectively on one side of the tubes but also trigger the deposition at both ends of the tubes that are modified on both sides with either the same material or two different materials (Figure 8.5f). The dissymmetrical dumbbell-like tubes were obtained in one experiment (Figure 8.12g), where pyrrole monomer was electrooxidized at one end of carbon tube to form PPy, and Cu^{2+} was reduced to Cu at the other end. Zhou et al. [76] have successfully obtained PEDOT fiber arrays on the surface of BPE by AC bipolar electrolysis without the use of any templates (Figure 8.12h–j). The mechanism for the PEDOT fiber array formation is showed in Figure 8.5h. First, the voltage between electrodes generates sufficient effect on the BPE surface (Figure 8.12h, stage 1). Then, PEDOT particles with several micrometer-sized diameter were formed on a surface of BPE by polymerization of EDOT, which is essential for inducting fiber growth (Figure 8.12h, stage 2). Once particles are formed, the deposited particles are considered as nodes to initiate the formation of PEDOT fiber (Figure 8.12h, stage 3).

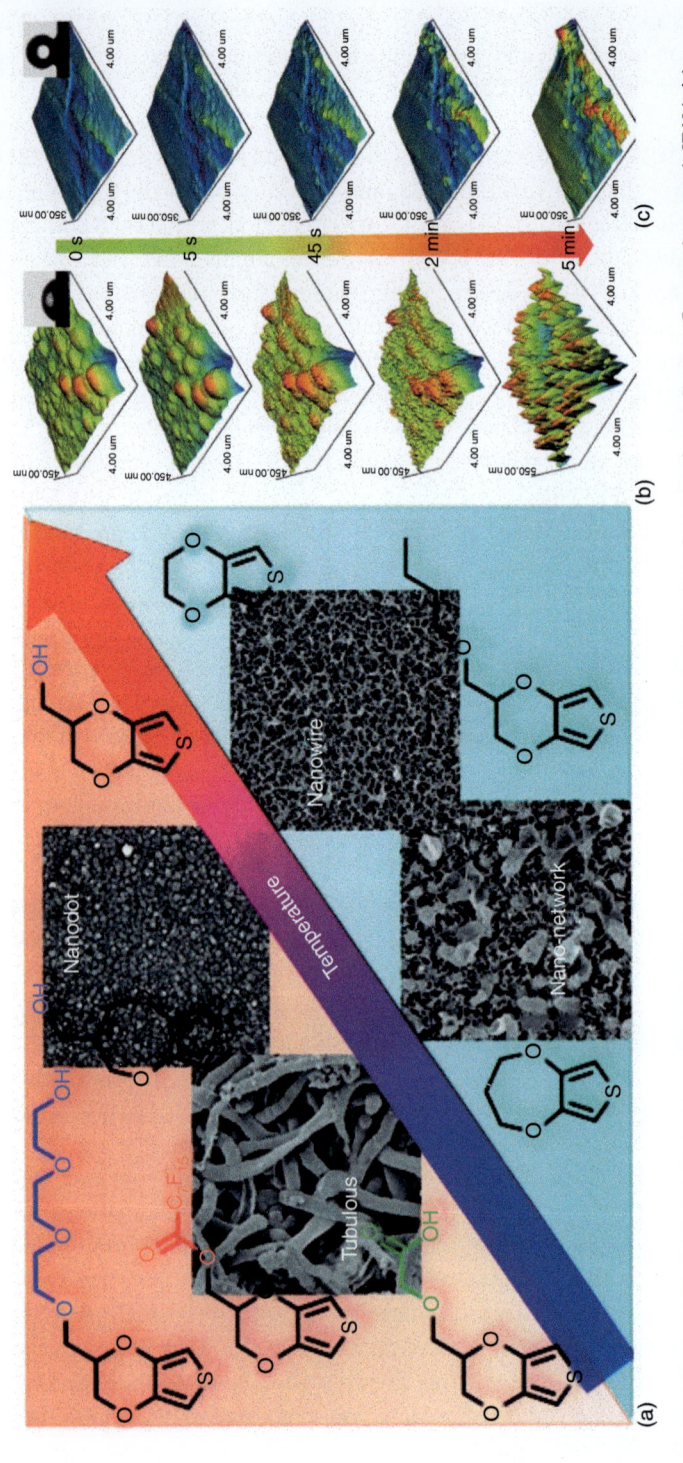

Figure 8.11 (a) The effects of temperature and functional groups on the nanomorphologies of conducting polymers. Source: Luo et al. [71]/with permission from American Chemical Society. (b,c) Height images of self-assembly fabrication process (0 seconds, 5 seconds, 45 seconds, 2 minutes, 5 minutes) of nanostructured conducting PPy on the modified Ti substrate (b) and Ti substrate (c) were recorded by *in situ* EC-AFM. Insert: spread of pyrrole droplets on the modified Ti and Ti substrates. Source: Liao et al. [72]. Reproduced with permission of American Chemical Society.

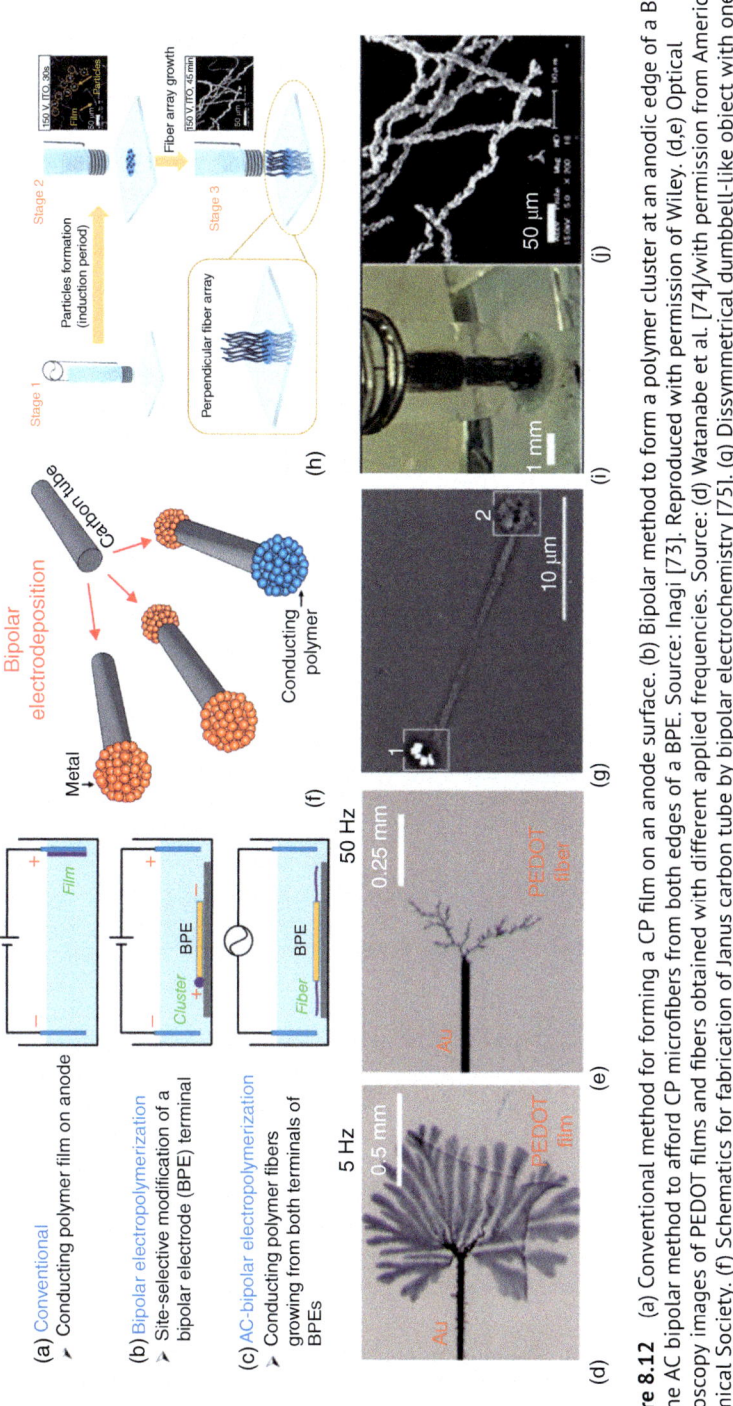

Figure 8.12 (a) Conventional method for forming a CP film on an anode surface. (b) Bipolar method to form a polymer cluster at an anodic edge of a BPE. (c) The AC bipolar method to afford CP microfibers from both edges of a BPE. Source: Inagi [73]. Reproduced with permission of Wiley. (d,e) Optical microscopy images of PEDOT films and fibers obtained with different applied frequencies. Source: (d) Watanabe et al. [74]/with permission from American Chemical Society. (f) Schematics for fabrication of Janus carbon tube by bipolar electrochemistry [75]. (g) Dissymmetrical dumbbell-like object with one Cu end (square 1) and one PPy end (square 2). Source: Loget et al. [75]/with permission from American Chemical Society. (h) Mechanism for the growth of a PEDOT fiber array. (g–j) PEDOT fiber arrays and SEM images of fibers. Source: (h, i) Zhou et al. [76]/with permission from Royal Society of Chemistry.

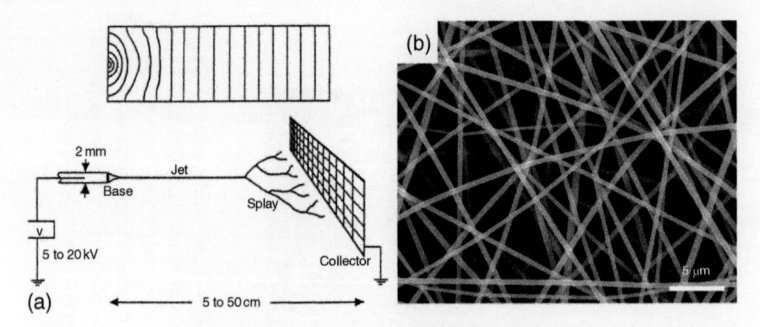

Figure 8.13 (a) Diagram of an electrospinning apparatus. Source: Reneker and Chun [77]. Reproduced with permission of IOP Publishing. (b) Electrospun polyester fibers. Source: Wei et al. [78]/with permission from Royal Society of Chemistry.

Electrospinning is an effective and versatile method for forming 1D micro- and nano-scaled fibers or tubes [77]. During electrospinning, a high voltage is employed to a solution of polymer to drive it toward a collector. As the solution travels through the air, the solvent evaporates, leaving behind an electrically charged thin fiber that can be collected by collector (Figure 8.13a). Figure 8.13b showed the polyester nanofibers made by electrospinning.

For example, Yu et al. [79] prepared PANI nanofibers doped with HCl or H_2SO_4 by electrospinning (Figure 8.14a,b). The diameter of acid-doped PANI fiber decreased;

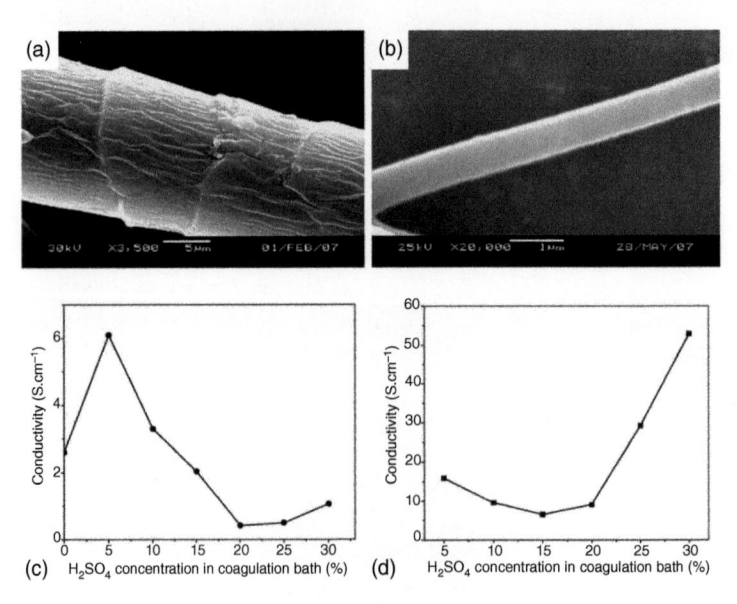

Figure 8.14 (a,b) SEM images of HCl-doped PANI fibers obtained from different diameters: (a) $d = 1.6\,\mu m$ and (b) $d = 1.4\,\mu m$, at 15% H_2SO_4 concentration in the coagulation bath. (c,d) Relationship between H_2SO_4 concentration in coagulation bath and the conductivity of H_2SO_4-doped PANI sub-micron: (c) single fiber with a diameter of 100 μm and (d) 100 μm fiber bundle collected by fibers with a diameter of 370 nm. Source: Yu et al. [79], Figure 04 [p. 073]/with permission from Elsevier.

then its electrical conductivity increased. When the ultrafine conductive fiber is formed into a fiber bundle, the electrical conductivity of the fiber bundle is much higher than that of a single fiber with the same diameter as the fiber bundle (Figure 8.14c,d). It indicates that reducing the diameter of the fiber can effectively improve the electrical conductivity of the nanofiber, because the fine fiber has a more compact, uniform, and smooth morphology. Also, the smaller the diameter, the more ordered the molecular chain on the fiber, and the higher the crystallinity. It is found that the conductivity is mainly affected by the doping level and the morphology of PANI sub-micron fibers. When the H_2SO_4 concentration in the coagulation bath is 30%, the conductivity of the H_2SO_4-doped PANI fiber bundle with a diameter of 100 μm increases from 1.06 to 52.9 S cm^{-1}, compared with single fiber with a diameter of 100 μm.

The development of polymer composites with high thermal conductivities and ideal chemical stabilities have become one of the hot topics in functional composites [80]. Yang et al. [80] prepared boron nitride nanosheets/polyvinyl alcohol (BNNS/PVA) thermally conductive composite films by electrospinning of mixture of BNNS and PVA, in which PVA could effectively realize the directional arrangement and BNNS increase the rigidity of the fiber (Figure 8.15). The thermal conductivity (λ) and thermal diffusivity (α) values of the BNNS/PVA-(I–III) composite films were enhanced with the increasing content of BNNS fillers. The λ and α in vertical direction ($\lambda_\perp$ and $\alpha_\perp$) of the BNNS/PVA-III films with 30 wt% BNNS were 1.031 W mK^{-1} and 0.888 mm^2 s^{-1}, respectively, higher than that of BNNS/PVA-I (0.573 W mK^{-1} and 0.494 mm^2 s^{-1}). Moreover, the corresponding λ and α in parallel direction ($\lambda_\parallel$ and $\alpha_\parallel$) were significantly enhanced to 18.630 W mK^{-1} and 9.706 mm^2 s^{-1}, higher than that of BNNS/PVA-(I–II) composite films. Through directional freeze-drying, the BNNS/PVA presented an ordered layered channel, which improved the possibility of an ordered heat conduction path (Figure 8.8), thereby improving the conductivity and thermal stability of the fiber. It is well known that ionic liquids (ILs) have good thermal stability. Trchová et al. [81] found that the combination of ILs (1-ethyl-3-methylimidazolium trifluoromethanesulfonate) and PANI can improve the conductivity and thermal stability of CPs, which showed the green color corresponding to conducting form of PANI. The results can be explained as follows, the "acidic" hydrogen atoms on the imidazole heterocyclic ring form proton-like bonds with the imine nitrogen atoms on the PANI chain. Also, the hydrogen atoms on the amino group of PANI form proton-like bonds with the oxygen atoms in the counterions of IL.

Kakunuri et al. [82] used a single-step nonconductive template assisted electrospinning process to obtain cellulose acetate nanofibers (Figure 8.16a). In this strategy, the nonconductive template placed in a high electric field is charged on the surface due to polarization, and the CP jets are attracted to the charged template and form a patterned surface according to the different charge distribution (Figure 8.16b–e), thereby changing the wettability of the CP fiber fabric. The CAs of micropatterned surfaces produced with the nylon meshes of opening sizes of 50, 100, and 200 μm were 137.7°, 123.7°, and 89.6°, respectively (Figure 8.16b′–d′), exhibiting hydrophobic behavior. However, CA on a non-patterned electrospun

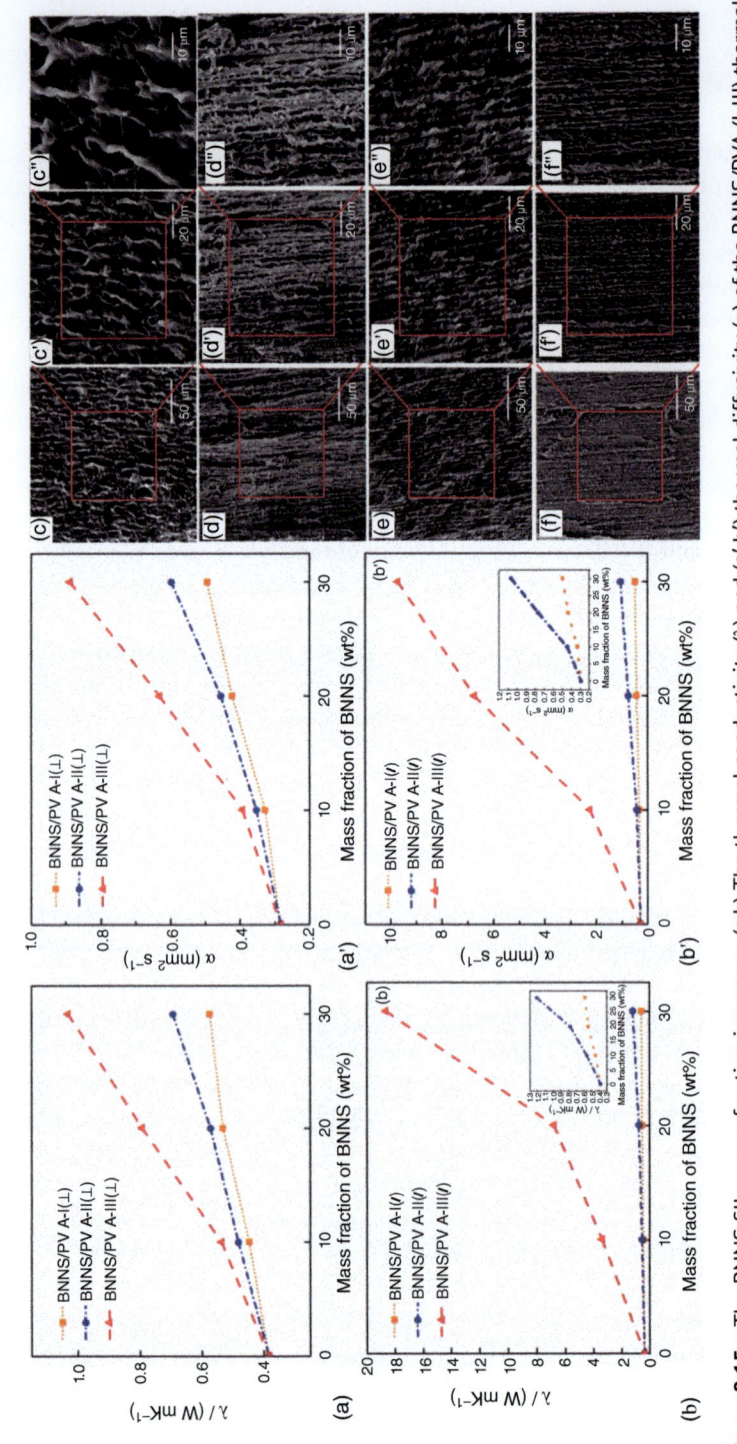

Figure 8.15 The BNNS filler mass fraction increases. (a,b) The thermal conductivity (λ) and (a',b') thermal diffusivity (α) of the BNNS/PVA (I–III) thermally conductive composite film in the vertical and parallel directions are improved [80]. SEM images of the PVA aerogel (c,c″) and BNNS/PVA aerogels with 10 wt% (d,d″), 20 wt% (e,e″), and 30 wt% (f–f″) BNNS fillers. Source: (c) Yang et al. [80]/with permission from Elsevier.

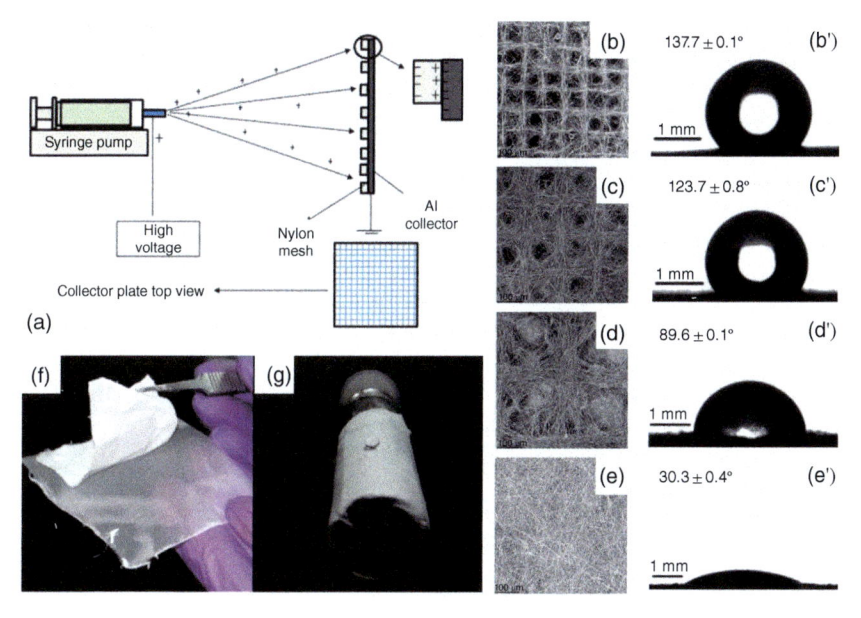

Figure 8.16 (a) Schematic of the electrospinning setup used for the production of the three-dimensional micropatterned nanofabric surfaces. (b–e) SEM images of Nylon meshes with opening sizes of 50, 100, and 200 μm and fiber mats. (b'–e') Images of water droplets on samples (a–d). (f) The peeling of a fiber mat from the mesh. (g) Water droplet sitting over a patterned fiber mat wrapped over a cylindrical glass vial. Source: Kakunuri et al. [82]. Reproduced with permission of Wiley.

nanofibrous mat was about 30°, showing a hydrophilic property. The hydrophobic film can be papered on the large scale, which may provide a promising application (Figure 8.16f,g).

Because of the low solubility of CPs, one of the most effective strategies for improving the spinnability of polymers is to mix with other spinnable polymers. Xu et al. [83] produced PVA/PEDOT ultrafine fibers with core–shell structure by electrospinning EDOT/PVA emulsion, followed by chemical polymerization. In the work, PEDOT concentration is as high as 45% (w/w, PVA), so this method provides a simple but effective route to obtain conductive composites with high PEDOT concentration. Interestingly, electrospinning EDOT/PVA emulsion can obtain bead-like PVA nanofibers with EDOT encapsulated within beads and fibers (Figure 8.17a). The uncovered PEDOT and the continuous PEDOT at the core of some fibers make a high and quick response to H_2O_2. However, this method requires a suitable solvent to dissolve the two mixed components, and this solvent is also suitable for electrospinning. In fact, there are few solvents and spinnable polymers that meet these two conditions at the same time. Therefore, two-fluid coaxial electrospinning (Figure 8.17b) has been developed as a promising strategy [84]. A spinneret composed of two coaxial capillaries is used to directly form a mixed jet in an electric field, which can prepare core–shell fibers, and subsequent measures can also be taken to remove the core to obtain uniform hollow nanofibers. Li et al. [84] blended poly[2-methoxy-5-(2-ethylhexyloxy)-1,4-phenylenevinylene] (MEH-PPV) dissolved

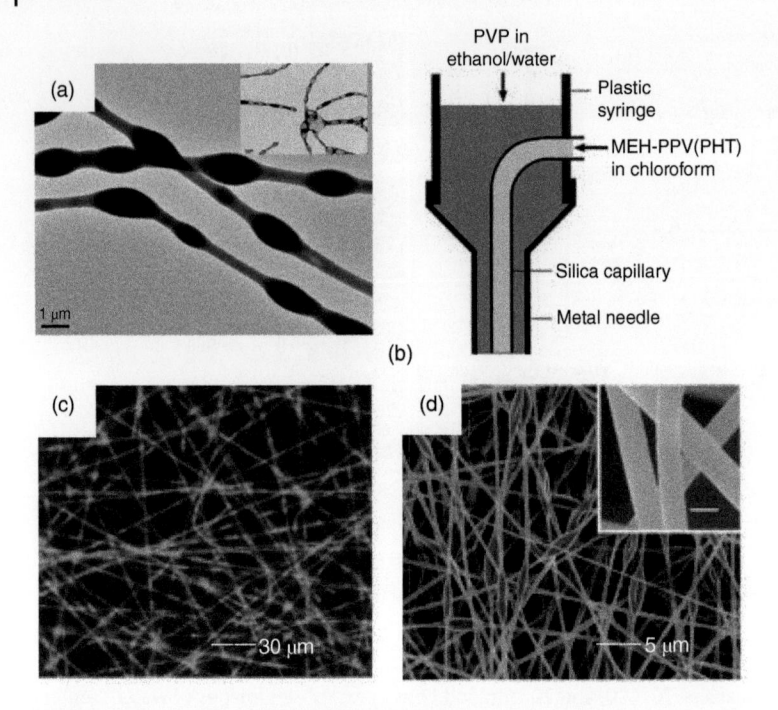

Figure 8.17 (a) TEM images of EDOT/PVA fibers with 15% EDOT. Source: Xu et al. [83]/with permission from Elsevier. (b) A schematic drawing of the spinneret constructed from two coaxial capillaries [84]. (c) Fluorescence microscopy image of as-spun MEH-PPV/PVP fibers. Source: Li et al. [84]/with permission from Wiley. (d) SEM image of as-spun MEH-PPV/PVP fibers. Source: Li et al. [84]. Reproduced with permission of Wiley.

in chloroform and PVP dissolved in ethanol/water, respectively, overcoming the problem of PPV insoluble in chloroform and obtaining nanofibers (Figure 8.17c,d).

The combination of electrospinning and impregnation method solved the lack of suitable solvents. Yuan et al. [85] electrospun multiwalled carbon nanotube/lauric acid/thermoplastic polyurethane (MWCNT/LA/PU) solution for the first time and then coated PEDOT:PSS by impregnation (Figure 8.18). PU as the substrate provides high elasticity for the composite fiber, LA has a phase transition temperature similar to that of the human body and can be used as a phase change material to store and convert thermal energy, and the outer coated PEDOT:PSS shows high thermal and electrical conductivity. For a comparison, the nanocomposite fibers were prepared $CLPF_0$, $CLPF_{0.5}$, $CLPF_1$, $CLPF_{1.5}$, and $CLPF_2$, corresponding to the weight ratios of LA (0, 0.5, 1, 1.5, and 2 times the amount of PU) and MWCNT (1 wt% of PU). The electric conductivity was 39.7, 29.3, 27.4, 18.6, and 13.3 S cm^{-1} for CSF_0, $CSF_{0.5}$, CSF_1, $CSF_{1.5}$, and CSF_2, respectively, indicating a decrease with increase of LA contents. Also, the melting enthalpies were 54.82, 73.95, 101.29, and 124.30 J g^{-1} for $CSF_{0.5}$, CSF_1, $CSF_{1.5}$, and CSF_2, respectively. As a result, the CSFs exhibited a rapid and unique response to environmental stimuli, such as electrical and photonic, with simultaneous variation of temperature. With the increase of LA content, the

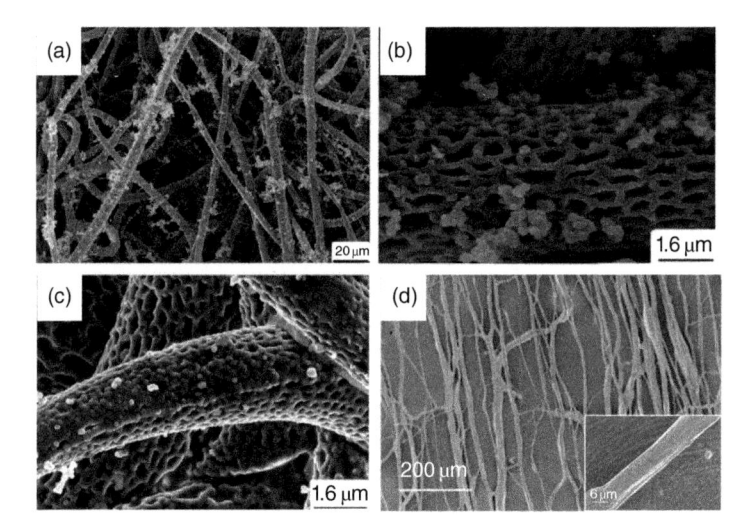

Figure 8.18 (a) Schematic description of the preparation processes of PEDOT:PSS-coated MWCNT/LA/PU nanocomposite smart fabrics. (b,c) SEM images of PEDOT:PSS-coated smart fabrics. Source: (a) Niu and Yuan [85]. Reproduced with permission of American Chemical Society, (b) Niu and Yuan [85], Figure 02 [p. 4510]/with permission from American Chemical Society.

highest temperature CSFs could decrease, for example, the highest temperature of 1% CSF_0 reached up to over 90 °C, as the input voltage was increased to 30 V.

Moreover, Jiao et al. [86] fabricated a highly interconnected 3D PP/GPN film (PP/GPNF) network by electrospinning of the glycerol-grafted polyacrylonitrile nanofibers (GPN), followed by PEDOT:PSS (PP) dip coating. The PP/GPNF film without any further treatment only showed a conductivity of around 170 S m^{-1}. The authors found that glycerol can be used to obtain hydrophilic glycerol-grafted PAN nanofibers (GPN) by electrospinning of mixture of glycerol and PAN solution. The hydrophilicity of GPN is attributed to glycerol molecules that can be easily linked to the nitrile group in PAN through the strong interaction. Compared with PP/GPNF, conductivity of PP/GPNF-EG increased to about 913 S m^{-1}. The average EMI SE values of PP/GPNF-EG films were measured to be about ~40 dB with a thickness of 260 nm.

In addition, Lan et al. [87] fabricated porous poly(L-lactic acid) (PLLA)/PPy composite micro/nanofiber films by combining electrospinning with *in situ* polymerization (Figure 8.19a–c). The main steps are as follows: PLLA fiber membrane is immersed in a H_2SO_4 solution containing pyrrole, and then ultrasonically treated, so that the pyrrole on the surface of the PLLA fiber membrane is saturated, and PPy is formed under the action of APS. The composite micro/nanofiber films had a dual multi-pore structure composed of pores both in the fibers and among the fibers. It was found that the conductivity of the optimized PLLA/PPy micro/nanofiber films was about 179.0 S cm^{-1} with the interstice size of about 250 μm, when the polymerization of pyrrole was up to 180 minutes. The mixing of CPs and other nonconductive polymers can improve the mechanical properties, but the electrical

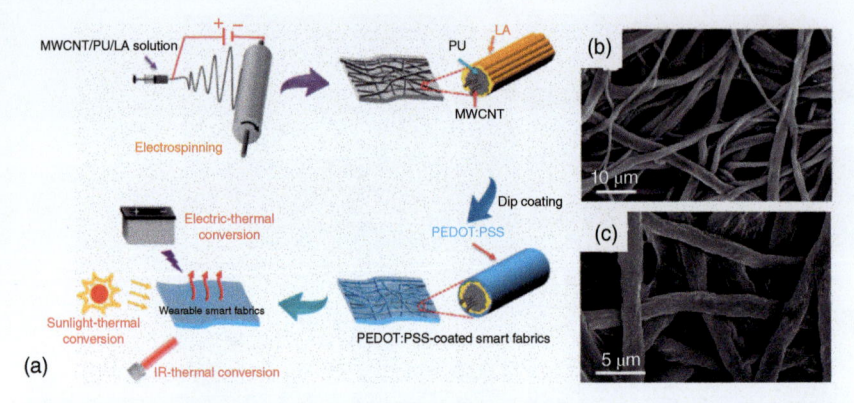

Figure 8.19 SEM images of PLLA/PPy composites when pyrrole *in situ* polymerization on the PLLA micro/nanofiber film for 300 minutes (a) at 20 °C, low magnification; (b) at 20 °C, high magnification; and (c) at 0 °C, high magnification. Pyrrole and the reaction solutions were dispersed by ultrasonic before polymerization or at initial reaction stage. Source: (a, c) Yu et al. [87]/with permission from Elsevier. SEM imaging of the as-spun fiber. Source: Bhattacharya et al. [88]/with permission from American Chemical Society.

conductivity will also be reduced to a certain extent, which can be adjusted by secondary doping. Bhattacharya et al. [88] applied the concept of secondary doping to electrospun fibers for the first time, using a rotating drum electrospinning design. PANI-poly(ethylene oxide) (PEO) mixed with PANI was electrospun to obtain composite nanofibers (Figure 8.19d), in which camphorsulfonic acid acts as the primary dopant and *m*-cresol as the secondary dopant, providing an one-step method to process highly aligned and highly branched electrospun fibers with high electrical conductivities. The conductivity of electrospun PANI-PEO fibers was measured to be $1.73\,\text{S}\,\text{cm}^{-1}$. These conductive fibers as electrodes for supercapacitors showed a specific capacitance up to $3121\,\text{F}\,\text{g}^{-1}$ at $0.1\,\text{A}\,\text{g}^{-1}$.

Recently, Zhao et al. [89] proposed for the first time the use of advanced 3D printing technology to prepare PEDOT:PSS nanomaterials. In general, CPs are processed in the form of monomers or polymer solutions, which cannot be directly used for printing because of their strong fluidity. Therefore, the key to this technology is to design a reasonable CP ink (Figure 8.20). In this work, the authors quickly frozen the PEDOT:PSS aqueous solution in nitrogen bath and dried it in low temperature and vacuum to obtain nanofiber foam structure. Then, they re-dispersed it with water/DMSO binary solvent by mechanical grinding and finally obtained a certain concentration of PEDOT:PSS ink. The appropriate viscosity and rheological properties can be obtained by adjusting the concentration, which can meet the needs of direct 3D printing. The electrical conductivity of the 3D-printed PEDOT:PSS fibers was about $155\,\text{S}\,\text{cm}^{-1}$ in the dry state and $28\,\text{S}\,\text{cm}^{-1}$ in the hydrogel state, potentially due to the shear-induced enhancements in the PEDOT:PSS nanofibril alignment.

In general, electrochemical method is a convenient and flexible way for the preparation and modification of CPs nanomaterials. Electrochemical preparation of CP nanomaterials is one of the important methods owing to its excellent conductivity, chemical stability, thermal stability, flexibility, and low cost and has great application value in industrial production.

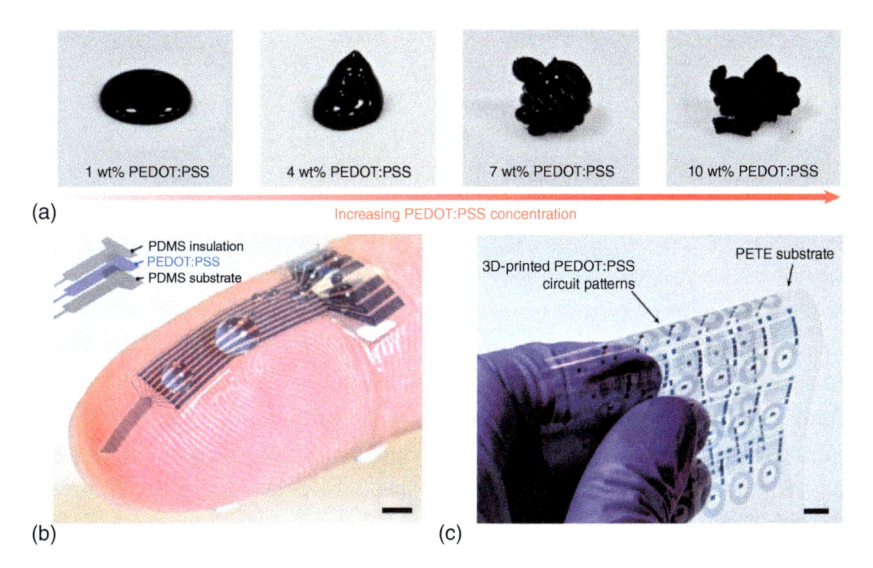

Figure 8.20 (a) Different concentrations of PEDOT:PSS printing ink. (b) Image of the 3D printed soft neural probe with 9 channels by the conducting polymer ink and the PDMS ink. (c) Bending of the 3D printed conducting polymer circuit without failure. Source: Yuk et al. [89]/Springer Nature/CC BY 4.0.

8.3 Summary

In summary, chemical and electrochemical polymerizations are the effective technologies for synthesizing CP nanomaterials, by which CP nanomaterials can be synthesized under moderate conditions, that is, at room temperature and ambient pressure. The properties of the resulting CP nanomaterials can be easily controlled by adjusting the reaction parameters and the amount and chemicals used for synthesis. In addition, electrospinning is a simple and versatile method for the preparation of CP nanofibers, which have potential applications in sensors, actuators, batteries, supercapacitors, electrochromic devices, etc.

References

1 Rasmussen, S.C. (2020). Conjugated and conducting organic polymers: the first 150 years. *ChemPlusChem* 85: 1412–1429. https://doi.org/10.1002/cplu.202000325.

2 Wolfart, F., Hryniewicz, B.M., Góes, M.S. et al. (2017). Conducting polymers revisited: applications in energy, electrochromism and molecular recognition. *J. Solid State Electrochem.* 21: 2489–2515. https://doi.org/10.1007/s10008-017-3556-9.

3 Shirakawa, H., Louis, E.J., Macdiarmid, A.G. et al. (1977). Synthesis of electrically conducting organic polymers: halogen derivatives of polyacetylene, $(CH)_x$. *J. Chem. Soc. Chem. Commun.* 16: 578–580. https://doi.org/10.1039/C39770000578.

4 Inzelt, G. (2011). Rise and rise of conducting polymers. *J. Solid State Electrochem.* 15: 1711–1718. https://doi.org/10.1007/s10008-011-1338-3.

5 Anonymous. 2000. *The Nobel Prize in Chemistry*. https://www.nobelprize.org/prizes/chemistry/2000/summary/ (accessed 9 December 2020).

6 Pnalwa, M.C. (2004). *Encyclopedia of Nanoscience and Nanotechnology*, vol. 2, 153–169. Los Angeles: American Scientific Publishers.

7 Bokbinder, D.C. and Wrighton, M.S. (1986). *Handbook of Conducting Polymers*, 543–659. New York: Marcel Dekker.

8 Su, W.P., Schrieffer, J.R., and Heeger, A.J. (1979). Solitons in polyacetylene. *Phys. Rev. Lett.* 42: 1698–1701. https://doi.org/10.1007/BFb0106888.

9 Brazovskii, S.A. and Kirova, N.N. (1980). Excitons, polarons, and bipolarons in conducting polymers. *JETP Lett.* 33: 4–8. https://doi.org/10.1007/BF00670555.

10 Furukawa, Y. (1996). Electronic absorption and vibrational spectroscopies of conjugated conducting polymers. *J. Phys. Chem.* 100: 15644–15653. https://doi.org/10.1021/jp960608n.

11 Sardar, S., Kar, P., Remita, H. et al. (2015). Enhanced charge separation and FRET at heterojunctions between semiconductor nanoparticles and conducting polymer nanofibers for efficient solar light harvesting. *Sci. Rep.* 5: 17313. https://doi.org/10.1038/srep17313.

12 Ahmad, S., Yum, J.H., Zhang, X. et al. (2010). Dye-sensitized solar cells based on poly(3,4-ethylenedioxythiophene) counter electrode derived from ionic liquids. *J. Mater. Chem.* 20: 1654–1658. https://doi.org/10.1039/b920210b.

13 Yang, X., Lin, Z., Zheng, J. et al. (2016). Facile template-free synthesis of vertically aligned polypyrrole nanosheets on nickel foams for flexible all-solid-state asymmetric supercapacitors. *Nanoscale* 8: 8650–8657. https://doi.org/10.1039/c6nr00468g.

14 Abidian, M.R., Kim, D.H., and Martin, D.C. (2006). Conducting-polymer nanotubes for controlled drug release. *Adv. Mater.* 18: 405–409. https://doi.org/10.1002/adma.200501726.

15 Li, L., Wang, Y., Pan, L. et al. (2015). Nanostructured conductive hydrogels-based biosensor platform for human metabolite detection. *Nano Lett.* 15: 1146–1151. https://doi.org/10.1021/nl504217p.

16 Hosseini, E., Arjmand, M., Sundararaj, U., and Karan, K. (2020). Filler-free conducting polymers as a new class of transparent electromagnetic interference shields. *ACS Appl. Mater. Interfaces* 12: 28596–28606. https://doi.org/10.1021/acsami.0c03544.

17 Guo, X. and Facchetti, A. (2020). The journey of conducting polymers from discovery to application. *Nat. Mater.* 19: 922–928. https://doi.org/10.1038/s41563-020-0778-5.

18 Strietzel, C., Oka, K., Strømme, M. et al. (2021). An alternative to carbon additives: the fabrication of conductive layers enabled by soluble conducting polymer precursors – a case study for organic batteries. *ACS Appl. Mater. Interfaces* 13: 5349–5356. https://doi.org/10.1021/acsami.0c22578.

19 Xue, Y., Chen, S., Yu, J. et al. (2020). Nanostructured conducting polymers and their composites: synthesis methodologies, morphologies and applications. *J. Mater. Chem. C* 8: 10136–10159. https://doi.org/10.1039/d0tc02152k.

20 Srabanti, G., Thandavarayan, M., and Rajendra, N.B. (2016). Nanostructured conducting polymers for energy applications: towards a sustainable platform. *Nanoscale* 8: 6921–6947. https://doi.org/10.1039/c5nr08803h.

21 Uppalapati, D., Boyd, B.J., Garg, S. et al. (2016). Conducting polymers with defined micro- or nanostructures for drug delivery. *Biomaterials* 111: 149–162. https://doi.org/10.1016/j.biomaterials.2016.09.021.

22 Lee, W. and Park, S.J. (2014). Porous anodic aluminum oxide: anodization and templated synthesis of functional nanostructures. *Chem. Rev.* 114: 7487–7556. https://doi.org/10.1021/cr500002z.

23 Jackowska, K., Bieguński, A.T., and Tagowska, M. (2008). Hard template synthesis of conducting polymers: a route to achieve nanostructures. *J. Solid State Electrochem.* 12: 437–443. https://doi.org/10.1007/s10008-007-0453-7.

24 Jang, J., Bae, J., and Park, E. (2006). Selective fabrication of poly(3,4-ethylenedioxythiophene) nanocapsules and mesocellular foams using surfactant-mediated interfacial polymerization. *Adv. Mater.* 18: 354–358. https://doi.org/10.1002/adma.200502060.

25 Ahn, K., Lee, Y., Choi, H. et al. (2015). Surfactant-templated synthesis of polypyrrole nanocages as redox mediators for efficient energy storage. *Sci. Rep.* 6: 14097. https://doi.org/10.1038/srep22501.

26 Zhu, Y., Hu, D., Wan, M. et al. (2007). Conducting and superhydrophobic rambutan-like hollow spheres of polyaniline. *Adv. Mater.* 19: 2092–2096. https://doi.org/10.1002/adma.200602135.

27 Huang, Z., Zhang, Y., Kotaki, M., and Ramakrishna, S. (2003). A review on polymer nanofibers by electrospinning and their applications in nanocomposites. *Compos. Sci. Technol.* 63: 2223–2253. https://doi.org/10.1016/S0266-3538(03)00178-7.

28 Ghosh, S., Datta, A., and Saha, A. (2009). Single step synthesis of highly stable good quality water soluble semiconductor/dendrimer nanocomposites through irradiation route. *Colloids Surf. A* 355: 130–138. https://doi.org/10.1016/j.colsurfa.2009.12.007.

29 Shirakawa, H., Louis, E.J., MacDiarmid, A.G. et al. (1977). Synthesis of electrically conducting organic polymers: halogen derivatives of polyacetylene, (CH). *J. Chem. Soc. Chem. Commun.* 578–580. https://doi.org/10.1039/C39770000578.

30 Branzoi, V., Branzoi, F., and Pilan, L. (2010). Electrochemical fabrication and capacitance of composite films of carbon nanotubes and polyaniline. *Surf. Interface Anal.* 42: 1266–1270. https://doi.org/10.1039/B418835G.

31 Lu, X., Zhang, W., Wang, C. et al. (2011). One-dimensional conducting polymer nanocomposites: synthesis, properties and applications. *Prog. Polym. Sci.* 36: 671–712. https://doi.org/10.1016/j.progpolymsci.2010.07.010.

32 Abdelhamid, M.E., O'Mullane, A.P., and Snook, G.A. (2015). Storing energy in plastics: a review on conducting polymers and their role in electrochemical energy storage. *RSC Adv.* 5: 11611–11626. https://doi.org/10.1039/C4RA15947K.

33 Gan, D., Shuai, T., Wang, X. et al. (2020). Mussel-inspired redox-active and hydrophilic conductive polymer nanoparticles for adhesive hydrogel bioelectronics. *Nano-Micro Lett.* 12: https://doi.org/10.1007/s40820-020-00507-0.

34 Zhang, M., Zhang, A.M., Chen, Y. et al. (2020). Polyoxovanadate-polymer hybrid electrolyte in solid state batteries. *Energy Storage Mater.* 29: 172–181. https://doi.org/10.1016/j.ensm.2020.04.017.

35 Zhang, T., Qi, H., Liao, Z. et al. (2019). Engineering crystalline quasi-two-dimensional polyaniline thin film with enhanced electrical and chemiresistive sensing performances. *Nat. Commun.* 10: 4225. https://www.ncbi.nlm.nih.gov/pubmed/31548543.

36 Chen, G., Rastak, R., Wang, Y. et al. (2019). Strain- and strain-rate-invariant conductance in a stretchable and compressible 3D conducting polymer foam. *Matter* 1: 205–218. https://doi.org/10.1016/j.matt.2019.03.011.

37 De Vito, S. and Martin, C.R. (1998). Toward colloidal dispersions of template-synthesized polypyrrole nanotubules. *Chem. Mater.* 10: 1738–1741. https://doi.org/10.1021/cm9801690.

38 Park, E., Kwon, O.S., Park, S.J. et al. (2012). One-pot synthesis of silver nanoparticles decorated poly(3,4-ethylenedioxythiophene) nanotubes for chemical sensor application. *J. Mater. Chem.* 22: 1521–1526. https://doi.org/10.1039/C1JM13237G.

39 Zhang, Z., Sui, J., Zhang, L. et al. (2005). Synthesis of polyaniline with a hollow, octahedral morphology by using a cuprous oxide template. *Adv. Mater.* 17: 2854–2857. https://doi.org/10.1002/adma.200501114.

40 Zhu, Y., Li, J., Wan, M. et al. (2007). A new route for the preparation of brain-like nanostructured polyaniline. *Macromol. Rapid Commun.* 28: 1339–1344. https://doi.org/10.1002/marc.200700073.

41 Choi, I.Y., Lee, J., Ahn, H. et al. (2015). High-conductivity two-dimensional polyaniline nanosheets developed on ice surfaces. *Angew. Chem. Int. Ed.* 54: 10497–10501. https://doi.org/10.1002/anie.201503332.

42 Yin, Z. and Zheng, Q. (2012). Controlled synthesis and energy applications of one-dimensional conducting polymer nanostructures: an overview. *Adv. Energy Mater.* 2: 179–218. https://doi.org/10.1002/aenm.201100560.

43 Zhou, C., Han, J., and Guo, R. (2009). Polyaniline fan-like architectures of rectangular sub-microtubes synthesized in dilute inorganic acid solution. *Macromol. Rapid Commun.* 30: 182–187. https://www.ncbi.nlm.nih.gov/pubmed/21706596.

44 Zhang, X., Zhang, J., Song, W., and Liu, Z. (2006). Controllable synthesis of conducting polypyrrole nanostructures. *J. Phys. Chem. B* 110: 1158–1165. https://doi.org/10.1021/jp054335k.

45 Yuan, R., Wang, H., Ji, T. et al. (2015). Superhydrophobic polyaniline hollow spheres with mesoporous brain-like convex-fold shell textures. *J. Mater. Chem. A* 3: 19299–19303. https://doi.org/10.1039/C5TA05614D.

46 Jang, J. and Yoon, H. (2003). Facile fabrication of polypyrrole nanotubes using reverse microemulsion polymerization. *Chem. Commun.* 720–721. https://www.ncbi.nlm.nih.gov/pubmed/12703790.

47 Liu, Z., Zhang, X., Poyraz, S. et al. (2010). Oxidative template for conducting polymer nanoclips. *J. Am. Chem. Soc.* 132: 13158–13159. https://doi.org/10.1021/ja105966c.

48 Park, H.-W., Kim, T., Huh, J. et al. (2012). Anisotropic growth control of polyaniline nanostructures and their morphology-dependent electrochemical characteristics. *ACS Nano* 6: 7624–7633. https://doi.org/10.1021/nn3033425.

49 Zhu, Y., Ren, G., Wan, M., and Jiang, L. (2009). 3D hollow microspheres assembled from 1D polyaniline nanowires through a cooperation reaction. *Macromol. Chem. Phys.* 210: 2046–2051. https://doi.org/10.1002/macp.200900317.

50 Zhu, Y., He, H., Wan, M., and Jiang, L. (2008). Rose-like microstructures of polyaniline by using a simplified template-free method under a high relative humidity. *Macromol. Rapid Commun.* 29: 1705–1710. https://doi.org/10.1002/marc.200800294.

51 Zhu, Y., Li, J., Wan, M., and Jiang, L. (2008). 3D-boxlike polyaniline microstructures with super-hydrophobic and high-crystalline properties. *Polymer* 49: 3419–3423. https://www.sciencedirect.com/science/article/pii/S0032386108005351.

52 Zhu, Y., Li, J., Wan, M., and Jiang, L. (2009). Electromagnetic functional urchin-like hollow carbon spheres carbonized by polyaniline micro/nanostructures containing FeCl3as a precursor. *Eur. J. Inorg. Chem.* 2009: 2860–2864.

53 Santino, L.M., Acharya, S., and D'Arcy, J.M. (2017). Low-temperature vapour phase polymerized polypyrrole nanobrushes for supercapacitors. *J. Mater. Chem. A* 5: 11772–11780. https://doi.org/10.1039/C7TA00369B.

54 Zhu, Y., Hu, D., Wan, M.X. et al. (2007). Conducting and superhydrophobic rambutan-like hollow spheres of polyaniline. *Adv. Mater.* 19: 2092–2096. https://doi.org/10.1002/adma.200602135.

55 Zhu, Y., Li, J., Wan, M., and Jiang, L. (2009). Electromagnetic functional urchin-like hollow carbon spheres carbonized by polyaniline micro/nanostructures containing FeCl3 as a precursor. *Eur. J. Inorg. Chem.* 2009: 2860–2864. https://doi.org/10.1002/ejic.200900040.

56 Huang, J. and Kaner, R.B. (2004). A general chemical route to polyaniline nanofibers. *J. Am. Chem. Soc.* 126: 851–855. https://doi.org/10.1021/ja0371754.

57 Su, K., Nuraje, N., Zhang, L. et al. (2007). Fast conductance switching in single-crystal organic nanoneedles prepared from an interfacial polymerization-crystallization of 3,4-ethylenedioxythiophene. *Adv. Mater.* 19: 669–672. https://doi.org/10.1002/adma.200602277.

58 Zhang, X., Goux, W.J., and Manohar, S.K. (2004). Synthesis of polyaniline nanofibers by "nanofiber seeding". *J. Am. Chem. Soc.* 126: 4502–4503. https://doi.org/10.1021/ja031867a.

59 Kerileng, P.M.N., Molapo, M., Ajayi, R.F. et al. (2012). Electronics of conjugated polymers (I): polyaniline. *Int. J. Electrochem. Sci.* 7: 11859–11875. http://www.electrochemsci.org/papers/vol7/71211859.pdf.

60 Lv, H., Pan, Q., Song, Y. et al. (2020). A review on nano-/microstructured materials constructed by electrochemical technologies for supercapacitors. *Nano-Micro Lett.* 12: 118. https://doi.org/10.1007/s40820-020-00451-z.

61 Diaz, A.F., Kanazawa, K.K., and Gardini, G.P. (1979). Electrochemical polymerization of pyrrole. *J. Chem. Soc. Chem. Commun.* 635–636. https://doi.org/10.1039/C39790000635.

62 Liu, Y., Goebl, J., and Yin, Y. (2013). Templated synthesis of nanostructured materials. *Chem. Soc. Rev.* 42: 2610–2653. https://doi.org/10.1039/C2CS35369E.

63 Cui, S., Zheng, Y., Liang, J., and Wang, D. (2016). Conducting polymer PPy nanowire-based triboelectric nanogenerator and its application for self-powered electrochemical cathodic protection. *Chem. Sci.* 7: 6477–6483. https://www.ncbi.nlm.nih.gov/pubmed/28451105.

64 Ma, Y., Liu, N., Xu, Z. et al. (2021). An ultrasensitive biosensor based on three-dimensional nanoporous conducting polymer decorated with gold nanoparticles for microRNA detection. *Microchem. J.* 161: 105780. https://www.sciencedirect.com/science/article/pii/S0026265X20331970.

65 Menaker, A., Syritski, V., Reut, J. et al. (2009). Electrosynthesized surface-imprinted conducting polymer microrods for selective protein recognition. *Adv. Mater.* 21: 2271–2275. https://doi.org/10.1002/adma.200803597.

66 Shao, Y., Jin, Y., and Dong, S. (2002). DNA-templated assembly and electropolymerization of aniline on gold surface. *Electrochem. Commun.* 4: 773–779. https://www.sciencedirect.com/science/article/pii/S1388248102004423.

67 Gonzalez-Vogel, A., Fogde, A., Crestini, C. et al. (2019). Molecularly imprinted conducting polymer for determination of a condensed lignin marker. *Sens. Actuator B-Chem.* 295: 186–193. https://www.sciencedirect.com/science/article/pii/S0925400519306999.

68 Fradin, C., Orange, F., Amigoni, S. et al. (2021). Micellar formation by soft template electropolymerization in organic solvents. *J. Colloid Interface Sci.* 590: 260–267. https://www.sciencedirect.com/science/article/pii/S0021979721000424.

69 Sow, S., Dramé, A., Thiam, E.H.Y. et al. (2020). Nanotubular structures via templateless electropolymerization using thieno[3,4-b]thiophene monomers with various substituents and polar linkers. *Prog. Org. Coat.* 138: 105382. https://www.sciencedirect.com/science/article/pii/S0300944019311725?via%3Dihub.

70 Khodja, M., El Kateb, M., Beji, M. et al. (2020). Tuning nanotubular structures by templateless electropolymerization with thieno[3,4-b]thiophene-based monomers with different substituents and water content. *J. Colloid Interface Sci.* 564: 19–27. https://www.ncbi.nlm.nih.gov/pubmed/31896424.

71 Luo, S.-C., Sekine, J., Zhu, B. et al. (2012). Polydioxythiophene nanodots, nanowires, nano-networks, and tubular structures: the effect of functional groups and temperature in template-free electropolymerization. *ACS Nano* 6: 3018–3026. https://doi.org/10.1021/nn300737e.

72 Liao, J., Wu, S., Yin, Z. et al. (2014). Surface-dependent self-assembly of conducting polypyrrole nanotube arrays in template-free electrochemical polymerization. *ACS Appl. Mater. Interfaces* 6: 10946–10951. https://doi.org/10.1021/am5017478.

73 Inagi, S. (2019). Site-selective anisotropic modification of conductive objects by bipolar electropolymerization. *Polym. J.* 51: 975–981. https://doi.org/10.1038/S41428-019-0223-2.

74 Watanabe, T., Ohira, M., Koizumi, Y. et al. (2018). In-plane growth of poly(3,4-ethylenedioxythiophene) films on a substrate surface by bipolar electropolymerization. *ACS Macro Lett.* 7: 551–555. https://doi.org/10.1021/acsmacrolett.8b00170.

75 Loget, G., Lapeyre, V.R., Garrigue, P. et al. (2011). Versatile procedure for synthesis of janus-type carbon tubes. *Chem. Mat.* 23: 2595–2599. https://pubs.acs.org/doi/10.1021/cm2001573.

76 Zhou, Y., Shida, N., Koizumi, Y. et al. (2019). Template-free perpendicular growth of a poly(3,4-ethylenedioxythiophene) fiber array by bipolar electrolysis under an iterative potential application. *J. Mater. Chem. C* 7: 14745–14751. https://doi.org/10.1039/c9tc04743c.

77 Reneker, D.H. and Chun, I. (1996). Nanometre diameter fibres of polymer, produced by electrospinning. *Nanotechnology* 7: 216–223. https://doi.org/10.1088/0957-4484/7/3/009.

78 Wei, J., Jiao, X., Wang, T., and Chen, D. (2018). Fast, simultaneous metal reduction/deposition on electrospun a-WO3/PAN nanofiber membranes and their potential applications for water purification and noble metal recovery. *J. Mater. Chem. A* 6: 14577–14586. https://doi.org/10.1039/C8TA03686A.

79 Yu, Q.-Z., Shi, M.-M., Deng, M. et al. (2008). Morphology and conductivity of polyaniline sub-micron fibers prepared by electrospinning. *Mater. Sci. Eng., B* 150: 70–76. http://www.sciencedirect.com/science/article/pii/S0921510708000950.

80 Yang, X., Guo, Y., Han, Y. et al. (2019). Significant improvement of thermal conductivities for BNNS/PVA composite films via electrospinning followed by hot-pressing technology. *Compos. Pt. B-Eng.* 175: https://doi.org/10.1016/J.COMPOSITESB.2019.107070.

81 Trchová, M., Šeděnková, I., Morávková, Z., and Stejskal, J. (2014). Conducting polymer and ionic liquid: improved thermal stability of the material – a spectroscopic study. *Polym. Degrad. Stab.* 109: 27–32. https://www.sciencedirect.com/science/article/pii/S0141391014002493.

82 Kakunuri, M., Wanasekara, N.D., Sharma, C.S. et al. (2017). Three-dimensional electrospun micropatterned cellulose acetate nanofiber surfaces with tunable wettability. *J. Appl. Polym. Sci.* 134: https://onlinelibrary.wiley.com/doi/abs/10.1002/app.44709.

83 Xu, Q., Li, Y., Feng, W., and Yuan, X. (2010). Fabrication and electrochemical properties of polyvinyl alcohol/poly(3,4-ethylenedioxythiophene) ultrafine fibers via electrospinning of EDOT monomers with subsequent in situ polymerization. *Synth. Met.* 160: 88–93. http://www.sciencedirect.com/science/article/pii/S0379677909005293.

84 Li, D., Babel, A., Jenekhe, S.A., and Xia, Y. (2004). Nanofibers of conjugated polymers prepared by electrospinning with a two-capillary spinneret. *Adv. Mater.* 16: 2062–2066. https://doi.org/10.1002/adma.200400606.

85 Niu, Z. and Yuan, W. (2021). Smart nanocomposite nonwoven wearable fabrics embedding phase change materials for highly efficient energy conversion-storage and use as a stretchable conductor. *ACS Appl. Mater. Interfaces* 13: 4508–4518. https://www.ncbi.nlm.nih.gov/pubmed/33439012.

86 Lai, H., Li, W., Xu, L. et al. (2020). Scalable fabrication of highly crosslinked conductive nanofibrous films and their applications in energy storage and electromagnetic interference shielding. *Chem. Eng. J.* 400: https://www.sciencedirect.com/science/article/abs/pii/S1385894720313140.

87 Yu, Q.-Z., Dai, Z.-w., and Lan, P. (2011). Fabrication of high conductivity dual multi-porous poly (l-lactic acid)/polypyrrole composite micro/nanofiber film. *Mater. Sci. Eng., B* 176: 913–920. https://www.sciencedirect.com/science/article/abs/pii/S0921510711002145.

88 Bhattacharya, S., Roy, I., Tice, A. et al. (2020). High-conductivity and high-capacitance electrospun fibers for supercapacitor applications. *ACS Appl. Mater. Interfaces* 12: 19369–19376. https://www.ncbi.nlm.nih.gov/pubmed/32275134.

89 Yuk, H., Lu, B., Lin, S. et al. (2020). 3D printing of conducting polymers. *Nat. Commun.* 11: 1604. https://www.ncbi.nlm.nih.gov/pubmed/32231216.

9

Conducting Polymer Nanomaterials for Electrochemical Energy Storage and Electrocatalysis

Mingwei Fang[1], Xingpu Wang[1], Xueyan Li[1], and Ying Zhu[1,2]

[1]*Beihang University, Key Laboratory of Bio-Inspired Smart Interfacial Science and Technology of Ministry of Education, School of Chemistry, 37 Xueyuan Road, Haidian District, Beijing, 100191, China*
[2]*Beihang University, Beijing Advanced Innovation Center for Biomedical Engineering, Beijing, 100191, China*

9.1 Introduction

Carbon papers (CPs) with highly π-conjugated polymeric chains have excellent electrical and optical properties similar to metals and inorganic semiconductors. They exhibit mechanical flexibility and processability associated with conventional polymers [1], including polyacetylene (PA), polypyrrole (PPy), polythiophenes (PTh), poly(*p*-phenylene) (PPP), poly(*p*-phenylenevinylene) (PPV), polyaniline (PANI), and their derivatives [2]. Conductivity is an important property for CPs, which is affected by chain structure that includes π-conjugated structure and length, crystalline and substituted grounds, and bounded fashion to the polymeric chain [3]. The conductivity of CPs could be tuned in a wide range from 10^{-10} up to 10^4 S cm^{-1} by using doping principles, endowing the conductive polymers to act as insulators, semiconductors, or conductors [4]. CPs are used in many fields, such as novel batteries, electrocatalysis, supercapacitors, electromagnetic shielding, etc. In addition, CPs can also be designed and prepared into composite materials, thus providing new exciting features, including tunable conductivities, flexibility, and mixed conductive mechanism, which lowers the interfacial impedance between electrodes and electrolytes [5]. This chapter will introduce the latest developments in applying CPs in electrochemical energy and material conversion.

9.2 Electrode Materials of Batteries

The increasingly rapid development of industry and massive consumption of fossil energies aggravate environmental pollution and global climate change, which has produced a common research theme about developing renewable energies, such as solar energy, tide energy, wind energy, and chemical energy [6]. Among the various energy storage systems, electrochemical energy conversion devices, especially

Functional Nanomaterials: Synthesis, Properties, and Applications, First Edition.
Edited by Wai-Yeung Wong and Qingchen Dong.
© 2022 WILEY-VCH GmbH. Published 2022 by WILEY-VCH GmbH.

various batteries, are dominant and have attracted substantial scientific and technological interests, owning to their high energy density [7]. To meet increasing energy requirements in various technologies, exploring new materials and methods to optimize the structure and properties of electrochemical materials is particularly important. CPs show several advantages, including good processibility, low cost, convenient molecular modification, and lightweight, making them candidates for battery electrodes [8]. However, poor stability during cycling and low conductivity in a reduced state inhibit their further applications. The adoption of nanostructured CPs can enhance stability during cycling and improve conductivity in a reduced state for battery applications due to their high surface and faster diffusion kinetics. Thus, this part aims to give a brief review of conducting polymer nanomaterials for battery applications.

9.2.1 Electrodes for Metal-Ion Batteries

9.2.1.1 Electrodes for Lithium-Ion Batteries

Lithium-ion batteries (LIBs) are extensively investigated as essential energy storage devices in various technological applications, such as portable electronic devices, electric vehicles, and grid-scale energy storage, due to their high theoretical capacity and relatively lightweight, enhanced safety, and low toxicity [9–11]. The CPs can play a positive role in the electrochemical properties of materials, such as PPy, PANI, and poly-3,4-ethylenedioxythiopene (PEDOT). For instance, Cheng et al. [12] synthesized ordered PANI nanotubes doped with $HClO_4$ and utilized them as cathodic materials, which showed better performance than the commercial PANI powders in LIBs (Figure 9.1a). The Li-PANI battery achieved a high practical discharge capacity of 75.7 mA h g^{-1} and retained 95.5% of the highest discharge capacity after 80 cycles. However, the capacity and power density are still relatively low, and the stability of organic materials remains a severe problem (Figure 9.1b). Traditional cathode materials, such as $LiCoO_2$, $LiFePO_4$, and $LiMn_2O_4$ with limited specific capacity (<170 mA h g^{-1}), barely meet the increasing requirements of energy storage systems and electric vehicles [14–16]. S. B. Schougaard and coworkers [13] successfully prepared free-standing poly(3,4-ethylenedioxythiophene) (PEDOT-LiFePO$_4$) composite films by dynamic three-phase interline electropolymerization (D3PIE), as shown in Figure 9.1c. These PEDOT-LiFePO$_4$ films were used without further modification as the cathode in standard LIBs. The PEDOT-LiFePO$_4$ composite film offers a discharge capacity of 75 mA h g^{-1} at C/10 rate and high capacity retention at C/2 rate. Its discharge capacity reached 160 mA h g^{-1}, close to the theoretical maximum value (170 mA h g^{-1}). This approach can produce a highly functional hybrid free-standing conducting polymer/active composite cathode with controllable size and structure. In addition, conducting polymer can also be utilized to produce cathodic binders. V. V. Kondratiev et al. [17] prepared eco-friendly water-based binder consisting of conducting polymer poly-3,4-ethylenedioxythiopene/polystyrene sulfonate (PEDOT:PSS) dispersion and carboxymethylcellulose (CMC) for the $LiMn_2O_4$ based cathode of LIBs. It was shown that the material with an optimal

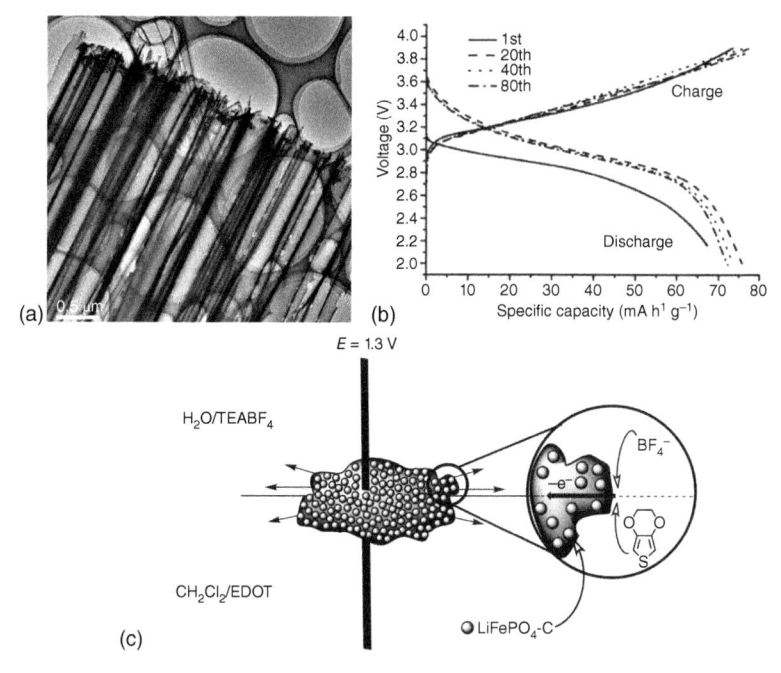

Figure 9.1 (a) TEM image of ordered HClO$_4$-doped PANI nanotubes. Source: Cheng et al. [12]/with permission from John Wiley & Sons, Inc. (b) Cycling characteristics of half cells fabricated by HClO$_4$-doped PANI nanotubes at a current density of 20 mA g^{-1}. Source: Cheng et al. [12]. Reproduced with permission of John Wiley and Sons. (c) Scheme of the D3PIE method in dynamic growth with LiFePO$_4$ at the water/dichloromethane interface. Source: Trinh et al. [13]. Reproduced with permission of Elsevier.

composition consisting of 86 wt% of LiMn$_2$O$_4$, 10 wt% of carbon black, and 4 wt% of conducting polymer binder exhibited good rate capability with discharge capacity 126 mA h g^{-1} (at 0.2 C, normalized by LiMn$_2$O$_4$ mass) and 75 mA h g^{-1} at 10 C and good cycling stability at 1 C (less than 5% decay after 200 cycles). These functional characteristics were better than those of poly(1,1-difluoroethylene) (PVDF)-bound LiMn$_2$O$_4$ materials.

The Ni-rich cathode has become popular due to its low cost, high capacity, and energy density [18]. Nevertheless, rapid capacity fading is a problem because of the direct contact of Ni-rich cathode with electrolytes, restraining its wide applications. W. Choi and coworkers [19] designed and prepared a uniform EDOT monomer layer on the LiNi$_{0.5}$Mn$_{1.5}$O$_4$ (LNMO), and chemical polymerization of the EDOT coating layer was then carried out to achieve PEDOT-coated cathode materials via a simple one-pot preparation process (Figure 9.2a). The electrochemical evaluation demonstrated that the PEDOT-coated LNMO showed better rate capability and cyclability, and the 2 wt% PEDOT-coated LNMO electrode exhibited superior power characteristics (Figure 9.2b). Z. Lu and coworkers [18] used poly(N-vinylpyrrolidone) (PVP) as an inductive agent to coat a uniform conductive PANI layer on LiNi$_{0.8}$Co$_{0.1}$Mn$_{0.1}$O$_2$ (NCM811) (Figure 9.2c). The coated PANI layer served as a rapid channel for electron conduction and prohibited direct

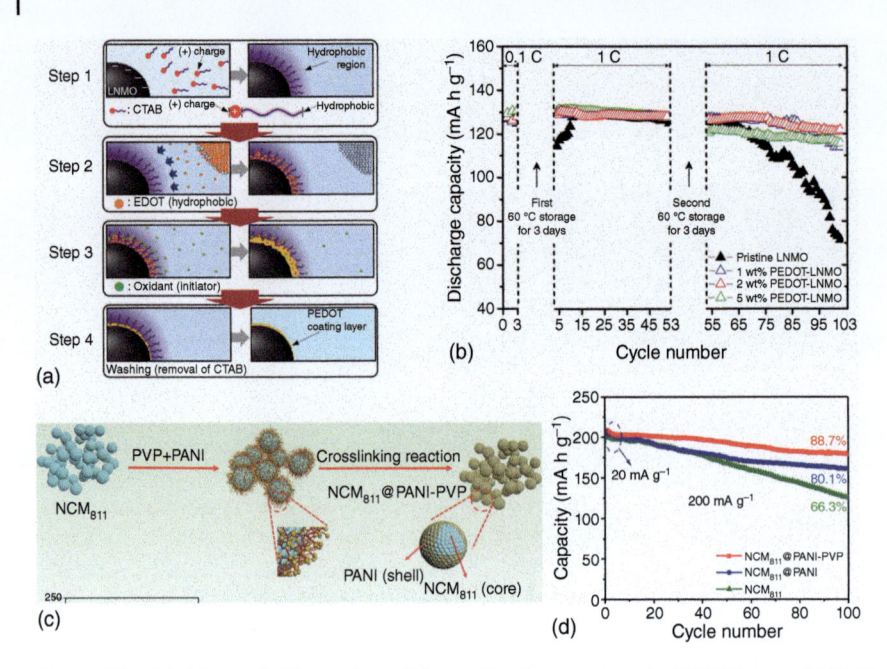

Figure 9.2 (a) Schematic illustration of the synthesis procedure for PEDOT-coated LNMO. Source: Kwon et al. [19]. (b) Comparison of cycle performances combined with the storage test for pristine LNMO and various levels of PEDOT coating on LNMO. The storage test at 60 °C for 3 days was performed after the charging process up to 4.9 V. Source: Kwon et al. [19]. Reproduced with permission of American Chemical Society. (c) Schematic illustration of the preparation of NCM811@PANI-PVP. Source: Gan et al. [18]. (d) Comparison of the cycling performance of NCM811, NCM811@PANI, and NCM811@PANI–PVP at 200 mA g^{-1}. Source: Gan et al. [18]. Reproduced with permission of American Chemical Society.

electrode contact with the electrolyte to effectively hinder the side reaction. The as-prepared NCM811@PANI-PVP exhibited excellent cyclability (88.7% after 100 cycles at 200 mA g^{-1}) and excellent rate performance (152 mA h g^{-1} at 1000 mA g^{-1}) (Figure 9.2d). This surfactant-modulated surface uniform coating strategy offers a new modification approach to stabilize Ni-rich cathode materials for LIBs.

Transitional metal sulfides became notable due to their relatively good electron conductivity compared to their oxide counterparts [20, 21], which could benefit charge transportation and promote rate capability. However, transitional metal sulfides have large crystal structures, which might offer unexpected opportunities for LIBs. K. Kang and coworkers [22] demonstrated that the electrochemical properties of $Li_3V_2(PO_4)_3$ can be improved by coating the conducting polymer PEDOT on it in a simple process (Figure 9.3a). The coating using oxidant-free polymerization at low temperature led to high power capability with outstanding cycle stability. The cathode using $Li_3V_2(PO_4)_3$/PEDOT composite delivered 110 mA h g^{-1} (more than 80% of its theoretical capacity) at a 30 C rate with 97% capacity retention after 100 cycles (at 10 C rate) without any carbon added. W. Fan and coworkers [23] prepared synthesized polymer-coated VS_4 submicrospheres as cathode material by solvothermal synthesizing VS_4 submicrospheres and coating three conductive

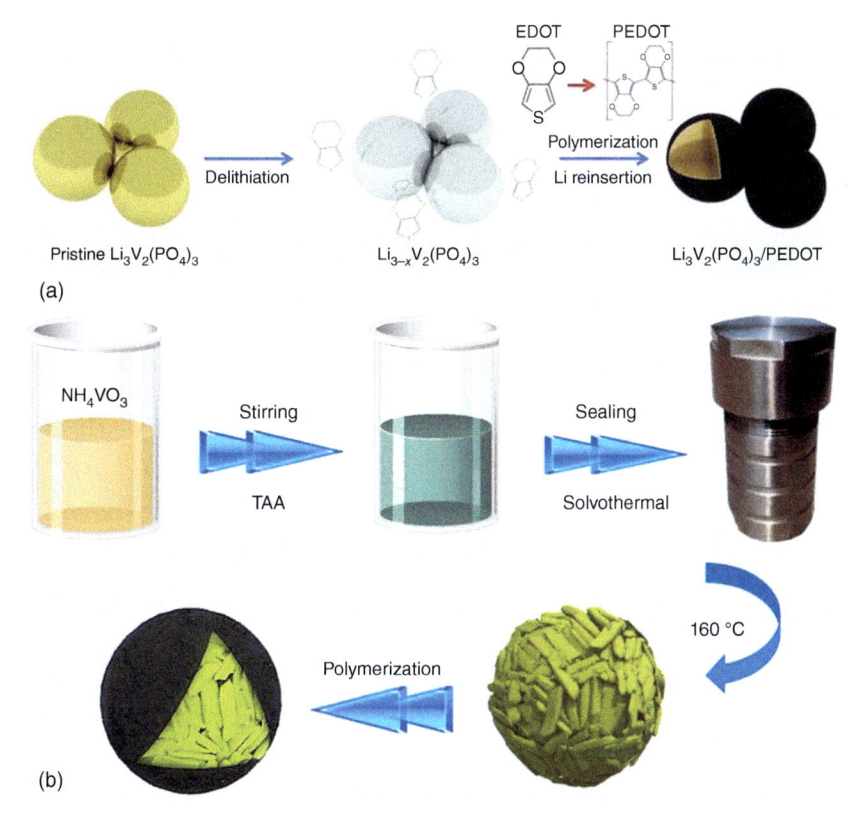

Figure 9.3 (a) Schematic illustration of the fabrication process of $Li_3V_2(PO_4)_3$/PEDOT composite. Source: Kim et al. [22]. Reproduced with permission of John Wiley and Sons. (b) Synthesis of polymer-coated VS_4 submicrospheres. Source: Zhou et al. [23], Figure 01 [p. 009]/with permission from American Chemical Society.

polymers, including PEDOT, PPy, and PANI, on the surface of VS_4 submicrospheres to improve the electron conductivity, suppress the diffusion of polysulfides and modify the interface between electrode/electrolyte (Figure 9.3b). PANI-coated VS_4 submicrospheres exhibited the best performance, which improved the Coulombic efficiency to 86% for the first cycle and kept the specific capacity at 755 mA h g^{-1} after 50 cycles, higher than the cases of naked VS_4 (100 mA h g^{-1}), VS_4@PEDOT (318 mA h g^{-1}), and VS_4@PPy (448 mA h g^{-1}).

The Al current collector is the most popular current collector for the cathode in lithium batteries because it has good price, conductivity, and availability. S.B. Schougaard and coworkers [24] modified conventional Al current collectors using a CP, decreasing contact resistance and improving battery performance. The oxidant Fe(III) p-toluenesulfonate (TOS) is aerosolized onto an aluminum foil current collector, followed by polymerization of 3,4-ethylenedioxythiophene (EDOT) via chemical vapor deposition (CVD) [24]. This fabrication technique should be scalable using roll-to-roll or similar fabrication methods. The coated foil was used as a cathode current collector in lithium batteries. The improvement of the current collector is determined by electrochemical tests in coin cells with C-LiFePO$_4$ (C-LFP) and lithium as active materials. At 15 C discharge rate, the new composite C-LFP-PEDOT-Al electrode provides a ~30% increase in discharge capacity compared to the standard C-LFP-Al electrode. The new composite has perfect stability over 50 cycles at C/2.5 rate.

Si has been considered a promising alternative anode for LIBs, but the commercial application of Si anodes is still limited due to their poor cyclability. The CPs have been employed to modify the Si anodes. X. Ai and coworkers [25] constructed a Si anode by embedding the nano-Si particles into a Li$^+$-conducting polymer PPP matrix to prevent the contact of the nano-Si surface with the electrolytes, suppressing the continual rupturing–reformation of solid electrolyte interphase (SEI) film on the Si surfaces (Figure 9.4a). The prepared Si/PPP composites demonstrated a high capacity of 3184 mA h g^{-1}, a high rate capability of 16 A g^{-1}, and a long-term

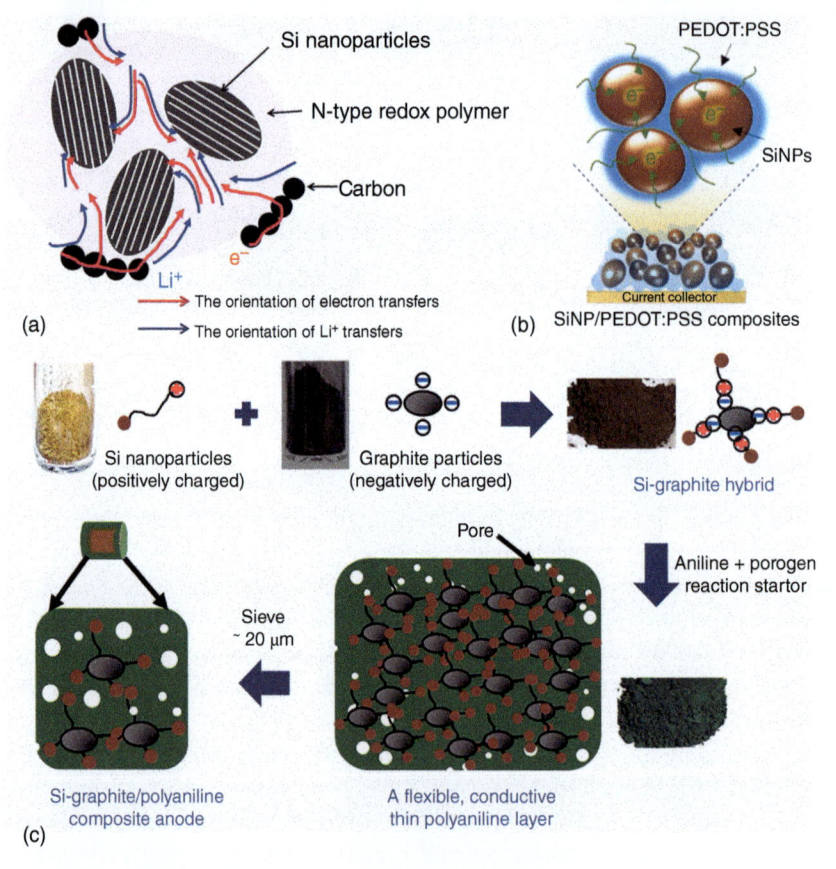

Figure 9.4 (a) Schematic illustration of electron transfer and Li$^+$ transfer for Si/PPP composites. Source: Chen et al. [25]. Reproduced with permission of American Chemical Society. (b) Scheme for the experimental PEDOT:PSS/SiNP electrode. Source: Higgins et al. [26]. Reproduced with permission of American Chemical Society. (c) Synthesis steps for porous Si/graphite/PANI composite. Source: Wiggers et al. [27], Figure 01 [p. 003]/with permission from Elsevier.

cyclability with 60% capacity retention over >400 cycles, which exceeded the electrochemical performances of conventional Si anodes. J.N. Coleman and coworkers [26] replaced the multiple non-active electrode additives (usually carbon black and an inert polymer binder) used in silicon nanoparticle-based LIB anodes with a single conducting polymer PEDOT:PSS (Figure 9.4b). Using an *in situ* secondary doping treatment of the PEDOT:PSS with small quantities of formic acid, electrodes containing 80 wt% SiNPs can be prepared with electrical conductivity as high as 4.2 S cm^{-1}. Even at a relatively high areal loading of 1 mg cm^{-2}, this system demonstrated a first cycle lithiation capacity of 3685 mA h g^{-1} (based on the SiNP mass) and a first cycle efficiency of ~78%. After 100 repeated cycles at 1 A g^{-1}, this electrode could still store an impressive 1950 mA h g^{-1} normalized to Si mass (~75% capacity retention), corresponding to 1542 mA h g^{-1} when the capacity is normalized total electrode mass. At the maximum electrode with a particular thickness (~1.5 mg cm^{-2}), a high areal capacity of 3 mA h g^{-1} was achieved. Additionally, scalable synthesis routes need to be developed to provide access to practically-relevant material quantities. H. Wiggers et al. [27] proposed a strategy for producing Si/graphite/PANI composites that addressed the challenges mentioned (Figure 9.4c). The Si/graphite/PANI nanocomposite showed favorable characteristics inherited from its three components. Si nanoparticles provided high capacity, and graphite acted as an electrical conductor and gave high Coulombic efficiency. The PANI coating further enhanced the electrical conductivity and protected the entire structure. Excellent Coulombic efficiency of 86.2% at the initial cycle was recorded for this nanocomposite material. Galvanostatic charge/discharge (GCD) tests demonstrated that this material delivered a discharge capacity of 2000 mA h g^{-1} with an excellent capacity retention of 76% after 500 cycles at a discharge rate of 0.5 C (1.25 A g^{-1}). The capacity is 870 mA h g^{-1} measured at 5 C (12.5 A g^{-1}).

Electrode binder engineering for Si anodes has become more recognized. Yinhua Zhou and coworkers [28] demonstrated a novel polymer binder that possessed high ion and electron conductivities, suitable for high-performance Si anodes. The binder was prepared by assembling ion-conducting polyethylene oxide (PEO) and polyethylenimine (PEI) onto the electron-conductive PEDOT:PSS chains via chemical crosslinking, chemical reduction, and electrostatic self-assembly due to the special chemical structure of the CPs (Figure 9.5a). The interactions between the polymer components (PEI and PEO) and lithium-ions made the binder improve the lithium-ion transport (Figure 9.5b). The interactions inside the binders (crosslinking between PEO and PSS, electrostatic interaction between PEI and PSS, and PEI chemically reducing PEDOT) contributed to the high ion and electron conductivities of the binder (Figure 9.5c). The polymer binder showed lithium-ion diffusivity and electron conductivity 14 and 90 times higher than the widely used carboxymethyl cellulose (with acetylene black) binder. The silicon anode with the polymer binder has a high reversible capacity of over 2000 mA h g^{-1} after 500 cycles at a current density of 1.0 A g^{-1}. It maintains a superior capacity of 1500 mA h g^{-1} at a high current density of 8.0 A g^{-1}. W. Liu and coworkers [29] reported a series of molecularly engineered conductive polymer binders, including star-like polyaniline (s-PANI), crosslinked polyaniline (c-PANI), and linear polyaniline (l-PANI). As a conductive

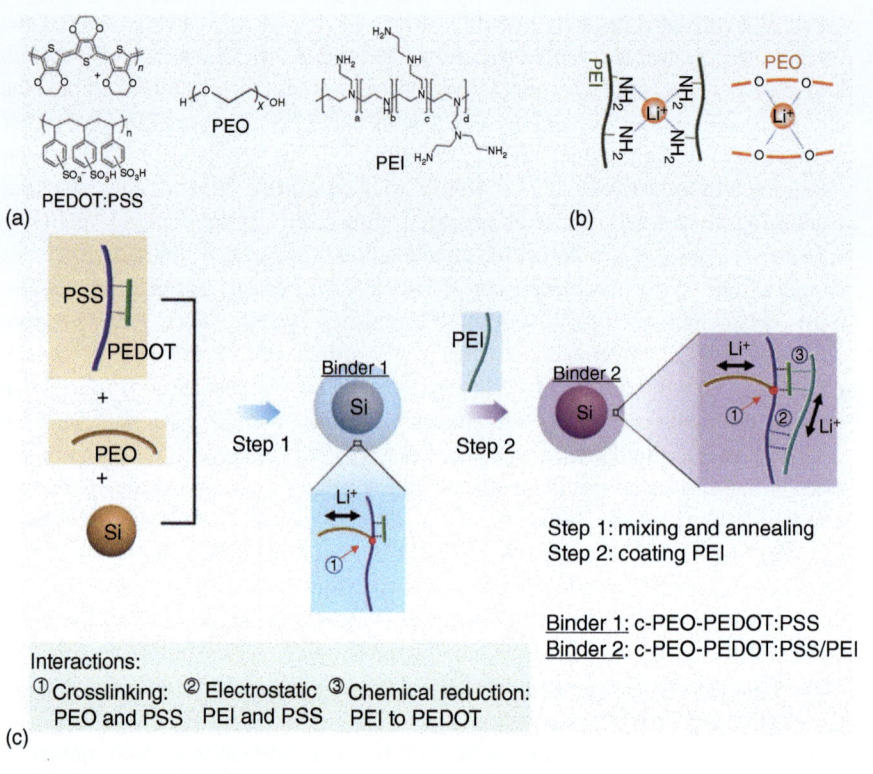

Figure 9.5 (a) Chemical structure of the polymers: PEDOT:PSS, PEO, and PEI used to prepare the binders. (b) Schematic diagram of the interactions between the polymer components (PEI and PEO) and lithium-ions that improve the lithium-ion transport. Source: (a, b) Zeng et al. [28]. (c) Schematic diagram of the interactions inside the binders with high ion and electron conductivities. (1) Crosslinking between PEO and PSS, (2) electrostatic interaction between PEI and PSS, (3) PEI chemically reducing PEDOT. Source: Zeng et al. [28]. Reproduced with permission of John Wiley and Sons.

binder, the molecular structure of PANI played a key role in determining the performance of a Si anode. The reversible capacity of 1776 mA h g^{-1} after 100 cycles at 500 mA g^{-1} was achieved using s-PANI as the conductive binder, superior to systems adopting c-PANI, l-PANI and conventional carboxymethyl cellulose binders.

New technologies demand materials with specially tailored structures and properties. Heterostructures consisting of inorganic oxides and organic CPs on metal support have drawn considerable interest in researching and developing many practical applications. M. Grzeszczuk and coworker [30] prepared nanocomposites of TiO$_2$ (anatase) with PPy or PEDOT via electrochemical routes. The deposition process of the conducting polymer films was performed in the presence of perchlorate, TOS, or bis(trifluoromethylsulfonyl)imide (TFSI) anions in propylene carbonate (PC). As a result of its interfacing with the polymers, improved lithium-ion intercalation/deintercalation properties of titanium (IV) oxide were observed. This effect depended on the thickness of the polymer layer and was closely related to the polymer facility for transporting lithium-ion. In contrast to the PEDOT case, PPy properties were susceptible to some substrate materials (Pt or Pt/TiO$_2$) during electropolymerization. PPy deposited on a rough surface exhibits an improvement in its ion exchange abilities. The impact of underlying TiO$_2$ layers on PPy properties has an indirect (synergic) influence on the effectiveness of lithium-ion intercalation into the oxide. Moreover, H. Hou and coworkers [31] prepared 1D chain-like Co$_3$O$_4$ through the thermal oxidation of the self-assembled rod-like Co-precursor, followed by *in situ* polymerization of pyrrole monomer. By particle size-tuning, 1D architecture-alteration, conducting PPy introduction could effectively broaden the ions' energy distribution, increase the speed of ions directional transfer, and improve the conductivity by protecting Co$_3$O$_4$/PPy with a stable capacity of 816.6 mA h g^{-1} at 1.0 A g^{-1} after 300 cycles. The full-cell capacity still delivered 526 mA h g^{-1} at 3.0 A g^{-1} after 50 loops. Changzhou Yuan and coworkers [32] fabricated bi-metal (Zn, Mn) metal–organic framework (MOF)-derived ZnMnO$_3$ (ZMO) micro-sheets (MSs) via bottom-up solvothermal method, which were further wrapped uniformly with flexible PPy via efficient gaseous polymerization. In the hybrid (denoted as PPy@ZMO), the conductive PPy, as a continuous electronic network, was dispersed throughout the porous ZMO MSs, which enhanced the structural stability and charge transfer of the hybrid anode. Due to compositional/structural advantages and intrinsic pseudocapacitive contributions, the resultant PPy@ZMO anode was endowed with a high-rate reversible capacity of 752 mA h g^{-1} at 2000 mA g^{-1} and desirable capacity retention with cycling (1037.6 mA h g^{-1} after 220 cycles at 500 mA g^{-1}). In addition, the PPy@ZMO-based full battery, along with remarkable cycling properties, exhibited an energy density of 206.2 W h kg^{-1} in the whole device, highlighting its promising application in advanced LIBs.

9.2.1.2 Electrodes for Other Metal-Ion Batteries

Sodium-ion batteries (SIBs) garnered attention in recent years. They have been widely developed as an alternative technology to LIBs, primarily due to lithium's much lower sodium price [33]. However, the radius of sodium ion is 34% bigger than that of lithium-ion, and sodium ions are also much heavier [34–36]. Hence, SIBs have relatively low capacity and poor reaction kinetics during charge/discharge, delaying commercialization. Shu-Lei Chou and coworkers [37] synthesized ClO$_4^-$-doped PPy coated Na$_{1+x}$MnFe(CN)$_6$ composite as a cathode material (NMHFC@PPy) for SIBs (Figure 9.6a). First, PPy served as a conductive coating layer that can increase the electronic conductivity of NMHFC to improve the rate capability. Second, PPy acted as a protective layer to reduce the dissolution of Mn in the electrolyte to improve the cycling performance. Finally, the PPy doped with ClO$_4^-$ can act as an active material to increase the capacity of the composite. NMHFC@PPy showed high energy density (428 Wh k g^{-1}), enhanced cycling performance (67% capacity retention after 200 cycles), and excellent rate capacity (46% capacity for 40 C rate). Zhongde Wang et al. [38] grew a string of nickel hexacyanoferrate (NiHCF) nanocubes coaxially on carbon nanotubes (CNTs)@bipolar conducting polymer derived from PANI and CNTs by a facile electrochemical route. It was used as a high-performance cathode material for SIBs. Upon the initial

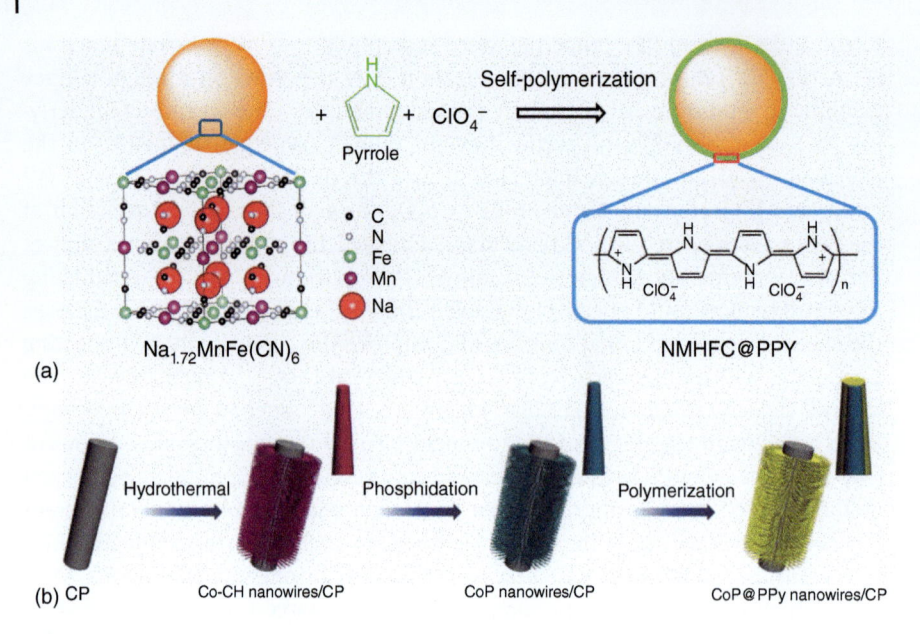

Figure 9.6 (a) Schematic illustration of the polymerization reaction between NMHFC particle to NMHFC@PPy. Source: Li et al. [37]. Reproduced with permission of Elsevier. (b) Schematic diagram of the fabrication process for CoP@PPy nanowires/CP electrode. Source: Zhang et al. [33]. Reproduced with permission of John Wiley and Sons.

discharge, the obtained cathode shows a high specific capacity of 194 mA h g^{-1}, good cycling performance, and excellent rate performance. The unique nanostructure of CPs effectively facilitated the electrode/electrolyte interaction and electronic and ionic transportation. It demonstrated a synergistic effect between the bipolar conducting polymer (BCP) and NiHCF nanocubes to trigger the electron and ion transport kinetics. Y.-M. Kang and coworkers [33] fabricated a CoP@PPy NWs/CP free-standing anode for SIBs (Figure 9.6b). They employed PPy encapsulating cobalt phosphide (CoP) nanowires (NWs) grown on carbon paper (CP) to realize 1D core–shell CoP@PPy NWs/CP. The CoP@PPy NWs/CP free-standing anode demonstrated superb electrochemical performance with a high areal capacity of 0.521 mA h cm^{-2} at 0.15 mA cm^{-2} after 100 cycles and 0.443 mA h cm^{-2} at 1.5 mA cm^{-2} even after 1000 cycles. Even at a high current density of 3 mA cm^{-2}, a significant areal discharge capacity reaching 0.285 mA h cm^{-2} was still maintained.

Aqueous zinc-ion batteries (AZIBs) have become a research focus because of their cost-effectiveness, high safety, and eco-friendliness [39–41]. Unfortunately, sluggish Zn^{2+} diffusion kinetics and poor cycling stability in cathode materials impede their large-scale application. X. Hu et al. [42] fabricated V$_2$O$_5$@PEDOT hybrid nanosheet arrays uniformly deposited on carbon cloth (CC) as a superior ZIB cathode (Figure 9.7a). The as-fabricated V$_2$O$_5$@PEDOT/CC electrode displayed a maximum capacity of 360 mA h g^{-1} at 0.1 A g^{-1}. Meanwhile, this hybrid array electrode also showed a high rate capability with a specific capacity of 232 mA h g^{-1} even at a large current density of 20 A g^{-1} and excellent cycling life with 97% retention after 600 cycles at 1 A g^{-1} and 89% retention after 1000 cycles at 5 A g^{-1}.

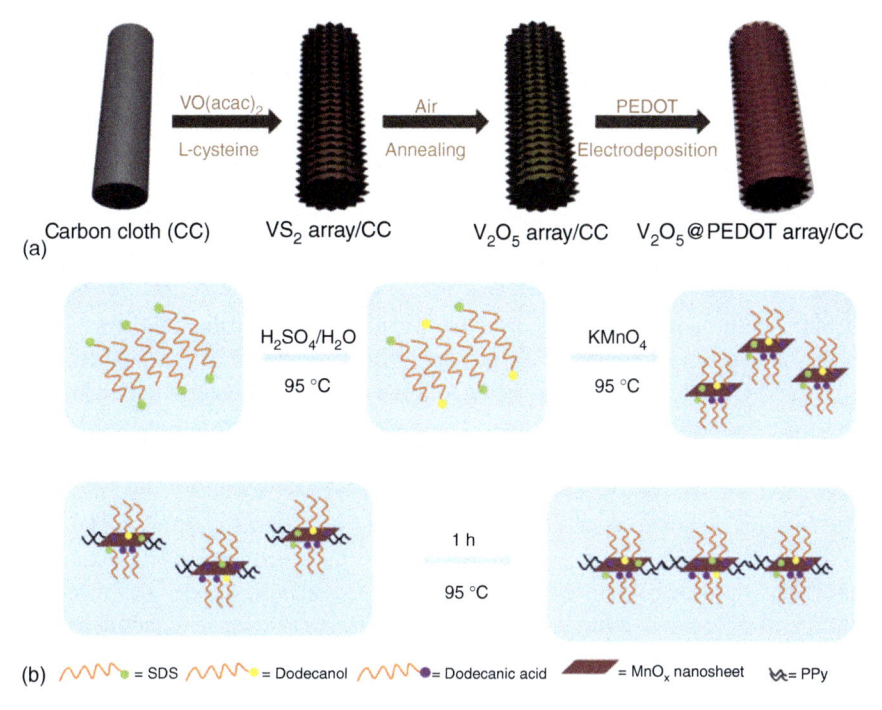

Figure 9.7 (a) Schematic illustration of the fabrication of V_2O_5@PEDOT/CC. Source: Xu et al. [42]. Reproduced with permission of John Wiley and Sons. (b) Synthesis process of pure MnO_x and MnO_x/PPy nanosheets. Source: Zang et al. [43]. Reproduced with permission of American Chemical Society.

The increased Zn-storage performance of the V_2O_5@PEDOT nanosheet arrays resulted from the synergistic effects of the two components: the V_2O_5 nanosheet arrays provide enough Zn-storage active sites, while the PEDOT coating shell increases zinc ion/electron transport kinetics and acts as a protective layer to restrain structural collapse during cycling. AZIB is a promising choice for future large-scale grid energy storage, in which Mn-based cathode materials have the advantages of low cost and sustainability [44]. However, the large bulk-induced incomplete zincation and structure pulverization limit their capacity delivery and cycling stability. M. Xue and coworkers [43] developed a strategy of epitaxial polymerization in the liquid phase to fabricate two-dimensional (2D) MnO_x/PPy nanosheets to enhance the zinc-ion storage by realizing the efficient utilization of active materials and improving the structural stability via a polymerized framework (Figure 9.7b). An ultrahigh capacity of 408 mA h g^{-1} was demonstrated at 1 C rate, and excellent capacity retention of 78% was realized after 2800 cycles at 5 C rate for the AZIB. Electrochemical and morphological characterizations revealed that the unique 2D structure contributed to the electron/ion conductivity and structural stability.

N.S. Hudak [45] demonstrated CPs as active materials in the positive electrodes of rechargeable aluminum-based batteries operating at room temperature. The battery chemistry is based on chloroaluminate ionic liquid (IL) electrolytes, allowing reversible stripping and plating aluminum metal at the negative electrode. Stable galvanostatic cycling of PPy and PTh cells was demonstrated, with electrode capacities near theoretical levels (30–100 mA h g^{-1}) and Coulombic efficiencies approaching 100%. The energy density of a sealed sandwich-type cell with polythiophene at the positive electrode was estimated to be 44 W h kg^{-1} relative to the total mass of active components.

9.2.2 Electrodes for Lithium–Sulfur Batteries

Lithium–sulfur batteries have become popular as promising electrochemical energy conversion and storage devices because of their high theoretical capacity, eco-friendliness, low operational costs, and the earth-abundance of sulfur [46, 47]. However, they still have difficulties to overcome to meet the requirements for many applications, such as grid storage and electric cars [48]. Two major drawbacks are associated with the use of sulfur as a cathode material. One of the obstacles is the low specific capacitance of sulfur due to its high electrical resistivity. The other is the shuttle effect, where polysulfide intermediates form during the charging/discharging process. When the polysulfide species reach the anode, they react with lithium and form insoluble Li_2S and Li_2S_2, leading to fast capacity fading [49–51]. Therefore, there is a demand for materials that can improve the entrapment of polysulfides while keeping conductive. Using CPs to help encapsulate sulfur was explored by Yi Cui and coworkers [52] In bare CMK-3/S particles, polysulfides diffuse out of the carbon matrix during lithiation/delithiation (Figure 9.8a). They used encapsulated CMK-3/S particles with PEDOT:PSS so that polysulfides could be confined within the carbon matrix and lithium-ions and electrons could move through this polymer layer (Figure 9.8b), which resulted in the reduction of polysulfide dissolution and improved the battery performance [52]. They reported an increase of ~10% in the initial discharge capacity (1140 mA h g^{-1}) compared with the non-coated carbon/sulfur particles. Furthermore, capacity retention was improved from ~60% per 100 cycles to ~85% per 100 cycles, and the coulombic efficiency increased from 93% to 97%. Chen et al. [53] successfully synthesized CPs CPs/graphene oxide (GO)@sulfur composites via a facile one-pot route, which were used as cathode materials for Li–S batteries. The PEDOT/GO and PANI/GO composites were prepared by interface polymerization of monomers on the surface of GO sheets. Then sulfur was *in situ* deposited on the CPs/GO composites in the same solution. The PEDOT/GO@S composites with the sulfur content of 66.2 wt% exhibited a reversible discharge capacity of 800.2 mA h g^{-1} after 200 cycles at 0.5 C, higher than that of PANI/GO@S composites (599.1 mA h g^{-1}) and PANI@S (407.2 mA h g^{-1}). Even at a high rate of 4 C, the PEDOT/GO@S composites still retained a high specific capacity of 632.4 mA h g^{-1}. Y. Xu and coworkers [54] prepared flexible and free-standing films with nanosulfur sandwiched between graphene and PEDOT:PSS (nanosulfur, graphene, and poly(3,4-ethylenedioxythiophene):poly(styrenesulfonate) [SGP]) via vacuum filtration. Benefiting from the ultrasmall sulfur particles, excellent mechanical and high conducting graphene and PEDOT:PSS, and strong interaction between lithium polysulfides

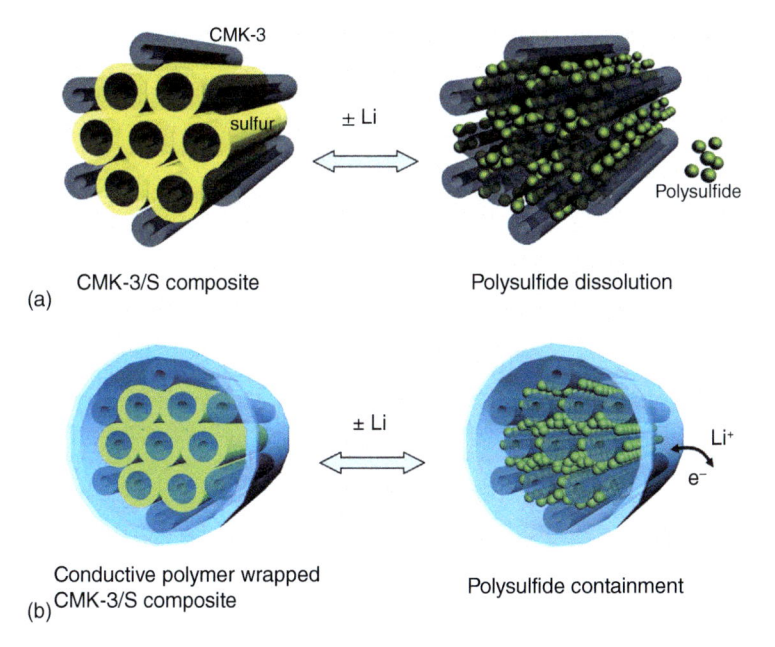

Figure 9.8 Scheme of PEDOT:PSS-coated CMK-3/sulfur composite for improving the cathode performance. Source: Yang et al. [52]. (a) In bare CMK-3/S particles (gray: CMK-3, yellow: sulfur), polysulfides (green color) still diffuse out of the carbon matrix during lithiation/delithiation. (b) Polysulfides could be confined within the carbon matrix with a conductive polymer coating layer (blue color). Lithium ions and electrons can move through this polymer layer. Source: Yang et al. [52]. Reproduced with permission of American Chemical Society.

and the functional groups in PEDOT:PSS or graphene, the compact layered SGP cathodes displayed a high initial capacity of 1584 mA h g^{-1} at 0.1 C, excellent rate performance with the capacity of 701 mA h g^{-1} at 4 C, and superior long-term cycling stability with 80% retention of the initial capacity after 500 cycles at 1 C. Meanwhile, with a high volumetric sulfur loading of ≈1 g cm^{-3}, the SGP cathodes show the highest volumetric capacity of 1432 A h l^{-1} among all reported Li–S battery cathodes. S. Chen and coworkers [55] prepared a sulfur cathode (S@DHPC) by incorporating sulfur into a 3D hierarchical porous carbon scaffold. Coating the S@DHPC electrode with PEDOT:PSS improved the electrochemical performance of the electrode. It was ascribed to the high porosity of the three-dimensional (3D) hierarchical porous carbon (DHPC) matrix that facilitated the encapsulation of sulfur. More importantly, the PEDOT:PSS capping layers impeded the diffusion/loss of polysulfides, leading to improved cycling stability of the electrode. With a high sulfur loading density of 5.8 mg cm^{-2}, the tri-layer electrode delivered a discharge capacity of 846 mA h g^{-1} in the first cycle. It retained 716 mA h g^{-1} after 500 cycles, corresponding to a fading rate of only 0.033% each cycle. These results suggest that a layered architecture of S@DHPC/PEDOT:POSS may be exploited to prepare effective cathode materials of high-performance Li–S batteries. The commercialization of Li$_2$S as a potential candidate for lithium–sulfur cathode material is hampered due to its low electronic conductivity, the "shuttle effect," and the initial energy barrier.

9.2.3 Electrodes for All-Polymer Batteries

All-polymer batteries mean that cathode and anode electrodes are all based on CPs. They are free of metal-containing electrodes, enabling a lightweight, flexible, environmentally benign, and sustainable battery system. This system utilizes capacity-carrying polymeric materials as both anodes and cathodes. Cations and/or anions move between electrodes in a rocking-chair motion to compensate for the charge [56]. In an ideal all-polymer battery, a p-type polymer with higher redox potentials functions as a cathode in combination with an n-type polymer anode [57]. However, limited n-type CPs are available due to their poor chemical stability and large impedance [58]. In 1997, Searson and coworkers [59] pioneered this field by coupling PTh electrodes bearing different fluorine-containing aromatic rings with the polymer gel electrolyte. The maximum n-doping level of functional PTh was correlated with the number of fluorine atoms on the phenyl ring. The assembled all-polymer dual-ion battery exhibited a discharge voltage of ~2.5 V and a capacity of 9.5–11.5 mA h g^{-1} in the electrolyte of tetrabutylammonium tetrafluoroborate. Then, several n-type conjugated redox polymers based on 5-alkylthieno[3,4-c]pyrrole-4,6-dione were developed [60]. The effect of the carbonyl group, alkyl chain, and comonomer unit on the electrochemical activity of these polymers has been demonstrated. Incorporating an electron-withdrawing or electron-donating group with the conjugated backbone of polymer impacted its redox potential.

An early attempt involved using PPy doped with a large polymeric anion PSS as a "pseudo-n-dopable" anode [61]. The charge compensation by Li$^+$ transport instead of PSS anions caused a negative shift in the oxidation potential, realizing a voltage separation between the PPy cathode at the reduced form and the anode at the oxidized form. This system exhibited a specific charge capacity of 22 mA h g^{-1} and no capacity loss over 100 cycles. Using PPy doped with different redox-active organic anions in an all-polymer battery system was demonstrated in 2006 [62]. The dopants used for cathode and anode materials were indigo carmine (IC) and 2,2′-azino-bis(3-ethylbenzothiazoline-6-sulfonate) (ABTS), respectively. Both PPy/IC and PPy/ABTS exhibited significant faradaic current response over the applied potential range of −0.1 to +0.8 V, like neat dopants. The potential difference generated between these two redox reactions allowed them to perform as a cathode and anode. The final full cell displayed a specific capacity of 15 mA h g^{-1}, equal to the estimated theoretical capacity. An all-polymer battery composed of free-standing electrodeposited PPy film electrodes has also been demonstrated, which provided a step toward developing a flexible all-polymer battery system [63]. PPy film doped with IC and para(toluene sulfonic acid) (pTS) functioned as anode and cathode, respectively. This battery system displayed an initial discharge capacity of 21 mA h g^{-1}, a capacity retention rate of 76% over 50 cycles. Like PPy,

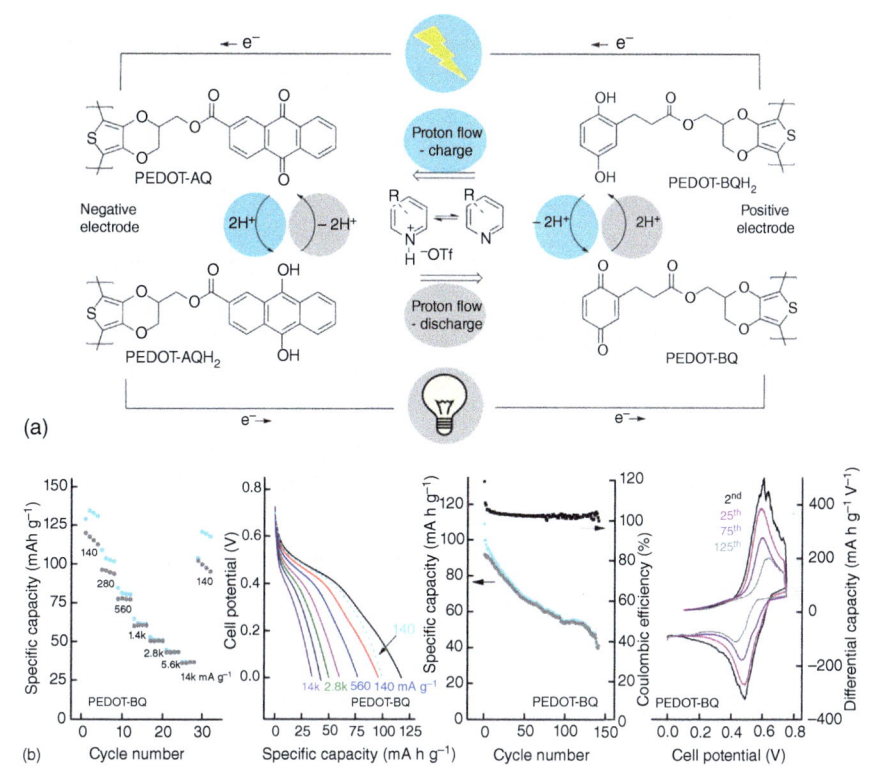

Figure 9.9 (a) Depiction of PEDOT-AQ, PEDOT-BQ, pyridine-based proton donors and acceptors (electrolyte), redox reactions occurred during charge (cyan) and discharge (gray) in an all-polymer battery [65]. (b) Cell characteristics measured in a two-electrode setup with PEDOT-BQ as limiting material. Source: Emanuelsson et al. [65]. Reproduced with permission of American Chemical Society.

a potential difference can also be generated between the neutral state and the oxidized state (p-doped) for PEDOT, enabling the realization of an all-polymer battery [64]. Sjödin and coworkers [65] recently demonstrated an all-polymer proton battery composed of PEDOT functionalized with anthraquinone (PEDOT-AQ) or benzoquinone (PEDOT-BQ) pendant groups as negative and positive electrode materials, respectively (Figure 9.9a). The electrolyte containing proton donor and acceptor allowed the $2e^-/2H^+$ quinone/hydroquinone redox reaction and enabled the PEDOT backbone to be conductive in the potential region of quinone redox reaction. This all-polymer proton battery delivered a capacity of $120\,\text{mA h g}^{-1}$ with an average cell potential of 0.5 V based on PEDOT-BQ, 75% of its theoretical specific capacity (Figure 9.9b). They further developed two conducting redox polymers (CRPs) based on PEDOT with a hydroquinone pendant group [66]. The hydroquinone pendant group underwent stable proton cycling. These polymers showed fast redox conversion, making them suitable for the aqueous rechargeable all-polymer battery.

9.2.4 Electrodes for Dye-Sensitized Solar Cell

Dye-sensitized solar cells (DSSCs) have become popular for various energy conversion applications. A typical DSSC consists of a transparent photo-anode with a dye-sensitized mesoporous thin film, a redox couple, and a counter electrode with a catalytic layer deposited on a fluorine-doped tin oxide (FTO) substrate [67–70]. Recently, CPs have been recognized as a relatively low-cost device for photovoltaic application due to their excellent electrocatalytic activity, high electrical conductivity, and good film-forming ability. For example, Kung et al. [71] examined a PEDOT hollow micro-flower array film as the catalytic material on the counter electrode of a DSSC. The PEDOT hollow micro-flower arrays contributed to high power conversion efficiency up to 7.20%, comparable to that of a cell with a sputtered Pt film (7.61%). Considering the low value of the Warburg diffusion resistance of the DSSC, the unique micro-flower structure would facilitate a hemispherical diffusion of ions for the electrocatalytic reaction in the electrolyte region. Lin et al. [69] reported the effect of PEDOT coating on the electrocatalytic activity of CNT/polypropylene plates. It was found that the PEDOT-coated CNT/polypropylene plate showed a Pt-like electrocatalytic activity for I^{3-} reduction. Consequently, a DSSC with the PEDOT-coated CNT/polypropylene counter electrode showed high power conversion efficiency up to 6.77%, even after the bending test. From numerous studies, it can be confirmed that the application of CP nanomaterials as potential substitutes for Pt counter electrodes in DSSCs is a recent trend. Nevertheless, a clear mechanism to describe the function of CP nanomaterials as counter electrode materials still needs clarification. It is anticipated that nanostructured CPs will open a new avenue for industrial fabrication of DSSCs due to their low cost and reasonable performance compared to Pt electrodes [70].

9.2.5 Electrodes for Bioelectric Batteries

Implantable medical devices (IMDs) have been widely used for therapies serving as functional devices to detect, prevent, and cure diseases [72–75]. A bioelectric battery (biobattery) with biocompatible electrodes that utilize body fluid electrolytes is safe for implantation [76]. It is a metal–air battery that consists of a cathode for oxygen reduction reaction and a sacrificial anode. The anode can be a biocompatible and bioresorbable metal, such as magnesium (Mg), zinc (Zn), and their alloys [77]. CPs with good electrical conductivity and electrocatalytic activity toward oxygen reduction have attracted attention as cathode materials in metal–air batteries [77]. In 2005, V.G. Khomenko et al. [78, 79] discussed the mechanism and feasibility of various CPs as an oxygen reduction catalyst in acid and neutral electrolytes. They proposed an "oxygen absorption" theory that the carbon atoms can absorb molecular oxygen on the polymer chain to form an "oxygen-conductive polymer" bridging complex, which could weaken the O—O bond of oxygen and lower the activation energy for reduction, leading to the electrocatalytic activity. In an Mg–air biobattery, the Mg anode is oxidized to Mg^{2+} and donates electrons when discharged. The CP cathode, taking oxidized PPy as an example, takes up electrons and is reduced concomitantly.

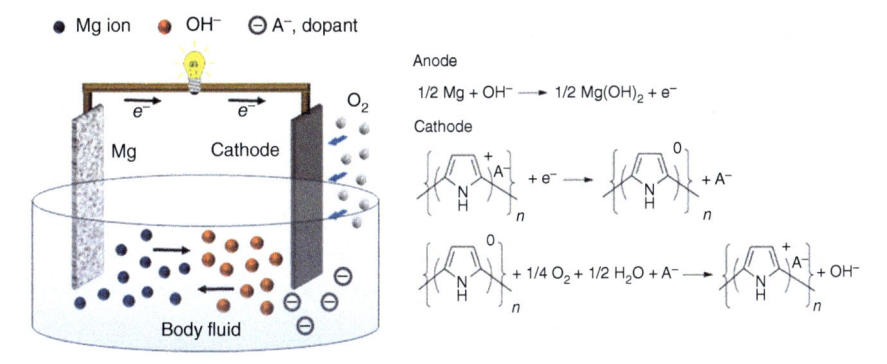

Figure 9.10 Schematic working principle and general redox chemistries of a Mg–air biobattery using PPy cathode when being discharged (A, represents mobile charge balance anions). Source: Jia et al. [80]. Reproduced with permission of American Chemical Society.

Then the reduced PPy is reoxidized ("recharged") by oxygen, as shown in Figure 9.10 [80]. The oxygen reduction mechanism in PEDOT has also been revealed using the density functional theory (DFT) [81]. Its catalytic activity is proposed to be related to the formation of polaronic states, which leads to the decreased highest occupied molecular orbital (HOMO)–lowest unoccupied molecular orbital (LUMO) gap and an enhanced reactivity. In a recent study, the doping of polymer by oxygen during oxygen reduction was done by decoupling the conductivity (intrinsic property) from electrocatalysis (an extrinsic phenomenon) [82]. Hence, the PEDOT electrode is electrochemically reduced in the voltage range of the oxygen reduction, while O_2 keeps it conducting, ensuring PEDOT acts as an electrode for the oxygen reduction.

Electrodes should be cyto- and tissue-compatible for applications in biobatteries. CPs, particularly PPy and PEDOT, are biocompatible with various cell types and tissues and have been thoroughly evaluated as biomaterials *in vitro* and *in vivo* [83]. They have demonstrated excellent cytocompatibility for supporting the *in vitro* adhesion, growth, and differentiation of many cell types such as bone, neural, glial, rat pheochromocytoma and endothelial cells, fibroblasts, keratinocytes, and mesenchymal stem cells [84]. It should be pointed out that the cytocompatibility of CPs depends on their surface charge, surface properties (i.e. roughness, morphology, wettability), and stiffness, which are closely related to the polymerization methods and conditions used [85]. The cytocompatibility can be improved by removing the impurities such as unused reactants and monomers and producing short oligomers. Additionally, the integration with bioactive molecules (polysaccharides, proteins, and even whole living cells) can further modulate the biological functionality of CPs, which can be achieved by physical (entrapping, blending) and chemical approaches (doping, covalent attaching) [86]. Dopants form a major constituent of CPs and affect their inherent chemical–biological properties. PPy doped with different biomolecules has demonstrated efficacy for biomedical applications, such as dermatan sulfate for increasing keratinocyte viability, heparin for increasing the proliferation of endothelial cells, and laminin-derived peptides to control neuron and astrocyte adhesion [87, 88]. The PPy cathode (doped with a biomolecule, dextran sulfate) coupled with a bioresorbable Mg alloy anode could maintain a cell voltage of 1.4 V for 19 hours at a current density of 30 mA g^{-1} in phosphate-buffered saline (PBS) electrolyte [89]. Incorporating a biocompatible dopant with an electrocatalytic ability to oxygen reduction can further improve performance. The molecule anthraquinone-1-sulfonate (AQS) with oxygen reduction capability was co-doped with reduced graphene oxide (RGO) into a PPy matrix [90]. The formed PPy/AQS/RGO composite delivered higher than that of PPy/pTS. The mechanical properties of CPs can also be manipulated to match that of body tissue (2–30 kPa). A two-component conducting polymer-based hydrogel was developed via electrodepositing PPy onto an ionically crosslinked PEDOT hydrogel with Mg^{2+} (Figure 9.11a) [91]. It demonstrated the conductivity and electroactivity of AQS and the "soft nature" of a hydrogel, making it a versatile platform for tissue engineering and energy storage applications (Figure 9.11b) [91]. This cyto-compatible PPy/PEDOT hydrogel coupled with an Mg anode provided a stable voltage of 0.7 V for 160 hours at a current density of 200 μA cm^{-2}, affording a capacity of 32 mA h cm^{-2} and an energy density of 22 mW h cm^{-2} in PBS electrolyte (Figure 9.11c) [91].

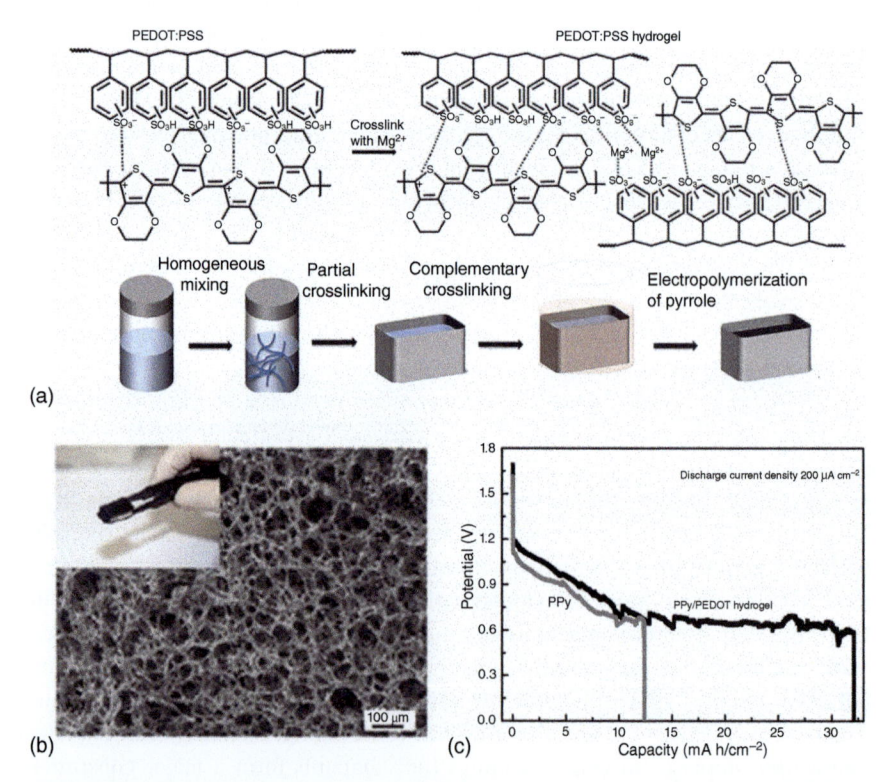

Figure 9.11 (a) Crosslinking process of PEDOT-PSS hydrogel and schematic procedures to fabricate PEDOT hydrogel and PPy/PEDOT hydrogel. (b) Cross-sectional and digital (inset) images of PPy/PEDOT hydrogel. Source: (b) Yu et al. [91]/with permission from John Wiley & Sons. (c) Discharge curves of a Mg biobattery with a PPy/PEDOT hydrogel or a PPy cathode at 200 μA cm^{-2} in PBS. Source: (a, c) Yu et al. [91]. Reproduced with permission of John Wiley and Sons.

A biodegradable battery system is an ideal choice to provide energy for emerging transient implants [92, 93]. They facilitate biological processes that enable/facilitate repair of damaged tissue by supporting biological processes in a controlled manner [94, 95]. However, CPs are stable under physiological conditions. Therefore, they need to be modified to attain biodegradability [96]. To achieve high conductivity and electrocatalytic activity while ensuring a large biodegradability, a bilayer structure for combining two components is an option [97]. Jia et al. [98] demonstrated a novel air cathode material by chemically depositing a thin continuous PPy layer only onto one side of a silk film polypyrrol (SF-PPy). This bilayer structure facilitated the enzyme attack the silk film with cleavage sites available on the opposite side. This film showed a conductivity of $\sim 1.1\,S\,cm^{-1}$. It degraded in a concentrated buffered protease XIV solution, with a weight loss of 82% after 15 days. Degradation of the silk substrate led to the fracture and disintegration of SF-PPy film, allowing elimination by renal excretion, phagocytosis, and/or endocytosis. The assembled Mg–air biobattery exhibited a discharge capacity up to $3.79\,mA\,h\,cm^{-2}$ at a current of $10\,\mu A\,cm^{-2}$, offering a specific energy density of $4.70\,mW\,h\,cm^{-2}$.

Research on batteries is driven by the electronic device market. With the urgent requirement for high-performance batteries, various conducting polymer nanomaterials for battery electrodes have been prepared, following the rapid development of synthetic strategies. These materials maintain attractive properties of CPs with efficient and reversible redox activity, lightweight and facile processability, and additional important features, including larger active surface areas and improved electrical, electrochemical, and mechanical properties for various battery types. The improvement of batteries and their unique properties (such as nontoxicity, lightweight renewability, facile processability, and tailorable electrochemical properties) make them ideal candidates for battery electrodes and many potential applications in other electronic devices, such as chemical sensors, light-emitting diodes, heat exchangers, and biosensors.

9.3 Electrocatalysis

Oxygen evolution reaction (OER), hydrogen evolution reaction (HER), and carbon dioxide reduction reaction (CO_2RR), are important electrochemical reactions in electrochemical energy and material conversion [99–101]. Generally, metals, such as Pt, Ag, and Pd, showed excellent performances for those electrochemical reactions, but their high price, low abundance, and low stability limit their commercial applications [102]. Due to their useful mechanical, optical, and electronic properties, CPs have been widely used in various fields [103, 104]. Recently, it was found that a thin layer of a conducting polymer deposited onto the surface of the substrate electrode can enhance the electrochemical kinetics of the electrode [105]. Their high conductivity and electrochemical properties make them suitable as catalysts for diverse redox reactions: CPs by themselves can have intrinsic electrocatalytic properties toward redox reactions [105, 106]. CPs have also been used to immobilize redox mediators for boosting the homogeneous electron transfer in the catalytic cycle [107]. It also uses heteroatoms of CPs to incorporate metals or metal oxide for generating heterogeneous electrocatalysts [108]. It is a convenient method to decrease the particle size and maximize the surface area and the metal loading, enhancing the electrocatalytic activity.

9.3.1 Oxygen Evolution Reaction (OER)

As a reaction involves four successive electron transfer steps, the OER is more complicated than other reduction reactions [102]. Moreover, OER suffers from slow kinetics and large overpotential, which is a hindrance in reaction efficiency. Low-cost, efficient catalysts for OER are crucial for the large-scale application of renewable energy technologies. As a result, noble metal-free or metal-free catalysts are anticipated. CPs are promising candidates for their tunable components via a simple processing strategy. Chinnappan et al. [106] fabricated well-defined PPy/IL nanoparticles with 24–44 nm diameters by chemical oxidation polymerization. PPy/IL nanoparticles contain 1-allyl-3-methylpyridinium cation and tetrachloronickelate anion complex, which acted as catalysts for OER based on the complexation of water-soluble polymers and metal cations (Figure 9.12a). The catalyst possessed a Tafel slope of 53 mV dec^{-1} and achieved a current density of 10 mA cm^{-2} at a potential of 583 mV that corresponded to an overpotential of 392 mV (Figure 9.12b–e).

Besides the direct use of CPs as catalysts, the polymer can also be a catalyst modifier and support. The CPs as catalyst supports were reported to be multi-functional. One example is that Liu et al. [109] developed an OER catalyst of PANI/MWCNT (MWCNT, multi-walled carbon nanotube) supported $CoFe_2O_4$ nanoparticles (NPs) (Figure 9.12f). The PANI was found to provide more active sites to attach $CoFe_2O_4$ NPs, improve the synergistic effect between the $CoFe_2O_4$ and MWCNT, and promote the electrical conductivity and stability of the catalyst (Figure 9.12g–j). The as-fabricated $CoFe_2O_4$/PANI-MWCNTs catalyst showed high OER activities at an overpotential of 314 mV at 10 mA cm^{-2} current density. Furthermore, the catalyst showed good long-term stability for at least 40 hours at 540 mV (vs. Ag/AgCl). Lattach et al. [110] electrodeposited IrO_2 nanoparticles into poly(pyrrole-alkylammonium)–iridium oxide (poly1-IrO$_x$) to fabricate nanocomposite anode OER catalyst. The electrocatalytic activity of as-prepared catalyst was well-maintained when incorporated in the polymer matrix and enhanced due to the nanostructure of the polymer matrix. The poly1-IrO$_x$ composite film modified electrode exhibited an overpotential for water oxidation varying from 0.4 to 0.5 V in 1–13 pH range. Moreover, in an acidic electrolyte (pH = 1), the poly1-IrO$_x$ composite material appeared more stable (decrease in the catalytic current: 15%, entry 5) than a pure oxide film (decrease in the catalytic current: 55%, entry 4).

Electrochemical CPs could be stable at a positive potential and showed good OER property. The incorporated noble or non-noble metal catalysts can further enhance the catalytic property. The interface of conducting polymer and metal particles played a crucial role in the performance of the composite catalysts. The metal-polymer composite catalysts were reported to be comparable to pure, noble

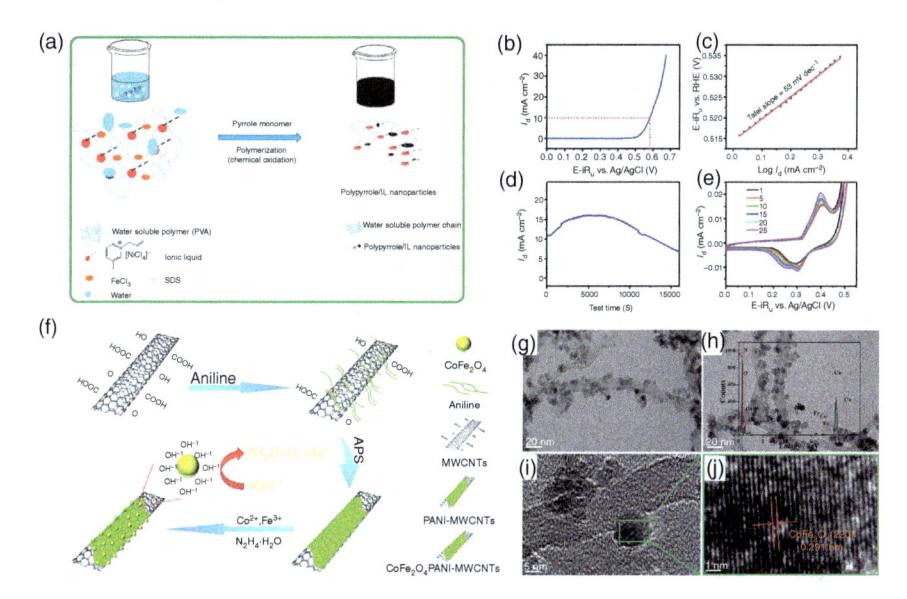

Figure 9.12 (a) Synthesis of PPy/IL nanoparticles. (b) Polarization curve obtained in 1 M KOH. (c) Tafel plot. (d) Potentiostatic analysis performed at a constant potential of 590 mV. (e) Cyclic voltammetry performed at the scan rate of 20 mV s^{-1}. Source: (a–e) Chinnappan et al. [106]. Reproduced with permission of Chemical Engineering Journal. (f) The process for the preparation of the CoFe$_2$O$_4$/PANI-MWCNTs and application in the oxygen evolution reaction. (g, h) TEM images of CoFe$_2$O$_4$/PANI-MWCNTs. (i, j) HRTEM images of CoFe$_2$O$_4$/PANI-MWCNTs. Source: (g) Liu et al. [109]/with permission of Royal Society of Chemistry.

catalysts, which provided an efficient substitute to reduce the cost of anode OER catalysts.

9.3.2 Hydrogen Evolution Reaction (HER)

Due to its high energy density and environment-friendly feature, hydrogen (H$_2$) is considered a promising energy carrier toward sustainable development [111]. The HER is an important step for electrochemical water splitting to produce hydrogen. Pt-based materials are the most popular electrocatalysts for HER, displaying an extremely high current density and a small Tafel slope, but usually suffering from rare materials and high prices [112]. There are numerous requirements for H$_2$ for economical and efficient HER catalysts. The conducting polymer catalysts are thought to be appropriate substitutes for noble metal catalysts. They are expected as ideal HER electrocatalysts, mainly due to their good conductivity, high stability, low price, easy access, and ability to store charges. CPs are considered to be a promising matrix that can stabilize the catalysts and enhance the conductivity. In addition, CPs could show HER activities without other components. For instance, Ng et al. [113] developed a poly(2,2′-bithiophene) (PBTh) catalyst for HER. Significant findings include the fourfold increase in performance by reducing thickness to facilitate charge transfer and the photo-catalysis of the HER at pH 11, with an onset of 0.14 V below E^0. Long-term stability testing shows the success of the catalyst over 12 days at neutral pH with corresponding turnover numbers exceeding 6×10^4.

The PPy is widely used in catalyst design for high conductivity, large electrochemical surface area, and good stability. For instance, Abrantes and Correia [114] reported a Ni–P deposited PPy film. Higher catalytic surface area played a similar role in improving electrocatalytic activity toward HER. The Ni–P loaded PPy composite was active for nitrite oxidation except HER, in which polymers worked as electron transfer channel and assistant inactive substrate. Wang et al. [115] reported a PPy/MoS$_x$ co-polymer film as HER catalyst fabricated by one-step electrochemical copolymerization (Figure 9.13a–d). Small Tafel slope of 29 mV dec^{-1}, positive onset potential of 0 V (vs. reversible hydrogen electrode [RHE]), and high exchange current density of 56 mA cm^{-2} were observed. The HER performances were comparable to commercial Pt/C catalysts (Figure 9.13e,f). The high HER activity of PPy/MoS$_x$ was attributed to the outstanding conductivity of PPy that offered a large electrochemical surface area, good interface of PPy/MoS$_x$ film to the substrate electrode, and high S ratio that provides more active S edge sites. Some works studied the role of PPy from another view. Mo et al. [117] analyzed the influence of different morphologies of PPy, and the electrode modified by PPy fibrils showed higher electrolysis current density (~230 mA cm^{-2} at 2.6 V vs. saturated calomel electrode [SCE]) than that of PPy cauliflower (~200 mA cm^{-2} at 2.6 V vs. SCE). Re-doping PPy fibrils with metal complex (Ppy(Ni-ethylene diamine tetraacetic acid [EDTA])) could further increase the hydrogen evolution current density (~600 mA cm^{-2} at 2.6 V vs. SCE). Jukic and Metikoš-Hukovic [116] prepared a PPy-coated GdNi$_4$Al intermetallic alloy catalyst, which showed a smaller Tafel slope ($\Delta b_c = -74$ mV) and better activity than pure alloy, as PPy coatings slowed the generation and growth of oxides to maintain metallic conductivity (Figure 9.13g). At overpotentials higher than -0.2 V, pure PPy and PPy-coated GdNi$_4$Al electrodes showed a considerable decrease in charge transfer resistance and increased pseudocapacitance of the surface process due to intense adsorption/absorption of hydrogen and its penetration in the subsurface structure resulting in brittle hydrogen.

CPs could be mixed with metallic catalysts simply or integrated at the molecular scale. Pandey and Lakshminarayanan [118] prepared Pd nanoparticle-dispersed PEDOT (Pd-PEDOT) thin film on the gold substrate as HER electrocatalyst in an acidic medium. The size of Pd-PEDOT clusters was around 200 nm, Pd was embedded in the PEDOT matrix, and the amount of Pd in Pd-PEDOT film was measured to be about 15% (Figure 9.14a). The low percent Pd catalyst showed improved catalytic activity than the Pd disk. The electron path provided by the PEDOT could be regarded as a major factor (Figure 9.14b). The measured onset potential of Pd-PEDOT (-0.28 V) is significantly lower than that of the Pd disk electrode (-0.50 V). At a potential of -0.45 V, the currents per unit area in the case of Pd electrode and Pd-PEDOT electrodes are 7.0 and 230 mA cm^{-2}, respectively, better by a factor of about 33. Navarro-Flores and Omanovic [119] developed a method that incorporated Ni into PPy and PANI to enhance HER activity. PPy and PANI worked as 3D matrices to control structures of Ni layers. The 3D Ni/conducting polymer layers with 3D HER zone showed higher HER activity than Ni electrodeposited on

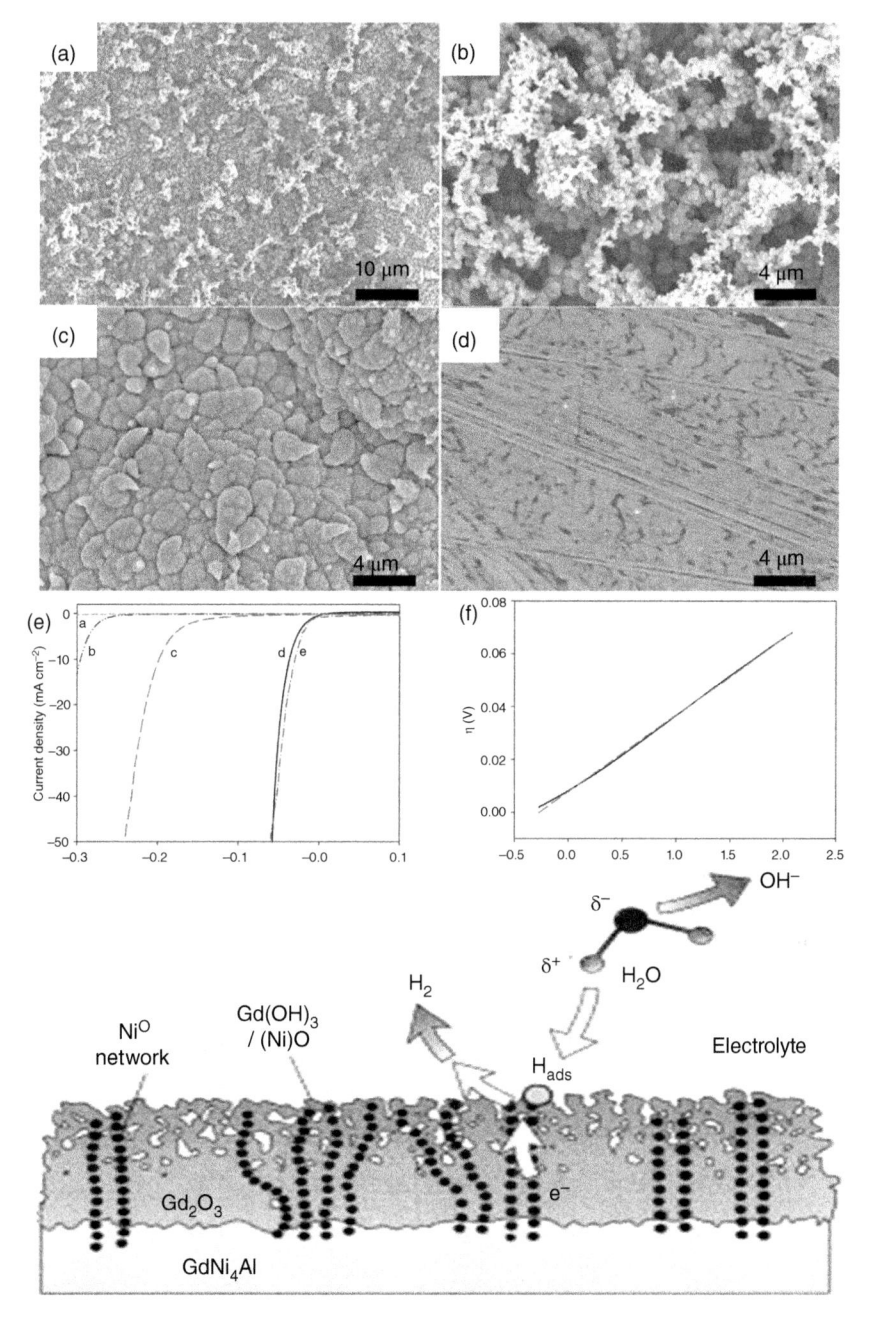

Figure 9.13 (a–d) SEM images of different electro-polymerized films. Source: (a) Wang et al. [115]/with permission of John Wiley and Sons. (a, b) PPy/MoS$_x$, (c) PPy, and (d) MoS$_x$. (e) Polarization curves of electrodes modified with: PPy (a), MoS$_x$ (b), (NH$_4$)$_2$MoS$_4$ (c), PPy/MoS$_x$ (d), and a commercial Pt/C electrode (e). (f) Tafel plot and the fitting curve (dash line) of the PPy/MoS$_x$ films in 0.5 M H$_2$SO$_4$ at a scan rate of 2 mV s^{-1}. Source: (e, f) Wang et al. [115]. Reproduced with permission of John Wiley and Sons. (g) Schematic representation of the structure of the surface film formed on the GdNi$_4$Al electrode in 1 M NaOH solution. Source: (g) Jukic et al. [116]. Reproduced with permission of Elsevier.

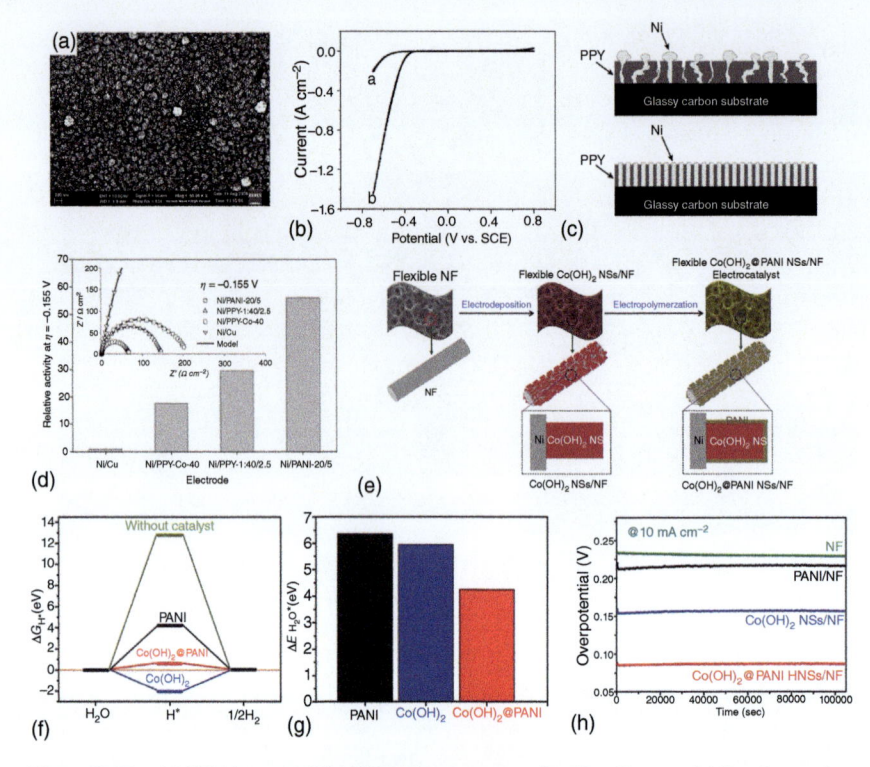

Figure 9.14 (a) SEM image of Pd-PEDOT nanocomposite film. Source: (a) Pandey and Lakshminarayanan [118]/with permission of American Chemical Society. (b) The cyclic voltammogram of the Pd disk electrode (a) and Pd-PEDOT nanocomposite-coated gold disk electrode (b) in 0.5 M H_2SO_4 at a scan rate of 100 mV s^{-1}. Source: (b) Pandey and Lakshminarayanan [118]. Reproduced with permission of American Chemical Society. (c) Schematic representation of Ni/CP catalyst layers produced by deposition of Ni on a pre-deposited PPY layer and co-deposition of Ni and PPY. (d) Relative electrocatalytic activity of the best three Ni/CP catalysts calculated at overpotential −0.155 V. Source: (c, d) Navarro-Flores and Omanovic [119]. Reproduced with permission of Elsevier. (e) The fabrication process of the Co(OH)$_2$@PANI HNSs/NF electrocatalysts. (f) HER free energy change for Co(OH)$_2$@PANI, Co(OH)$_2$, PANI, and without catalyst. (g) The histogram of H_2O adsorption energy change for Co(OH)$_2$@PANI, Co(OH)$_2$, and PANI. (h) Chronopotentiometric measurements of long-term stability of catalysts at 10 mA cm^{-2}. Source: (e–h) Feng et al. [120]. Reproduced with permission of John Wiley and Sons.

a 2D substrate. Further studies revealed that the small amounts of Ni with the thick pre-deposited PANI layer offered better catalytic activity, while Ni/PPy with lower porosity than Ni/PANI suffered from hydrogen accumulation at the Ni/PPy interface at high overpotentials (Figure 9.14c,d). The role of CPs in composite catalyst systems is controversial. CPs are less conductive materials than metal, and they show poor conductivity. Aydın and Köleli [121] demonstrated that PANI, PPy, and PANI-PPy coated Pt electrodes showed more negative potential values and higher activation energies in HER than Pt blank. For instance, the activation energies were c. 26 for PANI, 36.5 for PPy, 40.6 for PANI-PPy, and 20.6 kJ mol^{-1} for Pt. Moreover, metallic catalysts that linked to the polymer backbone also showed impressive activity and stability. Feng et al. [120] fabricated a nanosheet structured core–shell PANI-Co(OH)$_2$ on Ni foam (Co(OH)$_2$@PANI nanosheets/Ni foam) through electrodeposition and electropolymerized (Figure 9.14e). The Co(OH)$_2$@PANI nanosheets/Ni foam electrocatalyst possessed a large specific surface area, and the porous network accelerated the electrolyte diffusion. The Co(OH)$_2$ was bonded with PANI through synergistic coupling. The synergistic effects were shown by DFT to change the absorption energy of H atoms and accelerate the HER. Experiments further verified that the electron delocalization formed between the hydroxide d orbitals and the polymer π-conjugated ligand (Figure 9.14f,g). The shifted electronic state of Co(OH)$_2$ improved the activity of the catalyst in alkaline solution, which showed an onset potential of ~50 mV, a Tafel slope of 91.6 mV dec^{-1}, and a small overpotential of 90 mV at a current density of 10 mA cm^{-2}, and could maintain over 100 000 seconds (Figure 9.14h).

9.3.3 Carbon Dioxide Reduction Reaction (CO$_2$RR)

The most commonly explored electrocatalysts for CO$_2$RR are transition metal elements and their associated compounds [122]. These metals have vacant orbits and active d electrons, which energetically facilitate the bonding between the metal and the CO$_2$ molecule. Meanwhile, the organic components in the electrocatalysts were found to influence the catalytic performance like catalytic activity, product selectivity, and catalytic stability dramatically. CPs are generally used as catalysts directly or as support of the heterogeneous electrocatalysts of CO$_2$RR [105]. These polymers usually contain nitrogen functionalities that possess catalytic activity, complex metal cations, or anchor metal nanoparticles within the polymer backbone or the pendant chain [123, 124]. The CPs rich in nitrogen are the most studied polymer catalysts or catalyst supports for CO$_2$RR, including PANI and PPy. Aydin and Köleli [125] developed a PPy polymer catalyst and used it for electroreduction CO$_2$ under high pressure. The as-prepared catalyst showed a high CH$_3$COOH Faradic efficiency (FE) of 62.2% at an overpotential of −0.4 V (vs. Ag/AgCl) in methanol (CH$_3$OH) solution. The nitrogen-rich polymer PANI was also adopted as CO$_2$RR catalysts and showed excellent selectivity, mainly toward organic acid. For example, Köleli et al. [126] prepared a PANI catalyst, and the CO$_2$RR was carried out in CH$_3$OH solution at −0.4 V (vs. SCE). The reaction products were formic acid and acetic acid. The maximum FEs of the products were 12% for formic acid and 78% for acetic acid (Figure 9.15a). In these reports, the contained nitrogen functionalities were thought of as the active sites as they could adsorb and activate CO$_2$ molecules. Despite a certain CO$_2$RR ability on CPs, the catalytic property is unsatisfactory for limited product species and distribution. The metal–polymer catalysts combined the advantages of CPs and metal catalysts and were reported to be efficient in CO$_2$RR. Several electrochemical and non-electrochemical methods have been developed to produce metal–polymer catalysts, such as adsorption, solvent casting, chemical reduction deposition techniques, and electrochemical synthesizing dissolved monomers. The electrochemical synthesis is preferred for its controllable loading amount [130].

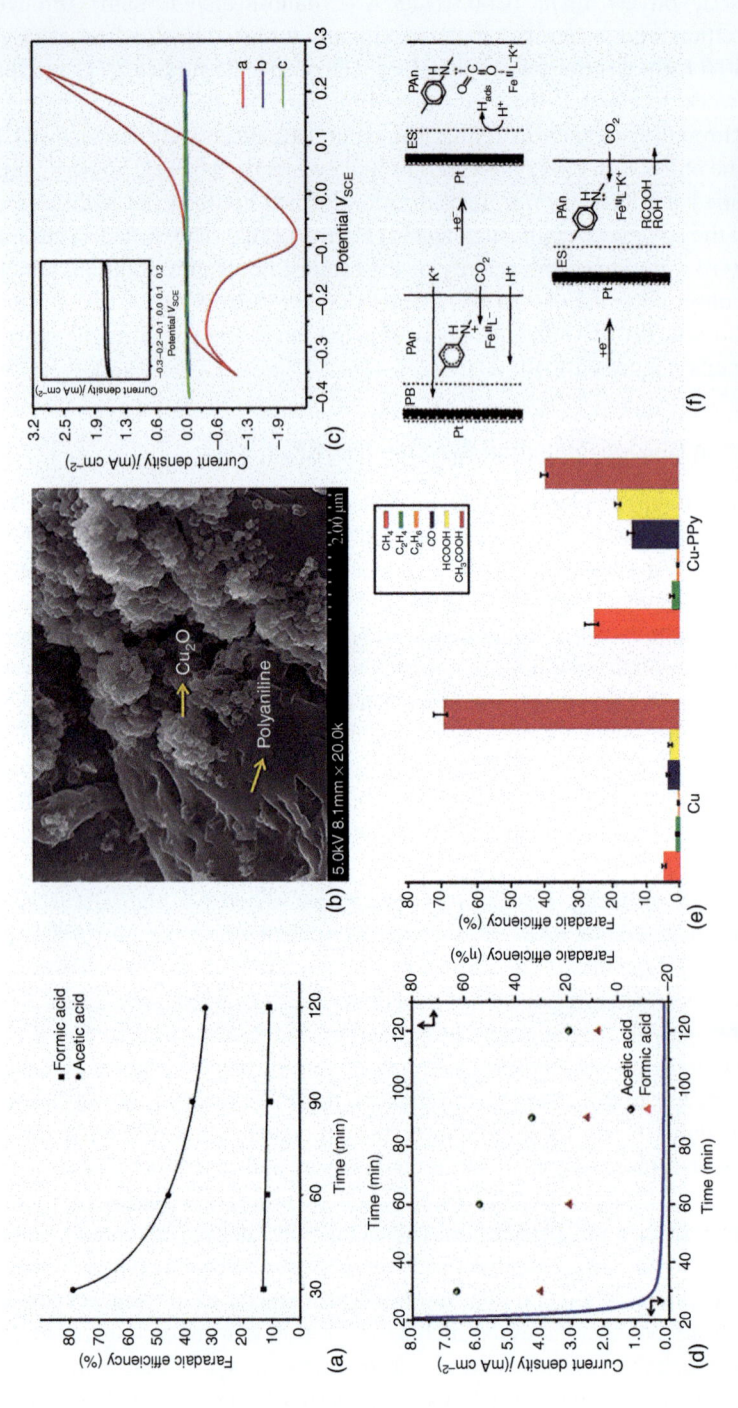

Figure 9.15 (a) Faradaic efficiency-time diagram of CO_2 reduction on PAn electrode in MeOH/0.25 M LiClO$_4$/H$^+$/H$_2$O. Source: Köleli et al. [126]. Reproduced with permission of Elsevier. (b) SEM images of PANI/Cu$_2$O films. Source: Grace et al. [127]/with permission from Elsevier. (c) Current-potential curves of CO_2 reduction on PANI/Cu$_2$O electrode in 0.1 M TBAP/methanol (a) CO_2 saturated (b) blank and (c) in N$_2$; scan rate = 50 mV s^{-1}. (d) Current density profile and Faradaic efficiencies of CO_2 reduction on PANI/Cu$_2$O electrode in TBAP/methanol system (−0.3 V vs. SCE). Source: (c, d) Grace et al. [127]. Reproduced with permission of Elsevier. (e) Faradaic efficiencies to formed products of CO_2 reduction on blank Cu and Cu-PPy electrocatalysts in CH$_3$OH/0.1 M LiClO$_4$/1.5 × 10^{-3} M H$^+$. Source: Aydin et al. [128]/with permission of Elsevier. (f) Schematic representation of the electrode reactions involving CO_2 at Pt/PB/PAn-Fe^{III}L. Source: Ogura et al. [129]. Reproduced with permission of IOP Publishing.

For example, Grace et al. [127] fabricated Cu_2O nanoparticles decorated PANI matrix (PANI/Cu_2O) catalyst using cyclic voltammetry (CV) and constant current technique (Figure 9.15b). Experiments showed that Cu(I) species were detected in the catalyst surface during the CO_2RR in 0.1 M tetrabutylammonium perchlorate and CH_3OH electrolyte. The adsorbed H atoms transferred to CO2 through the polymer film to form the products. The main products were HCOOH and CH_3COOH with FEs of 30.4% and 63.0% at a potential of −0.3 V (vs. SCE), respectively (Figure 9.15c,d). Aydın et al. [128] investigated the effects of PPy on Cu catalyst in a CH_3OH/0.1 M $LiClO_4$/H^+ electrolyte system. As-prepared catalysts achieved a CH_4 FE of 25% and HCOOH FE of 20% under optimum values (Figure 9.15e). The coating of PPy caused a shift in the product distribution of hydrocarbons, especially CH_4, which was different from the fixed product distribution for a specific catalyst mentioned in previous reports. Ogura et al. [129] designed a Pt [1,8-dihydroxynaphthalene-3,6-disulfonatoferrate(II)] complex immobilized PANI/Prussian blue-laminated catalyst and investigated the CO_2RR performance on it. Experiments showed that the major product was not CO but $C_3H_6O_3$, CH_3COOH, HCOOH, CH_3OH, and CH_3CH_2OH. The multi-carbon products were ascribed to the roles of PANI and the metal complex that render CO_2 active by combining the electrophilic C atom. Moreover, the activated CO_2 was hydrogenated and converted to organic acids and alcohols by reacting with H atoms (Figure 9.15f).

In summary, conducting polymer-based electrocatalysts are cheap, readily available, and scaled up for industry. They possess unique conjugated structures and excellent electrochemical properties, showing promising catalytic performances for electrocatalysis. Besides understanding the catalytic mechanisms and breakthroughs in synthesizing CPs with few structural defects and ideal nanostructures, conducting polymer-based electrocatalysts has a future with wide practical applications.

9.4 Supercapacitors

Supercapacitors (SC), also called ultracapacitors or electrochemical capacitors, show high power density, fast charge/discharge rate, low maintenance cost, safe operation, and long durability, representing one of the most promising energy storages and recycling [131]. According to their distinguished charge storage mechanisms, supercapacitors can be divided into electro double-layer capacitors (EDLC) and Faradic pseudocapacitors, depending on their distinguished charge storage mechanisms. The specific capacitance of EDLC, associated with the charge stored via electrostatic charge accumulation at the electrode/electrolyte interface, depends on the pore size distribution and the efficient surface area of the electrode materials. However, pseudo-capacitive possesses relatively high specific capacitances due to the fast and reversible redox reactions at or near the electrode surface [132]. CPs are pseudo-capacitive materials that can engage in electrochemical doping or redox reaction with anions and cations. The mechanism of storing charges of CPs can be described by the following two formulas: p-doping upon oxidization and n-doping upon reduction,

$$C \rightarrow CP^{n+}(A^-)_n + ne^- \text{(p-doping)} \quad (9.1)$$

$$C + ne^- \rightarrow (C^+)_n CP^{n-} \text{(n-doping)} \quad (9.2)$$

where A^- and C^+ represent the anion and cation in the electrolyte. Most CPs work as p-doped states, such as PANI and PPy, and only PTh materials can be n-doped in commonly used CPs. The counter ions with opposite charges will be entrapped or released from the polymer matrix during the reversible doping and de-doping. The conductivity of CPs can be tuned over about 13 orders of magnitude simply by adjusting the doping level. According to the design of electrode materials, CP-based electrode materials have been divided into conductive polymers and their composites. The former uses CPs as the active material without adding other materials, and the latter adopts CP composites with other electrode materials, such as transition metal oxides and carbon materials.

9.4.1 CP as the Active Material

Many CPs, including PANI [133], PPy [134], PTh [135], and derivatives [136], have been widely applied in electrode materials due to high capacitances, good flexibility, high conductivity, lightweight, and low cost. Park et al. [133] has synthesized three PANI nanostructures with nanospheres, nanorods, and nanofibers morphologies then employed a polymeric stabilizer of PVP. The diameters of the PANI nanostructures were similarly found to be around 50 nm, and the lengths of the nanorods and nanofibers were about 150 and 215 nm (Figure 9.16a–c), respectively. The authors explored the structure-property relationship of the CP nanostructures and the capacitive performances when the three nanostructures were used as the electrode-active materials for supercapacitors. The specific capacitance was determined from GCD measurements at a current density of $0.1\,A\,g^{-1}$, where the potential range was chosen by referring to the CV curves (Figure 9.16d). The specific capacitances were calculated from discharge curves to be 71, 133, and $192\,F\,g^{-1}$ for nanospheres, nanorods, and nanofibers, respectively (Figure 9.16e). The variations of the discharge capacitance with current density are plotted. Although the specific capacitances decreased at current densities of more than $0.1\,A\,g^{-1}$, the above tendency of the capacitance value remained unchanged over the entire current density range. The capacitors based on the three PANI nanostructures showed morphology-dependent capacitance values, and the nanofiber electrode had faster electrode kinetics and better capacitance than the nanorod and nanosphere electrodes. The discharging time increased in the order of nanospheres < nanorods < nanofibers over the same potential range, indicating that the nanofiber electrode has the highest specific discharge capacitance (Figure 9.16f).

Li et al. [137] synthesized uniform spherical PANI nanoparticles with the assistance of imidazolium-based ionic liquids. The PANI with different morphologies and sizes were successfully prepared through chemical polymerization in

Figure 9.16 SEM images of PANI nanostructures with different aspect ratios showing their size distribution. (a) Nanospheres, (b) nanorods, and (c) nanofibers. Source: (a) Park et al. [133]/with permission from American Chemical Society. Capacitances of PANI nanostructure electrodes measured at a current density of 0.1 A g^{-1} in a three-electrode cell: (d) GCD curves, (e) gravimetric discharge capacitances. Comparison of specific discharge capacitances at different current densities measured in a three-electrode cell (f) comparison of specific discharge capacitances at different current densities measured in a three-electrode cell. Source: (d–f) Park et al. [133]. Reproduced with permission of American Chemical Society.

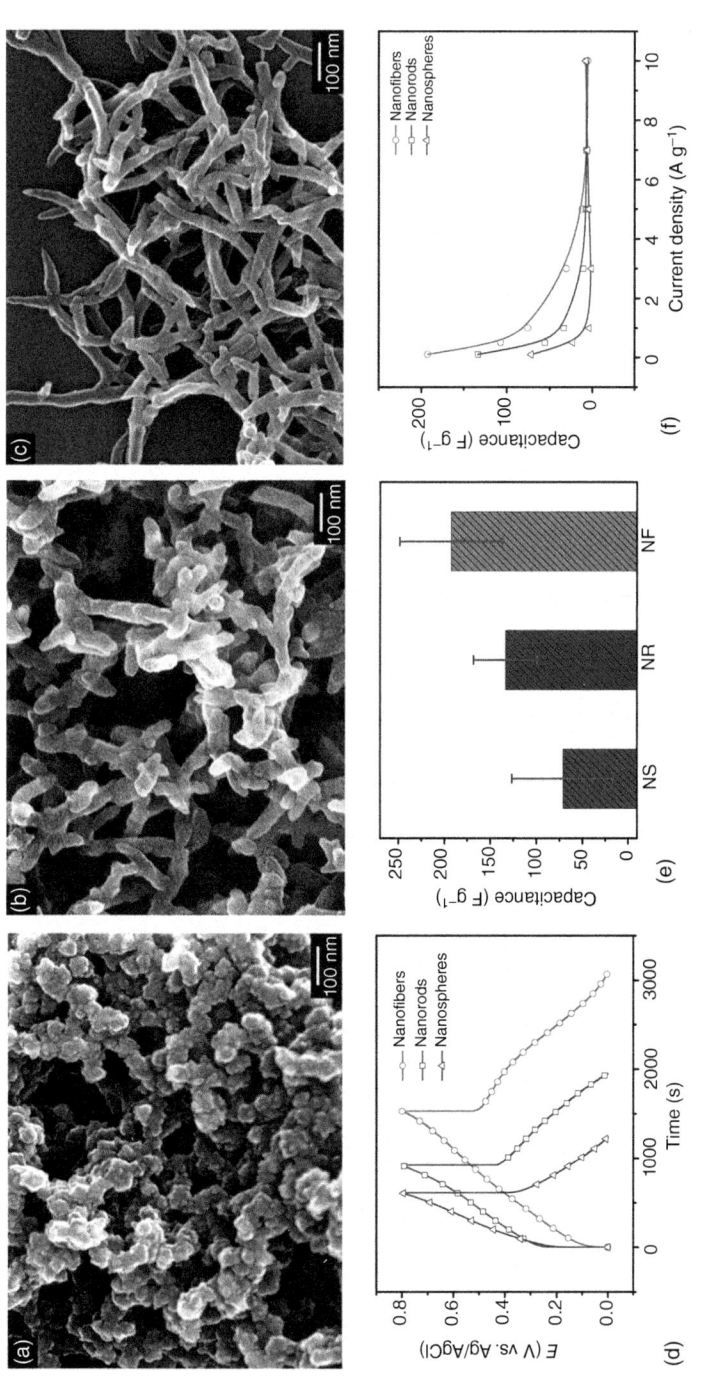

1-ethyl-3-methylimadozolium bromide ([emim][Br]), 1-ethyl-3-methylimadozolium tetrafluoroborate ([emim][BF$_4$]) and deionized water, denoted as PANI-1, PANI-2, and PANI-3, respectively. As shown in Figure 9.17a,b, PANI-1 and PANI-2 present spherical structures with different sizes ranging from 100 to 120 nm in [emim][Br] and ranging from 50 to 80 nm in [emim][BF$_4$]. However, the as-prepared PANI-3 using water as solvent presents random stacking nano-cudgel with size of 250–300 nm (Figure 9.17c). The results demonstrate that the imidazolium-based ILs with different anions as templates play an important role in forming PANI particles to improve supercapacitor performances, such as low charge transfer resistance, high apparent diffusion coefficient, large redox specific capacity, and high stability. As shown in Figure 9.17d,e, the PANI-2 achieved excellent electrochemical performance with an initial capacity of 625 F g^{-1} and retained an effective capacity of 565.8 F g^{-1} after 2000 cycles. Furthermore, the PANI-2 retain 90.4%, 87.5%, 85.0%, and 82.1% at current densities of 0.1, 0.2, 0.5, and 1.0 A g^{-1} after 2000 cycles, respectively (as shown in Figure 9.17f). The results reveal that the chemical polymerization in IL could be a new method in preparing specific-size nanoparticles for decreasing transfer resistance and increasing electronic conductivity.

Wang et al. [138] developed a 3D hierarchical PANI micro/nanostructure (H-PANI) as supercapacitor electrode material (Figure 9.18a–c). The H-PANI has excellent electrical conductivity and a unique porous structure, promoting rapid charge transfer and ion diffusion. The as-prepared H-PANI electrode shows a relatively high specific capacitance of 520 F g^{-1} at a current density of 0.5 A g^{-1} and exhibits excellent rate capability (65% specific capacitance retains from 0.5 to 50 A g^{-1}) (Figure 9.18d–f). The electrochemical performance of this hierarchically-structured electrode material showed a great improvement compared with the simple structured PANI. Notably, the H-PANI was delicately synthesized by adjusting the ionic strength and temperature of the regular dilute polymerization reaction system of PANI. No need for templates or organic structure-directing reagents, which made the preparation process simple and low-cost. This study will provide a new perspective for developing various PANI materials for supercapacitor applications.

Shi et al. [134] prepared a hierarchically 3D-nanostructured elastic PPy hydrogel by interfacial polymerization of pyrrole at the isopropanol alcohol/water interface using phytic acid as a crosslinker dopant and ammonium persulfate (APS) as an oxidant (Figure 9.19a,b). The simple synthesis method offers conductive hydrogel tunable nanostructures, electrochemical performance, and scalable processability. The pore size in the 3D porous hierarchical nanostructure can be tuned by adjusting the molar ratio of pyrrole to phytic acid. As shown in Figure 9.19c, the flexible symmetric PPy hydrogel supercapacitors showed a high areal capacitance of 6.4 F cm^{-2} at a mass loading of 20 mg cm^{-2} with good cycling stability (less than 7% capacitance degradation after 2000 cycles).

Among these nanomaterials, heterogeneous core–shell nanostructure represents a unique system for applications in electrochemical energy storage devices, having the advantages of both components and special synergistic effects. Li et al. [139] prepared PANI/PPy composite nanofibers with core–shell structures by covering

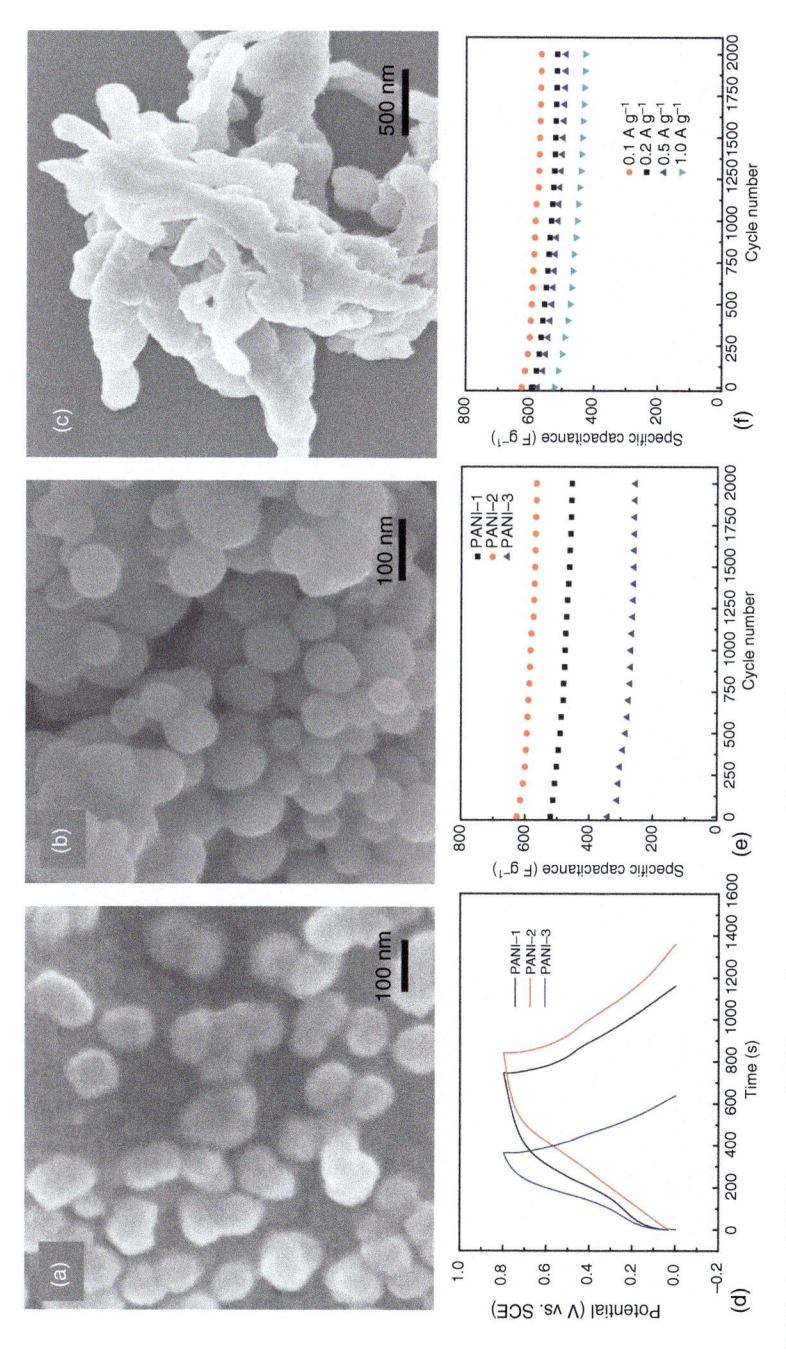

Figure 9.17 (a–c) SEM images of PANI samples were prepared in [emim][Br], [emim][BF$_4$], and deionized water, respectively. Source: (a) Li et al. [137]/with permission from Elsevier. (d) GCD of PANI-1, PANI-2, and PANI-3 at 0.1 A g^{-1} current density. Cyclic performance of (e) PANI-1, PANI-2, and PANI-3 electrode at 0.1 A g^{-1} current density. (f) PANI-2 electrode at different current densities. Source: (d–f) Li et al. [137]. Reproduced with permission of Elsevier.

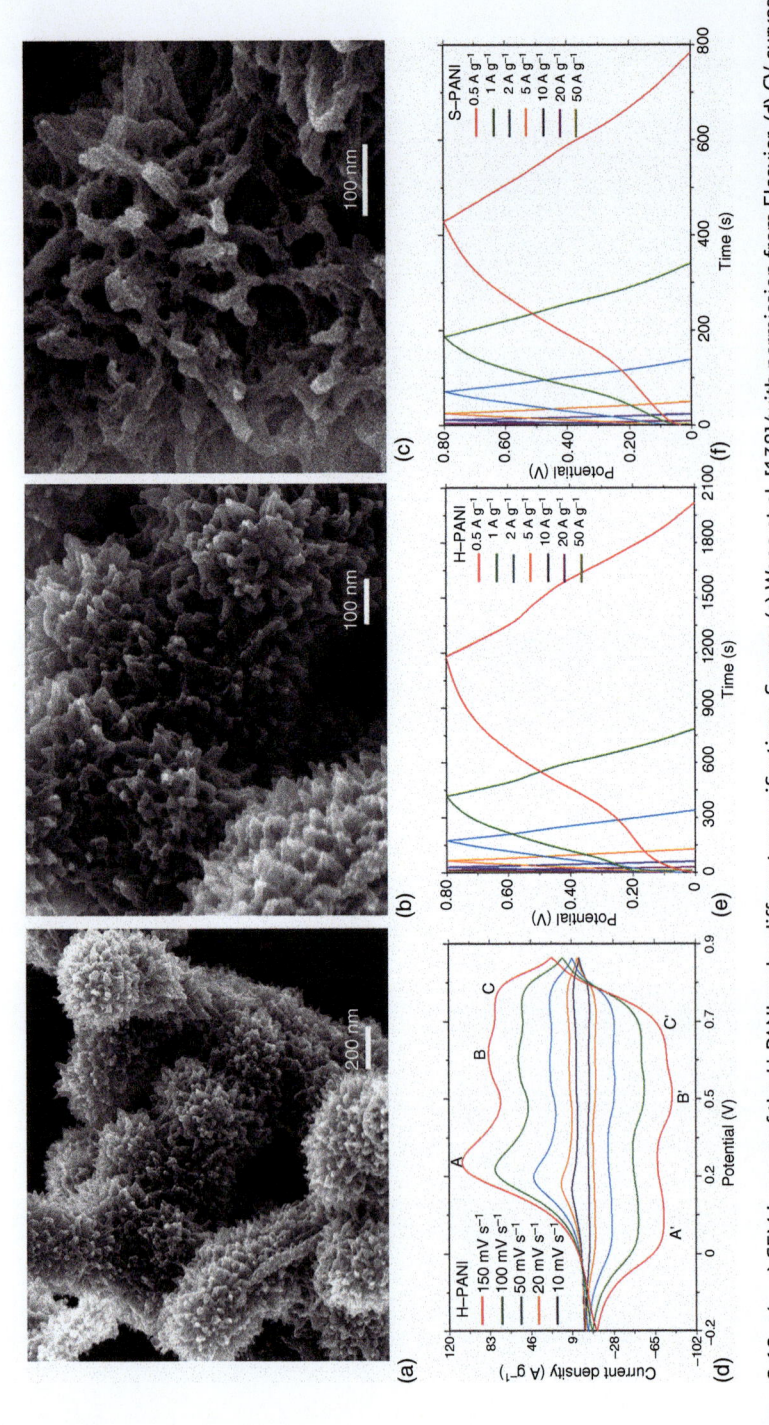

Figure 9.18 (a–c) SEM images of the H-PANI under different magnifications. Source: (a) Wang et al. [138]/with permission from Elsevier. (d) CV curves for the H-PANI electrode at varying scan rates from 10 to 150 mV s^{-1} within the potential window from −0.2 to 0.86 V. (e) GCD curves for the H-PANI electrode at varying current densities from 0.5 to 50 A g^{-1} within a potential window ranging from 0 to 0.8 V. (f) Galvanostatic charge–discharge curves for the S-PANI electrode at varying current densities from 0.5 to 50 A g^{-1} within a potential window ranging from 0 to 0.8 V. Source: (d–f) Wang et al. [138]. Reproduced with permission of Elsevier.

Figure 9.19 Microstructures and capacitance performance of PPy hydrogels with different ratios of pyrrole:phytic acid. (a, b) SEM images of the PPy hydrogel with Py:PA ratios of 5 : 1 and 10 : 1, respectively. Source: (a) Shi et al. [134]/with permission from Royal Society of Chemistry. (c) Specific capacitance vs. current density profiles of PPy hydrogel electrodes with Py:PA ratios of 5 : 1, 10 : 1, and 20 : 1. Source: (c) Shi et al. [134]. Reproduced with permission of Royal Society of Chemistry.

PPy thin layers on the surface of PANI nanofibers as high-performance electrode materials for supercapacitors in neutral aqueous electrolyte. The core part of PANI nanofibers was firstly synthesized through chemical oxidative polymerization of aniline monomers in the presence of phenylenediamine (PDA) isomers, including o-phenylenediamine (o-PDA), m-phenylenediamine (m-PDA), and p-phenylenediamine (p-PDA) without the assistance of any templates or usage of organic solvents. Then the shell part of PPy was fabricated by *in situ* chemical oxidative polymerization of pyrrole with the PANI as a seed (Figure 9.20a). Furthermore, electrochemical behaviors of PANI in H_2SO_4 and Na_2SO_4 electrolyte and corresponding composites in Na_2SO_4 electrolyte were tested by cyclic voltammetry and GCD techniques. As shown in Figure 9.20b–d, it is found that compared with pure PANI prepared without PDA isomers, the incorporation of o-PDA and p-PDA is helpful to improve the electrochemical property of PANI nanofibers. Especially for o-PDA, the resulting PANI nanofibers exhibited the largest specific capacitance of 1115.7 F g^{-1} at the scan rate of 5 mV s^{-1}, and 345.3 F g^{-1} at the specific current of 0.5 A g^{-1} in 1.0 M H_2SO_4 electrolyte. However, in 0.5 M Na_2SO_4

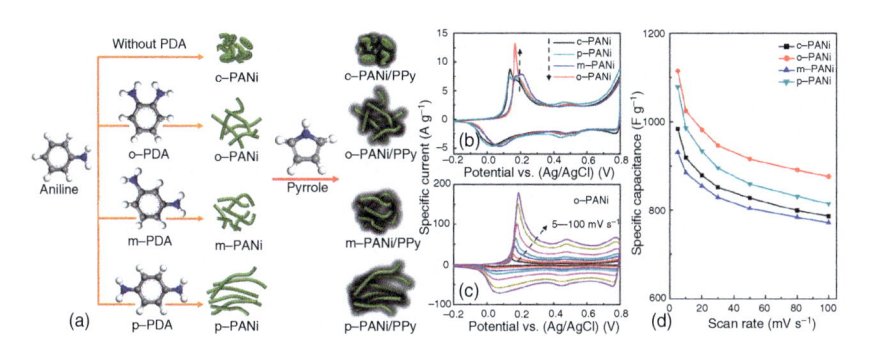

Figure 9.20 (a) Schematic illustration of the preparation of PANI and corresponding composites. (b) CV curves at the scan rate of 5 mV s^{-1}. (c) CV curves at various scan rates of 5–100 mV s^{-1}. (d) Plots of corresponding specific capacitance as a function of scan rate for as-prepared PANI in 1.0 M H_2SO_4. Source: Zhou et al. [139]. Reproduced with permission of Elsevier.

electrolyte, its specific capacitance decreased to 254.5 and 210.4 F g^{-1}, whereas 834.6 and 652.5 F A g^{-1} for PANI/PPy composites at the scan rate of 5 mV s^{-1}, respectively. Moreover, a significant improvement on cycling stability for the composites could be achieved due to the unique core–shell nanostructure and strong synergy effect between PANI and PPy.

Chen et al. [136] deposited PEDOT on the flexible paper by vapor phase polymerization (VPP), which can prepare electrodes on a large scale. The fabricated PEDOT/paper possesses a porous structure, facilitating sufficient contact between electrolyte and electrode material and benefiting ion diffusion (Figure 9.21a–c). The effect of different amounts of PEDOT on the electrochemical properties of the PEDOT/paper is investigated (Figure 9.21d–f). As expected, the as-prepared PEDOT/paper electrode exhibits a high areal capacitance of 639 mF cm^{-2} and good conductivity of 78 S cm^{-1}. It has remarkable stability under different bending angle conditions, which can be directly used as an electrode for flexible and lightweight energy storage devices.

9.4.2 CP Composites as the Active Materials

Recently, more intensive efforts have been devoted to designing hybrid supercapacitors based on CPs, combining the advantages of EDLC and pseudo capacitor and meeting the requirements of high energy density, high power density, and high cyclability, in which CPs are composited with carbon materials or transition metal complexes. Among them, carbon materials have high conductivity and good mechanical properties [140–143]. Transition metal oxide/hydroxide materials possess good electrochemical activity and high specific theoretical capacitance [144–151]. Composite materials consist of combining two or more materials. Each component exhibits its unique chemical, mechanical and physical properties. Therefore, with the adoption of composite-based electrode materials, the performance of supercapacitors can be improved by integrating the advantages of several materials.

Fu et al. [140] reported that polythiophene (PTh) was electropolymerized onto MWCNTs modified glassy carbon (GC) electrode in ionic liquid bmimPF$_6$ solution successfully (Figure 9.22a,b). The different electrochemical behavior of thiophene in bmimPF$_6$ solution was studied by cyclic voltammetry. As shown in Figure 9.22c, the CV test displays that PTh/MWCNT composites have better capacitive properties than pure MWCNT and pure PTh. Figure 9.22d depicts the specific capacitance of the PTh/MWCNT as a function of the mass ratio of PTh to MWCNT. It is found that the specific capacitance increases with increasing mass ratio. When the mass ratio rises to 60%, the specific capacitance approaches its maximum. As shown in Figure 9.22e, all the CV curves maintain a relatively rectangular shape even at a scan rate as high as 80 mV s^{-1}, indicating their good supercapacitor performances. It can be observed that the curve in Figure 9.22f behaves as triangular during the charge–discharge process within the whole potential range, which proves that the supercapacitor has good capacitive behavior and charge–discharge reversibility.

Dong et al. [141] synthesized an ionic liquid-modified RGO-IL/PANI composite (Figure 9.23a–c). The ionic liquid enlarged the interlayer distance of RGO

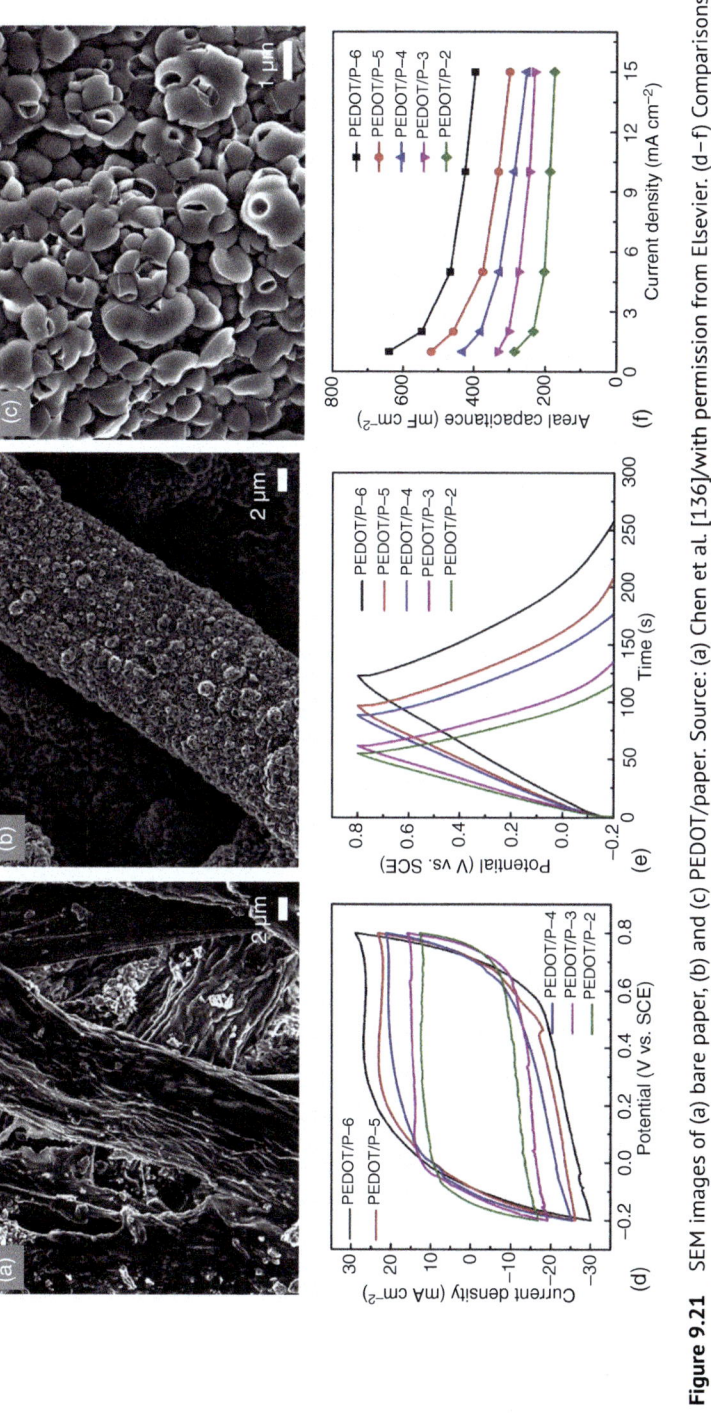

Figure 9.21 SEM images of (a) bare paper, (b) and (c) PEDOT/paper. Source: (a) Chen et al. [136]/with permission from Elsevier. (d–f) Comparisons of electrochemical performance of PEDOT/P-2, PEDOT/P-3, PEDOT/P-4, PEDOT/P-5, PEDOT/P-6: (d) CV curves at 50 mV s^{-1}, (e) GCD curves at 5 mA cm^{-2}, (f) areal capacitance at different current density. Source: (d–f) Chen et al. [136]. Reproduced with permission of Elsevier.

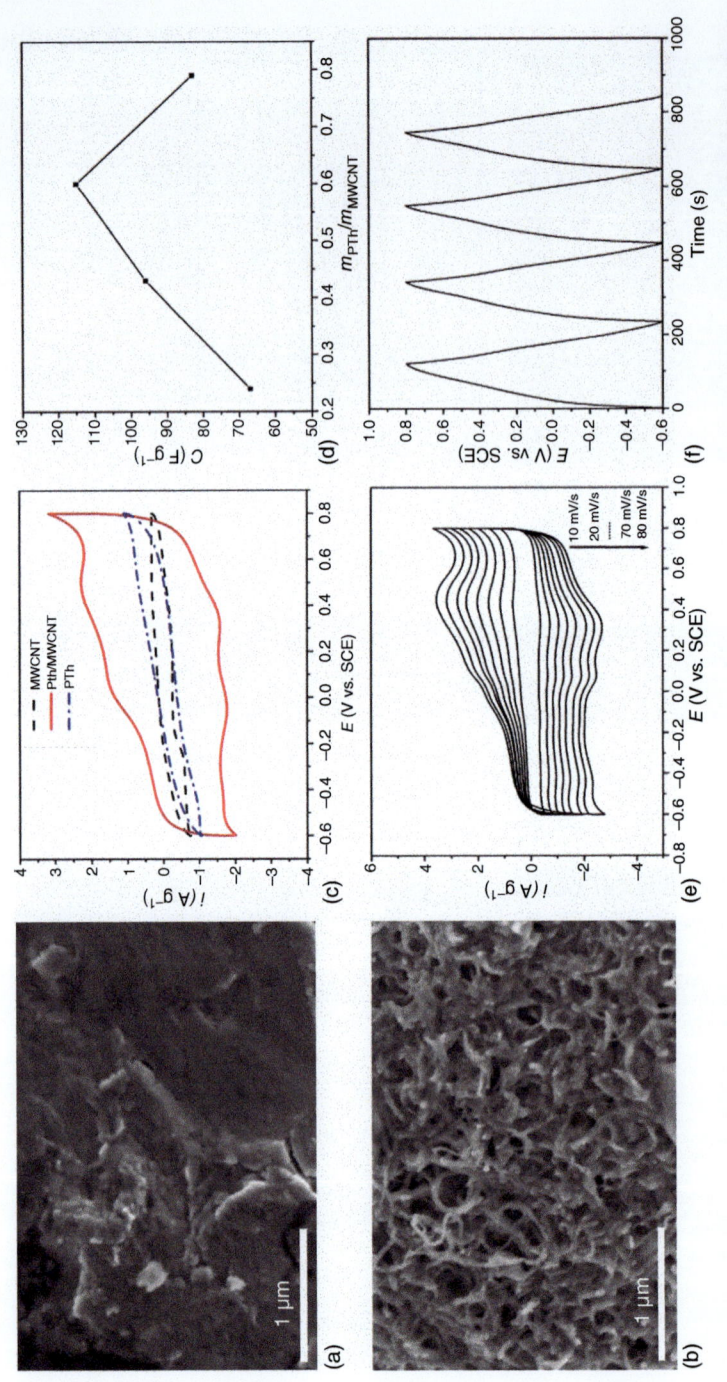

Figure 9.22 (a,b) SEM images of PTh and PTh/MWCNT composites. Source: (a) Fu et al. [140]/with permission from Elsevier. (c) CV of PTh/MWCNT (solid line), PTh (dash), and MWCNT (dash-dot) modified electrodes in 0.5 mol l^{-1} H$_2$SO$_4$ aqueous solution at the scan rate of 50 mV s^{-1}. (d) The specific capacitance of PTh/MWCNT as a function of the mass ratio of PTh to MWCNT. (e) CV curves of PTh/MWCNT composites electrode in 0.5 mol l^{-1} H$_2$SO$_4$ aqueous solution at different scan rates from 10 to 80 mV s^{-1}. (f) GCD curves of PTh/MWCNT composites in 0.5 mol l^{-1} H$_2$SO$_4$ aqueous solution at a constant current density of 1 A g^{-1}. Source: (c–f) Fu et al. [140]. Reproduced with permission of Elsevier.

Figure 9.23 (a–c) SEM images of RGO-IL/PANI at different magnifications. Source: (a) Dong et al. [141]/Royal Society of Chemistry/CC BY 4.0. (d) CV curves of different composite electrodes at 10 mV s^{-1}. (e) CV curves of RGO-IL/PANI at different scanning rates. (f) GCD curves of RGO-IL/PANI at current densities from 1 to 10 A g^{-1}. Source: (d–f) Dong et al. [141]. Reproduced with permission of Royal Society of Chemistry.

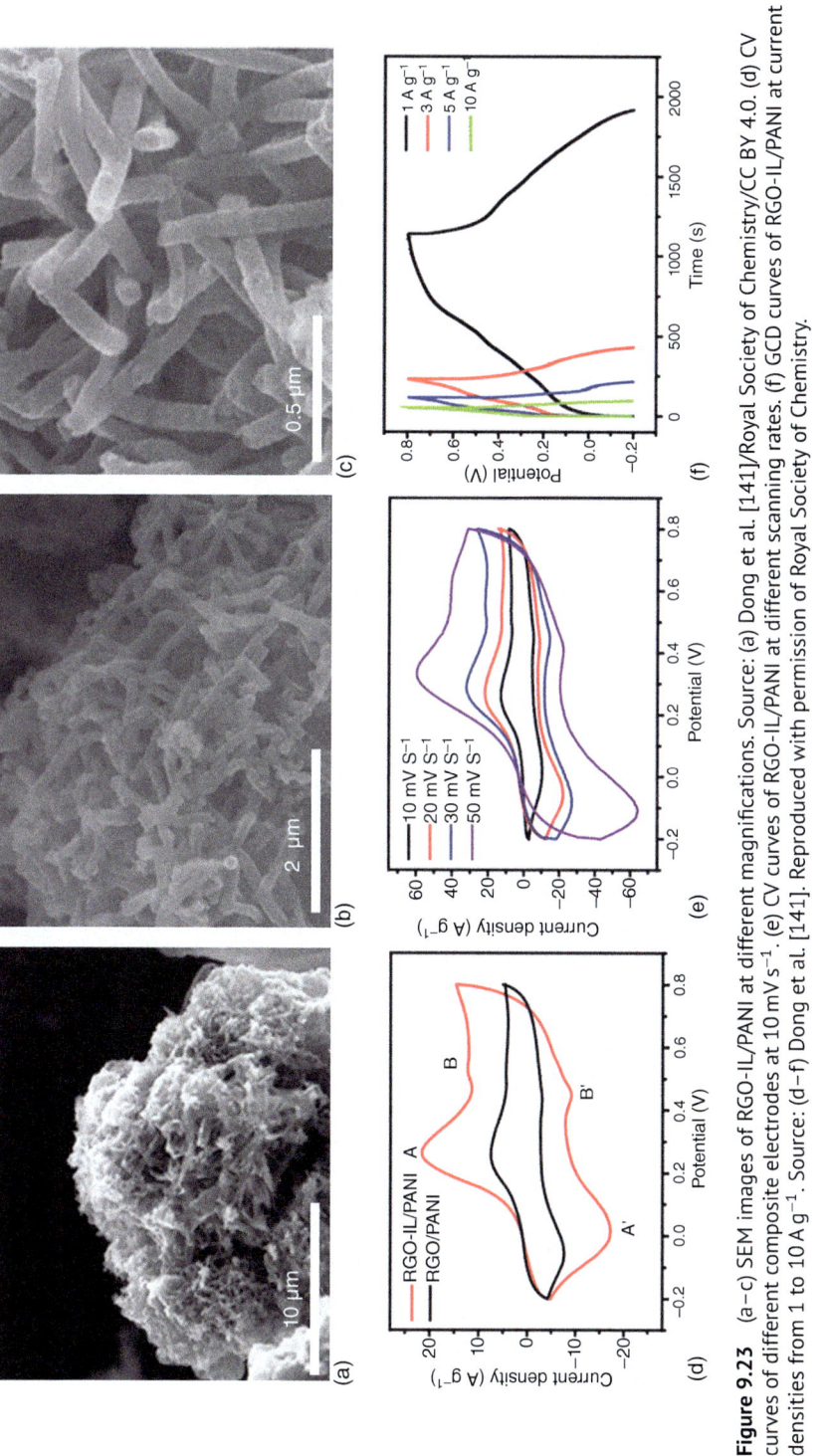

sheets and acted as a PANI dopant to improve the orderly establishment of PANI. Figure 9.23d shows the CV curves of RGO/PANI and RGO-IL/PANI at a scan rate of $10\,mV\,s^{-1}$. The curves for RGO/PANI and RGO-IL/PANI displayed two couples of apparent peaks (0.02/0.23 V and 0.42/0.54 V). When the scanning rate increased from 10 to $50\,mV\,s^{-1}$, the potential change corresponding to the redox peaks from RGO-IL/PANI (Figure 9.23e) was smaller than that from RGO/PANI, indicating that RGO-IL/PANI had better structural stability and rate capability. The quasitriangular shape of the charge–discharge curves in Figure 9.23f, which were not perfectly straight lines, implied that a faradaic transformation occurred. It confirmed the pseudocapacitive behavior in the composite materials over the range of current densities.

Ren et al. [142] fabricated highly stretchable electrodes based on a graphene foam/PPy composite through CVD and interfacial polymerization methods (Figure 9.24a–d). Because the tensile strain applied to the PPy network is shared by the structural deformation of the graphene foam, the graphene/PPy composite can accommodate the stretching deformation without a significant decay in conductivity. Benefiting from these superior features, an all-solid-state supercapacitor assembled with graphene/PPy compact film electrodes shows a high areal specific capacitance of $258\,mF\,cm^{-2}$, energy density of $22.9\,\mu Wh\,cm^{-2}$ at a power density of $0.56\,mW\,cm^{-2}$ (Figure 9.24e–g).

Kim and Shin [143] reported a method of fabricating PPy composites (CDPY) by CVD polymerization on flexible graphene substrate with the assistance of dopamine (Figure 9.25a). The dependency of initiators such as $FeCl_3$ and $CuCl_2$ for the polymerization of pyrrole is elucidated for preparing CDPY-Fe and the CDPY-Cu. The morphological changes and doping levels of PPy affect the capacitive behaviors of the nanocomposites. In addition, the assistance of dopamine between PPy and graphene improves the uniformity of PPy layers and enhances the supercapacitor performance. As shown in Figure 9.25b,c, the areal capacitance of the CDPY-Fe and the CDPY-Cu is 3.15 and $2.20\,mF\,cm^{-2}$, whose trend is identical to that from CVs curves.

Cai et al. [144] reported a novel hierarchical core/shell PEDOT@MoS_2 composite assembled by the electrochemical co-deposition of EDOT and MoS_2 submicron spheres. The transmission electron microscope (TEM) images of the as-obtained composite explicitly reveal the core/shell heterostructure of PEDOT@MoS_2 with a diameter of about 800 nm (Figure 9.26a–c). As shown in Figure 9.26d–f, the electrochemical characterizations show that the optimized PEDOT@MoS_2 composite possesses a high specific capacitance of $2540\,mF\,cm^{-2}$ at $1\,mA\,cm^{-2}$ and excellent capacitance retention of 98.5% after 5000 cycles at a high current density of $100\,mA\,cm^{-2}$. The assembled PEDOT@MoS_2/PEDOT asymmetric supercapacitor shows a high energy density of $937\,Wh\,m^{-2}$ at $6500\,W\,m^{-2}$ and outstanding cycling stability with capacitance retention of 100% after 5000 cycles. The enhanced performances are attributed to the robust hierarchical core/shell structures and the synergic effect between PEDOT and MoS_2, which is a promising candidate for applications in high-performance supercapacitors.

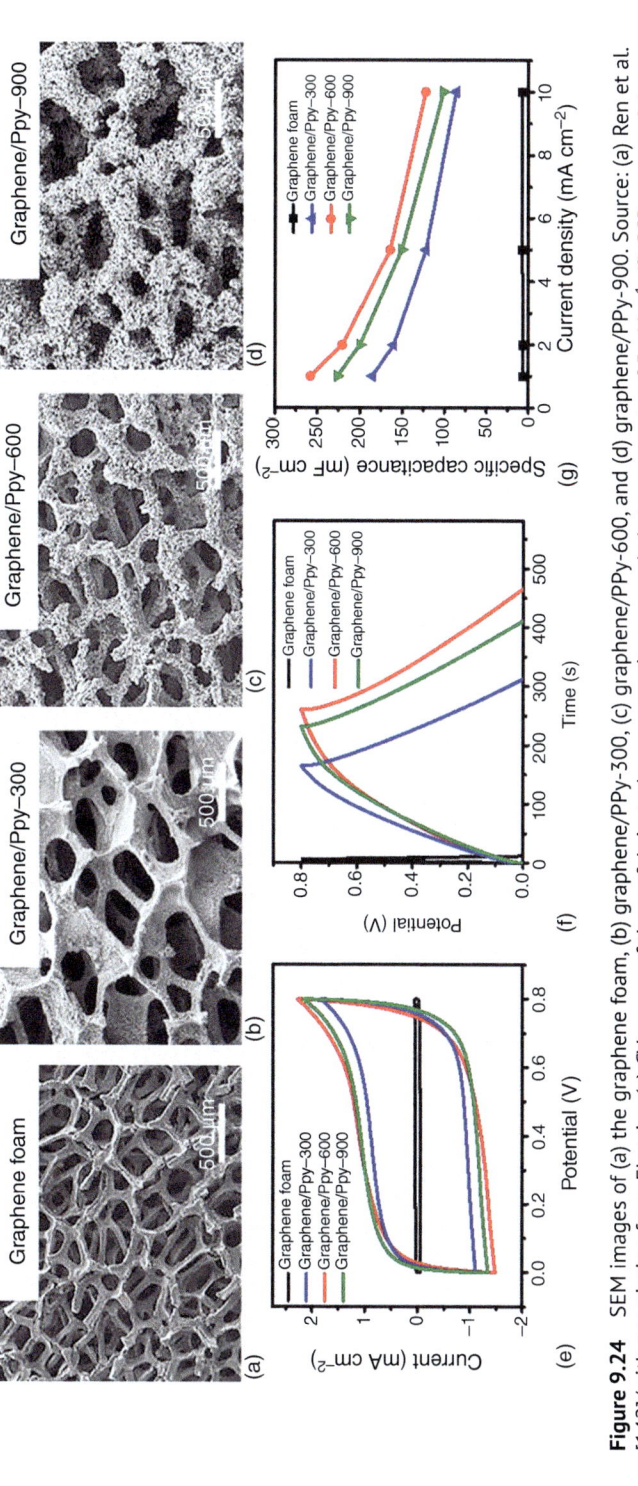

Figure 9.24 SEM images of (a) the graphene foam, (b) graphene/Ppy–300, (c) graphene/PPy–600, and (d) graphene/PPy–900. Source: (a) Ren et al. [142]/with permission from Elsevier. (e) CV curves of the as-fabricated supercapacitors recorded at a scan rate of 5 mV s^{-1}. (f) GCD curves of the supercapacitors at a current density of 1 mA cm^{-2}. (g) Dependence of the specific capacitance of the devices on the discharge current. Source: (e–g) Ren et al. [142]. Reproduced with permission of Elsevier.

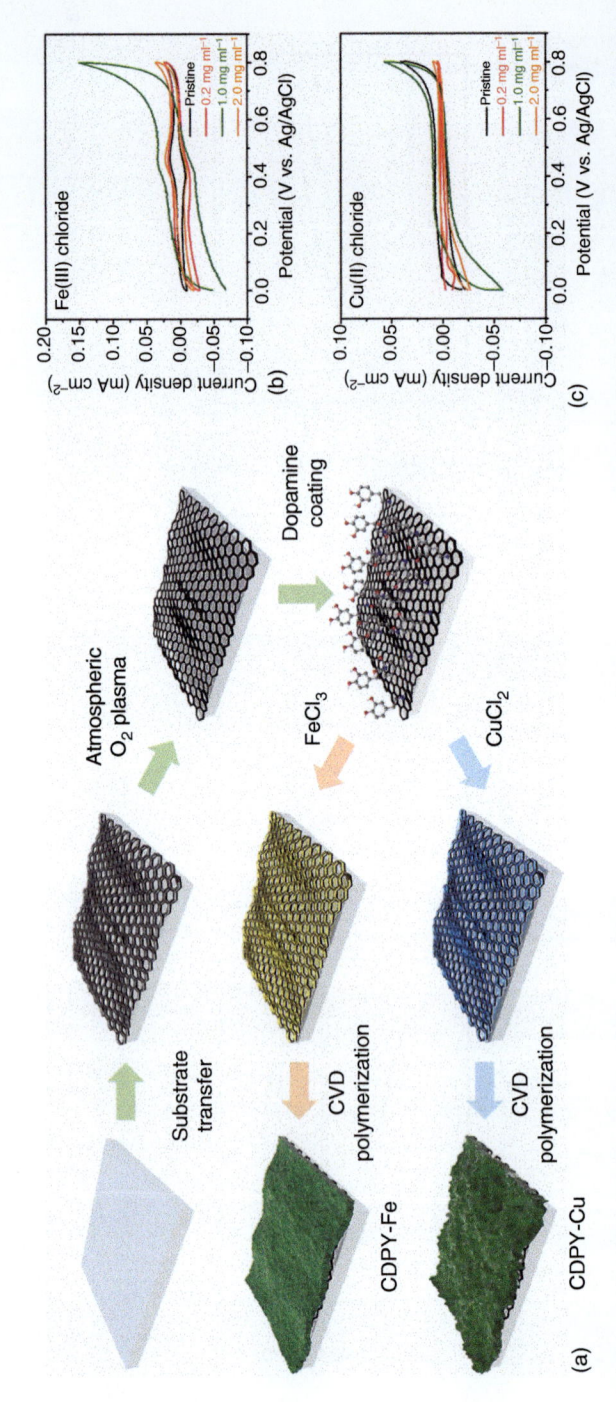

Figure 9.25 (a) A schematic illustration shows the fabrication of CDPY hybrid nanomaterials by introducing the dopamine layer onto the CVD graphene and CVD polymerization process. O_2 plasma treatment was conducted to modify the hydrophobic surface of CVD graphene to hydrophilic property. The dopamine was used as a dispersing agent, and $FeCl_3$ or $CuCl_2$ were utilized for initiator salts. (b, c) CV curves of CDPY-Fe and CDPY-Cu hybrid nanocomposites-based electrode as a function of the dopamine concentration. Source: Kim and Shin [143]. Reproduced with permission of Elsevier.

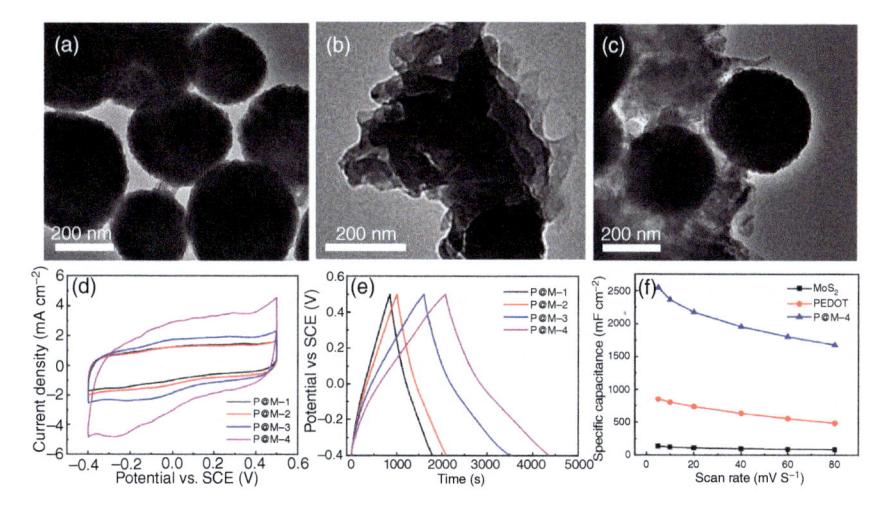

Figure 9.26 The TEM images of MoS$_2$ (a), PEDOT (b), and PEDOT@MoS$_2$/PEDOT (c) Source: (a) Cai et al. [144]/with permission from Elsevier. (d, e) CV and GCD curves of different PEDOT@MoS$_2$ composites at 5 mV s^{-1} and 1 mA cm^{-2}, respectively. (f) Specific capacitance of electrodes as a function of scan rates. Source: (d–f) Cai et al. [144]. Reproduced with permission of Elsevier.

Yuan et al. [145] reported spherical silver-incorporated conductive PPy (Ag/PPy) composite successfully prepared via oxidative polymerization of pyrrole in an aqueous Ag$^+$-containing solution in the presence of trisodium citrate (Figure 9.27a–d). Citrate produces stable and charged micelles, tailoring the facial polymerization of the pyrrole. Monodispersed submicron spheres of Ag/PPy composite are obtained by maintaining the molar ratio of citrate anion to silver cation at 3 : 1. Highly

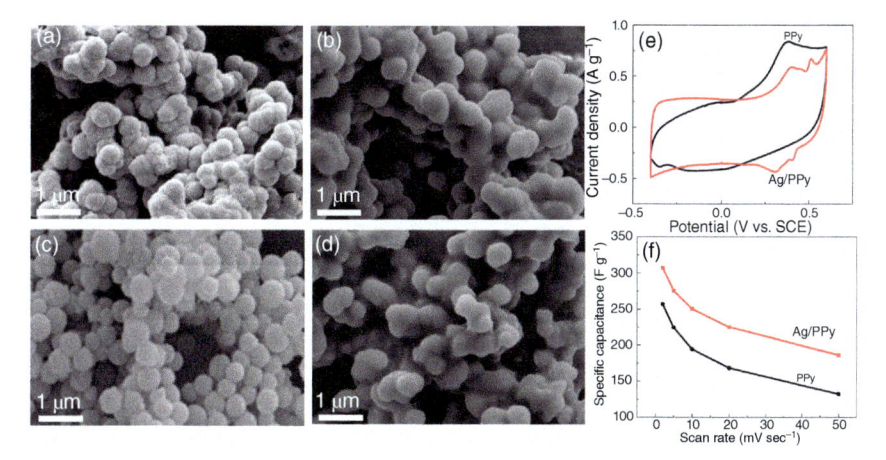

Figure 9.27 SEM images of the Ag/PPy composites obtained by controlling the molar ratio of citrate to Ag$^+$ at (a) 0, (b) 1.5 : 1, (c) 3 : 1, (d) 6 : 1. Source: (a) Yuan et al. [145]/with permission from Elsevier. (e) Cyclic voltammograms. (f) Variation of the specific capacitance to the scan rate of the PPy and Ag/PPy electrodes. Source: (e, f) Yuan et al. [145]. reproduced with permission of Elsevier.

dispersed metallic Ag is uniformly incorporated in the PPy matrix, increasing the charge transfer ability of PPy materials. The as-prepared Ag/PPy material exhibits ideal capacitive behavior and has excellent cyclic stability. As shown in Figure 9.27e,f, charge–discharge measurements demonstrate a maximum specific capacitance of 315.8 F g^{-1} for the Ag/PPy electrode, which is much higher than a pure PPy electrode (153.2 F g^{-1}). A high specific energy density of 43.9 Wh kg^{-1} can be achieved at 50 W kg^{-1} for PPy in the Ag/PPy electrode, and 30 Wh kg^{-1} remains at 500 W kg^{-1}. The facially synthesized Ag/PPy composite possesses promising properties as an electrode material for supercapacitors.

Zang et al. [146] prepared three-dimension hierarchical PPy nanotubes interconnected nickel cobalt-layered double hydroxide (PNT@NiCo-LDH) nanocages (Figure 9.28a), in which the formed networking hierarchical structure could facilitate the uniform distribution of ZIF-67-derived (ZIF, zeolitic imidazolate framework) NiCo-LDH nanocages (Figure 9.28b). Figure 9.28c,d shows the CV curves of NiCo-LDH and PNT@NiCo-LDH. Both of the CV curves exhibited apparent oxidation and reduction peaks, suggesting the presence of a faradic reaction. GCD curves of NiCo-LDH and PNT@NiCo-LDH showed prominent pseudocapacitive characteristics due to the existence of voltage plateau regions (Figure 9.28e,f). The PNT@NiCo-LDH achieved outstanding specific capacitance (1448.2 F g^{-1} at 1 A g^{-1}). Additionally, the asymmetric supercapacitor, assembled by positive PNT@NiCo-LDH and negative activated carbon, could achieve an energy density of 64.4 Wh kg^{-1} at 800 W kg^{-1}, superior to the electrochemical properties of the NiCo-LDH-dependent asymmetric supercapacitor, showing a promising prospect in high-performance supercapacitors.

Jabeen et al. [147] reported unique core–shell NiCo$_2$O$_4$@PANI nanorod arrays (NRAs), designed and employed as the electrode material for supercapacitors

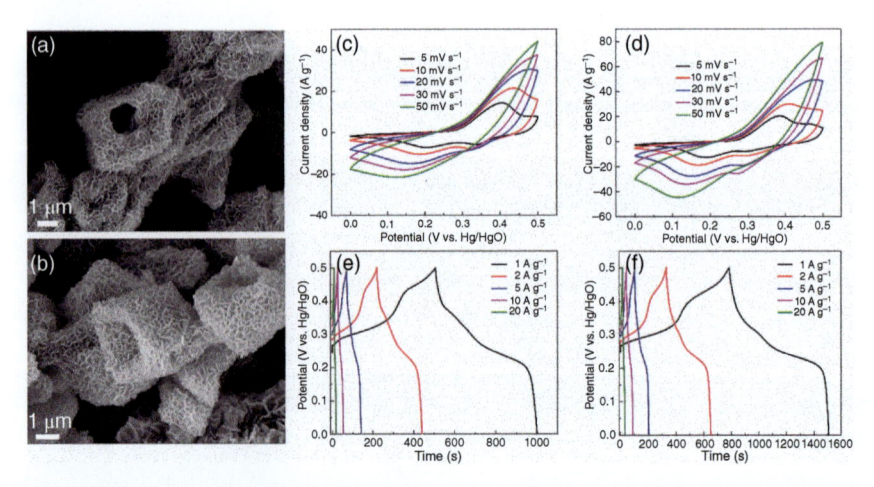

Figure 9.28 SEM images of (a) PNT@NiCo-LDH nanocages and (b) NiCo-LDH nanocages. Source: (a) Zang et al. [146]/with permission from American Chemical Society. CV curves of (c) NiCo-LDH and (d) PNT@NiCo-LDH nanocages in the range of 5–50 mV s^{-1}. GCD curves of (e) NiCo-LDH and (f) PNT@NiCo-LDH nanocages in the range of 1–20 A g^{-1}. Source: (c–f) Zang et al. [146]. Reproduced with permission of American Chemical Society.

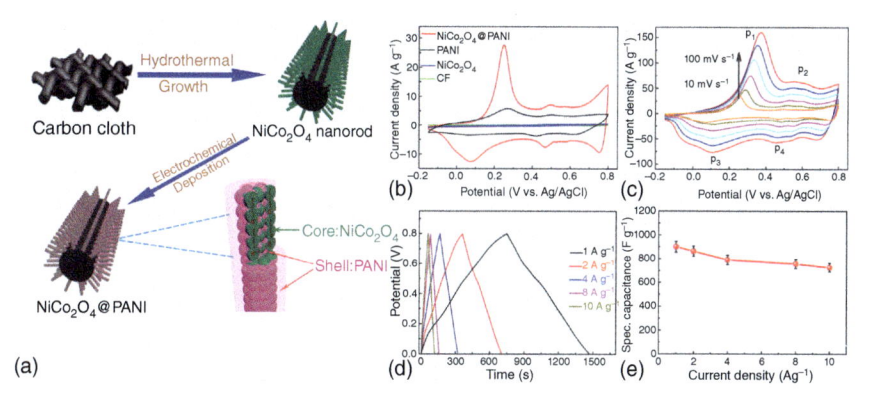

Figure 9.29 (a) Schematic illustration of the synthetic procedure for NiCo$_2$O$_4$@PANI NRAs. (b) A comparison of CV curves of the NiCo$_2$O$_4$@PANI NRAs, bare carbon cloth, NiCo$_2$O$_4$ NRAs, and pure PANI electrodes at a scan rate of 5 mV s^{-1}. (c) CV curves of NiCo$_2$O$_4$@PANI NRAs at different scan rates of 10, 20, 40, 60, 80, and 100 mV s^{-1}. (d) Charge and discharge curves of the NiCo$_2$O$_4$@PANI NRAs electrode at different current densities of 1, 2, 4, 8, and 10 A g^{-1}. (e) The plot of specific capacitances vs. current densities for the NiCo$_2$O$_4$@PANI NRAs electrode. Source: Jabeen et al. [147]. Reproduced with permission of American Chemical Society.

(Figure 9.29a). With highly porous NiCo$_2$O$_4$ as the conductive core and strain buffer support and nanoscale PANI layer as the electrochemically active component, a heterostructure achieves favorably high capacitance while maintaining good cycling stability and rate capability. As shown in Figure 9.29b–e, by adopting the uniform and intimate coating of PANI, the fabricated electrode exhibits a high specific capacitance of 901 F g^{-1} at 1 A g^{-1} in 1 M H$_2$SO$_4$ electrolyte and outstanding capacitance retention of 91% after 3000 cycles at a high current density of 10 A g^{-1}. The enhanced electrochemical performance demonstrates the complementary contributions of componential structures in the hybrid electrode design.

Ambade et al. [148] reported flexible polythiophene (e-PTh) on titania (Ti) wire electrode material (Figure 9.30a,b). To evaluate the performance of the optimized polymerization conditions of the f-WS all-solid-state symmetric SCs based on e-PTh/Ti wire with seven polymerization cycles, CV measurements were performed in a two-electrode configuration at various scan rates ranging from 5 to 100 mV s^{-1} with a maximum cell voltage window of 1.8 V (−0.8 V to 1 V). It is observed from Figure 9.30c that increasing the scan rate led the currents and the area under the CV curves to increase, indicating that the e-PTh electrodes show excellent capacitive behavior. The linear and symmetrical charge–discharge curves also indicate an excellent supercapacitor behavior (Figure 9.30d). As shown in Figure 9.30e,f, e-PTh/Ti wire show a high capacitive performance (1357.31 mF g^{-1} or 71.84 mF cm^{-2}). The robust f-WS all-solid-state symmetric SCs exhibit excellent mechanical flexibility with minimal capacitance changes upon bending at 360°. Furthermore, the SCs were implemented in the textile of a wearable/portable electronic device using a conventional weaving method, demonstrating a high potential for next-generation wearable textile electronic applications.

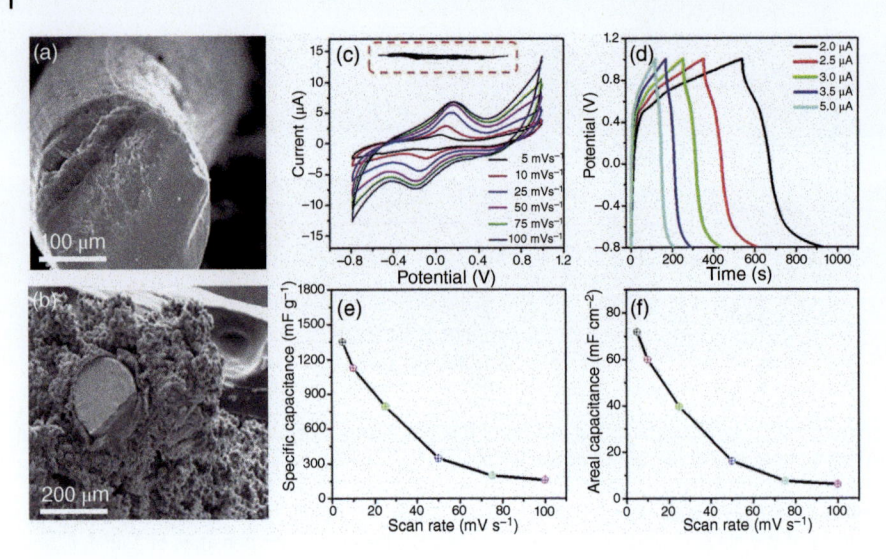

Figure 9.30 SEM images of the cross-section of bare Ti wire (a) cross-section of Ti–PTh at 7 cycles (b). Source: (a) Ambade et al. [148]/with permission from Royal Society of Chemistry. (c) CV curves at different sweep rates (inset is a digital photo of the device). (d) Charge–discharge at different current densities. (e) Specific capacitance. (f) Areal capacitance. Source: (c–f) Ambade et al. [148]. Reproduced with permission of Royal Society of Chemistry.

Bi et al. [149] reported the generating oxygen vacancies (Vö) in vanadium pentoxide (V_2O_5), which has been demonstrated as an effective approach to tailor its electrochemical properties. The present study investigates three kinds of CPs (PPy, PEDOT, and PANI) coated V_2O_5 nanofibers with oxygen vacancies generated at the interface during the polymerization process (Figure 9.31a–c). Surface Vö forms a local electric field and promotes the charge transfer kinetics of the resulting Vö-V_2O_5/CP nanocables. The accompanying V^{4+} and V^{3+} ions may also catalyze the redox reactions and improve the supercapacitor performance. The differences and similarities of the three CP coatings have been compared and discussed, dependent on their polymerization conditions and coating thickness. The distribution of Vö in the surface layer and in bulk has been elaborated, and the corresponding effects on the electrochemical properties and supercapacitor performance have also been investigated. As shown in Figure 9.31d–f, Vö-V_2O_5/PEDOT can deliver a high capacity of up to 614 F g^{-1} at a current rate of 0.5 A g^{-1}.

Moon et al. [150] reported a highly conductive 3D macroporous sponge fabricated by coating PEDOT:PSS/AgNWs on a commercial sponge using a simple and low-cost "immersion method," as shown in Figure 9.32a,b. The fabricated flexible 3D sponge conductor shows a high electrical conductivity of 3.94×10^{-4} S cm^{-1} with good stability in various environments and under bending deformation. To exploit the potential of the flexible 3D PEDOT:PSS/AgNW coating on the sponge as a current collector for energy-related applications, Co(OH)F arrays are directly grown on PEDOT:PSS/AgNW-on-sponge conductors (Co(OH)F/PEDOT:PSS/AgNW/polyurethane sponge [PUS]) for all-solid-state supercapacitors. As shown

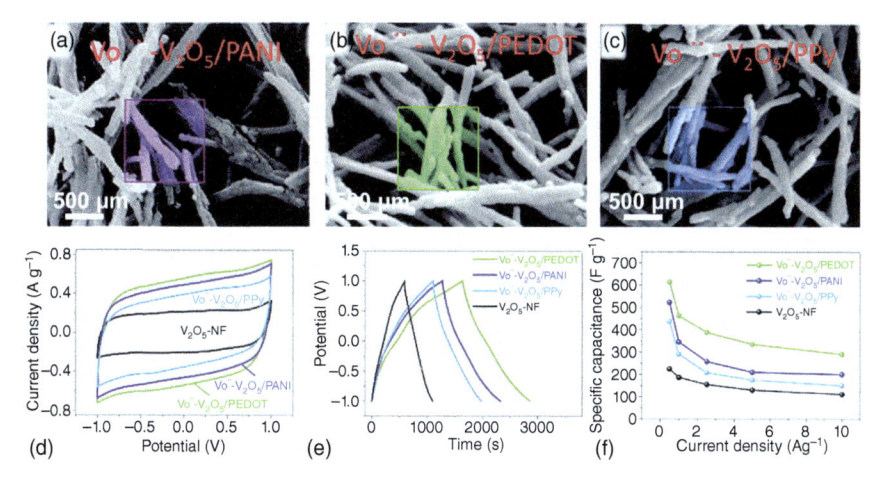

Figure 9.31 (a–c) SEM images of Vö-V$_2$O$_5$/CPs. (a) CV curves of Vö-V$_2$O$_5$/CPs and V$_2$O$_5$-NFs at a scan rate of 5 mV s^{-1}. Source: (a) Bi et al. [149]/with permission from Royal Society of Chemistry. (b) GCD curves of Vö-V$_2$O$_5$/CPs and V$_2$O$_5$-NFs at a current density of 0.5 A g^{-1}. (c) Specific capacitances of Vö-V$_2$O$_5$/CPs and V$_2$O$_5$-NFs at different current densities. Source: (d–f) Bi et al. [149]. Reproduced with permission of Royal Society of Chemistry.

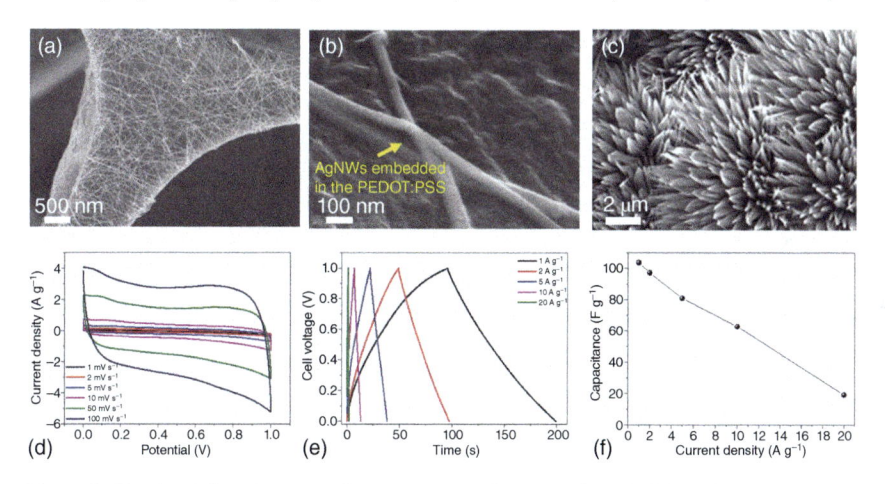

Figure 9.32 (a–c) SEM images of the obtained 3D AgNW/PUS, PEDOT:PSS/AgNW/PUS and Co(OH)F on PEDOT:PSS/AgNW/PUS electrode. Source: (a) Moon et al. [150]/with permission from John Wiley and Sons. (d) CV curves of symmetric Co(OH)F/PEDOT:PSS/AgNW/PUS supercapacitor device at different scan rates of 1–100 mV s^{-1} in the PVA-KOH gel electrolyte. (e) Galvanostatic charge–discharge curves at different current densities. (f) Specific capacitance (by account of the mass of Co(OH)F) as a function. Source: (d–f) Moon et al. [150]. Reproduced with permission of John Wiley and Sons.

in Figure 9.32d–f, the resulting symmetric all-solid-state supercapacitor exhibits a mass-specific capacitance of 103.7 F g^{-1} at a current density of 1 A g^{-1}. This 3D lightweight conductor fabrication can be adapted for mass production and new opportunities for flexible electronic applications.

Song et al. [151] constructed polyaniline/graphene/MnO$_2$ (PANI/G/MnO$_2$) paper using low-cost printing paper by a simple layer-by-layer *in situ* growth and vacuum

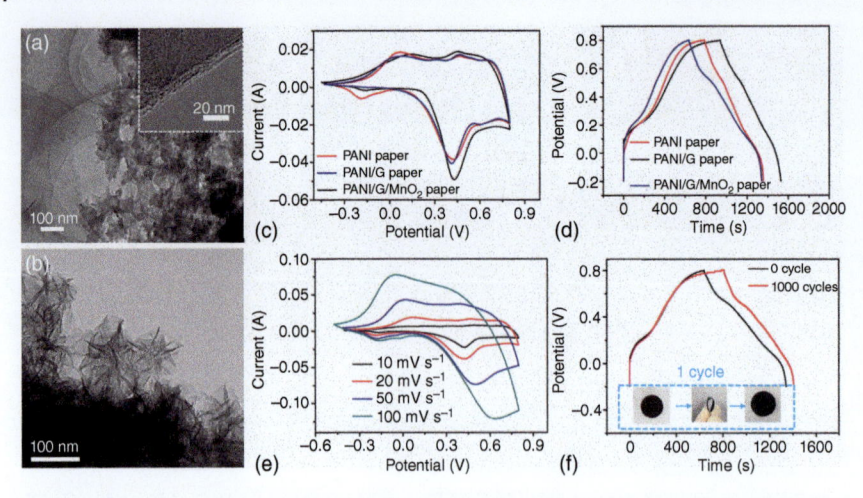

Figure 9.33 TEM images of PANI/G paper (a) (Source: (a) Song et al. [151]/with permission from Elsevier.) and PANI/G/MnO$_2$ paper (b). (c) CV curves of PANI paper, PANI/G paper, and PANI/G/MnO$_2$ paper at a scan rate of 20 mV s^{-1}. (d) The corresponding GCD curves at a current density of 5 mA cm^{-2}. (e) CV curves of PANI/G/MnO$_2$ paper at different scan rates. (f) GCD curves and photos of PANI/G/MnO$_2$ paper as original and bending 180° up to 1000 cycles. Source: (c–f) Song et al. [151]. Reproduced with permission of Elsevier.

filtration method (Figure 9.33a,b). The PANI nanofiber was polymerized on cellulose fibers. Then the pristine graphene sheet was introduced to generate the PANI/G paper, whose surface was subsequently scattered with the formed MnO$_2$ nanoflowers through oxidation of the cellulose fiber and graphene. Figure 9.33c shows the CV curves of PANI paper, PANI/G paper, and PANI/G/MnO$_2$ paper. The distinct redox peaks of these CV curves indicate the redox transition of PANI from leucoemeraldine to polaronic emeraldine form and from emeraldine to pernigraniline structure. The larger CV curve area of the PANI/G/MnO$_2$ paper was attributed to the introduction of MnO$_2$, which increased the capacitance. Figure 9.34d shows the corresponding GCD curves at a current density of 5 mA cm^{-2}. PANI/G/MnO$_2$ paper electrode shows a much longer discharge time than others, indicating PANI/G/MnO$_2$ has a higher columbic efficiency. The areal capacitance of PANI paper, PANI/G paper, and PANI/G/MnO$_2$ paper can be calculated as 2.84, 2.92, and 3.5 F cm^{-2}, respectively. The increased areal capacitance was ascribed to the faradic contribution of MnO$_2$. Also, the PANI/G/MnO$_2$ electrode showed a rapid current-voltage response and excellent electrochemical performance when the scan rate increased from 10 to 100 mV s^{-1} (Figure 9.33e). The electrode shows a sizeable areal capacitance of 3.5 F cm^{-2} at 5 mA cm^{-2} and excellent cycling stability of 90% after 1000 bending cycles (Figure 9.33f).

Porous materials, such as MOFs, possess features like large specific surface area, mesoporous structures, and multiple reaction sites that make them attractive for energy storage devices [152–154]. Further, integrating MOFs with suitable conducting agents is a promising strategy to introduce additional conductive pathways, facilitating fast charge transfer. Xu et al. [152] reported MOFs with high porosity and

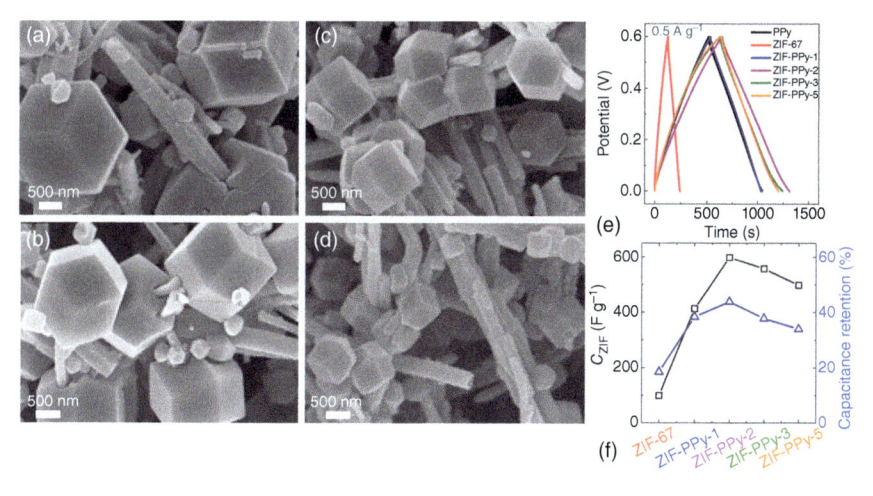

Figure 9.34 SEM images of (a) ZIF–PPy-1 (Source: (a) Xu et al. [152]/with permission from American Chemical Society.), (b) ZIF–PPy-2, (c) ZIF–PPy-3, and (d) ZIF–PPy-5. (e) GCD curves at 0.5 A g^{-1} of PPy, ZIF-67, and ZIF–PPy hybrids. (f) C ZIF and capacitance retention of ZIF-67 and ZIF–PPy hybrids. Source: (e, f) Xu et al. [152]. Reproduced with permission of American Chemical Society.

a regular porous structure have emerged as promising electrode material for supercapacitors. However, their poor electrical conductivity limits their utilization efficiency and capacitive performance. To increase the overall electrical conductivity and the efficiency of MOF particles, three-dimensional networked MOFs are developed using conductive PPy tubes to support *in situ* growth of MOF particles. This work selected the famous ZIF-67 as a typical example of MOFs, and the obtained hybrids are denoted as ZIF–PPy. As shown in (Figure 9.34a–d), ZIF–PPy samples obtained from 20, 40, 60, and 100 mg PPy tubes are denoted as ZIF–PPy-1, ZIF–PPy-2, ZIF–PPy-3, and ZIF–PPy-5, respectively. As a result, the highly conductive PPy tubes that run through the MOF particles increase the electron transfer between MOF particles and maintain the high effective porosity of the MOFs and endow the MOFs with flexibility. As shown in Figure 9.34e,f, promoted by elaborately designed MOF-PPy networks, the specific capacitance of MOF particles has been increased from 99.2 F g^{-1} for pristine ZIF-67 to 597.6 F g^{-1} for ZIF–PPy networks, indicating the importance of the design of the ZIF–PPy microstructure. Furthermore, a flexible supercapacitor device based on ZIF–PPy networks shows an outstanding areal capacitance of 225.8 mF cm^{-2}, better than other MOFs-based supercapacitors reported, confirming the significance of *in situ* synthetic chemistry and the importance of hybrid nanoscale materials.

Xu et al. [153] synthesized the composite of ZIF-67 and PANI by simply stirring the mixture of ZIF-67 and PANI (Figure 9.35a–d). Among them, PANI serves as a growth substrate of ZIF-67 for the transfer scaffold of electrons and works as a wire to connect nano-particles, improving the conductivity of ZIF-67. Furthermore, sulfur is introduced into ZIF-67/PANI by additional sulfurization (Co$_3$S$_4$/PANI). The analysis of GCD curves in Figure 9.35e also proved the above conclusion. The GCD

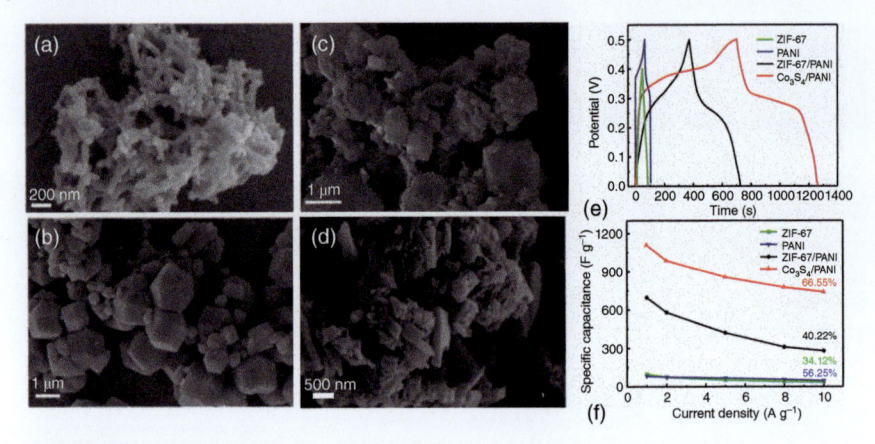

Figure 9.35 SEM images of PANI (a) (Source: (a) Xu et al. [153]/with permission from Elsevier.), ZIF-67 (b), ZIF-67/PANI (c), and Co_3S_4/PANI (d). (e) GCD curves of ZIF-67, PANI, ZIF-67/PANI, and Co_3S_4/PANI at 1 A g^{-1}. (f) Specific capacitance of ZIF-67, PANI, ZIF-67/PANI, and Co_3S_4/PANI at different current densities. Source: (e, f) Xu et al. [153]. Reproduced with permission of Elsevier.

curves were not ideal lines and further proved the pseudocapacitive behavior of these four-electrode materials. At low current densities, the electrolyte could contact well with the surface of the electrode, making the reaction more complete and leading to a higher specific capacitance. The specific capacitances of Co_3S_4/PANI, ZIF-67/PANI, PANI, and ZIF-67 at different current densities are shown in Figure 9.35f. It can be seen that Co_3S_4/PANI had the highest specific capacitance at any current density. Even if at a high current density of 10 A g^{-1}, the specific capacitance could still reach 736 F g^{-1}.

Shrivastav et al. [154] reported a heterostructure ZIF-67/PEDOT heterostructure composite (Figure 9.36a–d). The CV curves at different scan rates (5–100 mV s^{-1}) are presented in Figure 9.36e. These studies were carried out within a potential scan range of 0–0.5 V and took 1 M H_2SO_4 as an aqueous electrolyte. The area under the CV curves has been considered for computing the values of specific capacitance. The CV of the pristine ZIF-67 electrode has also been carried out to compare Figure 9.36f. The prepared ZIF-67/PEDOT composite has delivered a specific capacitance of 106.8 F g^{-1} at a current density of 1 A g^{-1} when analyzed for electrochemical performance in three-electrode systems. This value is significantly better than that with only pristine ZIF-67 electrode (34.75 F g^{-1} at 1 A g^{-1}). Extending the study, two ZIF-67/PEDOT electrodes have been used to assemble an all-solid-state symmetrical supercapacitor, which delivers excellent energy density values (∼11 Wh kg^{-1}) and power density (200 Wh kg^{-1}). This work paves the way for the potential ZIF-67/PEDOT composite application in developing next-generation supercapacitors.

Metal ion doping has also been applied to improve the capacitive performance of CPs. Wu et al. [155] reported Fe^{3+} doping used to enhance the capacity of PTh as

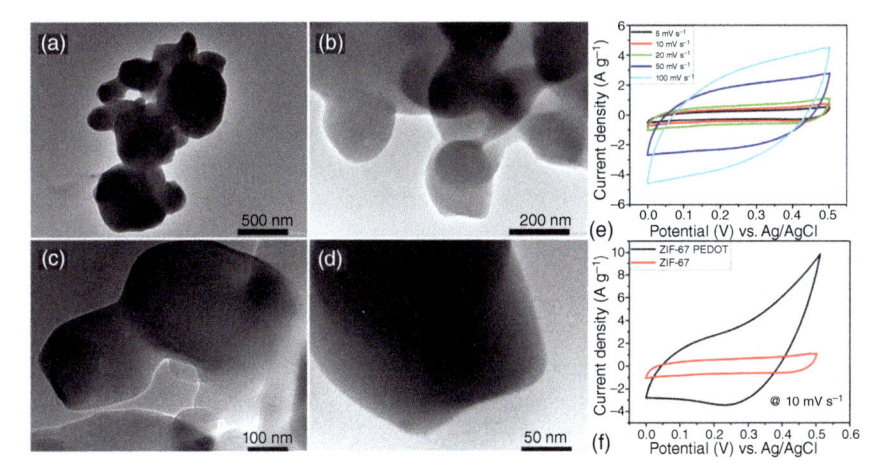

Figure 9.36 (a–d) TEM images of ZIF-67/PEDOT composite at different magnifications. Source: (a) Shrivastav et al. [154]/with permission from Elsevier. (e) CV curves for ZIF-67/PEDOTelectrode. (f) Comparative CV curves for ZIF-67 and ZIF-67/PEDOTelectrodes at a scan rate of 10 mV s^{-1}. Source: (e, f) Shrivastav et al. [154]. Reproduced with permission of Elsevier.

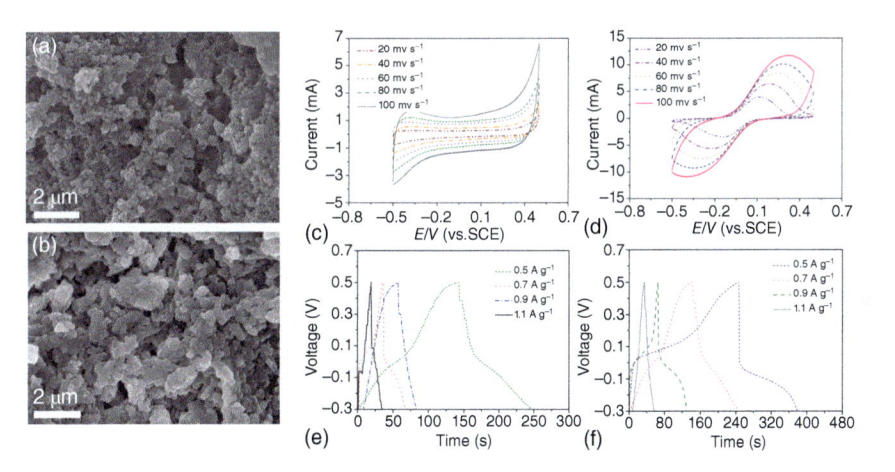

Figure 9.37 SEM images of (a) PTh (Source: Wu et al. [155]/with permission from Elsevier.) and (b) Fe^{3+}-doped PTh. Cyclic voltammograms of the (c) PTh, (d) Fe^{3+}-doped PTh at different scan rates. GCD curves of (e) PTh and (f) Fe^{3+}-doped PTh at different current densities. Source: (c–f) Wu et al. [155], Figure 01 [p. 012]/with permission from Elsevier.

a supercapacitor electrode, in which the flexible Fe^{3+} doped PTh electrode on carbon cloth was prepared by an electrochemical deposition method (Figure 9.37a,b). As shown in Figure 9.37c–f, the electrochemical performance of the PTh electrode is evaluated using CV and GCD technique. The pristine PTh exhibits a specific capacitance of 77.2 F g^{-1} at a current density of 0.5 A g^{-1}. After Fe^{3+} doping, the specific capacitance of Fe^{3+}-doped PTh is improved to 108.1 F g^{-1}.

9.5 Summary and Perspective

CPs have great application potential in electrochemical energy conversion and storage due to their high electrochemical activity, low cost, and easy synthesis. Moreover, CP micro/nanostructures were easily prepared using chemical and electrochemical methods, thus improving the performance of electrochemical devices. Importantly, CPs can be used as active electrodes for supercapacitors, and they easily combine with other electrochemically active materials for high-performance supercapacitors. CP composites also displayed good performance in shielding various electromagnetic interference due to their good environmental stability, high conductivity, and exceptional redox property. Thus, CPs and CP-based nanomaterials have offered tremendous advantages. They have emerged as ideal entrants for future electrochemical technology. Despite many hurdles, a remarkable development in the synthesis techniques and applications of nanomaterials is anticipated based on the promising results from current research. Further research is vital to overcome hurdles for CPs to be a forerunner in electrochemical energy conversion and storage. Further assessment of these results provides the basis for developing high-performance CPs. The potential of CPs in energy conversion and storage is incredibly prodigious.

References

1 Rasmussen, S.C. (2020). Conjugated and conducting organic polymers: the first 150 years. *ChemPlusChem.* 85: 1412–1429. https://doi.org/10.1002/cplu.202000325.
2 Pnalwa, M.C. (2004). *Encyclopedia of Nanoscience and Nanotechnology*, vol. 2, 153–169. American Scientific Publishers. https://doi.org/urn:isbn:9781588830012.
3 Brazovskii, S.A. and Kirova, N.N. (1980). Excitons, polarons, and bipolarons in conducting polymers. *JETP Lett.* 33: 4–8. https://doi.org/10.1007/BF00670555.
4 Furukawa, Y. (1996). Electronic absorption and vibrational spectroscopies of conjugated conducting polymers. *J. Phys. Chem.* 100: 15644–15653. https://doi.org/10.1021/jp960608n.
5 Xue, Y., Chen, S., Yu, J. et al. (2020). Nanostructured conducting polymers and their composites: synthesis methodologies, morphologies and applications. *J. Mater. Chem. C* 8: 10136–10159. https://doi.org/10.1039/D0TC02152K.
6 X. Hong, Y. Liu, Y. Li, X. Wang, J. Fu, X. Wang, Application progress of polyaniline, polypyrrole and polythiophene in lithium-sulfur batteries, *Polymers*, 12 (2020) 331, https://doi.org/10.3390/polym12020331
7 Shi, Y., Peng, L., Ding, Y. et al. (2015). Nanostructured conductive polymers for advanced energy storage. *Chem. Soc. Rev.* 44: 6684. https://doi.org/10.1039/C5CS00362H.
8 Q. Meng, K. Cai, Y. Chen, L. Chen, Research progress on conducting polymer based supercapacitor electrode materials, *Nano Energy*, 36 (2017) 268, https://doi.org/10.1016/j.nanoen.2017.04.040

9 D. A. Notter, M. Gauch, R. Widmer, P. Wager, A. Stamp, R. Zah, H.-J. Althaus, Contribution of Li-ion batteries to the environmental impact of electric vehicles, *Environ. Sci. Technol.*, 44 (2010) 6550, https://doi.org/10.1021/es903729a

10 K. Mizushima, P.C. Jones, P.J. Wiseman, J.B. Goodenough, Li_xCoO_2 ″(oless-thanxless-than-or-equal-To1) - a new cathode material for batteries of high-energy density, *Mater. Res. Bull.*, 15 (1980) 783, https://doi.org/10.1016/0025-5408(80)90012-4

11 Etacheri, V., Marom, R., Elazari, R. et al. (2011). Challenges in the development of advanced Li-ion batteries: a review. *Energy Environ. Sci.* 4: 3243. https://doi.org/10.1039/C1EE01598B.

12 Cheng, F., Tang, W., Li, C. et al. (2006). Conducting poly(aniline) nanotubes and nanofibers: controlled synthesis and application in lithium/poly(aniline) rechargeable batteries. *Chem. Eur. J.* 12: 3082. https://doi.org/10.1002/chem.200500883.

13 N. D. Trinh, M. Saulnier, D. Lepage, S.B. Schougaard, Conductive polymer film supporting $LiFePO_4$ as composite cathode for lithium-ion batteries, *J. Power Sources*, 221 (2013), 284, https://doi.org/10.1016/j.jpowsour.2012.08.006

14 Wang, J. and Sun, X. (2015). Olivine $LiFePO_4$: the remaining challenges for future energy storage. *Energy Environ. Sci.* 8: 1110. https://doi.org/10.1039/C4EE04016C.

15 Liu, J., Zhang, J.-G., Yang, Z. et al. (2013). Materials science and materials chemistry for large scale electrochemical energy storage: from transportation to electrical grid. *Adv. Funct. Mater.* 23: 929. https://doi.org/10.1002/adfm.201200690.

16 Dunn, B., Kamath, H., and Tarascon, J.-M. (2011). Electrical energy storage for the grid: a battery of choices. *Science* 334: 928. https://science-sciencemag-org-s.vpn.buaa.edu.cn:8118/content/334/6058/928.

17 K. A. Vorobeva, S. N. Eliseeva, R. V. Apraksin, M. A. Kamenskii, E. G. Tolstopjatova, V. V. Kondratiev, Improved electrochemical properties of cathode material $LiMn_2O_4$ with conducting polymer binder, *J. Alloys Compd.*, 766 (2018) 33, https://doi.org/10.1016/j.jallcom.2018.06.324

18 Gan, Q., Qin, N., Zhu, Y. et al. (2019). Polyvinylpyrrolidone-induced uniform surface-conductive polymer coating endows Ni-rich $LiNi_{0.8}Co_{0.1}Mn_{0.1}O_2$ with enhanced cyclability for lithium-ion batteries. *ACS Appl. Mater. Interfaces* 11: 12594. https://doi.org/10.1021/acsami.9b04050.

19 Kwon, Y., Lee, Y., Kim, S.O. et al. (2018). Conducting polymer coating on a high-voltage cathode based on soft chemistry approach toward improving battery performance. *ACS Appl. Mater. Interfaces* 10: 29457. https://doi.org/10.1021/acsami.8b08200.

20 J. W. Xiao, L. Wan, S. H. Yang, F. Xiao, S. Wang, Design hierarchical electrodes with highly conductive $NiCo_2S_4$ nanotube arrays grown on carbon fiber paper for high-performance pseudocapacitors, *Nano Lett.*, 14 (2014) 831, https://doi.org/10.1021/nl404199v

21 H. Liu, S. Bo, W. Cui, F. Li, C. Wang, Y. Xia, Nano-sized cobalt oxide/mesoporous carbon sphere composites as negative electrode material for lithium-ion batteries, *Electrochim. Acta*, 53 (2008) 6497, https://doi.org/10.1016/j.electacta.2008.04.030

22 J. Kim, J. K. Yoo, Y. Sikjung, K. Kang (2013). $Li_3V_2(PO_4)_3$/conducting polymer as a high power 4 V-class lithium battery electrode, *Adv. Energy Mater.*, 3: 1004, https://doi.org/10.1002/aenm.201300205

23 Y. Zhou, Y. Li, J. Yang, J. Tian, H. Xu, J. Yang, W. Fan, Conductive polymer-coated VS_4 submicrospheres as advanced electrode materials in lithium-ion batteries, *ACS Appl. Mater. Interfaces*, 8 (2016) 18797, https://doi.org/10.1021/acsami.6b04444

24 D. Lepage, L. Savignac, M. Saulnier, S. Gervais, S.B. Schougaar, Modification of aluminum current collectors with a conductive polymer for application in lithium batteries, *Electrochem. Commun.*, 102 (2019) 1, https://doi.org/10.1016/j.elecom.2019.03.009

25 Y. Chen, S. Zeng, J. Qian, Y. Wang, Y. Cao, H. Yang, and X. Ai, Li^+-Conductive polymer-embedded nano-Si particles as anode material for advanced Li-ion batteries, *ACS Appl. Mater. Interfaces*, 6 (2014) 3508, https://doi.org/10.1021/am4056672

26 T.M. Higgins, S.-H. Park, P.J. King, C. Zhang, N. MoEvoy, N.C. Berner, D. Daly, A. Shmeliov, U. Khan, G. Duesberg, V. Nicolosi, J.N. Coleman, A commercial conducting polymer as both binder and conductive additive for silicon nanoparticle-based lithium-ion battery negative electrodes, *ACS Nano*, 10 (2016) 3702, https://doi.org/10.1021/acsnano.6b00218

27 H. Wiggers, Y. H. Sehlleiera, F. Kunzec, L. Xiao, S. M. Schnurre, C. Schulz, Self-assembled nano-silicon/graphite hybrid embedded in a conductive polyaniline matrix for the performance enhancement of industrial applicable lithium-ion battery anodes, *Solid State Ionics*, 344 (2020) 115117, https://doi.org/10.1016/j.ssi.2019.115117

28 W. Zeng, L. Wang, X. Peng, T. Liu, Y. Jiang, F. Qin, L. Hu, P. K. Chu, K. Huo, Y. Zhou, Enhanced ion conductivity in conducting polymer binder for high-performance silicon anodes in advanced iithium-ion batteries, *Adv. Energy Mater.*, 8 (2018) 1702314, https://doi.org/10.1002/aenm.201702314

29 X. He, R. Han, P. Jiang, Y. Chen, W. Liu, Molecularly engineered conductive polymer binder enables stable lithium storage of Si, *Ind. Eng. Chem. Res.*, 59 (2020) 2680, https://doi.org/10.1021/acs.iecr.9b05838

30 P.M. Dziewoński, M. Grzeszczuk, Towards TiO_2-conducting polymer hybrid materials for lithium-ion batteries, *Electrochim. Acta*, 55 (2010) 3336, https://doi.org/10.1016/j.electacta.2010.01.043

31 P. Ge, S. Li, H. Shuai, T. Xu, Y. Yang, G. Zou, H. Hou, X. Ji, Engineering 1D chain-like architecture with conducting polymer towards ultra-fast and high-capacity energy storage by reinforced pseudo-capacitance, *Nano Energy*, 54 (2018) 26, https://doi.org/10.1016/j.nanoen.2018.09.062

32 X. Sun, Y. Zhang, J. Zhang, F. Zaman, L. Hou, C. Yuan, Bi-Metal (Zn, Mn) metal–organic framework–derived $ZnMnO_3$ micro-sheets wrapped uniformly with polypyrrole conductive network toward high-performance Li-ion batteries, *Energy Technol.* 2020, 8, 1901218, https://doi.org/10.1002/ente.201901218

33 J. Zhang, K. Zhang, J. Yang, G.-H. Lee, J. Shin, V. Wing-hei Lau, Y.-M. Kang, Bifunctional conducting polymer coated CoP core–shell nanowires on carbon

paper as a free-standing anode for sodium ion batteries, *Adv. Energy Mater.*, 8 (2018) 1800283, https://doi.org/10.1002/aenm.201800283

34 K. Zhang, M. Park, L. Zhou, G.-H. Lee, W. Li, Y.-M. Kang, J. Chen, Urchin-like $CoSe_2$ as a high-performance anode material for sodium-ion batteries, *Adv. Funct. Mater.*, 26 (2016) 6728, https://doi.org/10.1002/adfm.201602608

35 J. W. Choi, D. Aurbach, Promise and reality of post-lithium-ion batteries with high energy densities, *Nat. Rev. Mater.*, 1 (2016) 16013, https://doi.org/10.1038/natrevmats.2016.13

36 H. Hou, C. E. Banks, M. Jing, Y. Zhang, X. Ji, Carbon quantum dots and their derivative 3D porous carbon frameworks for sodium-ion batteries with ultralong cycle life, *Adv. Mater.*, 27 (2015) 7861, https://doi.org/10.1002/adma.201503816

37 W. Li, S.-L. Chou, J. Wang, J. Wang, Q. Gu, H. Liu, S. Dou, Multifunctional conducting polymer coated $Na_{1+x}MnFe(CN)_6$ cathode for sodium-ion batteries with superior performance via a facile and one-step chemistry approach, *Nano Energy*, 13 (2015) 200, https://doi.org/10.1016/j.nanoen.2015.02.019

38 Z. Wang, Y. Liu, Z. Wu, G. Guan, D. Zhang, H. Zheng, S. Xu, S. Liu, X. Hao, A string of nickel hexacyanoferrate nanocubes coaxially grown on a CNT@bipolar conducting polymer as a high-performance cathode material for sodium-ion batteries *Nanoscale*, 9 (2017) 823, https://doi.org/10.1039/C6NR08765E

39 Wang, X., Wang, F., Wang, L. et al. (2016). An aqueous rechargeable $Zn//Co_3O_4$ battery with high energy density and good cycling behavior. *Adv. Mater.* 28: 4904. https://doi.org/10.1002/aenm.201800283.

40 Pan, H., Shao, Y., Yan, P. et al. (2016). Reversible aqueous zinc/manganese oxide energy storage from conversion reactions. *Nat. Energy* 1: 16039. https://doi.org/10.1038/nenergy.2016.39.

41 C. Xu, B. Li, H. Du, F. Kang, Energetic zinc ion chemistry: the rechargeable zinc ion battery, *Angew. Chem.*, 124 (2012) 957, https://doi.org/10.1002/ange.201106307

42 D. Xu, H. Wang, Fu. Li, Z. Guan, R. Wang, B. He, Y. Gong, X. Hu, Conformal conducting polymer shells on V_2O_5 nanosheet arrays as a high-rate and stable zinc-ion battery cathode, *Adv. Mater. Interfaces*, 6 (2019) 1801506, https://doi.org/10.1002/admi.201801506

43 X. Zang, X. Wang, H. Liu, X. Ma, W. Wang, J. Ji, J. Chen, R. Li, M. Xue, Enhanced ion conduction via epitaxially polymerized two-dimensional conducting polymer for high-performance cathode in zinc-ion batteries, *ACS Appl. Mater. Interfaces*, 12 (2020) 9347, https://doi.org/10.1021/acsami.9b22470

44 F. Wan, L. Zhang, X. Dai, X. Wang, Z. Niu, J. Chen, Aqueous rechargeable zinc/sodium vanadate batteries with enhanced performance from simultaneous insertion of dual carriers, *Nat. Commun.*, 9 (2018) 1656, https://doi.org/10.1038/s41467-018-04060-8

45 N. S. Hudak, Chloroaluminate-doped CPs as positive electrodes in rechargeable aluminum batteries, *J. Phys. Chem. C*, 118 (2014) 5203, https://doi.org/10.1021/jp500593d

46 Borchardt, L., Oschatz, M., and Kaskel, S. (2016). Carbon materials for lithium sulfur batteries-ten critical questions. *Chem. Eur. J.* 22: 7324. https://doi.org/10.1002/chem.201600040.

47 M. A. Pope, I. A. Aksay, Structural design of cathodes for Li-S batteries, *Adv. Energy Mater.*, 5 (2015) 1500124, https://doi.org/10.1002/aenm.201500124

48 Chung, S.-Y., Bloking, J.T., and Chiang, Y.-M. (2002). Electronically conductive phospho-olivines as lithium storage electrodes. *Nat. Mater.* 1: 123. https://www.nature.com/articles/nmat732.

49 R. D. Rauh, F. S. Shuker, J. M. Marston and S. B. Brummer, Formation of lithium polysulfides in aprotic media, *J. Inorg. Nucl. Chem.*, 39 (1977) 1761, https://doi.org/10.1016/0022-1902(77)80198-X

50 Mikhaylik, Y.V. and Akridge, J.R. (2004). Polysulfide shuttle study in the Li/S battery system. *J. Electrochem. Soc.* 151: A1969. https://iopscience.iop.org/article/10.1149/1.1806394.

51 Cheon, S.E., Ko, K.S., Cho, J.H. et al. (2003). Rechargeable lithium sulfur battery II. Rate capability and cycle characteristics. *J. Electrochem. Soc.* 150: A800. https://iopscience.iop.org/article/10.1149/1.1571533.

52 Y. Yang, G. Yu, J. J. Cha, H. Wu, M. Vosgueritchian, Y. Yao, Z. Bao, Y. Cui, Improving the performance of lithium–sulfur batteries by conductive polymer coating, *ACS Nano*, 5 (2011) 9187, https://doi.org/10.1021/nn203436j

53 H. Chen, W. Dong, J. Ge, C. Wang, X. Wu, W. Lu, L. Chen, Ultrafine sulfur nanoparticles in conducting polymer shell as cathode materials for high performance lithium/sulfur batteries, *Sci. Rep.*, 3 (2013) 1910, https://doi.org/10.1038/srep01910

54 P. Xiao, F. Bu, G. Yang, Y. Zhang, Y. Xu, Integration of graphene, nano sulfur, and conducting polymer into compact, flexible lithium–sulfur battery cathodes with ultrahigh volumetric capacity and superior cycling stability for foldable devices, *Adv. Mater.*, 29 (2017) 1703324, https://doi.org/10.1002/adma.201703324

55 S. Zeng, X. Li, H. Zhong, S. Chen, Y. Mai, Layered electrodes based on 3D hierarchical porous carbon and CPs for high-performance lithium-sulfur batteries, small *Methods* (2019) 1900028, https://doi.org/10.1002/smtd.201900028

56 T. Suga, S. Sugita, H. Ohshiro, K. Oyaizu, H. Nishide, p and n-Type bipolar redox-active radical polymer: toward totally organic polymer-based rechargeable devices with variable configuration, *Adv. Mater.*, 23 (2011) 751, https://doi.org/10.1002/adma.201003525

57 G. A. Snook, P. Kao, A. S. Best, Conducting-polymer-based supercapacitor devices and electrodes, *J. Power Sources*, 196 (2011) 1, https://doi.org/10.1016/j.jpowsour.2010.06.084

58 G. Wang, L. Zhang, J. Zhang, A review of electrode materials for electrochemical supercapacitors, *Chem. Soc. Rev.*, 41 (2012) 797, https://doi.org/10.1039/C1CS15060J

59 Y. Gofer, H. Sarker, J. G. Killian, T. O. Poehler, P.C. Searson, An all-polymer charge storage device, *Appl. Phys. Lett.*, 71(1997) 1582, https://doi.org/10.1063/1.120074

60 A. Robitaille, A. Perea, D. Bélanger, M. Leclerc, Poly(5-alkyl-thieno[3,4-c]pyrrole-4,6-dione): a study of π-conjugated redox polymers as anode materials in lithium-ion batteries, *J. Mater. Chem. A*, 5 (2017) 18088, https://doi.org/10.1039/C7TA03786D

61 J. G. Killian, B. M. Coffey, F. Gao, T. O. Poehler, P. C. Searson, Polypyrrole composite electrodes in an all-polymer battery system, *J. Electrochem. Soc.*, 143 (1996) 9362, https://doi.org/10.1149/1.1836562

62 H. K. Song, G. T. R. Palmore, Redox-active polypyrrole: toward polymer-based batteries, *Adv. Mater.*, 18 (2006) 1764, https://doi.org/10.1002/adma.200600375

63 I. Sultana, M. M. Rahman, J. Wang, C. Wang, G. G. Wallace, H.-K. Liu, All-polymer battery system based on polypyrrole (PPy)/para(toluene sulfonic acid) (pTS) and polypyrrole (PPy)/indigo carmine (IC) free standing films, *Electrochim. Acta*, 83 (2012) 209, https://doi.org/10.1016/j.electacta.2012.08.043

64 Aradilla, D., Estrany, F., Casellas, F. et al. (2014). All-polythiophene rechargeable batteries. *Org. Electron.* 15: 40. https://doi.org/10.1016/j.orgel.2013.09.044.

65 R. Emanuelsson, M. Sterby, M. Strømme, M. Sjödin, An all-organic proton battery, *J. Am. Chem. Soc.*, 139 (2017) 4828, https://doi.org/10.1021/jacs.7b00159

66 M. Sterby, R. Emanuelsson, X. Huang, A. Gogoll, M. Strømme, M. Sjödin, Characterization of PEDOT-quinone conducting redox polymers for water based secondary batteries, *Electrochim. Acta*, 235 (2017) 35, https://doi.org/10.1016/j.electacta.2017.03.068

67 X. Zhang, S. Wang, S. Lu, J. Su, T. He, Influence of doping anions on structure and properties of electro-polymerized polypyrrole counter electrodes for use in dye-sensitized solar cells, *J. Power Sources*, 246 (2014) 49, https://doi.org/10.1016/j.jpowsour.2013.07.098

68 W. Jiang, D. Yu, Q. Zhang, K. Goh, L. Wei, Y. Yong, R. Jiang, J. Wei, Y. Chen, Ternary hybrids of amorphous nickel hydroxide-carbon nanotube-conducting polymer for supercapacitors with high energy density, excellent rate capability, and long cycle life, *Adv. Funct. Mater.*, 25 (2015) 1063, https://doi.org/10.1002/adfm.201403354

69 J.Y. Lin, W. Y. Wang, S.W. Chou, Flexible carbon nanotube/polypropylene composite plate decorated with poly(3,4-ethylenedioxythiophene) as efficient counter electrodes for dye-sensitized solar cells, *J. Power Sources*, 282 (2015) 348, https://doi.org/10.1016/j.jpowsour.2015.01.142

70 K. Saranya, M. Rameez, A. Subramania, Developments in conducting polymer based counter electrodes for dye-sensitized solar cells—an overview, *Eur. Polym. J.*, 66 (2015) 207, https://doi.org/10.1016/j.eurpolymj.2015.01.049

71 C. W. Kung, Y. H. Cheng, H. W. Chen, R. Vittal, K.C.Ho, Hollow microflower arrays of PEDOT and their application for the counter electrode of a dye-sensitized solar cell, *J. Mater. Chem. A*, 1 (2013) 10693, https://doi.org/10.1039/C3TA10803A

72 M. Southcott, K. MacVittie, J. Halámek, L. Halámková, W. D. Jemison, R. Lobel, E. Katz, A pacemaker powered by animplantable biofuel cell operating under conditions mimicking the human blood circulatory system – battery not included, *Phys. Chem. Chem. Phys.*, 15 (2013) 6278, https://doi.org/10.1039/C3CP50929J

73 R. Feiner, T. Dvir, Tissue–electronics interfaces: from implantable devices to engineered tissues, *Nat. Rev. Mater.*, 3 (2017) 17076, https://doi.org/10.1038/natrevmats.2017.76

74 S. Wang, J. Y. Oh, J. Xu, H. Tran, Z. Bao, Skin-inspired electronics: an emerging paradigm. *Acc. Chem. Res.*, 51 (2018) 1033, https://doi.org/10.1021/acs.accounts.8b00015

75 Tseung, A.C.C., King, W.J., and Wan, B.Y.C. (1971). An encapsulated, implantable metal-oxygen cell as a long-term power source for medical and biological applications. *Med. Biol. Eng.* 9: 175. https://doi.org/10.1007/BF02474813.

76 Y. F. Zheng, X. N. Gu, F. Witte, Biodegradable metals, *Mater. Sci. Eng., R*, 77 (2014) 1, https://doi.org/10.1016/j.mser.2014.01.001

77 Y. Cui, Z. Wen, X. Liang, Y. Lu, J. Jin, M. Wu, X. Wu, A tubular polypyrrole based air electrode with improved O_2 diffusivity for Li–O_2 batteries, *Energy Environ. Sci.*, 5 (2012) 7893, https://doi.org/10.1039/C2EE21638H

78 S. K. Singh, X. Crispin, I. V. Zozoulenko, Oxygen reduction reaction in conducting polymer PEDOT: density functional theory study, *J. Phys. Chem. C*, 121 (2017) 12270, https://doi.org/10.1021/acs.jpcc.7b03210

79 V. Khomenko, E. Frackowiak, F. Béguin, Determination of the specific capacitance of conducting polymer/nanotubes composite electrodes using different cell configurations, *Electrochim. Acta*, 50 (2005) 2499, https://doi.org/10.1016/j.electacta.2004.10.078

80 Jia, X., Yu, G., Liang, S. et al. (2019). Tunable conducting polymers: toward sustainable and versatile batteries. *ACS Sustainable Chem. Eng.* 7 (17): 14321–14340. https://pubs.acs.org/doi/10.1021/acssuschemeng.9b02315.

81 S. K. Singh, Crispin, X. I. V. Zozoulenko, oxygen reduction reaction in conducting polymer PEDOT: density functional theory study, *J. Phys. Chem. C*, 121 (2017) 12270, https://doi.org/10.1021/acs.jpcc.7b03210

82 E. Mitraka, M. J. Jafari, M. Vagin, X. Liu, M. Fahlman, T. Ederth, M. Berggren, M. P. Jonsson, X. Crispin, Oxygen-induced doping on reduced PEDOT, *J. Mater. Chem. A*, 5 (2017) 4404, https://doi.org/10.1039/C6TA10521A

83 J. G. Hardy, J. Y. Lee, C. E. Schmidt, Biomimetic conducting polymer-based tissue scaffolds, *Curr. Opin. Biotechnol.*, 24 (2013) 847, https://doi.org/10.1016/j.copbio.2013.03.011

84 R. Balint, N. J. Cassidy, S. H. Cartmell, Conductive polymers: towards a smart biomaterial for tissue engineering, *Acta Biomater.*, 10 2014 (6) 2341, https://doi.org/10.1016/j.actbio.2014.02.015

85 T. H. Qazi, R. Rai, A. R. Boccaccini, Tissue engineering of electrically responsive tissues using polyaniline based polymers: a review. *Biomaterials*, 35 (2014) 9068, https://doi.org/10.1016/j.biomaterials.2014.07.020

86 N. K. Guimard, N. Gomez, C. E. Schmidt, CPs in Biomedical engineering, *Prog. Polym. Sci.*, 32 (2007) 876, https://doi.org/10.1016/j.progpolymsci.2007.05.012

87 P. J. Molino, B. Zhang, G. G. Wallace, T. W. Hanks, Surface modification of polypyrrole/biopolymer composites for controlled protein and cellular adhesion, *Biofouling*, 29 (2013) 1155, https://doi.org/10.1080/08927014.2013.830110

88 W. R. Stauffer, X. T. Cui, Polypyrrole doped with 2 peptide sequences from laminin. *Biomaterials*, 27 (2006) 2405, https://doi.org/10.1016/j.biomaterials.2005.10.024

89 Y. Kong, C. Wang, Y. Yang, C. O. Too, G. G. Wallace, A battery composed of a polypyrrole cathode and a magnesium alloy anode-toward a bioelectric battery. *Synth. Met.*, 162 (2012) 584, https://doi.org/10.1016/j.synthmet.2012.01.021

90 Y. Yang, C. Wang, C. Zhang, D. Wang, D. He, G. G. Wallace, A novel codoping approach for enhancing the performance of polypyrrole cathode in a bioelectric battery, *Carbon*, 80 (2014) 691, https://doi.org/10.1016/j.carbon.2014.09.013

91 C. Yu, C. Wang, X. Liu, X. Jia, S. Naficy, K. Shu, M. Forsyth, G. G. Wallace, A cytocompatible robust hybrid conducting polymer hydrogel for use in a magnesium battery, *Adv. Mater.*, 28 (2016) 9349, https://doi.org/10.1002/adma.201601755

92 L. Yin, X. Huang, H. Xu, Y. Zhang, J. Lam, J. Cheng, J. A. Rogers, Materials, designs, and operational characteristics for fully biodegradable primary batteries, *Adv. Mater.*, 26 (2014) 3879, https://doi.org/10.1002/adma.201470152

93 X. Huang, D. Wang, Z. Yuan, W. Xie, Y. Wu, R. Li, Y. Zhao, D. Luo, L. Cen, B. Chen, H. Wu, H. Xu, X. Sheng, M. Zhang, L. Zhao, L. Yin, A fully biodegradable battery for self-powered transient implants, *Small*, 14 (2018) 1800994, https://doi.org/10.1002/smll.201800994

94 M. J. Tan, C. Owh, P. L. Chee, A. K. Kyaw, D. Kai, X. J. Loh, Biodegradable electronics: cornerstone for sustainable electronics and transient applications, *J. Mater. Chem. C*, 4 (2016) 5531, https://doi.org/10.1039/C6TC00678G

95 X. Jia, C. Wang, V. Ranganathan, B. Napier, C. Yu, Y. Chao, M. Forsyth, F. G. Omenetto, D. R. MacFarlane, G. G. Wallace, A biodegradable thin-film magnesium primary battery using silk fibroin–ionic liquid polymer electrolyte, *ACS Energy Lett.*, 2 (2017) 831, https://doi.org/10.1021/acsenergylett.7b00012

96 H. Zhu, W. Luo, P. N. Ciesielski, Z. Fang, J. Y. Zhu, G. Henriksson, M. E.Himmel, L. Hu, Wood-derived materials for green electronics, biological devices, and energy applications, *Chem. Rev.*, 116 (2016) 9305, https://doi.org/10.1021/acs.chemrev.6b00225

97 E. Steven, V. Lebedev, E. Laukhina, C. Rovira, V. Laukhin, J. S. Brooks, J. Veciana, Silk/molecular conductor bilayer thin-films: properties and sensing functions, *Mater. Horiz.*, 1 (2014) 522, https://doi.org/10.1039/C4MH00074A

98 X. Jia, C. Wang, C. Zhao, Y. Ge, G. G. Wallace, Toward biodegradable Mg–air bioelectric batteries composed of silk fibroin–polypyrrole film, *Adv. Funct. Mater.*, 26 (2016) 1454, https://doi.org/10.1002/adfm.201503498

99 Wang, Z.-L., Xu, D., Xu, J.-J., and Zhang, X.-B. (2014). Oxygen electrocatalysts in metal–air batteries: from aqueous to nonaqueous electrolytes. *Chem. Soc. Rev.* 43: 7746–7786. https://doi.org/10.1039/C3CS60248F.

100 Song, F., Li, W., Han, G., and Sun, Y. (2018). Electropolymerization of aniline on nickel-based electrocatalysts substantially enhances their performance for hydrogen evolution. *ACS Appl. Energy Mater.* 1: 3–8. https://doi.org/10.1021/acsaem.7b00005.

101 Blasco-Ahicart, M., Soriano-López, J., and Galán-Mascarós, J.R. (2017). Conducting organic polymer electrodes with embedded polyoxometalate

catalysts for water splitting. *ChemElectroChem* 4: 3296–3301. https://doi.org/10.1002/celc.201700696.

102 Spöri, C., Briois, P., Nong, H.N. et al. (2019). Experimental activity descriptors for iridium-based catalysts for the electrochemical oxygen evolution reaction (OER). *ACS Catal.* 9: 6653–6663. https://doi.org/10.1021/acscatal.9b00648.

103 Bullock, R.M., Das, A.K., and Appel, A.M. (2017). Surface immobilization of molecular electrocatalysts for energy conversion. *Chem. Eur. J.* 23: 7626–7641. https://doi.org/10.1002/chem.201605066.

104 Wang, L., Fan, K., Daniel, Q. et al. (2015). Electrochemical driven water oxidation by molecular catalysts in situ polymerized on the surface of graphite carbon electrode. *Chem. Commun.* 51: 7883–7886. https://doi.org/10.1039/C5CC00242G.

105 Zhou, Q. and Shi, G. (2016). Conducting polymer-based catalysts. *J. Am. Chem. Soc.* 138: 2868–2876. https://doi.org/10.1021/jacs.5b12474.

106 Chinnappan, A., Bandal, H., Ramakrishna, S., and Kim, H. (2018). Facile synthesis of polypyrrole/ionic liquid nanoparticles and use as an electrocatalyst for oxygen evolution reaction. *Chem. Eng. J.* 335: 215–220. http://www.sciencedirect.com/science/article/pii/S1385894717318235.

107 Tian, X.L., Xu, Y.Y., Zhang, W. et al. (2017). Unsupported platinum-based electrocatalysts for oxygen reduction reaction. *ACS Energy Lett.* 2: 2035–2043. https://doi.org/10.1021/acsenergylett.7b00593.

108 Dong, Y.-T., Feng, J.-X., and Li, G.-R. (2017). Transition metal ion-induced high electrocatalytic performance of conducting polymer for oxygen and hydrogen evolution reactions. *Macromol. Chem. Phys.* 218: 1700359. https://doi.org/10.1002/macp.201700359.

109 Liu, Y., Li, J., Li, F. et al. (2016). A facile preparation of $CoFe_2O_4$ nanoparticles on polyaniline-functionalised carbon nanotubes as enhanced catalysts for the oxygen evolution reaction. *J. Mater. Chem. A* 4: 4472–4478. https://doi.org/10.1039/C5TA10420C.

110 Lattach, Y., Rivera, J.F., Bamine, T. et al. (2014). Iridium oxide–polymer nanocomposite electrode materials for water oxidation. *ACS Appl. Mater. Interfaces* 6: 12852–12859. https://doi.org/10.1021/am5027852.

111 Chu, S. and Majumdar, A. (2012). Opportunities and challenges for a sustainable energy future. *Nature* 488: 294–303. https://doi.org/10.1038/nature11475.

112 Chen, W., Pei, J., He, C.-T. et al. (2018). Single tungsten atoms supported on MOF-derived N-doped carbon for robust electrochemical hydrogen evolution. *Adv. Mater.* 30: 1800396. https://doi.org/10.1002/adma.201800396.

113 Ng, C.H., Winther-Jensen, O., Ohlin, C.A., and Winther-Jensen, B. (2015). Exploration and optimisation of poly(2,2′-bithiophene) as a stable photo-electrocatalyst for hydrogen production. *J. Mater. Chem. A* 3: 11358–11366. https://doi.org/10.1039/C5TA00291E.

114 Abrantes, L.M. and Correia, J.P. (2000). Polypyrrole incorporating electroless nickel. *Electrochim. Acta* 45: 4179–4185. http://www.sciencedirect.com/science/article/pii/S0013468600005521.

115 Wang, T., Zhuo, J., Du, K. et al. (2014). Electrochemically fabricated polypyrrole and MoSx copolymer films as a highly active hydrogen evolution electrocatalyst. *Adv. Mater.* 26: 3761–3766. https://doi.org/10.1002/adma.201400265.

116 Jukic, A. and Metikoš-Hukovic, M. (2003). The hydrogen evolution reaction on pure and polypyrrole-coated $GdNi_4Al$ electrodes. *Electrochim. Acta* 48: 3929–3937. http://www.sciencedirect.com/science/article/pii/S0013468603005310.

117 Mo, X., Wang, J., Wang, Z., and Wang, S. (2004). The application of polypyrrole fibrils in hydrogen evolution reaction. *Synth. Met.* 142: 217–221. http://www.sciencedirect.com/science/article/pii/S0379677903004697.

118 Pandey, R.K. and Lakshminarayanan, V. (2010). Enhanced electrocatalytic activity of Pd-dispersed 3,4-polyethylenedioxythiophene film in hydrogen evolution and ethanol electro-oxidation reactions. *J. Phys. Chem. C* 114: 8507–8514. https://doi.org/10.1021/jp1014687.

119 Navarro-Flores, E. and Omanovic, S. (2005). Hydrogen evolution on nickel incorporated in three-dimensional conducting polymer layers. *J. Mol. Catal. A: Chem.* 242: 182–194. http://www.sciencedirect.com/science/article/pii/S1381116905005637.

120 Feng, J.-X., Ding, L.-X., Ye, S.-H. et al. (2015). $Co(OH)_2$@PANI hybrid nanosheets with 3D networks as high-performance electrocatalysts for hydrogen evolution reaction. *Adv. Mater.* 27: 7051–7057. https://doi.org/10.1002/adma.201503187.

121 Aydın, R. and Köleli, F. (2006). Hydrogen evolution on conducting polymer electrodes in acidic media. *Prog. Org. Coat.* 56: 76–80. https://www.sciencedirect.com/science/article/pii/S0300944006000348.

122 Wei, X., Yin, Z., Lyu, K. et al. (2020). Highly selective reduction of CO_2 to C_{2+} hydrocarbons at copper/polyaniline interfaces. *ACS Catal.* 10: 4103–4111. https://doi.org/10.1021/acscatal.0c00049.

123 Sinha, S., Berdichevsky, E.K., and Warren, J.J. (2017). Electrocatalytic CO_2 reduction using rhenium(I) complexes with modified 2-(2′-pyridyl)imidazole ligands. *Inorg. Chim. Acta* 460: 63–68. http://www.sciencedirect.com/science/article/pii/S0020169316305254.

124 Ponnurangam, S., Chernyshova, I.V., and Somasundaran, P. (2017). Nitrogen-containing polymers as a platform for CO_2 electroreduction. *Adv. Colloid Interface Sci.* 244: 184–198. http://www.sciencedirect.com/science/article/pii/S000186861630272X.

125 Aydin, R. and Köleli, F. (2004). Electrocatalytic conversion of CO_2 on a polypyrrole electrode under high pressure in methanol. *Synth. Met.* 144: 75–80. http://www.sciencedirect.com/science/article/pii/S0379677904000931.

126 Köleli, F., Röpke, T., and Hamann, C.H. (2004). The reduction of CO_2 on polyaniline electrode in a membrane cell. *Synth. Met.* 140: 65–68. http://www.sciencedirect.com/science/article/pii/S0379677903000213.

127 Grace, A.N., Choi, S.Y., Vinoba, M. et al. (2014). Electrochemical reduction of carbon dioxide at low overpotential on a polyaniline/Cu_2O nanocomposite based electrode. *ApEn* 120: 85–94. http://www.sciencedirect.com/science/article/pii/S0306261914000415.

128 Aydın, R., Doğan, H.Ö., and Köleli, F. (2013). Electrochemical reduction of carbondioxide on polypyrrole coated copper electro-catalyst under ambient and high pressure in methanol. *Appl. Catal., B* 140–141: 478–482. http://www.sciencedirect.com/science/article/pii/S0926337313002300.

129 Ogura, K., Endo, N., and Nakayama, M. (2019). Mechanistic studies of CO_2 reduction on a mediated electrode with conducting polymer and inorganic conductor films. *J. Electrochem. Soc.* 145: 3801–3809. https://doi.org/10.1149/1.1838877.

130 Santhosh, P., Gopalan, A., and Lee, K.-P. (2006). Gold nanoparticles dispersed polyaniline grafted multiwall carbon nanotubes as newer electrocatalysts: preparation and performances for methanol oxidation. *J. Catal.* 238: 177–185. http://www.sciencedirect.com/science/article/pii/S0021951705004902.

131 Wang, Y., Guo, X., and Yu, G. (2019). Conductive polymers for stretchable supercapacitors. *Nano Res.* 12: 1978. https://doi.org/10.1007/s12274-019-2296-9.

132 Y. Han, L. Dai, Conducting polymers for flexible supercapacitors, *Macromol. Chem. Phys.*, 220 (2018) 1800355. https://doi.org/10.1002/macp.201800355

133 H. Park, J. Huh, M. Kang, J. Lee, H Yoon, Anisotropic growth control of polyaniline nanostructures and their morphology-dependent electrochemical characteristics, *ACS Nano*, 6 (2012) 7624. https://doi.org/10.1021/nn3033425

134 Y. Shi, L. Pan, B. Liu, Y. Wang, Y. Cui, Z. Bao, G. Yu, Nanostructured conductive polypyrrole hydrogels as high-performance, flexible supercapacitor electrodes, *J. Mater. Chem. A*, 2 (2014) 6086. https://doi.org/10.1039/c4ta00484a

135 A. Laforgue, C. Sarrazin, J. Fauvarque, Polythiophene-based supercapacitors, *J. Power Sources*, 80 (1999) 142. https://doi.org/10.1016/s0378-7753(98)00258-4

136 Chen, X., Jiang, F., Jiang, Q. et al. (2020). Conductive and flexible PEDOT-decorated paper as high performance electrode fabricated by vapor phase polymerization for supercapacitor. *Colloids Surf., A* 603: 125173. https://doi.org/10.1016/j.colsurfa.2020.125173.

137 Li, X., Guo, W., Chen, J. et al. (2014). Synthesis of spherical PANI particles via chemical polymerization in ionic liquid for high-performance supercapacitors. *Electrochim. Acta* 135: 550. https://doi.org/10.1016/j.electacta.2014.05.051.

138 X. Wang, Y. Fu, S. Wang, T. Yang, K. Jiao, A highly conductive and hierarchical PANI micro/nanostructure and its supercapacitor application, *Electrochim. Acta*, 222 (2016) 701. https://doi.org/10.1016/j.electacta.2016.11.026

139 T. Li, Y. Zhou, Z. Dou, L. Ding, S. Dong, N. Liu, Z. Qin, Composite nanofibers by coating polypyrrole on the surface of polyaniline nanofibers formed in presence of phenylenediamine as electrode materials in neutral electrolyte, *Electrochim. Acta*, 243 (2017) 228. https://doi.org/10.1016/j.electacta.2017.05.087

140 Fu, C., Zhou, H., Liu, R. et al. (2012). Supercapacitor based on electropolymerized polythiophene and multi-walled carbon nanotubes composites. *Mater. Chem. Phys.* 135: 596. https://doi.org/10.1016/j.matchemphys.2011.11.074.

141 C. Dong, X. Zhang, Y. Yu, L. Huang, J. Li, Y. Wu, Z. Liu, An ionic liquid-modified RGO/polyaniline composite for high-performance flexible

all-solid-state supercapacitors. *Chem. Commun.*, 56 (2020) 11993. https://doi.org/10.1039/D0CC04691D

142 J. Ren, R. Ren, Y. Lv, Stretchable all-solid-state supercapacitors based on highly conductive polypyrrole-coated graphene foam, *Chem. Eng. J.*, 349 (2018) 111. https://doi.org/10.1016/j.cej.2018.05.075

143 Kim, Y. and Shin, K. (2021). Dopamine-assisted chemical vapour deposition of polypyrrole on graphene for flexible supercapacitor. *Appl. Surf. Sci.* 547: 149141. https://doi.org/10.1016/j.apsusc.2021.149141.

144 Y. Cai, L Xu, H. Kang, W. Zhou, J. Xu, X. Duan, X. Lu, Q. Xu, Electrochemical self-assembled core/shell PEDOT@MoS_2 composite with ultra-high areal capacitance for supercapacitor, *Electrochim. Acta*, 370 (2021) 137791. https://doi.org/10.1016/j.electacta.2021.137791

145 L. Yuan, C. Wan, X. Ye, F. Wu, Facial synthesis of silver-incorporated conductive polypyrrole submicron spheres for supercapacitors, *Electrochim. Acta*, 213 (2016) 115. https://doi.org/10.1016/j.electacta.2016.06.165

146 Y. Zang, H. Luo, H. Zhang, H. Xue, Polypyrrole nanotube-interconnected NiCo-LDH nanocages derived by ZIF-67 for supercapacitors, *ACS Appl. Energy Mater.*, 4 (2021) 1189, https://doi.org/10.1021/acsaem.0c02465

147 N. Jabeen, Q. Xia, M. Yang, H. Xia, Unique core−shell nanorod arrays with polyaniline deposited into mesoporous $NiCo_2O_4$ support for high-performance supercapacitor electrodes, *ACS Appl. Mater. Interfaces*, 8 (2016) 6093. https://doi.org/10.1021/acsami.6b00207

148 Ambade, R., Ambade, S., Salunkhe, R. et al. (2016). Flexible-wire shaped all-solid-state supercapacitors based on facile electropolymerization of polythiophene with ultrahigh energy density. *J. Mater. Chem. A* 4: 7406. https://doi.org/10.1039/c6ta00683c.

149 W. Bi, J. Huang, M. Wang, E. Jahrman, G. Seidler, J. Wang, Y. Wu, G. Gao, G. Wu, G. Cao, V_2O_5–Conductive polymer nanocables with builtin local electric field derived from interfacial oxygen vacancies for high energy density supercapacitors, *J. Mater. Chem. A*, 7 (2019) 17966. https://doi.org/10.1039/c9ta04264d

150 I. Moon, S. Yoon, J. Oh, 3D Highly conductive silver nanowire@PEDOT:PSS composite sponges for flexible conductors and their all-solid-state supercapacitor applications, *Adv. Mater. Interfaces*, 4 (2017) 1700860. https://doi.org/10.1002/admi.201700860

151 N. Song, Y. Wu, W. Wang, D. Xiao, H. Tan, Y. Zhao, Layer-by-layer in situ growth flexible polyaniline/graphene paper wrapped by MnO_2 nanoflowers for all-solid-state supercapacitor, *Mater. Res. Bull.*, 111 (2019) 267. https://doi.org/10.1016/j.materresbull.2018.11.024

152 X. Xu, J. Tang, H. Qian, S. Hou, Y. Bando, M. S. A. Hossain, L. Pan, Y. Yamauchi, Three-dimensional networked metal−organic frameworks with conductive polypyrrole tubes for flexible supercapacitors, *ACS Appl. Mater. Interfaces*, 9 (2017) 38737. https://doi.org/10.1021/acsami.7b09944.

153 M. Xu, H. Guo, T. Zhang, J. Zhang, X. Wang, W. Yang, High-performance zeolitic imidazolate frameworks derived three-dimensional Co_3S_4/polyaniline

nanocomposite for supercapacitors, *J. Energy Storage*, 35 (2021) 102303, https://doi.org/10.1016/j.est.2021.102303.

154 V. Shrivastav, S. Sundriyal, A. Kaur, U. K. Tiwari, S. Mishra, A. Deep, Conductive and porous ZIF-67/PEDOT hybrid composite as superior electrode for all-solid-state symmetrical supercapacitors, *J. Alloys Compd.*, 843 (2020) 155992. https://doi.org/10.1016/j.jallcom.2020.155992

155 Wu, K., Zhao, J., Wu, R. et al. (2018). The impact of Fe^{3+} doping on the flexible polythiophene electrodes for supercapacitors. *J. Electroanal. Chem.* 823: 527. https://doi.org/10.1016/j.jelechem.2018.06.052

10

Conducting Polymer Nanomaterials for Bioengineering Applications

Xiang Sun[1], Meiling Wang[1], You Liu[1], Xin Zhang[1], Yalan Chen[1], Shiying Li[1], and Ying Zhu[1,2]

[1] Beihang University, Key Laboratory of Bio-Inspired Smart Interfacial Science and Technology of Ministry of Education, School of Chemistry and Beijing Advanced Innovation Center for Biomedical Engineering, Beijing 100191, China
[2] Beihang University, Beijing Advanced Innovation Center for Biomedical Engineering, Beijing 100191, China

10.1 Introduction

Many functions of the human body are regulated by electrical signals, which require that the materials interfacing with electrically active tissues for bio-application purposes, such as tissue engineering or monitoring, must also be conductive to enhance the biological response to external stimuli [1]. Conducting polymers (CPs) showing electrical and optical properties such as polyaniline (PANI) [2], polypyrrole (PPy) [3], and poly(3,4-ethylenedioxythiophene) (PEDOT) [4], and their composites are attractive biomaterials due to their biocompatibility, facile synthesis, simple modification, and ability to electronically control a range of physical and chemical properties. The soft characteristics of organic conductive polymers provide better mechanical compatibility and structural tunability with cells and organs than conventional electronic inorganic and metal materials. These advantageous properties make them attractive in many biomedical applications, including drug delivery systems, electronic skins, biological actuators, biosensors, nervous systems, and tissue engineering [1, 5–7]. In this part, we focus on the application of CPs in bioengineering by virtue of their biocompatibility and conductivity, including tissue engineering, electrochemical sensors, and drug delivery systems.

10.2 Electronic Skin

Electronic skin (E-skin) is a flexible sensor that mimics the various functions of human skin for its tactile perception and temperature perception [8]. These sensors are mostly used in wearable electronic devices, bionic prostheses, and disease diagnosis and monitoring [9–11]. The popularization of E-skin will bring great convenience to people's lives. E-skin not only is required to be able to measure

Functional Nanomaterials: Synthesis, Properties, and Applications, First Edition.
Edited by Wai-Yeung Wong and Qingchen Dong.
© 2022 WILEY-VCH GmbH. Published 2022 by WILEY-VCH GmbH.

the spatial distribution of stress caused by a variety of mechanical stimuli (such as pressure, lateral strain and bending, etc.) but also has excellent stretchability to withstand long-term bending and twisting, even other mechanical deformations [12]. To realize their commercial applications, E-skin should possess comfort, low cost, and compatibility with large-area manufacturing. CPs, including PANI, PPy, PEDOT, and so on, are the most ideal polymer materials for E-skin due to their high conductivity, high chemical stability, flexibility, nontoxicity, lightweight, and low cost [13–16]. This section will summarize and discuss the current applications of CPs in E-skin.

10.2.1 Wearable Electronic Devices

Recently, wearable electronic devices have been attracting more and more attention due to their promising applications in fields such as portable, flexible/stretchable human-interactive sensors, displays, and energy devices. The functional materials provide promising solutions for wearable electronics, given their outstanding electronic, optical, and mechanical properties. Conducting polymer hydrogels (CPH) combined the unique advantages of hydrogels and organic conductors and received widespread attention due to their adjustable mechanical properties, biocompatibility, self-repair, hydrophilicity, and ease of preparation [17]. Because of its sensitivity to external stimuli, CPH has been widely used in various electronic devices, especially in the field of sensors. In particular, transparent conducting hydrogel is widely used in flexible electronic devices [18, 19]. Among CPs, PPy has biocompatibility, good stability, and excellent conductivity. For example, Han et al. [20] developed a transparent, stretchable, and self-adhesive hydrogel via *in situ* formation of polydopamine (PDA)-doped PPy nanofibrils in the polyacrylamide (PAM) network. The *in situ* formed PDA-PPy nanofibrils with good hydrophilicity were well-integrated with the hydrophilic polymer phase and interwoven into a nanomesh, which created a complete conductive path with a high intrinsic conductivity of $12\,S\,m^{-1}$ and allowed visible light to pass through for high transparency (70%). Moreover, these showed enhancement on the mechanical properties of the PDA-PPy-PAM hydrogel. The hydrogel was flexible and stretchable, allowing it to tolerate a large strain exceeding 2000%. It could be stretched to six times its original length and be able to recover to its original length in one second after being stretched. As an example, the hydrogels could be stably attached to the body, and they repeatedly detected movements even they went through large deformations. The corresponding time-dependent responses ($\Delta R/R_0$) of the hydrogels during body movement were recorded. The highest value of $\Delta R/R_0$ from the sensor on the wrist was 90%, while that of the knee joint was 80%. Moreover, the PDA-PPy-PAM hydrogel was also used as a self-adhesive electrode to detect biosignals, such as electrocardiograph (ECG) and magnetocardiography (EMG) signals, which were as accurate as those obtained by commercial electrodes.

Khalili et al. [21] proposed a constriction resistance model to capture the contact resistance change and the sensitivity of piezoresistive sensors on the basis of PPy based hydrogels with different ratios of incorporated electrically conducting

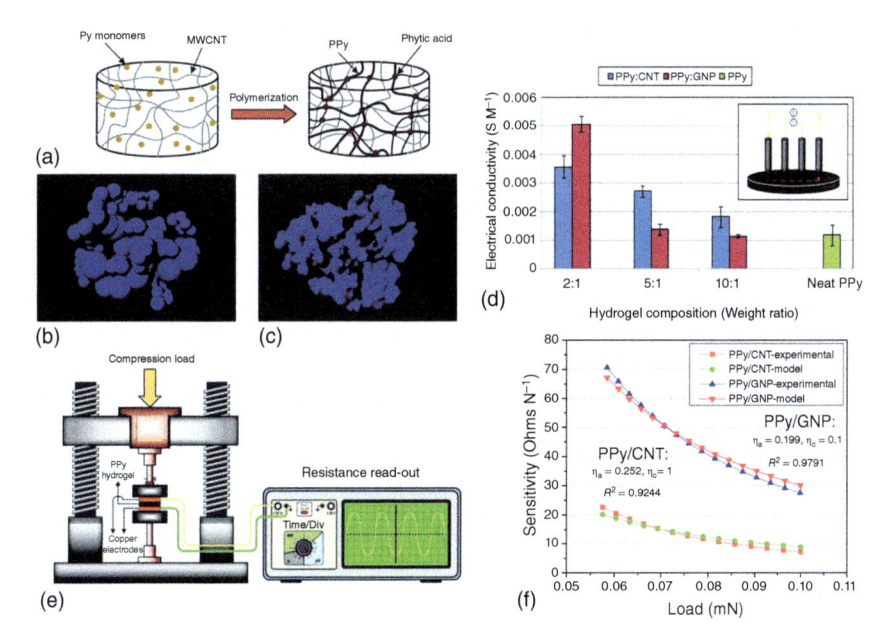

Figure 10.1 (a) Schematics of the polymerization process. (b) PPy:CNT hollow spherical structure. (c) The interconnected network of the hydrogels. (d) The measured electrical conductivity of the PPy-based hydrogels with different compositions. The inset photo depicts the schematic of a four-point probe measurement technique. (e) Schematic of the experimental setup to measure the piezoresistive response and time response of the sensors. (f) Experimental sensitivity curves vs. their modeling counterparts for different compositions and pressure ranges. Source: Khalili et al. [21]. Reproduced with permission of Royal Society of Chemistry.

particles. As-designed PPy-based hydrogels were prepared with incorporated carbon nanotubes (CNTs) and graphene nanoplatelets (GNPs) via a multiphase reaction (Figure 10.1a–c). It is shown that the electrical conductivity of the fabricated PPy-based piezoresistive sensors is enhanced to ~0.005 S m^{-1} for PPy:GNP and ~0.0035 S m^{-1} for PPy:CNT. As a result of adding conductive fillers, the PPy:GNP and PPy:CNT were employed as the sensors, for measuring the constriction resistance change, showing a higher sensitivity (Figure 10.1d,e). The response time of the sensors was also shown to be significantly low for both loading and unloading of the sensor (~70 ms). A constriction resistance model was further proposed to capture the contact resistance change and the sensitivity of piezoresistive sensors based on PPy hydrogels with different ratios of incorporated electrically conducting particles, which showed high degree of fit (R^2 (PPy:GNP) = 0.9791, R^2 (PPy:CNT) = 0.9244; Figure 10.1f).

To improve the mechanical properties and high stretchability of CPs sensors, the polymer matrix was applied to the E-skin. However, the mismatch of the mechanical properties of the matrix will result in severe interface delamination between the hard matrix and the soft human body [22–25]. Therefore, there is an urgent need for natural soft and biocompatible materials for interconnection that can transmit electrical signals. By combining CPs such as PANI, PPy, and PEDOT with the hydrated polymers, the functional E-skin devices have been produced. For instance, Lee et al. [26] used a gelation of (poly(3,4-ethylenedioxythiophene)/polystyrene sulfonate, PEDOT:PSS) dispersed EG solution with polyacrylamide (PAAm) and successfully obtained a uniformly distributed PEDOT:PSS-PAAm organogel. The as-manufactured PEDOT:PSS-PAAm organogel is conductive and highly stretchable. Even under a large tensile deformation of 300%, the organogel-based stretchable electronic conductor exhibits high electrical conductivity without an electrochemical reaction under DC voltage. It illustrates a stretchable light-emitting diode (LED) circuit where two LED lights are interconnected with the conductive PEDOT:PSS-PAAm organogel on an Ecoflex substrate. The LED is found to function well, even stretched above 200% strain, and it is elastically recovered after being released. Under a biaxial deformation on the stretchable 2×3 arrays of LEDs attached to an inflated balloon, the six LEDs remain in operation without mechanical and electrical failures. The resistance change during stretching is almost insensitive to strain, up to 50% strain, and remains unchanged even though it is deformed up to 1000 fatigue cycles. Moreover, the PEDOT:PSS-PAAm organogels can be stretched to 525% with the PEDOT:PSS weight ratio of 5.49%. The tensile strength of the organogels is in the range from 18 to 30 kPa, which are values mechanically compatible to human skin.

The flexibility and stretchability of electronic devices make it possible for smart electronic skins to be used in electronic consumers, robotics, and other fields. Such smart electronic devices include thin and compliant conductive materials and electronic devices that can withstand bending/torsion /stretching or can also adhere and conform to substrate surfaces. One example is that Oh et al. [27] fabricated piezoresistive pressure sensors by chemically grafting PPy on the surface of porous polydimethylsiloxane (PDMS) elastomer, which exhibited high uniformity and negligible hysteresis. The sensors had much lower coefficient of variance (2.43%) in relative resistance change than that of sensors made with randomly shaped pores (69.65%). The strong chemical bond between the PPy and the PDMS prevents their relative sliding and displacement, and the porosity of the elastomer enhances the elastic behavior, making it possible to accurately measure the pressure distribution over time. The degree of hysteresis of the sensor was 2% and was independent of the sweep rate or sweep range. The change in initial resistance after 1000 cycles at 50 kPa was minimal at 2.7%. Moreover, with eight sensors attached on a robotic hand gripper, it could be able to repeatedly grasp and release the objects. Resistance of the sensors changes in response to different grasp positions, through which applied pressure was deduced.

Zucca et al. [28] developed a roll-to-roll (R2R) technology to prepare PEDOT:PSS nanosheets on a large area to push the technology toward practical applications. The single-layer PEDOT:PSS nanosheet (thickness: ~110 nm) showed higher critical load (>50 000 N m^{-1}) than poly(D,L-lactic acid) (PDLLA) nanosheets (<45 000 N m^{-1}). As-prepared nanosheets were placed on the wrist and on the finger to assess the stability and evolution of electrical resistance of nanosheets. After 250 cycles, only negligible irreversible variations of electrical resistance were observed in the case of wrist (expansion/contraction on a +42/−42° range with respect to

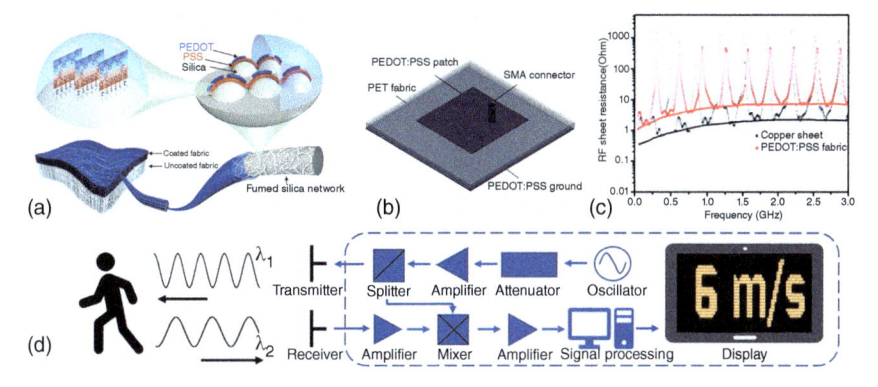

Figure 10.2 (a, b) Schematic diagram of the PEDOT:PSS-printed fabric, and the conductivity enhancement assisted by the silica nanoparticle network on the surface of the fabric fibers. (c) RF sheet resistance comparison between the PEDOT:PSS fabric and copper substrate. (d) Schematic diagram of the Doppler radar system with fabric patch antennas. Source: Zucca et al. [28]. Reproduced with permission of Royal Society of Chemistry.

relaxed state −0°) and minor variations (up to 20%) in the case of finger contraction, where nanosheet was placed across distal phalanges (flexion 70°). The R2R method enables continuous, high-yield, low-cost, and rapid implementation of industrialized/large-scale manufacturing technology on large-area rolls.

Next-generation wearable systems call for flexible, breathable, and skin-friendly wireless transmitters for realizing body area networks and the Internet of things. For instance, Li et al. [29] developed an all-organic flexible patch antenna based on a selected polyethylene terephthalate (PET) fabric substrate, screen-printed with PEDOT:PSS (Figure 10.2a,b). The fabricated patch antenna based on this conductive fabric shows an extremely low return loss of −50 dB and a satisfactory radiation efficiency of 28% at its resonant frequency of 2.35 GHz and preserves its performance characteristics when bended over a representative phantom. With the nanoparticle-assisted PEDOT:PSS phase segregation, the coated fabric benefited from the unique fiber-bundle structure to compensate skin effect and showed a low RF sheet resistance. At 2.4 GHz, the measured RF sheet resistance of PEDOT:PSS-printed fabric is 6.9 Ω, which is within the same order of magnitude of copper (Figure 10.2c). In addition, the Doppler radar system based on the fabric patch antennas demonstrates satisfactory speed and distance detection with high precision. As shown in Figure 10.2d, the volunteer walked ∼9 m away in 15 seconds, and a speed of 0.6 m s^{-1} can be obtained from the derivative of the distance. The RF signal transmitting capability of the fabric antennas is clearly demonstrated in this short-range Doppler radar system.

10.2.2 Self-Healing E-Skin

PEDOT:PSS has become one of the most successful organic conductive materials due to its high air stability, high conductivity, and biocompatibility. Zhang et al. [30] reported the first observation of the mechanical and electrical repairability

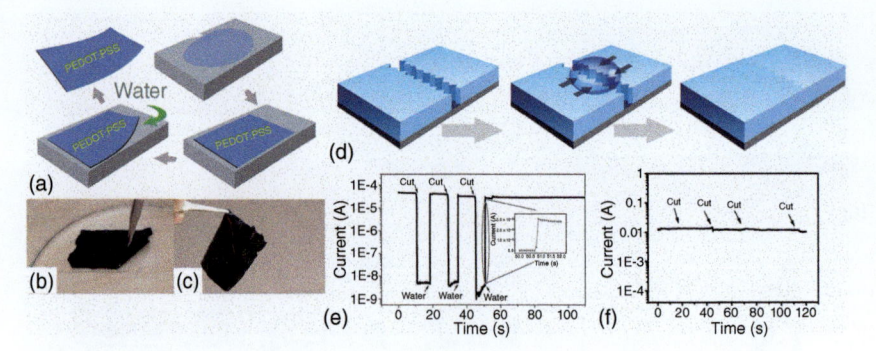

Figure 10.3 (a) Illustration of the water-assisted wedging method used to obtain freestanding PEDOT:PSS films. (b) The film is detached from the glass substrate with tweezers and (c) does not deteriorate during the process. (d) Schematic representation of water-induced mechanical and electrical healing of a PEDOT:PSS film. (e) Current vs. time measurements for PEDOT:PSS film showing the effect of damage and healing. (f) Current vs. time profile of wetted PEDOT:PSS film containing the conductivity enhancer glycerol after several cuts. Source: (a, d–f) Li et al. [29]. Reproduced with permission of Royal Society of Chemistry. (b, c) Li et al. [29] /with permission from Royal Society of Chemistry.

of PEDOT:PSS film with an electrical conductivity as high as 500 S cm^{-1}. Instantaneous electrical healing (about 150 ms) of pure PEDOT:PSS film after damage was observed by simply covering the damaged area of the films with a droplet of water. After being wetted, the films are transformed into autonomic self-healing materials without the need of external stimulation (Figure 10.3a–e). The film has long-term stability in air and water is independent and does not require substrate support. Thus, it can be directly laminated on regular or irregular objects. Moreover, the wet PEDOT:PSS films after soaking them into DI water for 5 seconds showed better autonomic self-healing properties. The current was totally unresponsive to repeated cuts in different regions of the film and maintained the initial value of ≈10 mA (Figure 10.3f). The healing effect can even be achieved by exposure to water vapor. After damage at a relative humidity (RH) of over 80%, the current recovered to its initial value within a few minutes.

Lu et al. [31] developed a new regenerative polymer complex composed of poly(2-acrylamido-2-methyl-1-propanesulfonic acid), PANI, and phytic acid (PAAMPSA/PANI/PA) as a skin-like electronic material. It can be attached directly on the skin to accurately detect and discriminate complex human motions. It exhibits ultra-high stretchability (1935%), quadratic response to strain ($R^2 > 0.9998$), and linear response to flexion bending ($R^2 > 0.9994$). Furthermore, the PAAMPSA/PANI/PA material exhibits a repeatable autonomous self-healing capability with high healing efficiencies (repeating healing efficiency >98%) and tunable healing time. Another noteworthy advantage of the PAAMPSA/PANI/PA strain sensor is its negligible resistance drift after the completion of motions. Notably, the sensor can detect bending angles of the finger knuckle with a high linearity. The coefficient of determination is shown to be 0.9935 with extremely low standard deviations (±1.1% for all bending angles). Similar to the finger knuckle motion detection, the sensor is capable of accurately detecting bending motions of knee joints with a linear response

($R^2 = 0.9914$). In addition, the present strain sensor can detect and discriminate complex motions. The knee motions of extending and flexing ($\Delta R/R_0 = 36$–40%), walking ($\Delta R/R_0 = 23$–27%), and squatting ($\Delta R/R_0 = 68$–70%) each show their own characteristic signal intensity, pattern, and frequency.

10.2.3 Energy-Saving E-Skin

Self-powered devices based on organic CPs have proven to be very promising for the next generation of fully integrated smart energy-saving skin applications, owing to their strong spin-charge interaction, ease of preparation, excellent stability, and nontoxicity [32–35]. In addition, PPy, PEDOT:PSS, and PANI have been successfully proven to be high-performance triboelectric materials, so it can be used for energy harvesting and sensing applications. Mule et al. [36] proposed PPy-based flexible and wearable triboelectric nanogenerators (TENGs), in which PPy was deposited on a cotton textile (PPy@CT) by *in situ* polymerization process. Afterward, PPy@CT was covered with PDMS and microtextures (PPy-WSEM-TENG) (Figure 10.4a). As-prepared PPy-WSEM-TENG device exhibited a maximum VOC, ISC, and power density values of ~200 V, ~6 μA, and ~82 μW cm^{-2}, respectively, under the pressing force of 10 N. Moreover, this device exhibited robust characteristics even after long-term cyclic operations and also generated an electrical output (VOC, ~180 V; ISC values, ~5.5 μA) by gently touching with the human hand (Figure 10.4b). For commercial applications, the as-fabricated PPy-WSEM-TENG was effectively employed as a self-powered source to drive portable electronic devices and LED (Figure 10.4c).

Ahmed et al. [37] proposed a new type of printed smart electronic skin (SES) with multi-sensing characteristics and energy conversion functions, which consisted of a flexible substrate of PDMS, electrode layer of PEDOT:PSS, and single active layer of

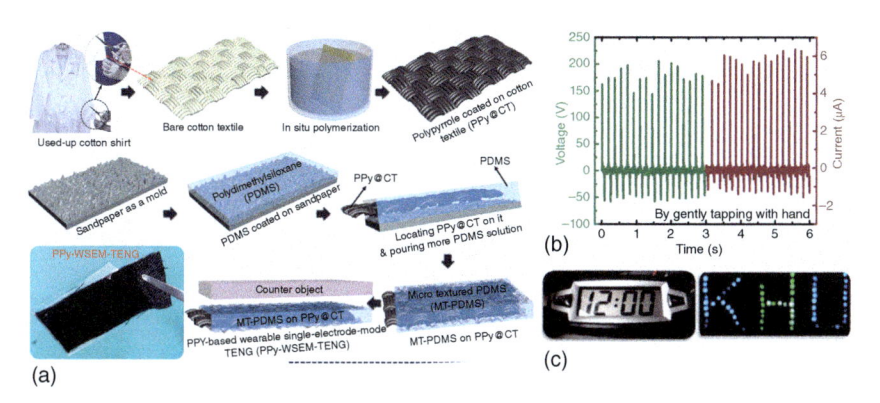

Figure 10.4 (a) Schematic diagram to illustrate the fabrication processes of a real PPy-based wearable single-electrode-mode TENG (PPy-WSEM-TENG) device. (b) VOC and ISC curves of PPy-WSEM-TENG by gently tapping with a human hand. (c) Commercial applications of PPy-WSEM-TENG. Source: Guan et al. [35]/with permission from American chemical Society.

poly(3-butylthiophene) blended with a poly[(9,9-dioctylfluorenyl-2,7-diyl)-co-(4,40-(N-(4-sec-butylphenyl)diphenyllam-ine)] (P3BT-TFB). It exhibits a good pressure sensitivity of 0.437 kPa^{-1} at a low detection limit of 0.1 kPa and high durability over 10 000 cycles of pressure loading and can detect the strain up to 30% with stable and reproducible performance. SES can be used as a human body monitoring device, which can detect various external stimuli, and can be used as a super-flexible biomechanical energy harvester.

CPs are similar to biomolecules in terms of chemical "naturalness," which can be engineered in many ways, including those with a Young's modulus similar to that of soft tissue and with ion conductivity hydrogels. The structure of organic materials can be adjusted through synthetic chemistry, and a variety of functionalization strategies can be used to control their biological properties. It can also be integrated with a variety of mechanical supports, resulting in a device with dimensions that can be integrated with biological systems [37]. The combination and integration of these CPs makes it widely used in electronic skins and wearable electronic devices, which is of great significance for detecting human health. We believe that in the near future, with people's in-depth research and exploration of CPs and printed E-skin, they can be widely used in the daily life, providing great convenience for our lives and medical health fields.

10.3 Bioengineering

The unique features of CPs, such as electronic-ion hybrid conductivity, mechanical flexibility, permeable porosity, and multifunctional chemical modification, make them very popular in a wide range of biomedical equipment applications. CPs have been used in different biomedical applications, including biosensors, drug delivery systems, artificial muscles, and so on [38–40]. This section focuses on the application research trends of CPs in tissue regeneration engineering, biosensors, and drug delivery systems. By summarizing and analyzing the current application status of CPs in various fields, it aims to further explore the application prospects and research directions of CPs in the future.

10.3.1 Tissue Regeneration Engineering

The application of CPs in the field of tissue engineering is inseparable from their good conductivity, redox, three-dimensional (3D) structure, and surface morphology. In particular, it benefits from the good biocompatibility of CPs. In this field, using scaffolds, biologically active molecules and living cells can restore or completely replace tissue functions [41]. The porous scaffold of CPs is made as a support template for cell growth and differentiation, which ultimately leads to the formation of new tissue. CPs can also regulate cell activities, including cell growth and migration through external stimulation [42]. As the control of cell behavior is essential for regeneration, CPs are attractive candidates as scaffold materials for regeneration engineering [43, 44].

Generally, the skin serves as the body's first line of defense against external hazards. Once the skin tissue is damaged, the infection will hinder the rapid recovery of the appearance and function of the skin tissue [45]. Therefore, infection control has become a research hotspot in wound healing. Wang and coworkers [45] synthesized PPy and Zn-functionalized chitosan molecules, which were cross-linked with polyvinyl alcohol (PVA) through dynamic glycol complexation, hydrogen bonds, and zinc-based coordination bonds to form PCPZ hydrogel (Figure 10.5a). The conductivity provided by PPy enables the hydrogel to sense strain and temperature, and the coordinated Zn significantly enhances the antibacterial activity of the hydrogel. The as-synthesized PCPZ hydrogel was used as a conductive dressing to treat chronic wounds in a rat model with electrical stimulation (ES). The results showed (Figure 10.5b,c) that four groups of full-thickness wounds were formed on each diabetic rat to produce chronic wounds. In the control group, the wound was not covered with a dressing, while in the Hydrosorb or PCPZ groups, the wound was treated with a commercial Hydrosorb dressing or PCPZ hydrogel, respectively. All wounds were inoculated with *Staphylococcus aureus* to cause infection. Figure 10.5b shows that the wound was purulent on day 1, indicating that the wound was infected. Compared with the control group, using only commercially available Hydrosorb dressings or PCPZ hydrogel alone to treat wounds did not accelerate wound healing, but stimulating the wound with a 3 V voltage through conductive PCPZ hydrogel would significantly promote wound closure. Using a diabetic rat model, it is proved that the conductive hydrogel can promote the healing of infected chronic wounds through ES.

Peripheral nerve injury is a common and serious disease. Currently, the most widely used technique to treat nerve injury is autologous nerve transplantation. However, autologous nerve transplantation itself has many defects, including the instability of autologous graft function, the lack of donor nerves, and the size mismatch of donor nerve tissue [47, 48]. Therefore, peripheral nerve regeneration has always been one of the central problems and difficulties of nervous system transplantation. Considering the structure and electrophysiological properties of nerve tissue, suitable scaffold materials should have specific electrical properties and morphological characteristics to guide cell behavior, stimulate neuronal activity, and guide nerve tissue repair [49, 50]. Feng and coworkers [50] used the hydrogen bonding and electrostatic adsorption between silk amino, reduced graphene oxide (RGO), and PANI to prepare a highly conductive braided silk (GO/PANI/silk) composite scaffold (Figure 10.6a). Compared with the traditional *in situ* polymerization of aniline, the conductive RGO/PANI/silk knitted scaffold prepared by the two-step electrostatic self-assembly method has more uniform PANI particles and lower resistance. The results show that when GO is $1\,\mathrm{g\,l^{-1}}$ and aniline is 0.4, 0.6, or $0.8\,\mathrm{mol\,l^{-1}}$, its conductivity is maintained between 0.62×10^{-3} and $1.72\times10^{-3}\,\mathrm{S\,cm^{-1}}$, indicating that the RGO/PANI/silk scaffold has better electrical property (Figure 10.6b,c). In addition, this polymer scaffold has good conductivity stability under different physical stresses and has good mechanical properties to provide good support. Park and coworkers [51] researched a method to produce hybrid nanocomposites by complexing gold nanoparticles with PANI

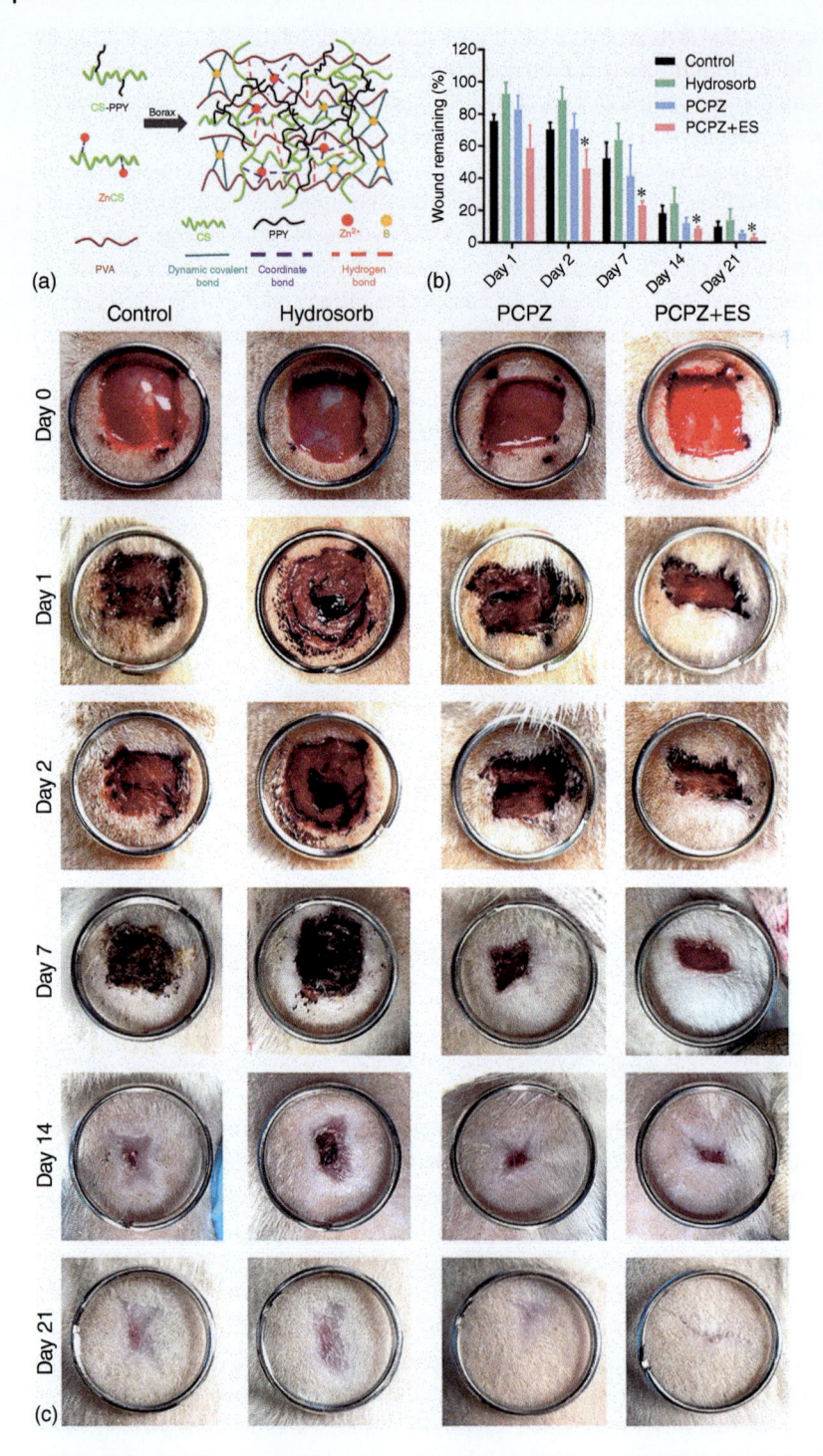

Figure 10.5 (a) Schematic illustration of the preparation of the PCPZ hydrogel. (b) Wound healing percentage of the different groups, indicating significant difference compared with the control group, $n = 3$. (c) Photos of the wounds of the different groups over 21 days. Source: Liang et al. [46]/with permission from American chemical society.

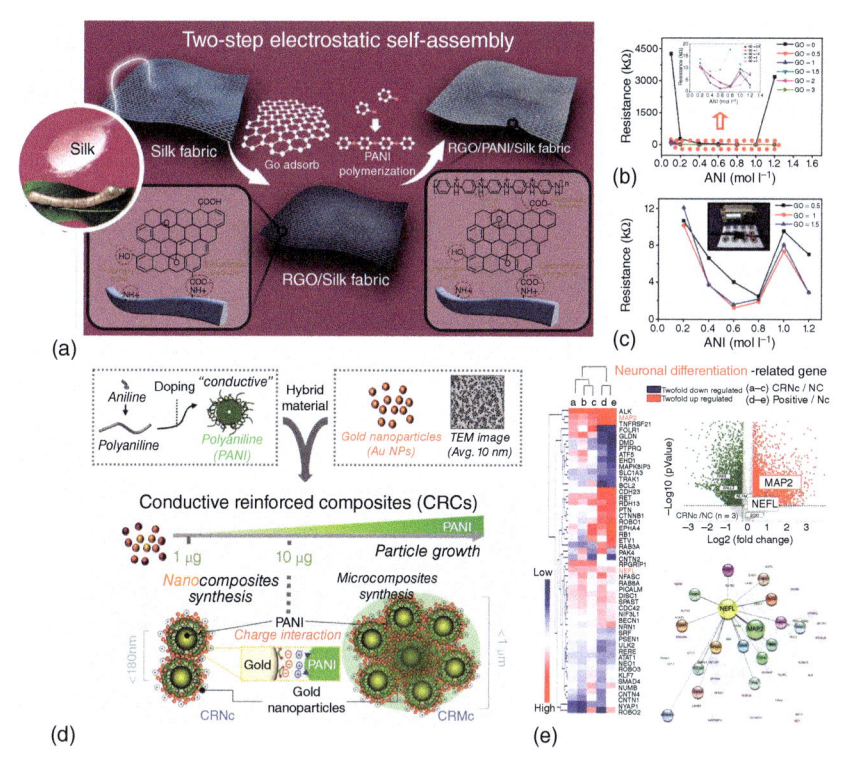

Figure 10.6 (a) Electrostatic self-assembly process of GO and ANI on silk fabrics. (b) Change of resistance with increase of GO and ANI concentration. (c) Change of resistance with GO = 0.5/1/1.5 g l^{-1} and ANI = 0.2–1.2 mol l^{-1}. Source: Kim et al. [51]/with permission from American chemical society. (d) CRC formation by the complexation of PANI with gold NPs. (e) Heat map of hierarchically clustered mRNA-array data in CRNc-treated ($n = 3$, a–c) and positive control samples ($n = 2$, d, e). Signal intensities were log$_2$-transformed, centered, and normalized. Source: Lu et al. [52]/with permission from American chemical society.

and tested the resulting composites in a nerve regeneration model (Figure 10.6d), which is named as conductivity enhanced nanocomposite (CRNc). When CRNc was delivered directly to cells, no cytotoxicity was observed. After delivering CRNc in the cell, an electroporator is used to electrically stimulate the stem cells. As a result of ES of CRNc internalized cells, followed by mRNA sequencing analysis, it was confirmed that CRNc internalized cells had a pattern similar to that of neuronal cells induced by positive groups. In particular, microtubule-associated protein 2 is more than twice that of the control group (negative control group), and neurofibrin is strongly expressed as in the positive control group (Figure 10.6e). These findings indicate that CRNc can be used to induce the formation of neuron-like cells by applying ES to stem cells.

In recent years, there has been an increasing need to use high-performance biomaterials for tissue reconstruction, including fracture and defect repair. However, traditional bone implants have poor anchoring properties to natural tissues, which may cause implant displacement accidents and local inflammation [53]. A number of studies have proven the effects of stimulation on cell activities, such as adhesion, proliferation, gene expression, and protein synthesis [54]. Inspired by the idea of improving bone healing through stimulation, the electrical properties of the stent can be adjusted by using biocompatible CPs. CPs with reversible redox, adjustable conductivity, and easy processing have attracted interest in biomedical applications. For example, Sun and coworkers [54]. reported a new type of immobilization method. This method self-assembles biotinylated mesenchymal stem cell-derived extracellular vesicles (MSC-EVs) on the surface of biotin-doped PPy titanium (Bio-Ppy-Ti) to improve its biological functions *in vivo* and *in vitro* (Figure 10.7a).

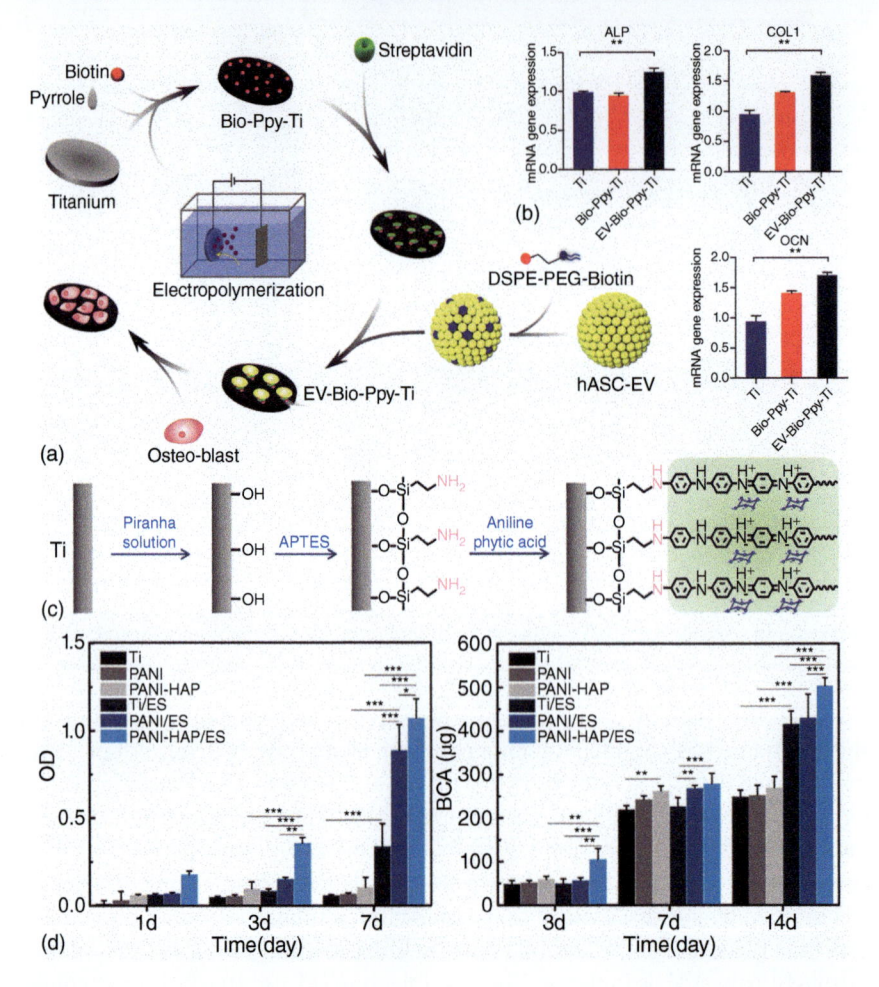

Figure 10.7 (a) Schematic representation of the functionalized titanium preparation and the entire experimental process [55]. (b) Expression of ALP, COL 1, and OCN genes in MG63 cells cultured on EV-Bio-PPy-Ti was significantly higher compared with that cultured on Ti (S–N–K test, **$p < 0.01$). Source: Wang et al. [55]. Reproduced with permission of American Chemical Society. (c) Schematic diagram of the experimental procedure for polymerizing PANI on titanium plates. (d) Activity of ALP and intracellular Ca ion content for MC3T3 cells cultured on different surfaces with/without ES. Source: Yan et al. [56]. Reproduced with permission of American Chemical Society.

Using this method, after 30 seconds of ultrasonic vibration, the amount of human adipose stem cell-EV anchored on the surface of Bio-Ppy-Ti was 185 times higher than that of pure Ti and remained stable within 30 seconds. For the *in vitro* assay, alkaline phosphatase (ALP), Collagen 1 (COL 1), and osteocalcin (OCN) mRNA levels were measured in MG63 cells cultured on Ti, Bio-Ppy-Ti, and EV-Bio-Ppy-Ti for 14 days. Compared with Ti, the expressions of ALP, COL 1, and OCN in cells cultured on EV-Bio-Ppy-Ti were significantly upregulated by 1.25, 1.6, and 1.7 times, respectively (Figure 10.7b). Therefore, this MSC-EV biotin immobilization method appears to be efficient and long-term stable for the modification of the biological activity of bone grafts. Li and his team [55] successfully synthesized phytic acid-doped PANI on medical titanium (Ti) tablets (Figure 10.7c). As phytic acid has abundant anodic phosphate groups, hydroxyapatite (HAP) nanocrystals are biomineralized on PANI to form PANI-HAP hybrid layer, which has good cell compatibility with MC3T3 cells. More importantly, the level of ALP expressed in each group was measured to assess the differentiation ability of osteoblasts. As shown in Figure 10.7d, the ALP activity showed a significant time dependence, slowly increasing after 3–14 days of incubation. After 7 days, PANI-HAP showed a clear advantage over PANI and Ti plates. In addition, it is worth noting that ES could amplify ALP levels compared with the corresponding unstimulated group. Therefore, the PANI-HAP/ES group has more ALP in all groups. After 14 days of incubation, the ALP activity measured on PANI-HAP/ES was 3 times higher than pure Ti without ES.

Poor adhesion between conventional implants and natural bone tissue may cause displacement, local inflammation, and unnecessary secondary operations. Therefore, conductive bioadhesives with strong adhesion property provide an effective method for the fixation and regeneration of comminuted fractures. Inspired by the chemistry of mussels, Zhang and coworkers [56] designed the conductive copolymers poly{[aniline tetramer methacrylamide]-*co*-[dopamine methacrylamide]-*co*-[poly(ethylene glycol) methyl ether methacrylate]} [poly-(ATMA-*co*-DOPAMA-*co*-PEGMA], AT/conductive aniline tetramer, DOPA/dopamine, and PEG/poly(ethylene glycol) with AT content of 3.0, 6.0, and 9.0 mol%, respectively (Figure 10.8). The tensile strength of the copolymer is enhanced during the stretching process and can reach 1.28 MPa at 6 mol% AT. In addition, conductive matrix and square wave electrical stimulation can synergistically enhance osteogenic differentiation, calcium deposition, and osteogenic gene expression. The ALP activity on the 14th day and the calcium deposition on the 28th day of the 9 mol% AT group were significantly higher than that of PLGA under electrical stimulation. In contrast, on the 7th day under electrical stimulation, compared with PLGA, the value of osteopontin was significantly increased by 5.9 times for the 9 mol% AT group. In general, CPs with strong adhesion can synergistically upregulate cell activity with electrical stimulation and may be a promising bioadhesive for plastic surgery and dental applications.

With the development of society, heart failure has become a key issue in international health. Myocardial infarction (MI) or heart attack is one of the main causes of death related to cardiovascular disease. At present, cardiac tissue engineering (CTE)

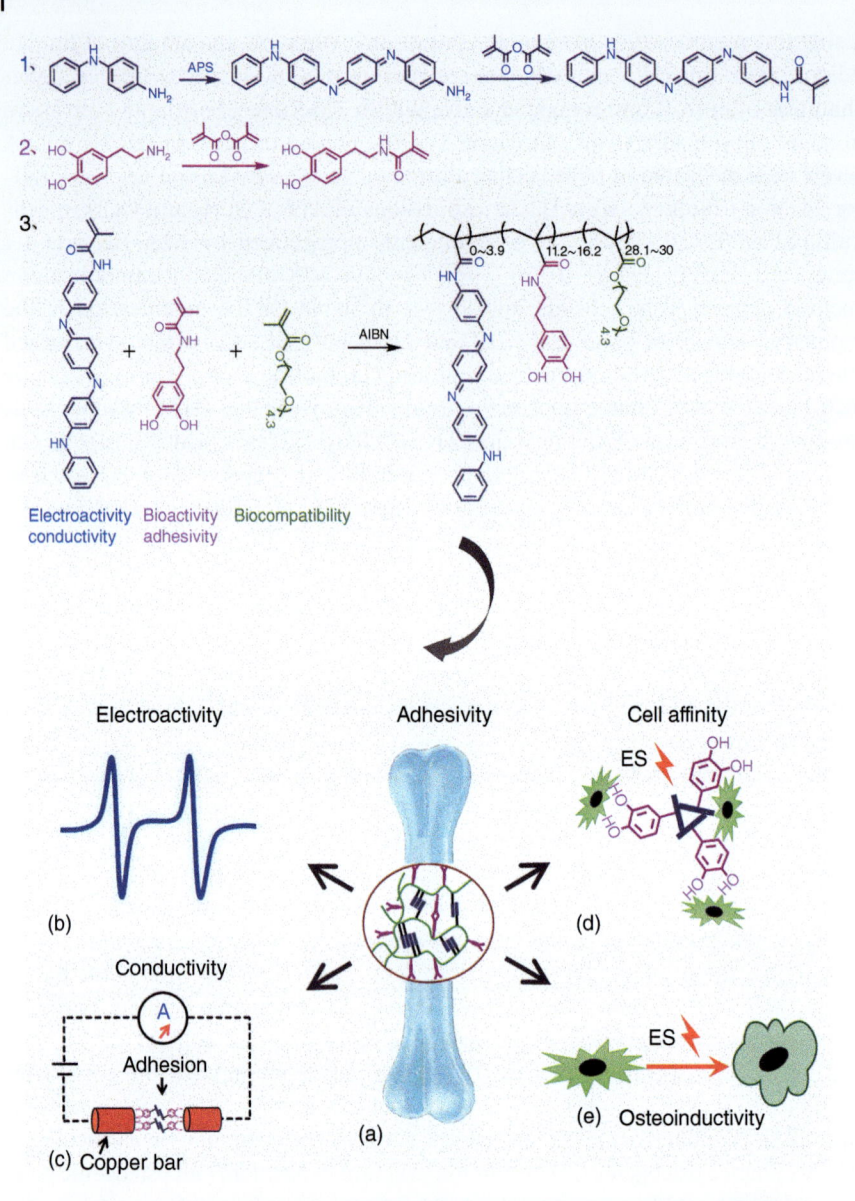

Figure 10.8 Synthetic route of conductivity mussel-inspired poly(ATMA-*co*-DOPAMA-*co*-PEGMA) bioadhesive and schematic illustration of the structure and function of bioadhesive that was (a) adhesive, (b) electroactive, (c) conductive, (d) cell affinitive, and (e) osteoinductive. Source: Ashtari et al. [57]. Reproduced with permission of American Chemical Society.

is a development method that aims to repair and regenerate damaged heart tissue through the application of cell transplantation and biomaterial 3D scaffolds [58]. The ideal stent for CTE should have conductivity and biocompatibility and have elasticity similar to that of natural myocardium. Damaged heart tissue can be regenerated by applying cell transplantation and 3D biomaterial scaffolds [1]. Choosing a 3D scaffold with conductive, biocompatible, and viscoelastic material design optimization can be functionalized with bioactive molecules or stem cells. Obviously, CPs are considered as potential candidates for enhancing the function of cardiac tissue engineering [59]. Qiu et al. [59] has developed thermoplastic (glycolic acid) surgical sutures and adheres mussel-inspired conductive particles to highly elastic, conductive, springlike coils. Biological springs coated with PPy act as electrodes that are assembled into solid supercapacitors (Figure 10.9a). After injection through a syringe needle (inner diameter of 0.33 mm), the tangled coils form an elastic and conductive 3D network to regulate heart function. In Figure 10.9b,

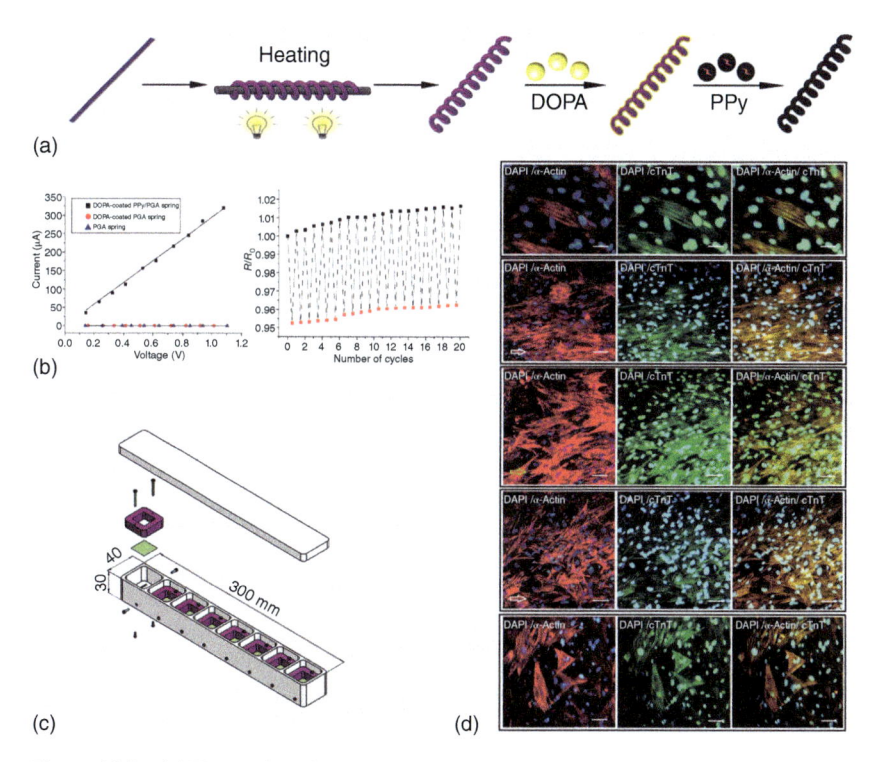

Figure 10.9 (a) The design of the conductive biospring *in vitro*. (b) *I–V* curve of PGA spring, DOPA-coated PGA spring, and DOPA-coated PPy/PGA spring. Source: Amirabad et al. [60]. Reproduced with permission of American Chemical Society. (c) Schematic design of the bioreactor using SolidWorks 3D CAD software. (d) Immunofluorescence staining of cardiac-specific structural including α-actin and cTnT in cardiomyocyte-like cell (zoom in view). The cells cultured on aligned and PANI/PES nanofibrous scaffolds with and without electrical stimulation. The cells cultured on random PANI/PES nanofibrous scaffolds with and without electrical stimulation. Source: Ulbrich et al. [61]/with permission from American Chemical Society.

the results show that the conductivity of the dopamine (DOPA)-coated PPy/PGA spring is 80.84 S m^{-1}, which is nearly 4 orders of magnitude higher than the conductivity of the PGA and DOPA-coated PGA spring coatings. In addition, the PPy/PGA spring coated with DOPA has very high electrochemical stability, and its resistance will only increase by 0.42% after 20 strain cycles under 35% strain. After injecting an elastic conductive biological spring into the area of MI, the left ventricle was shortened by about 12.6%, and the infarct size was reduced by about 34%. Interestingly, the spring can be used as a sensor to measure the contraction force of cardiomyocytes (CM), which is $1.57 \times 10^{-3} \pm 0.26 \times 10^{-3}$ mN (about 4.1×106 units). Therefore, this study emphasizes an injectable biological spring that can form a tangled conductive 3D network in the body to repair MI. Barzin and coworkers [60] prepared neatly arranged and random PANI/polysulfone (PANI/PES) nanofiber scaffolds doped with camphor-10-sulfonic acid (β-CPSA) by electrospinning and used to drive cardiomyocyte differentiation (Figure 10.9c). The electrical conductivity of the PES scaffold is within the range of electrically insulating materials, whereas the conductivity-aligned nanofibrous scaffolds were 15 times greater than random ones. However, the conductivity of both of them is within the range of semiconductor materials (10^2 to 10^{-6} S cm^{-1}), which is suitable for tissue engineering purposes. The bioreactor was subsequently engineered to apply electrical pulses to the cells cultured on the PANI/PES scaffold. The cardiovascular disease-specific induced pluripotent stem cells were seeded on the scaffold, cultured in cardiomyocyte inducing factor, and exposed to electric pulses in the bioreactor for 1 hour/day over a period of 15 days. Compared with multidirectional electrical stimulation using random fiber scaffolds, unidirectional electrical stimulation of cells significantly increased the number of cardiac troponin T (cTnT) cells (Figure 10.9d). The data shows that the unidirectional electrical stimulation obtained by aligned stents drives the cells more efficiently than multidirectional electrical stimulation generated by randomly oriented stents.

10.3.2 Drug Delivery

Traditional drug delivery methods (oral or injectable drug delivery) will cause the accumulation of drug/biomolecule concentration in the human body, and the peak concentration will gradually decrease to an ineffective level over time [62]. To reach the therapeutic level, there is an urgent need to develop a continuous and controlled drug delivery system to provide optimal drug delivery that regulates the drug level, so as to avoid insufficient and overdose drugs [63]. Recently, many studies have focused on the use of various stimulus-responsive biomaterials as "on–off" controllable drug carriers, in which drugs are released by adjusting the input of pH, temperature, or electrical energy [64, 65]. At present, CPs can be used as one of the more effective drug delivery platforms. For example, conductive hydrogels with biocompatibility have been designed to inherit the redox conversion and electrical properties of traditional CPs, which can promote drug release through local swelling and contraction of hydrogels. Alemán and his colleagues [65] prepared poly(ε-caprolactone) ultrafine fibers (PCL) loaded with

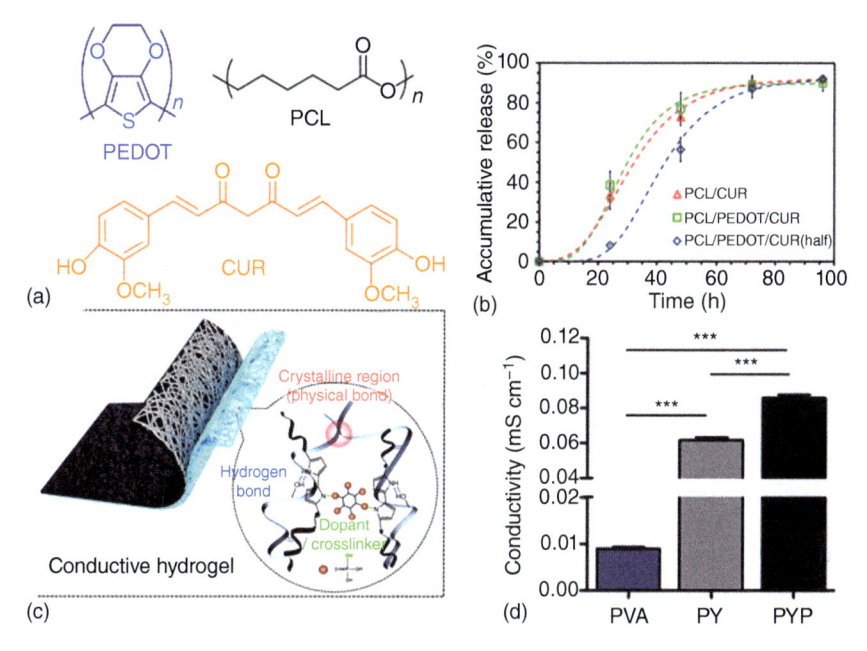

Figure 10.10 (a) Chemical structures of PEDOT, CUR, and PCL. (b) CUR release profiles from PCL/CUR and PCL/PEDOT/CUR MFs in PBS-EtOH. Source: An et al. [66]. Reproduced with permission of American Chemical Society. (c) Schematic diagram of the preparation of polyvinyl alcohol (PVA-PPy) conductive hydrogel with polypyrrole. (d) In the third FT, the conductivity of the hydrogels was calculated from the $I-V$ curve by the four-point probe method. Source: Boehler et al. [67]. Reproduced with permission of American Chemical Society.

PEDOT nanoparticles and curcumin (PCL/PEDOT/CUR) fibers by electrospinning as an electro-responsive drug delivery system (Figure 10.10a). The release of CUR is regulated by electrical stimulation. The results show that PCL/PEDOT/CUR acts as extracellular cell matrix, and their heterogeneity and roughness can promote cell spreading and enhance cell proliferation. The process of CUR release from PCL ultrafine fibers by simple diffusion is very slow, and external electrical stimulation is required to promote and regulate the delivery process of CUR (Figure 10.10b). When a well-defined potential pulse is applied, PEDOT NPs act as an electrical actuator, which promotes the release of CUR by influencing the structure of the PCL matrix. In addition, *in vitro* experiments proved that CUR still maintains anticancer activity after the electrospinning process. In general, PCL fibers loaded with electrically responsive PEDOT NP can quickly regulate the release of CUR by using pulsatile electrical stimulation. Hwang and coworkers [66] used a conductive hydrogel composed of PVA doped with PPy (PVA-PPy) (Figure 10.10c). At the same time, an iontophoretic skin drug delivery system was developed for effective delivery of electrokinetic drug nanocarriers (DNs). Compared with PVA hydrogel, the conductivity of PVA-PPy hydrogel with repeated freeze–thaw (FT) cycles has increased by more than 800% (Figure 10.10d). The FT cycle increases the conductivity of the PVA-PPy hydrogel, which may be due to the densification of the PVA macromolecular structure that promotes the closer contact between the PPy chains. DNs loaded with fluconazole or rosiglitazone can be functionalized with a charge-inducing agent, and charge-modified DNs promote transdermal transport through iontophoresis driven by repulsive reverse electrodialysis (RED). In addition, topical application and RED-driven iontophoresis of rosiglitazone DNs resulted in effective anti-obesity symptoms, manifested by weight loss, decreased glucose levels, and increased conversion of white adipose tissue to brown adipose tissue in the body.

Conductive PEDOT is a promising material with local drug delivery function, which can provide electrostatic bonding and release anionic molecules in thin film coating. C. Boehler et al. [67] prepared PEDOT/dexamethasone (PEDOT/Dex) and PEDOT/prednisolone phosphate (PEDOT/PreP) films by electrochemical deposition at deposition charges ranging from 50 to 300 mC cm^{-2} on either IrO_x or nanostructured platinum substrates, respectively. The thickness of these films can be adjusted by using different polymerization charges, including 25, 50, 100, 200, and 300 mC cm^{-2}. The authors found that cumulative Dex release can be actively controlled every day over a period of one week. The cumulative release of Dex from these samples revealed a clear correlation between the drug released content and the polymer bulk thickness with a slope of 13.74 ng per mC cm^{-2} at day 7 with R^2 of 0.998, which also can be identified between the polymerization charge and the cumulative drug release, as shown in Figure 10.11a,b. The work demonstrated the overall drug release for CP films scales linearly with the polymer thickness, thus making it possible to explore the polymer bulk to increase drug release capacity. David C. Martin and coworkers [68] synthesized a PEDOT nanotubes by combining electrospinning and electrochemical polymerization, which can be used to precisely control drug release. To produce the tubular structures, the biodegradable poly(L-lactide) (PLLA) or poly(lactide-co-glycolide) (PLGA) nanofibers were first electrospun onto the surface of a neural probe, followed by electrochemical polymerization of EDOT monomer around the electrospun nanofibers. Then, the fiber templates can be removed; to obtain nanotubes, dexamethasone was evenly dispersed within PEDOT film layers. Without applying electrical treatment, dexamethasone did not find significant diffusion though the PEDOT nanotube walls. However, a significant increase in the content of dexamethasone released was obtained after electrical excitation (Figure 10.11c). The authors precisely release individual drugs at desired points in time by using electrical stimulation of PEDOT nanotubes. The work provides a useful means for generating biologically active electrochemical coatings, which can facilitate the integration of high-performance electronic devices with living tissues. Cui and coworkers [69]. fabricated a dual-layer CP/acid functionalized carbon nanotube (fCNT) microelectrode containing the neurochemical 6,7-dinitroquinoxaline-2,3-dione (DNQX), where base layer was consisted of PEDOT/fCNT and the top layer was consisted of PPy/fCNT/DNQX. The incorporation of acid fCNT into PPy and PEDOT is a potential way to enhance conductivity and mechanical stability, which made the dual-layer microelectrode suitable for a loading and controlled delivery of NDQX. With the electrically triggered release of DNQX, sensory-induced neural activity is immediately (<1 second)

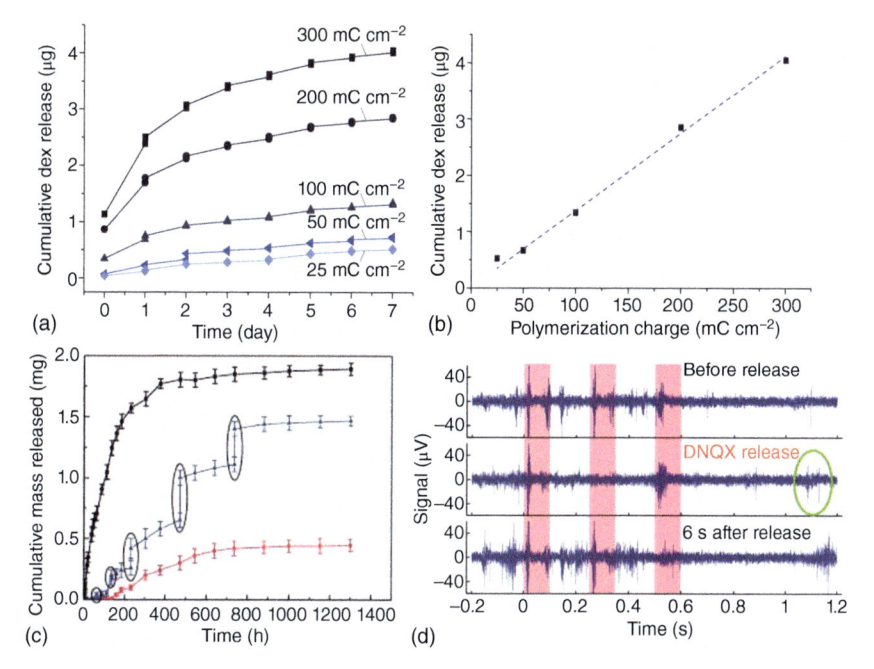

Figure 10.11 (a) Illustration of the cumulative Dex release over a period of one week with daily actively controlled drug expulsion. (b) The correlation between released drug amount and polymerization charge can be identified. Source: Abidian et al. [68]. Reproduced with permission of Elsevier. (c) Cumulative mass release of dexamethasone from PLGA nanoscale fibers (black squares), PEDOT-coated PLGA nanoscale fibers (red circles) without electrical stimulation, and PEDOT-coated PLGA nanoscale fibers with electrical stimulation of 1 V applied at the five specific times indicated by the circled data points (blue triangles). Source: Du et al. [69]. Reproduced with permission of John Wiley and Sons. (d) Representative air puff evoked sensory response (upper panel) and the drug release affected action potentials (middle panel) and recovered neural response 6 seconds after release (lower panel). Source: Krukiewicz et al. [70]. Reproduced with permission of John Wiley and Sons.

and locally (<446 µm) inhibited. The double-layer coating can induce effective nerve suppression at least 26 times without a significant decrease in efficacy. Figure 10.11d showed that the peak occurs about 1 second after the release of the drug and resumes 6 seconds after the release is triggered (middle panel, green circle), and the drug effect is almost completely washed away (lower panel). This *in vivo* DNQX clearance rate is much faster than *in vitro* conditions. Under *in vitro* conditions, CNQX release is still very effective 7.5 seconds after drug release. The results provided a great promising method for precise releasing the neurochemicals to modulate neural activity *in vivo*.

J.K. Zak and coworkers [70] also used PEDOP matrix as a reservoir of a model drug of IBU prepared by electrochemical polymerization of EDOP in the presence of IBU. The loading efficiency of PEDOP matrix is dependent on IBU concentration; for example, IBU concentration of 15 mM ensures the highest loading efficiency equal to $17.4 \pm 1.6\,\mu g\,cm^{-2}$. In addition, the amount of IBU released is highly dependent on the concentration of drug in solution (Figure 10.12a). Assuming

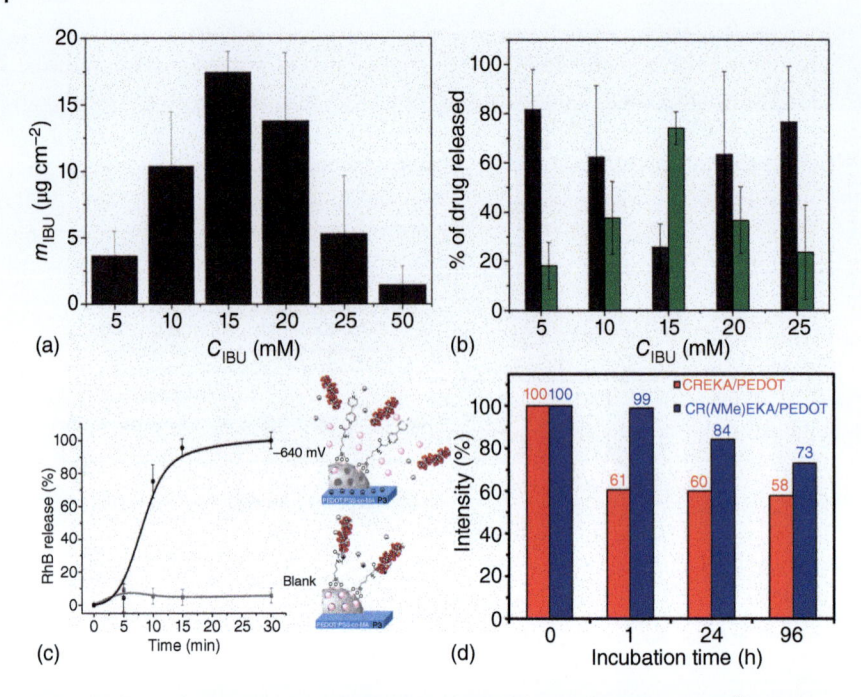

Figure 10.12 Cumulative mass of IBU released during 5 minutes from 1 cm^2 drug-loaded matrix via active mode as function of IBU concentration (a) and the percentage of drug released via passive (gray) and active (green) modes (b). Source: Garcia-Fernandez et al. [71]. Reproduced with permission of Elsevier. (c) Release profiles of Rh B from S3 nanoparticles in PBS at pH 7.5 alone and upon the application of a −640 mV vs. SCE potential. Source: Puiggali-Jou et al. [72]. Reproduced with permission of Elsevier. (d) Intensity ratio for CREKA/PEDOT and CR(NMe)EKA/PEDOT NPs before and after being incubated with proteinase K during different periods of time. Source: Stuart et al. [73]. Reproduced with permission of American Chemical Society.

that the percentage of drug released is 100%, the percentage of drug released can be determined through passive and active modes. Based on the data shown in Figure 10.12b, it can be concluded that the use of a 15 mM IBU solution results in a higher efficiency of electrically triggered release than spontaneous release. PEDOP loaded with IBU has been considered as a promising material for use in an implantable drug delivery device to achieve local treatment of inflammation. F. Sancenón and coworkers [71] reported the mesoporous silica nanoparticles (MSN) loaded with rhodamine B (Rh B), anchored to the PEDOT doped with poly[(4-styrenesulfonic acid)-co-(maleic acid)] (PEDOT:PSS-co-MA), functionalized with a bipyridinium derivative. Moreover, pores are capped with heparin (P3) by electrostatic interactions. The authors proved that delivery of Rh B was less than 5% after 30 minutes (Figure 10.12c) without the aid of potential. When a potential of −640 mV vs. SCE was applied for 15 minutes, on the contrary, Rh B released from the P3 film accounted for about 90% of the total dye. They also proved that release amount of scaffold with 2 μg of cargo cm^{-2} is approximately 2–4 times larger than the amounts delivered in reported CPs for drug release application.

Cys-Arg-Glu-Lys-Ala (CREKA) is an important fibrin pentapeptide, which has been widely used in diagnoses and therapies [73]. Therefore, C. Alemán and coworkers [72] fabricated PEDOT nanoparticles (PEDOT NPs) loaded with pentapeptides, CREKA and CR(NMe)EKA (denoted as CREKA/PEDOT and CR(NMe)EKA/PEDOT) by emulsion polymerization. Herein, the used CR(NMe)EKA is a CREKA analogue, where Glu was replaced by *N*-methyl-Glu to increase the resistance of the peptide against proteolysis that is one of the main drawbacks of therapeutic peptide delivery and improves the tumor-homing ability by stabilizing the bioactive conformation. Figure 10.12d indicated that the peptide signal for CR(NMe)EKA/PEDOT nanoparticles lasts longer at higher intensity, indicating that it needs more time to be digested by proteinase K. This shows that the loaded CR(NMe)EKA is more stable than loaded CREKA. Unsurprisingly, diffusion from those control samples did not significantly release CREKA after 24 hours, which proves that CREKA/PEDOT is an electronically controlled release system. After 50 CV cycles, $14.4 \pm 7.6\%$ of peptide was released from the initially loaded peptide at a scan rate of $100\,\mathrm{mV\,s^{-1}}$. While CR(NMe)EKA/PEDOT has significantly higher peptide release, reaching a total amount of $16.1 \pm 6.2\%$ after 50 cycles. Therefore, CREKA and CR(NMe)EKA combined with PEDOT NPs can achieve a modulated peptide release by fine-tuning the electrical stimuli.

In recent years, the hybridization of effective therapies has provided combination therapies with higher success rates, increased survival rates, and improved quality of life. Near-infrared (NIR)-induced photothermal therapy (PTT) is currently under extensive research and development to replace conventional cancer treatments due to its low invasiveness and high selectivity [74, 75]. The PTT agent absorbs NIR light and dissipates the absorbed energy by heating, thereby causing the local environment temperature to rise followed by irreversible cell damage. K. Sang and his groups [75] prepared paclitaxel-loaded polycaprolactone (PCL-PTX) mats by electrospinning and then used different concentrations of PPy to perform *in situ* surface functionalization as a customizable fiber-specific drug delivery platform. The manufacturing process of PPy-functionalized PCL-PTX membrane is shown in Figure 10.13. The results show that the obtained PPy-functionalized PCL-PTX mat exhibits excellent light stability and heating performance in response to NIR

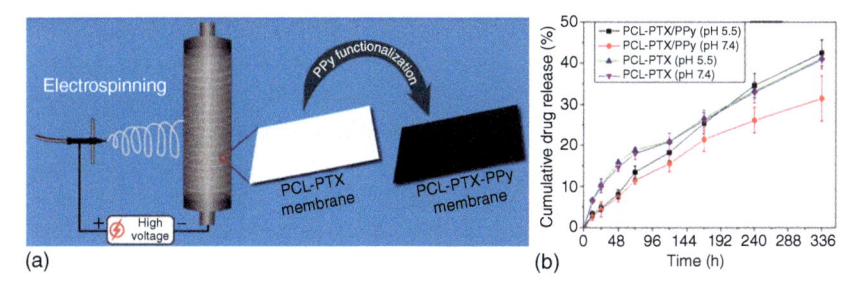

Figure 10.13 (a) Schematic representation of PPy-functionalized PCL-PTX membrane fabrication. (b) Drug release properties. PTX-release profile of PCL-PTX and PCL-PTX/PPy membranes in PBS solutions with different pH values. Source: Geng et al. [76]. Reproduced with permission of American Chemical Society.

exposure. Compared with the pH 7.4 environment, the PPy-coated PCL-PTX mats showed enhanced PTX release in the pH 5.5 environment. In both cases, the response to NIR further accelerated the release. However, compared with pH 7.4, superior release was observed at pH 5.5, indicating the existence of a dual stimulus response (pH and NIR) drug delivery platform. More importantly, compared with non-NIR or uncoated mats (PCL-PTX), PPy-coated PCL-PTX mats have significantly enhanced anticancer efficacy *in vitro* and *in vivo* under NIR radiation. Experiments have proved that the system has the function of pH and NIR responsive PPy, which aims to treat cancer through a combination of photothermal ablation and chemotherapy.

X. Yang and coworkers [76] developed a PNIPAM-based temperature-sensitive acidic triblock polymer poly(acrylic acid-*b*-*N*-isopropylamide-*b*-acrylic acid (PNA)) (D-PPy@PNAs) in the polymerization of PPy. The D-PPy@PNA was employed as thermosensitive PPy nanoplatform for synergistic photothermo-chemotherapy. Under the on–off cycle of NIR radiation, the pulse release of doxorubicin of D-PPy@PNA induced by NIR light was studied. Contrary to a continuous release without NIR irradiation, D-PPy@PNAs showed a sudden release (c. 13.2%) with NIR irradiation of $1.0\,W\,cm^{-2}$ in the switch-on period, followed by a slow release (c. 6.3%) in the switch-off period, suggesting that it indicates an NIR-induced pulsated release (Figure 10.14a). Moreover, D-PPy@PNAs also exhibited a pH-dependent releasing behavior, owing to the reduced the ionic bond between doxorubicin and carboxyl groups from PNA and the π–π stacking interactions between doxorubicin and PPy conjugated backbone when the pH decreased, thus promoting specific release in the acidic tumor microenvironment. Therefore, D-PPy@PNAs is a promising multifunctional nanoplatform for boosting photothermo-chemotherapy. Moreover, PPy was used as electrically responsive drug delivery systems typically prepared by electrochemical polymerization; however, the delivery amount of drug is generally low [77]. Therefore, D. Svirskis et al. produced PPy nanoparticles by chemical polymerization over drug-loaded micelles, which was used as drug carrier to realize electrically controlled release of dexamethasone base (Dex) and dexamethasone phosphate (DexP). When PPy particles were in its oxidized state, the slowest release rates of the DexP were achieved (Figure 10.14b). It is because that the positive charges of PPy backbone could bind to the anionic DexP in its oxidized state. On the contrary, DexP was quickly released in the PPy reduced state. DexP release was highest when a pulsed stimulus of $\pm 0.6\,V$ at 2 Hz was applied. The non-ionic Dex showed a similar release pattern to DexP (Figure 10.14c), but the reduction of PPy could not increase the release rates of drug beyond no stimulation, since Dex does not carry any changes. The work presented provides a reference to achieve controllable drug release by alternating the redox state of CPs. J.I. Amalvy and cowokers [78] prepared a conducting composite film based on PPy and poly(2-(diethylamino) ethyl methacrylate) hydrogel nanoparticles (PDEA-Dox/PPy), in which PDEA nanoparticles are deposited on a gold substrate by drop casting, followed by electrochemical synthesis of PPy through the hydrogel nanoparticles as a semi-interpenetrating polymer network. PPy releases H+ in the oxidized state, while PDEA nanoparticles reversibly swell at low pH value.

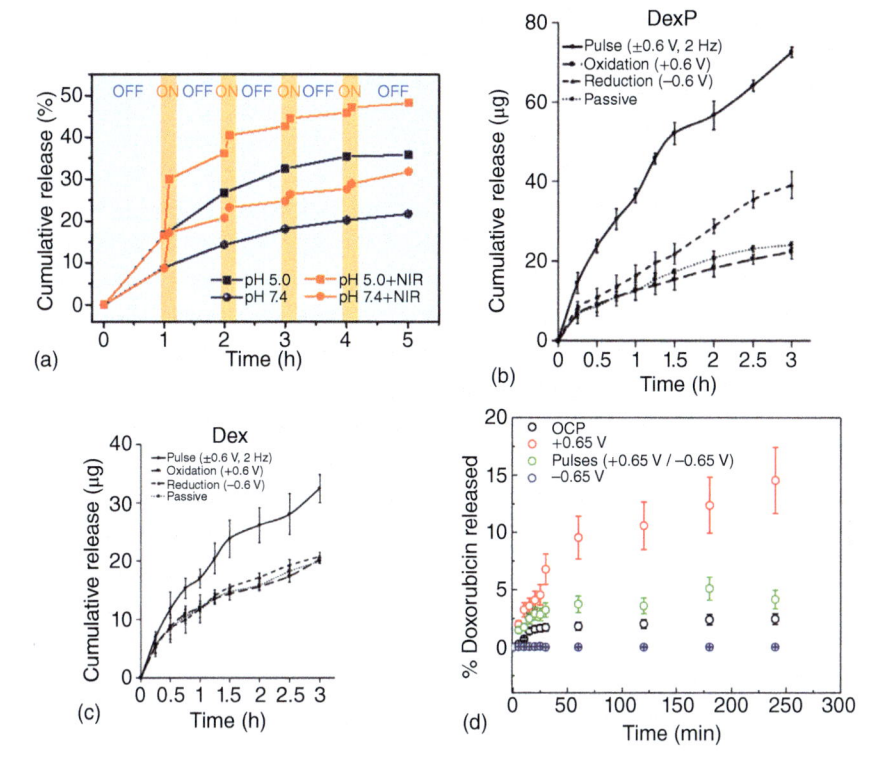

Figure 10.14 (a) Cumulative DOX release from D-PPy@PNA nanogels with or without 808 nm laser irradiation in PBS buffer of different pH values (1.0 W cm^{-2}, 5 minutes on/55 minutes off for every cycle). Source: Uppalapati et al. [77]. Reproduced with permission of American Chemical Society. Cumulative release profile of (b) DexP and (c) Dex on oxidation (+0.6 V), reduction (−0.6 V), and pulsed stimulation (±0.6 V) and under passive conditions with no stimulation. Data points represent mean values and error bars are SD ($n = 6$). Source: Gutiérrez-Pineda et al. [78]. Reproduced with permission of Elsevier. (d) Profiles of doxorubicin electrochemically released showing the influence of various applied potentials for the PDEA-Dox/PPy hybrid composite. Source: Woeppel et al. [79]. Reproduced with permission of John Wiley and Sons.

Therefore, the oxidation of PPy and decrease of pH could lead to the swelling of PDEA/PPy hybrid film under an electrical stimulation. Figure 10.14d proved that the largest amount release of Dox was about 15% of the loaded Dox at +0.65 V after 240 minutes, but a decrease release of Dox (around 1%) was achieved when the electrode potential was set at more negative potentials (−0.65 V) after 240 minutes of electrostimulation.

To address material limitations of biologically interfacing electrode, T. Cui and coworkers [79] prepared the sulfonate modified nanoparticle (SNP)-doped PEDOT for enhanced electrode coatings and drug delivery. The release amount of from PEDOT/SNP composite was quantitatively compared with a control group of PEDOT loaded with the dye directly and without the mesoporous SNP. Dyes were released under CV, but release was stopped. Compared with their respective control samples, a significant increase in dye release was observed in 9000 and 5000 stimuli, while the SNP-doped polymer continued to release up to 13 000 stimuli with a substantially higher release rate. Note that the dye release was increased for both fluorescein (6.4 times) and rhodamine (16.8 times) when SNPs were used as a reservoir, compared with PEDOT film that was polymerized only in the presence of the drug. This result indicated that SNP-doped PPy has completely improved the drug loading and release ability and broadened the range of drug choices, because most of the work on polymer release focused on negatively charged drugs.

10.3.3 Actuators

CPs, such as PPY, PAN, PA, polyalkylthiophene, etc., have been used as the electrical and mechanical stretching actuators studied through the electrochemical doping of electrolyte ions [80]. The driving mechanism of CPs is ion diffusion, originated from the change in the electric charge of the polymer chains. Under the applied voltage, ions with opposite charges are adsorbed on the surface of the CP chain, resulting in an increase in the volume and elongation of the CP, and then led to the actuator. Different types of CP and structures are used to improve the actuating stroke, stress, and working capacity. E. Choi and coworkers [81] developed an high-performance ionic soft actuator (CA-IL-GN) based on cellulose acetate (CA), graphene nanopowders (GN), ionic liquid (IL) as a plasticizer, and biofriendly flexible nonmetallic PEDOT:PSS as an electrode, which realized a large bending mechanical deformation and a fast response time. Figure 10.15a showed the tip displacement of the CA-IL-GN actuator at a 3 V DC-induced electrical step input. As a result, tip displacement and response time of the CA-IL-GN (0.2 wt%) actuators were 4.8 and 2.04 times larger than those of the pure CA-IL actuator, respectively, which was due to the higher electrical conductivity, lower stiffness, and proper control crystallinity. I. Enculescu and coworkers [82] developed electroactive polymer-coated microribbons composed of electrospun nylon-6/6 microribbons coated with a metallic layer and outmost layer of PPy by electrochemical deposition, which can undergo conformational changes in response to external physical and chemical parameters.

The actuator based on PPy-coated microribbons (≤40 μg) holds a small object (a piece of copper wire, 880 μg) with a weight 20 times higher than its weight by applying potential of −0.60 V for the specified time (Figure 10.15b). When the potential is shifted to +0.60 V, the actuator relaxes and releases the piece of metal (Figure 10.15c), which is attributed to the shrinking/swelling of the PPy film. After this process, the actuation properties are not affected, and no cracks appear, thus proving its capacity to manipulate millimeter size heavy objects while maintaining its motion characteristics. The actuator can be tailored according to the required biomedical application by depositing different metallic layers and CPs or adjusting its configurations. Y. Huang and coworkers [83] used the commercialization of PEDOT:PSS to prepare vapor-driven soft actuator by using H_2SO_4 treatment, which can undergo fast and reversible deformation in response to the ethanol vapor. The PEDOT:PSS films were used as a promising electrode material for constructing nanogenerator composed of a bottom PEDOT:PSS electrode and a top

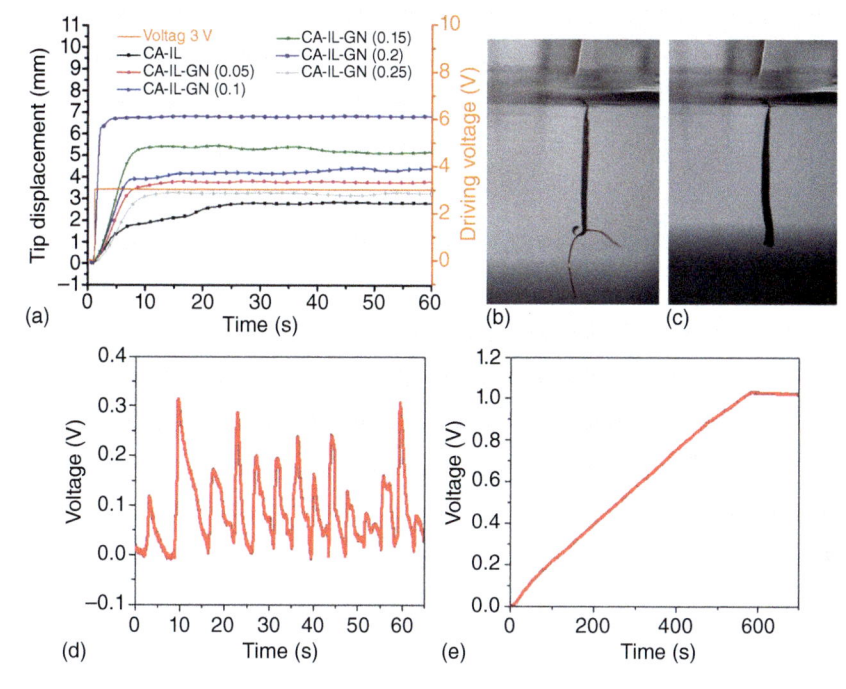

Figure 10.15 (a) Tip displacements under a DC excitation of 3 V voltage. Digital photos of a PPy-coated microribbon net strip. Source: Beregoi et al. [82]. Reproduced with permission of Elsevier. (b) Holding and (c) releasing a piece of copper wire in 1 M NaCl when the potential was switched from −0.60 to +0.60 V. Source: Xiong et al. [83]. Reproduced with permission of American Chemical Society. (d) The open-circuit voltage of this nanogenerator. (e) Voltage on a capacitance charged by the nanogenerator. Source: Khaldi et al. [84]. Reproduced with permission of John Wiley and Sons.

fluorinated ethylene propylene (FEP) film with a deposited Au layer on it, which can covert mechanical energy into electrical energy. When PEDOT:PSS electrode was exposed to ethanol vapor, the actuator deflects upward, and the gap distance between the FEP and PEDOT:PSS films decreases, generating an open-circuit voltage of 0.3 V with a 0.2 Hz frequency, controlled by the on–off state of ethanol vapor (Figure 10.15d). In addition, combined with a commercial full-wave bridge rectifier, the nanogenerator could charge a 10 μF supercapacitor to about 1.0 V in 10 minutes (Figure 10.15e). The authors anticipated that the excellent performance of nanogenerator based on PEDOT:PSS actuator could open a new avenue for the design of new generation of robots and smart sensors.

Khaldi et al. [84] displayed a individually controllable CP trilayer actuators by the microfabrication process that patterned firstly CP layer, followed by depositing a polymer electrolyte and a second patterning of the second CP layer. Also, the trilayer actuators are composed of polyvinylidenefluoride(PVDF) and poly(vinylidenefluoride-*co*-hexafluoropropylene) (PVDF-HFP) membranes as the solid polymer electrolyte to obtained the final devices Figure 10.16. A tip displacement of 100 μm was measured at 10 mm from the contact when ±1 V was applied at a frequency of 0.05 Hz. The development of bottom-up

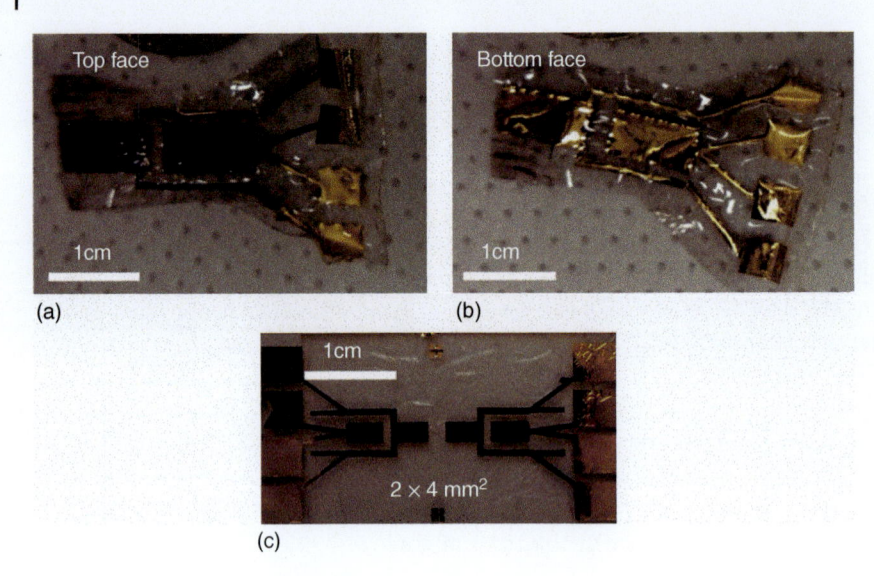

Figure 10.16 Bottom-up fabricated actuator devices (or "fingers") with individually addressable PPy trilayer actuators with 34 μm thick PVDF membrane manually cut with a scalpel: (a) top face and (b) bottom face of a device comprising two 5×10 mm^2 sized PPy actuators and (c) top face of two devices each comprising two 2×4 mm^2 sized PPy actuators. Source: Khaldi et al. [84]/with permission from Elsevier.

micromachining methods could provide an important way for the development of new micromanipulation tools.

In summary, CPs have a high degree of π-conjugated molecular backbone and universal side chains and have many excellent electrical conductivity, water dispersibility, and biocompatibility. Therefore, CPs are expected to develop more interesting applications in the future. However, so far, the practical biomedical and clinical applications of CPs still face many challenges, including their processability, cytotoxicity, huge gap between *in vivo* and *in vitro* results, and biodegradability. Obviously, the application of CPs in human biological tissues is still new. With the continuous advancement of material design, synthesis, and characterization technology, these challenges will be overcome in the near future.

10.4 Chemical Sensors and Biosensors

10.4.1 Chemical Sensors

The chemical sensor and biosensor employ capacitive readout cantilevers and electronics to analyze a transmitted signal for detecting a range of important chemical and biological targets, which have seen increasing importance in healthcare, medical science, agriculture, environment monitoring, food, and biosecurity. Owing to the unique redox properties and high electrical conductivity, CPs have emerged as potential candidates for electrochemical sensors. Mahmoudian et al. [85] reported polypyrrole coated on nanospherical platinum (Pt/PPy NSs) composites via the direct reduction of potassium tetrachloropalatinate (II) aqueous solution in the presence of pyrrole monomers in NaOH. The Pt/PPy NS composites showed a superior electrocatalytic activity toward the electrooxidation of 10 mM Hg^{2+} in a mixture of 0.33 M phosphate buffer solution. The estimated LOD (S/N = 3) of the sensor for the linear segment was 0.277 nM, and the sensitivity of this linear segment was 1.239 mA nM^{-1} cm^{-2}.

Heavy metal ions are a naturally occurring element that is toxic. The overuse of the heavy metal ions in the environment may have adverse effects on human health and the ecology. Therefore, the detection of metal ions is crucial, and many analytical methods are used to monitor them. Wei et al. [86] synthesized PPy/carbonaceous nanospheres (PPy/CNSs) on screen-printed electrode by dilute polymerization in the presence of CNSs and pyrrole monomer, in which CNSs were bond with PPy via electrostatic interaction and hydrogen bonding because of abundant hydroxyl groups on the surface (Figure 10.17a). PPy/CNS was exactly employed in the selective detection for Hg(II) and Pb(II) under ultralow concentration, which showed an excellent sensitivity with the detection limits of 0.113 μA nM^{-1} for Hg (II) and 0.501 A nM^{-1} for Pb(II), much lower than the guideline values in drinking water given by the World Health Organization. The work is expected to broaden our

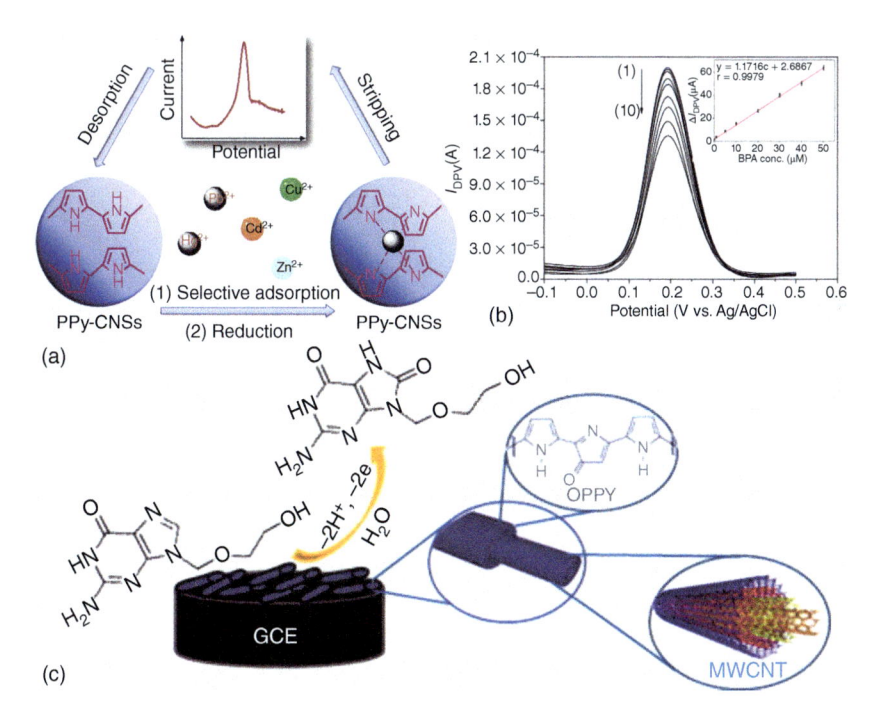

Figure 10.17 (a) Schematic representation for the selectively electrochemical detection of Hg(II) and Pb(II) by PPy–CNS nanocomposite. Source: Tan et al. [87]. Reproduced with permission of Elsevier. (b) DPV curves of 10 mM $K_3[Fe(CN)_6]$ at the MIPPy/GQDs electrode after incubation with different concentration BPA solutions. Source: Shahrokhian et al. [88]. Reproduced with permission of Elsevier. (c) Schematic diagram of the preparation process of OPPy/CNT/GCE electrode. Source: Liu et al. [89]. Reproduced with permission of Elsevier.

horizons of the selectivity of electrochemical sensing toxic metal ions. In addition, bisphenol A (BPA) is an important endocrine disrupter in the environment, for which sensitive and selective detection methods are highly necessary to carry out its recognition and quantification. Tan et al. [87] developed the molecularly imprinted polypyrrole/graphene quantum dots (MIPPy/GQDs) by the electropolymerization of pyrrole on a glassy carbon electrode with BPA as a template. The MIPPy/GQD used as a new electrochemical sensor to detect BPA in water samples resulted in the changes of peak currents of $K_3[Fe(CN)_6]$ at the MIPPy/GQD electrode in cyclic voltammetry and differential pulse voltammetry. It was found a linear relationship between BPA concentrations ranging from 0.1 µM to 50 M and response value (I_{DPV}) in DPV (Figure 10.17b), with a limit of detection of 0.04 µM (S/N = 3). The sensor displayed the recoveries of 94.5% and 93.11% for the detection of BPA in tap water and seawater samples, respectively. The method provides a rapid and sensitive detection of BPA in environmental samples. Furthermore, Shahrokhian et al. [88] prepared a thin film of multiwalled carbon nanotubes coated with an electropolymerized layer of tiron-doped PPy on glassy carbon electrode (OPPy/CNT/GCE) (Figure 10.17c). The OPPy/CNT/GCE was applied as electrochemical electrode to investigate the electrooxidation behavior of the compound to the drug of acyclovir (ACV). The synergistic effect originated from the unique properties of CNTs and PPy results in a considerable enhancement of the electrochemical activity and selectivity for sensing ACV, showing a wide linear dynamic range (0.03–10.0 µM) and a low detection limit (10.0 nM).

Liu et al. [89] fabricated the 3D macroporous PANI doped with PSS using the hard-template PS macroparticles (Figure 10.18a,b). The 3D macroporous PANI-modified electrode possesses large electroactive surface area and high conductivity, which acted as a new and sensitive electrochemical alpha-fetoprotein (AFP) immunosensor that achieved a wide linear range for AFP from 0.01 to 1000 pg ml^{-1}, with a detection limit of 3.11 fg ml^{-1}. Moreover, the electrochemical signal of this biosensor was obtained by measuring the inherent redox signal change of PANI, without the addition of other redox reagents, which may provide great potential in practical applications. Robertson et al. [90] have proved that PANI functionalized with homopolymer poly(3-aminobenzoic acid) (P3ABA) acted as novel antimicrobial agents that are active against a broad range of bacteria. In the work, PANI could induce production of hydrogen peroxide, which can promote formation of hydroxyl radicals, leading to the microorganism destruction and, subsequently, cell death. P3ABA causes an increase in intracellular free iron, which involved in perturbation of metabolic enzymes and promoted production of reactive oxygen species. The authors found that PANI was more active against *Escherichia coli* in aerobic condition, compared with the anaerobic one. PANI will be incorporated into surfaces for infection control and food safety applications. Yang and coworkers [91] prepared PANI on two spatially resolved areas of the ITO electrode for attaching AuNPs by electropolymerization of aniline. Then, human mucin1 protein (MUC1) aptamer was immobilized onto AuNPs for capturing MUC1-positive MCF-7 cells (Figure 10.18c). The quantification of MCF-7 cells on the two spatially resolved areas could be achieved over the linear range from 10^2 to

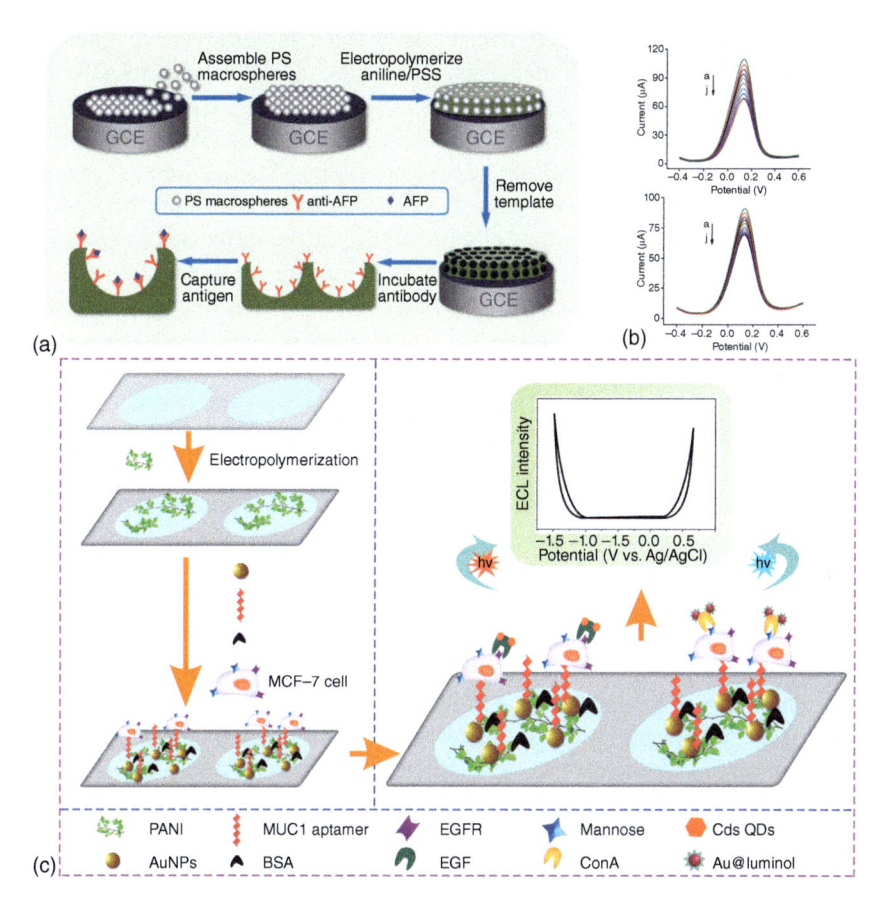

Figure 10.18 (a) Fabrication process of the AFP immunosensor based on macroporous PANI. Source: Robertson et al. [90]. Reproduced with permission of Elsevier. (b) DPV responses of porous PANI/GCE and planar PANI/GCE toward different concentrations of AFP. Source: Zhou et al. [91]. Reproduced with permission of Elsevier. (c) Preparation and schematic illustration of the dual electrochemiluminescence signal system. Source: Sharma et al. [92]. Reproduced with permission of American Chemical Society.

1.0×10^6 cells ml^{-1} with a detection limit of 20 cells ml^{-1}. Therefore, the composite has been developed as a cytosensor that simultaneously detects the expression levels of mannose and epidermal growth factor receptor on MCF-7 cells; their average numbers were 1.2×10^6 and 0.86×10^5 with the relative standard deviation of 5.3% and 4.2%, respectively. The proposed PANI cytosensor could be used to the detect of multiple cell-surface receptors, which may provide a promising disease diagnosis tool.

10.4.2 Biosensors

Biosensors are analysis devices that connect biologically active materials such as enzymes, cells, and DNA as a detector or sensor. After the biomolecules respond, the detector/sensor can convert the biochemical reaction as a biological signal into a calculable electrical signal [93]. In recent years, biosensors have played an important role in diagnosis, monitoring, and maintenance of biomolecules, and their applicability has received extensive attention. To improve the performance of biosensors, there is an urgent need for a simple and fast electron transfer process between the analyte and the electrode surface. Despite many efforts, biosensors based on direct electron transfer still show poor stability and conductivity [94]. In this regard, CPs provide a variety of solutions for the development of biosensors. The superior electrical conductivity, mechanical elasticity, and biocompatibility of CPs are the most ideal characteristics required to improve the sensing performance and analytical performance of the biosensors [95, 96]. Biosensors based on CPs have shown increased conductivity, selectivity, and sensitivity. It can be seen that the emergence and development of CPs are expected to become a new generation of biosensors. Biosensors based on CPs have been widely used in enzyme detection, such as the assessment of blood glucose levels that are very important in the diagnosis of home care. Yu and coworkers [96] prepared a high-sensitivity glucosidase sensor based on Pt nanoparticle (Pt NP)-PANI hydrogel heterostructure (Figure 10.19a). The glucose sensor based on the heterostructure of Pt NP/PANI hydrogel synergizes the advantages of conductive hydrogel and nanoparticle catalyst. The porous structure of PANI hydrogel facilitates the high-density immobilization of enzymes and the penetration of water-soluble molecules, thereby helping to effectively catalyze the oxidation of glucose. In addition, Pt NPs catalyze the decomposition of hydrogen peroxide produced during enzymatic reactions. The results indicate that the glucose enzyme sensor based on this heterogeneous structure shows unprecedented sensitivity up to 96.1 μA mM^{-1} cm^{-2}, response time as high as 3 seconds, linear range from 0.01 to 8 mM, and detection limit as low as 0.7 μM. In addition to enzyme biosensors, CPs are also used to detect other biological molecules. For example, Mak and coworkers [97] introduced a facile approach to modulate the carboxylate functional groups on the PEDOT interface through a systematic evaluation on the effect of a series of carboxylate-containing molecules as counterion dopant integrated into the PEDOT backbone, including acetate as monocarboxylate (mono-COO$^-$), malate as dicarboxylate (di-COO$^-$), citrate as tricarboxylate (tri-COO$^-$), and poly(acrylamide-co-acrylate) as polycarboxylate (poly-COO$^-$) bearing different amounts of molecular carboxylate moieties to create tunable PEDOT:COO$^-$ interfaces with improved polymerization efficiency. The experiment proved that the modulation of PEDOT and COO$^-$ interface has various particle structures with different size ranging from 0.33 to 0.11 μm and adjustable surface carboxylate density from 0.56 to 3.6 μM cm^{-2}, which could improve electrochemical kinetics and cycle stability. In addition, simple covalent chemistry further proved the effective and stable coupling of the enzyme model lactate dehydrogenase (LDH) with the optimized PEDOT:poly-COO$^-$ interface and developed a biofunctionalized PEDOT as a lactate biosensor (Figure 10.19c,d). The results showed that the LDH-PEDOT biosensor was used to detect lactic acid in spiked serum samples, with a high recovery rate of 91–96% and relative standard deviation of between 2.1% and 3.1%.

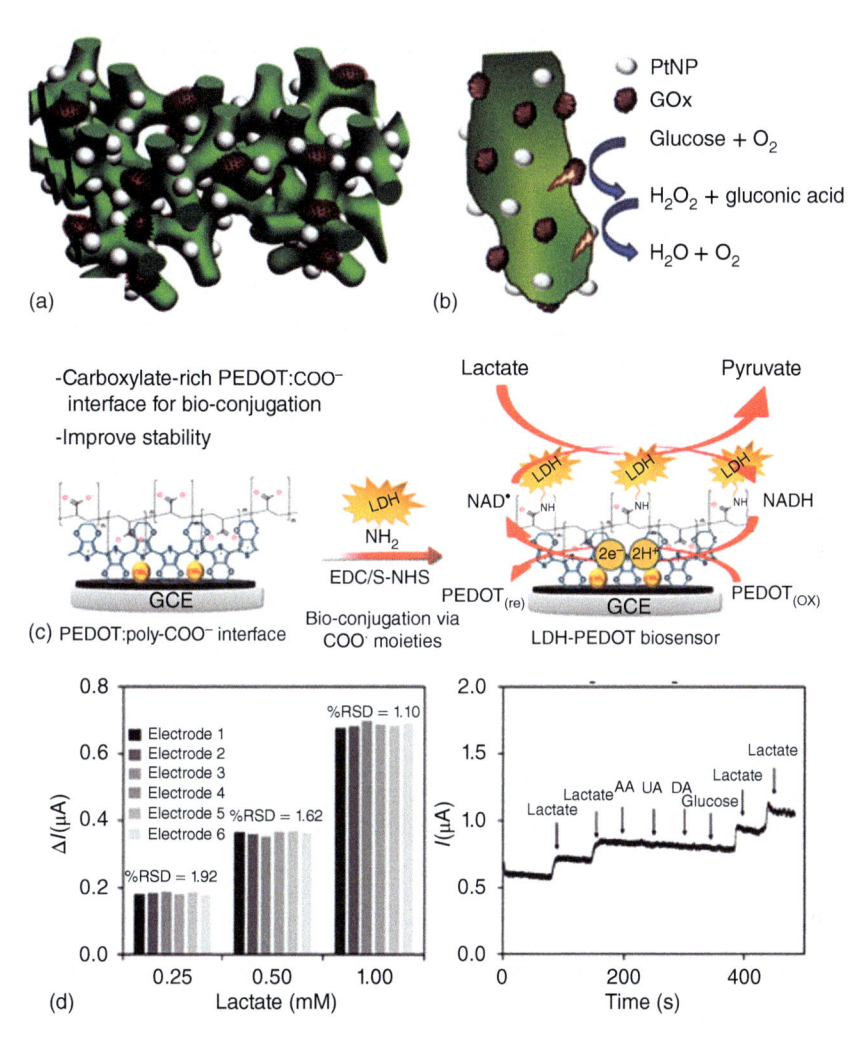

Figure 10.19 (a) Schematic representation of the 3D heterostructure of the PtNP/PAni hydrogel, in which the PANI hydrogel acts as a matrix for the immobilization of the GOx enzyme and homogeneous loading of Pt NPs. (b) A scheme showing the reaction mechanism of the glucose sensor based on the PtNP/PANI hydrogel heterostructure. Source: Promsuwan et al. [97]. Reproduced with permission of American Chemical Society. (c) Lactate biosensing mechanism at the LDH-PEDOT interface. (d) Six kinds of LDH/PEDOT. The reproducibility and selectivity of poly-COO$^-$/GCE to 0.25, 0.5, and 1.0 mM lactic acid. Source: Zhang et al. [98]. Reproduced with permission of American Chemical Society.

Zhang et al. [98] reported that a molecularly imprinted polymer (MIP)-modified SnO$_2$ electrode was fabricated by electropolymerization of pyrrole on SnO$_2$ electrode using BPA as the template (Figure 10.20a). It was found that photoelectrochemical oxidation of BPA is mediated by [Ru(bpy)$_3$]$^{2+}$; photocurrent enhancement was observed in the presence of BPA upon light excitation, which therefore served as photoelectrochemical sensor for BPA detection. This sensor is highly sensitive,

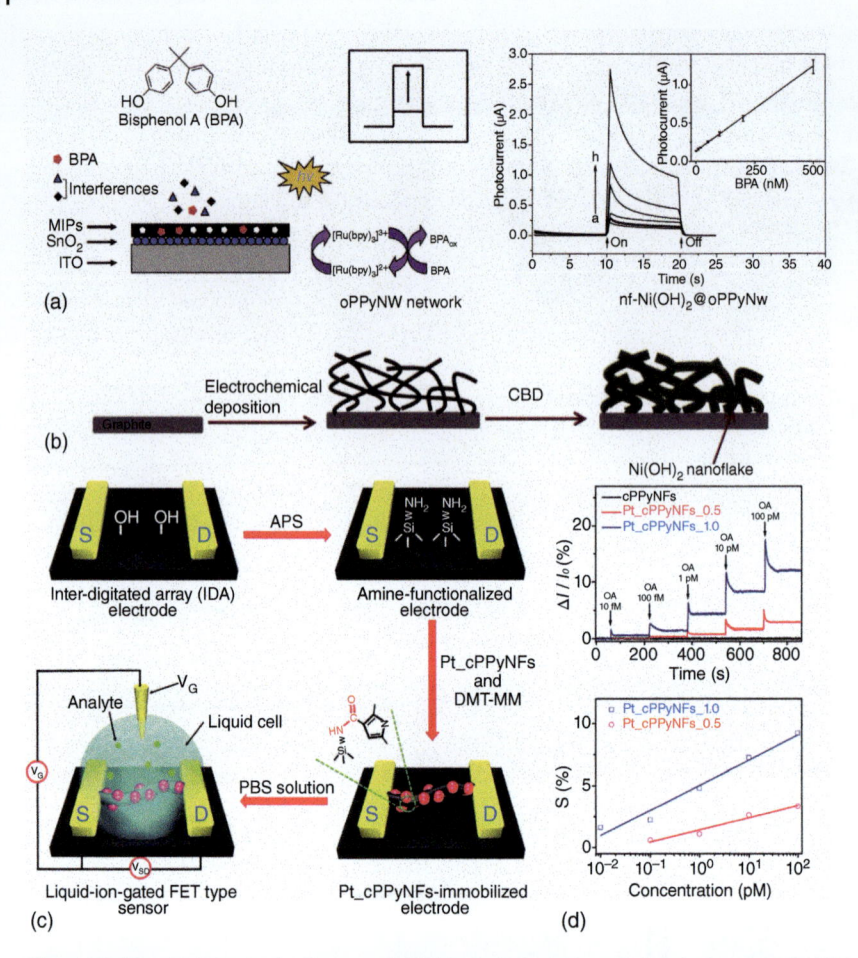

Figure 10.20 (a) Schematic diagram of MIP-based photoelectrochemical sensor for BPA detection. Source: Yang et al. [99]. Reproduced with permission of Elsevier. (b) Scheme of two-step synthesis of nf-Ni(OH)$_2$@oPPyNW on graphite electrode. Source: Kim et al. [100]. Reproduced with permission of Elsevier. (c) Schematic diagram of fabrication steps for liquid-ion-gated FET-type sensor fabrication. (d) Real-time response of the FET-type sensor in normalized current change ($I/I_0 = (I - I_0)/I_0$, where I_0 is the initial current and I is the instantaneous current). Source: Tertis et al. [101]. Reproduced with permission of Royal Society of Chemistry.

selective, and low cost for BPA detection with a detection limit of 1.2 nM with a linear range of 2–500 nM. Yang et al. [99] synthesized a PPy peroxide nanowire modified with nickel hydroxide nanosheet (nf-Ni(OH)$_2$@oPPyNW) composite material. The preparation of nf-Ni(OH)$_2$@oPPyNW was as follows (Figure 10.20b): electrodeposit oPPyNW on the graphite electrode by chronoamperometry, and then grow Ni(OH)$_2$ nanosheets on the PPyNW by chemical method. The prepared nf-Ni(OH)$_2$@oPPyNW-modified electrode possesses a high surface area and fast ion/electron transfer and then induces high sensitivity to glucose detection. Therefore, the modified electrode is used as sensor for detecting glucose, which showed a high sensitivity of 1049.2 μA mM^{-1} cm^{-2} and a low detection limit of 0.3 μM in a wide dynamic detection range of 0.001–4.863 mM. The work illustrated that nf-Ni(OH)$_2$@oPPyNW-modified electrode can be developed a fast, sensitive, selective, and stable electrochemical sensor for early diabetes diagnosis. Kim et al. [100] synthesized platinum nanoparticle-polypyrrole-3-carboxylated nanofiber composites (Pt_cPPyNFs), as electrode materials for non-enzyme field-effect transistor (FET) oxalic acid (OA) sensors. For achieving uniform decoration of Pt on carboxylated polypyrrole nanofibers (cPPyNFs), ultrasonication and chemical reduction were introduced as a Pt immobilization process. The prepared Pt_cPPyNFs were assembled using an interdigitated array electrode to fabricate the FET-type oxalic acid non-enzyme sensor via covalent bonding (Figure 10.20c). The resulting Pt_cPPyNF sensor exhibited high sensitivity (10 pM) toward oxalic acid (10–14 M), which was attributed to the uniformity of the Pt-nanoparticle and distinct properties of the FET configuration (Figure 10.20d). The work provides a simple and effective method to assemble a non-enzyme FET-type sensor.

Tertis et al. [101] synthesized PPy nanoparticles decorated with gold nanoparticle (AuNPs@PPyNPs) on graphite-based screen-printed electrodes (GSPEs) by electrochemical polymerization of pyrrole and electrochemical reduction of HAuCl$_4$ (Figure 10.21a). AuNPs@PPyNPs nanocomposite was used for the highly selective and sensitive electrochemical detection of serotonin in serum samples, even in

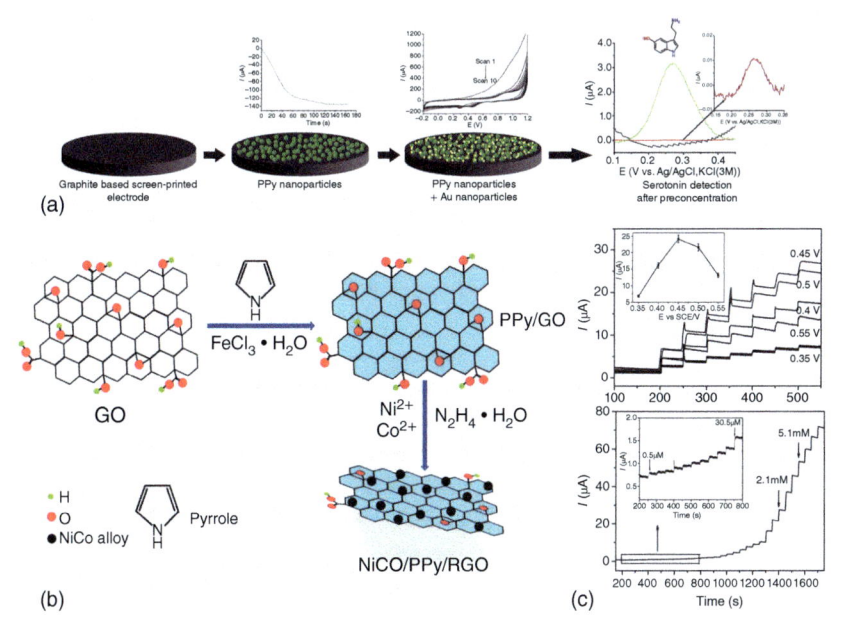

Figure 10.21 (a) AuNPs@PPyNPs nanocomposite preparation process and sensitive electrochemical detection. Source: Sheng et al. [102]. Reproduced with permission of Elsevier. (b) Schematic representation of the formation processes of NiCo/PPy/RGO nanocomposites. (c) Amperometric response of NiCo/PPy/RGO/GCE with dropwise addition of 0.1 mol l^{-1} glucose into the 0.1 M NaOH at different applied potentials. Source: Hui and Wang [103]. Reproduced with permission of Royal Society of Chemistry.

the presence of its common interferents. The analytical performance was 320 times better than that of the bare electrode, owing to an enhanced active surface area resulting from a synergistic effect of PPy and AuNPs. In addition, Sheng et al. [102] employed a two-step *in situ* chemical polymerization and reduction method for preparation of the NiCo alloy nanoparticles anchored on polypyrrole/reduction of graphene oxide nanocomposites (NiCo/PPy/RGO) complexes, in which the supporting substrate composed of PPy and RGO provides a larger surface area to load metal NPs (Figure 10.21b,c). Serving as a non-enzymatic glucose sensor, the NiCo/PPy/RGO nanocomposite-modified electrode exhibited remarkable electrocatalytic activity toward glucose oxidation, with detection limit of 0.17 μM, linear range from 0.5 μM to 4.1 mM, and sensitivity of 153.5 μA mM^{-1} cm. The work may provide a promising material candidate for the development of enzyme-free glucose sensors. Moreover, Wang and coworker [103] synthesized honeycomb-like cobalt nanostructure-modified graphene oxide-doped polypyrrole (CoNS/RGO/PPy) nanocomposites by a simple two-step electrochemical synthesis route. The CoNS/RGO/PPy nanocomposites was employed as enzyme-free glucose sensor that exhibited outstanding electrocatalytic performance for in alkaline media, possessing with a high sensitivity of 297.73 μA mM^{-1} cm^{-2}, a very low detection limit of 29 nM (at S/N = 3), and a superfast response time of 1 second under optimum conditions.

Qian et al. [104] synthesized oxygen-containing PPy-modified CNT composite (PPy/CNTs-MIPs) by MIP, which is used as electrochemical sensor for *in vivo* detection of dopamine (DA) through by hydrogen bonding between DA and PPy amino used to construct the DA. The preparation process of PPy/CNTs-MIPs is shown in Figure 10.22a. The electrochemical sensor exhibits an excellent sensitivity toward DA with a line arrange of 5.0×10^{-11} to 5.0×10^{-6} M and limit of detection as low as 1×10^{-11} M in the detection of DA. The authors proposed that the plenty cavities for binding DA through π–π stacking between aromatic rings and hydrogen bonds play an important role in the detection of DA. Cui et al. [105] synthesized a novel class of molecular tags, nanogold–polyaniline–nanogold microspheres (GPGs) functionalized with horseradish peroxidase-conjugated thyroid-stimulating hormone antibody (HRP-Ab2) for sensitive electrochemical immunoassay of thyroid-stimulating hormone (TSH) (Figure 10.22b,c). The results of the study showed that the strong combination of HRP-Ab2 and GPG resulted in good repeatability and moderate accuracy as low as 7%. The range of dynamic concentration is 0.01–20 μIU ml^{-1}, and the limit of detection (LOD) is 0.005 μIU ml^{-1} TSH under the $3s_B$ standard. Importantly, in the analysis of 0.05 spiked serum samples, there was no significant difference between the developed electrochemical immunoassay and the commercially available enzyme-linked immunosorbent assay (ELISA) to measure TSH.

The ever-changing world system not only provides us with a relaxed life but also brings many diseases that threaten human health (such as cancer, genetic diseases, and infectious diseases). To identify and diagnose these diseases, it needs to rely on electrochemical biosensors to detect their genetic material. Therefore, electrochemical DNA (E-DNA) sensors are very important for use in various

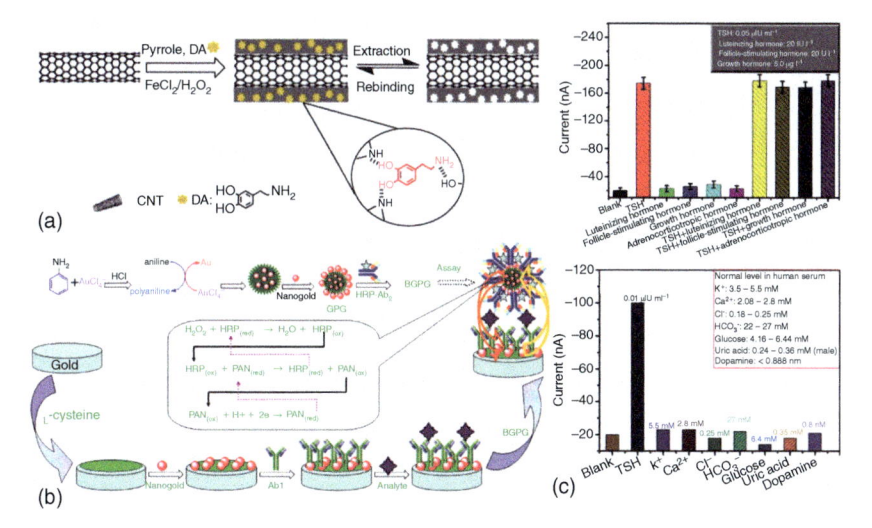

Figure 10.22 (a) The chemical route to the preparation of PPy/CNTs-MIPs. Source: Qian et al. [104]. Reproduced with permission of Elsevier. (b) Fabrication process of the BGPG bionanolabels and measurement protocol of the sandwich-type electrochemical immunoassay. (c) The interfering effects of other hormones and sample matrix components on the currents of the electrochemical immunoassays. Source: (b, c) Cui et al. [105]. Reproduced with permission of Elsevier.

fields such as cancer diagnosis, infectious diseases, chronic diseases, genetic diseases, tissue incompatibility, pathogen detection, and forensic medicine [106]. Panahi and coworkers [107] developed an electrochemical DNA (E-DNA) sensor based on hollow carbon spheres (HCS) decorated with PANI and used it for the sensitive detection of HBV DNA markers. Figure 10.23a shows that this method uses gold nanoparticles electrodeposited on HCS-PANI to immobilize thiolated 21-mer oligonucleotides with HBV DNA characteristics. The results show that the proposed biosensor allows the detection of HBV DNA target sequences at a concentration as low as 10 fM (i.e. 109 DNA copies ml^{-1}) (Figure 10.23b). In addition, the HCS/PANI-based E-DNA sensor can detect HBV DNA in real samples with high sensitivity. Kıranşan and Topçu [108] prepared a flexible, conductive PEDOT:PSS/RGO composite sponge from a composite dispersion containing GO and PEDOT:PSS by using simple hydrothermal and freeze-drying methods. The preparation process is shown in Figure 10.23c. The single-stranded DNA (ssDNA) and methylene blue (MB)-labeled ssDNA probes were immobilized on the PEDOT:PSS/RGO composite sponge under optimal conditions. The detection limit of the sensor is as low as 17 fM, and the linear range is from 50 fM to 2 μM, which can be used to determine hybrid DNA.

Biofouling arising from nonspecific adsorption is a substantial outstanding challenge in diagnostics and disease monitoring. Hui et al. [109] prepared a grafting of polyethylene glycol polymer onto polyaniline (PANI/PEG) nanofibers by using electrochemical polymerization of aniline, followed by grafting 4-armed PEG onto the PANI nanofiber surfaces (Figure 10.24a). PANI/PEG nanofibers have a large

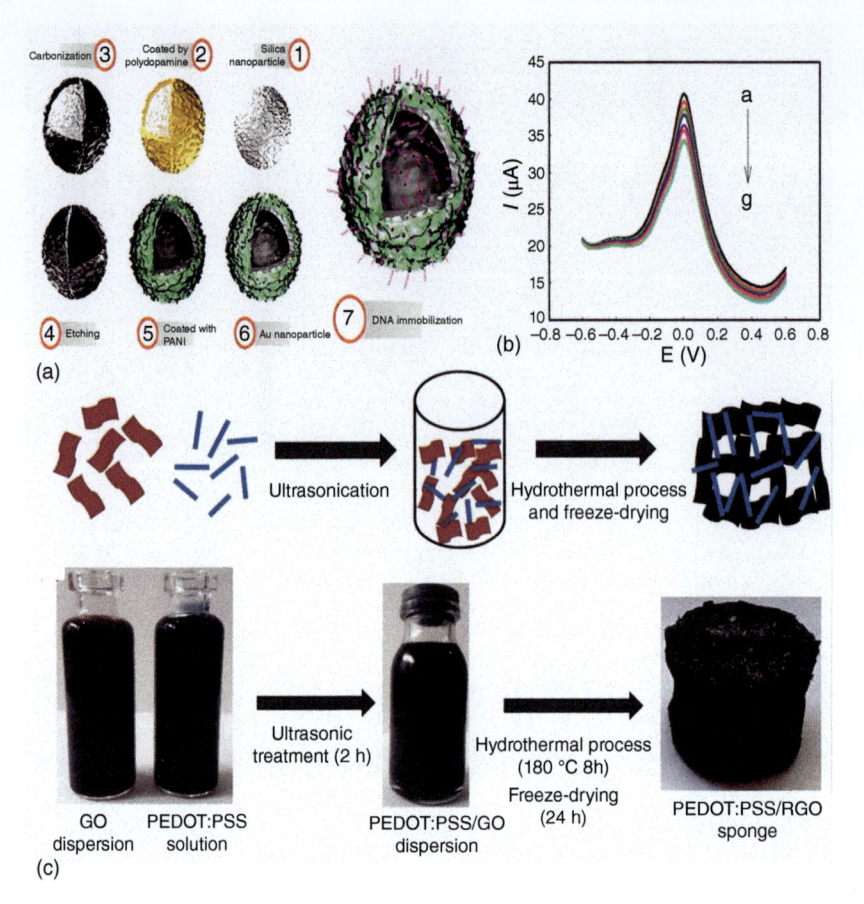

Figure 10.23 (a) The preparation flow chart of electrodepositing gold nanoparticles on hollow carbon spheres decorated with polyaniline (PANI). (b) Background-subtracted differential pulse voltammograms of biosensor after exposure to different concentrations of HBV from 0.01 pM to 1 nM (from (a) to (g), (a) is a control sample). Source: (a, b) Salimian et al. [107]. Reproduced with permission of American Chemical Society. (c) Schematic illustration and digital camera photograph of the preparation of PEDOT:PSS/RGO composite sponge material. Source: Kıranşan and Topçu [108]/with permission from American Chemical Society.

surface area and high conductivity. At the same time, these displayed an excellent antifouling performance in single protein solutions and complex human serum samples. Importantly, the attachment of DNA probes to the PANI/PEG nanofibers can be easily constructed antifouling biosensor for the breast cancer susceptibility gene (BRCA1). The biosensor showed a very high sensitivity and selectivity to target BRCA1 with a linear range from 0.01 pM to 1 nM, indicating great potential of this novel biomaterial for application in biosensors and bioelectronics. Shaikh et al. [110] developed a the carbon working electrode modified with polyaniline and electrodeposited gold nanocrystals (PANI/AuNCs) by electropolymerization of aniline monomer in acidic media followed by the electrodeposition of AuNCs, in which

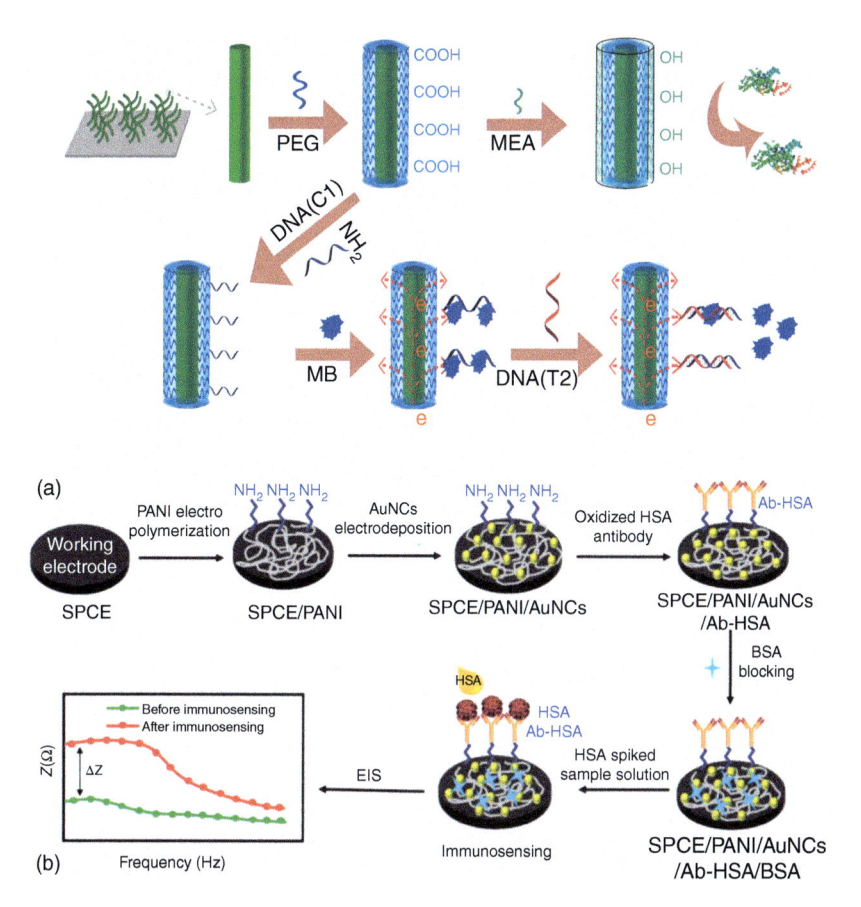

Figure 10.24 (a) Illustration of the construction of the PANI/PEG nanofiber antifouling interface and its application in ultrasensitive and low fouling DNA sensors. Source: Hui et al. [109]. Reproduced with permission of American Chemical Society. (b) Schematic of the systematic protocol for SPCE surface modification and immunosensing. Source: Shaikh et al. [110]. Reproduced with permission of American Chemical Society.

PANI matrix serves as an interconnected nanostructured scaffold for homogeneous distribution of AuNCs. The resulting PANI/AuNC nanocomposite was developed an impedance immunosensor for the sensitive, specific, and label-free detection of human serum albumin (HSA) (a valuable clinical biomarker for early detection of chronic kidney disease) in urine (Figure 10.24b). The normalized impedance variation during immunosensing increased linearly with HSA concentration in the range of 3–300 μg ml^{-1}, and a highly repeatable response was observed for each concentration. Furthermore, the immunosensor displayed high specificity when tested using spiked sample solutions containing different concentrations of actin protein and J82 cell lysate, a complex fluid containing a multitude of interfering proteins.

In summary, the unique conductivity and electrochemical stability of CPs will become a promising choice for manufacturing sensors. This section highlights the application of CPs in electrochemical sensors and emphasizes the versatility of CP-based sensors. It is not difficult to imagine that with the latest developments in universal, inexpensive, and reliable technologies for manufacturing sensors, CPs may promote the development of advanced sensors and apply them to the food industry, clinical and immediate diagnosis, and design of biofuel cells.

10.5 Summary and Perspective

In summary, this chapter summarizes the application of CPs in the field of bioengineering, including biosensors, electronic skins, drug delivery devices, and tissue engineering scaffolds. CPs show many positive properties, such as their biocompatibility, adjustable conductivity, ease of synthesis, and simple modification. CPs have shown biocompatibility *in vitro*, and significant progress has been made in the synthesis and functionalization of these beneficial polymer-based biomaterials. However, a systematic study of its biocompatibility and biodegradability *in vivo* is still needed. Extensive *in vivo* experiments on long-term cytotoxicity and biodegradability are necessary to ensure the nontoxicity and biodegradability of conductive biomaterials. In addition, durability and stability are one of the main challenges that must be overcome, for long-term applications, such as equipment for continuous monitoring. For biomedical applications related to tissue regeneration, the development of biodegradable conductive polymers remains a huge challenge, although several examples have been proven. Recently, due to the emergence of nanotechnology and 3D printing technology, some predetermined limitations including magnetic and mechanical properties are being resolved. Obviously, the application of CPs in human bioengineering is still in its infancy; it is hopeful that it will break through the bottleneck in this field.

References

1 Nezakati, T., Seifalian, A., Tan, A., and Seifalian, A.M. (2018). Conductive polymers: opportunities and challenges in biomedical applications. *Chem. Rev.* 118: 6766–6843. https://doi.org/10.1021/acs.chemrev.6b00275.

2 Wang, J.S. and Hui, N. (2019). Zwitterionic poly(carboxybetaine) functionalized conducting polymer polyaniline nanowires for the electrochemical detection of carcinoembryonic antigen in undiluted blood serum. *Bioelectrochemistry* 125: 90–96. https://doi.org/10.1016/j.bioelechem.2018.09.006.

3 Lee, J.Y., Bashur, C.A., Goldstein, A.S., and Schmidt, C.E. (2009). Polypyrrole-coated electrospun PLGA nanofibers for neural tissue applications. *Biomaterials* 30: 4325–4335. https://doi.org/10.1016/j.biomaterials.2009.04.042.

4 Liu, X.M., Xiao, T.F., Wu, F. et al. (2017). Ultrathin cell-membrane-mimic phosphorylcholine polymer film coating enables large improvements for in vivo electrochemical detection. *Angew. Chem. Int. Ed.* 56: 11802–11806. https://doi.org/10.1002/anie.201705900.

5 Guo, B. and Ma, P.X. (2018). Conducting polymers for tissue engineering. *Biomacromolecules* 19: 1764–1782. https://doi.org/10.1021/acs.biomac.8b00276.

6 Khan, M.A., Cantu, E., Tonello, S. et al. (2019). A review on biomaterials for 3D conductive scaffolds for stimulating and monitoring cellular activities. *Appl. Sci.* 9: 961. https://doi.org/10.3390/app9050961.

7 Zeglio, E., Rutz, A.L., Winkler, T.E. et al. (2019). Conjugated polymers for assessing and controlling biological functions. *Adv. Mater.* 31: 1806712. https://doi.org/10.1002/adma.201806712.

8 S. J. Benight, C. Wang, B. H. Tok, Z. N. Bao, Stretchable and self-healing polymers and devices for electronic skin, *Prog. Polym. Sci.*, 38 (2013) 1961. https://doi.org/10.1016/j.progpolymsci.2013.08.001

9 Xu, S., Zhang, Y.H., Jia, L. et al. (2014). Soft microfluidic assemblies of sensors, circuits, and radios for the skin. *Science* 344: 70. https://doi.org/10.1126/science.1250169.

10 Wu, W.Z., Wen, X.-N., and Wang, Z.-L. (2013). Taxel-addressable matrix of vertical-nanowire piezotronic transistors for active and adaptive tactile imaging. *Science* 340: 952. https://doi.org/10.1126/science.1237145.

11 Z. W. Xu, C. X. Wu, F. S. Li, W. Chen, T. L. Guo, T. W. Kim, Triboelectric electronic-skin based on graphene quantum dots for application in self-powered, smart, artificial fingers, *Nano Energy*, 49 (2018) 274. https://doi.org/10.1016/j.nanoen.2018.04.059

12 M. Hammock, A. Chortos, C. K. Tee, B. H. Tok, Z. N. Bao, 25th Anniversary article: the evolution of electronic skin (e-skin): a brief history, design considerations, and recent progress, *Adv. Mater.*, 25 (2013): 5997. https://doi.org/10.1002/adma.201302240

13 Yu, P., Li, Y., Zhao, X. et al. (2014). Graphene-wrapped polyaniline nanowire arrays on nitrogen-doped carbon fabric as novel flexible hybrid electrode materials for high-performance supercapacitor. *Langmuir* 30: 5306. https://doi.org/10.1021/la404765z.

14 Benoudjit, A., Bader, M.M., and Salim, W.W.A.W. (2018). Study of electropolymerized PEDOT:PSS transducers for application as electrochemical sensors in aqueous media. *Sens. Bio-Sens. Res.* 17: 18. https://doi.org/10.1016/j.sbsr.2018.01.001.

15 Ma, L., Liu, R., Niu, H. et al. (2016). Flexible and freestanding electrode based on polypyrrole/graphene/bacterial cellulose paper for supercapacitor. *Compos. Sci. Technol.* 137: 87. https://doi.org/10.1016/j.compscitech.2016.10.027.

16 Shrestha, B.K., Ahmad, R., Shrestha, S. et al. (2017). Globular shaped polypyrrole doped well-dispersed functionalized multiwall carbon nanotubes/nafion composite for enzymatic glucose biosensor application. *Sci. Rep.* 7: 16191. https://doi.org/10.1038/s41598-017-16541-9.

17 Hao, G.P., Hippauf, F., Oschatz, M. et al. (2014). Stretchable and semitransparent conductive hybrid hydrogels for flexible supercapacitors. *ACS Nano* 8: 7138. https://doi.org/10.1021/nn502065u.

18 G. Kaur, R. Adhikari, P. Cass, M. Bown, P. Gunatillake, Electrically conductive polymers and composites for biomedical applications, *RSC Adv.*, 5 (2015) 37553. https://doi.org/10.1039/c5ra01851j

19 Wang, Y., Zhu, C., Pfattner, R. et al. (2017). A highly stretchable, transparent, and conductive polymer. *Sci. Adv.* 3: e1602076. https://doi.org/10.1039/C5RA01851J.

20 Han, L., Yan, L.W., Wang, M.H. et al. (2018). Transparent, adhesive, and conductive hydrogel for soft bioelectronics based on light-transmitting polydopamine-doped polypyrrole nanofibrils. *Chem. Mater.* 30: 5561. https://doi.org/10.1021/acs.chemmater.8b01446.

21 Khalili, N., Naguib, H.E., and Kwon, R.H. (2016). A constriction resistance model of conjugated polymer based piezoresistive sensors for electronic skin applications. *Soft Matter* 12: 4180. https://doi.org/10.1039/C6SM00204H.

22 Baumberger, T., Caroli, C., and Martina, D. (2006). Solvent control of crack dynamics in a reversible hydrogel. *Nat. Mater.* 5: 552. https://doi.org/10.1038/nmat1666.

23 Henderson, K.J., Zhou, T., Otim, K.J., and Shull, K.R. (2010). Ionically cross-linked triblock copolymer hydrogels with high strength. *Macromolecules* 43: 6193. https://doi.org/10.1021/ma100963m.

24 Sun, J., Zhao, X., Illeperuma, W. et al. (2012). Highly stretchable and tough hydrogels. *Nature* 489: 151. https://doi.org/10.1038/nature11409.

25 Sun, T., Kurokawa, T., Kuroda, S. et al. (2013). Physical hydrogels composed of polyampholytes demonstrate high toughness and viscoelasticity. *Nat. Mater.* 12: 060. https://doi.org/10.1038/nmat3713.

26 Lee, Y.Y., Kang, H.Y., Gwon, S.H. et al. (2016). A strain-insensitive stretchable electronic conductor: PEDOT: PSS /acrylamide organogels. *Adv. Mater.* 28: 1636. https://doi.org/10.1002/adma.201504606.

27 Oh, J., Kim, J.O., Kim, Y. et al. (2019). Highly uniform and low hysteresis piezoresistive pressure sensors based on chemical grafting of polypyrrole on elastomer template with uniform pore size. *Small* 15: 1901744. https://doi.org/10.1002/smll.201901744.

28 Zucca, A., Yamagishi, K., Fujie, T. et al. (2015). Roll to roll processing of ultraconformable conducting polymer nanosheets. *J. Mater. Chem. C* 3: 6539. https://doi.org/10.1039/C5TC00750J.

29 Li, Z.Z., Sinha, S.K., Treich, G.M. et al. (2020). All-organic flexible fabric antenna for wearable electronics. *J. Mater. Chem. C* 8: 5662. https://doi.org/10.1039/D0TC00691B.

30 Zhang, S.M. and Cicoira, F. (2017). Water-enabled healing of conducting polymer films. *Adv. Mater.* 29: 1703098. https://doi.org/10.1002/adma.201703098.

31 Lu, Y., Liu, Z.Q., Yan, H.M. et al. (2019). Ultrastretchable conductive polymer complex as a strain sensor with a repeatable autonomous self-healing ability. *ACS Appl. Mater. Interfaces* 11: 22. https://doi.org/10.1021/acsami.9b05464.

32 Zhu, Z., Li, R., and Pan, T. (2018). Imperceptible epidermal–iontronic interface for wearable sensing. *Adv. Mater.* 30: 1705122. https://doi.org/10.1002/adma.201705122.

33 Wang, S.H., Xu, J., Wang, W.C. et al. (2018). Skin electronics from scalable fabrication of an intrinsically stretchable transistor array. *Nature* 555: 83. https://doi.org/10.1038/nature25494.

34 Xu, B., Chakraborty, H., Remsing, R.C. et al. (2017). A freestanding molecular spin–charge converter for ubiquitous magnetic-energy harvesting and sensing. *Adv. Mater.* 29: 1605150. https://doi.org/10.1002/adma.201605150.

35 Guan, Y.S., Zhang, Z., Tang, Y. et al. (2018). Kirigami-inspired nanoconfined polymer conducting nanosheets with 2000% stretchability. *Adv. Mater.* 30: 1706390. https://doi.org/10.1002/adma.201706390.

36 Mule, R., Dudem, B., Patnam, H. et al. (2019). Wearable single -electrode-mode triboelectric nanogenerator via conductive polymer-coated textiles for self-power electronics. *ACS Sustainable Chem. Eng.* 7: 16450. https://doi.org/10.1021/acssuschemeng.9b03629.

37 Ahmed, A., Guan, Y.S., Hassan, I. et al. (2020). Multifunctional smart electronic skin fabricated from two-dimensional like polymer film. *Nano Energy* 75: 105044. https://doi.org/10.1016/j.nanoen.2020.105044.

38 Inal, S., Rivnay, J., Suiu, A.-O. et al. (2018). Conjugated polymers in bioelectronics. *Acc. Chem. Res.* 51: 1368. https://doi.org/10.1021/acs.accounts.7b00624.

39 Ding, R., Krikstolaityte, V., and Lisak, G. (2019). Inorganic salt modified paper substrates utilized in paper based microfluidic sampling for potentiometric determination of heavy metals. *Sens. Actuators, B* 290: 347. https://doi.org/10.1016/j.snb.2019.03.079.

40 Joon, N., He, N., Ruzgas, T. et al. (2019). PVC-Based ion-selective electrodes with a silicone rubber outer coating with improved analytical performance. *Anal. Chem.* 91: 10524. https://doi.org/10.1021/acs.analchem.9b01490.

41 Long, Y., Li, M., Gu, C. et al. (2011). Recent advances in synthesis, physical properties and applications of conducting polymer nanotubes and nanofibers. *Prog. Polym. Sci.* 36: 1415. https://doi.org/10.1016/j.bios.2017.08.030.

42 Talikowska, M., Fu, X., and Lisak, G. (2019). Application of conducting polymers to wound care and skin tissue engineering: a review. *Biosens. Bioelectron.* 135: 50. https://doi.org/10.1016/j.bios.2019.04.001.

43 Gumus, J., Califano, A., Wan, J. et al. (2010). Control of cell migration using a conducting polymer device. *Soft Matter* 6: 5138. https://doi.org/10.1039/B923064E.

44 Proksch, E., Brandner, J., and Jensen, J. (2008). The skin: an indispensable barrier. *Exp. Dermatol.* 17: 1063. https://doi.org/10.1111/j.1600-0625.2008.00786.x.

45 Zhang, J., Wu, C., Xu, Y. et al. (2020). Highly stretchable and conductive self-healing hydrogels for temperature and strain sensing and chronic wound treatment. *ACS Appl. Mater. Interfaces* 12: 40990. https://doi.org/10.1021/acsami.0c08291.

46 Liang, Y., Chen, B., Li, M. et al. (2020). Injectable antimicrobial conductive hydrogels for wound disinfection and infectious wound healing. *Biomacromolecules* 21: 1841. https://doi.org/10.1021/acs.biomac.9b01732.

47 Dedkov, E., Kostrominova, T., Borisov, A., and Carlson, B. (2002). Survival of Schwann cells in chronically denervated skeletal muscles. *Acta Neuropathol.* 103: 565. https://doi.org/10.1007/s00401-001-0504-6.

48 Zhang, L. and Webster, T. (2009). Nanotechnology and nanomaterials: promises for improved tissue regeneration. *Nano Today* 4: 66. https://doi.org/10.1016/j.nantod.2008.10.014.

49 Prabhakaran, M., Mobarakeh, L., Jin, G., and Ramakrishna, S. (2011). Electrospun conducting polymer nanofibers and electrical stimulation of nerve stem cells. *J. Biosci. Bioeng.* 112: 501. https://doi.org/10.1016/j.jbiosc.2011.07.010.

50 Meng, C., Jiang, W., Huang, Z. et al. (2020). Fabrication of a highly conductive silk knitted composite scaffold by two-step electrostatic self-assembly for potential peripheral nerve regeneration. *ACS Appl. Mater. Interfaces* 12: 12317. https://doi.org/10.1021/acsami.9b22088.

51 Kim, H., Lee, J., Park, J. et al. (2020). Fabrication of nanocomposites complexed with gold nanoparticles on polyaniline and application to their nerve regeneration. *ACS Appl. Mater. Interfaces* 12: 30750. https://doi.org/10.1021/acsami.0c05286.

52 Lu, D., Wang, H., Li, T. et al. (2017). Mussel-inspired thermoresponsive polypeptide–pluronic copolymers for versatile surgical adhesives and hemostasis. *ACS Appl. Mater. Interfaces* 9: 16756. https://doi.org/10.1021/acsami.6b16575.

53 Meng, S., Rouabhia, M., and Zhang, Z. (2013). Electrical stimulation modulates osteoblast proliferation and bone protein production through heparin-bioactivated conductive scaffolds. *Bioelectromagnetics* 34: 189. https://doi.org/10.1002/bem.21766.

54 Chen, L., Mou, S., Li, F. et al. (2019). Self-assembled human adipose-derived stem cell-derived extracellular vesicle-functionalized biotin-doped polypyrrole titanium with long-term stability and potential osteoinductive ability. *ACS Appl. Mater. Interfaces* 11: 46183. https://doi.org/10.1021/acsami.9b17015.

55 Wang, Q., Wu, M., Xu, X. et al. (2020). Direct current stimulation for improved osteogenesis of MC3T3 cells using mineralized conductive polyaniline. *ACS Biomater. Sci. Eng.* https://doi.org/10.1021/acsbiomaterials.9b01821.

56 Yan, H., Li, L., Wang, Z. et al. (2020). Mussel-inspired conducting copolymer with aniline tetramer as intelligent biological adhesive for bone tissue engineering. *ACS Biomater. Sci. Eng.* 6: 634. https://doi.org/10.1021/acsbiomaterials.9b01601.

57 Ashtari, K., Nazari, H., Ko, H. et al. (2019). Electrically conductive nanomaterials for cardiac tissue engineering. *Adv. Drug Delivery Rev.* 144: 162. https://doi.org/10.1016/j.addr.2019.06.001.

58 Zia, S., Mozafari, M., Natasha, G. et al. (2016). Hearts beating through decellularized scaffolds: whole-organ engineering for cardiac regeneration and transplantation. *Crit. Rev. Biotechnol.* 36: 705. https://doi.org/10.3109/07388551.2015.1007495.

59 Song, C., Zhang, X., Wang, L. et al. (2019). An injectable conductive three-dimensional elastic network by tangled surgical-suture spring for heart repair. *ACS Nano* 13: 14122. https://doi.org/10.1021/acsnano.9b06761.

60 Amirabad, L., Massumi, M., Shamsara, M. et al. (2017). Enhanced cardiac differentiation of human cardiovascular disease patient-specific induced pluripotent stem cells by applying unidirectional electrical pulses using aligned electroactive nanofibrous scaffolds. *ACS Appl. Mater. Interfaces* 9: 6849. https://doi.org/10.1021/acsami.6b15271.

61 Ulbrich, K., Holá, K., Šubr, V. et al. (2016). Targeted drug delivery with polymers and magnetic nanoparticles: covalent and noncovalent approaches, release control, and clinical studies. *Chem. Rev.* 116: 5338. https://doi.org/10.1021/acs.chemrev.5b00589.

62 Tibbitt, M., Dahlman, J., and Langer, R. (2016). Emerging frontiers in drug delivery. *J. Am. Chem. Soc.* 138: 704. https://doi.org/10.1021/jacs.5b09974.

63 Stuart, M.A.C., Huck, W.T.S., Genzer, J. et al. (2010). Emerging applications of stimuli-responsive polymer materials. *Nat. Mater.* 9: 101. https://doi.org/10.1038/nmat2614.

64 Kang, H., Trondoli, A., Zhu, G. et al. (2011). Near-infrared light-responsive core–shell nanogels for targeted drug delivery. *ACS Nano* 5: 5094. https://doi.org/10.1021/nn201171r.

65 Anna, P., Cejudo, A., Valle, L., and Alemán, C. (2018). Smart drug delivery from electrospun fibers through electroresponsive polymeric nanoparticles. *ACS Appl. Bio Mater.* 1: 1594. https://doi.org/10.1021/acsabm.8b00459.

66 An, Y., Lee, J., Son, D. et al. (2020). Facilitated transdermal drug delivery using nanocarriers-embedded electroconductive hydrogel coupled with reverse electrodialysis-driven iontophoresis. *ACS Nano* 14: 4523. https://doi.org/10.1021/acsnano.0c00007.

67 Boehler, C., Oberueber, F., and Asplund, M. (2019). Tuning drug delivery from conducting polymer films for accurately controlled release of charged molecules. *J. Controlled Release* 304: 173. https://www.ncbi.nlm.nih.gov/pubmed/31096016.

68 Abidian, M.R., Kim, D.H., and Martin, D.C. (2006). Conducting-polymer nanotubes for controlled drug release. *Adv. Mater.* 18: 405. https://www.ncbi.nlm.nih.gov/pubmed/21552389.

69 Du, Z.J., Bi, G.Q., and Cui, X.T. (2018). Electrically controlled neurochemical release from dual-layer conducting polymer films for precise modulation of neural network activity in rat barrel cortex. *Adv. Funct. Mater.* 28: 1703988. https://www.ncbi.nlm.nih.gov/pubmed/30467460.

70 Krukiewicz, K., Zawisza, P., Herman, A.P. et al. (2016). An electrically controlled drug delivery system based on conducting poly(3,4-ethylenedioxypyrrole) matrix. *Bioelectrochemistry* 108: 13. https://www.ncbi.nlm.nih.gov/pubmed/26606716.

71 Garcia-Fernandez, A., Lozano-Torres, B., Blandez, J.F. et al. (2020). Electro-responsive films containing voltage responsive gated mesoporous silica nanoparticles grafted onto PEDOT-based conducting polymer. *J. Controlled Release* 323: 421. https://www.ncbi.nlm.nih.gov/pubmed/32371265.

72 Puiggali-Jou, A., Del Valle, L.J., and Aleman, C. (2020). Encapsulation and storage of therapeutic fibrin-homing peptides using conducting polymer nanoparticles for programmed release by electrical stimulation. *ACS Biomater. Sci. Eng.* 6: 2135. https://www.ncbi.nlm.nih.gov/pubmed/33455313.

73 Stuart, M., Huck, W., Genzer, J. et al. (2016). Near-infrared (NIR)-absorbing conjugated polymer dots as highly effective photothermal materials for in vivo cancer therapy. *Chem. Mater.* 28: 8669. https://doi.org/10.1021/acs.chemmater.6b03738.

74 Huang, X., Jain, P., El-Sayed, I., and El-Sayed, M. (2007). Plasmonic photothermal therapy (PPTT) using gold nanoparticles. *Laser Med. Sci.* 23: 217. https://doi.org/10.1007/s10103-007-0470-x.

75 Prasad, T., In, H., Jung-Mi, O. et al. (2018). pH/NIR-Responsive polypyrrole-functionalized fibrous localized drug-delivery platform for synergistic cancer therapy. *ACS Appl. Mater. Interfaces* 10: 20256. https://doi.org/10.1021/acsami.7b17664.

76 Geng, S., Zhao, H., Zhan, G. et al. (2020). Injectable in situ forming hydrogels of thermosensitive polypyrrole nanoplatforms for precisely synergistic photothermo-chemotherapy. *ACS Appl. Mater. Interfaces* 12: 7995. https://www.ncbi.nlm.nih.gov/pubmed/32013384.

77 Uppalapati, D., Sharma, M., Aqrawe, Z. et al. (2018). Micelle directed chemical polymerization of polypyrrole particles for the electrically triggered release of dexamethasone base and dexamethasone phosphate. *Int. J. Pharm.* 543: 38. https://www.ncbi.nlm.nih.gov/pubmed/29581065.

78 E Gutiérrez-Pineda, P. R. Cáceres-Vélez, M. J. Rodríguez-Presa, S. E. Moya, C. A. Gervasi, J. I. Amalvy, Hybrid conducting composite films based on polypyrrole and poly(2-(diethylamino)ethyl methacrylate) hydrogel nanoparticles for electrochemically controlled drug delivery, *Adv. Mater. Interfaces*, 5 (2018) 1800968. https://doi.org/10.1002/admi.201800968

79 Woeppel, K.M., Zheng, X.S., Schulte, Z.M. et al. (2019). Nanoparticle doped PEDOT for enhanced electrode coatings and drug delivery. *Adv. Healthc. Mater.* 8: 1900622. https://www.ncbi.nlm.nih.gov/pubmed/31583857.

80 Zou, M., Li, S., Hu, X. et al. (2021). Progresses in tensile, torsional, and multifunctional soft actuators. *Adv. Funct. Mater.* 2007437. https://doi.org/10.1002/adfm.202007437.

81 Nan, M., Wang, F., Kim, S. et al. (2019). Ecofriendly high-performance ionic soft actuators based on graphene-mediated cellulose acetate. *Sens. Actuators, B* 301: 127127. https://doi.org/10.1016/j.snb.2019.127127.

82 Beregoi, M., Evanghelidis, A., Diculescu, V.C. et al. (2017). Polypyrrole actuator based on electrospun microribbons. *ACS Appl. Mater. Interfaces* 9: 38068. https://www.ncbi.nlm.nih.gov/pubmed/28976177.

83 Xiong, L., Jin, H., Lu, Y. et al. (2020). A solvent molecule driven pure PEDOT:PSS actuator. *Macromol. Mater. Eng.* 305: 2000327. https://doi.org/10.1002/mame.202000327.

84 Khaldi, A., Maziz, A., Alici, G. et al. (2016). Bottom-up microfabrication process for individually controlled conjugated polymer actuators. *Sens. Actuators, B* 230: 818. http://dx.doi.org/doi:10.1016/j.snb.2016.02.140.

85 Mahmoudian, M.R., Basirun, W.J., and Alias, Y. (2016). A sensitive electrochemical Hg^{2+} ions sensor based on polypyrrole coated nanospherical platinum. *RSC Adv.* 6: 36459. https://doi.org/10.1039/c6ra03878f.

86 Wei, Y., Yang, R., Liu, J.H., and Huang, X.J. (2013). Selective detection toward Hg(II) and Pb(II) using polypyrrole/carbonaceous nanospheres modified screen-printed electrode. *Electrochim. Acta* 105: 218. https://doi.org/10.1016/j.electacta.2013.05.004.

87 Tan, F., Cong, L., Li, X. et al. (2016). An electrochemical sensor based on molecularly imprinted polypyrrole/graphene quantum dots composite for detection of bisphenol A in water samples. *Sens. Actuators, B* 233: 599. https://doi.org/10.1016/j.snb.2016.04.146.

88 Shahrokhian, S., Azimzadeh, M., and Ainini, M.K. (2015). Modification of glassy carbon electrode with a bilayer of multiwalled carbon nanotube/tiron-doped polypyrrole: application to sensitive voltammetric determination of acyclovir. *Mater. Sci.* 53: 134. https://doi.org/10.1016/j.msec.2015.04.030.

89 Liu, S., Ma, Y., Cui, M., and Luo, X. (2017). Enhanced electrochemical biosensing of alpha-fetoprotein based on three-dimensional macroporous conducting polymer polyaniline. *J. Mol. Struct.* 225: 2568. https://doi.org/10.1016/j.snb.2017.09.062.

90 Robertson, J., Gizdavic-Nikolaidis, M., Nieuwoudt, M.K., and Swift, S. (2018). The antimicrobial action of polyaniline involves production of oxidative stress while functionalisation of polyaniline introduces additional mechanisms. *PeerJ. J. Med. Chem.* 6: 5135. https://doi.org/10.7717/peerj.5135.

91 Zhou, B., Qiu, Y., Wen, Q. et al. (2016). A dual electrochemiluminescence signal system for in-situ and simultaneous evaluation of multiple cell-surface receptors. *ACS Appl. Mater. Interfaces* 3: 2074. https://doi.org/10.1021/acsami.6b12411.

92 Sharma, A., Kumar, B., Singh, S.K. et al. (2018). In-vitro and in-vivo pharmacokinetic evaluation of guar gum-eudragit® S100 based colon-targeted spheroids of sulfasalazine co-administered with probiotics. *Curr. Drug Delivery* 15, 367: https://doi.org/10.2174/1567201815666171207165059.

93 Zamani, F., Moulahoum, H., Ak, M. et al. (2019). Current trends in the development of conducting polymers-based biosensors. *TrAC, Trends Anal. Chem.* 118: 264. https://doi.org/10.1016/j.trac.2019.05.031.

94 Wang, Y., Yu, H., Li, Y. et al. (2019). Facile preparation of highly conductive poly(amide-imide) composite films beyond $1000\,S\,m^{-1}$ through ternary blend strategy. *Polymer* 11: 546. https://doi.org/10.3390/polym11030546.

95 Fang, Y., Yu, H., Wang, Y. et al. (2020). Simultaneous improvement of mechanical and conductive properties of poly(amide-imide) composites using carbon nano-materials with different morphologies. *J. Polym. Eng.* 40: 806. https://doi.org/10.1515/polyeng-2020-0091.

96 Zhai, D., Liu, B., Shi, Y. et al. (2013). Highly sensitive glucose sensor based on Pt nanoparticle/polyaniline hydrogel heterostructures. *ACS Nano* 7: 3540. https://doi.org/10.1021/nn400482d.

97 Promsuwan, K., Meng, L., Suklim, P. et al. (2020). Bio-PEDOT: Modulating carboxyl moieties in poly(3,4-ethylenedioxythiophene) for enzyme-coupled bioelectronic interfaces. *ACS Appl. Mater. Interfaces* 12: 39841. https://doi.org/10.1021/acsami.0c10270.

98 Zhang, B., Lu, L., Huang, F., and Lin, Z. (2015). [Ru(bpy)$_3$]$^{2+}$-mediated photoelectrochemical detection of bisphenol A on a molecularly imprinted polypyrrole modified SnO$_2$ electrode. *Anal. Chim. Acta* 887: 59. https://doi.org/10.1016/j.aca.2015.05.051.

99 Yang, J., Cho, M., Pang, C., and Lee, Y. (2015). Highly sensitive non-enzymatic glucose sensor based on over-oxidized polypyrrole nanowires modified with Ni(OH)$_2$ nanoflakes. *Sens. Actuators, B* 211: 93. https://doi.org/10.1016/j.snb.2015.01.045.

100 Kim, W., Lee, J.S., Shin, D.H., and Jang, J. (2018). Platinum nanoparticles immobilized on polypyrrole nanofibers for non-enzyme oxalic acid sensor. *J. Mater. Chem. B* 6: 1272. https://doi.org/10.1039/c7tb00629b.

101 Tertis, M., Cemat, A., Lacatis, D. et al. (2017). Highly selective electrochemical detection of serotonin on polypyrrole and gold nanoparticles-based 3D architecture. *Electrochem. Commun.* 75: 43. https://doi.org/10.1016/j.elecom.2016.12.015.

102 Sheng, Q., Liu, D., and Zheng, J. (2016). NiCo alloy nanoparticles anchored on polypyrrole\reduced graphene oxide nanocomposites for nonenzymatic glucose sensing. *New J. Chem.* 40: 6658. https://doi.org/10.1039/c6nj01264g.

103 Hui, N. and Wang, J. (2017). Electrodeposited honeycomb-like cobalt nanostructures on graphene oxide doped polypyrrole nanocomposite for high performance enzymeless glucose sensing. *J. Electroanal. Chem.* 798: 9. https://doi.org/10.1016/j.jelechem.2017.05.021.

104 Qian, T., Yu, C., Zhou, X. et al. (2014). Ultrasensitive dopamine sensor based on novel molecularly imprinted polypyrrole coated carbon nanotubes. *Biosens. Bioelectron.* 58: 237. https://doi.org/10.1016/j.bios.2014.02.081.

105 Cui, Y., Chen, H., Hou, L. et al. (2012). Nanogold–polyaniline–nanogold microspheres-functionalized molecular tags for sensitive electrochemical immunoassay of thyroid-stimulating hormone. *Sens. Actuators, B* 738: 76. https://doi.org/10.1016/j.aca.2012.06.013.

106 Lima, D., Hacke, A., Inaba, J. et al. (2020). Electrochemical detection of specific interactions between apolipoprotein E isoforms and DNA sequences related to Alzheimer's disease. *Bioelectrochemistry* 133: 107447. https://doi.org/10.1016/j.bioelechem.2019.107447.

107 Salimian, R., Shahrokhian, S., and Panahi, S. (2019). Enhanced electrochemical activity of a hollow carbon sphere/polyaniline-based electrochemical biosensor for HBV DNA marker detection. *ACS Biomater. Sci. Eng.* 5: 2587. https://doi.org/10.1021/acsbiomaterials.8b01520.

108 Kıranşan, K. and Topçu, E. (2020). Conducting polymer-reduced graphene oxide sponge electrode for electrochemical detection based on DNA hybridization. *ACS Appl. Nano Mater.* 3: 5449. https://doi.org/10.1021/acsanm.0c00782.

109 Hui, N., Sun, X., Niu, S., and Luo, X. (2017). PEGylated polyaniline nanofibers: antifouling and conducting biomaterial for electrochemical DNA sensing. *ACS Appl. Mater. Interfaces* 9: 2914. https://doi.org/10.1021/acsami.6b11682.

110 Shaikh, M.O., Srikanth, B., Zhu, P.Y., and Chuang, C.H. (2019). Impedimetric immunosensor utilizing polyaniline/gold nanocomposite-modified screen-printed electrodes for early detection of chronic kidney disease. *Sensors* 19: 3990. https://doi.org/10.1021/acsami.6b12411.

11

Methods for Synthesizing Polymer Nanocomposites and Their Applications

Muwei Ji[1], Jintao Huang[2], and Caizhen Zhu[1]

[1] Shenzhen University, School of Chemistry and Environmental Engineering, No. 3688, Nanhai Avenue, Nanshan District, Shenzhen, Guangdong 518060, China
[2] Guangdong University of Technology, School of Materials and Energy, Department of Polymeric Materials and Engineering, No. 100, Outer Ring West Road, Panyu District, Guangzhou, Guangdong 510006, China

As an important nanoscale material, polymer nanocomposites were considered significant candidates for electronic, catalytic, and biological applications, enhancing toughness, drug delivery, cell imaging, and exploration of new powerful devices [1–6]. The methods and technologies for the controlled preparation of polymer nanocomposites were the basis for the new devices and applications. This chapter reviewed the traditional methods for preparing polymer nanocomposites and the recently developed technologies for synthesizing polymer nanocomposites with controlled sizes, surfaces, and morphology. In the past decade, the study of polymer composites and nanocomposites has gained interest from scientists and engineers. As shown in Figure 11.1a, the published papers on composites and nanocomposites have increased rapidly since last year. The percentage of publications on polymer nanocomposites increased up to 25% (Figure 11.1b). Thus, the study of polymer nanocomposites is becoming popular in science and technology.

The methods for preparing polymer nanocomposites are essential as they are the basis of the design and application. Popular methods are emulsion polymerization, dispersion polymerization or copolymerization, self-assembly, melting, and *in situ* polymerization. By using these methods, various polymer nanocomposites can be prepared in facile ways. For example, the emulation polymerization and dispersing polymerization methods can obtain the composite in nanoscale. The particles with a larger size can be prepared by using precipitate polymerization. In recent years, many works have been reported on molecule design, controlling size, and dispersing, which enhance polymerization properties in the application. This chapter discusses the reviewed methods for preparing polymer nanocomposites.

Functional Nanomaterials: Synthesis, Properties, and Applications, First Edition.
Edited by Wai-Yeung Wong and Qingchen Dong.
© 2022 WILEY-VCH GmbH. Published 2022 by WILEY-VCH GmbH.

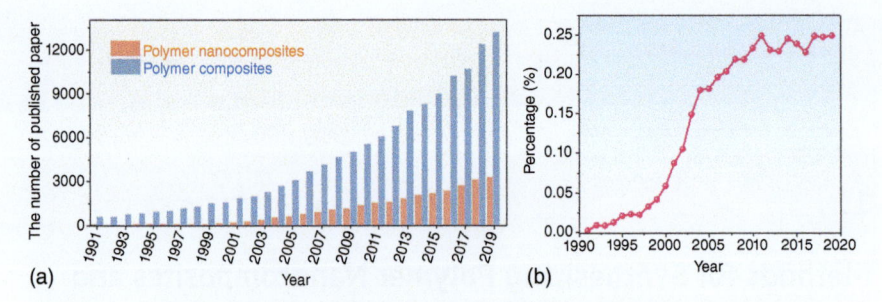

Figure 11.1 (a) The published paper increasing in the recent 20 years. The data was collected from the Web of Science. (b) The percentage of the publications of polymer nanocomposites in the reported studies on polymer composites. The data were collected from Web of Science.

11.1 Factors for Synthesizing Polymer Nanocomposites

For synthesizing polymers nanocomposites, several factors should be seriously considered such as thermal conductivity, solubility, etc., which play an important role in the wide applications. The more details are included in the controlled preparation process, the more possibility for designing and constructing polymer nanocomposites. For example, in the emulsion polymerization process, the hydrophilic properties of the composite surface of inorganic nanoparticles, the size of the polymer compositions, and the surfactant of emulation should be considered. More factors should be included when using polymer composites for devices, or complicated structures, such as 3D printing [7, 8]. For instance, polyvinylidene difluoride (PVDF) was an important material to prepare polymer composite for electronic studies [9–11].

In the solution or emulsion, pH is another important factor for synthesizing polymer nanocomposites. For example, Cai and coworkers reported that the pH could be used in preparing polymer nanomaterials with tunable surfaces [12]. In a close natural condition, the nanotubes with 4% and 6% N-2-aminoethylacrylamide hydrochloride (AEAM) (A-1, B-1 in Figure 11.2) were porous. The tubular pores were closed when the pH was adjusted to 7.8 (A-2, B-2 in Figure 11.2). The nanotubes were disrupted as the pH changed to 3.0 (A-3, B-3 in Figure 11.2). The results were attributed to the deionization of NH_3^+-motifs and the quantitative ionization dependent on the pH. Furthermore, the spherical vesicles at 4% AEAM (a-1 in Figure 11.2) coarsened and perforated at higher content (6%, b-1). When the conditions were adjusted to neutral, the vesicular thickness and vesicular pores remained (a-2, b-2 in Figure 11.2). However, the polymerization system was adjusted to pH 3.0. As the result of the quantitative ionization of NH_2-motifs, the pores became more pronounced, and the thickness decreased (a-3, b-3 in Figure 11.2). Hence, pH was an important factor because many groups were ionized during polymerization, especially in an aqueous solution.

The solubility of the polymer precursors or composition is also important because the polymerization reaction depends on the concentration [13, 14]. An

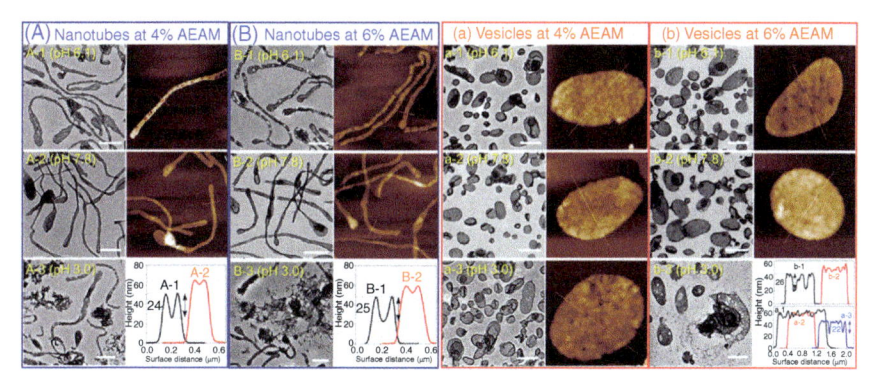

Figure 11.2 TEM and AFM images of the AEAM-inserted objects by diluting the final dispersions in water to 0.05 mg ml^{-1} at pH 6.1 (top) and followed by deionization at pH 7.8 (middle) and ionization at pH 3.0 (bottom). (A) Nanotube at 4% AEAM, (B) Nanotube at 6% AEAM, (a) Vesicle at 4% AEAM, (b) Vesicle at 6% AEAM. Scale bar: 1 μm. Source: Gao et al. [12], Figure 4 [p. 1330]/with permission from American Chemical Society.

and coworkers reported that the solubility of [2-(methacryloyloxy)ethyltrimethylammonium salt ([META$^+$]) affected the polymer nanocomposites' growth. As the [META$^+$][PF$_6^-$] was introduced, the polymerization was fast, and about 3% conversion occurred in 2.5 minutes by using equiv 2,2′-azobis(2-methylpropionamidine) dihydrochloride (V-50) as the initiator (Figure 11.3) [13]. Combining with the degree of polymerization (DP$_n$) tailoring, polymer nanocomposites could be controlled. The water-soluble macromolecular chain transfer agents were obtained to generate nano-objects consisting of a neutral stabilizer block (Figure 11.3c–e). For higher solubility, the mixed solution was used. Simultaneously, the mixed solution was also important in self-assembly due to its polarity. Inorganic salt was added to obtain high polarity in ionized polymerization [15].

The DP$_n$ of polymer compounds is also essential for synthesizing polymer nanocomposites [13]. Using polymer p(MAA-co-PEOMA) with different degrees of polymerization resulted in different morphologies [16]. Only nanoparticles could be obtained using low DP$_n$ (DP$_n$ = 80), and the morphologies transformed from nanoparticles to nanowires as the DP$_n$ increases. The morphologies of the obtained samples were nanowires without any particles as the DP$_n$ reached 330. As the DP$_n$ became higher, the morphologies re-transformed into nanosphere with large sizes. The polyion complexes (PIC) also revealed the morphologies of polymer nanocomposites depending on the DP$_n$. Ionic histamine acrylamide hydrochloride (HisAM), 2-acrylamido-2-methylpropanesulfonate (AMPS) monomers, and nonionic poly2-hydroxypropylmethacrylamide (PHPMA) were mixed to form different polymer composites with different DP$_n$. It illustrated that the PIC were nanoparticles whose sizes were smaller than 100 nm. When the DP$_n$ was controlled in the range of 35 and 55, the obtained PICs were ribbon. As the DP$_n$ went higher than 60, some PIC vesicles were observed, reducing the film density. The size of the well-defined vesicles increased till the range of the DP$_n$ reached around 85 and 100. They aggregated and reformed into irregular bulk as the DP$_n$ constant rose. The DP$_n$

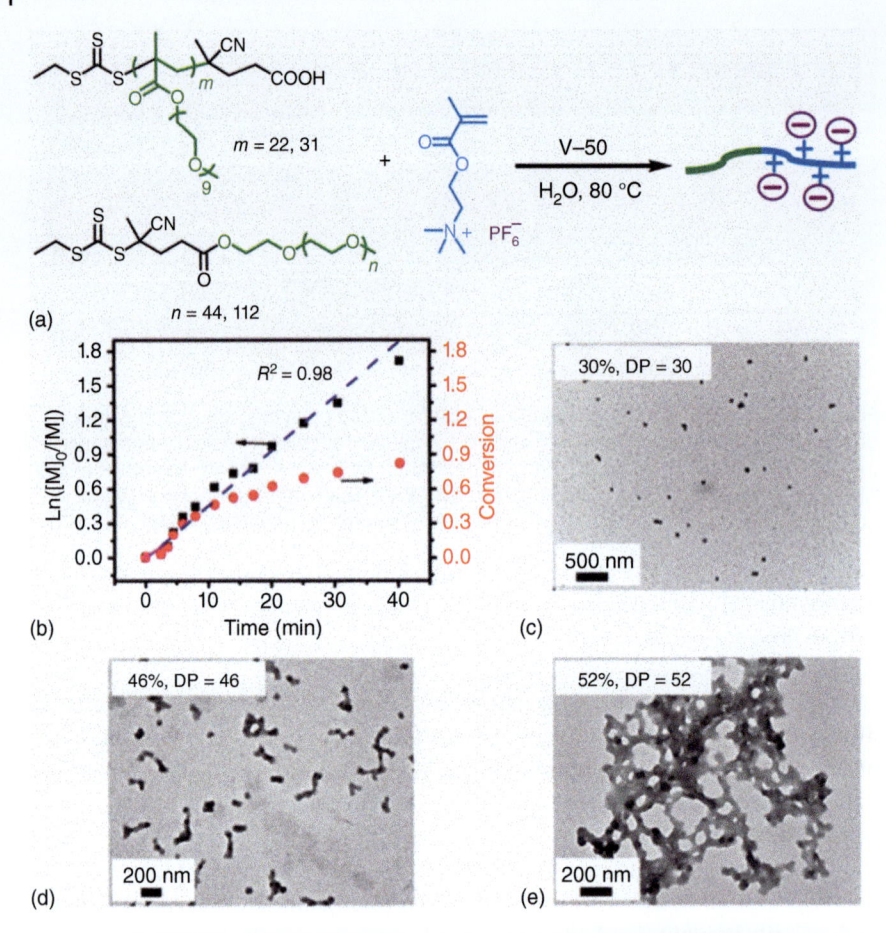

Figure 11.3 (a) The schematic of polymerization of ionic monomer to produce Neutral-polyelectrolyte block. (b) The kinetic results of polymerization. (c–e) The TEM images of as-prepared polymer nanocomposites. Source: Zhang et al. [13], Figure 1 [p. 226]/with permission from American Chemical Society.

was not the only main factor for polymer nanocomposite synthesis. Deng et al. found that by using the 2-hydroxypropyl methacrylate (HPMA) with 170 or 240 of DP_n, the well-defined membrane-forming PHPMA block could be obtained [17]. Therefore, the effect of DP_n was more complex because it also depended on the polymerization system and reaction conditions, such as solubility, temperature, and concentration.

Reaction time is also an important factor for controlled polymer nanocomposites. For instance, Gao et al. prepared PEG-b-PS/PS-b-PEG-b-PS via polymerization-induced self-assembly (PISA) methods (Figure 11.4) [14]. By controlling the reaction time, they successfully tailored the morphology and structure (Figure 11.4). As presented in Figure 11.4b, PEG_{45}-b-PS/PS-b-PEG_{45}-b-PS obtained after eight hours of polymerizing were nanoparticles, and the nanosheet hollow structure was obtained by prolonging the polymerizing time. When the reaction time was prolonged to 14–24 hours, pores were found on the thin shell of as-prepared PEG_{45}-b-PS/PS-b-PEG_{45}-b-PS.

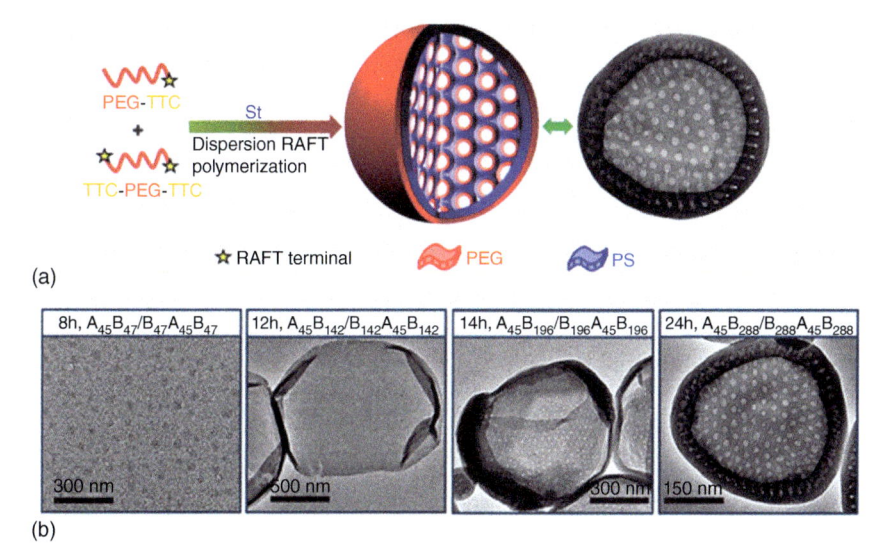

Figure 11.4 (a) Schematic of preparation of PEG_{45}-b-PS/PS-b-PEG_{45}-b-PS. (b) the TEM images of PEG_{45}-b-PS/PS-b-PEG_{45}-b-PS are controlled by the reaction time. Source: Wang and An [14]/with permission from Wiley-VCH.

The feed molar ratio of polymers or monomers can also be used for tailoring polymer nanocomposites [18, 19]. Pan and coworker used trithiocarbonate-terminated poly(4-vinylpyridine) (P4VP-TC), 2,20-azobisisobutyronitrile (AIBN), and styrene (St) in methanol to prepare copolymers P4VP-TC/St/AIBN [18]. By tailoring the composition molar ratio, the copolymer nanocomposites produced different morphologies. Using the feed molar ratio of P4VP-TC/St/AIB = 10 : 200 000 : 1, the amount of methanol resulted in the morphology variation from chain to sphere. Furthermore, varying the amount of St during the polymerization also tailored the morphologies [18]. Besides reaction time, DP_n, feed molar ratio, and pH, hydrogen bonding should also be considered in the dispersing system, especially in an aqueous system [20–23]. In other words, polymerization was a complex process, and besides the factors mentioned, other factors also affected the preparation of polymer nanocomposites. Therefore, the preparation of polymer nanocomposites needed suitable conditions and reasonable designs. Furthermore, some polymerizations were affected by multiple factors, and the synergetic effect from components and reactions should be tailored for special combinations to form polymer nanocomposites. In summary, more factors should be further studied for tailoring polymer nanocomposites with multiple dimensions.

11.2 Solution Mixing

Solution mixing is a broad method for preparing polymers/inorganic nanocomposites. This strategy is usually employed in fields such as electronics, thermal conductivity, and so on [24]. In the PVDF-based polymer nanocomposites, solution mixing is a popular and facile way to obtain various nanocomposites. However, surface modification is usually needed during the combination to ensure the tight interaction between two components. For instance, $BaTiO_3$ nanocrystals combined with PVDF would enhance the polar properties of PVDF, energy storage and electronics, and other properties.

Hu and coworkers used the dopamine coated $BaTiO_3$@TiO_2 coaxial nanofibers, and PVDF was dispersed in DMF under ultrasonication for two hours. After stirring for 12 hours, the resulted mixture was cast on a clean glass plate and dried at 45 °C. After modifying the surface with dopamine, the $BaTiO_3$@TiO_2 coaxial nanofibers were dispersed into N,N'-dimethylformamide (DMF) and formed a uniform film (Figure 11.5a,b). The in-plane and cross-section morphologies of the as-prepared $BaTiO_3$@TiO_2/PVDF polymer nanocomposites illustrated that no aggregation of the $BaTiO_3$@TiO_2 coaxial nanofibers occurred and confirmed the dispersing and combining with PVDF (Figure 11.5c,d) [25]. By utilizing the surface modifying technology, $BaTiO_3$ nanocrystals could be further connected with multiple composites. He and coworkers dealt with $BaTiO_3$ nanocrystals with H_2O_2 and a titanate coupling agent DN-101. The $BaTiO_3$ was successfully combined with PVDF by further treating the surface because of the formed –OH group and hydrophilic surface. The resulting $BaTiO_3$/PVDF polymer nanocomposites performed higher dielectrics permittivity. In the preparation process, the $BaTiO_3$ nanocrystals were treated in H_2O_2 for several hours and then collected by centrifugation. After purification, the $BaTiO_3$ nanocrystals were dispersed in isopropanol under sonication for 30 minutes. Next, DN-101 was added, and the resulting mixture was stirred at 70 °C for two hours. The treated $BaTiO_3$ nanocrystals were then purified and redispersed into isopropanol and centrifugated. The treated $BaTiO_3$ nanocrystals dispersed in N,N-dimethylacetamide (DMAC) were mixed with PVDF/DMAC solution with sonication and stirring to form homogeneous nanocomposite (Figure 11.5e). The nanocomposites were cast on a glass plate at 120 °C, and the solvent was removed at 80 °C for 12 hours. $BaTiO_3$/PVDF polymer nanocomposites were further treated at high temperatures to enhance the crystalline (Figure 11.5e).

With more direct surface modification, Dang and his coworkers prepared the $BaTiO_3$ (BT)/PVDF polymer nanocomposite in solution. The PVDF was successfully combined by modifying the surface with H_2O_2 to produce OH groups (Figure 11.6) [9]. The BT nanoparticles were dispersed in the aqueous solution of H_2O_2 (35%) at 106 °C for six hours. After drying, the BT nanoparticles were dispersed into DMF solution with ultrasonication for 30 minutes, then PVDF was added at 70 °C with stirring for two hours. Finally, the resulting mixture was cast on the glass plate at 80 °C in an oven to remove the solvent. During the process, the hydrophilic surface of BT nanoparticles was critical to combining with PVDF. Considering the functional application of PVDF, the method was valuable for designing and fabricating devices [9, 10, 27–29].

Carbon nanotubes (CNTs) were also considered an important functional material, and combining with a polymer to form polymer nanocomposites was widely reported (Figure 11.7) [30]. Before combining, the surface of a multiple-walled nanotube (MWNT) should be modified to achieve excellent dispersity. Firstly,

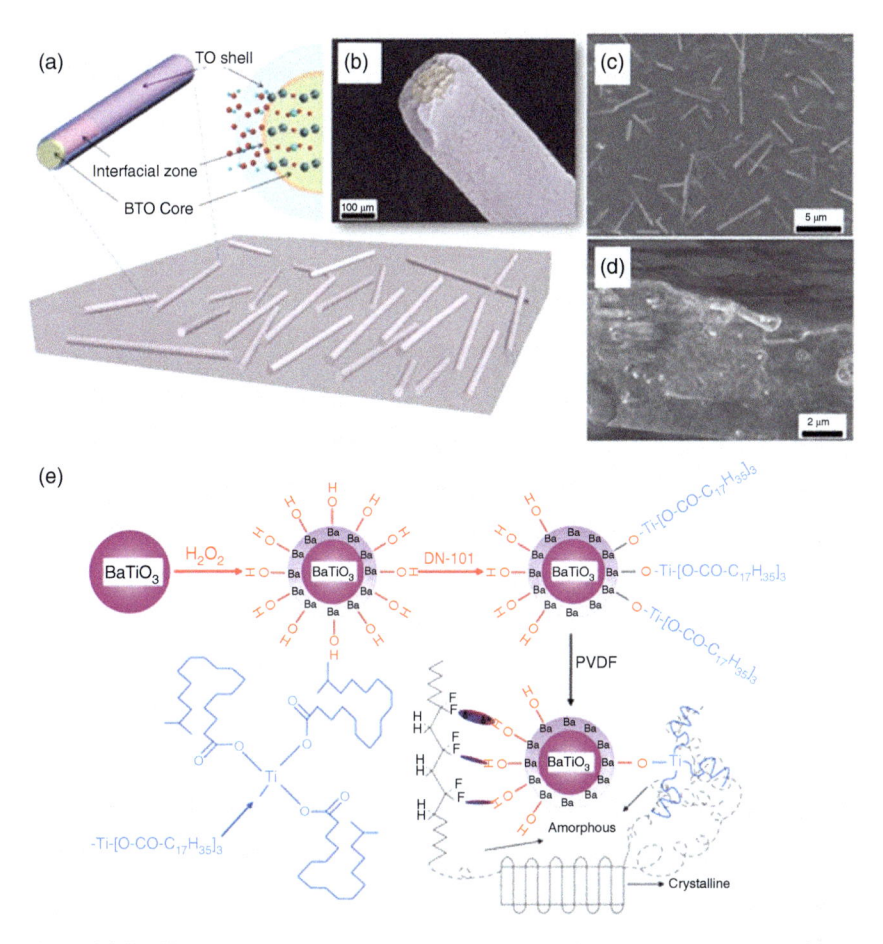

Figure 11.5 The polymer/inorganic nanoparticle composites were prepared by using solution mixing. (a) The structure schematic of PVDF/BaTiO$_3$@TiO$_2$ nanocrystals nanocomposites. (b) SEM images of the BaTiO$_3$@TiO$_2$ hetero-structure. (c,d) The in-plane (c) and cross-section (d) morphologies of the BaTiO$_3$@TiO$_2$/PVDF polymer nanocomposites. Source: Lin et al. [25], Figure 3 [p. 4]/with permission from Royal Society of Chemistry. (e) The illustration of surface modification of BaTiO$_3$ nanoparticles and combining with VDF to form BaTiO$_3$/PVDF polymer nanocomposites. Source: Gao et al. [26]. Copyright 2014 American Chemical Society.

the CNTs were dispersed in HNO$_3$ and H$_2$SO$_4$ solution, and the oxidation produced –COOH groups on the surface of the CNT. The thionyl chloride (SOCl$_2$) was added to transform the –COOH into –COCl via refluxing, and the obtained CNTs-COCl$_2$ were dispersed in N-methyl-2-pyrrolidone (NMP) via ultrasonication. Then the phthalazinone-containing diamine (DHPZDA) was added with stirring at 130 and 100 °C for 72 hours. The modified CNTs could be used for preparing multi-walls carbon nanotubes-phthalainone-containing diamine (MWNT-DHPZDA)/poly(phthalazinone ether sulfoneketone)s (PPESK) via solution mixing methods. The obtained MWNT-DHPZDA was dispersed in NMP by ultrasonication, and the PPESK was added under strong mechanical stirring. The

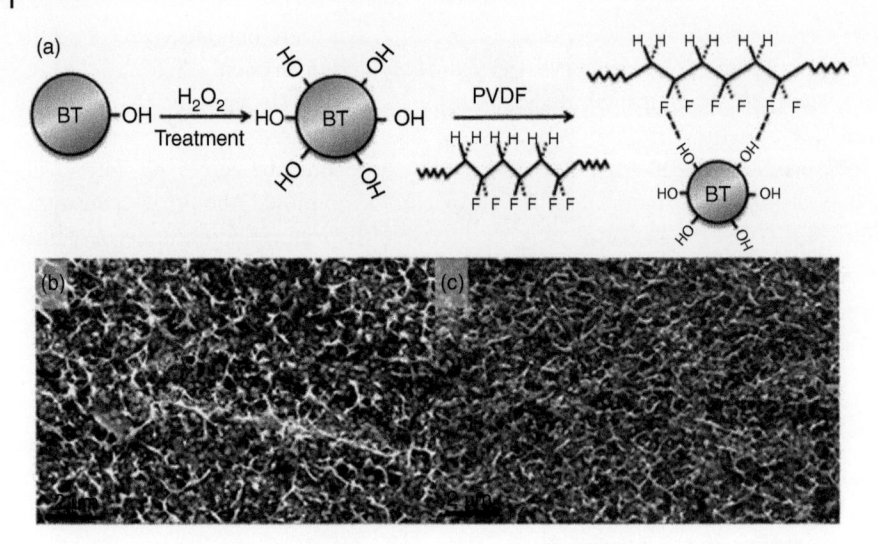

Figure 11.6 The BaTiO$_3$/PVDF composites prepared by using solution mixing: (a) The schematic of BT surface modification and combined with PVDF. (b,c) The SEM images of the fractured cross-surface of c-BT/PVDF (b) and h-BT/PVDF (c) nanocomposites with the filler concentration of 30%. Source: Zhou et al. [9]. Copyright 2011 American Chemical Society.

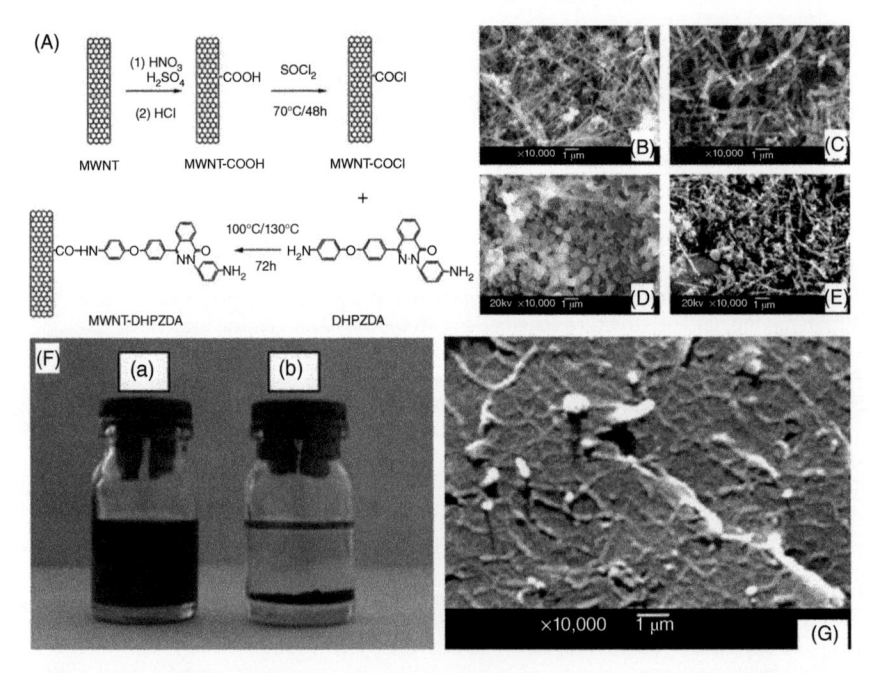

Figure 11.7 (A) The schematic of modification of multiple walled carbon nanotubes with DHPZDA. (B-E) The SEM images of carbon nanotubes during the preparation: (B) the pristine carbon nanotubes, (C) the acid treated carbon nanotubes, (D) the reaction medium and (E) the as-obtained MWNT-DHPZDA. (F) The photographs of dispersed in NMP: (a) MWNT-DHPZDA, (b) MWNT. (G) The SEM images of obtained MWNT-DHPZDA/PPESK polymer nanocomposites. Source: Feng et al. [30]. Copyright 2009 Wiley-VCH.

MWNT-DHPZDA/PPESK polymer nanocomposites were obtained by casting the resulting solution on clean glass plates and drying it in an oven.

Similarly, N-bromosucinimide (NBS), 2,2-aobissobutyrinitrile (AIBN), and poly(2,6-dimethyl-1,4-phenylene oxide) (PPO) were mixed under ultraviolet (UV) light irradiation to form a PPO-Br chain which could be combined with CNTs (Figure 11.8a). Although the surface of CNT was modified, the process was quite different. The PPO-Br was dissolved in toluene to form a solution under Ar atmosphere. Then CNT, CuBr, and 1,1,4,7,7-pentamethyldiethyltriamine (PMDETA) were added. After de-gassing, the resulting mixture was reacted at 70 °C for 24 hours, then poured into tetrahydrofuran (THF) and hot water to remove the physically adsorbed compounds. The obtained product was dried in an oven and then mixed with a solution of PPO in chloroform with ultrasonication for one hour. The polymer composites were obtained by casting on a clean glass plate and drying at ambient conditions for 24 hours. The transmission electron microscopy (TEM) images of CNT and CNT-PPO (Figure 11.8b,c) showed that the PPO coated the CNT with 4.5 nm thickness. Furthermore, by using solution mixing, the CNTs was uniformly dispersed in the PPO film, and no aggregation of CNTs was found during combination (Figure 11.8d,e) [31].

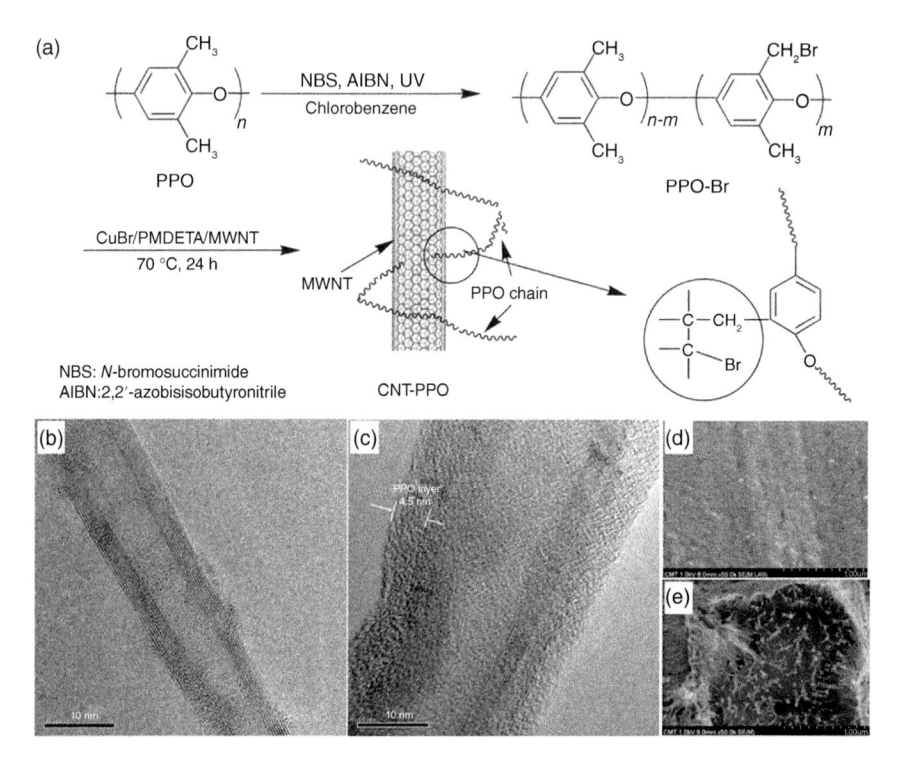

Figure 11.8 (a) The schematic of preparation CNT-PPO. The TEM images of CNT (b) and CNT-PPO polymer nanocomposites (c). (d,e) The SEM images of pristine PPO film (d) and PPO-CNT (e). Source: Liu et al. [31]. Copyright 2008 Elsevier.

Solution mixing was widely used in CNTs-based polymer nanocomposites to enhance the electronic or thermal conductivity and mechanical properties. During the preparation, the CNTs were usually modified with acid or other treatments to obtain suitable surface. Therefore, to reach good dispersity in the solvent, one of the main factors included surface modification of nanocomposites, such as inorganic nanoparticles, hetero-nanostructure, and CNTs. The solution mixing strategy also provides a general method for preparing polymer nanocomposites.

11.3 Emulsion Polymerization

Emulsion polymerization is another widely employed method for preparing polymer composites. The polymers and polymer composites prepared through this method were used in medicine, biotechnology, industry, and so on. The free radical addition polymerization can be driven in water/oil (W/O) and oil/water (O/W) systems using an aqueous solution as median. By taking advantage of the shape of the emulsion, the morphology of polymerization can be tailored, and the structure of the polymer nanocomposites can be designed and constructed. In general, the process of emulsion polymerization can be concluded in three steps: (i) nucleation, (ii) the growth of emulsoid particles, and (iii) the formation of polymer nanocrystals or composites. During the period of these steps, the compositions or the conditions such as pH, buffer solution etc., can be adjusted to obtain the designed structure. In recent years, the reversible addition-fragmentation chain transfer (RAFT) was considered a unique technology to prepare polymer nanocomposite in aqueous solutions and emulation systems [32–35].

Jan C.M. van Hest and coworkers found that polystyrene-b-poly(acrylic acid) (PS-b-PAA) prepared in polymersome pickering emulsions (PPE) was pH-responsive [36]. The PPE system was prepared by mixing with dilute HCl solution and ethyl acetate. Next, the polymers and polymer/metal-polymer nanocomposites were prepared in the PPE system. For example, the crosslinked polymersomes were prepared in the PPE system. In detail, PAA_{30}-b-$P(S_{135}$-co-$4VBA_{15})$ block copolymer was dissolved in THF, and 1,4-dioxane mixed solvent and the ultrapure water was added under stirring. Then, 4,7,10,13,16-pentaoxanonadeca-1,18-diyne, copper sulfate, sodium ascorbate, and bathophenanthrolinedisulfonic acid disodium salt hydrate were added into the suspension to obtain polymersome crosslinking.

Luzinov and coworkers reported that the poly(styrene-$block$-2-vinylpyridine-$block$-ethylene oxide) (P(S-b-2VP-b-EO)) was prepared by emulsion polymerization (Figure 11.9) [37]. In the process, the silica nanoparticles were dispersed in toluene by ultrasonication. 11-Bromoundodeciltrimethoxisilane (BTMS) was added into the particle dispersion under vigorous stirring for a certain time. After purifying the as-prepared polymer nanocomposites by centrifugation, it was redispersed in toluene. Finally, the particles were redispersed in nitromethane, and the nitromethane suspension of BTMS-functionalized particles was added in a flask containing the P(S-b-2VP-b-EO) solution in nitromethane and stirred at 60 °C for 68–72 hours [37].

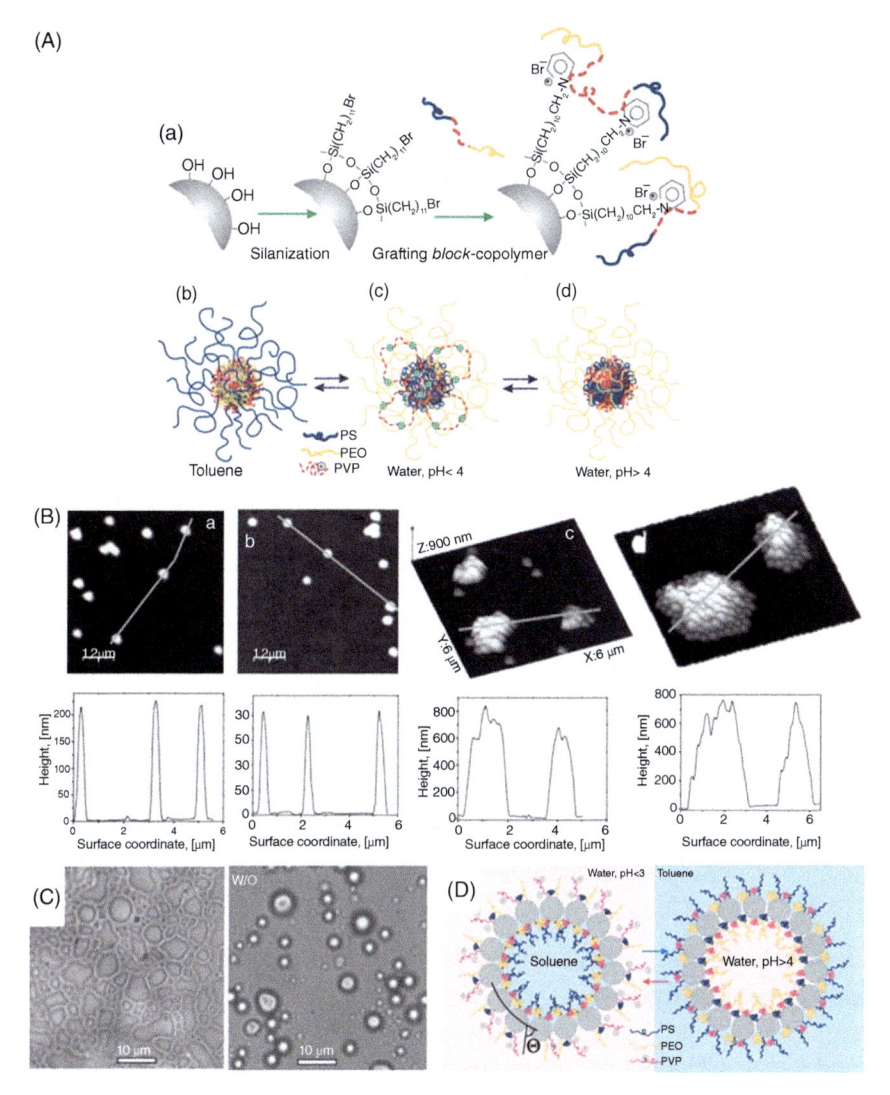

Figure 11.9 (A) The illustration of synthesizing P(S-*b*-2VP-*b*-EO). (B) The AFM images of the responsive nanoparticles. (C) Optical microscopy images of o/w and w/o emulsions stabilized by nanoparticles. (D) The Schematics of structure of o/w and w/o emulsions stabilized by smart particles. Source: Motornov et al. [37]. Copyright 2007 Wiley-VCH.

Zhu and coworkers utilized emulsion polymerization to prepare poly((diethylamino)-ethyl methacrylate)-*co*-sodium methacrylate (PDEAEMA-*co*-SMA) polymer nanocomposites which were switched to respond to CO_2 and N_2 [38]. In the emulsion polymerization process, methacrylic acid (MAA), deionized water, and NaOH aqueous solution were mixed, followed by adding the 2-(diethylamino)ethyl methacrylate (DEAEMA) and *N*,*N*′-methylene bis(acrylamide) (MBA) under mechanical stirring. The resulting mixture was heated at 70 °C then injected with the potassium persulfate (KPS) initiator solution. After five hours of polymerization,

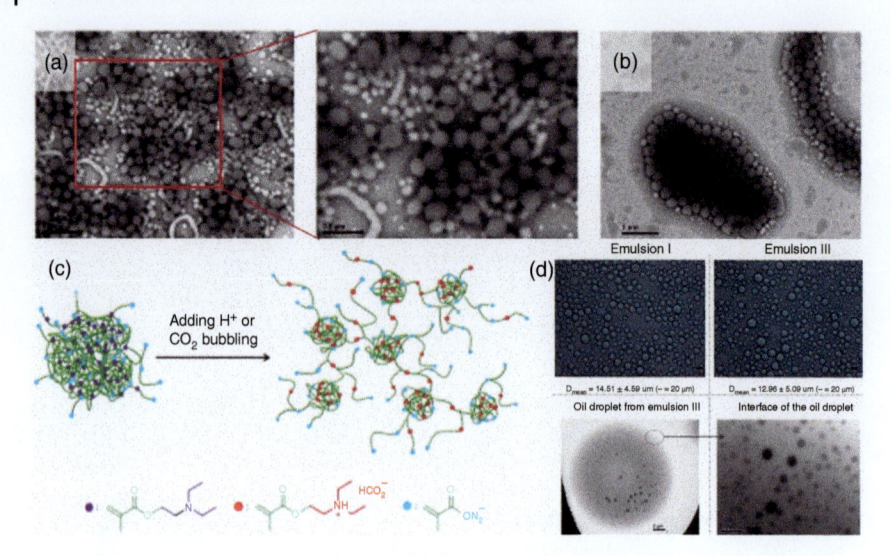

Figure 11.10 (a) The TEM images of PDEAEMA-co-SMA prepared emulsion polymerization with SMA, MBA and DEAEMA. (b) The TEM images of PDEAEMA-co-SMA after dialysis. (c) The illustration of structural change of PDEAEMA-co-SMA in the process of adding H$^+$ or CO$_2$. (d) Optical microscopy image (top) and TEM images (bottom) of Pickering emulsions before and after stabilized by PDEAEMA-co-SMA. Source: Liu et al. [38]. Copyright 2014 American Chemical Society.

the obtained PDEAEMA-co-SMA polymer composites were purified for the lyophilization of corresponding latex solutions. The TEM characterization showed the sphere shape of as-prepared PDEAEMA-co-SMA (Figure 11.10a), and the structure changed in response to CO$_2$/N$_2$ (Figure 11.10c,d).

11.4 Dispersion Polymerization and Dispersion Copolymerization

Dispersion polymerization is driven by dispersing the monomers, stabilizers, and initiators into a solution or medium to form a nucleus. Then the polymerization reaction carries out on the surface of the nuclear-resulting polymer chain or frameworks. The stabilizers absorb the formed nucleus or nanoparticles and make the reaction system more stable. Therefore, by tuning the employed amount of stabilizer, the size of the polymer nanocomposite can be controlled. In particular, dispersion copolymerization can be driven by using the monomers with macromolecules. The macromolecules stabilize the formed particles, prepare polymer composites with a larger size and promote the properties in other fields. Based on this approach, size, morphologies, and surface components of polymer nanocomposites can be controlled by copolymerization conditions.

By using irradiation, polymer nanocomposites also could be prepared in a dispersed system with radical control [39]. In the past years, several mediates and photoinitiators were explored. They were employed in the system including the Cu-mediated system in the absence of photoinitiators, Fe-mediated photo atom transfer radical polymerization (photoATRP), Ir-mediated photoATRP, Ru-mediated photoATRP, Au-mediated photoATRP, Metal-free ATRP, and so on [39].

The RAFT dispersed polymerization prepared polymer nanocomposites [23]. Tailoring the polarity of the solution was a facile path to the controllable synthesis of polymer nanocomposites. The poly(ethylene oxide) monomethyl ether methacrylate end-capped by a trithiocarbonate group (P(MMA-co-POEMA)-b-P(BzMA)) block copolymer nanoparticles could be synthesized via dispersed polymerization. Furthermore, their sizes and morphologies were synthesized by using solvent molar ratios. The nanowires with some nanoparticles were obtained using 95 : 5 of ethanol:water ratio. As the water concentration increased, the polymer nanoparticles were observed to increase. As the water concentration was up to 50%, monodispersed polymer nanoparticles were obtained. By substituting ethanol with dioxane, the nanoparticles were reduced, and the nanofibers were observed. As the dioxane:water ratio increased to 80 : 20, the diameter decreased, and a few nanoparticles were found [16].

Ni^{2+} and carboxymethyl guar gum (CMGG) polymers were mixed and reduced to form polymer nanocomposites (Figure 11.11). In the process, a $Ni(NO_3)_2$ aqueous solution, oleic acid, and methanolic solution were mixed, followed by adding CMGG. After adding sodium borohydride, the resulting mixture turned black

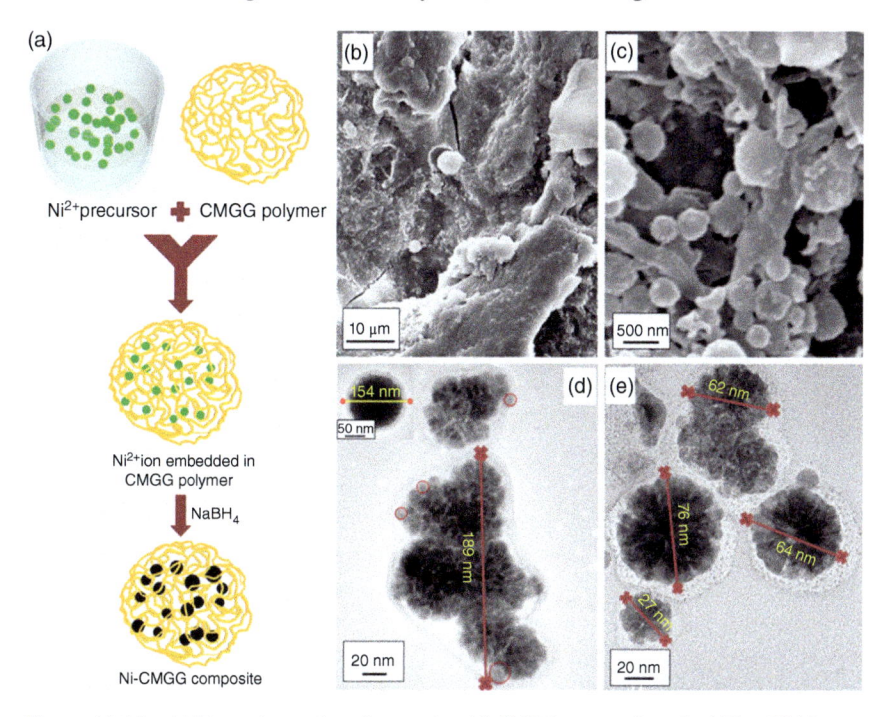

Figure 11.11 (a) The schematics of preparing Ni-CMGG composites. (b,c) The SEM image of as-prepared Ni-CMGG composites. (d,e) The TEM images of as-prepared Ni-CMGG composites. Source: Sardar et al. [40]. Copyright 2017 Elsevier.

due to the Ni^{2+} reducing to Ni^0 (Figure 11.11a). The polymer composites could be tailored with the repeated operation. Figure 11.11b–d showed the Ni-CMGG polymer nanocomposites with different sizes and the single Ni nanoparticle presented in the CMGG without aggregation [40]. By using different polymers as coordinated ligands, cations could be used to produce the polymer/metal nanocrystal nanocomposites. Besides using cations, inorganic nanoparticles with modified surfaces could be substituted with cation to prepare polymer nanocomposites. Zheng and coworkers reported that the reduced graphene oxide (rGO) reacted with polymers and studied their π–π interaction (Figure 11.12) [41]. The polymer was

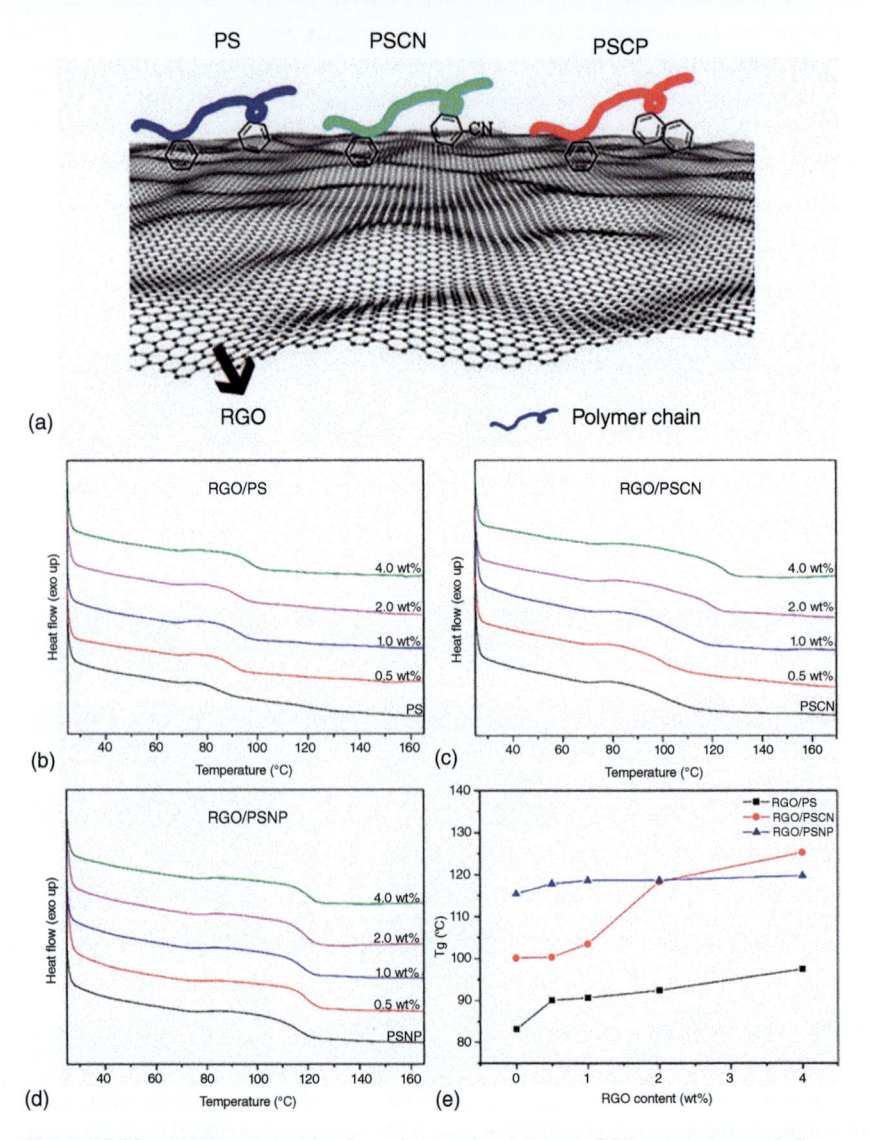

Figure 11.12 (a) The schematic of π–π interaction between RGO and polymer chains. (b–d) The thermal properties of the polymer nanocomposites. Source: Zhao et al. [41]. Copyright 2018 MDPI.

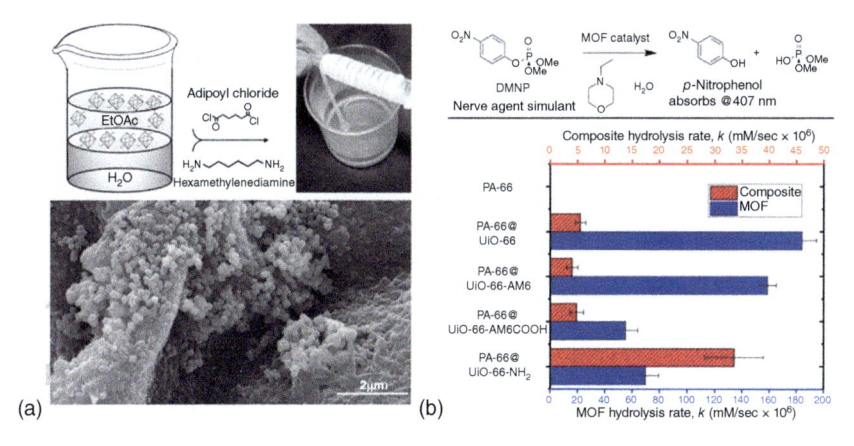

Figure 11.13 (a) The schematic for preparing the PA-66@UiO-66-NH$_2$. (b) The catalysis performance on the DMNP degradation reaction and the corresponding rate. Source: Kalaj et al. [47]. Copyright 2019 Wiley-VCH.

synthesized before the combination of rGO and copolymers. The rGO was added to chloroform and dispersed under sonication, followed by adding the copolymer in chloroform. The mixture was then sonicated to form polymer nanocomposites. Due to the existence of the π–π interaction between RGO and copolymers, the copolymer would adhere to the rGO to form the polymer nanocomposites. As presented in Figure 11.12b–d, after combining rGO and copolymers, the thermal properties were enhanced depending on the amount of rGO. The study on the π–π interaction between RGO and copolymers also revealed the importance of the interactions.

Metal–organic frameworks (MOFs) were considered an important material in inorganic or coordination chemistry. Their properties and derivatives were widely applied in Li-battery, gas storage, electrocatalysis, and so on [42–46]. By using the dispersing method, MOFs could be combined with a polymer to form polymer/MOFs nanocomposites. Figure 11.13a showed the process and the SEM images of PA-66@UiO-66-NH$_2$, which could be applied to the catalysis performance on the DMNP degradation reaction (Figure 11.13) [47].

Dispersion polymerization and dispersion copolymerization were the facile methods for combined polymer compositions or polymer/inorganic compounds. The polymerization process would be tailored by many parameters and conditions in the aqueous or organic solution. Hence, using suitable conditions, dispersion polymerization and dispersion copolymerization were proven effective paths to prepare polymer nanocomposites.

11.5 Self-Assembly

Self-assembly is also an effective path to synthesize polymer/polymer composites. Self-assembling polymer nanocomposite employs the interactions among molecules to form the order structure. The interactions are hydrogen bonding, electrostatic attraction, π–π interaction, and so on. By using these methods, many polymer composites were reported with various morphologies, such as nanorods, nanowires, nanorings, nanotubes, core–shell structures with multiple shells, nanoflowers, multiple-layers membranes, and so on. The polymer nanocomposites with multiple composites and structures provide new applications in medicine, drug-controlled releasing, targeted drug delivery, catalysis, and so on.

Wang and coworkers used a puzzle approach to prepare the organic nanocrystals and construct the topologic hetero-structure using self-assembly method [48]. Wang

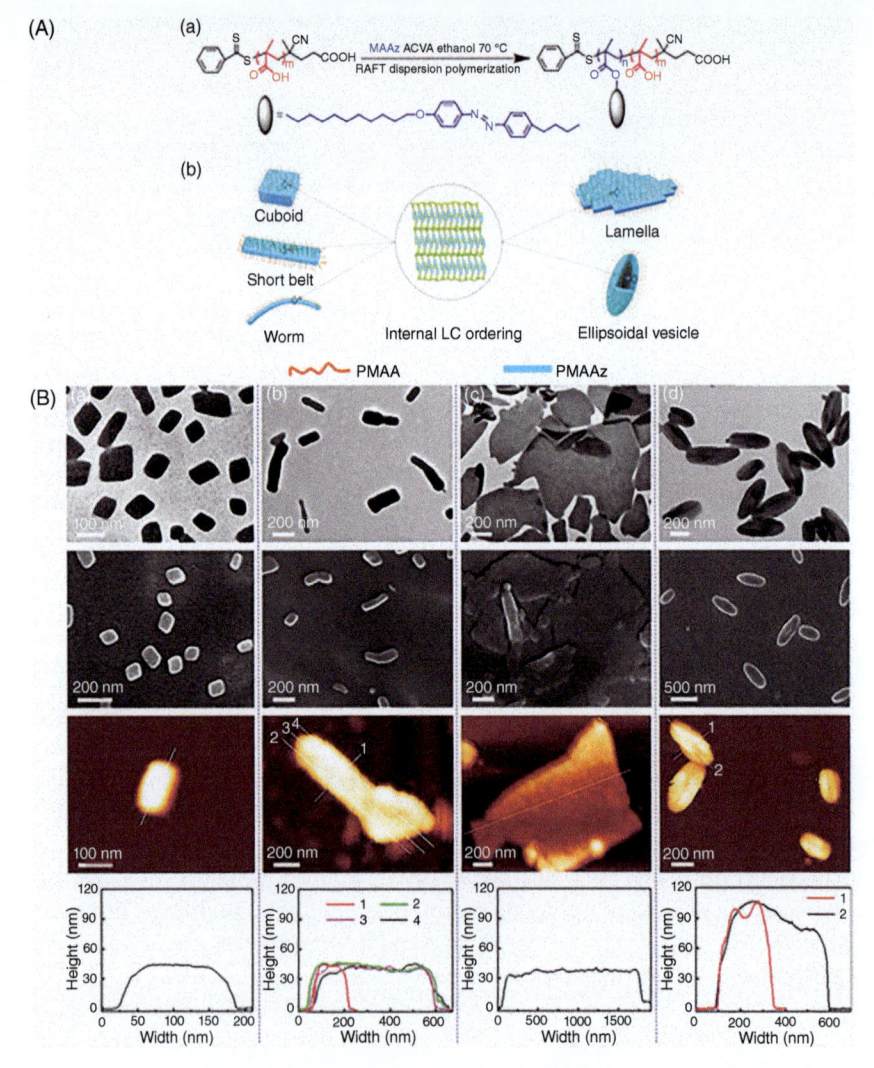

Figure 11.14 (a) Scheme of preparing the PMAA-b-PMAAz BCP NPs via RAFT dispersion polymerization in ethanol and the tailoring for the constructing of anisotropic morphologies. (B) The TEM, SEM, AFM images, and the corresponding height profiles of as-prepared PMAA$_{112}$-b-PMAAz$_n$ NPs with different DPPMAAz$_n$. (a) PMAA$_{112}$-b-PMAAz$_{66}$, (b) PMAA$_{112}$-b-PMAAz$_{89}$, (c) PMAA$_{112}$-b-PMAAz$_{115}$, (d) PMAA$_{112}$-b-PMAAz$_{142}$. Source: Guan et al. [49]. Copyright 2018 American Chemical Society.

and his coworkers reported benzo[ghi]perylene (BGP) grow to form a 2D microplate, and the 1,2,4,5-tetracyanobenzene (TCNB), tetrafluoroterephthalonitrile (TFP), and octafluoronaphthalene (OFN) grow to form 1D microwires. Combining the lattices of these organic compounds, the self-assembly method was employed for the crystal puzzle, forming the polymer nanocomposites.

Polymerization based on self-assembly was also an important technique for preparing polymer nanocomposites. Nowadays, RAFT emulation polymerization is a valuable self-assembly technique for synthesizing polymer nanocomposites with different morphologies and sizes [14]. In particular, PISA was a technique for producing polymeric nanomaterials.

By using the self-assembly method, it is possible to construct anisotropic morphologies. For instance, 4-cyanopentanoic acid dithiobenzoate (CPADB) was taken for preparing the poly(methylacrylic acid) (PMAA) macromolecular chain then to PMAA-b-PMAAz$_n$ nanoparticles via PISA (Figure 11.14) [49]. The corresponding TEM, SEM, and atomic force microscopy (AFM) images of PMAA$_{112}$-b-PMAAz$_n$ with different DP$_n$ of PMAAz show that the resulting morphologies of polymer composites depended on the DP$_n$ of PMAAz$_n$ (Figure 11.14). Using the PMAAz with larger DP$_n$, the anisotropic growth became more apparent. Although the PMAA-b-PMAAz$_n$ grew along three axes, the thickness changes slowly while the long and short axis grows quickly.

Self-assembly was also widely used in inorganic nanocrystals for device preparation. For example, polymer nanocomposite could be synthesized with a special pattern for device fabrication using lattice matching by combining the polymer self-assembly. Compared to other polymerizations, self-assembly for polymer nanocomposites provide a new way to fabricate devices with designed patterns.

11.6 Melting

Polymers were melted and coated on the inorganic nanoparticles to form polymers/inorganic NP composites by making some inorganic nanoparticles as a template or substance. Most polymers, such as polyethylene (PE), polypropylene (PP), polyamides, polyesters, polyurethane, and polystyrene, were also employed to prepare polymer nanocomposites [28, 50–53]. The inorganic nanocrystals acted as functional compositions as polymer supports performed important roles in the device preparation [54].

Liu and coworkers used cellulose nanocrystals, poly((butylene adipate-co-terephthalate) (PBAT), and SiO$_2$ nanocrystals to produce polymer nanocomposites [55]. PBAT was blended with different amounts of modified SiO$_2$ nanoparticles by stirring at 100–170 °C following a temperature profile (Figure 11.15). The obtained polymer composites were used to produce cellulose nanofibers and prepare polymer nanocomposites. Figure 11.15b showed that the TEM images of the PBAT/SiO$_2$-EO/CNF-NH$_2$ and the SiO$_2$ nanoparticles were uniformly dispersed in the polymer compositions. The extensive bonding between CNF-NH$_2$ and SiO$_2$-EO nanocrystals resulted in the stable structure of formed polymer nanocomposites.

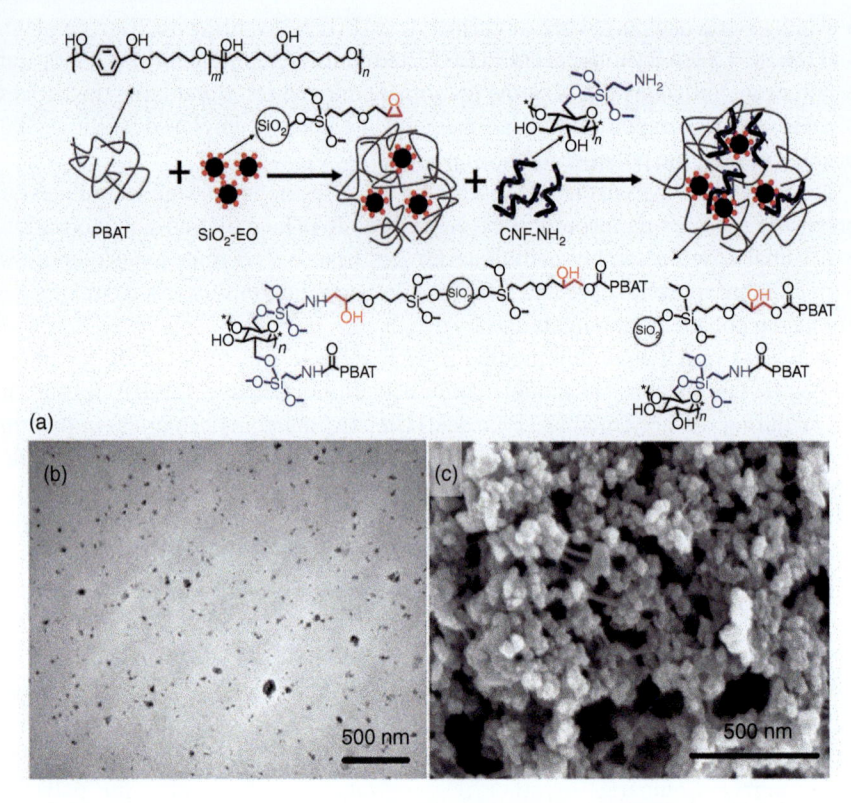

Figure 11.15 (a) The schematic of preparation of PBAT/SiO$_2$-EO/CNF-NH$_2$. (b) The TEM images of PBAT/SiO$_2$-EO/CNF-NH$_2$. (c) The SEM images of as-prepared PBAT/SiO$_2$-EO/CNF-NH$_2$. Source: Lai et al. [55]. Copyright 2020 Elsevier.

For functional materials, thermoplastic vulcanizates (TPV) were necessary for the industry due to their biodegradability, renewability, and other excellent properties. Using the melting method, cellulose nanocrystals could enhance mechanical and shape memory properties. The poly(lactic acid) (PLA) was melted at 155 °C with rotor speed in the process. Then the epoxidized natural rubber (ENR) and cellulose nanocrystals were added and blended with PLA melt for another three minutes. After adding dicumyl peroxide (DCP) as a curing agent, the reaction continued for four minutes. Finally, the surface of the as-obtained TPV could be tailored by etching with dichloromethane. The cellulose nanocrystals were found affecting their properties [54].

Based on the function of multiple-walled carbon nanotubes (MWCNTs), polymer/CNTs polymer nanocomposites were expected to fabricate a device with excellent capacity. There were many methods for synthesizing polymer/MWCNTs composites, including solution mixing, emulation polymerization, melting, *in situ* polymerization, and so on [56]. Wang and coworkers used the poly(L-lactide) (PLLA) to combine with MWCNTs. PLLA and MWCNTs with different loadings were melt-mixed in a rotational mixer at 170 °C to obtain the L-PLANT master batches. H-PLLA granules and the as-prepared L-PLANT master bathes

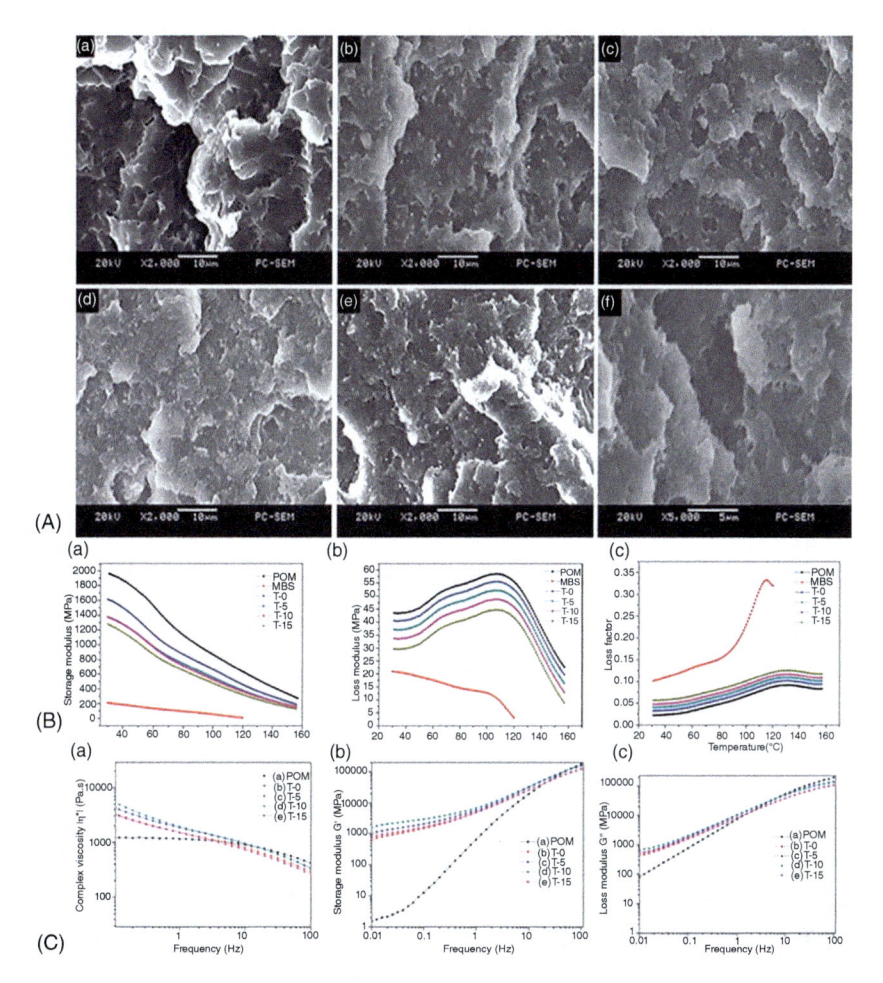

Figure 11.16 (A) The SEM images of the impact-fractured surface of the obtained POM/MBS/TPU blends samples; (B) the storage modulus (a), loss modulus (b), and loss factor (c) vs. temperature curve of various POM/MBS/TPU blends; (C) time-temperature superposition of complex viscosity (a), G' (b), and G'' (c) of POM, POM/MBS blends and POM/MBS/TPU blends. Source: Yang et al. [57]. Copyright 2019 Wiley-VCH.

were melt-mixed in the rotational mixer at 140 °C to obtain L-PLANT/ H-PLLA. Finally, the targeted PLLA/MWCNT polymer nanocomposites were obtained via pressing [50]. The electrical conductivity and electromagnetic interference shielding were tailored using different feed amounts (Figure 11.16b,c).

Different polymer compositions could also prepare polymer nanocomposites via melting because of the low melting points or other factors. For example, Wang and his co-authors applied the melting method to prepare methacrylate–butadiene–styrene (MBS) to enhance the materials' toughness [57]. During the preparation process, different weight ratios were employed to prepare ternary blends using a co-rotating twin-screw extruder under 110–118 °C (Figure 11.16). As a result, the enhanced toughness of the polymer composites was apparent, and this path could be used to prepare the various porous multifunctional nanocomposites.

In summary, the melting method was a traditional path for preparing polymer composites, and it was changed to apply in the nanosize.

11.7 *In situ* Polymerization

Besides the melting methods, polymer nanocomposites could also be prepared via *in situ* polymerization to enhance and strengthen toughness. Graphene and its derivatives combined with polymers via *in situ* polymerization would enhance the toughness of the polymer [58]. Nowadays, new polymer nanocomposites were synthesized for toughness materials, thermoelectric materials, dielectric materials, electromagnetic absorption materials, and so on [58–65].

In situ polymerization is a high technology for surface and interface control of polymer nanocomposites [66–69]. Figure 11.17 showed that PVA *in situ* grew on the surface of cellulose crystal reinforcing nanofibers to produce coaxial nanocomposite fibers. The aqueous PVA and cellulose crystal solution were mixed *in situ* with mechanical stirring at 80 °C for 12 minutes.

Polymerization was quenched by cooling. The cellulose crystal was dispersed in the PVA aqueous solution during the process. The concentration of PVA would be important in controlling the polymerization.

For reinforcing the nanocomposite, graphene was usually applied as a composition to form polymer nanocomposites [58, 68, 71, 72]. By using *in situ* polymerization, graphene surfaces were usually modified prior to synthesizing

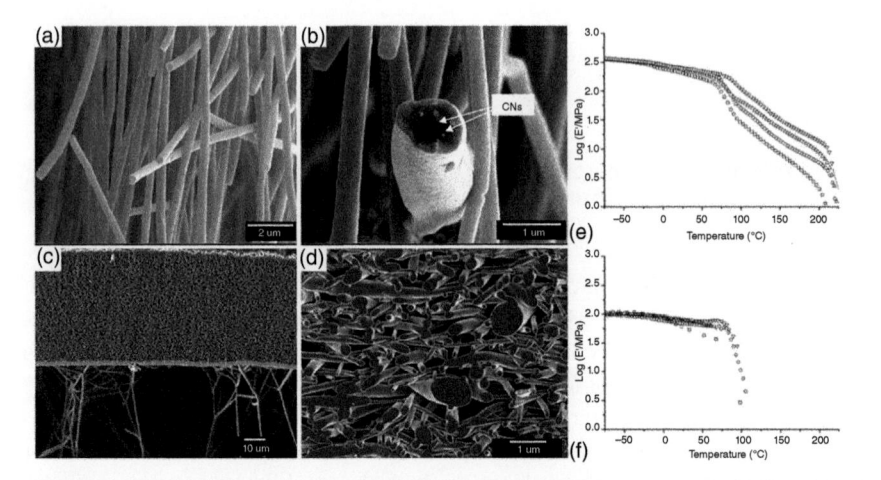

Figure 11.17 (a–d) the SEM images of PVA/cellulose polymer nanocomposites: (a,b) The Cryo-SEM images and (c,d) high resolution field-emission SEM images. (e,f) Storage modulus of electrospun nanofiber mats vs temperature for fully hydrolyzed PVA-98 (e) and partially hydrolyzed PVA-88 (f) with different loadings of cellulose nanocrystals. Source: Peresin et al. [70]. Copyright 2010 American Chemical Society.

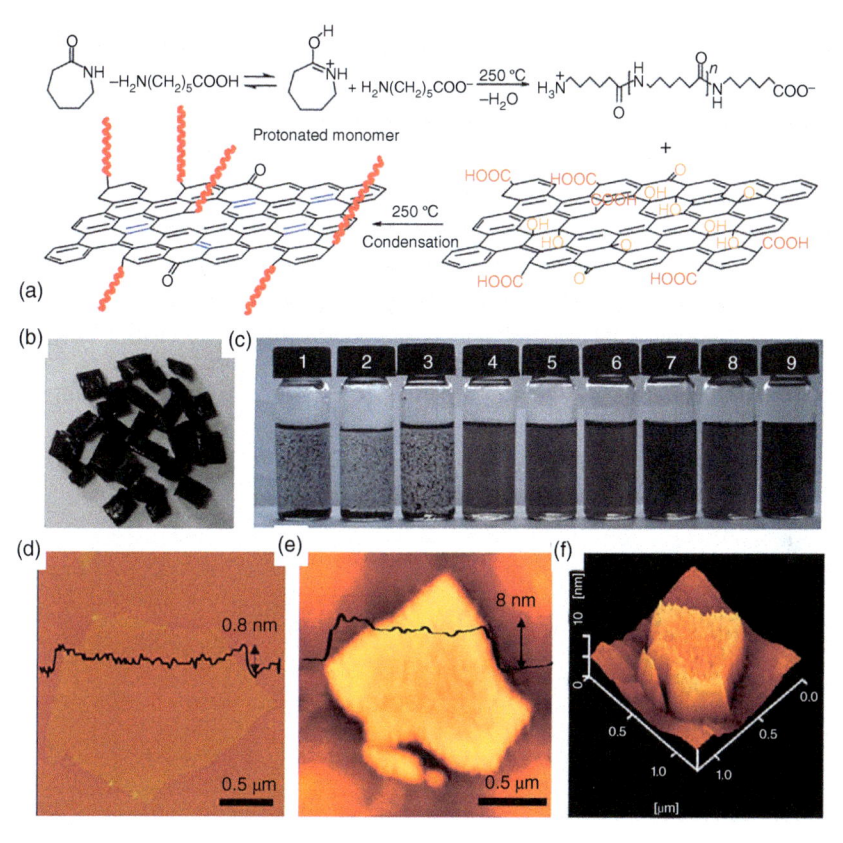

Figure 11.18 (a) The schematic of in-situ ring-open polymerization for combining with graphene to form polymer nanocomposites. (b) The photograph of Nylon-6-graphene. (c) the reduced graphene oxide (1–3), nylon-6 (4–6) and polymer nanocomposites (7–9) dispersed in sulfuric acid and formic acid and *m*-cresol respectively. (d) The AFM image of graphene and (e,f) nylon/graphene polymer nanocomposites. Source: Xu and Gao [66]. Copyright 2010 American Chemical Society.

polymer nanocomposites. Graphene oxides could also be utilized due to the function group reacting with the monomers. Figure 11.18 showed the Nylon-6-graphene polymer nanocomposite by *in situ* ring-open polymerization to combine with graphene via reaction with –COOH or –OH [66]. For synthesizing the polymer nanocomposites, graphene oxide and caprolactam were mixed with sonication at 80 °C for two hours to form the homogenous solution. After adding aminocaproic acid, the polymerization was done at 180 °C for one hour and at 250 °C for nine hours with stirring. The raw polymer nanocomposites were purified by washing them in boiled water for five hours three times. The *in situ* polymerization process was closed to that of the preparation of PVA/cellulose polymer nanocomposites, which illustrated that the substrate was important to trap the monomer. Then the polymerizations were initiated in some conditions. Besides carbon-based nanomaterials, other kinds of materials, such as MOFs also could be used for preparing polymer nanocomposites [47].

For multiple compositions of nanocrystals, *in situ* polymerization can also be applied. The BiOBr nanoplates were prepared via hydrothermal methods, and the Ag nanoparticles were loaded by photoreduction. By taking BiOBr/Ag hetero-structure as a template, PPy was *in situ* polymerized by oxidation of pyrrole. The as-prepared BiOBr/Ag nanocrystals were dispersed in pyrrole aqueous monomer solution with sonication. $FeCl_3$ aqueous solution was employed for initiating oxidative polymerization under stirring for 12 hours at room temperature (Figure 11.19a) [73]. The raw BiOBr/Ag/PPy nanocomposites were purified by washing with deionized water repeatedly and dried. The TEM images of the as-prepared samples showed the PPy layer growing on the surface of the nanocrystals, and the corresponding HRTEM images showed the Ag nanoparticles' lattice and the BiOBr nanoplates (Figure 11.19b,c). The clear interface of nanocrystals and PPy illustrated the application of *in situ* polymerization method for preparing polymer nanocomposites.

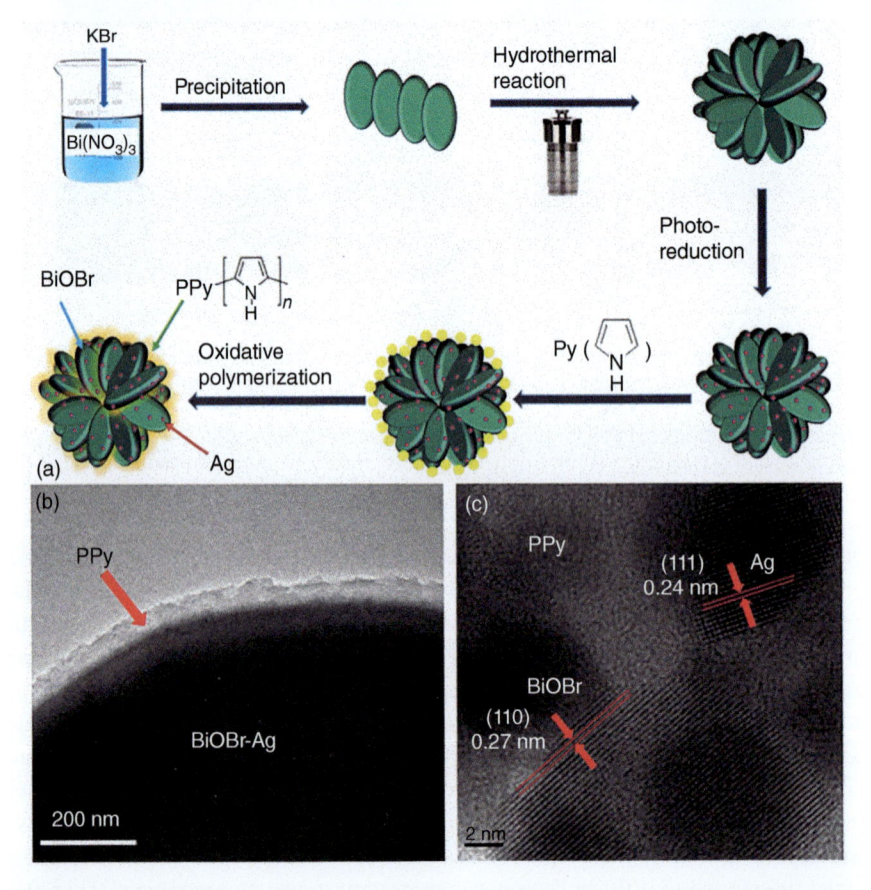

Figure 11.19 (a) The schematic of preparation of indirect Z-scheme BiOBr/Ag/PPy system and (b,c) the according to TEM images (b) and HRTEM images (c). Source: Liu and Li [73]. Copyright 2018 Elsevier.

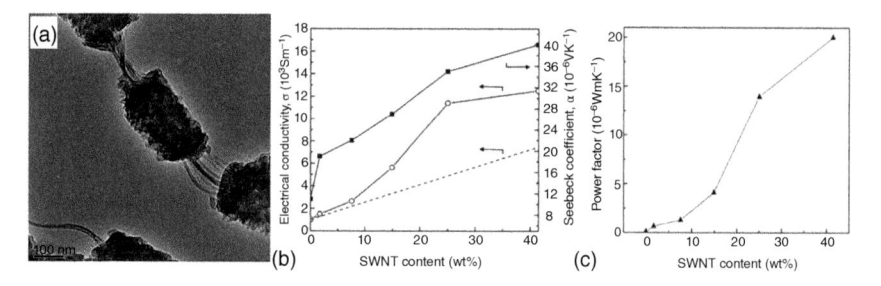

Figure 11.20 (a) TEM images of SWNT/PANI composites. (b) The Seebeck coefficient (closed squares) and electrical conductivity (open circles) of SWNT/PANI composites with different SWNT content. (c) The power factor (closed triangle) of SWNT/PANI content. Source: Yao et al. [75]. Copyright 2010 American Chemistry Society.

Single walls carbon nanotubes (SWCNTs) coated with polyaniline (PANI) were prepared by *in situ* polymerization, and the resulting SWCNTs/PANI polymer nanocomposite performed a higher thermoelectric performance [53, 74]. SWCNTs were dispersed in the HCl solution for surface modifications, and then the aniline monomer was initiated by the ammonium peroxidisulfate (APS) to form the polymer nanocomposites. As shown in Figure 11.20, as the content of SWCNTs went up to 25%, the Seebeck coefficient and the electrical conductivity were not promoted. In other words, the combination of SWCNTs and PANI would reduce the number of SWCNTs in the thermoelectric devices, and it would be a possible way to reduce the cost of fabricated devices [75].

The polymers/metal nanoparticles could also be synthesized by anchoring the catch metal cations with polymer via coordinating reaction. The metal cations were *in situ* reduced into metal nanoparticles to form the polymer/metal nanocomposites. Before catching metal cations, the polymers could be modified or copolymerized with other chain fragments. For the poly high internal phase emulsions (PHIPE)-poly glycidyl methacrylate (PGMA)-triethylenetetramine (TETA)/Au polymer nanocomposites preparation, the copolymers were synthesized before adsorbing and reducing the Au^{3+} (Figure 11.21). Emulation polymerization provides a way to generate pores, and the *in situ* reductions ensure maintaining pores. In the preparation process, an aqueous solution of potassium peroxodisulfate ($K_2S_2O_8$) was added into the mixture of styrene, divinylbenzene, 2-acryloxyethyl-2′-bromoisobutyrate, and Span 80 surfactant to form PHIPE-Br in the emulsion. After purification by washing and drying, the PHIPE-Br was mixed with GMA, THF, CuBr, and PMDETA. The mixture was stirred at 65 °C in Ar atmosphere to get PHIPE-PGMA. The residue catalysts and monomers were removed by extraction. The obtained PHIPE-PGMA was merged into the TETA solution in the presence of Na_2CO_3 solution, and the reaction was carried out at 70 °C for two hours under stirring. The monoliths were obtained by separating from solution and drying. The PHIPE-PGMA-TETA/Au nanoparticles were obtained by merging the as-prepared PHIPE-PGMA-TETA into $HAuCl_4$ solution and reducing

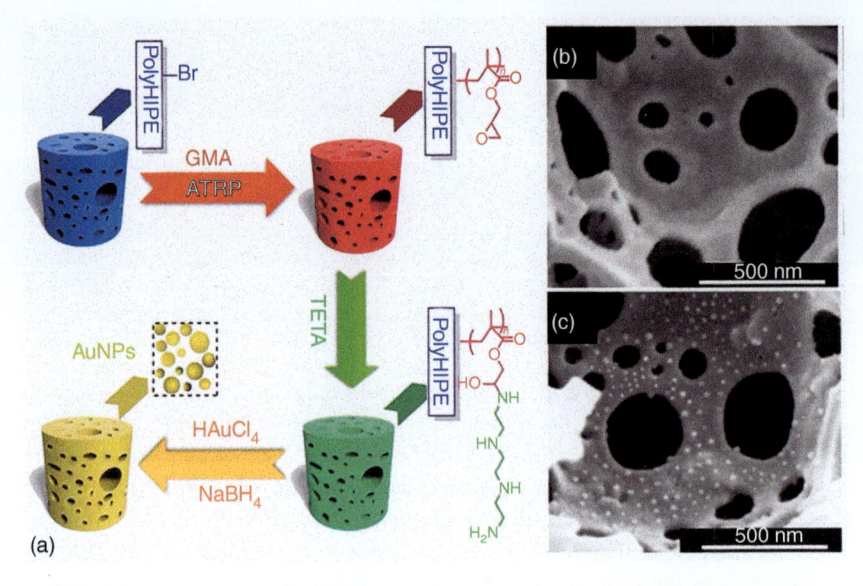

Figure 11.21 (a) The schematic of *in-situ* polymerization for preparing the PHIPE-PGMA-TETA/Au and the SEM images of (b) PHIPE-Br and (c) PHIPE-PGMA-TETA/Au NPs. Source: Yuan et al. [76]. Copyright 2018 Royal Society of Chemistry.

the Au^{3+} by $NaBH_4$. After obtaining PHIPE, other composites were combined via *in situ* polymerization and reduction. Therefore, the morphologies and pores remained (Figure 11.21b,c).

Compared with the supported polymer of nanocomposites, some inorganic materials could also be used as support. The functional group of polymers could be used to trap cation in solution, which was expected to apply to water treatment. Poly(acrylic acid) (PAA) *in situ* grew on the mesoporous alumina and then used in Pb(II) adsorption for environmental studies [63]. For the preparation of PAA/alumina nanocomposites, a PAA solution was typically loaded on the alumina and repeatedly rinsed several times, followed by drying overnight. After adding alkaline solution, the *in situ* polymerization was carried out at 65 °C for 12 hours at N_2 atmosphere and quenched with ice water. Such samples were purified with water and dried at 60 °C overnight [63].

The polymer/polymer nanocomposites were also prepared by using *in situ* methods. For example, PVDF and PANI were combined for strain sensors [11]. First, the PVDF membrane was prepared via electrospinning with patterns and then immersed in the ammonium persulfate, aniline, and sulfosalicylic acid mixed solution at room temperature for 12 hours. The polymerization was monitored by the color-changing and purified by washing with deionized water and drying at room temperature. Due to the *in situ* polymerization path, the PVDF membrane could be designed with patterns resulting in the unique application of the polymer nanocomposites sensors.

The *in situ* polymerization of polymer nanocomposites was listed above. The main objectives of this method were inducing the monomer to the surface

11.8 Tailoring of Polymers Nanocomposite

In general, the polymerization systems and the reaction conditions were the factors to control or tailor polymer nanocomposites. In the reported studies, the polymer nanocomposites presented that controlling the growth of polymers or polymerization enabled tailoring of the nanocomposites, such as pH, temperature, polarity of system, and so on [77–79]. In the aqueous solution, the ionized polymers were affected by pH. Consequently, the ionization and deionization by pH could be utilized to tailor the porosity, thickness, and so on.

The polymer nanocomposites with different structures were obtained by tailoring the nanoparticles' concentration and the solvent's polarity. Park and coworkers found that the different solvents could result in the self-assembly of the block polymer PAA-b-PS and iron oxides [19]. Figure 11.22a showed the schematic illustration of the formation of PAA-b-PS/iron oxides polymer nanocomposites with different structures. According to their reports, in the DMF/THF (96.8 : 3.2) system, magneto-core shells were obtained, and the iron oxides, while using THF solvent, only magneto-micelles structures were generated. When the DMF was substituted with dioxane, the hollow structure was obtained. Figure 11.2b showed the corresponding STEM images of the three structures (a–c) in the different systems, and the Fe intensity line further confirmed the structures. In addition, the mass presence of iron oxide nanoparticles also affected their structures (Figure 11.22c). Figure 11.22 showed that the employment of 10% of nanoparticles resulted in magneto-core shells while using a large presence of nanoparticles (35.8%) generated magneto-micelles structure. The polarity of the solvent and the feed concentration were the conditions and tailored factors of polymerization. Utilizing these variations was an effective way to control the structure or morphologies of polymer nanocomposites.

The polymer monomers ratio tailoring was also an effective path for synthesizing polymer nanocomposites. In aqueous solution, poly-[2-(methacryloyloxy)ethyl phosphorylcholine]-b-poly[2-(dimethylamino)ethyl methacrylate]-b-poly(2-hydroxypropylmethacrylate) (PMPC-b-PDMA-b-PHPMA$_z$) were prepared via dispersion polymerization. The polymerization under the various solid concentrations (Table 11.1) was carried out. The results showed that the sizes and morphologies of the PMPC-b-PDMA-b-PHPMA$_z$ depended on the feed amounts and DP$_n$ (Figure 11.23) [80]. The entry 1–3 in Table 11.1 found nanospheres, and the morphologies varied from worms to jellyfish, worms/vesicles, and vesicles (Table 11.1 and Figure 11.23d) [80].

The factors, such as reaction time, the polarity of solvents, DP$_n$, pH, emulsion system, and so on, affecting polymerization could also be used to tailor polymer nanocomposites. However, the complexities of polymerization also increased the difficulties. Nevertheless, the tailoring of nanocomposites was expected to

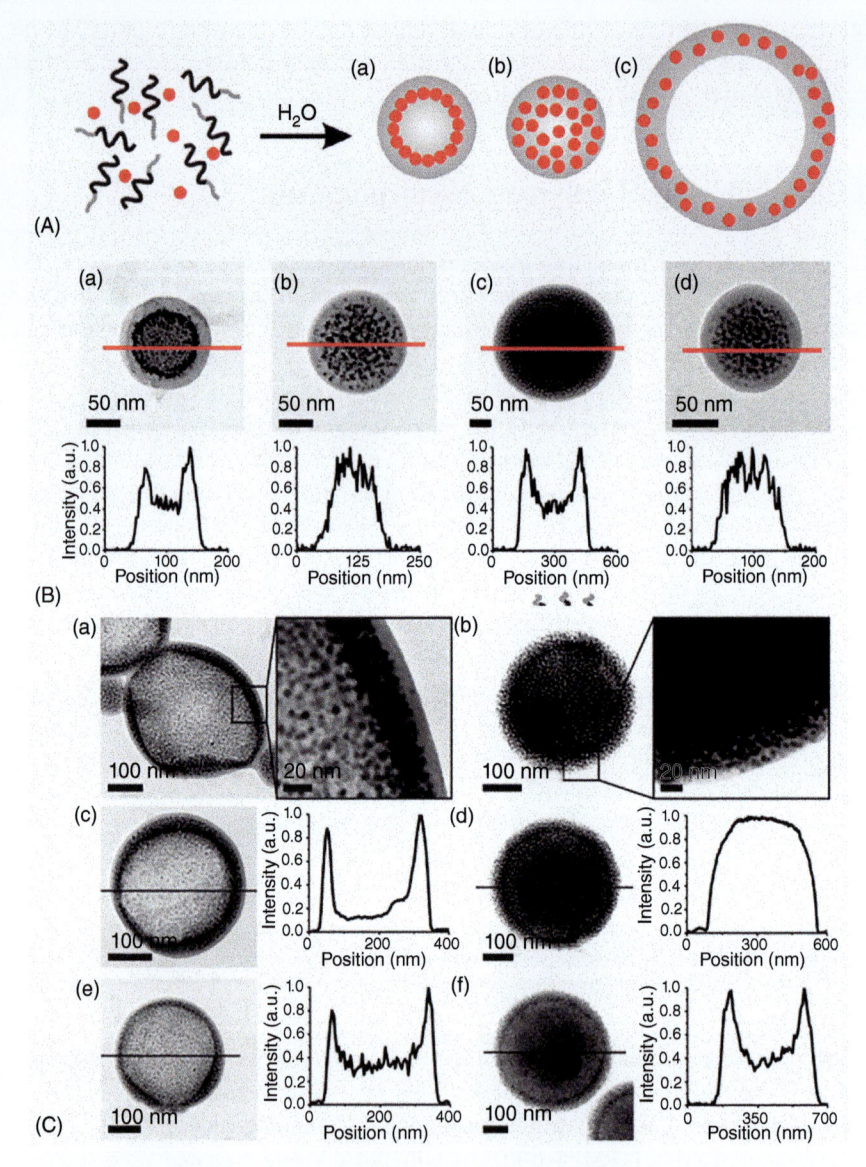

Figure 11.22 (A) The Schematic illustration for the controlling iron oxides nanoparticles PAA-b-PS core/shell structure. (B) STEM images of the iron oxide nanoparticles/PAA-b-PS magneto-core shell and the corresponding Fe intensity line scans. (C) The structural characterization of magneto-poly mersomes with different nanoparticle mass percents. (a) Nanoparticle mass percent of 10.0% and (b) nanoparticle mass percent of 35.8%. (c, d) EELS relative height profiles for assemblies formed at a (c) Nanoparticle mass percent of 10.0% and at a (d) Nanoparticle mass percent of 35.8%. (e, f) EDS Fe intensity profiles for assemblies formed at a (e) Nanoparticle mass percent of 10.0% and at a (f) Nanoparticle mass percent of 35.8%. Source: Hickey et al. [19]. Copyright 2011 American Chemical Society.

Table 11.1 The condition for preparing of PMPC-b-PDMA-bPHPMA$_z$ by RAFT aqueous dispersion polymerization.

Entry	Polymer structure[b]	Targeted DP of HPMA[c]	Solids[d] (w/w %)	M_n[e] (kg mol^{-1})	M_w/M_n[e]	D_h[f] (nm)	PDI[f]	Morphology[g]
1	PMPC$_{25}$-b-PDMA$_4$-b-PHPMA$_{100}$	100	25	2.07	1.08	25	0.06	Spheres
2	PMPC$_{25}$-b-PDMA$_4$-b-PHPMA$_{152}$	150	25	3.10	1.08	33	0.13	Spheres
3	PMPC$_{25}$-b-PDMA$_4$-b-PHPMA$_{201}$	200	25	3.97	1.09	57	0.28	Spheres
4	PMPC$_{25}$-b-PDMA$_4$-b-PHPMA$_{251}$	250	25	4.59	1.09	309	0.28	Worms
5	PMPC$_{25}$-b-PDMA$_4$-b-PHPMA$_{300}$	300	25	5.27	1.12	589	0.30	Worms
6	PMPC$_{25}$-b-PDMA$_4$-b-PHPMA$_{328}$	325	25	5.50	1.15	753	0.29	Jellyfish
7	PMPC$_{25}$-b-PDMA$_4$-b-PHPMA$_{350}$	350	25	5.85	1.15	359	0.20	Worms/vesicles
8	PMPC$_{25}$-b-PDMA$_4$-b-PHPMA$_{400}$	400	25	6.08	1.15	117	0.08	Vesicles
9	PMPC$_{25}$-b-PDMA$_4$-b-PHPMA$_{400}$	400	20	6.25	1.17	268	0.26	Worms
10	PMPC$_{25}$-b-PDMA$_4$-b-PHPMA$_{400}$	400	15	6.00	1.20	51	0.08	Spheres

a) [PMPC$_{25}$-b-PDMA$_4$]$_0$/[V-501]$_0$/[HPMA]$_0$ molar ratio = 1/0.4/100–400, total solids concentration = 25.0% (w/w).
b) The segment DP in the formula was determined by ^{1}H NMR spectroscopy, based on PMPC$_{25}$-b-PDMA$_4$ macro-CTA.
c) [HPMA]$_0$/[PMPC$_{25}$-b-PDMA$_4$]$_0$.
d) 100 × [PMPC^{25}-b-PDMA$_4$ (g) + HPMA (g)]/[all reaction mixtures (g)].
e) Determined by GPC [poly(methyl methacrylate) (PMMA) standards, 3 : 1 CHCl$_3$/methanol eluent with 2 mM LiBr].
f) Determined by DLS measurement at 25 °C.
g) By DM-AFM.
Source: Sugihara et al. [80] Copyright 2020 American Chemical Society.

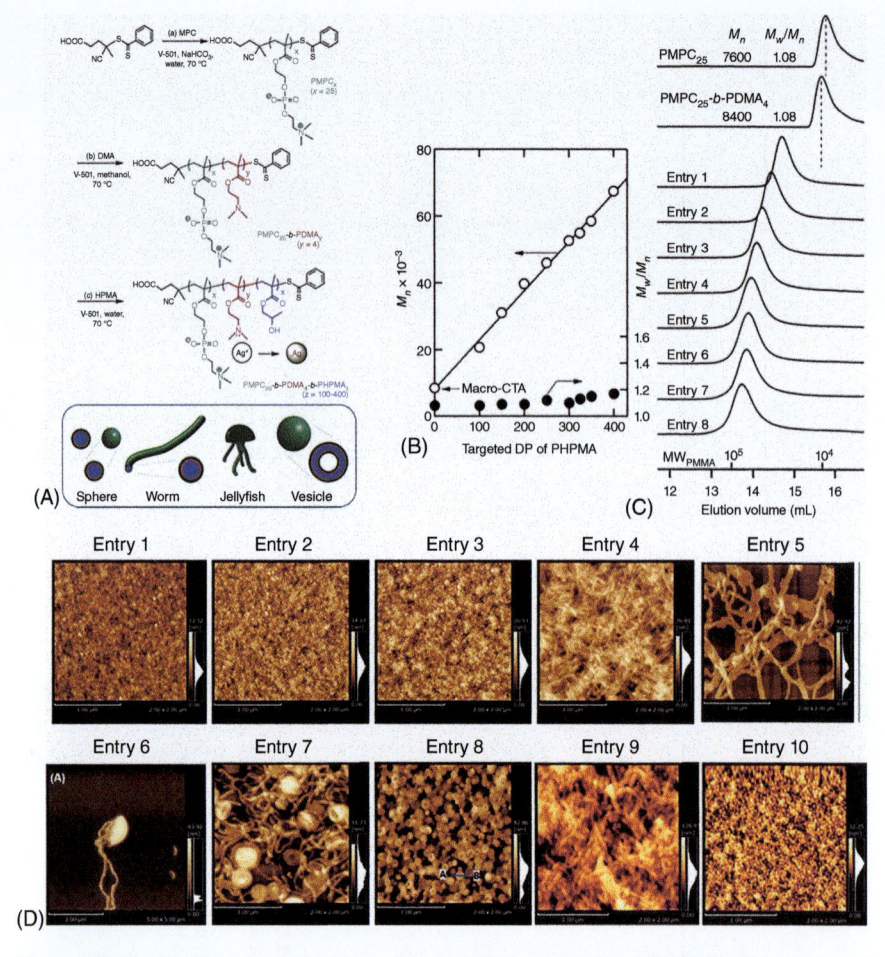

Figure 11.23 (A) The schematic illustration of PMPC$_{25}$-b-PDMA$_4$-b-PHPMA. (B) Variation of Mn and Mw/Mn with targeted DPn of HPMA for the resulting PMPC$_{25}$-b-PDMA$_4$-b-PHPMAz (z = 100–400) triblock copolymer assemblies (entries 1–8) and (C) MWDs for aqueous dispersion polymerization of HPMA at 70 °C (entries 1–8) and for the related PMPC$_{25}$ and PMPC$_{25}$-b-PDMA$_4$ macro-CTAs:[PMPC$_{25}$-b-PDMA4]$_0$/[V-501]$_0$/[HPMA]$_0$. (D) Representative DM-AFM (height) images of PMPC$_{25}$-b-PDMA$_4$-b-PHPMAz assemblies for entries 1-10. Source: Sugihara et al. [80]. Copyright: American Chemical Society, 2019.

provide new applications with better designs and control. Hence, it was a great opportunity and challenge to explore new methods to synthesize and tailor polymer nanocomposites.

11.9 Application of Polymer Nanocomposites

Polymer nanocomposites are synthesized to obtain multiple functions or enhance properties such as electronics, photonics, magnetics, and mechanical properties [8].

Preparation and tailoring of these polymer composites with multiple functions provide a way to design a novel device.

In biology, polymer nanocomposites were used for drug delivery, imaging, delivery of large therapeutics, worm gels for cell culture, blood cryopreservation, and so on [81–83]. For electronics and magnetics applications, aggregation of nanoparticles was a significant obstacle, and combining with polymer to form polymer nanocomposites was a possible path to utilize magnetic nanoparticles. The role of the polymer was only to be the substance or interface shield. Zhan et al. used the natural rubber to combine with rGO/Fe_3O_4 via sonication, *in situ* reducing, and thermal treatment to form natural rubber/rGO/Fe_3O_4 (NMGO) (Figure 11.24a). After combining, the blending-release process of the as-obtained composites was proven (Figure 11.24b), and TEM characterization also showed the uniformly dispersed magnetic nanoparticles (Figure 11.24c,d). The electronic and magnetic properties performed well after combining. Therefore, the strategy effectively utilized magnetic nanoparticles by transforming the magnetic nanoparticles into polymer nanocomposites.

Nanocrystals were considered necessary materials in biology [85–90], and toxicity was one of the main limits. Combining with polymer to form polymer nanocomposites was one of the effective ways to reduce the toxicity and remain as excellent nanocrystals. For example, Khimyak and coworkers used the trimethoxysilane propyl methacrylate (MPTS) to coat oligoperoxide metal complex (MOC) treated SiO_2 nanoparticles via copolymerization (Figure 11.25) [91]. After copolymerization, Rhodamine could be loaded on the surface of nanoparticles to promote luminescence, providing an application for identifying lymphocytic leukemia L1210 cells (Figure 11.25b,c).

Park and coworkers prepared uniform, high areal density Pt nanocrystals supported by conducting PANI [92]. The Pt nanoparticles were loaded on the PANI nanosheets on the air and water solution interface. The PANI absorbed Pt^{4+} and *in situ* reduced to PANI/Pt polymer nanoparticles (Figure 11.26a). The corresponding TEM images showed that the Pt nanocrystals grow on the PANI nanosheets, providing a large catalytic special area. The electrocatalysis of PANI/Pt on ethanol oxidation performed a high activity and durability (Figure 11.26c,d).

Besides supporting the excellent conductivity of the conducting polymer, some polymer/metal nanocomposites could be employed for electrocatalysis on organic molecular oxidation. Srabanti Ghosh et al. reported [93] that the multimetallic palladium alloy nanoparticles/polypyrrole (Ppy) nanofibers were synthesized via *in situ* nucleation and colloidal radiolytic path. The as-prepared nanoalloy/Ppy performed high electrochemical activity on ethanol oxidation (Figure 11.27), which hinted that the synergetic effect of polymers and inorganic nanomaterials could be reached using a suitable combining method.

As supporting inorganic semiconductors or metal nanocrystals, polymer/inorganic nanocomposites were also taken as photocatalysts [94]. Au nanoparticles were considered a catalyst with high activity. However, the aggregation usually occurred due to the high surface energy. The polymer with functional groups could anchor the metal cations so that such polymers could support catalysts to prevent them from aggregating. For instance, 2,4,6-tris(thiophen-2-yl) aniline was

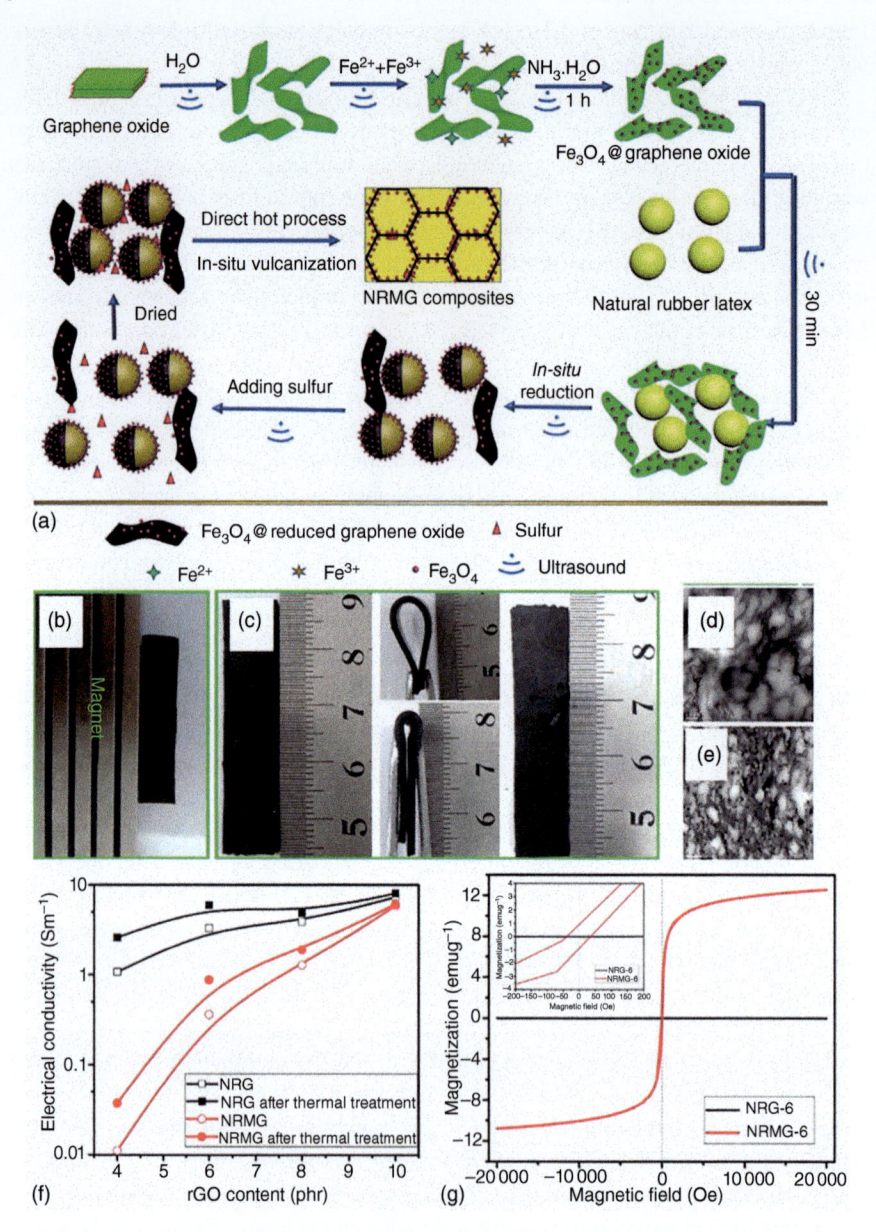

Figure 11.24 (a) The schematic of preparation of NRMG composites. (b,c) The photograph of the NRMG and the bending-release process. (d,e) The TEM images of the NRMG. (f) The electrical conductivity effect by thermal treatment and rGO content. (g) The magnetic properties of NRG and NRMG. Source: Zhan et al. [84]. Copyright 2018 Elsevier.

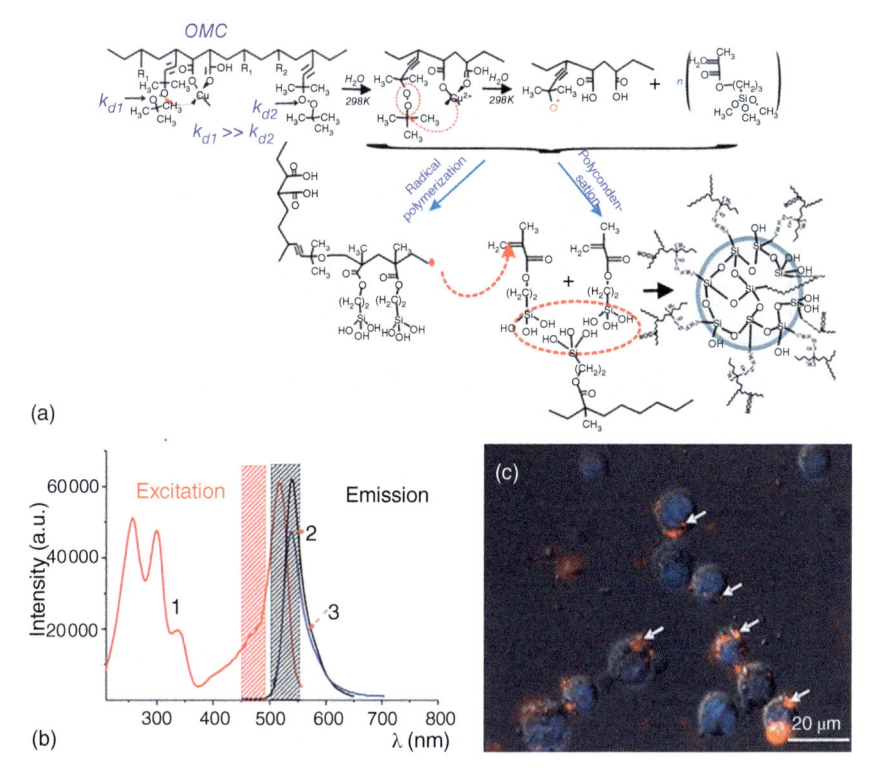

Figure 11.25 (a) Scheme of the formation of OMC coated SiO_2 NPs. (b) Spectra of excitation and luminescence of SiO_2 NPs with Rhodamine 6G and SiO_2 NPs of "core–shell" type with the grafted spacers poly(NVP-co-GMA). (c) Murine lymphocytic leukemia L1210 cells were treated for 24 hours with Rhodamine G-conjugated NPs. Source: Cropper et al. [91]. Copyright 2018 Elsevier.

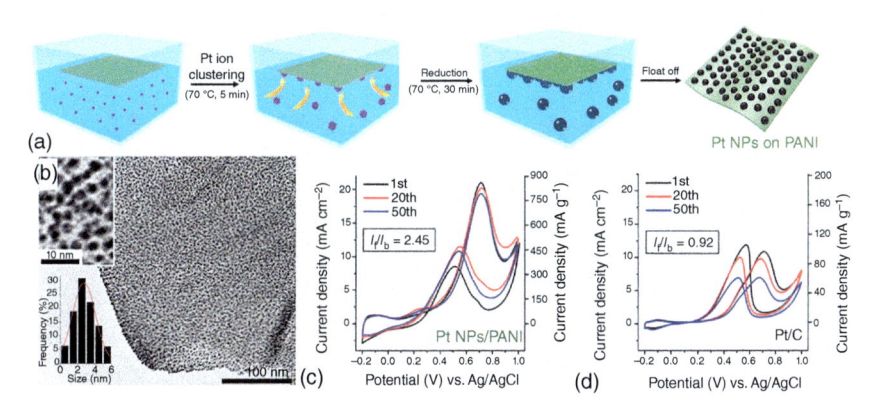

Figure 11.26 (a) The schematic of preparation (b) The TEM image of Pt. (c,d) The electrocatalysis on methanol oxidation. Source: Kim et al. [92]. Copyright 2017 American Chemical Society.

Figure 11.27 (a–c) The TEM images of prepared Ppy (a) and the multimetallic nanoalloy/Ppy (b,c). (d) The TEM images of as-prepared $Pd_{89}Pt_{11}$/Ppy nanohybrid and the corresponding element mapping. (e) The TEM images of as-prepared $Pd_{30}Pt_{29}Au_{41}$/Ppy nanohybrid and the corresponding element mapping. (f–i) The electrocatalysis performance on electrochemical oxidation of ethanol of as-prepared multimetallic nanoalloy/Ppy. Source: Ghosh et al. [93]. Copyright 2017 American Chemical Society.

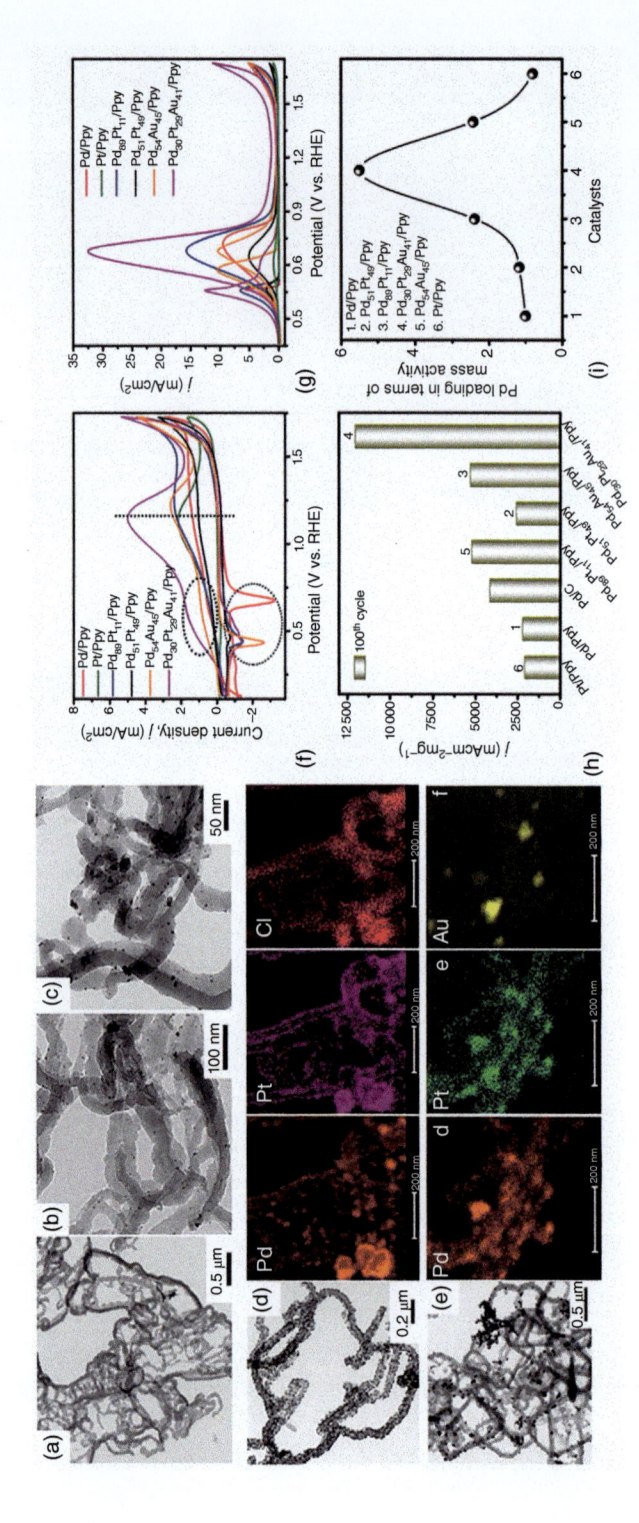

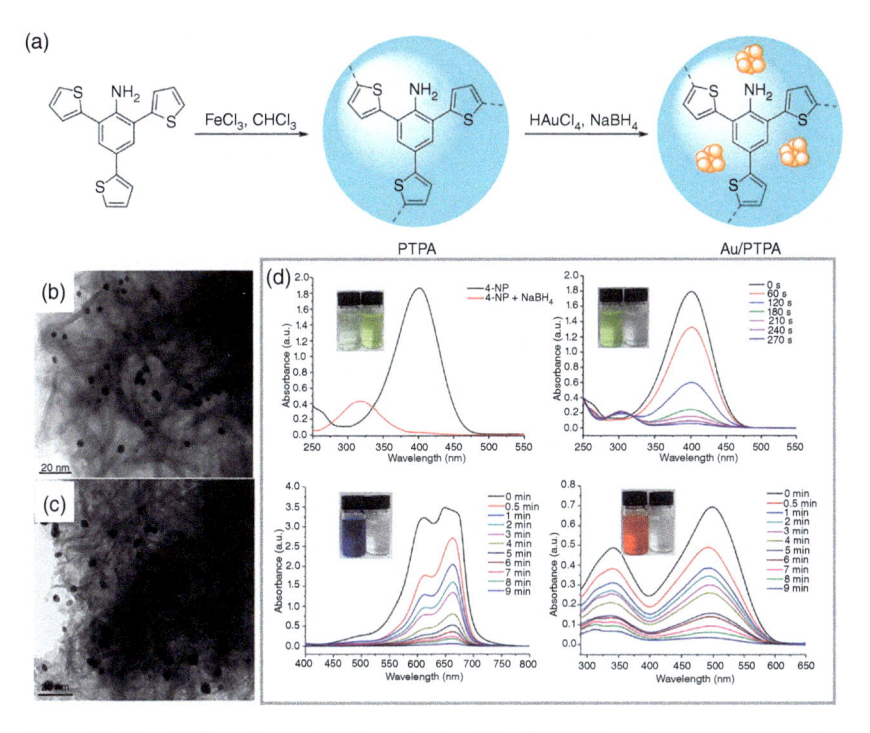

Figure 11.28 (a) The schematics of synthesis of Au NPs/PTPA polymer nanocomposites. (b) The TEM image of Au NPs/PTPA polymer nanocomposites and (c) the according to TEM images after catalysis for six hours. (d) The reduction catalysis of 4-nitrophenol, MB, and CR. Source: Wei et al. [94]. Copyright 2018 Elsevier.

employed as a monomer to produce a porous polythiophene polymer (PTPA), which could capture Au^{3+} with coordination (Figure 11.28a). After reduction, Au nanoparticles were loaded on the PTPA and acted as catalysts (Figure 11.28b,c). The results showed that the as-prepared Au NPs/PTPA polymer nanocomposites performed activity on reduction on 4-nitrophenol and dyes by using $NaBH_4$ as a reductant (Figure 11.28d) [94]. As supported by PTPA, the Au nanoparticles effectively remained, and the aggregation was limited (Figure 11.28c). The polymer nanocomposites with multiple compositions could also be prepared for catalysis by using a similar strategy [95]. Lü and coworkers reported that graphene oxide (GO)–Fe_3O_4 was combined. Then the polydopamine and poly(N-isopropylacrylamide-co-episulfide) (P(NIPAM-co-ETMA)) were coated on the hetero-structure by *in situ* polymerization, followed by capturing Au^{3+} to grow Au nanoparticles. The as-obtained polymer nanocomposites performed high catalytic activity with magnetic property for catalyst reusing. Due to the low critical solution temperature (LCST) of P(NIPAM-co-ETMA), the catalytic reaction rate of Au nanoparticles could be tailored. Hence, utilizing the suitable path was critical for the multifunction of polymer nanocomposites and their applications.

For organic synthesis and industrial catalysis, polymers supporting novel metal nanocrystal catalysts also played important roles. For instance, Pd was an important

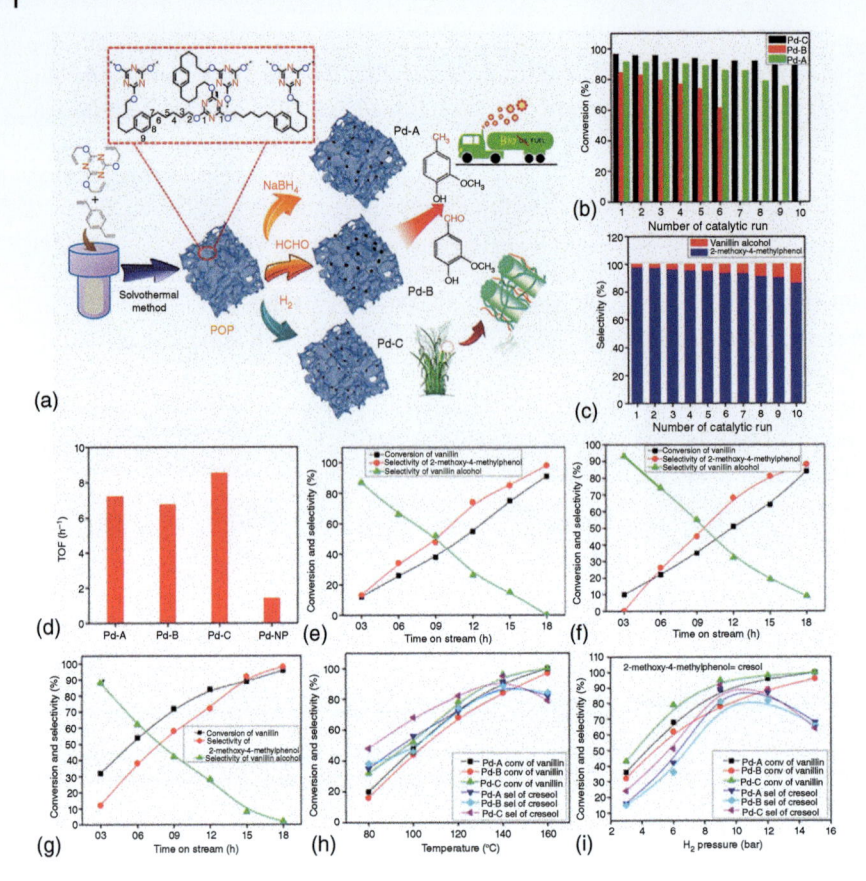

Figure 11.29 (a) Schematic of the synthetic strategy of Pd nanoparticles/POP polymer nanocomposites and its selective catalysis in biofuel upgrade reactions. (b) Recycling diagram for catalytic the hydrodeoxygenation of vanillin with Pd-A, Pd-B, and Pd-C catalysts. (c) Evolution of reactant and product distributions in recyclability test for Pd-C catalyst. (d) Comparison of the catalytic activity with respect to TOF (turnover frequency) for the vanillin hydrodeoxygenation reaction with Pd-based catalysts. Evolution of reactant and product distributions as a function of time over (e) Pd-A, (f) Pd-B, and (g) Pd-C catalysts. (h) Influence of reaction temperature with different Pd-POP catalysts in the vanillin HDO reaction. (i) Effect of H_2 pressure on conversion and product selectivity. Source: Singuru et al. [97]. Copyright 2017 Wiley-VCH.

catalyst in the organic synthesis and chemical industry. Nevertheless, solving the aggregation of Pd nanocrystals is still a challenge for their catalysis applications [96]. By using polyether polyol (POP) as the support for Pd nanoparticles, the formed POP/Pd polymer nanocomposites performed high catalytic activity and stability on the hydrodeoxygenation of vanillin (Figure 11.29) [97].

A further study found that the Pd nanoparticles loaded on the POP supports became more stable during the catalysis (Figure 11.29b,c). The study proved that their activity and selectivity remained after 10 cycles [97]. Furthermore, their turnover frequency (TOF) was higher than that of Pd nanoparticles (Figure 11.29d), and the conversion and selectivity measurements showed the potential applications of polymer nanocomposites. Because the polymers acted as supports and prevented the metallic nanoparticles from aggregating [96, 98], the polymers/metallic nanoparticle composites could perform at high activity. Another sample was reported that the poly(triphenylimidazole) decorated with palladium nanoparticles, and their works showed that the polymer nanocomposites performed high activity and selectivity on cyanation of aryl iodides. Metallic nanoparticles, such as Pt, Pd, Au, Cu, and so on, were widely employed in the catalysis of organic reactions. Their catalytic activities were enhanced greatly in the nanoscale due to the special surface area and the high index facets. Using suitable polymer supports was proven a good strategy to prevent aggregation of nanocrystals and maintain the catalysis capacity. Further studies could also explore the synergetic effect of polymer supports and metallic nanocrystals to enhance the selectivity or chiral controlling of organic synthesis. Considering the conductivity of polymer nanocomposites with special structures, they can also be employed in photovoltaic transformation to generate green energy [99].

The examples provided illustrated that polymer nanocomposites have promising applications. The methods for designing and controlled synthesizing polymer nanocomposites will be a challenge for a long time.

11.10 Outlook

In the future, polymer nanocomposites will combine with different materials to bring new functions and applications. Polymer composites and nanocomposites for multifunctional applications require synthesis methods considering cost, sustainability, and efficiency. Meanwhile, the challenges have existed for a long time, and the difficulties are still to be overcome. For instance, the nanoparticles obtained by the emulation polymerization and dispersing polymerization methods are limited to biotechnology and medicine due to the difficulty of removing surfactant on the nanoparticles. Using precipitate polymerization, the obtained nanoparticles with large size cannot be dispersed uniformly, which are unstable for application. The new fabricating technologies are also needed to prepare polymer nanocomposites with multiple structures and compositions. The surface controlling still needs more research for tailoring the interface between different compositions. The polymerization mechanism should also be further discovered.

In summary, the changes and challenges coexist in polymer nanocomposites. Polymer nanocomposites are derived from polymer composites and cover polymer materials and nanomaterials, bringing more possibilities for designing and fabricating novel materials with excellent properties and functions.

List of Abbreviations

AEAM	*N*-2-aminoethylacrylamide
AIBN	2,20-azobisisobutyronitrile

AMPS	2-acrylamido-2-methylpropanesulfonate
BT	BaTiO$_3$
BzMA	benzyl methacrylate
PADB	4-cyanopentanoic acid dithiobenzone
CNT	carbon nanotube
DHPZDA	phthalazinone-containing diamine
DMAC	N,N-dimethylacetamide
DMF	N,N-dimethylformamide
DP$_n$	degree of polymerization
GO	graphene oxide
HisAM	histamine acrylamide hydrochloride
HPMA	2-hydroxypropyl methacrylate
LCST	lower critical solution temperature
MBS	methacrylate–butadiene–styrene
MOFs	metal–organic frameworks
MWCNT	multi-walls carbon nanotubes
P4VP-TC	poly(4-vinylpyridine)
PAA	poly(acrylic acid)
PANI	polyaniline
PBAT	poly(butylene adipate-co-terephthalate)
P(BzMA)	poly(benzyl methacrylate)
PDEAEMA-co-SMA	poly((diethylamino)-ethyl methacrylate)-co-sodium methacrylate
PE	polyethylene
PEOMA	poly(ethylene oxide) monomethyl ether methanrylate
PGMA	poly glycidyl methacrylate
PHIPE	poly high internal phase emulsions
PhotoATRP	photo atom transfer radical polymerization
PHPMA	poly2-hydroxypropylmethacrylamide
PIC	polyion complexes
PISA	polymerization-induced self-assembly
PLA	poly(lactic acid)
PMAA	poly(methylacrylic acid)
P(MMA-co-POEMA)-b-P(BzMA)	poly(ethylene oxide) monomethyl ether methacrylate end-capped by a trithiocarbonate group
PMPC-b-PDMA-b-PHPMA$_z$	poly-[2-(methacryloyloxy)ethyl phosphorylcholine]-b-poly[2-(dimethylamino)ethyl methacrylate]-b-poly(2-hydroxypropyl-methacrylate)
PMDETA	1,1,4,7,7-pentamethyldiethyltriamine
P(NIPAM-co-ETMA)	poly(N-isopropylacrylamide-co-episulfide)
POP	polyether polyol
PP	polypropylene

PPO	poly(2,6-dimethyl-1,4-phenylene oxide)
PPESK	poly(phthalazinone ether sulfoneketone)
Ppy	polypyrrole
P(S-*b*-2VP-*b*-EO)	poly(styrene-*block*-2-vinylpyridine-*block*-ethylene oxide)
PS	polystyrene
PTPA	polythiophene polymer
PVA	polyvinyl alcohol
RAFT	reversible addition-fragmentation chain transfer
SMA	sodium methacrylate
SWCNTs	single walls carbon nanotubes
SEM	scanning electron microscopy
St	styrene
TEM	transmission electron microscopy
TETA	triethylenetetramine
TPVs	thermoplastic vulcanizates.

References

1 Gaitzsch, J., Huang, X., and Voit, B. (2016). Engineering functional polymer capsules toward smart nanoreactors. *Chem. Rev.* 116 (3): 1053–1093.

2 Thomas, B., Raj, M.C., Athira, K.B. et al. (2018). Nanocellulose, a versatile green platform: from biosources to materials and their applications. *Chem. Rev.* 118 (24): 11575–11625.

3 Wu, D.C., Xu, F., Sun, B. et al. (2012). Design and preparation of porous polymers. *Chem. Rev.* 112 (7): 3959–4015.

4 Hu, J., Zhou, S.X., Sun, Y.Y. et al. (2012). Fabrication, properties and applications of Janus particles. *Chem. Soc. Rev.* 41 (11): 4356–4378.

5 Luo, H., Zhou, X.F., Ellingford, C. et al. (2019). Interface design for high energy density polymer nanocomposites. *Chem. Soc. Rev.* 48 (16): 4424–4465.

6 Wu, S.H., Mou, C.Y., and Lin, H.P. (2013). Synthesis of mesoporous silica nanoparticles. *Chem. Soc. Rev.* 42 (9): 3862–3875.

7 Ngo, T.D., Kashani, A., Imbalzano, G. et al. (2018). Additive manufacturing (3D printing): a review of materials, methods, applications and challenges. *Composites Part B* 143: 172–196.

8 Deng, Z., Hu, T., Lei, Q. et al. (2019). Stimuli-responsive conductive nanocomposite hydrogels with high stretchability, self-healing, adhesiveness, and 3D printability for human motion sensing. *ACS Appl. Mater. Interfaces* 11 (7): 6796–6808.

9 Zhou, T., Zha, J.-W., Cui, R.-Y. et al. (2011). Improving dielectric properties of $BaTiO_3$/ferroelectric polymer composites by employing surface hydroxylated $BaTiO_3$ nanoparticles. *ACS Appl. Mater. Interfaces* 3 (7): 2184–2188.

10 Santos, J.P.F., Carvalho, B.D., and Bretas, R.E.S. (2019). Remarkable change in the broadband electrical behavior of poly(vinylidene fluoride)-multiwalled carbon nanotube nanocomposites with the use of different processing routes. *J. Appl. Polym. Sci.* 136 (17): 47409.

11 Yu, G.F., Yan, X., Yu, M. et al. (2016). Patterned, highly stretchable and conductive nanofibrous PANI/PVDF strain sensors based on electrospinning and in situ polymerization. *Nanoscale* 8 (5): 2944–2950.

12 Gao, P., Cao, H., Ding, Y. et al. (2016). Synthesis of hydrogen-bonded pore-switchable cylindrical vesicles via visible-light-mediated RAFT room-temperature aqueous dispersion polymerization. *ACS Macro Lett.* 5 (12): 1327–1331.

13 Zhang, B., Lv, X., and An, Z. (2017). Modular monomers with tunable solubility: synthesis of highly incompatible block copolymer nano-objects via RAFT aqueous dispersion polymerization. *ACS Macro Lett.* 6 (3): 224–228.

14 Gao, C., Wu, J., Zhou, H. et al. (2016). Self-assembled blends of AB/BAB block copolymers prepared through dispersion RAFT polymerization. *Macromolecules* 49 (12): 4490–4500.

15 Cai, M., Ding, Y., Wang, L. et al. (2018). Synthesis of one-component nanostructured polyion complexes via polymerization-induced electrostatic self-assembly. *ACS Macro Lett.* 7: 208–212.

16 Zhang, X., Rieger, J., and Charleux, B. (2012). Effect of the solvent composition on the morphology of nano-objects synthesized via RAFT polymerization of benzyl methacrylate in dispersed systems. *Polym. Chem.* 3 (6): 1502–1509.

17 Deng, R., Derry, M.J.D., Mable, C.J. et al. (2017). Using dynamic covalent chemistry to drive morphological transitions: controlled release of encapsulated nanoparticles from block copolymer vesicles. *J. Am. Chem. Soc.* 22: 7616–7623.

18 Wan, W.-M. and Pan, C.-Y. (2010). One-pot synthesis of polymeric nanomaterials via RAFT dispersion polymerization induced self-assembly and re-organization. *Polym. Chem.* 1 (9): 1475–1484.

19 Hickey, R.J., Haynes, A.S., Kikkawa, J.M., and Park, S.J. (2011). Controlling the self-assembly structure of magnetic nanoparticles and amphiphilic block-copolymers: from micelles to vesicles. *J. Am. Chem. Soc.* 133 (5): 1517–1525.

20 Habibi, Y., Lucia, L.A., and Rojas, O.J. (2010). Cellulose nanocrystals: chemistry, self-assembly, and applications. *Chem. Rev.* 110 (6): 3479–3500.

21 Jamekhorshid, A., Sadrameli, S.M., and Farid, M. (2014). A review of microencapsulation methods of phase change materials (PCMs) as a thermal energy storage (TES) medium. *Renewable Sustainable Energy Rev.* 31: 531–542.

22 Jessop, P.G., Mercer, S.M., and Heldebrant, D.J. (2012). CO_2-Triggered switchable solvents, surfactants, and other materials. *Energy Environ. Sci.* 5 (6): 7240–7253.

23 Khor, S.Y., Quinn, J.F., Whittaker, M.R. et al. (2019). Controlling nanomaterial size and shape for biomedical applications via polymerization-induced self-assembly. *Macromol. Rapid Commun.* 40 (2): 1800438.

24 Giannelis, E.P. (1996). Polymer layered silicate nanocomposites. *Adv. Mater.* 8 (1): 29–35.

25 Lin, X., Hu, P., Jia, Z., and Gao, S. (2016). Enhanced electric displacement induces large energy density in polymer nanocomposites containing core–shell structured $BaTiO_3@TiO_2$ nanofibers. *J. Mater. Chem. A* 4 (6): 2314–2320.

26 Gao, L., He, J., Hu, J., and Li, Y. (2014). Large enhancement in polarization response and energy storage properties of poly(vinylidene fluoride) by improving the interface effect in nanocomposites. *J. Phys. Chem. C* 118 (2): 831–838.

27 Yuan, J.K., Yao, S.H., Dang, Z.M. et al. (2011). Giant dielectric permittivity nanocomposites: realizing true potential of pristine carbon nanotubes in polyvinylidene fluoride matrix through an enhanced interfacial interaction. *J. Phys. Chem. C* 115 (13): 5515–5521.

28 Uyor, U.O., Popoola, A.P.I., Popoola, O.M., and Aigbodion, V.S. (2020). Enhanced dielectric performance and energy storage density of polymer/graphene nanocomposites prepared by dual fabrication. *J. Thermoplast. Compos. Mater.* 33 (2): 270–285.

29 Dun, C., Hewitt, C.A., Huang, H. et al. (2015). Layered Bi_2Se_3 nanoplate/polyvinylidene fluoride composite based n-type thermoelectric fabrics. *ACS Appl. Mater. Interfaces* 7 (13): 7054–7059.

30 Feng, X., Liao, G., He, W. et al. (2009). Preparation and characterization of functionalized carbon nanotubes/poly(phthalazinone ether sulfone ketone)s composites. *Polym. Compos.* 30 (4): 365–373.

31 Liu, Y.-L., Chang, Y.-H., and Liang, M. (2008). Poly(2,6-dimethyl-1,4-phenylene oxide) (PPO) multi-bonded carbon nanotube (CNT): preparation and formation of PPO/CNT nanocomposites. *Polymer* 49 (25): 5405–5409.

32 D'Agosto, F., Rieger, J., and Lansalot, M. (2020). RAFT-mediated polymerization-induced self-assembly. *Angew. Chem. Int. Ed.* 59 (22): 8368–8392.

33 Khan, M.U., Reddy, K.R., Snguanwongchai, T. et al. (2016). Polymer brush synthesis on surface modified carbon nanotubes via in situ emulsion polymerization. *Colloid. Polym. Sci.* 294 (10): 1599–1610.

34 Moad, G. (2017). RAFT polymerization to form stimuli-responsive polymers. *Polym. Chem.* 8 (1): 177–219.

35 Sun, J.T., Hong, C.Y., and Pan, C.Y. (2013). Recent advances in RAFT dispersion polymerization for preparation of block copolymer aggregates. *Polym. Chem.* 4 (4): 873–881.

36 Wang, Z., Rutjes, F.P.J.T., and van Hest, J.C.M. (2014). pH responsive polymersome Pickering emulsion for simple and efficient Janus polymersome fabrication. *Chem. Commun.* 50 (93): 14550–14553.

37 Motornov, M., Sheparovych, R., Lupitskyy, R. et al. (2006). Stimuli-responsive colloidal systems from mixed brush-coated nanoparticles. *Adv. Funct. Mater.* 17: 2307–2314.

38 Liu, P., Lu, W., Wang, W.J. et al. (2014). Highly CO_2/N_2-switchable zwitterionic surfactant for pickering emulsions at ambient temperature. *Langmuir* 30 (34): 10248–10255.

39 Pan, X.C., Tasdelen, M.A., Laun, J. et al. (2016). Photomediated controlled radical polymerization. *Prog. Polym. Sci.* 62: 73–125.

40 Sardar, D., Sengupta, M., Bordoloi, A. et al. (2017). Multiple functionalities of Ni nanoparticles embedded in carboxymethyl guar gum polymer: catalytic activity and superparamagnetism. *Appl. Surf. Sci.* 405: 231–239.

41 Zhao, D., Zhu, G.D., Ding, Y., and Zheng, J.P. (2018). Construction of a different polymer chain structure to study pi-pi interaction between polymer and reduced graphene oxide. *Polymers* 10 (7): 716.

42 Zhang, M., Zhang, E., Hu, C. et al. (2020). Controlled synthesis of Co@N-doped carbon by pyrolysis of ZIF with 2-aminobenzimidazole ligand for enhancing oxygen reduction reaction and the application in Zn–air battery. *ACS Appl. Mater. Interfaces* 12 (10): 11693–11701.

43 Li, R., Hu, J., Deng, M. et al. (2014). Integration of an inorganic semiconductor with a metal–organic framework: a platform for enhanced gaseous photocatalytic reactions. *Adv. Mater.* 26 (28): 4783–4788.

44 Liu, L., Diaz, U., Arenal, R. et al. (2017). Generation of subnanometric platinum with high stability during transformation of a 2D zeolite into 3D. *Nat. Mater.* 16: 132–138.

45 Zhao, S., Wang, Y., Dong, J. et al. (2016). Ultrathin metal–organic framework nanosheets for electrocatalytic oxygen evolution. *Nat. Energy* 1: 16184.

46 Su, J., Ge, R., Jiang, K. et al. (2018). Assembling ultrasmall copper-doped ruthenium oxide nanocrystals into hollow porous polyhedra: highly robust electrocatalysts for oxygen evolution in acidic media. *Adv. Mater.* 30: e1801351.

47 Kalaj, M., Denny, M.S., Bentz, K.C. et al. (2019). Nylon-MOF composites through postsynthetic polymerization. *Angew. Chem. Int. Ed.* 58 (8): 2336–2340.

48 Zhuo, M.-P., He, G.-P., Yuan, Y. et al. (2021). Super-stacking self-assembly of organic topological heterostructures. *CCS Chemistry* 3 (1): 413–424.

49 Guan, S., Zhang, C., Wen, W. et al. (2018). Formation of anisotropic liquid crystalline nanoparticles via polymerization-induced hierarchical self-assembly. *ACS Macro Lett.* 7 (3): 358–363.

50 Zhang, K., Li, G.H., Feng, L.M. et al. (2017). Ultralow percolation threshold and enhanced electromagnetic interference shielding in poly(L-lactide)/multi-walled carbon nanotube nanocomposites with electrically conductive segregated networks. *J. Mater. Chem. C* 5 (36): 9359–9369.

51 Niu, L., Xu, J.L., Yang, W.L. et al. (2019). Synergistic effect between nano-Sb_2O_3 and brominated epoxy resin on the flame retardancy of poly(butylene terephthalate). *Sci. Adv. Mater.* 11 (4): 466–475.

52 Shi, Y.Q., Liu, C., Liu, L. et al. (2019). Strengthening, toughing and thermally stable ultra-thin MXene nanosheets/polypropylene nanocomposites via nanoconfinement. *Chem. Eng. J.* 378: 122267.

53 Han, Z. and Fina, A. (2011). Thermal conductivity of carbon nanotubes and their polymer nanocomposites: a review. *Prog. Polym. Sci.* 36 (7): 914–944.

54 Cao, L.M., Liu, C., Zou, D.J. et al. (2020). Using cellulose nanocrystals as sustainable additive to enhance mechanical and shape memory properties of PLA/ENR thermoplastic vulcanizates. *Carbohydr. Polym.* 230: 115618.

55 Lai, L., Li, J.X., Liu, P.W. et al. (2020). Mechanically reinforced biodegradable poly(butylene adipate-*co*-terephthalate) with interactive nanoinclusions. *Polymer* 197: 122518.

56 Bhattacharya, S. and Samanta, S.K. (2016). Soft-nanocomposites of nanoparticles and nanocarbons with supramolecular and polymer gels and their applications. *Chem. Rev.* 116 (19): 11967–12028.

57 Yang, J., Yang, W.Q., Wang, X.L. et al. (2019). Synergistically toughening polyoxymethylene by methyl methacrylate-butadiene-styrene copolymer and thermoplastic polyurethane. *Macromol. Chem. Phys.* 220 (12): 1800567.

58 Wang, J.F., Jin, X.X., Li, C.H. et al. (2019). Graphene and graphene derivatives toughening polymers: toward high toughness and strength. *Chem. Eng. J.* 370: 831–854.

59 Wang, Y., Zhang, W.Z., Wu, X.M. et al. (2017). Conducting polymer coated metal–organic framework nanoparticles: facile synthesis and enhanced electromagnetic absorption properties. *Synth. Met.* 228: 18–24.

60 Kim, H., Miura, Y., and Macosko, C.W. (2010). Graphene/polyurethane nanocomposites for improved gas barrier and electrical conductivity. *Chem. Mater.* 22 (11): 3441–3450.

61 Dai, H.X., Wang, N., Wang, D.L. et al. (2016). An electrochemical sensor based on phytic acid functionalized polypyrrole/graphene oxide nanocomposites for simultaneous determination of Cd(II) and Pb(II). *Chem. Eng. J.* 299: 150–155.

62 Gu, J.W., Lv, Z.Y., Wu, Y.L. et al. (2017). Dielectric thermally conductive boron nitride/polyimide composites with outstanding thermal stabilities via in-situ polymerization-electrospinning-hot press method. *Composites Part A* 94: 209–216.

63 Wang, Y.P., Zhou, P., Luo, S.Z. et al. (2018). Controllable synthesis of monolayer poly(acrylic acid) on the channel surface of mesoporous alumina for Pb(II) adsorption. *Langmuir* 34 (26): 7859–7868.

64 Yang, X.T., Guo, Y.Q., Luo, X. et al. (2018). Self-healing, recoverable epoxy elastomers and their composites with desirable thermal conductivities by incorporating BN fillers via in-situ polymerization. *Compos. Sci. Technol.* 164: 59–64.

65 Cheng, C.B., Fan, R.H., Fan, G.H. et al. (2019). Tunable negative permittivity and magnetic performance of yttrium iron garnet/polypyrrole metacomposites at the RF frequency. *J. Mater. Chem. C* 7 (11): 3160–3167.

66 Xu, Z. and Gao, C. (2010). In situ polymerization approach to graphene-reinforced nylon-6 composites. *Macromolecules* 43 (16): 6716–6723.

67 Peresin, M.S., Habibi, Y., Zoppe, J.O. et al. (2010). Nanofiber composites of polyvinyl alcohol and cellulose nanocrystals: manufacture and characterization. *Biomacromolecules* 11 (3): 674–681.

68 Karakassides, A., Ganguly, A., Tsirka, K. et al. (2020). Radially grown graphene nanoflakes on carbon fibers as reinforcing interface for polymer composites. *ACS Appl. Nano Mater.* 3 (3): 2402–2413.

69 Zhang, L.Q., Wang, R., Wang, J.L. et al. (2019). Mechanically robust nanocomposites from screen-printable polymer/graphene nanosheet pastes. *Nanoscale* 11 (5): 2343–2354.

70 Peresin, M.S., Habibi, Y., Zoppe, J.O., and Rojas, O.J. (2010). Nanofiber composites of polyvinyl alcohol and cellulose nanocrystals: manufacture and characterization. *Biomacromolecules* 11: 674–681.

71 Layek, R.K. and Nandi, A.K. (2013). A review on synthesis and properties of polymer functionalized graphene. *Polymer* 54 (19): 5087–5103.

72 Punetha, V.D., Rana, S., Yoo, H.J. et al. (2017). Functionalization of carbon nanomaterials for advanced polymer nanocomposites: a comparison study between CNT and graphene. *Prog. Polym. Sci.* 67: 1–47.

73 Liu, X. and Li, C. (2018). Novel indirect Z-scheme photocatalyst of Ag nanoparticles and polymer polypyrrole co-modified biobr for photocatalytic decomposition of organic pollutant. *Appl. Surf. Sci.* 445: 242–254.

74 Qin, Y., Chen, L.C., Zhang, W. et al. (2010). Enhanced thermoelectric performance of single-walled carbon nanotubes/polyaniline hybrid nanocomposites. *ACS Nano* 4: 2445–2451.

75 Yao, Q., Chen, L., Zhang, W. et al. (2010). Enhanced thermoelectric performance of single-walled carbon nanotubes/polyaniline hybrid nanocomposites. *ACS Nano* 4 (4): 2445–2451.

76 Yuan, W., Chen, X., Xu, Y. et al. (2018). Preparation and recyclable catalysis performance of functional macroporous polyHIPE immobilized with gold nanoparticles on its surface. *RSC Adv.* 8: 5912–5919.

77 Yang, H.C., Cai, Z.N., Liu, H.T. et al. (2020). Tailoring the surface of attapulgite by combining redox-initiated RAFT polymerization with alkynyl-thiol click reaction for polycarbonate nanocomposites: effect of polymer brush chain length on mechanical, thermal and rheological properties. *Mater. Chem. Phys.* 241: 122334.

78 Wang, Y.X., Huang, X.Y., Li, T. et al. (2019). Polymer-based gate dielectrics for organic field-effect transistors. *Chem. Mater.* 31 (7): 2212–2240.

79 Allasia, M., Passeggi, M.C.G., Gugliotta, L.M., and Minari, R.J. (2019). Waterborne hybrid acrylic/protein nanocomposites with enhanced hydrophobicity by incorporating a water repelling protein. *Ind. Eng. Chem. Res.* 58 (46): 21070–21079.

80 Sugihara, S., Sudo, M., and Maeda, Y. (2018). Synthesis and nano-object assembly of biomimetic block copolymers for catalytic silver nanoparticles. *Langmuir* 35 (5): 1346–1356.

81 Naskar, S., Panda, A.K., Jana, A. et al. (2020). UHMWPE-MWCNT-nHA based hybrid trilayer nanobiocomposite: processing approach, physical properties, stem/bone cell functionality, and blood compatibility. *J. Biomed. Mater. Res. Part B: Appl. Biomater.* 108 (5): 2320–2343.

82 Tang, Z.H., He, C.L., Tian, H.Y. et al. (2016). Polymeric nanostructured materials for biomedical applications. *Prog. Polym. Sci.* 60: 86–128.

83 Armentano, I., Dottori, M., Fortunati, E. et al. (2010). Biodegradable polymer matrix nanocomposites for tissue engineering: a review. *Polym. Degrad. Stab.* 95 (11): 2126–2146.

84 Zhan, Y.H., Wang, J., Zhang, K.Y. et al. (2018). Fabrication of a flexible electromagnetic interference shielding Fe_3O_4@ reduced graphene oxide/natural rubber composite with segregated network. *Chem. Eng. J.* 344: 184–193.

85 Ji, M., Xu, M., Zhang, W. et al. (2016). Structurally well-defined Au@Cu_{2-x}S core–shell nanocrystals for improved cancer treatment based on enhanced photothermal efficiency. *Adv. Mater.* 28 (16): 3094–3101.

86 Xia, Y., Li, W., Cobley, C.M. et al. (2011). Gold nanocages: from synthesis to theranostic applications. *Acc. Chem. Res.* 44 (10): 914–924.

87 Zou, Q., Abbas, M., Zhao, L. et al. (2017). Biological photothermal nanodots based on self-assembly of peptide-porphyrin conjugates for antitumor therapy. *J. Am. Chem. Soc.* 139 (5): 1921–1927.

88 Li, B., Wang, Q., Zou, R. et al. (2014). $Cu_{7.2}S_4$ nanocrystals: a novel photothermal agent with a 56.7% photothermal conversion efficiency for photothermal therapy of cancer cells. *Nanoscale* 6 (6): 3274–3282.

89 Qin, J., Peng, Z., Li, B. et al. (2015). Gold nanorods as a theranostic platform for in vitro and in vivo imaging and photothermal therapy of inflammatory macrophages. *Nanoscale* 7 (33): 13991–14001.

90 Zhang, W.J., Hong, C.Y., and Pan, C.Y. (2019). Polymerization-induced self-assembly of functionalized block copolymer nanoparticles and their application in drug delivery. *Macromol. Rapid Commun.* 40 (2).

91 Cropper, C., Mitina, N., Klyuchivska, O. et al. (2018). Luminescent SiO_2 nanoparticles for cell labelling: combined water dispersion polymerization and 3D condensation controlled by oligoperoxide surfactant-initiator. *Eur. Polym. J.* 103: 282–292.

92 Kim, K., Ahn, H., and Park, M.J. (2017). Highly catalytic Pt nanoparticles grown in two-dimensional conducting polymers at the air–water interface. *ACS Appl. Mater. Interfaces* 9 (36): 30278–30282.

93 Ghosh, S., Bera, S., Bysakh, S., and Basu, R.N. (2017). Highly active multimetallic palladium nanoalloys embedded in conducting polymer as anode catalyst for electrooxidation of ethanol. *ACS Appl. Mater. Interfaces* 9 (39): 33775–33790.

94 Wei, F., Lu, C., Wang, F. et al. (2018). A novel functionalized porous polythiophene polymer network for Au catalyst deposition. *Mater. Lett.* 212: 251–255.

95 Wang, D., Duan, H., Lü, J., and Lü, C. (2017). Fabrication of thermo-responsive polymer functionalized reduced graphene oxide@Fe_3O_4@Au magnetic nanocomposites for enhanced catalytic applications. *J. Mater. Chem. A* 5 (10): 5088–5097.

96 Molla, R., Bhanja, P., Ghosh, K. et al. (2017). Pd Nanoparticles decorated on hypercrosslinked microporous polymer: a highly efficient catalyst for the formylation of amines through carbon dioxide fixation. *ChemCatChem* 9: 1939–1946.

97 Singuru, R., Dhanalaxmi, K., Shit, S.C. et al. (2017). Palladium nanoparticles encaged in a nitrogen-rich porous organic polymer: constructing a promising robust nanoarchitecture for catalytic biofuel upgrading. *ChemCatChem* 9: 2550–2564.

98 Yu, H., Xu, S., Liu, Y. et al. (2018). A porous organic poly(triphenylimidazole) decorated with palladium nanoparticles for the cyanation of aryl iodides. *Chem. Asian J.* 13 (18): 2708–2713.

99 Zoromba, M.S. and Al-Hossainy, A.F. (2020). Doped poly(*o*-phenylenediamine-*co*-*p*-toluidine) fibers for polymer solar cells applications. *Sol. Energy* 195: 194–209.

12

Spin-Related Electrode Reactions in Nanomaterials

Shengnan Sun and Yanglong Hou

Beijing Innovation Centre for Engineering Science and Advanced Technology (BIC-ESAT), Beijing Key Laboratory for Magnetoelectric Materials and Devices (BKLMMD), College of Engineering, Department of Materials Science and Engineering, Peking University, Beijing 100871, China

12.1 Introduction

Spin-related electrode reactions in nanomaterials attract much attention in recent years. People are trying to establish the link between the behavior of electrodes and the spin states/magnetic properties of the electrochemical system and to use the external magnetic field to regulate the electrochemical performance to meet our needs. Insofar as the research areas we are involved in, we summarize the recent process about spin-related electrode reactions from the fundamental principle to the application and present the endeavor people are making in the understanding of the mechanism and the observation of experimental phenomenon. The stimulus to writing this chapter is that people are realizing the spin states and magnetic properties play important roles in the electrode reactions, making efforts to looking for the guidance of utilizing spin and magnetism to design the electrochemical system. In this chapter, we will discuss the progress on the spin-related electrode reactions from the following aspects including the possible factors that influence the electrochemical behavior like the force in electrochemical system and spin state of the materials, and the related applications such as electrodeposition, and the catalytic reactions like hydrogen evolution reaction (HER), oxygen evolution reaction (OER), oxygen reduction reaction (ORR), and so on. The influence of spin on batteries and some other fields is also discussed briefly.

Till now, three mechanisms by the interaction of the magnetic field (or spin states) and chemical systems have been established. They arise from the magnetomechanic, magnetohydrodynamic (MHD), and electronic interactions [1]. The magnetomechanic mechanism applies to two types of magnetic fields, namely, the homogeneous field and the high gradient field. The homogeneous field induces the orientation of the materials with magnetically anisotropism or the movement with ferro/ferrimagnetic or paramagnetic properties. When applying the magnetic field, the force will orientate or pull the electrolyte to the specific direction [2]. High gradient fields cause the movement of paramagnetic species to the high gradient

Functional Nanomaterials: Synthesis, Properties, and Applications, First Edition.
Edited by Wai-Yeung Wong and Qingchen Dong.
© 2022 WILEY-VCH GmbH. Published 2022 by WILEY-VCH GmbH.

directions. Such behavior can also be regarded as MHD effect. The MHD mechanism arises from the interaction of magnetic fields with the flow of electrolytes, since the force is induced when the electrolyte flow in the solution under magnetic field [3]. The electronic interaction mechanism involves electron transfer via radical pair intermediates in the magnetic field [4], spin channel [5], and superexchange interactions [6]. In our discussion in this chapter, we mainly focus on the effect from MHD and electronic interaction on the electrochemical system.

12.2 Factors Influencing the Electrochemical System

12.2.1 Forces Caused by Magnetic Fields in Aqueous Solution

In magnetism-related electrochemistry in aqueous solution, different types of forces originating from magnetic fields and their importance have been intensively investigated. Generally, there are five possible forces caused by magnetic fields in aqueous electrochemical cells, namely, Lorentz force ($\mathbf{F_L}$), magnetic gradient force ($\mathbf{F_B}$), paramagnetic gradient force/concentration gradient force ($\mathbf{F_P}$), electrokinetic force ($\mathbf{F_E}$), and magnetic damping force ($\mathbf{F_M}$) [7]. When the temperature is 298 K, concentration is 1 mol L^{-1}, magnetic induction strength is 1 T, and magnetic gradient is 1 T m^{-1}, the typical values of the forces associated with magnetic fields and other forces from migration, diffusion, and convection can be found in Table 12.1.

Among them, the Lorentz force can be described according to the equation

$$\mathbf{F} = \mathbf{j} \times \mathbf{B},$$

where $\mathbf{j}$ and $\mathbf{B}$ are the current density (C cm^{-2} s^{-1}) and magnetic induction (T), respectively. The Lorentz force can change the distribution of the charged species and create the convection close to the electrode surface like a local stirring, and thus accelerate the mass transport of ions toward to the electrode surface. When the directions of $\mathbf{j}$ and $\mathbf{B}$ are parallel, the Lorentz force is zero, and when the directions of $\mathbf{j}$ and $\mathbf{B}$ are perpendicular, the Lorentz force is maximum [1, 3, 8]. In the real electrochemical system, two types of convections caused by the Lotentz force may be involved caused. When the direction of magnetic field is parallel to the electrode surface, the magnetohydrodynamic convection (MHDC) in the bulk solution has the same magnitude with that in natural convection between the anode and the cathode; when the homogeneous magnetic field is perpendicular to the electrode surface, the Lorentz force also arises due to the distortion of the electric field distortion (Figure 12.1) [9]. For example, in the case of the existence of gas bubbles adhering at the electrode surface, MHDC arises within the scale of single bubble [10].

When the magnetic field is not uniform, a nonuniform magnetization in the electrolyte solution is induced, which exerts a force that drives the movement of any paramagnetic species (Figure 12.2) [1, 3, 8]. The force can be defined as magnetic gradient force ($\mathbf{F_B}$) that influences electrode reactions involving paramagnetic ions in solution. This force can be expressed as

$$\mathbf{F_B} = c\chi_m \nabla \mathbf{B}^2 / 2\mu_0,$$

Table 12.1 Typical forces acting in aqueous electrolytes.

Force	Typical value (N m^{-3})
Lorentz force	10^3
Magnetic gradient force	10^1
Paramagnetic force	10^4
Magnetic damping force	10^1
Electrokinetic force	10^3
Viscous drag	10^1
Driving force for diffusion	10^{10}
Driving force for electromigration	10^{10}
Driving force for forced convection	10^5
Driving force for natural convection	10^3

Source: Modified from Hinds et al. [7].

where μ_0 is the magnetic constant and c and χ_m are the concentration and the molar susceptibility of a species in the electrolyte, respectively [11]. The field gradient force may become significant in nonuniform magnetic fields, whether on the scale of the cell or on a microscopic scale at the surface of ferromagnetic electrodes [7]. Under typical experimental conditions, $\mathbf{F_B}$ is the order of $10\,\mathrm{N\,m^{-3}}$. When a field gradient $\nabla \mathbf{B} \gg 1\,\mathrm{T\,m^{-1}}$ is imposed deliberately, $\mathbf{F_B}$ can become dominant [12].

The paramagnetic gradient force or concentration gradient force is defined as

$$\mathbf{F_p} = \chi_m \mathbf{B}^2 \nabla c / 2\mu_0.$$

$\mathbf{F_p}$ arises from the concentration gradient of paramagnetic ions. Only in the diffusion layer, $\mathbf{F_p}$ becomes significant. The electrokinetic force $\mathbf{F_E}$ is the force acting on the charge carriers in diffuse double layer under the influence of a non-electrostatic field parallel to the electrode surface and is expressed as

$$\mathbf{F_E} = \sigma_d \mathbf{E}_\| / \delta_0,$$

where $\mathbf{E}_\|$ is the non-electrostatic field parallel to the electrode surface and σ_d is the charge density in the diffuse layer.

The magnetic damping force ($\mathbf{F_M}$) can be expressed as

$$\mathbf{F_M} = \sigma \mathbf{v} \times \mathbf{B} \times \mathbf{B},$$

where $\mathbf{v} \times \mathbf{B}$ is non-electrostatic field due to the applied magnetic field. The five forces have different strength order, $\mathbf{F_L}\,10^3$, $\mathbf{F_B}\,10^1$, $\mathbf{F_p}\,10^4$, $\mathbf{F_E}\,10^3$, $\mathbf{F_M}\,10^1$, with the dimension of $\mathrm{N\,m^{-3}}$ [7]. In our discussion, we just focus on the influence of Lorentz force ($\mathbf{F_L}$) and magnetic field gradient ($\mathbf{F_B}$) to the electrochemical system.

In an electrochemical cell under a magnetic field, generally both the Lorentz force ($\mathbf{F_L}$) and the magnetic gradient force ($\mathbf{F_B}$) are involved. It is not always clear which one contributes to the effects observed in the magneto-electrochemical experiment. Though the convection caused by the Lorentz force can be minimized when the

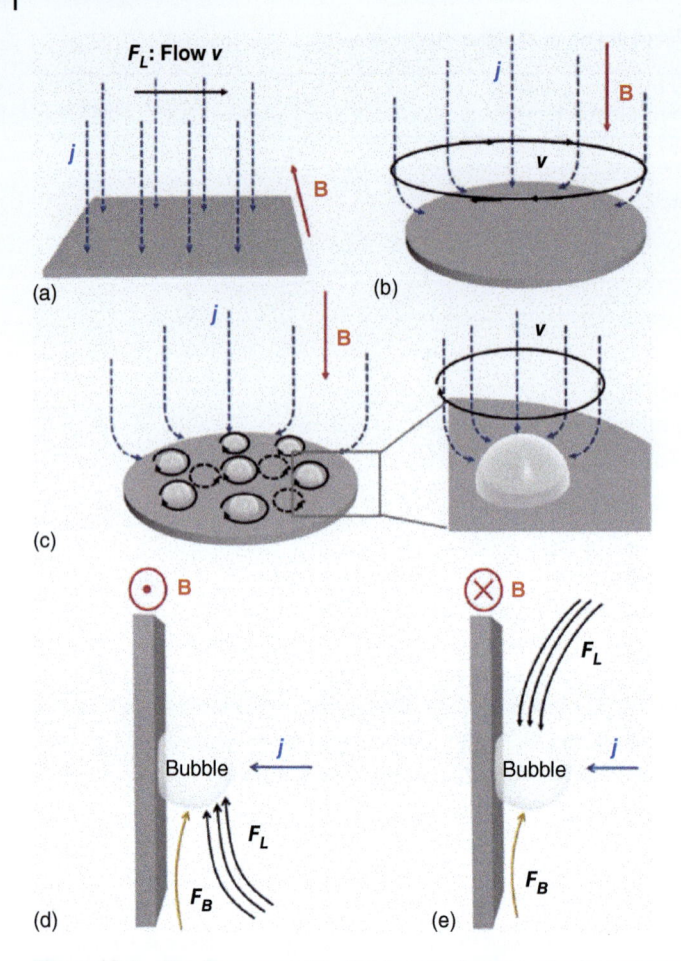

Figure 12.1 The flow caused by the Lorentz forces under homogeneous magnetic field. The magnetic field is perpendicular to the current and parallel to the electrode surface (a), parallel to the current and perpendicular to the electrode surface (b), convection around the bubbles (c), and the buoyancy force and Lorentz force exerted on the bubbles, F_B and F_L in the same (d) and different (e) directions. Source: Adapted with permission from Zhang et al. [9]. Copyright (2018) American Chemical Society.

applied magnetic field is perpendicular to the electrode according to the Lorentz force definition, the dimension is also an important factor that affect the convection. For the large vertical electrode, it is known that natural convection can affect the current direction [13]. If the dimension of the electrode is small or micro, radical components of the current can cause the azimuthal flow of the electrolyte near the edge of the electrode [11, 14]. Similar azimuthal flow is created above large electrodes that are magnetically structured. The electrolyte flow driven by the Lorentz force brings about the increase of limiting current. For example, the increase of the current has been observed in terms of the nitrobenzene reduction in tetrabutylammonium hexafluorophosphate in acetonitrile by using the paired and bonded

Figure 12.2 Magnetic gradient force under the magnetic field parallel (a) with or perpendicular (b) to the current. Magnetic gradient force is responsible for the movement of electrogenerated paramagnetic species into the magnetic field and independent of the current direction. Source: Adapted with permission from Ragsdale et al. [8]. Copyright (1998) American Chemical Society.

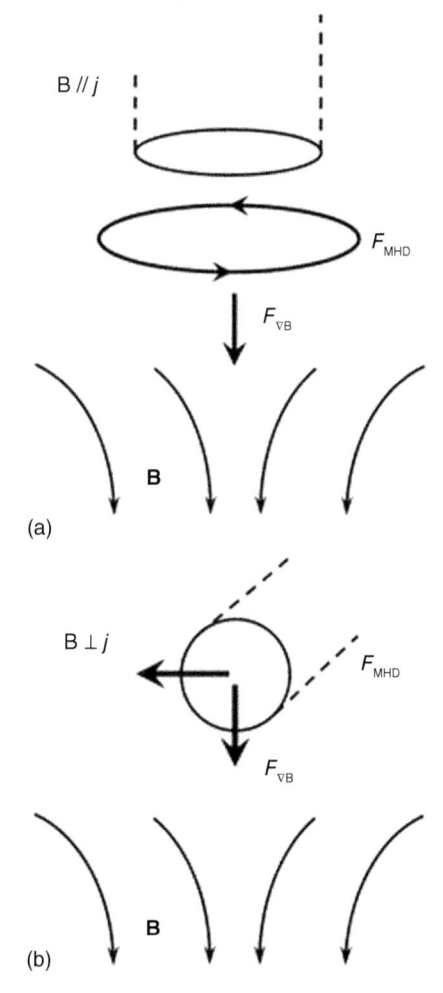

NdFeB magnets [15]. In addition to the static magnetic field, rotating magnetic field is another approach for the augmentation of the process intensity with the similar effect like mechanical mixing [16], for example, in controlling horseradish peroxidase activity and enhancing the decolorization process of the anthraquinone synthetic dye enzymatic [17].

12.2.2 Spin States of Electrocatalysts

Except for the force caused by the interaction between the charge current and magnetic field, the electronic state presented by the spin orientation is another key factor to influence the catalytic activity. This factor can be improved by applying magnetic field. Different from the current convection leading to the enhancement of the catalytic current, i.e. limiting current, the effect from electronic states presents the aspect in kinetics. Recent examples are the oxygen electrochemistry.

Changing the spin orientation of materials increases the reaction rate and adjusts the selectivity of products.

12.3 Spin-Related Electrode Reactions

Here, we choose some typical electrochemical process to exhibit the spin-related electrochemistry. The electrochemical processes are electrodeposition, HER, ORR, OER, CO_2 reduction, organic compound oxidation, battery reactions, and so on. We will combine the three factors such as attraction force, convection caused by the charge current and the magnetic field, and spin states of the electrocatalysts to discuss the electrochemical processes above.

12.3.1 Electrodeposition of Metals or Alloys

Electrodeposition is a method to prepared the metals or alloys by reducing the metal cations to the substrate in solution. Applying an external magnetic field has been employed to electrodeposition process. The effect of external magnetic field can merely apply to the paramagnetic cations, and it does not work for diamagnetic cations [18]. For example, Tschulik et al. showed a new method and employed a local magnetic gradient field for structured electrodeposition. They electrodeposited the paramagnetic Cu^{2+} ions with the magnetic susceptibility $\sim 16 \times 10^{-9}$ m^3 mol^{-1} by means of the application of magnetic field gradients and found a direct correlation of the distribution of magnetic induction **B** at the electrode and the thickness and morphology of the deposit. In contrast, as for the electrodeposition of Bi from dimagnetic Bi^{3+} ions, the influence from the magnetic gradient field on the deposit structure was not observed (Figure 12.3). This result indicates that the structuring effect is mainly attributed to the magnetic gradient force [18]. Based on experimental results and numerical simulation analysis, Murschke et al. showed the formation of microstructured copper by electrodeposition is mainly attributed to the magnetic gradient force and the deposition structure is not due to the action of the Lorentz force. It also stresses that the direction of the magnetic gradient force does not have the direct relationship with the bulk concentration of the ions with magnetic moment in solution and the direction of the magnetic gradient force direction does not change when the direction of magnetic field is inverted. Meanwhile, the susceptibility that determines the direction of the magnetic gradient force is not the total susceptibility in the bulk solution but the susceptibility of the deposited species [3]. Efforts have been made to apply the gradient magnetic field to the deposition of thick coatings. The thick (>100 μm) copper films with microstructure features by electrodeposition under gradient magnetic field were showed by Murdoch et al. in 2018. Investigations found that gradient magnetic field affects the grain size and twin boundary fraction; lower plating thicknesses and larger grain size form due to the depletion of Cu^{2+} ions at the substrate caused by the magnetic gradient force. In contrast, when the direction of gradient magnetic field and

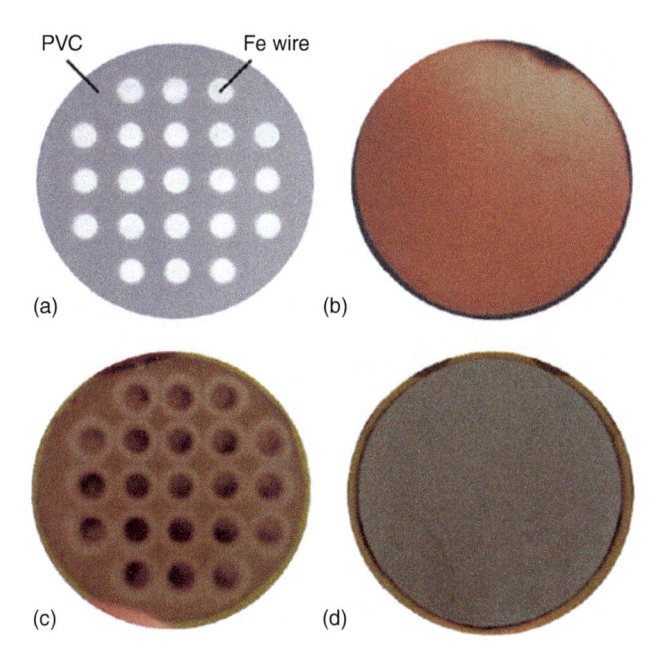

Figure 12.3 The optical images of the magnetic field template (a), the Cu electrodeposition in the absence (b) and presence (c) of the magnetic field template, and the Bi electrodeposition in the presence of magnetic field template (d). Source: Tschulik et al. [18]/with permission from Elsevier.

current is antiparallel, little effect on the thickness and grain size can be observed under gradient magnetic field. In both cases substantially higher twin boundary fraction forms because of the increased local convective flow [19]. In addition to the pure metal, external magnetic field can also affect the electrodeposition of alloys. Aaboubi and Msellak prepared the CoNiMo alloy by electrodeposition and noticed that the grains became more refined and the atom distribution became more homogeneous under the external magnetic field (Figure 12.4). It is worth noting that though the standard potential of Co^{2+}/Co (−0.280 V vs. standard hydrogen electrode (SHE)) is more negative than Ni^{2+}/Ni (−0.257 V vs. SHE), the presence of magnetic field makes the Co content increase and Ni decrease in the product [20]. It suggests that the external magnetic field can change the electrodeposition rate and trend for those metal ions with different magnetic moments. Besides, in the study by Karnbach et al., the magnetic gradient fields were used to electrodeposit the structured Co and Fe metals and the industrially highly relevant magnetic CoFe alloys. Meanwhile, the current efficiency for alloy deposition is significantly improved with respect to that for hydrogen reduction under magnetic gradient field by electrochemical quartz crystal microbalance study [21]. Magnetic fields are also employed for the metal recycle. Aboelazm et al. electrodeposited the cobalt element from the used lithium-ion battery cathode under magnetic fields, and results show that Co_3O_4 grew along the specific direction with a double yield (Figure 12.5) [22].

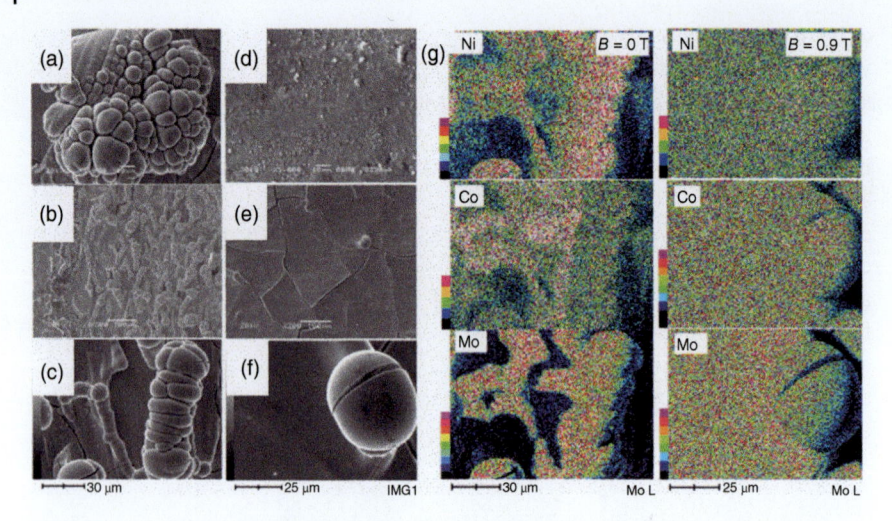

Figure 12.4 SEM images for a CoNiMo alloy electrodeposited from bath deposition. $E = -1.0$ V vs. SCE (a and d) and at $E = -1.20$ V vs. SCE (b, c, e, and f). Deposition time $t = 30$ minutes. (a–c) $B = 0$ T and (d–f) $B = 0.9$ T. Characteristic X-ray maps around a grain nodular. $E = -1.20$ V vs. SCE. CoNiMo alloy electrodeposited from bath deposition (g). Source: Aaboubi and Msellak [20]/with permission from Elsevier.

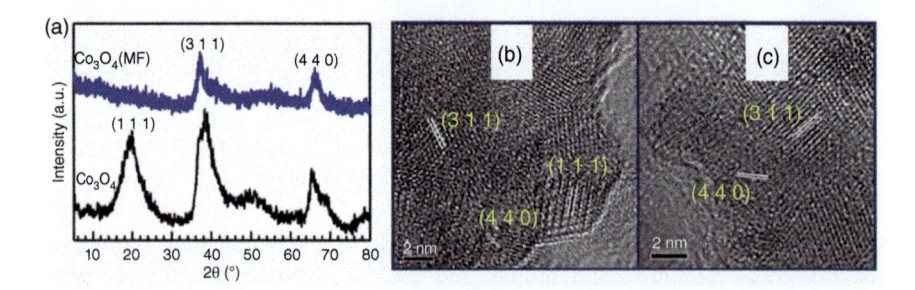

Figure 12.5 XRD (a) and HRTEM patterns of Co_3O_4 in the absence (b) and the presence (c) of magnetic field (MF). Source: Adapted with permission from Eslam et al. [22]. Copyright (2018) American Chemical Society.

12.3.2 Hydrogen Evolution Reaction

The electrochemical reaction accompanied with gas evolution is a very hot topic in the area of the industrial production [23], energy usage in space [24], and nanoscience [25]. In the area of hydrogen evolution by electrolysis under magnetic fields, two major aspects become the major investigation objects: one of which is the bubble management by means of magnetic field application and the other is the activity enhanced by the spin state manipulation in the absence or presence of magnetic field. Hydrogen gas is the desired product. However, in the electrolysis process since the nonconducting hydrogen bubbles appear, the ohmic voltage drops, and electrode overpotential increases [26, 27]. As for the efficiency of energy conversion in electrochemical hydrogen evolution, the behavior and management of hydrogen bubbles is one of the most important issues [28]. In industrial electrolysis, the electrodes are generally tall and installed in vertically that at high current situation, there will be a self-pumping effect to drive the bubble dispersion upward to enhance the mass transfer [29–31]. However, as for HER in water electrolysis, it is the considerably larger overpotential that is required for high current density. The energy is also consumed substantially in electrolyzer due to the resistance of electrolyte [32]. In addition, increasing the void fraction in the solution and the covered parts in the electrodes with the bubbles caused by the high current situation acts as a barrier for HER [33, 34]. Applying a magnetic field is a successful approach for enhancing the HER [35]. The reason is that the force caused by magnetic field accelerates the floating dispersion flow in the electrochemical cell including the gas bubbles [36–38]. As for the half-cell reaction, hydrogen evolution process in water electrolysis, the electrode potential E, is expressed by the following equation:

$$E = E_{iR} + \eta_s + \eta_{cs} + \eta_{ci}$$

where E_{iR} and η_s are the ohmic resistance between the working and reference electrodes and the surface overpotential for the charge transfer reaction, respectively, and are the overpotential from concentration in the vicinity of the electrode surface, η_{cs} is the supersaturation of the dissolved gas, and η_{ci} is the depletion of the reacting ions. The overpotential components η_{cs} and η_{ci} are small enough to be neglected according to the study of Leistra and Sides that in the case of the hydrogen evolution in 0.5 M H_2SO_4, hydrogen supersaturation is dominant [39]. Therefore, the electrode potential E can be simplified as follows:

$$E = E_{iR} + \eta_s$$

To investigate the effect from the convection caused by the Lorentz and eliminate the contribution from the magnetic gradient force, a uniform magnetic field is commonly used. Studies have found that the solution resistance E_{iR} can be influenced by via the convection caused by the Lorentz force. Matsushima et al. applied a uniform magnetic field to investigate the electrode potential of hydrogen evolution in 0.5 M H_2SO_4. They found that the ohmic resistance between working and reference electrodes can be reduced by means of the current interrupter method. Meanwhile, the convection is found to restrict the increase of supersolubility considerably and promote the mass transfer of the dissolved gas. They also found that at intermediate current density region with respect to the mass transfer coefficient, the supersolubility is affected to a larger content [40]. The convection caused by the Lorentz force is also found to influence the desorption of adsorbed H_2 bubbles. Koza et al. applied a uniform magnetic field perpendicular to the electrode to investigate the effect of on the hydrogen evolution and found that desorption of hydrogen is facilitated [38]. It is noted that though the magnetic field is perpendicular to electrode, which suggests that the Lorentz force is zero according to the Lorentz force definition, here Lorentz force arises as a result of the distorted current distribution around H_2 bubbles [41]. Koza et al. also found and proved that when the directions of the magnetic field and the primary current density lines are parallel, the desorption of hydrogen can be enhanced at a drastically reduced bubble size [42]. In the following investigation, Koza et al. found that the desorption of hydrogen is enhanced by the presence of uniform magnetic field irrespective of the magnetic field direction. The magnetic field can reduce the mean bubble size and narrow the bubble size distribution. Furthermore, the magnetic field strongly retards the fractional bubble coverage, and thus more active sites are available for hydrogen evolution [43]. As for the issue of hydrogen bubble desorption from the electrode, Diao et al. recorded the noise spectrum of overpotential fluctuations of hydrogen evolution on a platinum electrode in an aqueous electrolyte at a constant density under a static magnetic field. In addition to that the overpotential for hydrogen evolution decreasing with applied field, they also observed $1/f^2$ variation of the power spectrum characteristic of droplet coalescence at the frequency higher than 10 Hz. There exists a threshold strength 0.5 T beyond which the released bubbles approximately have the double size, leading to the enhanced coalescence of small bubbles across the electrode surface induced by the convection from Lorentz force. The effect is similar like the mechanical agitation field that improves the hydrogen evolution activity [44]. Liu et al. investigated the behavior of hydrogen bubbles under electrode perpendicular magnetic field by using the electrode partially covered with hydrophobic materials. Under the magnetic field, the anchored bubbles are observed at hydrophobic islets observed and release randomly, and the convection caused by magnetic field exactly facilitates the pair of bubbles to release from neighboring hydrophobic islands (Figures 12.6 and 12.7) [45]. However, the situation is different if the electrode is the microelectrode. The convection caused by magnetic field has the effect of stabilizing the bubbles, and consequently there will be larger bubbles that release from the electrode surface [46]. Besides, foam electrodes are also used to investigate water splitting under

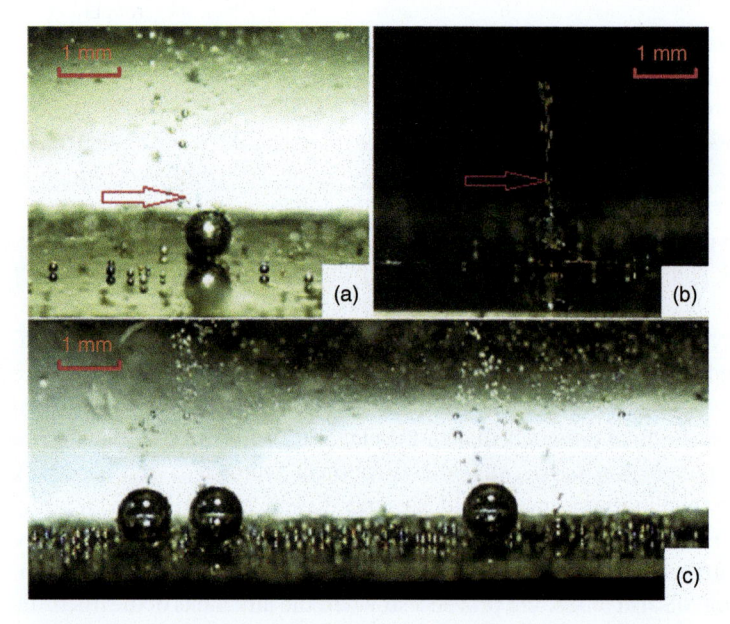

Figure 12.6 Hydrogen bubble growth at single hydrophobic islet with magnetic field (a) and without magnetic field (b), pair of bubbles evolution from neighboring hydrophobic islets (Ld = 1.1 mm) under 0.9 T field (c). Source: Liu et al. [45]/with permission from Elsevier.

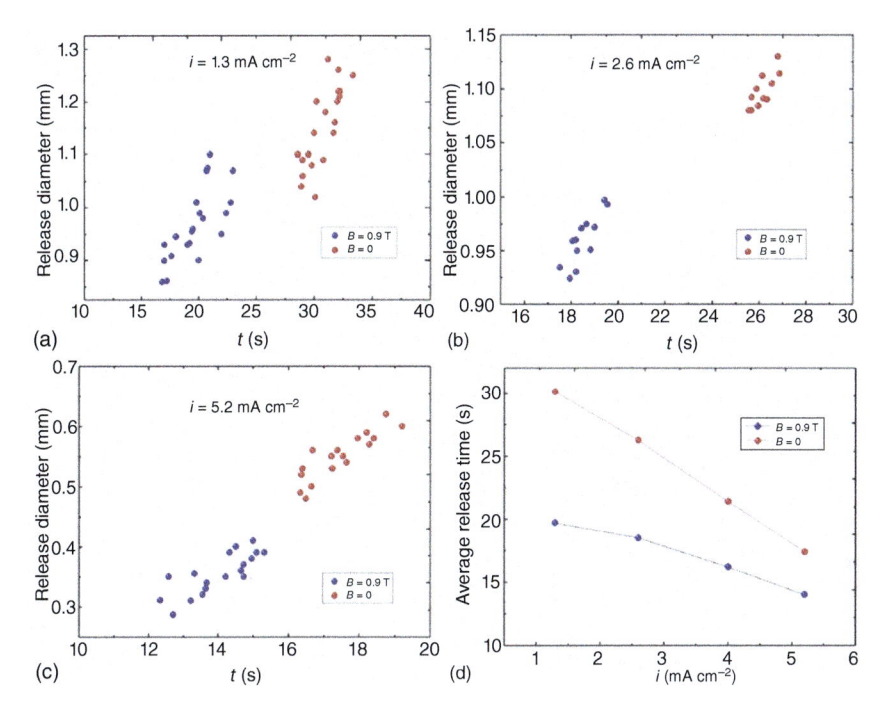

Figure 12.7 Bubbles release from the pair of neighboring hydrophobic islets with 1.1 mm distance between each other under various current conditions (a, 1.3 mA cm^{-2}; b, 2.6 mA cm^{-2}; c, 5.2 mA cm^{-2}) with ($B = 0.9$ T) and without magnetic field ($B = 0$) and the relationship between average release time and current (d). Source: Reprinted from Liu et al. [45]. Copyright (2019), with permission from Elsevier.

the magnetic field. Foam electrodes have their advantages that there is more space for the reaction because internal surfaces can be utilized and active areas are not limited on the outer surface. Especially, when the inner bubbles become large enough, they will dash out to the solution at a quite high speed, and the solution will be agitated intensely. This effect can be strengthened by magnetic field. For example, Liu et al. used the copper foams as the cathode and nickel foams as the anode for water splitting in alkaline. The reduction of about 3.4% energy consumption is observed under the magnetic field. Moreover, uneven distributed Lorentz force will induce the convection scouring the inner space of the foam electrodes [47].

Except for that the force caused by magnetic fields has an obvious effect on hydrogen evolution, the spin state of the catalyst itself has been reported to affect the HER activity. Here, we will use the metal dichalcogenide family MS$_2$ (M = Mo, W) that intensively investigated in HER area to demonstrate the recent progress and discuss the HER activity correlation with the spin states of catalysts. When talking about the spin states of MS$_2$ (M = Mo, W), it is necessary to consider the crystal structure. Taking MoS$_2$ as an example, the Mo atom is sixfold coordinated and hexagonally packed between the S atomic layers. There are two distinct typical phases depending on the S atom arrangement. One is the 2H phase (D_{3h}, semiconductor, band gap

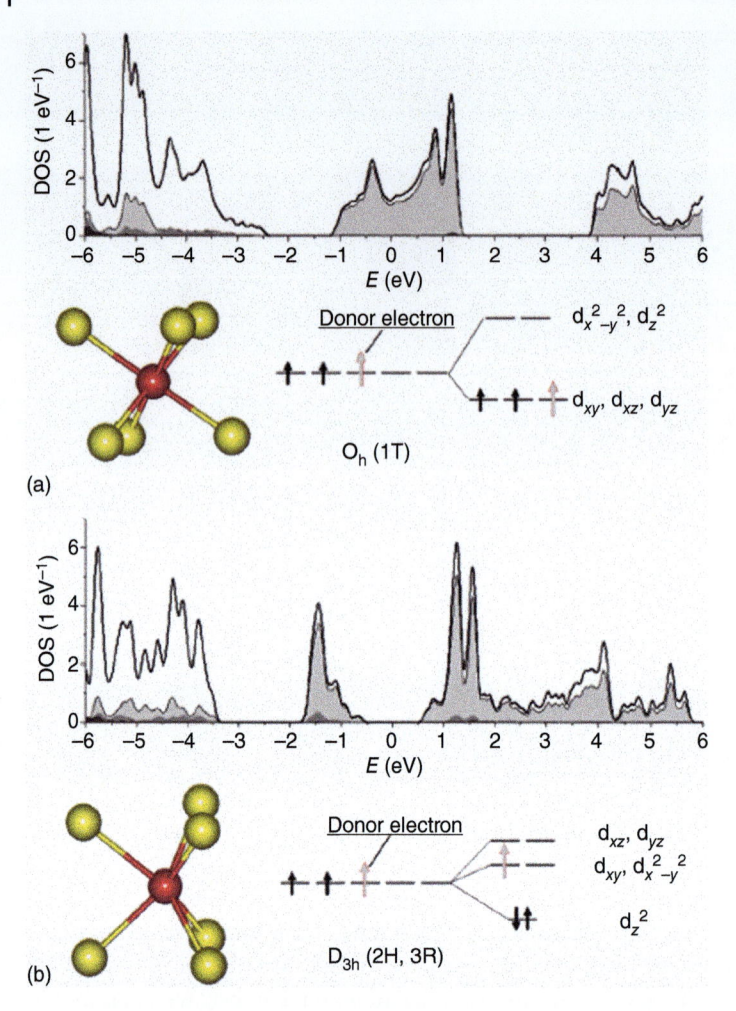

Figure 12.8 Total (full line) and partial Mo 4d densities of states for 1T (a) and 2H (b) MoS$_2$ and their d orbital splitting according to the crystal field theory. Source: Adapted with permission from Andrey et al. [48]. Copyright (2011) American Chemical Society.

~2.2 eV) with two S–Mo–S layers built from edge-sharing MoS$_6$ trigonal prisms, and the other one is the 1T phase (O$_h$, metal) with a single S–Mo–S edge-sharing MoS$_6$ octahedra layer with octahedral O$_h$ phases. These two phases can be converted mutually. The 2H phase is thermodynamically stable, and the 1T phase is not naturally found. The MoS$_2$ electronic structure strongly depends on the phase. According to the crystal field theory, octahedrally coordinated Mo atoms form degenerate d$_{xy}$, d$_{yz}$, d$_{zx}$, and d$_{x^2-y^2}$, d$_{z^2}$ orbitals. On the other hand, Mo atoms in trigonal prismatic coordination form three groups: d$_{z^2}$; degenerate d$_{x^2-y^2}$, d$_{xy}$; and degenerate d$_{xz}$, d$_{yz}$ (Figure 12.8) [48–50]. The HER activity is highly dependent on the phase of MS$_2$ (M = Mo, W), one of which with 1T phase has a higher HER activity than that with 2H phase (Figure 12.9) [51]. Chemical exfoliation is one method to

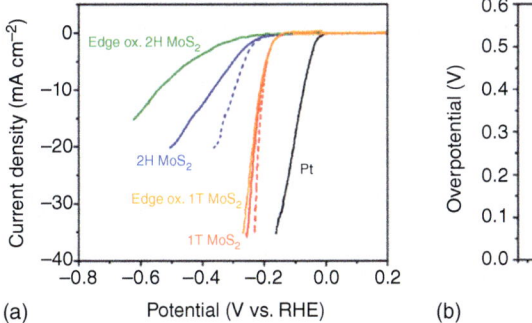

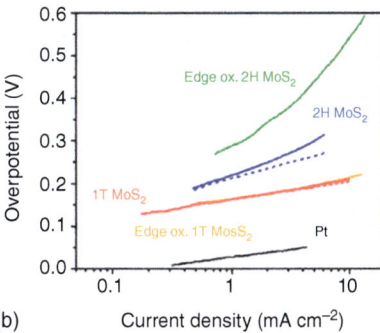

Figure 12.9 Polarization curves (a) and the corresponding Tafel plots (b) (dash line, after iR correction) of 2H and 1T MoS$_2$ before and after edge oxidation and Pt. Source: Adapted with permission from Voiry et al. [51]. Copyright (2013) American Chemical Society.

prepare the 1T MoS$_2$ from 2H MoS$_2$ to enhance the HER activity. For example, Mark A. Lukowski et al. grew 2H MoS$_2$ on the graphite and use the lithium intercalation to exfoliate for the 1T MoS$_2$ and found the HER activity was dramatically enhanced [52]. Voiry et al. prepared chemically exfoliated monolayered WS$_2$ nanosheets and found in the as-exfoliated nanosheets that the high concentration of strained 1T phase contributes to the enhanced HER activity [53]. In addition, the active sites of 2H- and 1T-phase for HER are reported to be different. In 2007, Jaramillo et al. correlated the HER activity of 2H-MoS$_2$ with the edge site number and found a linear relationship by experimental results [54]. Hinnemann et al. reported that the edge of MoS$_2$ contributes to the catalytic activity and the basal surfaces are catalytic inert by calculation [55]. Voiry et al. partially oxidized 2H-phase MoS$_2$ and noticed the significantly reduced HER activity, while the HER activity in 1T-phase MoS$_2$ remains unaffected after partially oxidation, suggesting that edges are not the main active sites in 1T-phase (Figure 12.9) [51]. However, it could not give an explanation why after oxidation the HER activity of 1T phase MoS$_2$ is not affected. To enhance the HER activity of MoS$_2$, carbon material combination is a good way. For instance, the introduction of reduced graphene oxides makes the MoS$_2$ have an abundance of exposed edge and thus improve the HER activity increase, which is a contrast to large aggregated MoS$_2$ [56]. It is worth noting that Zhou et al. [57] employed the external magnetic field to enhance the HER performance on bowl-like ferromagnetic MoS$_2$ with a magnetic moment ($\sim$2.4 μ_B) (Figure 12.10). The origin of magnetic properties of bowl-like MoS$_2$ is the symmetry destroyed, and there are unpaired spins on the edge terminations, in particular Mo terminations [58, 59]. In contrast, the bilayer MoS$_2$ is nonmagnetic due to the symmetry of spin-up and spin-down bands. Thus, under external vertical magnetic field H, the bowl-like MoS$_2$ could be more sensitive to obtain extra energy than nonmagnetic bilayer MoS$_2$, and the electrons in the bowl-like MoS$_2$ structure are more ready to transport than those in bilayer MoS$_2$. As a result, under the external magnetic field, the electrons in ferromagnetic MoS$_2$ are stimulated easier and transfer from the substrate to the active sites. Moreover, the ferromagnetic MoS$_2$ exhibits an obvious HER enhancement (Figure 12.10) [60]. It is worth emphasizing that the electron transfer in intralayer and between interlayers

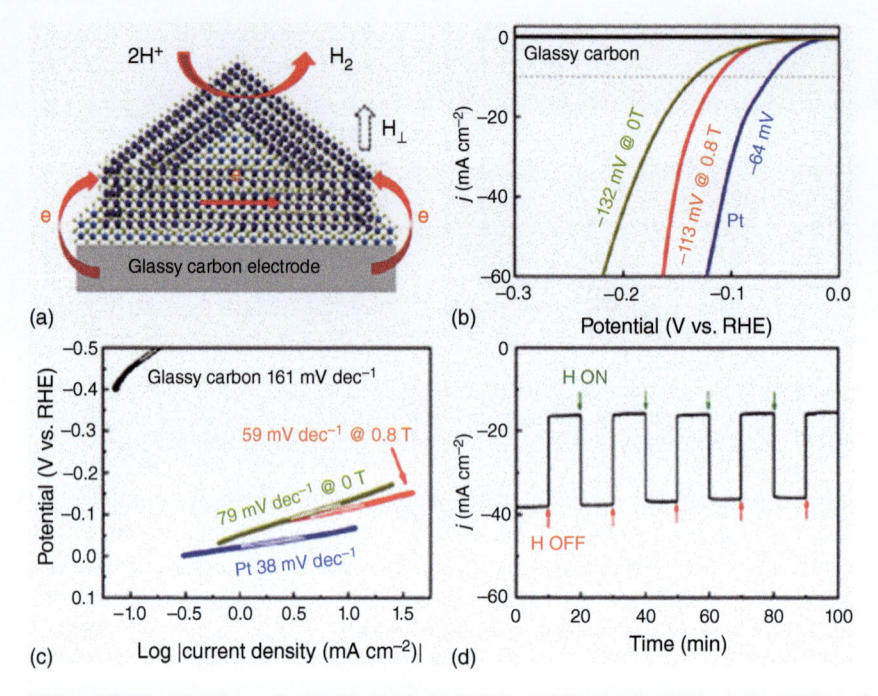

Figure 12.10 (a) Schematic of electron transfer in bowl-like MoS$_2$ flakes during HER under the vertical magnetic field. (b) HER curves of Pt and bowl-like MoS$_2$ flakes in the absence and presence of the external vertical magnetic field. (c) Tafel plots corresponding to the (b). (d) j–t curves of bowl-like MoS$_2$ flakes at −150 mV vs. RHE under pulsed magnetic field. All measurements were carried out in Ar saturated 0.5 M H$_2$SO$_4$ solution. Source: Adapted with permission from Zhou et al. [57]. Copyright (2020) American Chemical Society.

coexists in MoS$_2$ during HER process. The resistivity between the interlayers is 2 orders of magnitude higher than that of in the intralayer [60, 61]. As a result, the electron transfer between interlayers dominates the HER activity in layered MoS$_2$ [62], and increasing the electron transfer between interlayers is a promising strategy to improve the HER activity. Besides, creating anion vacancy is one approach for promoting HER activity. For example, Wang et al. prepared the MoS$_2$ with S vacancy and optimized the 12.11% single S-vacancies, having the highest HER activity. Meanwhile, MoS$_2$ with single S-vacancies is better than that with agglomerate S-vacancies [63]. Except for 2D materials, the perovskite oxide with the superexchange interaction exhibits excellent HER activity. Dai et al. synthesized the SrTi$_{0.7}$Ru$_{0.3}$O$_{3-\delta}$ perovskite oxide with a better OER activity than SrTiO$_3$ and SrRuO$_3$ in alkaline media, in which Ti, Ru, and O sites function to dissociate water, OH* desorption, and H* adsorption and H$_2$ desorption, respectively. In SrTi$_{0.7}$Ru$_{0.3}$O$_{3-\delta}$, charge redistributes between titanium and ruthenium cations from Ti^{4+} and Ru^{4+} to Ti^{3+} and Ru^{5+} [64].

12.3.3 Oxygen Evolution Reaction

OER is an important gas-evolving reaction widely applied in the water splitting and metal–air battery. Designing of the catalyst is playing an important role, and

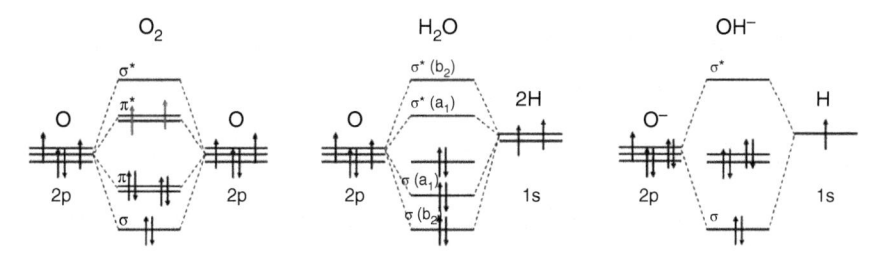

Figure 12.11 Molecular orbital diagram of O_2, H_2O, and OH^-. Source: Reproduced with permission from Sun et al. [65]. Copyright (2020) Wiley-VCH.

in recent years, more and more attention focuses in the spin states of the catalysts. The basic principle of oxygen catalyst design is the difference of the spin states of oxygen species. Considering the process from water oxidation to oxygen, the background states of water and oxygen are singlet and triplet, respectively. The other oxygen species like dioxo $(O^{2-})_2$ and peroxo (O_2^{2-}) are singlets (Figure 12.11) [65]. Therefore, the difference of oxygen species in spin states causes the importance of developing spin catalysts. Here, we take OER on metal oxides in alkaline as an example to illustrate the OER process. First, the metal adsorbed by the hydroxyl (M–OH) deprotonates and loses one electron to form M–O, and then M–O combines with OH^- in solution together with losing one electron to form M–OOH. The M–OOH continues deprotonating and losing one electron to M–OO, and then O_2 evolves from the M–OO followed by OH^- adsorption [5]. Among these step, the process from M–OO to oxygen release involves the change of spin states. O_2^{2-} has a molecular orbital $\sigma(2)\pi(4)\pi^*(4)\sigma^*(2)$. To form O_2 with the molecule orbital $\sigma(2)\pi(4)\pi^*(4)\sigma^*(2)$, the removal of two electrons with opposite spin in the O–O σ^* orbital is the first step, and further oxidation to O_2 requires to remove the two electrons with the same spin in the O–O π^* orbitals. In 1999, McGrady and Stranger calculated O—O σ and π bonds in a peroxo-bridged manganese dimer and pointed out that the coupling between the metal centers, ferromagnetic or antiferromagnetic, influences the stability of the intermediate species. Moreover, oxidative formation of π component of the O—O bond is favored by ferromagnetic coupling [66, 67]. In magnetic metals, electrical resistance is crucially affected by the scattering of electrons on the magnetic lattice of the crystal. Scattering is weaker when the spin direction of the traveling electrons is opposite to the magnetization of the lattice than when it is parallel [68–70]. The selective removal of spin-oriented electrons from the active sites, at the applied potential, will allow the appropriate magnetic moment for oxygen molecules to evolve [71].

So far, there is a lot of studies about the catalysts with various spin states having different OER activity. It should be emphasized at first the OER activity is not expected to be quantitated to the specific descriptor because there are many important aspects relevant to OER activity like the number of charge carrier, the adsorption energy of oxygen intermediates, the conductivity of the catalyst, and the surface thermodynamics. Pioneering experimental results by Shao-Horn's group have revealed that the decrease in Co—O bond strength and the increase in spin states can highly affect the oxygen surface exchange in $LaCoO_3$ model [72]. Guo et al. reported that change the $Ca_{0.9}Yb_{0.1}MnO_{3-\delta}$ electronic state by using hydrogen treatment can tune the OER activity. The OER activity has the same order with the effective magnetic moment in $Ca_{0.9}Yb_{0.1}MnO_{3-\delta}$ series [73]. In 2017, Xu's group doped less amount of Fe into $LaCoO_3$ for the OER enhancement. High spin state of Co^{3+} was induced by proper Fe doping, and the percentage of high spin of Co^{3+} is high for those catalysts with high activity (Figure 12.12) [74]. They also substituted Fe for partial Co in $ZnCo_2O_4$ to realize the OER activity enhancement, and the trend of valence state of Co from 3+ to 4+ was observed, which means the electronic state Co^{3+} partially changes from $t_{2g}^6 e_g^0$ to $t_{2g}^5 e_g^0$. Some other possibilities cannot be ruled out [75]. Vanadium was also introduced to $LiCo_2O_4$ to obtain $LiCoVO_4$ to stabilize high spin Co^{2+} ($t_{2g}^5 e_g^2$, $s = 3/2$) for OER enhancement [76]. The spin-selective electron transfer and orbital symmetry are also emphasized to explain the OER activity. It is well known that Mn element is nature-selected in oxygen evolution complex. In the observation of Xu's group, they noticed the OER activity of $ZnMn_2O_4$ is lowered than that of $ZnCo_2O_4$ [5]. The e_g filling number of Mn is one in $ZnMn_2O_4$, whereas the number of Co is zero in $ZnCo_2O_4$. If the e_g descriptor established in perovskite oxides was suitable in spinel oxides, $ZnMn_2O_4$ would have a higher OER activity than $ZnCo_2O_4$. To explain this unexpected phenomenon that $ZnCo_2O_4$ has higher OER activity than $ZnMn_2O_4$, Xu's group found that the rate-determining step gradually changes from the first step and that deprotonation of OH^- adsorbed on $ZnMn_2O_4$ to the second step, O—O bond formation, by comparing the Tafel slope of $ZnCo_xMn_{2-x}O_4$ ($x = 0, 0.2, 0.4, 0.6, 0.8, 1.0, 1.2, 1.4, 1.6, 1.8, 2.0$) systematically. By the density of state analysis, the electron transfer close to Fermi level is limited by the spin selection and has a higher band gap from the energy state of metal adsorption sites to the conduction band, whereas it does not happen in $ZnCo_2O_4$. It indicates that the lower OER activity in $ZnMn_2O_4$ can be attributed to the spin selection in the first step (Figure 12.13) [5]. In the study on $Zn_xCo_{3-x}O_4$ for OER by Ramsundar et al., they found that the OER activity of $Zn_xCo_{2-x}O_4$ is directly correlated to number of high spin Co^{3+} and at the same time optimized the Zn ratio, $Zn_{0.8}Co_{2.2}O_4$ [77]. However, it still suggests that spin state increases. In 2018, a noteworthy work by Jiang et al. [78] noticed that the overall spin state could not be completely applied for estimating OER activity, and the local spin state of metal element should be considered. They prepared $La_2NiMnO_{6-\delta}$ annealed at different annealing temperatures. The valence states of Mn and Ni change from 4+ to 3+ and from 2+ to 3+, respectively, with the increase of annealing temperature. At low annealing temperature, Mn and Ni exist in the electron configuration $t_{2g}^3 e_g^0$ (Mn^{4+}) and $t_{2g}^6 e_g^2$ (Ni^{2+}), respectively, while at high annealing temperature Mn exists in $t_{2g}^3 e_g^1$ and Ni in $t_{2g}^6 e_g^1$. The t_{2g} orbital and the average e_g number are not changed, and according to the report by Shao-Horn et al., the OER activity should be same. However, experimental results show that $La_2NiMnO_{6-\delta}$ annealed at high temperature has a high OER activity, where Ni and Mn have the electron configuration of $t_{2g}^3 e_g^1$ and Ni in $t_{2g}^6 e_g^1$. They proposed the average deviation from unity e_g occupation is a suitable activity descriptor. This work suggests the situation that only one electron in e_g orbitals of each metals is preferable for OER activity.

At the same year, another excellent work on La$_2$NiMnO$_6$ for OER was reported by Tong et al. They stressed the superexchange interaction in La$_2$NiMnO$_6$ has a very important role in promoting surface reconstruction and thus enhanced OER activity (Figure 12.14). When the particle size of La$_2$NiMnO$_6$ decreases from bulk to about 33 nm, the electronic states of nickel and manganese change from Ni^{2+}–O–Mn^{4+} to Ni^{3+}–O–Mn^{3+}. The electron configuration of Ni^{3+} ($t_{2g}^6 e_g^1$) and Mn^{3+} ($t_{2g}^3 e_g^1$) causes strong Jahn–Teller distortion of NiO$_6$ and MnO$_6$ octahedra, where the elongated Ni—O and Mn—O bond would facilitate the formation of Ni and Mn oxides or hydroxides. The facilitating process was confirmed by double-layer capacitance [79].

It is worth noting that OER is a surface reaction, and the surface spin state is also a key factor except for the bulk spin state. Investigating the magnetic properties of the materials with relative highly OER activity is also one approach for linking the activity and spin states. RuO$_2$ is one of the best OER catalyst, which is not magnetic, while RuO$_2$ (110) surface is magnetic [80]. IrO$_2$ is also reported to have a weak ferromagnetic coupling [81]. The surface spin state of nanosized LaCoO$_3$ also was changed from low spin to high spin by reducing the nanoparticle size, and meanwhile, the OER activity is enhanced [82]. The moderate OER activity material Co$_3$O$_4$ is paramagnetic at room temperature, and the (110)-B surface is magnetic [83]. Experimentally, Ma et al. noticed that high spin state in Co$_3$O$_4$ can promote OER in neutral media by comparing the OER activity of synthesized Co$_3$O$_4$ quantum dot and Co$_3$O$_4$ nanocubes. The Co$_3$O$_4$ quantum dots have a higher content of high spin Co^{2+} and Co^{3+} [84]. Co$_3$S$_4$ nanosheets was reported to have a superior OER activity than bulk materials. It can be ascribed to the exposition of the active octahedra with Jahn–Teller distortions. As the thickness of nanosheets decreases to atomic scale, Co^{3+} changes from the low spin states to the high spin states [85]. The different orientation of surface is also investigated to illustrate the spin influence on OER activity. Tong et al. synthesized the LaCoO$_3$ with different surface orientation exposures (100), (110), and (111) faces. The different orientation caused different distortion degrees in CoO$_6$ octahedrons, which induce the transition of spin state from low spin $t_{2g}^6 e_g^0$ to intermediate spin state $t_{2g}^5 e_g^1$. The (100), (110), and (111) surfaces have 87%, 48%, and 31% intermediate spin states, respectively, and the OER activity has the same trend. The surface with a more intermediate spin state contributes to a higher OER activity [86]. Also, ultrathin nanosheets of NiSe$_2$ and NiS$_2$ exhibit promoted OER activity compared with the bulk ones, and this reason is attributed to the delocalized spin of nickel atoms in ultrathin structure, which provide both high electrical conductivity and low adsorption of reaction intermediates [87]. The van der Waals layer epitaxy structure based on the spin-coupled interface is also used to promote the OER activity. Tiwari et al. activated the inactive basal planes of MoS$_2$ by incorporating Mo$_2$C to realize MoS$_2$ lattice strain. The spin coupling exists between MoS$_2$ and Mo$_2$C epitaxy along (100) facets, which provides an easy electron transfer for enhancing OER activity [88]. Tkalych et al. indicated that the magnetic states of the β-NiOOH are a ferromagnetic state on the surface in contrast with the antiferromagnetic state in bulk (Figure 12.15) [89].

Except for single-metal hydroxides or oxyhydroxides, layered double hydroxides (LDHs), belonging to hydrotalcite-like compounds $\left([M^{2+}_{1-x} M^{3+}_x (OH)_2]^{x+} \left[A^{m-}_{x/m} \right] \cdot nH_2O \right)$

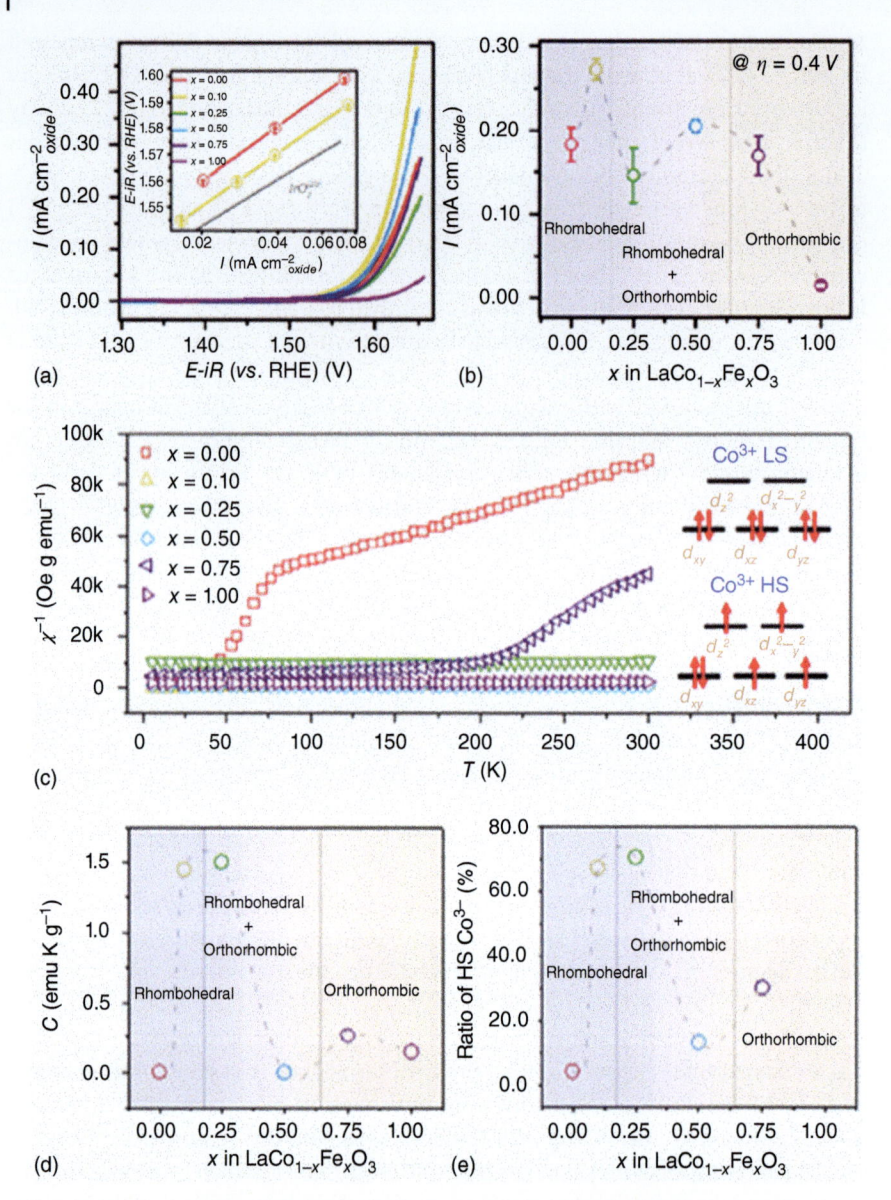

Figure 12.12 (a) Representative cyclic voltammetry (CV) curves (the second cycle) of oxygen evolution reaction in 1.0 M KOH. The inset of panel a shows the Tafel plot of $LaCo_{0.90}Fe_{0.10}O_3$ (yellow line) for comparison with unsubstituted $LaCoO_3$ (red line). (b) Specific current density for $LaCo_{1-x}Fe_xO_3$ at 1.63 V vs. RHE. (c) Spin structure analyses of $LaCo_{1-x}Fe_xO_3$ (x = 0.00, 0.10, 0.25, 0.50, 0.75, 1.00). (a) The temperature-dependent inverse susceptibilities for $LaCo_{1-x}Fe_xO_3$ samples. Inset: the plots for Co^{3+} LS ($t_{2g}^6 e_g^0$, S = 0) and Co^{3+} HS ($t_{2g}^4 e_g^2$, S = 2) in the octahedral ligand field. (d) Curie constant for $LaCo_{1-x}Fe_xO_3$ (x = 0.00, 0.10, 0.25, 0.50, 0.75, 1.00). (e) The ratio of HS Co^{3+} in $LaCo_{1-x}Fe_xO_3$ with x = 0.00, 0.10, 0.25, 0.50, 0.75. Source: Adapted with permission from Duan et al. [74]. Copyright (2017) American Chemical Society.

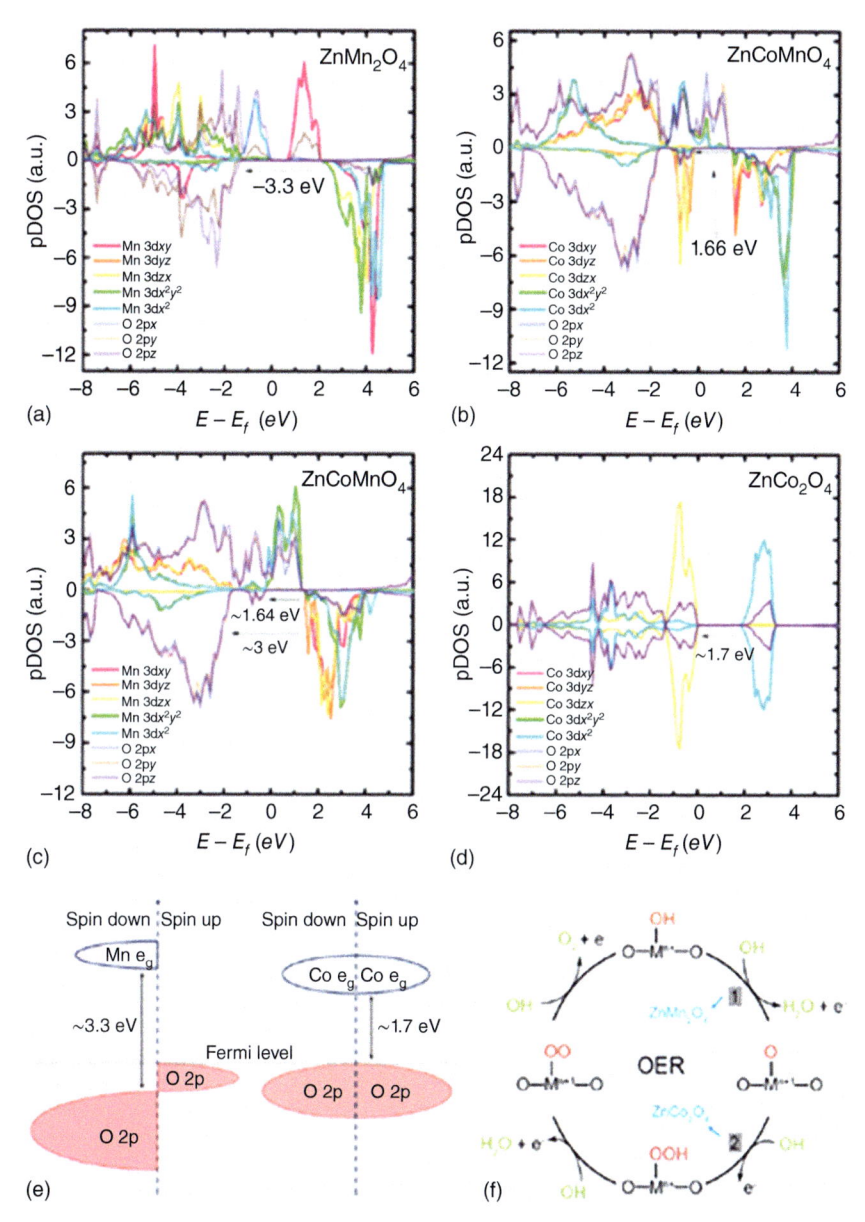

Figure 12.13 Projected DOS (pDOS) for O 2p orbital and Mn 3d orbital of $ZnMn_2O_4$ (a). pDOS for O 2p orbital and Co 3d orbital of $ZnCoMnO_4$ (b). pDOS for O 2p orbital and Mn 3d orbital of $ZnCoMnO_4$ (c). pDOS for O 2p orbital and Co 3d orbital of $ZnCo_2O_4$ (d). Illustration of spin selection and energy band gap in $ZnMn_2O_4$ and $ZnCo_2O_4$ (e). The contributions from metal orbitals below Fermi level and oxygen orbitals above Fermi level are not shown for better displaying the electron transfer between the M−OH_{ad} and its neighboring lattice oxygen because of the shorter distance and stronger orbital overlap of M−O than M−M. Proposed OER RDSs on spinel oxides (step 1 for $ZnMn_2O_4$ and step 2 for $ZnCo_2O_4$) (f). Source: Adapted with permission from Sun et al. [5]. Copyright (2019) American Chemical Society.

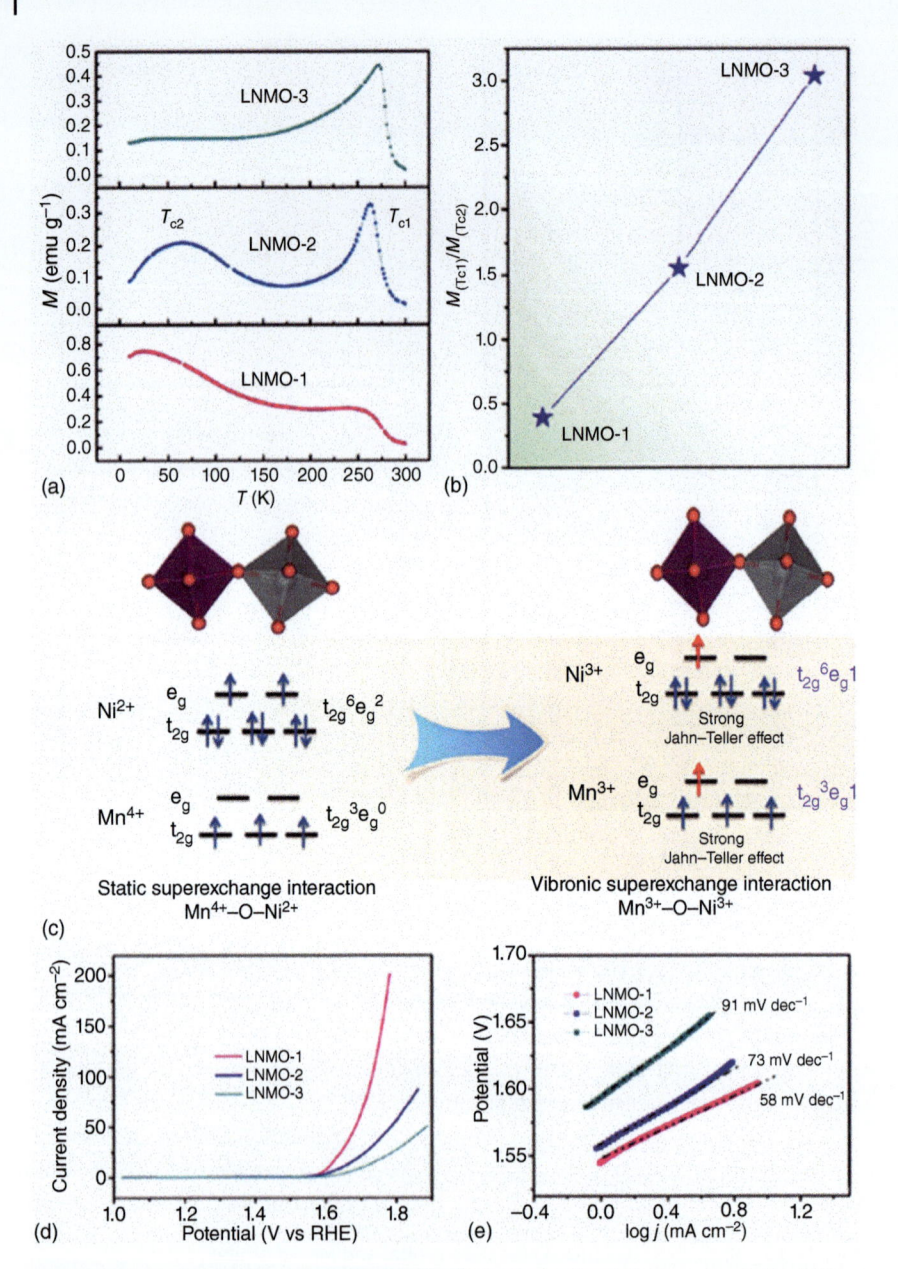

Figure 12.14 (a) Temperature dependence of magnetization (M–T) curves under ZFC. (b) Magnetization ratio of T_{c1}/T_{c2} for LNMO samples with different particle sizes. (c) Schematic for the deformation of MnO_6 and NiO_6 octahedra. (d) OER polarization curves of LNMO samples. (e) Corresponding Tafel slopes of LNMO samples. Source: Adapted with permission from Tong et al. [79]. Copyright (2018) American Chemical Society.

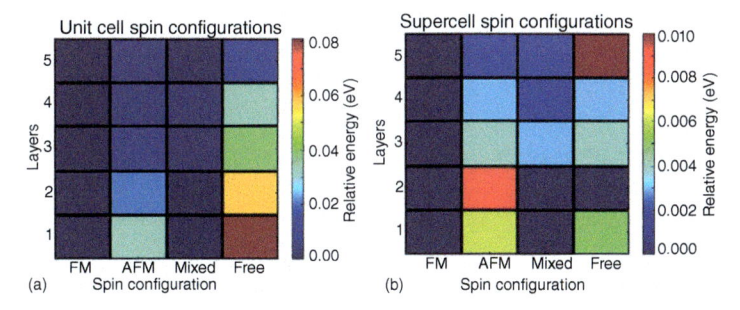

Figure 12.15 Spin configuration relative stabilities for unit-cell (a) and supercell (b) structures. Source: Adapted with permission from Alexander et al. [89]. Copyright (2017) American Chemical Society.

or the anionic clays, are a family for OER catalysis. LDHs are built by stacking the brucite-like octahedral layers with the positive charge, in which part of M^{2+} cations are replaced by M^{3+} ones. The charge is compensated by accommodated solvated anions (A^{m-}) between the layers [90]. Charge transport can take place in LDHs along the layers and between interlayers under applying potential, which is attributed to M^{II}/M^{III} redox reaction and a migration of anions between the layers for compensating the extra charge [91, 92]. Among LDHs, Ni–Fe LDHs have a good OER performance [93–95]. The study by Oliver-Tolentino et al. indicated that Fe plays an important role in magnetic properties of LDHs, and there exists a superexchange interaction among metal centers through the OH bridges across the cationic sheets favored by ferromagnetic interactions in Ni–Fe LDHs. Such a superexchange interaction enhances OER performance in alkaline media [96]. Dai et al. compared $SrCo_{0.9}Ru_{0.1}O_{3-\delta}$ with $SrCoO_{3-\delta}$ and $SrRuO_3$ for their OER activity. There exists Co^{3+}–O–Ru^{5+} superexchange interactions in $SrCo_{0.9}Ru_{0.1}O_{3-\delta}$, which induces abundant Co^{3+}/Co^{4+}, Ru^{5+}, and O_2^{2-}/O^- OER active species, metal-oxygen covalency, and high electrical conductivity [97]. In another promising OER catalyst Co–Pi with reasonable efficiency and stability in neutral media [98–100], the influence of spin state is also observed. In the study of Co–Pi catalyst with different loading amounts, Co–Pi is found to have high spin states of Co^{2+} ($s = 3/2$) and a $3d^8L$ ($s = 1$) charge transfer state, and the high loading of Co–Pi has additional low spin state Co^{3+} ($s = 0$) and a $3d^7L$ state ($s = 1/2$). The ability to have the Co^{2+} and Co^{3+} is significant for Co–Pi catalysts. The low spin state of Co^{3+} should facilitate the water adsorption and oxidation. The high spin state of Co^{2+} facilitates oxygen desorption because of the spin forbidden of back reaction [101]. Shichuan Chen et al. introduced structural distortion in the atomic layers to form the delocalize spin states in ultrathin nanosheets of $NiSe_2$. The delocalized spin states for Ni atoms can provide not only high electrical conductivity but also low adsorption energy between the actives and reaction intermediates for system. As a result, it led to a dramatic enhanced OER activity compared with their corresponding bulk samples [87]. In 2019, an excellent work on external magnetic field controlled OER activity has been reported by Galan-Mascaros's group in 2019. They employed a magnetic field less than 450 mT to promote the OER activity above 100% for highly magnetic catalysts, like the NiZnFe$_4$O$_x$. They found that the current enhancement is strong in high pH environment and becomes weaker at the low pH environment [102].

Another category of electrocatalyst related to the spin transition is the chiral molecule. Naaman's group has done a very comprehensive study in this area and focused on photoelectrochemical catalysis to regulate the conversion among water ($s = 0$), hydrogen peroxide ($s = 0$), and oxygen ($s = 1$) by using chiral and achiral molecules as a bridge, where chiral molecules are used to create spin channel to facilitate the conversion between oxygen and water, whereas achiral molecules cannot work like spin channel [103–108]. The generally principle is that only those electrons with one spin state, spin up or spin down, can pass chiral molecules. After excitation, the left electron has an opposite spin orientation with the transferred one, whereas the spin-selected electron transfer does not happen in achiral molecules. This principle arises from the experimental result of Naaman's group that the spin specificity of electrons transferred through chiral molecules is the origin of a more efficient oxidation process and is further applied to oxygen electrochemistry. In 2015, Naaman's group found the hydrogen production on the photocathode Pt is not the same when the chiral and achiral molecules work to link CdSe to the photoanode TiO$_2$ in Na$_2$SO$_3$ and Na$_2$S mixed solution. Chiral molecule-coated photocathodes induce a higher hydrogen production on photoanode than achiral molecule-coated photoanodes. Under photoexcitation, an electron is transferred from CdSe to TiO$_2$ photoanode, and a hole was left, which can oxidize S^{2-} to S$_2^{2-}$ and eventually to S$_2$ ($s = 1$) [103]. Later, in 2017, they used the chiral organic semiconductors from helically aggregated dyes as sensitizers to coat the photoanode and impose the spin selectivity to suppress hydrogen peroxide formation. It can be regarded as a method to enhance photoelectrode stability and efficiency in water splitting [104]. Considering the spin selectivity of electron transfer through chiral molecules, they also designed and fabricated an electrochemical device based on the Hall effect that can provide the spin selectivity information in electrochemical process, where the chiral molecules served as the working electrode. The magnetic field was provided by the spin in the transferred electron, rather than the external magnetic field [105]. Except for the experimental work, some calculation work has also reported on spin catalysis during OER process. Zhang et al. used chiral molecule to decorate Fe$_3$O$_4$ as oxygen evolution anode based on the principle that the formation of H$_2$O$_2$ is suppressed and the triplet oxygen evolution is enhanced. Adsorbed chiral molecules can induced the assembly chirality, and the spin filter properties of the chiral Fe$_3$O$_4$ nanoparticles induce spin alignment of electrons in generated hydroxyl radicals, thus effectively inhibiting the formation of hydrogen peroxide when promoting the production of triplet oxygen (Figure 12.16) [106].

In addition to the abundant experimental work, there are some excellent calculation work on spin-selected OER in recent years. Among them, Gracia and coworkers did a lot systematically theoretical work. They, for the first time, connected OER activity with magnetic ordering in the bulk perovskite as a model [71]. The most OER active catalysts were found to possess the conduction channels with spin selection at the ground states. Like the chiral molecule catalysts, the active sites can extract electrons with one spin state from the singlet reactants, leaving the opposite

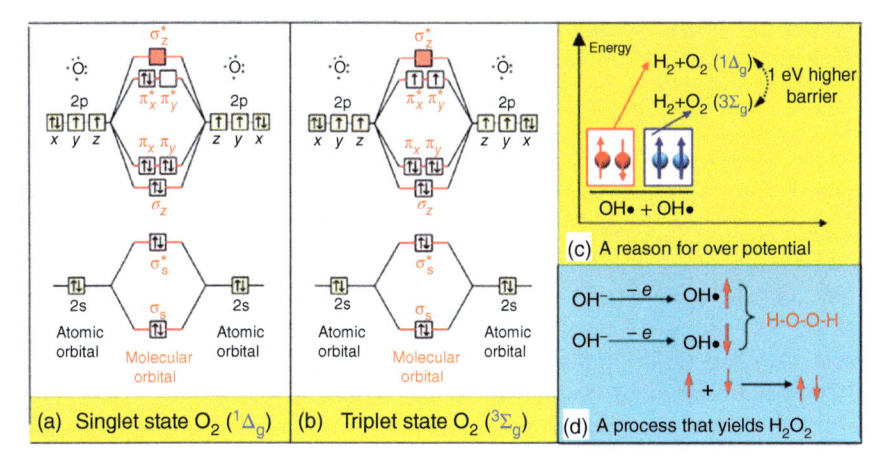

Figure 12.16 (a) Scheme of a molecular orbit of a singlet-state oxygen molecule. (b) Scheme of a molecular orbit of a triplet-state oxygen molecule. (c) Energy scheme of different mechanistic pathways for spin-parallel (blue) and spin-antiparallel (red) photogenerated holes (or ·OH radicals) when water is oxidized to produce oxygen and hydrogen. (d) Scheme of H_2O_2 production in the water splitting reaction. Source: Adapted with permission from Zhang et al. [106]. Copyright (2018) American Chemical Society.

spin orientation. They noted that the magnetic moment accumulation on the conduction plane atoms could be a promising OER descriptor in bulk perovskites. With the increase of the magnetic polarization on the conduction atoms, charge conduction is gradually coupled with spin transport. The spin accumulation and transport are the key to determine OER activity but not the overall magnetism [71]. The spin-polarized conduction states are observed in the perovskites $CaCu_3Fe_4O_{12}$ and $Ba_{0.5}Sr_{0.5}Co_{0.8}Fe_{0.2}O_{3-\delta}$, where a considerable number of unpaired electrons are localized in the inner d shell with high-spin configurations. Meanwhile, they found the magnetization average fluctuation in the metal atoms could linearly be linked to onset potential for OER in the studied system [109]. Gracia et al. also found that the best catalysts show metal sites with electron delocalization and localized spins, and there exists a relationship of the rate constant and spin-dependent electron mobility. Magnetic potentials in optimum catalysts act as selective gates to enhance the transport of local spin currents [110]. They continued to link ferromagnetic exchange interactions and antibonding orbitals of catalysts and the OER activity. They indicated that delocalizing spin potentials facilitates the coherent propagation of electrons at the covalent magnetic interfaces [111]. In 2017, they investigated the electron transfer step between individual atoms of $CaMn_4O_5$ in oxygen evolution complex and found that the oxygen evolution complex is an exchange coupled electron and spin acceptor with mixed valence state [112].

12.3.4 Oxygen Reduction Reaction

As the reverse reaction of the OER, the oxygen reduction reaction plays an important role in the field of industries and life sciences. As mentioned in the OER part, one oxygen molecule has two unpaired electrons with the parallel spins in the π^* orbitals. Owing to this electron configuration, oxygen is forced to move toward a stronger magnetic field. This force, $\mathbf{F}_m$, is expressed as follows:

$$\mathbf{F}_m = (\chi_{O_2}/\mu_0)\mathbf{B} \cdot (d\mathbf{B}/dx)$$

where μ_0 is the absolute magnetic permeability of vacuum ($4\pi \times 10^{-7}$ H m^{-1}), χ is the volume magnetic susceptibility (for oxygen gas, 1.91×10^{-6}), H is the magnetic induction, and $d\mathbf{B}/dx$ is the magnetic field gradient. The positive or negative of χ determines the force direction. The force is attractive for $\chi > 0$ while force is repulsive for $\chi < 0$ [113]. Since the mass transfer of O_2 that influences the oxygen concentration is a key part to ORR process, the effect of external magnetic field on the mass transport and concentration of dissolved oxygen gas has attracted much attention, considering that O_2 has a much greater value than any other common gas in the magnetic susceptibility. As early in 1946, an instrument for measuring the concentration of O_2 was invented by Pauling et al. [114]. The concentration and dissolution rate of O_2 also become the research subject under magnetic fields. In the study by Hirota et al., the magnetic field up to 10 T did not affect the equilibrium of oxygen gas but significantly accelerated the dissolution rate. During the dissolution of oxygen into the water phase, there exists a convection induced magnetically, and the maximum acceleration position is at the maximum position $\mathbf{B} \cdot d\mathbf{B}/dx$. In contrast, the dissolution of CO_2 was not accelerated by the magnetic field as it is diamagnetic [115]. Pt is a typical ORR catalyst, and applying magnetic field gradient to Pt electrodes can promote ORR process in H_2SO_4 media, where the rate-determining step is magnetic O_2 transport [113]. To utilize the magnetic field gradient force, Okada et al. incorporated the magnetic powders into the cathode layers and then placed a magnet behind it and investigated the ORR activity in H_2SO_4. Under magnetic field, O_2 was forced to the direction of the increase of magnetic field gradient, and the current increased with the increase of the magnetic field. When incorporating the Nd–Fe–B magnetic powder into the cathode layers, the ORR current was higher [116]. Except for the magnetic convection effect in aqueous solution, the O_2/O_2^- magnetic convection in acetonitrile was detected under the magnetic field. With the increase of the magnetic field, due to the convection of O_2 under the magnetic force, the decreased reduction peaks and steady state currents were observed in CV curves [117]. The external magnetic field is not needed sometimes. Coey and coworkers incorporated CoPt nanowires to alumina membranes to generate a 25 mT magnetic field within 100 nm of the electrode surface to create a gradient of 10^5 T m^{-1} to investigate the ORR activity in alkaline media [11]. When CoPt nanowires are magnetized, the ORR enhancement can be observed. They thought the ORR enhancement is attributed to the increase of HO_2^- concentration close to the electrode surface, rather than the O_2 movement to the electrode surface under the magnetic field by comparing the relatively small magnetic gradient force ($\sim$400 N m^{-3}) and strong effective driving force for diffusion ($\sim 10^7$ N m^{-3}) applied on O_2. Moreover, the other effect of the field gradient force is to inhibit convection. The electroactive HO_2^- can reside close to the electrode surface for a longer time due to the inhibiting convection of the paramagnetic HO_2^- radical [11, 118]. The effect from material choices (Zn, Co, Fe) on ORR activity under the magnetic field is also investigated. Under magnetic field, only 3% increase in the maximum ORR current was observed for Zn, and this slight increase was attributed to Lorentz force caused by the current and external magnetic field. Co and Fe have significant enhancement, ~12% and 8%, respectively. Such significant increase is attributed to enhanced convection caused by magnetic field gradient force due to the ferromagnetic properties of Co and Fe [119]. The difference in saturated magnetization of materials also affects the change in ORR current under the applied magnetic field. High saturated magnetization may inhibit the increase in ORR current under magnetic field. Taking Fe_3O_4, γ-Fe_2O_3, and Fe–N–C, among which Fe_3O_4 and γ-Fe_2O_3 are relatively higher in saturated magnetization, as example, Wang et al. found that though increasing the applied magnetic field could increase the ORR current and the direction of applied magnetic field did not affect the oxygen transport, the difference in the increase of ORR current is not the same and has the order Fe–N–C > γ-Fe_2O_3 > Fe_3O_4. They explained that when the applied magnetic field was stronger than the saturation magnetization of Fe–N–C and fully saturated Fe–N–C magnetically, all the tiny magnetic moments in Fe–N–C have the same direction as the magnetic field. In contrast, Fe_3O_4 and γ-Fe_2O_3 have high saturation magnetization, higher than the strength of the applied magnetic field. Under applied magnetic field, some magnetic moments in Fe_3O_4 and γ-Fe_2O_3 samples have different direction from the applied magnetic field. Hence, under the applied magnetic field, the magnetic force promotes O_2 transport, which causes the ORR current increase on Fe–N–C. In contrast, the acceleration of O_2 transport on Fe_3O_4 and γ-Fe_2O_3 was inhibited [120]. Besides, a porous Ag@Fe_3O_4 nanoassembly was developed by Tufa et al., and applying magnetic field can give a 1.13-fold high ORR efficiency [121]. The iron/nitrogen/sulfur-co-doped carbon gel (Fe–N–C–S) was synthesized by Kicinski et al. for oxygen reduction reaction under magnetic field (140 mT). No matter in acid or alkaline media, applying magnetic field can increase four-electron step and enhance the ORR process. Different from the previous report, the magnetic field direction has an effect to the ORR current increase. When the magnetic field is parallel to the electrode surface, the increase of the current is at least ca. 27%, and when the magnetic field has an angle with the electrode surface with 60°, the increase is ca. 48%. They also mentioned exerting an external magnetic field to the magnetic catalysts can cause a strong gradient field formation near the electrode surface [122]. The application of the magnetic field also influences the synthesis of the catalysts. For example, Jia's group employed the intense magnetic field to develop porous Pt–Ni alloys, predominating in (111) facets. Such alloys exhibited a superior ORR catalytic activity and stability [123].

In addition to the effect from the force caused by magnetic field, the spin state of the materials and the change in electronic states under the magnetic field also play a vital role in ORR process [124]. The typical model is iron porphyrin. Kim and Kim investigated the spin state of Fe^{II}-N_4 in structure to indicate the spin state importance in tuning the oxophilicity of the Fe–N_4. Fe^{II} (d^6) center has three possible spin states: a singlet state (low spin, $s = 0$), a triplet state (intermediate spin, $s = 1$), and a quintet state (high spin, $s = 2$). The study shows the ground state of O_2 bound to Fe^{II}-N_4 (O_2^*@Fe^{II}-N_4) has the open-shell singlet state, the

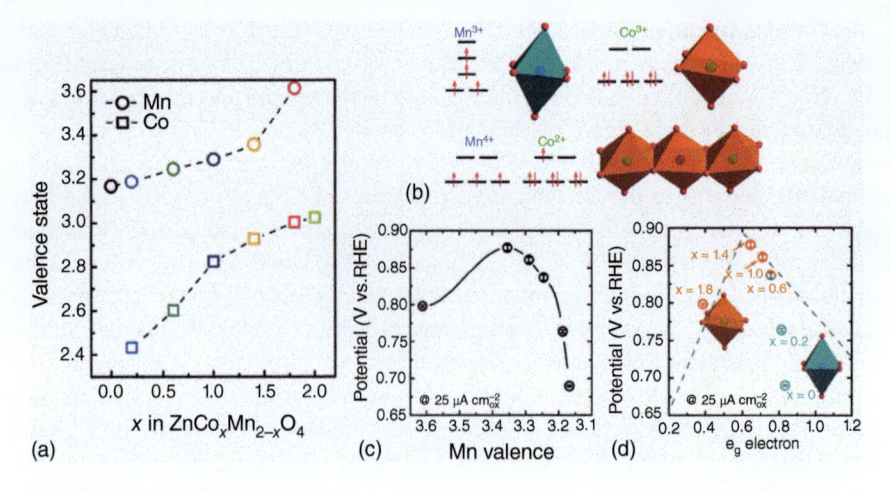

Figure 12.17 (a) The quantified Mn and Co valence states obtained from XANES analysis. (b) Charge redistribution via $Mn^{3+} + Co^{3+} \rightarrow Mn^{4+} + Co^{2+}$. (c) Correlation between Mn valence and ORR potentials at 25 $\mu A\ cm_{ox}^{-2}$. (d) ORR potentials at 25 $\mu A\ cm_{ox}^{-2}$ as a function of e_g electron in $ZnCo_xMn_{2-x}O_4$. Source: Reproduced with permission Zhou et al. [6]. Copyright (2018), Wiley-VCH.

open-shell triplet state, and closed-shell singlet state, among which the open-shell triplet is a spin-forbidden process [125]. In 2013, Jeon et al. proposed a new mechanism from the aspect of electron spin to explain the enhanced ORR activity of sulfurized graphene nanoplatelets compared with graphite by the results from experiments and calculations. After oxidation, the spin density of the sulfurized graphene nanoplatelets increased, and the ORR activity further increased [126]. Xu's group substituted Co for Mn in $ZnMn_2O_4$ to induce the superexchange interaction between Mn and Co, and the change of valence state takes place as follows: $Co^{3+} + Mn^{3+} \rightarrow Co^{2+} + Mn^{4+}$. This superexchange can tune the electron number of e_g orbitals to optimize the ORR activity (Figure 12.17) [6]. Besides, in the study of Wei's group, applied magnetic field was found to increase the electron exchange number of ORR on Co_3O_4/carbon fiber, which can be attributed to the unpaired electrons in Co_3O_4 under applying a magnetic field. The magnetic field polarization and electron energy degeneracy were induced based on the unpaired electron spin of Co_3O_4 under the magnetic field, facilitating the oxygen reaction rate [127]. Lu et al. focused on the intrinsic magnetic characteristic of hard-magnet $L1_0$-PtFe alloy for its ORR activity and noticed the activity of the magnetized $L1_0$-PtFe alloy has a fivefold enhancement compared with that of the unmagnetized one (Figure 12.18). The investigation shows that the magnetic field of catalysts controls the Pt d_{yz}–O_2 π^* coupling, primary dominating the chemisorbed oxygen coverage on catalyst surface, which is key for ORR activity [128]. The present study shows that magnetic promotion of the electrochemical oxygen reduction is feasible and thus can be successfully applied to improve the performance of low-temperature fuel cells such as polymer electrolyte fuel cell. Because a permanent magnet can induce a steep magnetic field gradient at the vicinity of the platinum catalyst, this will enable the

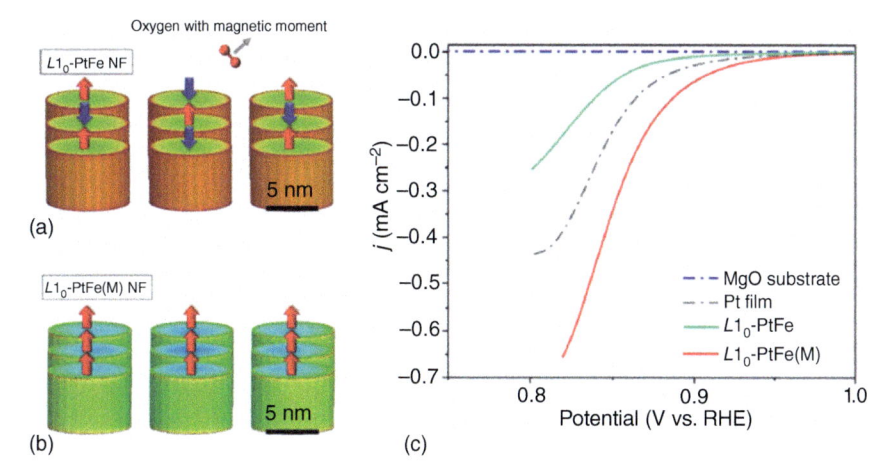

Figure 12.18 Simulated surface magnetic field induced by (a) $L1_0$-PtFe NF and (b) $L1_0$-PtFe(M) NF. (c) ORR polarization curve of the MgO substrate, Pt film, $L1_0$-PtFe NF, and $L1_0$-PtFe(M) NF. Source: Reprinted from Lu et al. [128], with permission from Elsevier.

design of a novel cathode catalyst layer in which magnetic particles are dispersed with a carbon supported catalysts in the gas diffusion electrode of fuel cells.

12.3.5 Other Catalytic Reactions

Except for the HER, OER, and ORR, some other catalytic reactions like CO_2 reduction, oxalate oxidation, and alcohol oxidation also exhibit the involvement of spin state or applied magnetic field. Taking CO_2 reduction as an example, as the reactants of CO_2 reduction, the solubility of CO_2 is affected under the magnetic field. With the increase of magnetic field, the solubility of CO_2 in seawater increases while such an effect is not observed in distilled water or glucose solution. It suggests that the dielectric boundary layers are necessary [129]. The optimization of CO_2 reduction can be realized by regulating the radical pair spin states. Like the effect of spin state in OER catalyst, the radical pair spin states play a key role in electrochemical processes due to the significance of spin states in defining the final product [130, 131]. Pan et al. observed that applying the magnetic field can increase the number of singlet radical pairs, facilitate the transition from triplet to singlet spin evolution, and thus increase the overall electrochemical CO_2 reduction reaction. They used tin nanoparticles as catalysts to investigate the catalytic activity and the yield of formate/formic acid in the absence or presence of magnetic field. Magnetic field can boost the catalytic activity and increase the formic acid yield [132]. As for the explanation of magnetic field-induced yield increase in formic acid proposed by Pan et al., Player and Hore raised a different opinion. They pointed out that the difference in the electron Zeeman interaction of the two radicals, is not enough to explain the observed magnetic field induced current. It is the anisotropy of the g-tensor of CO_2^- that gives rise to the field-dependent spin relaxation, combined with isotropic hyperfine and Zeeman interactions giving rise to the coherent singlet–triplet interconversion that accounts

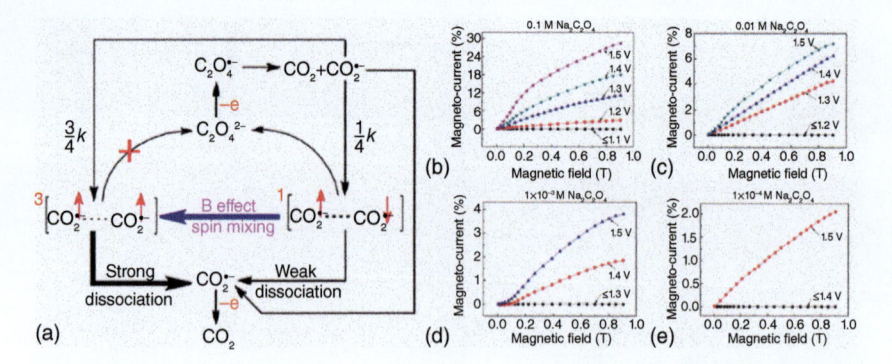

Figure 12.19 (a) Schematic diagram to show spin-dependent oxidation of oxalate-induced positive MC in aqueous solutions. The magneto-current is shown at different applied electrode potentials for different concentrations of $Na_2C_2O_4$ in water, from 0.1 M (b), 0.01 M (c), 1×10^{-3} M (d), and 1×10^{-4} M (e). Source: Adapted with permission from Pan et al. [134]. Copyright (2018) American Chemical Society.

for the observed magnetic field effects [133]. In addition to CO_2 reduction reaction, Pan et al. applied the external magnetic field to the electrochemical oxalate oxidation in 0.1 M NaH_2PO_4 aqueous solution, and a nearly 30% current increase was observed. They hold the opinion that according to the Pauli exclusion principle, the dissociation of the triplet radical pair has a larger probability than that of the singlet radical pair. The application of magnetic field accelerates the spin evolution of radical pair from a singlet to triplet and thus leads to the increase in the oxalate oxidation rate and generates a larger current (Figure 12.19) [134]. Fe_2Pt nanocrystals were synthesized by Wang and Song in the absence or presence of magnetic field for the ethanol oxidation. Applying magnetic field can induce more Pt^{2+} reduction to Pt^0 and cause bigger nanocrystals and more unpaired electrons on the surface. The Fe_2Pt nanocrystals with more unpaired electrons exhibited a better methanol and ethanol oxidation activity in $HClO_4$ [135].

12.3.6 Battery

In the battery field, the magnetic properties and spin states of electrode materials and the effects of external magnetic field are attracting much attention [136]. Take graphite as an example. Till now, graphite is a state-of-the-art negative electrode material. However, graphite flakes are easy to stack together, which impede the charge carrier transport and lower the lithiation and delithiation kinetics. To solve the issue, application of magnetic field can promote the battery performance by aligning the electrode materials. Zhang et al. aligned the graphite flakes by using a magnetic field and found the vertically aligned electrode has a 4.5 times higher specific capacity than the control one (Figure 12.20) [137]. Besides, Wang et al. applied a magnetic field to the wrinkled Fe_3O_4/reduced graphene oxide in Li-ion batteries to induce the alignment of the reduced graphene oxide along the magnetic field and thus promoted the Li^+ transport and electron transfer [138]. The magnetic field is also used to induce the formation of 1D Fe_3O_4/C microrods, storing more Li^+ than

Figure 12.20 Transmission path of Li⁺ (Fuchsia ball) in (a) reference and (b) aligned graphite electrode. (c) Evolution of the specific charge at 2C as a function of angle between graphite flakes and current collector. Source: Adapted with permission from Zhang et al. [137]. Copyright (2019) American Chemical Society.

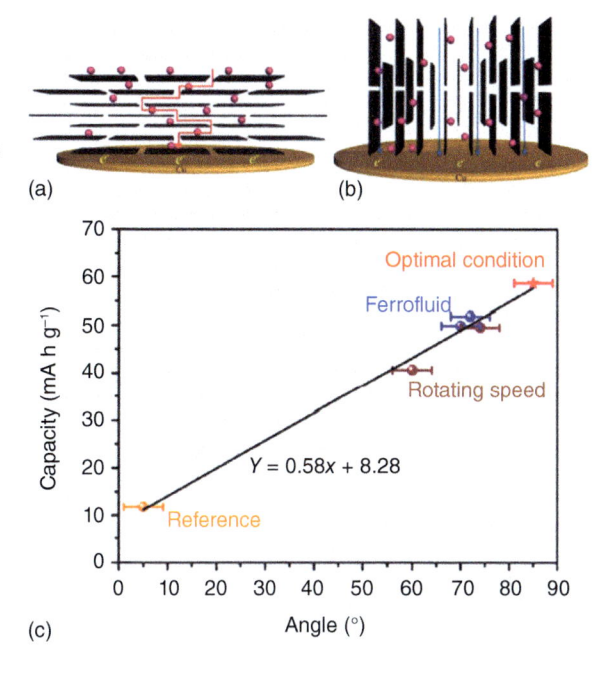

Fe_3O_4 nanospheres [139]. Applying a magnetic field to the ferromagnetic α-Fe_2O_3 nanoparticle/N-doped carbon anode incremented about twofold in specific capacity due to the enhanced the diffusion and convection [140]. The magnetic control is also used to increase Li-S battery performance. Cui's group raise a biphasic solution containing the magnetic nanoparticles (superparamagnetic γ-Fe_2O_3) and lithium polysulfide. When applying the magnetic field, most of the polysulfide and magnetic nanoparticles can be pulled to current collector for a high-concentration polysulfide phase (Figure 12.21) [141]. In the storage capacity, a lot of promising transition metal oxides has an extra value exceeding their theoretical values. In 2020, Li et al. found the presence of spin-polarized surface capacitance of Fe_3O_4/Li and revealed the extra storage capacity by investigating the transition metal/Li_2O electronic state evolution. They demonstrated the Fe nanoparticles appeared during low-voltage discharge process due to the electrochemical reduction, and these Fe nanoparticles lead to extra capacity due to the storage of a number of the spin-polarized electrons and interface magnetization. Such capacitance is also found in CoO, NiO, FeF_2, and Fe_2N electrode materials [142]. Electrodeposition of Co_3O_4 from spent lithium-ion batteries is enhanced by the magnetic field effect, and the products are reused in the storage field. A higher electroactive surface area and hierarchy of Co_3O_4 nanostructures with higher electroactive surface areas are formed during the electrodeposition process under a magnetic field compared with the agglomerated sheetlike structure in the control one. Co_3O_4 formed under the magnetic field has a specific capacitance of 1273 F g⁻¹ compared with the control one (315 F g⁻¹) [22]. Besides the anode alignment, the lithium dendrite is another crucial factor to affect the battery performance. Applying the external magnetic field to the battery has been found to have an influence on the growth of lithium dendrites. Till now, there are two contrary opinions

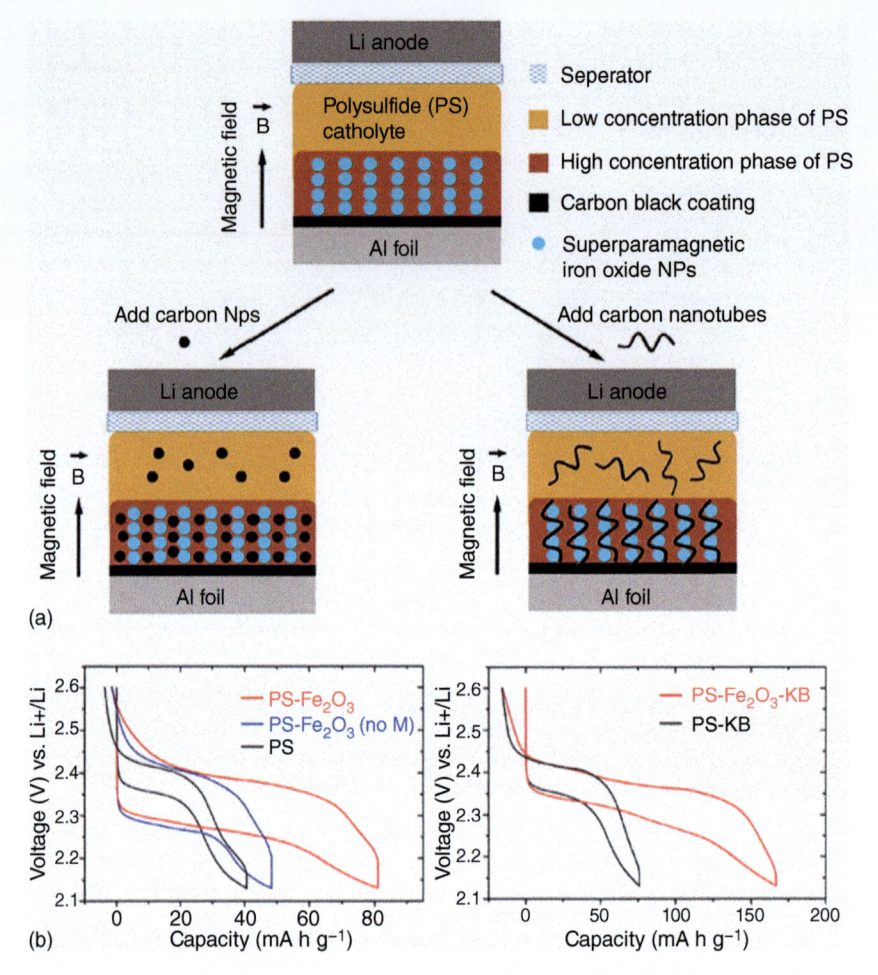

Figure 12.21 (a) Schematic illustrating the proof-of-concept magnetic field-controlled semiliquid battery. The catholyte is composed of two phases, where the polysulfide and the superparamagnetic iron oxide nanoparticles can be extracted together to a phase in close contact with the current collector under the influence of applied magnetic field. (b) Discharge–charge voltage profiles during the second cycle of cells made from the biphasic magnetic polysulfide catholyte containing the γ-Fe_2O_3 NPs with (PS-Fe_2O_3, red line) and without the applied magnetic field (PS-Fe_2O_3, no M, where M refers to magnetic field; blue line), measured at a current density of 0.2 °C. Source: Adapted with permission from Li et al. [141]. Copyright (2015) American Chemical Society.

that magnetic field inhibits and promotes the growth of lithium dendrite based on the experimental results. Shen et al. supposed that Li$^+$ is forced spiral motion under the Lorentz force, assisting in obtaining the uniform and compact lithium deposition to suppress the dendrite growth by means of promoting the Li$^+$ transport. The copper coated with the ferromagnetic NiCo alloy is more responsive to the magnetic field [143]. Similar phenomenon is also be observed by other groups [144, 145]. Besides, Huang et al. reported Li electrodeposition under a parallel (**B** // **j**) and perpendicular

($\mathbf{B} \perp \mathbf{j}$) magnetic field. The parallel magnetic field facilitated denser and more uniform lithium growth than the perpendicular one. They explained the Lorentz force on Li$^+$ for $\mathbf{B} \perp \mathbf{j}$ becomes weak as the velocity direction of Li$^+$ approaches to the magnetic field direction and cannot enable uniform ion distribution. In contrast, for the $\mathbf{B}//\mathbf{j}$ case, the Li$^+$ is under the force perpendicular to magnetic field considering the velocity direction of Li$^+$, and the magnetic field direction is not strictly parallel to each other. Li$^+$ is driven to rotate around the nuclei or deposits, like a stirring bar, uniform Li$^+$ distribution. The current collectors with different magnetic properties are investigated under the parallel magnetic field; uniform Li deposits could not be associated with the magnetic property of the current collectors. For example, on the diamagnetic Cu collectors, the Li dendrites disappear under a magnetic field. At the same capacity, with the increase of magnetic field strength, the diameter of lithium deposits increases, suggesting more uniform Li$^+$ deposition. Such a trend is also observed on the ferromagnetic stainless steel collectors. They also estimated the force in this case that Lorentz force (1.4×10^{-25} sinα N) and electric field force (3.2×10^{-26} N) have no obvious difference, implying the Lorentz force could change Li$^+$ transport direction (Figures 12.22 and 12.23) [146]. In the most recent report by Xu's group, the overall MHD effect is negative. Uneven current distribution is considered as the domain MHD effect, which shortens the lithium anode cycling life

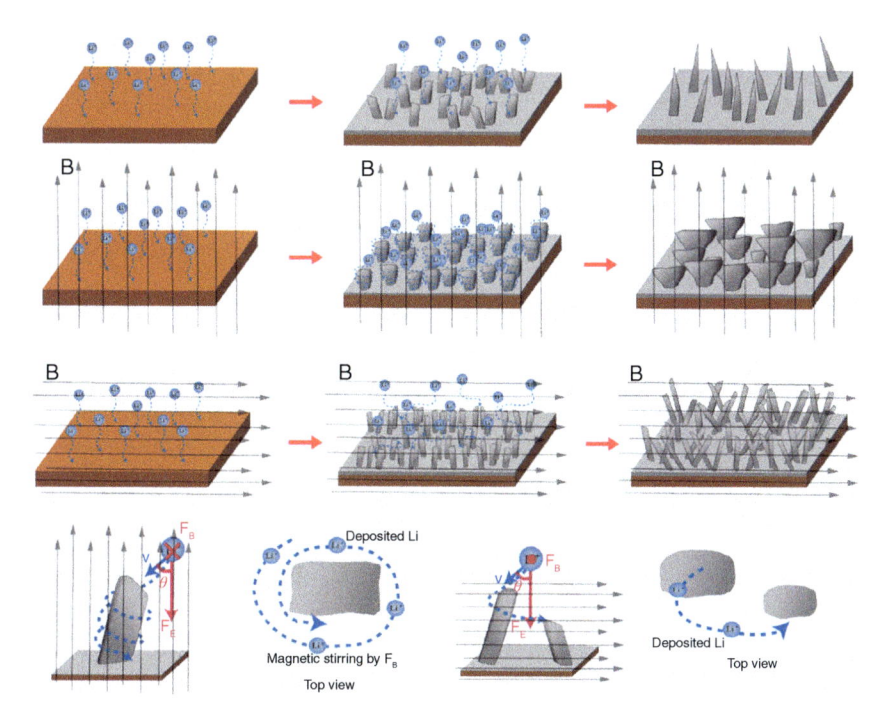

Figure 12.22 The mechanism of lithium electrodeposition on metallic substrates under magnetic field with perpendicular and parallel to current direction. The perpendicular magnetic field can control Li-ions deposited more uniformly than the parallel magnetic field and the magnetic field-free. Source: Reprinted from Huang et al. [146], with permission from Elsevier.

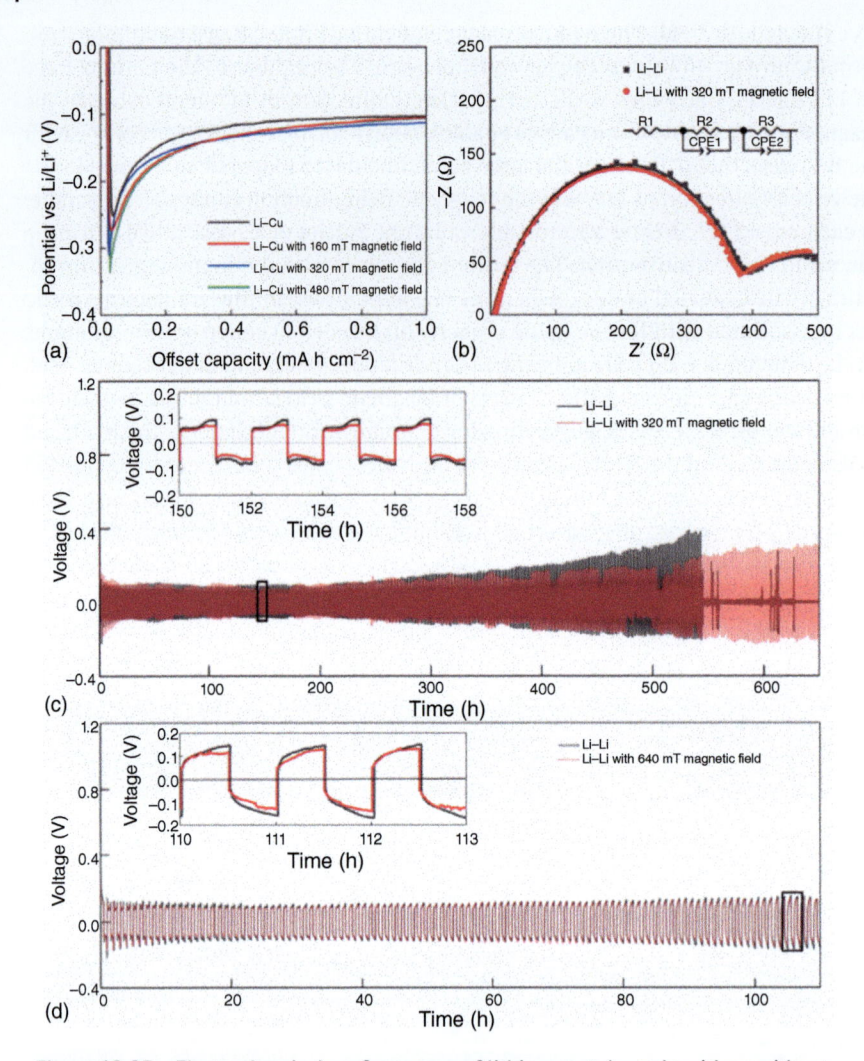

Figure 12.23 Electrochemical performances of lithium metal anode with or without parallel magnetic field using carbonate electrolyte. (a) Voltage profiles of Li platting on Cu current collector without magnetic field and with 480 mT magnetic field under 2.0 mA cm^{-2}. (b) Nyquist plots of the Cu|Li cells with and without magnetic field. The cycling performance of Li|Li symmetrical cells with and without (c) 320 mT magnetic field and (d) 640 mT magnetic field at the deposited capacity of 1.0 mA h cm^{-2} under the current density of 2.0 mA cm^{-2}. Source: Reprinted from Huang et al. [146], with permission from Elsevier.

through a higher decomposition rate of the electrolyte, and the separator destroyed by the growth of lithium dendrites was observed [147].

12.3.7 Others

Except for the effect above we stress, the effects of magnetic field in electrochemistry are exhibited in other aspects. For example, the attraction force caused by magnetism is utilized for stabilizing the electrode materials. Yang and Liang synthesized the self-magnetic-attracted supported $Ni_{0.7}Fe_{0.3}@Ni_{0.7}Fe_{0.3}O$/nickel foam OER electrode, and the superparamagnetic properties and the size effect of $Ni_xFe_{1-x}@Ni_xFe_{1-x}O$ nanoparticles gave rise to the conglutination of nanoparticles on the nickel foam and exhibited excellent activity and stability. Moreover, the adhesion to nickel foam was improved after conversion to NiFe-LDH [148]. The magnetic field force is also used for immobilizing biomolecules even though no external magnetic field is applied. For example, Barsan et al. used the magnetic Ni nanoparticles as the biomolecule carriers to immobilize the biomolecule onto the Ni electrode [149]. The magnetic field can also influence the form of individual nanoparticles in solution. For instance, the form of Fe_3O_4 nanoparticles is different with or without the external magnetic field. Under magnetic field gradient force, the released Fe^{2+} ions close to the surface can be trapped, and their mass transport was hindered. As a result, the dissolution of Fe_3O_4 was inhibited [150].

12.4 Conclusion and Outlook

In summary, the force arising from the gradient magnetic field and the interaction of current and magnetic field influences the liquid-phase electrochemical system. In our opinion, the design of electrochemical cell and the emergence of certain gradient magnetic field are very promising for different reaction systems. In the area related to spin state, it has been focused on catalysis, and there will be a lot potential opportunities in the energy storage.

References

1 Waskaas, M. and Kharkats, Y.I. (1999). Magnetoconvection phenomena: a mechanism for influence of magnetic fields on electrochemical processes. *J. Phys. Chem. B* 103: 4876–4883.
2 Sun, S., Wei, C., Zhu, Z. et al. (2014). Magnetic iron oxide nanoparticles: synthesis and surface coating techniques for biomedical applications. *Chin. Phys. B* 23: 037503.
3 Mutschke, G., Tschulik, K., Weier, T. et al. (2010). On the action of magnetic gradient forces in micro-structured copper deposition. *Electrochim. Acta* 55: 9060–9066.
4 Pan, H., Wang, M., Shen, Y., and Hu, B. (2018). Large magneto-current effect in the electrochemical detection of oxalate in aqueous solution. *J. Phys. Chem.* 122: 19880–19885.
5 Sun, S., Sun, Y., Zhou, Y. et al. (2019). Switch of the rate-determining step of water oxidation by spin-selected electron transfer in spinel oxides. *Chem. Mater.* 31: 8106–8111.
6 Zhou, Y., Sun, S., Xi, S. et al. (2018). Superexchange effects on oxygen reduction activity of edge-sharing [$Co_xMn_{1-x}O_6$] octahedra in spinel oxide. *Adv. Mater.* 30: 1705407.

7 Hinds, G., Coey, J.M.D., and Lyons, M.E.G. (2001). Influence of magnetic forces on electrochemical mass transport. *Electrochem. Commun.* 3: 215–218.

8 Ragsdale, S.R., Grant, K.M., and White, H.S. (1998). Electrochemically generated magnetic forces. Enhanced transport of a paramagnetic redox species in large, nonuniform magnetic fields. *J. Am. Chem. Soc.* 120: 13461–13468.

9 Zhang, Y., Liang, C., Wu, J. et al. (2020). Recent advances in magnetic field-enhanced electrocatalysis. *ACS Appl. Energy Mater.* 3: 10303–10316.

10 Monzon, L.M.A. and Coey, J.M.D. (2014). Magnetic fields in electrochemistry: the Lorentz force. A mini-review. *Electrochem. Commun.* 42: 38–41.

11 Chaure, N.B., Rhen, F.M.F., Hilton, J., and Coey, J.M.D. (2007). Design and application of a magnetic field gradient electrode. *Electrochem. Commun.* 9: 155–158.

12 Grant, K.M., Hemmert, J.W., and White, H.S. (1999). Magnetic focusing of redox molecules at ferromagnetic electrodes. *Electrochem. Commun.* 1: 319–323.

13 Mutschke, G. and Bund, A. (2008). On the 3D character of the magnetohydrodynamic effect during metal electrodeposition in cuboid cells. *Electrochem. Commun.* 10: 597–601.

14 Grant, K.M., Hemmert, J.W., and White, H.S. (2002). Magnetic field-controlled microfluidic transport. *J. Am. Chem. Soc.* 124 (3): 462–467.

15 Arumugam, P.U., Clark, E.A., and Fritsch, I. (2005). Use of paired, bonded NdFeB magnets in redox magnetohydrodynamics. *Anal. Chem.* 77: 1167–1171.

16 Rakoczy, R. and Masiuk, S. (2011). Studies of a mixing process induced by a transverse rotating magnetic field. *Chem. Eng. Sci.* 66 (11): 2298–2308.

17 Wasak, A., Drozd, R., Jankowiak, D., and Rakoczy, R. (2019). The influence of rotating magnetic field on bio-catalytic dye degradation using the horseradish peroxidase. *Biochem. Eng. J.* 147: 81–88.

18 Tschulik, K., Koza, J.A., Uhlemann, M. et al. (2009). Effect of well-defined magnetic field gradients on the electrodeposition of copper and bismuth. *Electrochem. Commun.* 11: 2241–2244.

19 Murdoch, H.A., Yin, D., Hernandez-Rivera, E., and Giri, A.K. (2018). Effect of applied magnetic field on microstructure of electrodeposited copper. *Electrochem. Commun.* 97: 11–15.

20 Aaboubi, O. and Msellak, K. (2017). Magnetic field effects on the electrodeposition of CoNiMo alloys. *Appl. Surf. Sci.* 396: 375–383.

21 Karnbach, F., Uhlemann, M., Gebert, A. et al. (2014). Magnetic field templated patterning of the soft magnetic alloy CoFe. *Eletrochim. Acta* 123: 477–484.

22 Aboelazm, E.A.A., Ali, G.A.M., Algarni, H. et al. (2018). Magnetic electrodeposition of the hierarchical cobalt oxide nanostructure from spent lithium-ion batteries: its application as a supercapacitor electrode. *J. Phys. Chem. C* 122 (23): 12200–12206.

23 Kuwertz, R., Gonzalez, I., Vidakovic-Koch, T. et al. (2016). Material development and process optimization for gas-phase hydrogen chloride with oxygen depolarized cathode. *J. Appl. Electrochem.* 46: 755–767.

24 Matsushima, H., Nishida, T., Konishi, Y. et al. (2003). Water electrolysis under microgravity Part 1. Experimental technique. *Electrochim. Acta* 48: 4119–4125.

25 Cheng, Y., Song, H., Wu, H. et al. (2020). Defects enhance the electrocatalytic hydrogen evolution properties of MoS_2-based materials. *Chem. Asian J.* 15: 3123–3134.

26 Liu, H., Pan, L.-m., Huang, H. et al. (2015). Hydrogen bubble growth at micro-electrode under magnetic field. *J. Electroanal. Chem.* 754: 22–29.

27 Wang, M., Wang, Z., Gong, X., and Guo, Z. (2014). The intensification technologies to water electrolysis for hydrogen production-a review. *Renewable Sustainable Energy Rev.* 29: 573–588.

28 Sakuma, G., Fukunaka, Y., and Matsushima, H. (2014). Nucleation and growth of electrolytic gas bubbles under microgravity. *Int. J. Hydrogen Energy* 39: 7638–7645.

29 Fukunaka, Y., Suzuki, K., Ueda, A., and Kondo, Y. (1989). Mass-transfer rate on a plane vertical cathode with hydrogen gas evolution. *J. Electrochem. Soc.* 136: 1002–1009.

30 Hine, F. and Murakami, K. (1980). Bubble effects on the solution IR drop in a vertical electrolyzer under free and forced convection. *J. Electrochem. Soc.* 127: 292–297.

31 Rousar, I. (1987). Pumping effect due to gas evolution in flow-through electrolysers. *J. Appl. Electrochem.* 17: 134–146.

32 Cheng, H., Scott, K., and Ramshaw, C. (2002). Intensification of water electrolysis in a centrifugal field. *J. Electrochem. Soc.* 149 (11): D172–D177.

33 Perron, A.L., Kiss, L.I., and Poncsak, S. (2007). Mathematical model to evaluate the ohmic resistance caused by the presence of a large number of bubbles in Hall-Heroult cells. *J. Appl. Electrochem.* 37: 303–310.

34 Kiuchi, D., Matsushima, H., Fukunaka, Y., and Kuribayashi, K. (2006). Ohmic resistance measurement of bubble froth layer in water electrolysis under microgravity. *J. Electrochem. Soc.* 153: E138–E143.

35 Iida, T., Matsushima, H., and Fukunaka, Y. (2007). Water electrolysis under a magnetic field. *J. Electrochem. Soc.* 154 (8): E112–E115.

36 Diao, Z., Dunne, P.A., Zangari, G., and Coey, J.M.D. (2009). Electrochemical noise analysis of the effects of a magnetic field on cathodic hydrogen evolution. *Electrochem. Commun.* 11: 740–743.

37 Chaure, N.B. and Coey, J.M.D. (2009). Enhanced oxygen reduction at composite electrodes producing a large magnetic gradient. *J. Electrochem. Soc.* 156 (3): F39–F46.

38 Koza, J.A., Uhlemann, M., Gebert, A., and Schultz, L. (2008). Desorption of hydrogen from the electrode surface under influence of an external magnetic field. *Electrochem. Commun.* 10: 1330–1333.

39 Leistra, J.A. and Sides, P.J. (1987). Voltage components at gas evolving electrodes. *J. Electrochem. Soc.* 134: 2442–2446.

40 Matsushima, H., Kiuchi, D., and Fukunaka, Y. (2009). Measurement of dissolved hydrogen supersaturation during water electrolysis in a magnetic field. *Electrochim. Acta* 54: 5858–5862.

41 Shinohara, K., Hashimoto, K., and Aogaki, R. (2002). Macroscopic fluid motion accompanying copper corrosion in nitric acid under a vertical magnetic field. *Electrochemistry* 70 (10): 772–778.

42 Koza, J.A., Mühlenhoff, S., Uhlemann, M. et al. (2009). Desorption of hydrogen from an electrode surface under influence of an external magnetic field-in-situ microscopic observations. *Electrochem. Commun.* 11: 425–429.

43 Koza, J.A., Mühlenhoff, S., Żabiński, P. et al. (2011). Hydrogen evolution under the influence of a magnetic field. *Electrochim. Acta* 56: 2665–2675.

44 Diao, Z., Dunne, P.A., Zangari, G., and Coey, J.M.D. (2009). Electrochemical noise analysis of the effects of magnetic field on cathodic hydrogen evolution. *Electrochem. Commun.* 11: 740–743.

45 Liu, H.-b., Hu, Q., Pan, L.-m. et al. (2019). Electrode-normal magnetic field facilitating neighbouring electrochemical bubble release from hydrophobic islets. *Electrochim. Acta* 306: 350–359.

46 Fernández, D., Martine, M., Meagher, A. et al. (2012). Stabilizing effect of a magnetic field on a gas bubble produced at a microelectrode. *Electrochem. Commun.* 18: 28–32.

47 Liu, Y., Pan, L.-m., Liu, H. et al. (2019). Effects of magnetic field on water electrolysis using foam electrodes. *Int. J. Hydrogen Energy* 44: 1352–1358.

48 Enyashin, A.N., Yadgarov, L., Houben, L. et al. (2011). New route for stabilization of 1T-WS_2 and MoS_2 phases. *J. Phys. Chem. C* 115: 24586–24591.

49 Zhu, J., Wang, Z., Yu, H. et al. (2017). Argon plasma induced phase transition in monolayer MoS_2. *J. Am. Chem. Soc.* 139: 10216–10219.

50 Chhowalla, M., Shin, H.S., Eda, G. et al. (2013). The chemistry of two-dimensional layered transition metal dichalcogenide nanosheets. *Nat. Chem.* 5: 263–275.

51 Voiry, D., Salehi, M., Silva, R. et al. (2013). Conducting MoS_2 nanosheets as catalysts for hydrogen evolution reaction. *Nano Lett.* 13: 6222–6227.

52 Lukowski, M.A., Daniel, A.S., Meng, F. et al. (2013). Enhanced hydrogen evolution catalysis from chemical exfoliated metallic MoS_2 nanosheets. *J. Am. Chem. Soc.* 135: 10274–10277.

53 Voiry, D., Yamaguchi, H., Li, J. et al. (2013). Enhanced catalytic activity in strained chemically exfoliated WS_2 nanosheets for hydrogen evolution. *Nat. Mater.* 12: 850–855.

54 Jaramillo, T.F., Jørgensen, K.P., Bonde, J. et al. (2007). Identification of active edge sites for electrochemical H_2 evolution from MoS_2 nanocatalysts. *Science* 317 (5834): 100–102.

55 Hinnemann, B., Moses, P.G., Bonde, J. et al. (2005). Biomimetic hydrogen evolution MoS_2 nanoparticles as catalyst for hydrogen evolution. *J. Am. Chem. Soc.* 127: 5308–5309.

56 Li, Y., Wang, H., Xie, L. et al. (2011). MoS_2 nanoparticles grown on graphene: an advanced catalyst for the hydrogen evolution reaction. *J. Am. Chem. Soc.* 133 (19): 7296–7299.

57 Zhou, W., Chen, M., Guo, M. et al. (2020). Magnetic enhancement for hydrogen evolution reaction on ferromagnetic MoS_2 catalyst. *Nano Lett.* 20: 2923–2930.

58 Zhang, Z., Zou, X., Crespi, V.H., and Yakobson, B.I. (2013). Intrinsic magnetism of grain boundaries in two-dimensional metal dichalcogenides. *ACS Nano* 7 (12): 10475–10481.

59 Cai, L., He, J., Liu, Q. et al. (2015). Vacancy-induced ferromagnetism of MoS_2 nanosheets. *J. Am. Chem. Soc.* 137: 2622–2627.

60 Kong, D., Wang, H., Cha, J.J. et al. (2013). Synthesis of MoS_2 and $MoSe_2$ films with vertically aligned layers. *Nano Lett.* 13 (3): 1341–1347.

61 Yu, Y., Huang, S.-Y., Li, Y. et al. (2014). Layer-dependent electrocatalysis of MoS_2 for hydrogen evolution. *Nano Lett.* 14 (2): 553–558.

62 Jiang, Z., Zhou, W., Hong, A. et al. (2019). MoS_2 Moire superlattice for hydrogen evolution reaction. *ACS Energy Lett.* 4: 2830–2835.

63 Wang, X., Zhang, Y., Si, H. et al. (2020). Single-atom vacancy defect to trigger high-efficiency hydrogen evolution of MoS_2. *J. Am. Chem. Soc.* 142: 4298–4308.

64 Dai, J., Zhu, Y., Tahini, H.A. et al. (2020). Single-phase perovskite oxide with super-exchange induced atomic-scale synergistic active centers enables ultrafast hydrogen evolution. *Nat. Commun.* 11: 5657.

65 Sun, Y., Sun, S., Yang, H. et al. (2020). Spin-related electron transfer and orbital interactions in oxygen electrocatalysis. *Adv. Mater.* 32: 2003297.

66 Nishida, Y. (1988). On the chemical mechanism of O_2-evolution in photosystem II. *Inorg. Chim. Acta* 152 (2): 73–74.

67 McGrady, J.E. and Stranger, R. (1999). Redox-induced formation and cleavage of O-O σ and π bonds in a peroxo-bridged manganese dimer: a density functional study. *Inorg. Chem.* 38: 550–558.

68 Wang, J.T., Tang, F., Brown, W.D., and Bagayoko, D. (1998). Spin-dependence of the electron scattering cross section by a magnetic layer system and the magneto-resistance. *Int. J. Mod. Phys. B* 12: 3376–3380.

69 Chappert, C., Fert, A., and Van Dau, F.N. (2007). The emergence of spin electronics in data storage. *Nat. Mater.* 6: 813–823.

70 Binasch, G., Griinberg, P., Saurenbach, F., and Zinn, W. (1989). Enhanced magnetoresistance in layered magnetic structures with antiferromagnetic interlayer exchange. *Phys. Rev. B* 93: 4828–4830.

71 Lim, T., Niemantsverdriet, J.W. (Hans), and Gracia, J. (2016). Layer antiferromagnetic ordering in the most active perovskite catalysts for the oxygen evolution reaction. *ChemCatChem* 8: 2968–2974.

72 Hong, W.T., Gadre, M., Lee, Y.-L. et al. (2013). Tuning the spin state in $LaCoO_3$ thin films for enhanced high-temperature oxygen electrocatalysis. *J. Phys. Chem. Lett.* 4: 2493–2499.

73 Guo, Y., Tong, Y., Chen, P. et al. (2015). Engineering the electronic state of a perovskite electrocatalyst for synergistically enhanced oxygen evolution reaction. *Adv. Mater.* 27: 5989–5994.

74 Duan, Y., Sun, S., Xi, S. et al. (2017). Tailoring the Co 3d-O 2p covalency in $LaCoO_3$ by Fe substitution to promote oxygen evolution reaction. *Chem. Mater.* 29: 10534–10541.

75 Zhou, Y., Sun, S., Song, J. et al. (2018). Enlarged Co-O covalency in octahedral sites leading to highly efficient spinel oxides for oxygen evolution reaction. *Adv. Mater.* 30: 1802912.

76 Chen, R.R., Sun, Y., Ong, S.J.H. et al. (2020). Antiferromagnetic inverse spinel oxide LiCoVO$_4$ with spin-polarized channel for water oxidation. *Adv. Mater.* 32: 1907976.

77 Ramsundar, R.M., Pillai, V.K., and Joy, P.A. (2018). Spin state engineered Zn$_x$Co$_{3-x}$O$_4$ as an efficient oxygen evolution electrocatalyst. *Phys. Chem. Chem. Phys.* 20: 29452–29461.

78 Jiang, M., Li, J., Zhao, Y. et al. (2018). Double perovskites as model bifunctional catalysts toward rational design: the correlation between electrocatalytic activity and complex spin configuration. *ACS Appl. Mater. Interfaces* 10: 19746–19754.

79 Tong, Y., Wu, J., Chen, P. et al. (2018). Vibronic superexchange in double perovskite electrocatalyst for efficient electrocatalytic oxygen evolution. *J. Am. Chem. Soc.* 140: 11165–11169.

80 Torun, E., Fang, C.M., de Wijs, G.A., and de Groot, R.A. (2013). Role of magnetism in catalysis: RuO$_2$ (100) surface. *J. Phys. Chem. C* 117: 6353–6357.

81 Ping, Y., Galli, G., and Goddard, W.A. III, (2015). Electronic structure of IrO$_2$: the role of the metal d orbitals. *J. Phys. Chem. C* 119: 11570–11577.

82 Zhou, S., Miao, X., Zhao, X. et al. (2016). Engineering electrocatalytic activity in nanosized perovskite cobaltite through surface spin-state transition. *Nat. Commun.* 7: 11510.

83 Chen, J. and Selloni, A. (2012). Electronic states and magnetic structure at the Co$_3$O$_4$ (110) surface: A first-principles study. *Phys. Rev. B* 85: 085306.

84 Ma, L., Hung, S.-F., Zhang, L. et al. (2018). High spin state promotes water oxidation catalysis at neutral pH in spinel cobalt oxide. *Ind. Eng. Chem. Res.* 57: 1441–1445.

85 Liu, Y., Xiao, C., Lyu, M. et al. (2015). Ultrathin Co$_3$S$_4$ nanosheets that synergistically engineer spin states and exposed polyhedral that promote water oxidation under neutral conditions. *Angew. Chem.* 127: 11383–11387.

86 Tong, Y., Guo, Y., Chen, P. et al. (2017). Spin-state regulation of perovskite cobaltite to realize enhanced oxygen evolution activity. *Chem* 3: 812–821.

87 Chen, S., Kang, Z., Hu, X. et al. (2017). Delocalized spin states in 2D atomic layers realizing enhanced electrocatalytic oxygen evolution. *Adv. Mater.* 29: 1701687.

88 Tiwari, A.P., Yoon, Y., Novak, T.G. et al. (2019). Lattice strain formation through spin-coupled shells of MoS$_2$ on Mo$_2$C for bifunctional oxygen reduction and oxygen evolution reaction electrocatalysts. *Adv. Mater. Interfaces* 6: 1900948.

89 Tkalych, A.J., Zhuang, H.L., and Carter, E.A. (2017). A density functional + U assessment of oxygen evolution reaction mechanism on β-NiOOH. *ACS Catal.* 7: 5329–5339.

90 Miyata, S. (1983). Anion-exchange properties of hydrotalcite-like compounds. *Clays Clay Miner.* 31: 305–311.

91 Scavetta, E., Berrettoni, M., Giorgetti, M., and Tonelli, D. (2002). Electrochemical characterization of Ni/Al-X hydrotalcites and their electrocatalytic behavior. *Electrochim. Acta* 2451–2461.

92 Aguilar-Vargas, V., Valente, J.S., and González, I. (2013). Electrochemical characterization of carbon paste electrodes modified with MgZnGa and ZnGaAl hydrotalcite-like compounds. *J. Solid State Electrochem.* 17: 3145–3152.

93 Corrigan, D.A. (1987). The catalysis of the oxygen evolution reaction by iron impurities in thin film nickel oxide electrodes. *J. Electrochem. Soc.* 134: 377–384.

94 Luo, J., Im, J.-H., Mayer, M.T. et al. (2014). Water photolysis at 12.3% efficiency via perovskite photovoltaics and earth-abundant catalysts. *Science* 345: 1593–1596.

95 Dionigi, F. and Strasser, P. (2016). NiFe-based (oxy)hydroxide catalysts for oxygen evolution reaction in non-acidic electrolytes. *Adv. Energy Mater.* 6: 1600621.

96 Oliver-Tolentino, M.A., Vázquez-Samperio, J., Manzo-Robledo, A. et al. (2014). An approach to understanding the electrocatalytic activity enhancement by superexchange interaction toward OER in alkaline media of Ni-Fe LDH. *J. Phys. Chem. C* 118: 22432–22438.

97 Dai, J., Zhu, Y., Yin, Y. et al. (2019). Super-exchange interaction induced overall optimization in ferromagnetic perovskite oxides enables ultrafast water oxidation. *Small* 15: 1903120.

98 Kanan, M.W. and Nocera, D.G. (2008). In situ formation of an oxygen-evolving catalyst in neutral water containing phosphate and Co^{2+}. *Science* 321: 1072–1075.

99 Shao, Y., Xiao, X., Zhu, Y.-P., and Ma, T.-Y. (2019). Single-crystal cobalt phosphate nanosheets for biomimetic oxygen evolution in neutral electrolytes. *Angew. Chem.* 131: 14741–14746.

100 Esswein, A.J., Surendranath, Y., Reece, S.Y., and Nocera, D.G. (2011). Highly active cobalt phosphate and borate based oxygen evolving catalysts operation in neutral and natural waters. *Energy Environ. Sci.* 4: 499–504.

101 Richter, M. and Schmeißer, D. (2013). Spin states in Co-Pi catalyst. *Appl. Phys. Lett.* 102: 253904.

102 Garcés-Pineda, F.A., Blasco-Ahicart, M., Nieto-Castro, D. et al. (2019). Direct magnetic enhancement of electrocatalytic water oxidation in alkaline media. *Nat. Energy* 4: 519–525.

103 Mtangi, W., Kiran, V., Fontanesi, C., and Naaman, R. (2015). Role of the electron spin polarization in water splitting. *J. Phys. Chem. Lett.* 5: 4916–4922.

104 Mtangi, W., Tassinari, F., Vankayala, K. et al. (2017). Control of electrons' spin eliminates hydrogen peroxide formation during water splitting. *J. Am. Chem. Soc.* 139: 2794–2798.

105 Kumar, A., Capua, E., Vankayala, K. et al. (2007). Magnetless device for conducting three-dimensional spin-specific electrochemistry. *Angew. Chem. Int. Ed.* 56: 14587–14590.

106 Zhang, W., Banerjee-Ghosh, K., Tassinari, F., and Naaman, R. (2018). Enhanced electrochemical water splitting with chiral molecule-coated Fe_3O_4 nanoparticles. *ACS Energy Lett.* 3: 2308–2313.

107 Mondal, P.C., Fontanesi, C., Waldeck, D.H., and Naaman, R. (2016). Spin-dependent transport through chiral molecules studied by spin-dependent electrochemistry. *Acc. Chem. Res.* 49: 2560–2568.

108 Tassinari, F., Banerjee-Ghosh, K., Parenti, F. et al. (2017). Enhanced hydrogen production with chiral conductive polymer-based electrodes. *J. Phys. Chem. C* 121: 15777–15783.

109 Sharpe, R., Lim, T., Jiao, Y. et al. (2016). Oxygen evolution reaction on perovskite electrocatalysts with localized spins and orbital rotation symmetry. *ChemCatChem* 8: 3762–3768.

110 Gracia, J. (2017). Spin dependent interactions catalyse the oxygen electrochemistry. *Phys. Chem. Chem. Phys.* 19: 20451–20456.

111 Gracia, J., Sharpe, R., and Munarriz, J. (2018). Principles determining the activity of magnetic oxides for electron transfer reactions. *J. Catal.* 361: 331–338.

112 Jiao, Y., Sharpe, R., Lim, T. et al. (2017). Photosystem II acts as a spin-controlled electron gate during oxygen formation and evolution. *J. Am. Chem. Soc.* 139: 16604–16608.

113 Wakayama, N.I., Okada, T., Okano, J.-i., and Ozawa, T. (2001). Magnetic promotion of oxygen reduction reaction with Pt catalyst in sulfuric acid solution. *Jpn. J. Appl. Phys.* 40: L269–L271.

114 Pauling, L., Wood, R.E., and Sturdivant, J.H. (1946). An instrument for determining the partial pressure of oxygen in a gas. *J. Am. Chem. Soc.* 68 (5): 795–798.

115 Hirota, N., Ikezoe, Y., Uetake, H. et al. (2000). Magnetic field effect on the kinetics of oxygen dissolution into water. *Mater. Trans., JIM* 41 (8): 976–980.

116 Okada, T., Wakayama, N.I., Wang, L. et al. (2003). The effect of magnetic field on the oxygen reduction reaction and its application in polymer electrolyte fuel cells. *Electrochim. Acta* 48: 531–539.

117 Kishioka, S.-y., Yamada, A., and Aogaki, R. (2001). Electrochemical detection of magnetic convection: effect of high magnetic fields on O_2/O_2^- electrode reaction in acetonitrile. *Electrolysis* 13 (14): 1161–1164.

118 Saito, E. and Bielski, B.H.J. (1961). The electron paramagnetic resonance spectrum of the HO_2 radical in aqueous solution. *J. Am. Chem. Soc.* 83 (21): 4467–4468.

119 Monzon, L.M.A., Rode, K., Venkatesan, M., and Coey, J.M.D. (2012). Electrosynthesis of iron, cobalt, and zinc microcrystals and magnetic enhancement of the oxygen reduction reaction. *Chem. Mater.* 24: 3878–3885.

120 Wang, L., Yang, H., Yang, J. et al. (2016). The effect of the internal magnetism of ferromagnetic catalysts on their catalytic activity toward oxygen reduction reaction. *Ionics* 22: 2195–2202.

121 Tufa, L.T., Jeong, K.-J., Tran, V.T., and Lee, J. (2020). Magnetic-field-induced electrochemical performance of a porous magnetoplasmonic Ag@Fe_3O_4 nanoassembly. *Appl. Mater. Interfaces* 12: 6598–6606.

122 Kiciński, W., Sęk, J.P., Matysiak-Brynda, E. et al. (2019). Enhancement of PGM-free oxygen reduction electrocatalyst performance for conventional and enzymatic fuel cells: the influence of an external magnetic field. *Appl. Catal., B* 258: 117955.

123 Lyu, X., Zhang, W., Liu, S. et al. (2021). A magnetic field strategy to porous Pt-Ni nanoparticles with predominant (111) facets for enhanced electrocatalytic oxygen reduction. *J. Energy Chem.* 53: 192–196.

124 Li, X., Cao, C.-S., Hung, S.-F. et al. (2020). Identification of the electronic and structural dynamics of catalytic centers in single-Fe-atom material. *Chem* 6: 3440–3454.

125 Kim, S. and Kim, H. (2017). Oxygen reduction reaction at porphyrin-based electrochemical catalysts: mechanistic effects of pH and spin states studied by density functional theory. *Catal. Today* 295: 119–124.

126 Jeon, I.-Y., Zhang, S., Zhang, L. et al. (2013). Edge-selectivity sulfurized graphene nanoplatelets as efficient metal-free electrocatalysts for oxygen reduction reaction: the electron spin effect. *Adv. Mater.* 25: 6138–6145.

127 Zeng, Z., Zhang, T., Liu, Y. et al. (2018). Magnetic field-enhanced 4-electron pathway for well-aligned Co_3O_4/electrospun carbon nanofibers in the oxygen reduction reaction. *ChemSusChem* 11: 580–588.

128 Lu, F., Wang, J., Li, J. et al. (2020). Regulation of oxygen reduction reaction by the magnetic effect of $L1_0$-PtFe alloy. *Appl. Catal., B* 278: 119332.

129 Pazur, A. and Winklhofer, M. (2008). Magnetic effect on CO_2 solubility in seawater: a possible link between geomagnetic field variations and climate. *Geophys. Res. Lett.* 35: L16170.

130 Pan, H., Xiao, X., Hu, B. et al. (2017). Generating huge magnetocurrent by using spin-dependent dehydrogenation based on electrochemical system. *J. Phys. Chem. C* 121: 28420–28424.

131 Kumar, A., Capua, E., Vankayala, K. et al. (2017). Magnetless device for conduction three-dimensional spin-specific electrochemistry. *Angew. Chem. Int. Ed.* 56: 14587–14590.

132 Pan, H., Jiang, X.X., Wang, X. et al. (2020). Effective magnetic field regulation of the radical pair spin states in electrocatalytic CO_2 reduction. *J. Phys. Chem. Lett.* 11: 48–53.

133 Player, T.C. and Hore, P.J. (2020). Source of magnetic field effects on the electrocatalytic reduction of CO_2. *J. Chem. Phys.* 153: 084303.

134 Pan, H., Wang, M., Shen, Y., and Hu, B. (2018). Large magneto-current in the electrochemical detection of oxalate in aqueous solution. *J. Phys. Chem. C* 122: 19880–19885.

135 Wang, J. and Song, Y. (2020). Magnetic field coupling microfluidic synthesis of Fe_2Pt/C nanocatalysts for enhanced electrochemical catalytic oxidation of alcohol. *Int. J. Hydrogen Energy* 45: 13035–13044.

136 Nguyen, H. and Clement, R.J. (2020). Rechargeable batteries from the perspective of the electron spin. *ACS Energy Lett.* 5: 3848–3859.

137 Zhang, L., Zeng, M., Wu, D., and Yan, X. (2019). Magnetic field regulating the graphite electrode for excellent lithium-ion batteries performance. *ACS Sustainable Chem. Eng.* 7: 6152–6160.

138 Wang, H., Xie, J., Follette, M. et al. (2016). Magnetic field-induced fabrication of Fe_3O_4/graphene nanocomposites for enhanced electrode performance in lithium-ion batteries. *RSC Adv.* 6 (86): 83117–83125.

139 Wang, Y., Zhang, L., Gao, X. et al. (2014). One-pot magnetic field induced formation of Fe_3O_4/C composite microrods with enhanced lithium storage capability. *Small* 10 (14): 2815–2819.

140 Ganguly, D., V.S., A.P., Ghosh, A., and Ramaprabhu, S. (2020). Magnetic field assisted high capacity durable Li-ion battery using magnetic α-Fe_2O_3. 10: 9945.

141 Li, W., Liang, Z., Lu, Z. et al. (2015). Magnetic field-controlled lithium polysulfide semiliquid battery with ferrofluidic properties. *Nano Lett.* 15: 7394–7399.

142 Li, Q., Li, H., Xia, Q. et al. (2020). Extra storage capacity in transition metal oxide lithium-ion batteries revealed by in situ magnetometry. *Nat. Mater.* https://doi.org/10.1038/s41563-020-0756-y.

143 Shen, K., Wang, Z., Bi, X. et al. (2019). Magnetic field-suppressed lithium dendrite growth for stable lithium-metal batteries. *Adv. Energy Mater.* 9: 1900260.

144 Dong, J., Dai, H., Wang, C., and Lai, C. (2019). Uniform lithium deposition driven by vertical magnetic field for stable lithium anodes. *Solid State Ionics* 341: 115033.

145 Wang, A., Deng, Q., Deng, L. et al. (2019). Eliminating tip dendrite growth by Lorentz force for stable lithium metal anodes. *Adv. Funct. Mater.* 29: 1902630.

146 Huang, Y., Wu, X., Nie, L. et al. (2020). Mechanism of lithium electrodeposition in a magnetic field. *Solid State Ionics* 345: 115171.

147 Yu, L., Wang, J., and Xu, Z.J. (2020). A perspective on the behavior of lithium anodes under a magnetic field. *Small Struct.* 2000043.

148 Yang, Z. and Liang, X. (2020). Self-magnetic-attracted $Ni_xFe_{(1-x)}$@$Ni_xFe_{(1-x)}$O nanoparticles on nickel foam as highly active and stable electrocatalysts towards alkaline oxygen evolution reaction. *Nano Res.* 13 (2): 461–466.

149 Barsan, M.M., Enache, T.A., Preda, N. et al. (2019). Direct immobilization of biomolecules through magnetic forces on Ni electrodes via Ni nanoparticles: applications in electrochemical biosensors. *ACS Appl. Mater. Interfaces* 11 (22): 19867–19877.

150 Lu, Z., Shoji, T., and Yang, W. (2010). Anomalous surface morphology of iron generated after anodic dissolution under magnetic fields. *Corros. Sci.* 2680–2686.

Index

a
ABX$_3$ perovskite   228, 229
acidic conditions   2, 5, 6, 9, 24, 30, 75, 100, 108, 306
2-acrylamido-2-methylpropanesulfonate (AMPS) monomers   449
activated carbon (AC)   97, 116, 378
actuators   422–424
acyclovir, electrooxidation behavior of   426
adriamycin (DOX)   169
adsorbate evolution mechanism (AEM)   16, 17
Al current collectors   341
alkaline conditions   2–4, 6, 48, 74–76
alkaline electrolysis   4, 5
alkaline fuel cell, PEMFC   121–124
all-inorganic perovskite materials   232–234
all-organic flexible patch antenna   403
alpha-fetoprotein (AFP) immunosensor   426
alternating current (AC)   319
N-2-aminoethylacrylamide hydrochloride (AEAM)   448
ammonium peroxidisulfate (APS)   469
ammonium persulfate (APS)   100, 101, 306, 366, 470
aniline   101, 109, 308–311, 313–315, 317, 369, 407, 411, 426, 433, 434, 469, 470, 475
anion exchange membrane fuel cells (AEMFCs)   57
anodic aluminum oxide (AAO)   306
antifouling biosensor   434
2,2′-azobisisobutyronitrile (AIBN)   455
aqueous electrolytes   369, 384, 493, 500

aqueous zinc-ion batteries (AZIBs)   346, 347
atomic force microscopy (AFM)   208, 213, 309, 463
atomic layer deposition (ALD)   9
Austin Model 1 (AM1)   213
autologous nerve transplantation   407

b
Barrett–Joyner–Halenda (BJH) model   102
battery field   518
benzenedithiol (BDT)   79
benzo[ghi]perylene (BGP)   463
bioelectric batteries, electrodes for   352–355
bioengineering
   actuators   422–424
   drug delivery methods   414–422
   tissue regeneration engineering   406–414
biofouling   433
biological imaging   165, 168, 182, 184, 189–190, 258
biologically active electrochemical coatings   416
biologically interfacing electrode   421
biosensors   172, 304, 305, 315, 316, 355, 399, 406, 424–436
bipolar electrodes (BPEs)   319
bis(dipyrrinato)zinc complex nanosheet   218
bis(dipyrrinato)zinc(II) complex nanosheet   212, 218, 219
bis(terpyridine)metal complex nanosheet   215
bisphenol A (BPA) detection   426

Functional Nanomaterials: Synthesis, Properties, and Applications, First Edition.
Edited by Wai-Yeung Wong and Qingchen Dong.
© 2022 WILEY-VCH GmbH. Published 2022 by WILEY-VCH GmbH.

black phosphorus (BP)   275
boron-doped carbon nanotubes (BCNTs)   77
boron nitride nanosheets/polyvinyl alcohol (BNNS/PVA)   323
bottom-up fabricated actuator devices   424
Brewster angle microscopy (BAM)   214
11-bromoundodeciltrimethoxisilane (BTMS)   456
Brunauer–Emmett–Teller (BET)   95
(1-butyl-3-methylimidazolium bis(trifluoromethyl-sulfonyl)imide (BMITFSI)   121

c

cadmium sulphide (CdS) nanowire   289
CA-IL-GN actuator   422
calcium titanate ($CaTiO_3$)   225
camphor-10-sulfonic acid (β-CPSA)   414
capacitors based on PANI nanostructures   364
carbazole-decorated covalent triazine framework (CTF-CSU1)   118
carbon-based organic framework   117–119
carbon black (CB)   87, 122–124, 339, 343
carbon dioxide reduction reaction (CO_2RR)   355, 361–363
carbon nanocages (CNCs)   13, 101, 117
carbon nanotubes (CNTs)   8, 12, 75, 76–80, 108, 121, 123, 124, 345, 356, 401, 452–455, 464
carbon spheres (CS)   116, 186
carboxymethyl guar gum (CMGG) polymers   459
cardiac tissue engineering (CTE)   411, 413
catalytic reactions   9, 42, 65, 67, 479, 491, 517–518
cathode ray tubes (CRT)   184
CDPY hybrid nanomaterials   376
CeO_2-based electrocatalysts   42, 44, 48, 55, 60–63
CeO_2-based nanomaterials   42, 64, 67
CeO_2 composites   47, 61
CeO_2-embedded NiO (Ce–NiO–E)   54
CeO_x nanoparticles (NPs)   53
ceria-based hybrid electrocatalysts   55
cerium-based electrocatalysts   49, 63
    CeO_2-based electrocatalysts   60
    earth-abundant electrocatalysts   59
    for HER   44–49
    for OER   49–57
    for ORR   57–63
    for other electrochemical reactions   63–65
cerium-doped electrocatalysts   44–45, 50
cetyltrimethylammonium bromide (CTAB)   109, 309
CGHNs   107
Channelrhodopsin-2 (ChR2)   170
chemical oxidation method   306
chemical sensors   355, 424–436
chemical vapor deposition (CVD)   77, 287, 341
citric acid (CA)   180–182, 308
C-LFP-PEDOT-Al electrode   342
CMK-3   105, 348, 349
cobalt chalcogenides   14
Co-based oxides/hydroxides   22–23
$CoFe_2O_4$/PANI-MWCNTs OER catalyst   356
computer-assisted materials discovery   28–29
concerted proton–electron transfer (CPET)   17
conducting polymer-based electrocatalysts   363
conducting polymer hydrogels (CPH)   400
conducting polymer nanomaterials
    chemical synthesis and properties   306–314
    electrochemical synthesis and properties   314–329
conducting polymers (CPs)   303
    advantages of   338
    anode surface   321
    applications of   305
    challenges of   424
    conductivity of   337
    for electrocatalysis   355–356
    electrochemical properties   355
    for electrochemical sensors   424
    molecular structure   304
    sulfur composites   348
    trilayer actuators   423
conductivity enhanced nanocomposite (CRNc)   409
confocal laser scanning microscope (CLSM)   190

CoNS/RGO/PPy nanocomposites 432
constriction resistance model 400, 401
contact angles (CAs) 309, 323
continuous-wave (CW) 258
Co_3O_4 (spinel oxides) 51
cooperative upconversion (CU) 144
coprecipitation method 149, 178, 183
coprecipitation synthesis 178
CoPt nanowires 514
CO_2 reduction reaction (CO_2RR) 41, 517, 518
Co_3S_4/PANI, specific capacitance of 384
PANI nanosheets/Ni foam electrocatalyst 361
Co-terpyridine nanosheets 217
coulombic efficiency 215, 216, 341, 343, 348
covalent organic framework (COF) 75
CR(NMe)EKA/PEDOT 418, 419
cross-linked liquid crystal polymer (CLCP) 175
cross relaxation (CR) 144, 160
crystal field theory 502
$CsPbX_3$ crystal structures 234
Cu based oxides/hydroxides 24
Cu–Ce–O oxide 60
CuO/MoS_2-based heterostructure flexible photodetector 290
Cu_2O nanoparticles decorated PANI matrix (PANI/Cu_2O) catalyst 363
4-cyanopentanoic acid dithiobenzoate (CPADB) 463
cyclic voltammograms (CVs) 81, 83, 216, 312, 313, 360, 377, 385
Cys-Arg-Glu-Lys-Ala (CREKA) 419

d

decyltrimethylammonium bromide (DTAB) 309
defective carbon nanomaterials 116
deionized (DI) 313, 366, 367, 457, 468, 470
density functional theory (DFT) 9, 42, 44, 114, 208, 236, 255, 353
dexamethasone phosphate (DexP) release pattern 420
DHPZDA 453–455
dicyandiamide (DCDA) 103
2-(diethylamino)ethyl methacrylate (DEAEMA) 457
drift current 278, 279

2,5-dihydroxyterephthalaldehyde (DHTA)-COF 103
diisooctylphosphinic acid (DIOP) 243
N,N-dimethylacetamide (DMAC) 452
dimethylammonium (DMA^+) 228, 232
dimethyl sulfoxide (DMSO) 103, 245
dioctylamine (DOAm) 243
diphenyl diselenide (DDS) 106
4,4′-dipyridyl disulfide (DPDS) 211
dipyrrin ligand molecule 212
direct piezoelectric effect 277
dispersion copolymerization 458–461
dispersion polymerization 447, 458–462, 471, 473, 474
dithienylethene (DTE) molecules 175
docetaxel (Dtxl) therapy 170
dopamine (DOPA)-coated PPy/PGA spring 414
dopamine (DA) detection 432
downconversion nanoparticles (DCNPs) 142
  $LaPO_4:Ce^{3+}, Tb^{3+}$ 187–189
  $LnVO_4:Ln^{3+}$ (Ln = La, Gd, Y; Ln^{3+} = $Eu^{3+}, Dy^{3+}, Sm^{3+}$) 186–187
  $SrAl_2O_4:Eu^{2+}, Dy^{3+}$ 182–184
  $Y_2O_3:Eu^{3+}$ 184–186
  $Y_3Al_5O_{12}$:RE (RE = Ce^{3+}, Tb^{3+}) 178–182
dual-layer conducting polymer/acid functionalized carbon nanotube (fCNT) microelectrode 416
dye-sensitized solar cells (DSSCs), electrodes for 352

e

earth-abundant electrocatalysts 59–60
earth-abundant metallic nanomaterials
  hydrogen evolution reaction 6–16
  oxygen evolution reaction 16–27
edge-halogenated graphene nanoplatelets (XGnPs) 89, 90
electrical healing, of pure PEDOT:PSS film 404
electrocatalytic activity 41, 44, 45, 53, 60, 76, 82, 85, 87, 89, 103, 111, 125, 352, 355, 356, 358, 360, 425, 432
electrochemical atomic force microscopy (EC-AFM) 319
electrochemical capacitors 363
electrochemical DNA (E-DNA) sensors 432, 433

electrochemical impedance spectroscopy (EIS)   107
electrochemical reaction   76, 121, 355, 402, 498
   cerium-based electrocatalysts   63–65
electrochemical surface area (ECSA)   16, 30, 358
electrochemical system
   in aqueous solution   492–495
   spin states of electrocatalysts   495–496
electrochemical water splitting
   current techniques   4–5
   overpotential and Tafel slope   3–4
   principle   1–3
electrodeposition   45, 361, 434, 491, 496–498, 519–521
electrodes
   for all-polymer batteries   350–351
   for aqueous zinc-ion batteries   346
   for bioelectric batteries   352–355
   for dye-sensitized solar cells   352
   for lithium–sulfur batteries   348–350
   for sodium ion batteries   345
electrodes for lithium-ion batteries
   $HClO_4$-doped PANI nanotubes   339
   PEDOT-$LiFePO_4$ films   338
   Si anodes   342
electro double-layer capacitor (EDLC)   363
electrokinetic drug nanocarriers, delivery   415
electroluminescence (EL)   254, 255
electromagnetic interference (EMI)   304, 386, 465
electron beam lithography (EBL)   291
electronic interaction mechanism   492
electronic skin (E-skin)   399
   energy-saving   405–406
   self-healing   403–405
   wearable electronic devices   400–403
electron spin resonance (ESR)   54
electron transportation   289
electropolymerization   315, 317–319, 338, 345, 426, 429, 434
electroproperties   215–217
electrospinning   52, 183–184, 305, 322, 323, 325–329, 414–416, 419, 470
emulsion polymerization   310, 419, 447, 448, 456–458
energy-dispersive X-ray spectrum (EDS)   94
energy migration-mediated upconversion (EMU)   145

energy-saving E-skin   405–406
energy transfer upconversion (ETU)   143–144
epoxidized natural rubber (ENR)   464
ethylenediamine tetraacetic acid ester (EDTA)   149, 185
3,4-ethylenedioxythiophene (EDOT)   306, 319, 325
excited-state absorption (ESA)   143
exciton binding energy   240, 241, 258

f

$FAPbI_3$ 228
Faradaic efficiencies (FEs)   4, 63, 64, 66, 362
Faraday's constant   1
Faradic pseudocapacitors   363
fast crystallization-deposition (FDC) method   246
$[Fe(acac_2\text{-}trien)][MnCr(Br_2\text{-}An)_3] \cdot (CH_3CN)_2$ 210
(Fe)-based MOFs   175
Fe-based oxides/hydroxides   24
Fe-diterephthalate grid   209
Fe-doped β-$Ni(OH)_2$ porous   22
Fe^{3+} doped PTh electrode on carbon cloth   385
Fermi level pinning   282
Fe-terpyridine nanosheet   217
few-layer oxidized graphdiyne (FLGDYO)   96
field emission displays (FEDs)   184
field-emission scanning electron microscope (FESEM)   94, 95
flexible polythiophene (e-PTh) on titania (Ti) wire electrode material   379
flexible symmetric PPy hydrogel supercapacitors   366
4-fluorophenylmethylammonium-trifluoroacetate (FPMATFA)   255
fluorescence resonance energy transfer (FRET)   162, 171–172
fluorescent ink   191–192
formamidinium (FA)   228, 239
free-standing electrodeposited PPy film electrodes   350
fuel cells, application in   120–129

g

gas/liquid interfacial synthesis   213–214
Ge-based perovskites   235, 236

Gibbs free energy 1, 9, 10, 17, 18, 22, 51, 54
GO/PANI/silk composite scaffold 407
Goldschmidt tolerance factor 235
grafted PAN nanofibers (GPN) 327
graphdiyne 94–97, 121, 122, 125, 127, 129
graphene foam/PPy composite stretchable electrodes 374
graphite oxide composite 91
grazing incidence X-ray diffraction (GIXRD) 213
ground state absorption (GSA) 144
guanidinium (GA^+) 228, 232

h

Haber–Bosch method 63
halogen ions 238
$HClO_4$-doped PANI nanotubes 338
HCS/PANI-based E-DNA sensor 433
hematoporphyrin (HP) 168, 170
heteroatom-doped carbon-based materials 75, 105, 129
heteroatom-doped carbon nanotubes 76–80
heteroatom-doped composite materials 105–108
heteroatom-doped graphdiyne 94–97
heteroatom-doped graphenes 80–94
heteroatom-doped nanocarbon materials 105, 129
heteroatom-doped porous carbon nanomaterials 97–105, 117, 118
heteroatom-substituted CNTs 78, 79
hierarchically porous carbon materials (HPCMs) 99, 100
highly ordered pyrolytic graphite (HOPG) 208
highly oriented pyrolytic graphite (HOPG) 105
high-resolution TEM (HRTEM) 45, 88, 103, 106, 211
histamine acrylamide hydrochloride (HisAM) 449
homogeneous field 491
homojunction 285, 286
host matrix screening 157–158
hot injection method 242–244, 255
H_2TCPP molecular structures 213
humidity sensing 256, 257
hydrogen evolution reaction (HER) 1, 41, 357, 491, 498, 504
CeO_2 composites 45
cerium-based electrocatalysts 44
energy diagram 45
mechanism 6–7
metal (M^0) nanoparticles 7–8
metal (M^0) single-atom catalysts 8–11
metal carbides 15
metal chalcogenides 12–14
metal nitrides 14–15
metal oxides/(oxy)hydroxides 15–16
metal phosphides 11–12
metal single-atom electrocatalyst 9
hydrogen gas 6, 498
hydrogen oxidation reaction (HOR) 73
Hydrosorb dressings, for wound healing 407
hydrothermal method 15, 149, 182–184, 191, 290, 293, 468
2-hydroxypropyl methacrylate (HPMA) 450
hypophosphorous acid 235

i

imidazolium tetrafluoroborate ($IMBF_4$) 237–238
immobilization method 410, 411
indium selenide (InSe) 283
indium tin oxide (ITO) 215, 318
infrared radiation (IR) 290
intermediate-spin state (IS) 60, 507
interparticle energy transfer 158
inverse opal carbon (IOC) 61
ionic liquid-modified (RGO-IL/PANI) composite 370, 374
iontophoretic skin drug delivery system 415
isopropanol (IPA) 181, 188, 247, 366, 452

k

KillerRed-UCNPs 169
Koutecky–Levich (K–L) equation 87

l

lactate dehydrogenase (LDH) 428
Langmuir isotherm 214
lanthanide-doped upconversion nanocrystals 147
lanthanide ions 141, 156, 158, 159, 162
$LaPO_4$:Ce^{3+}, Tb^{3+} 187–189
lattice oxygen mechanism (LOM) 16, 17
layered double hydroxide (LDH) 16, 44, 378, 507

lead-based perovskite materials 234
Lewis bases 105, 248, 249, 251
ligand-assisted reprecipitation (LARP) method 242, 244, 245
light emitting diodes (LEDs) 238, 251, 254–256, 278, 355, 402
Li-ion batteries 518
limited anion-exchange reaction (LAER) 189
linear conjugated polymers (LCPs) 127
linear sweep voltammetry (LSV) 47, 78
Li-PANI battery 338
liquid-phase exfoliation 209–211, 276
liquid–solid-solution (LSS) 149
lithium electrodeposition 521
lithium-ion batteries (LIBs)
  applications 338
  Ni-rich cathode materials 340
  Si anodes for 342
  ZMO anode 345
lithium metal anode 522
lithium–sulfur battery (Li-S battery) 93, 348–350
$Li_3V_2(PO_4)_3$/PEDOT composite 340
$LnVO_4:Ln^{3+}$ (Ln = La, Gd, Y; Ln^{3+} = Eu^{3+}, Dy^{3+}, Sm^{3+}) 186–187
Lorentz force 492–494, 496, 499–501, 515, 520, 521
low critical solution temperature (LCST) 479
low graphitic carbon (LGC) 65
low-lead perovskite material 234–238
low-temperature magnetic force microscopy (LT-MFM) 220

m

magnesium oxide (MgO) 99
magnetic field template 497
magnetohydrodynamic (MHD) 491
magnetohydrodynamic convention (MHDC) 492
magnetomechanic mechanism 491
magnetoproperties 219–221
$MAPbI_3$ 228–230, 232, 234, 236, 238–240, 245, 247, 252, 253, 258
mechanical exfoliation 208–209, 289
melamine fiber (MF) 99, 101
membrane electrode assembly (MEA) 121
metal carbides 15, 207
metal chalcogenides/nitrides/phosphides 12, 26–27
metal complex nanosheets
  electroproperties 215
  gas/liquid interfacial synthesis 213–214
  liquid/liquid interfacial synthesis 211–213
  liquid-phase exfoliation 209–211
  magnetoproperties 219–221
  mechanical exfoliation 208–209
  photoproperties 217–219
  vacuum phase fabrication 208
metal halide perovskite (MHP) 225–258
  all-inorganic perovskite materials 232–234
  crystal structure and phase 225–227
  excellent charge transport performance 240
  high absorption coefficient 239
  hot injection method 243–244
  lead-free perovskite materials 234–238
  ligand-assisted reprecipitation method 244
  optoelectronic devices 258
  organic–inorganic hybrid perovskite materials 228–232
  photodetector 257
  photoluminescence properties 240–242
  sensing 256–257
  solution deposition methods 244–245
  tunable bandgap 238–239
metal (oxy)hydroxides 16
metal-ion batteries 338–348
metallic (M^0) nanoparticles 19
metal (M^0) nanoparticles 7–8
metal nitrides 14–15, 26, 48
metal–organic frameworks (MOFs) 9, 27, 53, 99, 117, 175, 207, 210, 345, 382, 383, 461
metal oxides/(oxy)hydroxides 15–16
metal oxides/hydroxides 19
  Co-based oxides/hydroxides 22–23
  Fe-based oxides/hydroxides 24
  Ni-based oxides/hydroxides 19–22
metal phosphides 11–12, 26, 27, 47, 48
metal–semiconductor junction 282–284
metal/semiconductor/metal (MSM) 258
metal–semiconductor (M–S) system 282

metal (M^0) single-atom catalysts 8–11
metal (Mn^+) single-atom catalysts 25–26
metal-to-ligand charge transfer (MLCT) 215
methacrylic acid (MAA) 457
methylammonium (MA) 228, 232, 238, 239, 256
[2-(methacryloyloxy)ethyltrimethylammonium salt ([META$^+$]) 449
N,N'-methylene bis(acrylamide) (MBA) 457
2,2'-methylenebis(2-methoxy-4-methylphenol) 317
Mg–air biobattery 352, 353, 355
micromanipulation tools 424
microporous polycarbonate filters 306
microwave-assisted method 150, 183
mix-dimensional (MD) heterojunctions 286
mixed A-site cation perovskite 229–232
mixed cellulose ester filter films (MCEFs) 97
Mn-based cathode materials, for ZIBs 347
molecularly engineered conductive polymer binders 343
molecularly imprinted polymer (MIP)-modified SnO_2 electrode 429
molecularly imprinted polypyrrole/graphene quantum dots (MIPPy/GQDs) as electrochemical sensor 426
molecule-based nanosheets 207, 214
molybdenum disulfide (MoS_2) 280
M–OOH, linear scaling relation of 18
MoS_2 nanosheet 290
m-phenylenediamine (m-PDA) 369
MSC-EV biotin immobilization method 411
multiple-walled carbon nanotubes (MWCNTs) 122, 464
multiple-walled nanotube (MWNT) 452
multiwalled carbon nanotubes/lauric acid/thermoplastic polyurethane (MWCNTs/LA/PU) 326
mussel-inspired poly(ATMA-co-DOPAMA-co-PEGMA) bioadhesive 412
MUV-1-Cl 219, 220
MWNT-DHPZDA 453–455
MXenes 207

n

N and S dual-doped carbon (NSC) 61
nanoclusters (NCs) 46
nanowires arrays (NWAs) electrodes 62
1,4,5,8-naphthalenetetracarboxylic dianhydride (NTCDA) 118
naphtho[2,3-b]thieno[3,4-e][1,4]dioxine (NaphDOT) 317
natural rubber/rGO/Fe_3O_4 (NMGO) 475
N-bromosucinimide (NBS) 455
ND-GLC 97, 98, 126
N-doped carbon foam (CF_N) 99
N-doped carbon nanocages (NCNCs) 101
N-doped carbon nanoribbon (NCNR) 101
N-doped carbon nanosheets (NDCNs) 109
N-doped fullerene-like carbon shell (NDCS) 97, 99
N-doped graphdiyne (N-GD) 94–97
N-doped graphdiyne (N-GDY) 94–97
N-doped graphene/CNT composite (N-G-CNT) 122
N-doped graphene framework (NGF) 85
N-doped graphene quantum dots (N-GQDs) 83, 84
N-doped graphene structure 87
N-doped mesoporous carbon nanospheres (NMCNs) 100
N-doped porous carbon (NPC) electrocatalysts 55
N-doped reduced graphene oxide (N-rGO) 107
N-dual-doped graphene 85
near-infrared (NIR) 156, 174, 176, 191, 236, 238, 254, 419
next-generation wearable systems 403
N/F co-doped graphdiyne (NFGD) 122
Ni-based oxides/hydroxides 19–22
Ni–C electrocatalyst 9
nickel foam (NF) substrate 45
NiCo-LDH nanocages 378
NiCo/PPy/RGO nanocomposite 431, 432
NiDI 215, 216
NiFe layered double hydroxide 45
Nile Red Derivative (NRD) 172
Ni–P deposited PPy film 358
Ni-rich cathode 339, 340
nitrogen-doped carbon materials 10, 75, 98

nitrogen-doped carbon nanotubes (N-CNT)   76, 107
nitrogen-doped graphene   76
nitrogen-rich polymer PANI catalyst   361
N-methyl-2-pyrrolidone (NMP)   453
noble metals   5, 8, 19, 25, 26, 30, 41, 42, 46, 57–59, 61, 62, 65, 77, 80, 190, 356, 357
N,P-co-doped CGHNs (N,P-CGHNs)   107
N/P co-doped hierarchical carbon (NPHC)   125
N_2 reduction reaction (NRR)   41
N-rGO-CNT-0.2   107
N/S-co-doped carbon aerogel (NSCA)   110
N/S-co-doped graphene (NSG)   90–92
nuclear magnetic resonance (NMR) technology   153

o

1-octadecene (OD)   146
octafluoronaphthalene (OFN)   463
octyltrimethylammonium bromide (OTAB)   309
oleamine (OM)   146
oleic acid (OA)   146, 153, 154, 169, 186, 243, 459
oleylamine (OAm)   243
1D cesium lead bromide ($CsPbBr_3$) nanowires   287
one-dimensional (1D) cylindrical giant molecules   76
1D–2D homojunction photodetectors   286–289
o-phenylenediamine (o-PDA)   124, 369
OPPy/CNT/GCE electrochemical electrode   425, 426
optical microscopy (OM)   211, 321, 457, 458
optical sensing   256, 257
optogenetics   170–171
organic–inorganic hybrid perovskite materials   228
   $MAPbI_3$ 228
   mixed A-site cation perovskite   229–232
   $FAPbI_3$ 228–229
organic light-emitting diode (OLED)   254
organic molecular beam epitaxy (OMBE)   208
ORR-active carbon nanocages (CNCs)   117
oxidation template assembly (OTA)   311
oxygen evolution reaction (OER)   1, 41, 356, 491, 504, 512, 513
   catalyst design   17, 29
   CeO_2 composites   50
   computer-assisted materials discovery   28
   mechanism   16–19
   metal chalcogenides/nitrides/phosphides   26
   metal (Mn^+) single-atom catalysts   25–26
   metal oxides/hydroxides   19
oxygen reduction reaction (ORR)   356–357, 504–513
   acidic medium   108–111
   electrocatalysts   57–58, 88
   electrochemical experiment   112
   fuel cells   120–121
   heteroatom-doped graphenes   80–94
oxygen vacancies (OVs)   15, 23, 42, 43, 47, 50, 51, 53–55, 61–65, 380

p

PAAMPSA/PANI/PA
   regenerative polymer complex   404
   strain sensor   404
paclitaxel-loaded polycaprolactone (PCL-PTX) mats   419
PANI-poly(ethylene oxide) (PEO)   328
PANI/PES nanofibrous scaffolds   413
PANI-PPy coated Pt electrodes   360
PANI/PPy composite nanofibers   366, 370
paramagnetic gradient force   492, 493
partial density of states (PDOS)   52
particle track-etched membranes (PTM)   306
Pb-based perovskite   235–237
PCL/PEDOT/CUR   415
PCPZ hydrogel   407, 408
PDA-PPy-PAM hydrogel   400
Pd electrode   358
Pd nanoparticle-dispersed PEDOT (Pd-PEDOT)
   CeO_2 model   46
   thin film on gold substrate as HER catalyst   358
P-doped graphene (PG)   86

PEDOT-coated cathode materials   339
PEDOT-coated CNT/polypropylene plate electrode   352
PEDOT-coated LNMO   339, 340
PEDOT/dexamethasone (PEDOT/Dex) films   416
PEDOT functionalized with anthraquinone (PEDOT-AQ)   351
PEDOT functionalized with benzoquinone (PEDOT-BQ)   351
PEDOT-LiFePO$_4$ composite film   338
PEDOP loaded with IBU, in implantable drug delivery device   418
PEDOT/paper   370, 371
PEDOT/prednisolone phosphate (PEDOT/PreP) films   416
PEDOT PSS-coated CMK-3/sulfur composite   349
PEDOT PSS-PAAm organogels   402
PEDOT PSS/RGO composite sponge   433, 434
1,1,4,7,7-pentamethyldiethyltriamine (PMDETA)   455
perfluorooctane sulfonic acid (PFOSA)   312
perfluorooctanoic acid (PFOA)   309
perovskite films   245
perovskite light-emitting diode (PeLED)   254–256
perovskite nanoparticles (PeNPs)   255
perovskite oxides   60, 504, 506
perovskite quantum dots (PQDs)   163, 242, 257
perovskite solar cells (PSCs)   177, 178, 238, 247, 249–253
phase-transition-assisted (PTA) strategy   99
phenylethylammonium iodide (PEAI)   253, 255
phenylpropylammonium iodide (PPAI)   256
photoacoustic (PA)   168, 191
photo atom transfer radical polymerization (photoATRP)   459
photocatalysis   42, 173–175, 304, 305
photo(electro)catalytic activity   24
photodynamic therapy (PDT)   168, 170, 175
photoelectrochemical sensor, for BPA detection   429, 430
photoisomerization   175

photoluminescence properties   240–242
photoluminescence quantum yield (PLQY)   240
photon avalanche (PA)   144
photoproperties   217–219
physical vapor deposition (PVD)   289
phytic acid (PA)   103, 366, 404
phytic acid-doped PANI on medical titanium (Ti) tablets   411
piezoelectricity   275, 277, 278, 284, 291
piezo-phototronics   275
   fundamental physics   277–278
   metal–semiconductor junction   282–284
   P–N junction   278–281
piezoresistive pressure sensors   402
piezotronics   275, 277, 278, 282, 292, 295
platinum nanoparticle-polypyrrole-3-carboxylated nanofiber composites (Pt&uscore;cPPyNFs)   431
PLLA/PPy composites   328
PNIPAM-based temperature-sensitive acidic triblock polymer   420
P–N junction photodetectors   278
   based on 1D–2D homojunction   286–289
   based on 2D homojunction   285–286
   based on 2D–2D homojunction   289–293
   based on 3D–2D homojunction   293–294
P(NIPAM-co-ETMA)   479
polyacetylene (PA)   303, 306
poly(acrylic acid-b-N-isopropylamide-b-acrylic acid) (PNA)   420, 421
polyacrylonitrile (PAN)   110, 327
poly[3-amino-5-mercapto-1,2,4-triazole]   91
polyaniline (PANI)   303, 309, 469, 470
   cytosensor   427
   nanofibers   313, 314
   nanostructures   364, 365
   poly(3-aminobenzoic acid)   426
polyaniline/graphene/MnO$_2$ (PANI/G/MnO$_2$) paper   381, 382
poly-ATMA-co-DOPAMA-co-PEGMA conductive copolymers   411
poly(2,2′-bithiophene) (PBTh) catalyst, for HER   357
poly((butylene adipate-co-terephthalate) (PBAT)   463

polycarbonate membrane (PCM) 315
poly((diethylamino)-ethyl methacrylate)-*co*-sodium methacrylate (PDEAEMA-co-SMA) 457
poly(2,6-dimethyl-1,4-phenylene oxide) (PPO) 455
polydimethylsiloxane (PDMS) 283, 402
polydopamine (PDA) 97, 102, 109, 400, 479
poly-3,4-ethylenedioxythiopene/polystyrene sulfonate (PEDOT:PSS)
  hydrogel, crosslinking process of 354
  nanosheets 402
  printed fabric 403
  printing ink 329
poly(3,4-ethylenedioxythiophene) (PEDOT) 315, 319
  concentration 325
  hydrogel 354
  MoS_2/PEDOT asymmetric supercapacitor 374
  nanoneedles 313, 314
  NTs 307
polyethylene glycol (PEG) 167, 433
polyethylene glycol polymer onto polyaniline (PANI/PEG) nanofibers 433–435
polyethyleneimine (PEI) 149, 165
polyethylene terephthalate (PET) 283, 403
poly glycidyl methacrylate (PGMA) 469
poly2-hydroxypropylmethacrylamide (PHPMA) 449
polyion complexes (PIC) 449
poly1-IrO_x composite film modified electrode 356
poly(lactic-*co*-glycolic acid) 171
poly(L-lactide) (PLLA) 416, 464
polymer composites 255, 323, 356, 447, 448, 449, 455, 456, 458, 460, 461, 463, 465, 466, 475, 481
polymerization-induced self-assembly (PISA) 450, 463
polymer precursors 448
polymers/metal nanoparticles 469
polymers nanocomposites
  application 474–481
  emulsion polymerization 456–458
  in situ polymerization 466–471
  synthesis 448–451
  solution mixing 451–456
  tailoring 471–474
polymersome pickering emulsion (PPE) 456
polymer sulfonate (PSS) 307, 315
poly-[2-(methacryloyloxy)ethyl phosphorylcholine]-b-poly[2-(dimethylamino)ethyl methacrylate]-b-poly-(2-hydroxypropylmethacrylate) (PMPC-b-PDMA-bPHPMA$_z$) 471
poly[2-methoxy-5-(2-ethylhexyloxy)-1, 4-phenylenevinylene] (MEH-PPV) 325
poly(methylacrylic acid) (PMAA) 463
poly(*N*-vinylpyrrolidone) (PVP) 310, 339
polyoxovanadate (POV) 307
poly(phthalazinone ether sulfoneketone)s (PPESK) 453
poly(*p*-phenylene) (PPP) 303, 304, 337
poly(*p*-phenylenevinylene) (PPV) 303, 304, 337
poly(3,4-propylenedioxythiophene) (PProDOT) 318
polypyrrole coated on nanospherical platinum (Pt/PPy NSs) composites 424
poly(3,4-propylenedioxythiophene) (PProDOT) 318, 319
poly(styrene-block-2-vinylpyridine-block-ethylene oxide) (P(S-b-2VP-b-EO)) 456
polystyrene-*b*-poly(acrylic acid) (PS-b-PAA) 456
polythiophene polymer (PTPA) 479
polythiophene (PTh) 303, 304, 337, 348, 370, 379, 479
poly(triphenylimidazole) 481
polyvinylidene difluoride (PVDF) 448
polyvinylidene fluoride (PVDF) 315
poly(4-vinylpyridine) 451
polyvinylpyrrolidone (PVP) 149, 190
porous carbon materials 99, 100, 103, 117
porous nanocarbon materials 124
porous N-doped carbon 97, 100, 101, 124
poly(L-lactic acid) (PLLA) 327
positron emission tomography (PET) technique 168
potassium peroxodisulfate ($K_2S_2O_8$) 469
potassium persulfate (KPS) 457

p-phenylenediamine (p-PDA)   369
PPy/ABTS   350
PPy/AQS/RGO composite   354
PPy-based piezoresistive sensors, electrical conductivity of   401
PPy-based wearable single-electrode-mode TENG (PPy-WSEM-TENG) device   405
PPy/carbonaceous nanospheres (PPy/CNSs)   425
PPy/CNTs-MIPs preparation   432, 433
PPy-coated $GdNi_4Al$ intermetallic alloy catalyst   358
PPy-coated microribbons   422, 423
PPy-functionalized PCL-PTX membrane fabrication   419
PPy hydrogels   366, 369, 401, 415
PPy/IC materials   350
PPy/IL nanoparticles   356, 357
PPy/MoS_x co-polymer film   358, 359
PPyNPs nanocomposite   431
PPy, ZMO anode, for LIBs   345
printed smart electronic skin   405
projected DOS (pDOS)   509
proton exchange membrane (PEM) electrolysis   4, 108
proton exchange membrane fuel cell (PEMFC)   57, 108, 123, 124
Pt-based electrocatalysts   44, 75
Pt/C electrocatalyst   93
Pt_cPPyNF sensors   431
Pt [1,8-dihydroxynaphthalene-3,6-disulfonatoferrate(II)] complex immobilized PANI/Prussian blue-laminated catalyst   363
PTh/MWCNT composites   370, 372
Pt nanoparticle (Pt NP)-PANI hydrogel based glucose sensor   428
PVA-PPy hydrogel, conductivity of   415
pyridine analogue of graphdiyne (PyN-GDY)   127
pyridinium chlorochromate (PCC)   287

q

quantum yields (QYs)   155, 158, 182, 185, 189, 218, 240, 244

r

radiative energy transfer (RET)   146, 162, 187
radiotherapy (RT)   170

rare earth based DCNPs
  $LaPO_4$:Ce^{3+}, Tb^{3+} 187–189
  $LnVO_4$:Ln^{3+} (Ln = La, Gd, Y; Ln^{3+} = Eu^{3+}, Dy^{3+}, Sm^{3+})   186–187
  upconversion material development   142–143
  upconversion mechanism
    cooperative upconversion   144
    cross relaxation   144
    energy migration-mediated upconversion   145
    energy transfer upconversion   143–144
    excited-state absorption   143
    photon avalanche   144
  Y_2O_3:Eu^{3+} 184–186
  $Y_3Al_5O_{12}$:RE (RE = Ce^{3+},Tb^{3+})   178–182
rare earth luminescent material   141
rate-determining step (RDS)   6, 45, 50, 506, 514
RbCsMAFA-based perovskite   232
reactive oxygen species (ROS)   168, 190, 426
redox reactions   27, 215–217, 350, 351, 355, 363, 380, 511
reduced graphene oxide (RGO)   12, 87, 107, 354, 407, 460, 467, 503, 518
reduced graphite oxides (rGOs)   63, 127
relative humidity (RH)   124, 255, 257, 311, 404
reverse electrodialysis (RED)-driven iontophoresis   416
reversible addition-fragmentation chain transfer (RAFT)   456, 463
reversible hydrogen electrode (RHE)   63, 73, 75, 358
RF sheet resistance   403
roll-to-roll (R2R)   250, 341, 402
rotating magnetic field technique (RMF)   249
rotating ring-disk electrode (RRDE)   83

s

Sabatier principle   6
salicylic acid (SA)   311, 470
scanning electron microscopy (SEM)   47, 88, 152, 211
scanning transmission electron microscopy (STEM)   52, 152
Schottky barrier height (SBH)   277, 282

Schottky contact   276, 280, 282, 283, 285
Schottky junction   283, 284
Scotch tape   80, 276
S-doped graphene (SG)   86, 90, 91, 93
second harmonic generation (SHG)   283
selected area electron diffraction (SAED)   51, 151, 213
selenium-doped CNT/graphene composites (Se-CNT)   106
self-assembly method   100, 186, 187, 207, 305, 309, 311, 312, 319, 320, 343, 407, 409, 447, 449, 461–463, 471
self-healing E-skin   403–405
Shockley–Queisser model   237
Si-doped nanosphere (Si-CNS)   80
Si/graphite/PANI nanocomposites   342, 343
Si/N-co-doped nanotube (SiN-CNT)   80
Si/PPP composites   342
silicon nanoparticles-based LIB anodes   343
silver nanoparticles (Ag NPs)   306
single-atom catalysts (SACs)   8–11, 25–26
single photon emission computed tomography (SPECT)   168
singlet-state oxygen molecule   513
single-walled CNT   77
single walls carbon nanotubes (SWCNTs)   469
slot-die coating   249, 250
Sn-based perovskites   235–238
S/N-co-doped CNTs (SN-CNTs)   109
sodium bis(2-ethylhexyl) sulfosuccinate   310
sodium dodecyl sulfate (SDS)   309
sodium ion batteries (SIBs)   345
   bipolar conducting polymer   345
   ClO_4^--doped PPy coated $Na_{1+x}MnFe(CN)_6$ composite cathode for   345
   PPy NWs/CP free-standing anode   346
sol–gel method   141, 147, 150, 178–181, 183, 185
solid-state laser (SSL)   184
solution deposition method   244
   blade-coating process   249–250
   one-step method   245–247
   two-step method   247–249
solution mixing   451–456, 464
solvothermal reactions   118, 181

spherical silver-incorporated conductive PPy (Ag/PPy) composite   377
spin related electrode reactions   491
   battery   518–522
   catalytic reactions   517–518
   electrochemical system
      in aqueous solution   492–495
      spin states of electrocatalysts   495–496
   electrodeposition of metals or alloys   496–498
   hydrogen evolution reaction   498–504
   oxygen evolution reaction   504–513
   oxygen reduction reaction   513–517
spin states of electrocatalysts   495–496
sp-N-doped few-layer graphdiyne (sp-NFLGDY)   96, 97
$SrAl_2O_4:Eu^{2+}, Dy^{3+}$ 182–184
stimulus-responsive biomaterials   414
sulfonate modified nanoparticle (SNP)-doped PEDOT composite   421
sulfur-doped carbon nanotubes (SCNTs)   79
sulfur-doped graphene nanoplatelets (SOGnP)   89
supercapacitors (SC)   363
   CP as active material   364–385
   MOFs-based   383
Superconducting Quantum Interference Device (SQUID)   156
surfactant molecules   310
synthetic metals   303

t

Tafel slope   3, 4, 6, 9, 10, 14, 19, 22, 23, 25, 44, 45, 49, 56, 57, 61, 356–358, 361, 506, 510
tailoring, polymers nanocomposites   471–474
TCNTs   114
temperature programmed desorption (TPD)   54, 105
temperature sensing   256, 257
terephthalic acid (TPA)   208
terpyridine ligand molecule   214
4-*tert*-butylpyridine (TBP)   249
1,2,4,5-tetracyanobenzene (TCNB)   463
tetra-ethyl orthosilicate (TEOS)   121

tetrabutylammonium hexafluorophosphate (TBAHFP) 494
tetrabutylammonium perchlorate 83, 363
tetrafluoroterephthalonitrile (TFP) 463
tetrahydrofuran (THF) 308, 455
5,10,15,20-tetrakis(4-carboxyphenyl)-porphyrin 210, 213
3D heterostructures 286, 287, 429
3D hierarchical PANI micro/nanostructure (H-PANI) supercapacitor electrode material 366, 368
3D macroporous PANI modified electrode 426
3D macroporous sponge 380
3D-nanostructured elastic PPy hydrogel 366
3D–2D homojunction photodetectors 293–294
3D-printed PEDOT PSS fibers 328
thermal decomposition 141, 146–149
thermoplastic (glycolic acid) surgical sutures 413
thermoplastic vulcanizates (TPV) 464
Ti based oxides/hydroxides 24
tissue reconstruction, high-performance biomaterials for 409
transition metal-based electrocatalysts 50
transition metal chalcogenides (TMCs) 12
transition metal dichalcogenides (TMDs) 207, 275, 280
transition metal (TM)-LDH electrocatalysts 50
transition metal nitrides 14
transition metal phosphides (TMPs) 11, 47
transition metal sites (TM) 11
transitional metal sulfides 340
transmission electron microscopy (TEM) 45, 52, 152, 181, 211, 213, 374, 455, 458
transparent conducting hydrogel 400
triarylphosphine (TPP) 103
4,4′,4″-(1,3,5-triazine-2,4,6-triyl)-trianiline(TAPT) 103
triboelectric nanogenerator (TENG) 315, 405

triethylenetetramine (TETA) 469
trimethoxysilane propyl methacrylate 475
1,3,5-trimethylbenzene (TMB) 102
triphenylborane (TPB) 77
triphenylphosphine (TPP) 86, 107
1,3,5-tris(4-aminophenyl)benzene (TAPB) 118
tris(4-bromophenyl)amine 118
tumor treatment 190–191
tunable bandgap 238–239, 256
tuning UC emission
  chemical composition 156–157
  core/shell structures 160
  cross-relaxation processes 160
  through energy migration 158–159
  external stimulus 165
  FRET or RETU 162–165
  host matrix screening 157–158
  interparticle energy transfer 158
  size- and shape-induced surface effects 160–162
turnover frequency (TOF) 12, 51, 480
two-dimensional (2D) carbon allotrope 80
2D homojunction photodetector 285–286
2D layered piezoelectric semiconductors 283
(2D) MnO_x/PPy nanosheets 347
2D PANI nanosheets 308
2D–2D homojunction photodetectors 289–293

u

UC-based nanothermometer 172
ultracapacitors 363
ultra-high vacuum (UHV) 208
undoped carbon nanomaterials 111
  edge as defect 112–114
  intrinsic/topological defects 114–117
unidirectional electrical stimulation, cell 414
upconversion (UC) 142
upconversion nanoparticles (UCNPs) 142
  activators 145–146
  based drug delivery system 169
  bioimaging 165–168
  composition determination 154–155

upconversion nanoparticles (UCNPs) (*contd.*)
  crystal structure identification 151–152
  magnetic properties measurement 156
  mediated molecular switches 175–176
  optical properties measurement 155
  optogenetics 170–171
  sensing and detection 171–173
  sensing platform 172
  size and morphology 152–153
  surface moieties 153–154
  synthesis
    coprecipitation method 149
    hydro/solvothermal synthesis 149
    microwave-assisted synthesis 150
    sol–gel synthesis 150
  therapy 168–170

v

VA-BCN nanotubes 78
van de Waals (vdW) interactions 276
  heterostructure photodetector 289
vapor-assisted solution process (VASP) 247, 248
V-based oxides/hydroxides 25
Vegard's law 236
ventral tegmental area (VTA) 171
vertically aligned nitrogen-doped CNTs (VA-NCNTs) 76, 83
vibrating sample magnetometer (VSM) 156
volatile organic compounds (VOCs) 42, 256
Volmer–Heyrovsky mechanism 6, 10, 45
Vö–V_2O_5/CP nanocables 380

w

Warburg diffusion resistance, of DSSC 352
water-assisted wedging method, for PEDOT 404
water splitting to hydrogen and oxygen 1
wearable electronic devices 399–403, 406
wet chemistry method 9

x

X-ray absorption fine structure (XAFS) 59
X-ray absorption near-edge structure (XANES) 53, 96, 156
X-ray absorption spectroscopy (XAS) 55, 151
X-ray diffraction (XRD) 151, 213, 229–231
X-ray photoelectron spectroscopy (XPS) 53, 82

y

Y_2O_3:Eu^{3+} 184–186
$Y_3Al_5O_{12}$:RE (RE = Ce^{3+},Tb^{3+}) 178–182
YAG nanoparticles 178, 180, 182

z

Ziegler–Natta polymerization method 306
ZIF-67 383
  NiCo-LDH nanocages 378
  PANI composites 383
  PEDOT heterostructure composite 384, 385
ZIF–PPy samples 383
zigzag-edged graphene ribbons 122
zinc–air (Zn–air)
  battery 75, 124–130
  fuel cell 120–122, 124, 126
ZnO wurtzite crystal 293
Zn-terpyridine nanosheet 217, 218
Zr-doped CeO_2 (ZDC) 61